Electromagnetism

Electromagnetism

Maxwell Equations, Wave Propagation and Emission

Tamer Bécherrawy

ISTE

WILEY

First published 2012 in Great Britain and the United States by ISTE Ltd and John Wiley & Sons, Inc.

ISTE Ltd
27-37 St George's Road
London SW19 4EU
UK

www.iste.co.uk

John Wiley & Sons, Inc.
111 River Street
Hoboken, NJ 07030
USA

www.wiley.com

Library of Congress Cataloging-in-Publication Data

Bécherrawy, Tamer.
Electromagnetism : Maxwell equations, wave propagation, and emission / Tamer Bécherrawy.
p. cm.
Includes bibliographical references and index.
ISBN 978-1-84821-355-5
1. Electromagnetism. 2. Maxwell equations. 3. Electromagnetic waves. 4. Field emission. I. Title.
QC670.B37 2012
537--dc23

2012009826

British Library Cataloguing-in-Publication Data
A CIP record for this book is available from the British Library
ISBN: 978-1-84821-355-5

Table of Contents

Preface

The scientific study of electric and magnetic forces started as two distinct sciences during the second half of the 18th Century. The concepts of electric and magnetic fields were introduced as independent constructs to facilitate the calculation of forces. However, after the discovery by Oersted in 1819 that an electric current produces a magnetic field, and the discovery by Faraday in 1831 that a variable magnetic field induces currents, it became clear that electric and magnetic fields are related and that they are very important physical concepts. In 1873, Maxwell unified electricity and magnetism in a single theory, called *electromagnetism*, based on four fundamental equations. An important prediction of this theory was the existence of electromagnetic waves that propagate with the speed of light. This prediction was confirmed experimentally by Hertz in 1887.

Thanks to the discovery of induction, the large-scale production of electricity became possible, opening the door to a new technological era in the second half of the 19th Century. The discovery of electromagnetic waves and the development of electronics generated a real revolution in telecommunications in the 20th Century with considerable economical, social, cultural and political impact.

The electromagnetic field, which is an association of the electric and magnetic fields, is a real physical object with energy, momentum, and angular momentum, which may be static or propagating as waves exactly like sound, elastic waves, or even particles. This is the first example of *field theories* in modern physics. It was followed by the discovery of the gravitational field in the framework of General Relativity and quantum fields in the framework of Quantum Electrodynamics and Quantum Chromodynamics. On the other hand, Maxwell's theory solved the very long-standing problem of the nature of light; it is an electromagnetic wave of short wavelength. Thus, Maxwell's work unified electricity, magnetism and optics in a single theory. Electromagnetic theory is in such complete agreement with experiments that any theory in conflict with it should be modified or abandoned.

The formulation of the electromagnetic theory was a major event in the history of physics in its incessant search to explain the maximum of phenomena with the minimum of basic principles. Furthermore, electromagnetism is the prototype of the so-called *gauge theories* in modern physics. They include the unification of electromagnetic and weak interactions by Glashow, Salam and Weinberg around 1967, Quantum Chromodynamics around 1973 and the so-called Grand Unification Theories that try to unify all interactions in Nature.

The electromagnetic theory posed two challenging problems, which produced real revolutions in physics and even in philosophy at the beginning of the 20th Century. The first was the disagreement of the propagation of light with the Galilean transformation, which is one of the basic principles of Classical Mechanics. This was shown by several experiments (namely Michelson's historical experiment) and it is fundamental since Maxwell's equations, which are obeyed by light as electromagnetic waves, are not covariant in the Galilean transformation. This contradiction was solved by the Special Theory of Relativity that modified the Galilean transformation, and had far-reaching consequences. The second problem was the understanding of the black body radiation and the discrete emission spectrum of atoms, which contradict both Classical Mechanics and the electromagnetic theory. Its solution led to the formulation of Quantum Theory. At present, the interaction of electromagnetic radiations with matter remains a very important subject both in theoretical physics and in various domains of applied physics.

Electromagnetism plays an important part in almost all branches of physics: atomic physics, molecular physics, solid-state physics, astrophysics, atmospheric physics, etc., and it even intervenes in chemistry and biology. In fact, almost all properties of matter are fundamentally electromagnetic on both the macroscopic scale and the atomic and molecular microscopic scale. On the other hand, electromagnetic waves play a fundamental part in the transfer of energy and information. Thus, a good understanding of electromagnetism is essential in any scientific activity and in the training of future physicists and engineers.

The purpose in writing this book is to study electromagnetism at the upper undergraduate level following teaching experience of several years. The goal is to understand the concept of electromagnetic fields, to obtain Maxwell's equations and to analyze some of their consequences regarding the propagation and emission of radiation.

Writing a book on electromagnetism is not an easy task for two reasons: the first is that the subject is so well established and so many excellent books already exist that one can expect originality only in didactical details: selection of topics, clear presentation of the material, choice of exercises, etc. The second is that electromagnetism is very connected to other subjects, namely quantum theory, relativity, properties of matter, and it has countless applications. Thus, it is hard to set the limits of the text.

Some authors prefer to start with Maxwell's equations as basic equations and then study time-independent phenomena and time-dependent phenomena. This approach is similar to starting Classical Mechanics with Newton's principles or, at a higher level, starting with Hamilton's principle and Lagrange equations. I think that the traditional approach, starting with the time-independent phenomena, is more pedagogical because of the mathematical complexity of the fields as functions of space and time, and the complexity of Maxwell's equations as partial differential equations for vector quantities. Thus, this text may be divided into four parts:

– The first part of seven chapters studies the time-independent electric and magnetic phenomena. This study goes beyond introductory electricity and magnetism by the use of vector calculus, differential and partial differential equations, etc. In this part, the basic concepts of electric and magnetic fields, energy and symmetries are analyzed, as well as the properties of dielectrics and magnetic matter. Conduction in solids is introduced, but we do not develop circuit analysis. In Chapter 5, some useful mathematical techniques (Legendre polynomials, Bessel's functions and multipole expansion) are introduced.

– The second part studies the time-dependent phenomena. It includes a detailed study of induction with some of its applications in Chapter 8 and the formulation of Maxwell's equations in Chapter 9.

– The third part studies the propagation effects. It includes a detailed study of electromagnetic waves in Chapter 10 (including propagation in dielectrics, in conductors and in plasmas, the quantization of radiation and its emission), reflection, interference, diffraction and diffusion in Chapter 11, and guided waves in Chapter 12.

– The fourth part includes Chapter 13 on the Special Theory of Relativity (including its applications to mechanics and electrodynamics), the motion of charged particles in electromagnetic fields (both non-relativistic and relativistic) in Chapter 14, and the emission of electromagnetic waves by antennas and particles in Chapter 15. The chapter on the Special Theory of Relativity is necessary as an introduction to the subject and for a better understanding of the electromagnetic theory.

Electromagnetism if one of the first physics courses in which vector calculus and partial differential equations are extensively used. The electromagnetic theory in vacuum requires one electric field and one magnetic field, and the electromagnetic theory in matter requires two more fields. All of them are vector fields. They may be represented by their 12 components measured with respect to convenient Cartesian axes. The four Maxwell's equations couple these components to the charge and current densities. It is unthinkable to handle these equations and analyze their consequences without the use of vector calculus. Only this analysis allows us to study electromagnetism independently of the used frame and to use curvilinear coordinates, which are very often more convenient to solve the equations. Thus, some knowledge of mathematical analysis (both real and complex) and vector calculus are assumed. The required mathematical techniques are introduced as the need arises. Appendix A summarizes the principal mathematical formulas, integrals and vector analysis.

I have tried to use clear notations by assigning similar symbols for the various physical quantities: a boldfaced symbol for a vector quantity, an italic symbol for a scalar quantity or a component of a vector quantity, an underlined symbol for a complex quantity, and script symbol for a curve, a surface, a volume and some special quantities. Physical quantities of the same type are referred to by symbols with different indexes: for instance, $\mathbf{F}_E$, $\mathbf{F}_M$, $f_{(ex)}$, etc., for the different types of force. The charge densities, per unit volume, per unit surface and per unit length are respectively q_v, q_s and q_L. To avoid confusion with the components of the electric field $\mathbf{E}$, the energy is designated by U (U_K for the kinetic energy, U_E for the electric energy, etc.). The frequency is represented by $\tilde{\nu}$, instead of the usual Greek symbol ν, to avoid confusing it with the velocity v.

A unit vector is often represented by $\mathbf{e}$, while the unit vectors of the axes are $\mathbf{e}_x$, $\mathbf{e}_y$ and $\mathbf{e}_x$. In order to write summations in a condensed form, the Cartesian coordinates x, y and z are sometimes designated by x_1, x_2 and x_3 respectively, and the components of a vector $\mathbf{V}$ by $V_1 \equiv V_x$, $V_2 \equiv V_y$ and $V_3 \equiv V_z$. The partial derivatives of $u(x, y, z, t)$ are represented by $\partial_x u$ for $\partial u/\partial x$, $\partial^2_{xt}u$ for $\partial^2 u/\partial x\,\partial t$, etc. We also use the notation $\partial_\alpha u$ for $\partial u/\partial x_\alpha$ and $\partial_\beta V_\alpha$ for $\partial V_\alpha/\partial x_\beta$ (α and $\beta = 1, 2, 3$) and occasionally $\dot{u}$ for $\partial u/\partial t$ or du/dt and $\ddot{u}$ for $\partial^2 u/\partial t^2$.

Some sections, indicated by an asterisk (*), have some difficulty and may be omitted without loss of continuity. At the end of each chapter, I have included numerous problems, which are ordered according to the sections of the chapter. The answers to most of the problems are given in a special addendum entitled *Answers to Some Problems*, which enables the student to check the results.

I hope that this text makes the subject more accessible for students, and that it is utilized as a good teaching tool for professors.

T. Bécherrawy
May 2012

List of Symbols

A	mass number
$\mathbf{A}$	vector potential
$\overline{\mathbf{A}}$	four-vector
$\mathbf{A}_{\mathrm{m}}$	magnetization vector potential
A_{μ}	four-vector potential
b	impact parameter in scattering
$\mathbf{B}$	magnetic induction field
$\mathbf{B}_{l}$	local magnetic field
$\mathbf{B}_{\mathrm{m}}$	magnetization magnetic field
B_{r}	remanent field
c	speed of light in vacuum
C	Curie constant
C	capacitance
$\mathcal{C}$	contrast or visibility factor
$\mathrm{C_i}(x)$	cosine integral
C_{ik}	coefficient of electric influence
C_{l}	capacitance per unit length
$\mathbf{D}$	electric displacement field
e	elementary charge
$\mathbf{e}$	unit vector
$\mathbf{e}_{\mathrm{p}}$	polarization direction
$\mathbf{E}$	electric field
$\mathcal{E}$	electromotive force (emf)
$\mathbf{E}_{\perp}$	normal component of $\mathbf{E}$
$\mathbf{E}_{//}$	tangential or parallel component of $\mathbf{E}$
$\mathbf{E}^{(+)}$	right-handed circularly polarized wave
$\mathbf{E}^{(-)}$	left-handed circularly polarized wave
E_{f}	Fermi energy
$\mathbf{E}_{\mathrm{g}}$	generating field
$\mathbf{E}_{\mathrm{H}}$	Hall electric field
$\mathbf{E}_{l}$	local field
$\mathcal{E}_{\mathrm{M}}$	magnetomotive force
E_{n}	energy levels
$\mathbf{E}_{\mathbf{p}}$	polarization field
$f(\theta)$	scattering amplitude
$\mathbf{f}_{\mathrm{g}}$	generating force
f_{R}	energy reflection factor
f_{T}	energy transmission factor
F	flux of particles
$\overline{\mathbf{F}}$	four-vector force
$\mathcal{F}_{\mathrm{dg}}(\phi)$	diffraction grating function
$\mathbf{F}_{\mathrm{M}}$	magnetic force
g	gyromagnetic ratio
h	Planck's constant
$\hbar$	Planck's reduced constant
G	gravitational constant
H	Hubble constant
$\mathbf{H}$	magnetic field (or magnetic excitation)
H_{c}	coercive field
i_{L}	critical angle (or limiting angle)
I	current intensity

$\mathcal{I}$ intensity of a wave
$\mathcal{Im}\ \underline{z}$ imaginary part of $\underline{z}$
$I_m(u)$ modified Bessel function
$\mathbf{j}'$ volume magnetization current density
$\mathbf{j}(\mathbf{r})$ current density
$\mathbf{j}_s$ surface current density
$\mathbf{j}'_s$ surface magnetization current density
J_μ four-vector current density
$\mathbf{j}_c$ conduction current density
$\mathbf{J}$ total angular momentum
$J_m(u)$ Bessel function of the first kind
k wave number
k_B Boltzmann's constant
$\mathbf{k}$ wave vector
K_o Coulomb force constant
l mean free path
L self-inductance (or inductance)
$L(x)$ Langevin function
$L_{\mu\alpha}$ Lorentz transformation matrix
$\mathbf{L}_a$ angular momentum of the atom
L_l inductance per unit length
$\mathbf{L}_v$ density of orbital angular momentum
m_e electron mass
$\mathbf{M}$ intensity of magnetization
M_{12} mutual inductance
$\mathcal{M}$ magnetic moment
$\mathcal{M}_a$ magnetic moment of the atom
M_r remanent magnetization
$\mathcal{M}_s$ intrinsic magnetic moment
n index of refraction
$\mathbf{n}$ normal unit vector
N number of neutrons in atoms
N_A Avogadro number
$N_m(u)$ Bessel function of the second kind (or Neumann function)
$\mathbf{p}$ dipole moment
p_r radiation pressure
P power
$\mathcal{P}$ degree of polarization
$\mathbf{P}$ polarization density (or polarization)
$\mathbf{P}$ relativistic momentum
$P_{(ED)}$ emitted power by an electric dipole
$\mathbf{P}_{EM,v}$ density of electromagnetic momentum
$P_{(MD)}$ emitted power by a magnetic dipole
$P_{(rad)}$ radiated power
P_μ four-vector momentum-energy
P_J Joule heat
$P_l(u)$ Legendre polynomials
$\mathbf{P}_{rad}$ momentum of the radiation
$P_l^{|m|}(u)$ associate Legendre functions
Q quality factor
q, Q charge
q'_s surface polarization charge density
q_v volume charge density
q'_v volume polarization charge density
$Q_{\alpha\beta}$ electric quadrupole moment
R resistance
$\widetilde{R}$ transposed matrix of R
$\mathcal{R}$ reflection coefficient
$\mathcal{R}$ reluctance
$R_{(ED)}$ electric dipole radiation resistance
$R_{(MD)}$ magnetic dipole radiation resistance
$\mathcal{Re}\ \underline{z}$ real part of $\underline{z}$
R_H Rydberg constant
R_{ji} matrix elements of R
r_n radius of Bohr orbits
$\mathcal{R}_p$ resolving power
R^{-1} inverse matrix of R
$\mathbf{s}$ intrinsic angular momentum or spin
$\mathbf{S}$ Poynting vector

$S_i(x)$	sine integral
$\mathbf{s}_V$	density of intrinsic angular momentum
T	period
$\mathcal{T}$	transmission coefficient
T_c	Curie temperature (or critical temperature)
T_{sc}	superconductivity critical temperature
T_W	Weiss temperature
$\mathcal{T}$	coefficient of transmission
u	wave function
U_{EM}	electromagnetic energy
U_{12}	interaction energy
U_E	electric potential energy
U_K	kinetic energy
U_M	magnetic energy
U_P	potential energy
V	electrostatic potential
V_M	scalar magnetic potential
v	speed of propagation
$v_{(p)}$	phase velocity
$\mathbf{v}_{(g)}$	group velocity
$\mathbf{v}_d$	drift velocity
V_μ	four-vector velocity
V_H	Hall potential
$V_\mathbf{p}$	polarization potential
W	work
W	relativistic energy
W_o	rest energy
$\overline{\mathbf{x}}$	four-coordinates
$x'_{m,j}$	j^{th} zero of the function $J'_m(x) \equiv dJ_m/dx$
$x_{m,j}$	j^{th} zero of the Bessel function $J_m(x)$
$Y_l^m(\theta, \varphi)$	spherical harmonics
Z	atomic number
Z	impedance
Z_o	impedance of vacuum

α_e	electronic polarizability
α_o	orientation polarizability
α_o	temperature coefficient of resistivity
α_{molar}	molar polarization
α_M	magnetic polarizability
β	velocity in unit of c
γ	time dilation factor
γ	Euler's constant
$\mathbf{\Gamma}$	moment of forces
δ	skin depth or penetration depth
$\delta_{m\,n}$	Kronecker symbols
$\delta(z - z_o)$	Dirac delta-function
$\delta^3(\mathbf{r} - \mathbf{r}_o)$	three-dimensional Dirac delta-function
Δ	Laplacian
ΔS^2	four-dimensional interval
Δt	time duration of a signal
Δx	space interval
$\Delta\omega$	frequency band
ε	electric permittivity
ε_o	permittivity of vacuum
ε_r	relative electric permittivity
η	attenuation coefficient
η	inclination factor in diffraction
θ	angle of incidence
θ'	angle of reflection
θ''	angle of refraction
θ_B	Brewster angle
λ	wavelength
μ	magnetic permeability
μ	absorption coefficient
μ_B	Bohr magneton
μ_N	nuclear magneton
μ_o	magnetic permeability vacuum
μ_r	relative magnetic permeability
$\tilde{\nu}$	frequency
Π	probability
ρ	resistivity
σ	conductivity
σ	cross section
$\sigma(\Omega)$	differential cross-section
τ	relaxation time
τ	proper time

τ	collision time
τ_c	characteristic time
$\tau_{\alpha\beta}$	Maxwell's tensor
ϕ	phase
Φ	flux
χ_E	electric susceptibility
χ_M	magnetic susceptibility
$\chi_{\alpha\beta}$	susceptibility tensor
$\Psi(\mathbf{r},t)$	wave function
ω	angular frequency
ω_c	characteristic angular frequency
ω_c	cyclotron frequency
ω_L	Larmor frequency
ω_p	plasma angular frequency
Ω	solid angle

$\Box$	d'Alembertian
∇	del, nabla or derivation operator
∇f	gradient of a scalar field f
$\nabla.\mathbf{V}$	divergence of a vector field **V**
$\nabla\times\mathbf{V}$	curl of a vector field **V**
$\partial_x f$	derivative with to *z* or partial derivative
$<f>$	average value of a function $f(x)$

Chapter 1

Prologue

Most physical phenomena are fundamentally electromagnetic. This makes electromagnetism a basic theory in many branches of physics (solid state physics, electronics, atomic and molecular physics, relativity, atmospheric physics, etc.) also in some other sciences and most technologies.

Although physics is an experimental science, it uses mathematical language to formulate its theories and its laws and analyze their consequences. Electromagnetism is a typical theory that is impossible to formulate without extensive use of vector analysis, differential equations, complex analysis, etc. The use of mathematics can even lead to the prediction of new physical laws and new phenomena (the discovery of the electromagnetic waves by Maxwell is a typical example). However, only experiments can decide whether a particular solution or prediction and even the whole theory is acceptable. Until now, no experiment has contradicted electromagnetic theory, both on the macroscopic scale and the microscopic scale (nuclear, atomic or molecular).

Although permanent magnets and electrification by rubbing were known in antiquity, scientific observations of magnetism began around 1270 with the French army engineer Pierre de Marincourt. The observation of electric effects began much later with the French botanist C. Dufay around 1734. Contrary to the gravitational interaction between masses, the large majority of objects around us are globally neutral and, if they become charged, they discharge rapidly in the surrounding air. The scientific study of electricity started with Franklin (1706-1790), Priestley (1733-1804), Cavendish (1731-1810), Coulomb (1736-1806), Laplace (1749-1827), Ampère (1775-1836), Gauss (1777-1855), and Poisson (1781-1840) who formulated the laws of electricity and magnetism. Faraday (1791-1867) introduced the notions of *influence* and *fields* and discovered *electric induction*, which allowed the large-scale production of electricity. Electricity and magnetism were unified in a single theory by Maxwell in 1864. This long itinerary led to the present technological era with the considerable influence of electromagnetism and its consequences on our industrial, economical and cultural environment.

In this chapter, we introduce some basic mathematical methods and some general invariances and symmetries that we use in the formulation of any theory and especially electromagnetic theory.

1.1. Scalars and vectors

The basic elementary concepts in the formulation of physical theories are *position* and *time*. The position is specified by the coordinates with respect to a reference frame *Oxyz*, supported by a material body and represented by an origin *O* and three mutually orthogonal axes. Although these concepts seem to be simple, their analysis poses deep practical and philosophical questions even in classical mechanics. In modern physics, their analysis has been one of the corner-stones of the special theory of relativity (see Chapter 13), general relativity, and quantum theory.

Some physical quantities are determined by a single algebraic quantity with no characteristic orientation. Mass, time, temperature, and electric charge are examples of such quantities; these are *scalar* quantities. They may be strictly positive (mass, pressure, etc.), positive or negative (position along an axis, potential energy, electric charge, etc.), or even complex (wave function, impedance, etc.). Other physical quantities **A** are specified, each one by a positive *magnitude A* and an *orientation*; these are said to be *vector quantities*. Displacement, velocity, acceleration, force, electric field, magnetic field, etc., are examples of vector quantities. A more precise definition of a vector quantity is given in section 1.2.

A vector **A** is conveniently specified by its Cartesian components A_x, A_y and A_z with respect to a frame *Oxyz* (Figure 1.1a). We may write $\mathbf{A} = A_x\mathbf{e}_x + A_y\mathbf{e}_y + A_z\mathbf{e}_z$, where $\mathbf{e}_x$, $\mathbf{e}_y$ and $\mathbf{e}_z$ are the unit vectors of the axes *Ox*, *Oy* and *Oz*; they are the *basis vectors* of the reference frame *Oxyz*. To simplify the writing of summations, we use the numbers 1, 2 and 3 instead of *x*, *y* and *z* to label the components and we write

$$\mathbf{A} = A_1\mathbf{e}_1 + A_2\mathbf{e}_2 + A_3\mathbf{e}_3 = \Sigma_\alpha A_\alpha \mathbf{e}_\alpha \qquad \alpha = 1, 2, \text{and } 3 \qquad [1.1]$$

The component A_1, for instance, is the projection of **A** on the axis *Ox*. It is well known that the decomposition [1.1] is unique.

The *product* $k\mathbf{A}$ of a scalar k and a vector $\mathbf{A}$ is the vector $k\mathbf{A}$ parallel to $\mathbf{A}$ and of magnitude k times the magnitude of $\mathbf{A}$. The components of $k\mathbf{A}$ are simply those of $\mathbf{A}$ multiplied by k. The *resultant* (or *sum*) $(\mathbf{A} + \mathbf{B})$ of two vectors $\mathbf{A}$ and $\mathbf{B}$ is defined by the usual parallelogram rule (Figure 1.1b). The components of $(\mathbf{A} + \mathbf{B})$ is simply the sum of the corresponding components of $\mathbf{A}$ and $\mathbf{B}$:

$$k\mathbf{A} = \Sigma_\alpha\, kA_\alpha\, \mathbf{e}_\alpha \qquad \text{and} \qquad \mathbf{A} + \mathbf{B} = \Sigma_\alpha\, (A_\alpha + B_\alpha)\, \mathbf{e}_\alpha\,. \qquad [1.2]$$

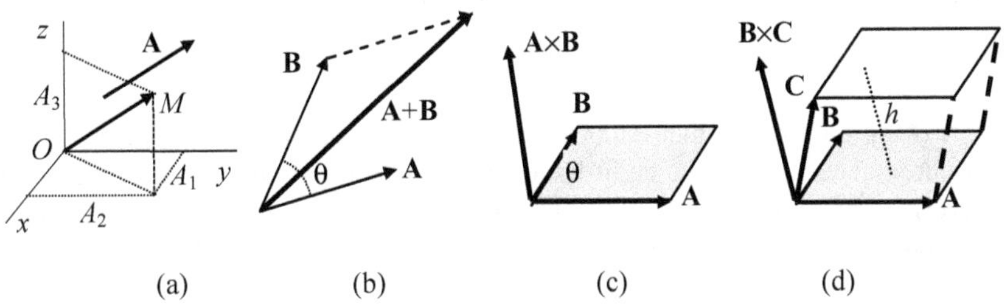

Figure 1.1. *a) Cartesian components of a vector. b) Sum of two vectors* **A** *and* **B**. *c) The cross product* $\mathbf{A} \times \mathbf{B}$. *d) The triple scalar product* $(\mathbf{A}{\times}\mathbf{B}).\mathbf{C}$

Scalar product

The *scalar product* (or *dot product*) of two vectors $\mathbf{A}$ and $\mathbf{B}$, written as $\mathbf{A.B}$, is the product of their magnitudes and the cosine of their angle θ. Thus, the scalar product of a vector $\mathbf{A}$ by itself, written as $\mathbf{A}^2$, is the square of its magnitude, $\mathbf{A}^2 = A^2$. We note that the scalar product is linear in $\mathbf{A}$ and $\mathbf{B}$. In the case of the basis vectors, we have $\mathbf{e}_\alpha{}^2 = 1$ and $\mathbf{e}_\alpha.\mathbf{e}_\beta = 0$ if $\alpha \neq \beta$. Using the *Kronecker* symbols $\delta_{\alpha\beta}$, we may write:

$$\mathbf{e}_\alpha.\mathbf{e}_\beta = \delta_{\alpha\beta}, \qquad \text{where } \delta_{\alpha\beta} = 1 \text{ if } \alpha = \beta \text{ and } \delta_{\alpha\beta} = 0 \text{ if } \alpha \neq \beta. \qquad [1.3]$$

This allows us to write the scalar product of $\mathbf{A}$ and $\mathbf{B}$ in terms of their components:

$$\mathbf{A.B} = BA\cos\theta = (\Sigma_\alpha\, A_\alpha \mathbf{e}_\alpha).(\Sigma_\beta B_\beta \mathbf{e}_\beta) = \Sigma_{\alpha\beta}\, A_\alpha B_\beta\, (\mathbf{e}_\alpha.\mathbf{e}_\beta) = \Sigma_{\alpha\beta}\, A_\alpha B_\beta\, \delta_{\alpha\beta}$$
$$= \Sigma_\alpha\, A_\alpha B_\alpha. \qquad [1.4]$$

The unitary vector $\mathbf{e}_\mathrm{B}$ in the direction of a vector $\mathbf{B}$ is obtained by dividing $\mathbf{B}$ by its magnitude

$$\mathbf{e}_\mathrm{B} = \mathbf{B}/B\,, \qquad \text{i.e.} \qquad \mathbf{B} = B\,\mathbf{e}_\mathrm{B}. \qquad [1.5]$$

If a vector $\mathbf{A}$ forms an angle θ with $\mathbf{B}$, the projection of $\mathbf{A}$ on $\mathbf{B}$ is

$$A_B = (\mathbf{A}.\mathbf{e}_B) = A \cos\theta = (\mathbf{A}.\mathbf{B})/B. \quad [1.6]$$

A may be written as $\mathbf{A} = \mathbf{A}_{/\!/} + \mathbf{A}_\perp$, where $\mathbf{A}_{/\!/}$ is parallel to **B** and $\mathbf{A}_\perp$ is normal to **B**:

$$\mathbf{A}_{/\!/} = A_B\,\mathbf{e}_B = (\mathbf{A}.\mathbf{B})\,\mathbf{B}/B^2 \quad \text{and} \quad \mathbf{A}_\perp = \mathbf{A} - \mathbf{A}_{/\!/} = \mathbf{A} - (\mathbf{A}.\mathbf{B})\,\mathbf{B}/B^2. \quad [1.7]$$

Cross product

The *cross product* (or *vector product*), designated by $\mathbf{A} \times \mathbf{B}$, is the vector

$$\mathbf{V} = \mathbf{A} \times \mathbf{B} = AB \sin\theta\ \mathbf{n}, \quad [1.8]$$

where **n** is the unit vector that is normal to the plane containing the vectors **A** and **B** and oriented according to the right-hand rule: if the thumb and the forefinger are in the directions of **A** and **B**, respectively, the middle finger points in the direction of $\mathbf{A}\times\mathbf{B}$ (Figure 1.1c). Note that the area of the parallelogram of sides **A** and **B** is just the magnitude of $\mathbf{A} \times \mathbf{B}$.

Contrary to the scalar product, the cross product $\mathbf{A} \times \mathbf{B}$ is not commutative: it is odd in the exchange of the vectors: $(\mathbf{A} \times \mathbf{B}) = -(\mathbf{B} \times \mathbf{A})$. The cross product of two parallel (or antiparallel) vectors is equal to zero because $\theta = 0$ (or $\theta = \pi$). It may be verified that $\mathbf{e}_\alpha \times \mathbf{e}_\beta = \mathbf{e}_\gamma$, where (α, β, γ) is a circular permutation of (1, 2, 3), that is

$$\mathbf{e}_1 \times \mathbf{e}_2 = \mathbf{e}_3, \quad \mathbf{e}_2 \times \mathbf{e}_3 = \mathbf{e}_1, \quad \mathbf{e}_3 \times \mathbf{e}_1 = \mathbf{e}_2. \quad [1.9]$$

This allows us to write the components $(\mathbf{A}\times\mathbf{B})_\alpha = A_\beta B_\gamma - A_\gamma B_\beta$, that is,

$$(\mathbf{A}\times\mathbf{B})_1 = A_2B_3 - A_3B_2, \quad (\mathbf{A}\times\mathbf{B})_2 = A_3B_1 - A_1B_3, \quad (\mathbf{A}\times\mathbf{B})_3 = A_1B_2 - A_2B_1. \quad [1.10]$$

We may also write the cross product as a determinant

$$(\mathbf{A} \times \mathbf{B}) = \begin{bmatrix} \mathbf{e}_1 & \mathbf{e}_2 & \mathbf{e}_3 \\ A_1 & A_2 & A_3 \\ B_1 & B_2 & B_3 \end{bmatrix} \quad [1.11]$$

Triple scalar product

The so-called *triple scalar product* of three vectors is defined by $U = (\mathbf{A} \times \mathbf{B}).\mathbf{C}$. It is invariant in a circular permutation of the vectors and odd in the exchange of any two vectors. It can be interpreted as the volume of the parallelepiped of sides **A**, **B**, and **C** with a positive sign if the trihedron **A**, **B**, **C**, taken in this order, is right-handed and a negative sign otherwise (Figure 1.1d). It may be expressed as the determinant of the components

$$(\mathbf{A}\times\mathbf{B})\,.\,\mathbf{C} = \begin{bmatrix} A_1 & A_2 & A_3 \\ B_1 & B_2 & B_3 \\ C_1 & C_2 & C_3 \end{bmatrix}. \qquad [1.12]$$

Differentiation of vectors

The differentiation rules for sums and products of functions hold for vectors. To simplify the notation, the derivatives $\partial f/\partial t$, $\partial^2 f/\partial x\partial t$, etc., are written as $\partial_t f$, $\partial^2{}_{xt} f$, etc. If a vector $\mathbf{A}$ depends on time, the components A_α depend on time. Thus, if the basis $\mathbf{e}_\alpha$ is time-independent, the differential of $\mathbf{A}$ and its derivative with respect to time are

$$d\mathbf{A} = \Sigma_\alpha\, dA_\alpha\, \mathbf{e}_\alpha = \Sigma_\alpha\, \partial_t A_\alpha\, dt\, \mathbf{e}_\alpha \qquad \text{and} \qquad d\mathbf{A}/dt = \Sigma_\alpha\, \partial_t A_\alpha\, \mathbf{e}_\alpha. \qquad [1.13]$$

If the basis vectors depend on time, we must write

$$d\mathbf{A}/dt = \Sigma_\alpha\, (\partial_t A_\alpha)\, \mathbf{e}_\alpha + \Sigma_\alpha\, A_\alpha\, \partial_t \mathbf{e}_\alpha. \qquad [1.14]$$

1.2. Effect of rotations on scalars and vectors

The choice of the origin O and the orientation of the axes of reference are completely arbitrary and observers in different places and different times often use different reference frames, different origins of time and even moving frames, relative to each other. Although these observers may find different coordinates and different time for any given event, it is evident that they must find the same laws for any physical phenomenon (otherwise, physics would not be a science at all). This is known as the *relativity principle*. Thus, it is necessary to know how physical quantities are related in different frames $Oxyz$ and $O'x'y'z'$. A physical quantity that depends on position $\overrightarrow{OM} \equiv \mathbf{r}$ and time t in $Oxyz$ is a *field*, which we write as $f(\mathbf{r}, t)$ or $f(x_\alpha, t)$, where x_α is a shorthand notation for the coordinates x, y and z of $\mathbf{r}$.

We consider two parallel frames $Oxyz$ and $O'x'y'z'$, such that the origin O' has a fixed position $\overrightarrow{OO'} \equiv \mathbf{r}_o$ (of coordinates $x_{o,\alpha}$ with respect to $Oxyz$), the position of an event $\overrightarrow{O'M} \equiv \mathbf{r}'$ with respect to $O'x'y'z'$ is related to its position $\mathbf{r}$ with respect to $Oxyz$ by the equation

$$\overrightarrow{OM} = \overrightarrow{OO'} + \overrightarrow{O'M}, \qquad \text{thus} \quad \mathbf{r} = \mathbf{r}' + \mathbf{r}_o, \quad (\text{i.e. } x_\alpha = x'_\alpha + x_{o,\alpha})\,. \qquad [1.15]$$

This is a simple translation in space. Any field, whatever its nature, must be specified by equal values $f(\mathbf{r}, t)$ and $f'(\mathbf{r}', t)$ in these frames, thus

$$f'(\mathbf{r}', t) = f(\mathbf{r}, t), \qquad \text{i.e. } f'(\mathbf{r}', t) = f(\mathbf{r}' + \mathbf{r}_o, t). \qquad [1.16]$$

We consider now the more interesting case of reference frames related by rotations. The basis vectors $\mathbf{e}'_\beta$ of the frame $O'x'y'z'$ are related to the basis $\mathbf{e}_\alpha$ of the frame $Oxyz$ by a linear transformation

$$\mathbf{e}_\alpha = \Sigma_\beta R_{\beta\alpha} \mathbf{e}'_\beta, \qquad \text{and} \qquad \mathbf{e}'_\alpha = \Sigma_\beta R^{-1}{}_{\beta\alpha} \mathbf{e}_\beta, \qquad [1.17]$$

where R is a 3×3 matrix and R^{-1} is its inverse. Writing $\mathbf{A} = \Sigma_\alpha A_\alpha \mathbf{e}_\alpha$ and expressing the $\mathbf{e}_\alpha$ in terms of the $\mathbf{e}'_\beta$ by using [1.17], we find

$$\mathbf{A} = \Sigma_{\alpha\beta} A_\alpha R_{\beta\alpha} \mathbf{e}'_\beta. \qquad [1.18]$$

Comparing with $\mathbf{A} = \Sigma_\beta A'_\beta \mathbf{e}'_\beta$, we deduce that

$$A'_\beta = \Sigma_\alpha R_{\beta\alpha} A_\alpha, \qquad \text{and} \qquad A_\beta = \Sigma_\alpha R^{-1}{}_{\beta\alpha} A'_\alpha. \qquad [1.19]$$

In particular, these transformations hold for the coordinates that are the components of the vector $\mathbf{r}$. Using vector notation, we write

$$\mathbf{r}' = R\,\mathbf{r}, \qquad \mathbf{A}' = R\mathbf{A}, \qquad \text{and} \qquad \mathbf{r} = R^{-1}\mathbf{r}', \quad \mathbf{A} = R^{-1}\mathbf{A}'. \qquad [1.20]$$

The transformation R conserves the scalar products (and in particular the magnitude of vectors) if it is *orthogonal* (that is, its transposed $\tilde{R}$ is equal to its inverse). In other words, it verifies the condition

$$\tilde{R}R = R\tilde{R} = I, \qquad \text{i.e. } R_{\beta\alpha} R_{\beta\gamma} = \delta_{\alpha\gamma}, \qquad [1.21]$$

where I is the unit matrix (that is, it has $\delta_{\alpha\gamma}$ as matrix elements).

A physical quantity f is a *scalar* if it is invariant in any rotation R. If it is a *scalar field*, it must verify the condition

$$f(\mathbf{r}) = f'(\mathbf{r}'), \qquad \text{where} \quad \mathbf{r}' = R\,\mathbf{r}. \qquad [1.22]$$

This is the case of $\mathbf{r}^2$ or any scalar function of $\mathbf{r}^2$ (i.e. $r = \sqrt{\mathbf{r}^2}$).

The three quantities A_α are the components of a *vector* $\mathbf{A}$ if they transform according to [1.19], exactly as the coordinates x_α in any rotation R. The functions $A_\alpha(\mathbf{r})$ are the components of a *vector field* $\mathbf{A}(\mathbf{r})$, if they transform according to

$$A'_\alpha(\mathbf{r}') = R_{\alpha\beta} A_\beta(\mathbf{r}), \qquad \text{where} \quad \mathbf{r}' = R\,\mathbf{r}. \qquad [1.23]$$

1.3. Integrals involving vectors

Circulation of a vector field

The circulation of a vector field **E** in a displacement $d\mathbf{r} = dx\,\mathbf{e}_x + dy\,\mathbf{e}_y + dz\,\mathbf{e}_z$ is $\mathbf{E}.d\mathbf{r}$ (Figure 1.2a). The work of a force is a typical example of circulation. The circulation of **E** along a curve $\mathcal{C}$ going from **r** to $\mathbf{r}_o$ is the *line integral*

$$C = \int_{\mathcal{C}} d\mathbf{r}.\mathbf{E} = \int_{\mathbf{r}}^{\mathbf{r}_o} d\mathbf{r}.\mathbf{E} = \int_{\mathbf{r}}^{\mathbf{r}_o} dr\, E_{//} = \int_{\mathbf{r}}^{\mathbf{r}_o} [dx\, E_1 + dy\, E_2 + dz\, E_3], \qquad [1.24]$$

where $d\mathbf{r}$ is the infinitesimal displacement along the path $\mathcal{C}$ and $E_{//}$ is the tangential component of **E**. The circulation is a scalar quantity defined as the limit of the sum of the scalar products $d\mathbf{r}_n.\mathbf{E}_n$ of the infinitesimal elements $d\mathbf{r}_n$ of $\mathcal{C}$ and the fields $\mathbf{E}_n$ at these elements. To calculate the integral in the general case, a parametric representation of the curve $x = x(u)$, $y = y(u)$ and $z = z(u)$ may be used, where u is any parameter with u and u_o corresponding to the extreme positions **r** and $\mathbf{r}_o$. The components E_α become functions of u and the circulation becomes an integral over u

$$C = \int_{\mathcal{C}} d\mathbf{r}.\mathbf{E} = \int_{u}^{u_o} du\, [\,(dx/du)\, E_1(u) + (dy/du)\, E_2(u) + (dz/du)\, E_3(u)\,]. \qquad [1.25]$$

If the field has a uniform tangential component $E_{//} = E$ along the path $\mathcal{C}$, its circulation is $C = EL_{\mathcal{C}}$, where $L_{\mathcal{C}}$ is the length of the path. On the other hand, if $\mathbf{E} = E\,\mathbf{e}_z$ is uniform in the direction Oz, its circulation is $\int_{\mathcal{C}} d\mathbf{r}.\mathbf{E} = \int_z^{z_o} dz\, E = (z_o - z)E$.

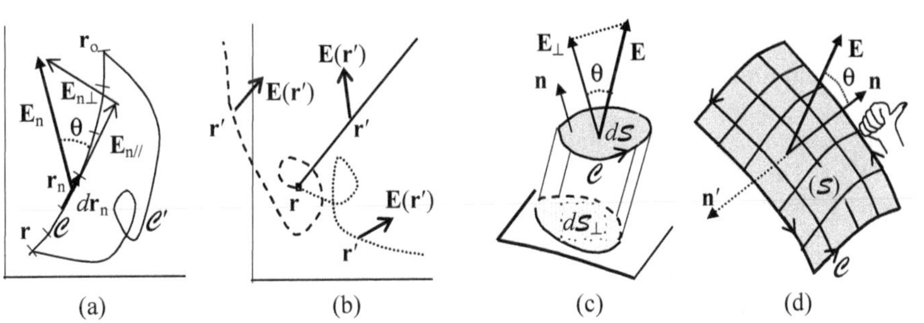

Figure 1.2. *a) Circulation of* **E** *along a path* $\mathcal{C}$ *going from* **r** *to* $\mathbf{r}_o$. If **E** *is conservative, this circulation is equal to* $V(\mathbf{r}) - V(\mathbf{r}_o)$ for any $\mathcal{C}$. *b) Setting* $V(\infty) = 0$, $V(\mathbf{r})$ *is the circulation of* **E** *along an arbitrary path going from* **r** *to infinity. c) The flux of* **E** *through an infinitesimal surface* $d\mathcal{S}$. *d) The flux through an open surface* $\mathcal{S}$ *bounded by an oriented contour* $\mathcal{C}$

Flux of a vector field

Consider the integral over a surface $\mathcal{S}$

$$\Phi = \iint_{\mathcal{S}} d\mathcal{S}\, \mathbf{n}.\mathbf{E}(\mathbf{r}) = \iint_{\mathcal{S}} d\mathcal{S}\, E(\mathbf{r}) \cos\theta = \iint_{\mathcal{S}} d\mathcal{S}\, E_{\perp}(\mathbf{r}), \qquad [1.26]$$

where $\mathbf{E}(\mathbf{r})$ is a vector field, $\mathbf{n}$ is the unit vector normal to the surface $\mathcal{S}$ at the running point $\mathbf{r}$, θ is the angle of $\mathbf{E}$ with $\mathbf{n}$ and $E_{\perp}$ is the component of $\mathbf{E}$ in the direction of $\mathbf{n}$. This integral is the *flux* of $\mathbf{E}$ through $\mathcal{S}$. The flux $d\Phi = d\mathcal{S}\, \mathbf{n}.\mathbf{E}(\mathbf{r})$ through the infinitesimal area $d\mathcal{S}$ is a scalar quantity and so is the flux. Note that we may write $d\Phi = d\mathcal{S}\, E_{\perp}(\mathbf{r})$, where $E_{\perp}(\mathbf{r})$ is the normal component of $\mathbf{E}$, or $d\Phi = d\mathcal{S}_{\perp}\, E(\mathbf{r})$, where $d\mathcal{S}_{\perp}$ is the projection of $d\mathcal{S}$ on the normal plane to $\mathbf{E}$ (Figure 1.2c). $d\Phi$ is positive or negative, depending on whether θ is acute or obtuse, and it vanishes if $\mathbf{E}$ is tangent to $\mathcal{S}$. Note also that $\mathbf{n}$ has two possible orientations; by changing the direction of $\mathbf{n}$, we change the sign of Φ. In the case of an open surface, which is bounded by an oriented closed curve $\mathcal{C}$, we choose $\mathbf{n}$ according to the right-hand rule (Figure 1.2d). In the case of a closed surface $\mathcal{S}$, we choose $\mathbf{n}$ oriented outward; Φ is then the outgoing flux.

The flux is additive both for the vector field and for the area. In the particular case of a field having a uniform component in the direction of $\mathbf{n}$, its flux is $\Phi = B_{\perp}\mathcal{S}$. Another physically interesting case is that of a radial field $\mathbf{E} = Kq\mathbf{r}/r^3$ of a charge q. Its flux through a closed surface $\mathcal{S}$ is $\Phi = Kq \iint_{\mathcal{S}} d\mathcal{S}\, \mathbf{n}.\mathbf{r}/r^3 = Kq\Omega$, where Ω is the solid angle of the cone, whose apex is at q and which is subtended by $\mathcal{S}$: it is equal to 4π if q is inside $\mathcal{S}$ and equal to 0 if q is outside $\mathcal{S}$.

1.4. Gradient and curl, conservative field and scalar potential

The work of a force $\mathbf{F}$ acting on a particle of mass m in a displacement $d\mathbf{r}$ is $dW = \mathbf{F}.d\mathbf{r}$. This work is transformed into kinetic energy if no other force acts on the particle. Conversely, to displace the particle without acquiring kinetic energy dU_K, an external agent must exert a force $\mathbf{F}' = -\mathbf{F}$ and supply a work $dW' = -\mathbf{F}.d\mathbf{r}$. If the force is *conservative*, this work is transformed into potential energy dU_P of the particle in the field of force $\mathbf{F}$. This analysis can be repeated for any vector field $\mathbf{E}$. Its circulation along a path $\mathcal{C}$ going from $\mathbf{r}$ to $\mathbf{r}_o$ depends in general on $\mathbf{r}$ and $\mathbf{r}_o$ and also on the path $\mathcal{C}$. Its circulation on a closed path is not necessarily equal to zero. The differential form $dx\, E_1 + dy\, E_2 + dz\, E_3$ is a total differential if the components E_α are the partial derivatives of a scalar function $-V$ where V is called the *scalar potential* corresponding to the field of force $\mathbf{F}$. Then, we have $E_1 = -\partial_1 V$, $E_2 = -\partial_2 V$, and $E_3 = -\partial_3 V$, which we write in the vector form

$$\mathbf{E} = -\nabla V, \qquad \text{where} \qquad \nabla = \mathbf{e}_1\,\partial_1 + \mathbf{e}_2\partial_2 + \mathbf{e}_3\partial_3. \qquad [1.27]$$

The vector differential operator ∇ is called *nabla* or *del* and ∇V is the gradient of V. It may be shown that the gradient of any scalar function V is a vector. In this case the circulation [1.24] becomes

$$C = \int_{\mathcal{C}} d\mathbf{r}.\mathbf{E} = \int_{\mathbf{r}}^{\mathbf{r}_o} d\mathbf{r}.\mathbf{E} = -\int_{\mathbf{r}}^{\mathbf{r}_o} d\mathbf{r}.\nabla V = -\int_{\mathbf{r}}^{\mathbf{r}_o} dV = V(\mathbf{r}) - V(\mathbf{r}_o). \qquad [1.28]$$

In this special case, the circulation between two points is equal to the drop of the potential. It depends only on the points $\mathbf{r}$ and $\mathbf{r}_o$ for any path $\mathcal{C}$ connecting these points (Figure 1.2a). In the case of a closed path $\mathcal{C}$ ($\mathbf{r} \equiv \mathbf{r}_o$), the circulation vanishes. We say that the field $\mathbf{E}$ is *conservative.* For instance, in the case of a uniform field $\mathbf{E}$, the potential is $V = -\mathbf{E}.\mathbf{r} + V_o$ and in the case $\mathbf{E} = Kq\mathbf{r}/r^3$, $V = Kq/r + V_o$, where V_o is an arbitrary constant. In the last case it is convenient to assume that V vanishes at infinity, hence $V_o = 0$, and we may interpret $V(\mathbf{r})$ as the circulation of $\mathbf{E}$ along an arbitrary path going from $\mathbf{r}$ to infinity (Figure 1.2b). In the case where $\mathbf{E}$ is a conservative field of force $\mathbf{F}$, we may write $\mathbf{F} = -\nabla U_P$, where U_P is the *potential energy*. The work of $\mathbf{F}$ along a path $\mathcal{C}$ going from $\mathbf{r}$ to $\mathbf{r}_o$ is $\int_{\mathbf{r}}^{\mathbf{r}_o} d\mathbf{r}.\mathbf{F} = U_P(\mathbf{r}) - U_P(\mathbf{r}_o)$ and the work of $\mathbf{F}$ along a closed path vanishes.

To know whether a vector field $\mathbf{E}$ is conservative, we do not have to evaluate the circulation on all imaginable paths. We may use the important property that the partial derivatives of a function are independent of the order of differentiation. If $\mathbf{E}$ is conservative (that is, $\mathbf{E} = -\nabla V$), the equation $\partial_\alpha\partial_\beta V = \partial_\beta\partial_\alpha V$ may be written as $\partial_\alpha E_\beta - \partial_\beta E_\alpha = 0$. Using the differential vector operator ∇, we define the vector

$$\text{curl}\,\mathbf{E} \equiv \nabla\times\mathbf{E} = \begin{bmatrix} \mathbf{e}_1 & \mathbf{e}_2 & \mathbf{e}_3 \\ \partial_1 & \partial_2 & \partial_3 \\ E_1 & E_2 & E_3 \end{bmatrix}$$
$$= (\partial_2 E_3 - \partial_3 E_2)\,\mathbf{e}_1 + (\partial_3 E_1 - \partial_1 E_3)\,\mathbf{e}_2 + (\partial_1 E_2 - \partial_2 E_1)\,\mathbf{e}_3. \qquad [1.29]$$

A vector field $\mathbf{E}$ is conservative if its curl is identically equal to 0, and it may be shown that the converse is true: if $\nabla \times \mathbf{E} = 0$, $\mathbf{E}$ is conservative. In this case, we may define a potential V at each point $\mathbf{r}$ (see section A.7 in Appendix A)

Even if a vector field $\mathbf{A}$ is non-conservative, *Stokes' theorem* (see section A.8 of Appendix A) allows the expression of the circulation of $\mathbf{A}$ along a closed path $\mathcal{C}$ as the flux of $\nabla\times\mathbf{A}$ through any surface $\mathcal{S}$ bounded by $\mathcal{C}$

$$\int_{\mathcal{C}} d\mathbf{r}.\mathbf{A} = \iint_{\mathcal{S}} d\mathcal{S}\,\mathbf{n}.(\nabla\times\mathbf{A}). \qquad [1.30]$$

Note that the normal **n** is oriented according to the right-hand rule. We see from this theorem that, in the special case of a conservative field ($\nabla \times \mathbf{A} = 0$), its circulation along any closed path $\mathcal{C}$ vanishes and this is the definition of a conservative field.

1.5. Divergence, conservative flux, and vector potential

In general, the flux of a vector field **B** through the surfaces $\mathcal{S}$ bounded by a given closed contour $\mathcal{C}$ depends on the special choice of $\mathcal{S}$ and the flux through a closed surface is not necessarily equal to zero. We define the *divergence* of **B** as

$$\nabla.\mathbf{B} = \partial_1 B_1 + \partial_2 B_2 + \partial_3 B_3. \qquad [1.31]$$

It may be shown (see section A.7 of Appendix A) that, if $\mathbf{B} = \nabla \times \mathbf{A}$, its divergence vanishes ($\nabla.\mathbf{B} = 0$) and conversely, if $\nabla.\mathbf{B} = 0$, we may write

$$\mathbf{B} = \nabla \times \mathbf{A}. \qquad [1.32]$$

A is the *vector potential.* In fact, there are an infinite number of vectors **A** that correspond to the same **B**. They differ by a gradient term

$$\mathbf{A}' = \mathbf{A} + \nabla f \qquad [1.33]$$

because $\nabla \times \nabla f \equiv 0$. The relation [1.33] is called *gauge transformation.*

Gauss-Ostrogradsky's theorem (see section A.9 of Appendix A) allows the expression of the flux of any vector field **B** through a closed surface $\mathcal{S}$ as the integral of $\nabla.\mathbf{B}$ over the volume $\mathcal{V}$ enclosed by $\mathcal{S}$

$$\iint_{\mathcal{S}} d\mathcal{S}\, \mathbf{n}.\mathbf{B} = \iiint_{\mathcal{V}} d\mathcal{V}\; \nabla.\mathbf{B}. \qquad [1.34]$$

Note that to apply this theorem, the unit normal vector **n** must point outward $\mathcal{S}$. We deduce that, if $\nabla.\mathbf{B} = 0$, the flux of **B** through any closed surface $\mathcal{S}$ vanishes. We say that **B** has a *conservative flux*. On the other hand, if $\nabla.\mathbf{B} > 0$ at a point M, the flux of **B** outgoing from any surface surrounding M is positive; thus, the field is divergent from M. On the contrary, if $\nabla.\mathbf{B} < 0$, this flux is negative and **B** is convergent at M.

1.6. Other properties of the vector differential operator

Here are some useful properties of the operator ∇ acting on scalar fields $f(\mathbf{r})$ and $g(\mathbf{r})$ and on vector fields $\mathbf{A}(\mathbf{r})$ and $\mathbf{B}(\mathbf{r})$:

$$\nabla (f+g) = \nabla f + \nabla g, \qquad [1.35]$$
$$\nabla (fg) = g\,\nabla f + f\,\nabla g, \qquad [1.36]$$
$$\nabla (\mathbf{A} . \mathbf{B}) = (\mathbf{A} .\nabla)\,\mathbf{B} + (\mathbf{B} .\nabla)\,\mathbf{A} + \mathbf{A} \times (\nabla \times \mathbf{B}) + \mathbf{B} \times (\nabla \times \mathbf{A}), \qquad [1.37]$$
$$\nabla .(f\,\mathbf{A}) = f\,(\nabla . \mathbf{A}) + \mathbf{A} .\nabla f, \qquad [1.38]$$
$$\nabla .(\mathbf{A} \times \mathbf{B}) = \mathbf{B}.(\nabla \times \mathbf{A}) - \mathbf{A} .(\nabla \times \mathbf{B}), \qquad [1.39]$$
$$\nabla \times (f\mathbf{A}) = f\,(\nabla \times \mathbf{A}) + (\nabla f) \times \mathbf{A}, \qquad [1.40]$$
$$\nabla \times (\mathbf{A} \times \mathbf{B}) = \mathbf{A}\,(\nabla .\mathbf{B}) - \mathbf{B}\,(\nabla . \mathbf{A}) + (\mathbf{B} .\nabla)\,\mathbf{A} - (\mathbf{A} .\nabla)\,\mathbf{B}, \qquad [1.41]$$
$$\Delta (f g) = f\,\Delta g + 2\,(\nabla f).(\nabla g) + g\,\Delta f, \qquad [1.42]$$

where $(\mathbf{A}.\nabla) = \Sigma_\alpha A_\alpha \partial_\alpha = A_1 \partial_1 + A_2 \partial_2 + A_2 \partial_3$ is a scalar operator.

The successive application of ∇ on scalar and vector fields is very useful in physics. In Cartesian coordinates, if we evaluate the divergence of the gradient of a scalar function, we find

$$\nabla .(\nabla f) = \Sigma_\alpha\, \mathbf{e}_\alpha\, \partial_\alpha .[\Sigma_\beta\, \mathbf{e}_\beta\, \partial_\beta f\,] = \Sigma_{\alpha\beta}\, (\mathbf{e}_\alpha .\mathbf{e}_\beta) \partial^2{}_{\alpha\beta} f = \Sigma_\alpha\, \partial^2{}_{\alpha\alpha} f \equiv \Delta f, \qquad [1.43]$$

where the operator Δ, called *Laplacian*, is defined by

$$\Delta \equiv \nabla^2 = \partial^2{}_{xx} + \partial^2{}_{yy} + \partial^2{}_{zz}. \qquad [1.44]$$

As ∇ is a vector operator, the Laplacian is a scalar operator. Acting on a scalar field, it gives a scalar field and, acting on a vector field, it gives the vector field

$$\Delta \mathbf{A} = \Sigma_{\alpha\beta}\, \nabla_\alpha \nabla_\alpha (A_\beta\, \mathbf{e}_\beta) = \Sigma_\beta\, (\Delta A_\beta)\, \mathbf{e}_\beta. \qquad [1.45]$$

Thus, in Cartesian coordinates (and only in these coordinates), the components of $\Delta \mathbf{A}$ are simply ΔA_α. Other useful relationships may be obtained by successive applications of ∇:

$$\text{curl (grad } f) = \nabla \times (\nabla f) = 0, \qquad [1.46]$$
$$\text{div (curl } \mathbf{A}) = \nabla .(\nabla \times \mathbf{A}) = 0, \qquad [1.47]$$
$$\text{curl (curl } \mathbf{A}) = \nabla\,(\nabla .\mathbf{A}) - \Delta \mathbf{A}. \qquad [1.48]$$

1.7. Invariance and physical laws

By *transformation*, we mean a change of the coordinates or the variables of a system. A transformation is said to be *continuous* if it depends on parameters taking continuous values (as in the case of translations and rotations), otherwise it is said to be *discrete* (as in the case of reflections). A physical system is *invariant* in a transformation if it remains unchanged in the transformation (for instance, an infinite homogeneous medium is invariant in translations and a cone is invariant in rotations about its axis). A physical theory is *invariant* if it remains valid in the transformation (for instance, classical mechanics is invariant in the translation of time) and a physical quantity is *invariant* if it is unchanged in the transformation. An equation is said to be *covariant* in a transformation if it remains valid in the transformation (although the value of its terms may change in the transformation).

The invariance of a physical theory imposes some restrictions on the mathematical formulation of the laws and the allowed processes. A general principle, which was formulated by Noether, associates a conserved physical quantity with each invariance in a continuous transformation.

a) Invariance in geometrical transformations

Geometrical transformations are those of spatial coordinates and time, which conserve distances and intervals of time in classical physics. They include translations, rotations, and reflections. In a transformation $(\mathbf{r}, t) \to (\mathbf{r}', t')$, a physical quantity (or a field) $f(\mathbf{r}, t)$ becomes $f'(\mathbf{r}', t')$.

It is evident that physical laws do not depend on the origin of coordinates. In other words, an isolated system evolves in the same way, whatever its position in space (we say that the space is *homogeneous*). Mathematically, any physical law should not be modified if the positions $\mathbf{r}_{(k)}$ of all the particles (k) of the system are modified by the same translation $\mathbf{r}'_{(k)} = \mathbf{r}_{(k)} + \mathbf{a}$. For instance, the interaction energy U_{12} of two particles located at $\mathbf{r}_1$ and $\mathbf{r}_2$ is invariant in the translations if U_{12} depends on $\mathbf{R} = \mathbf{r}_2 - \mathbf{r}_1$ and not on $\mathbf{r}_1$ and $\mathbf{r}_2$ separately. Thus, we must have $U_{12} = U(\mathbf{R})$. Consequently, the forces that act on the particles are $\mathbf{F}_{1\to 2} = -\boldsymbol{\nabla}_2 U_E = -\boldsymbol{\nabla}_R U$ and $\mathbf{F}_{2\to 1} = -\boldsymbol{\nabla}_1 U_E = \boldsymbol{\nabla}_R U$, where $\boldsymbol{\nabla}_R$ means the vector differential operator with respect to the components of $\mathbf{R}$. Thus, the invariance in translations implies that $\mathbf{F}_{12} = -\mathbf{F}_{21}$ and, consequently, the conservation of the total momentum of a system of interacting particles such as electrically charged particles.

On the other hand, the physical laws do not depend on the orientation of the axes of coordinates. In other words, the space is *isotropic*. Mathematically, any physical law should not be modified if the reference frame is rotated. This requires, for instance, that the interaction energy of two particles is $U_{12} = U(R)$, i.e. a function of the magnitude of $\mathbf{R}$ and not its direction. Consequently, the force $\mathbf{F}_{1\to 2}$ may be written as $\mathbf{F}_{1\to 2} = -\boldsymbol{\nabla}_R U = -(\partial U/\partial R)\, \mathbf{R}/R$. Thus, it is oriented along the line that joins the two particles. This implies that the total angular momentum of an isolated system of particles $\mathbf{L} = \Sigma_i\, m_i\, \mathbf{r}_i \times \mathbf{v}_i$ is conserved.

Physical laws obey another important invariance law: they do not depend on the choice of the origin of time. In other words, if an experiment is repeated in time, the result should be the same. Mathematically, any physical law should not be modified in translations of time $t' = t + t_o$. This invariance requires that the potential energy of two bodies does not depend explicitly on time. Thus, the total energy of an isolated system is conserved if there are no dissipative forces.

To implement these invariance laws, the physical quantities must have well-defined transformation laws: they must be scalars, vectors, or other types of mathematical objects. A typical vector is the position **r**. Quantities, such as the force **F** and the electric field $\mathbf{E}(x, y, z)$, that transform exactly like **r** in rotations are *vectors*. For instance, in a rotation through an angle φ about Oz, they transform according to:

$$x' = x\cos\varphi + y\sin\varphi, \qquad y' = -x\sin\varphi + y\cos\varphi, \qquad z' = z, \qquad \text{(for } \mathbf{r}\text{)}$$
$$F'_x = F_x\cos\varphi + F_y\sin\varphi, \qquad F'_y = -F_x\sin\varphi + F_y\cos\varphi, \qquad F'_x = F_z. \qquad \text{(for } \mathbf{F}\text{)}$$
$$E'_x = E_x\cos\varphi + E_y\sin\varphi, \qquad E'_y = -E_x\sin\varphi + E_y\cos\varphi, \qquad E'_x = E_z. \qquad \text{(for } \mathbf{E}\text{)}$$

Here, the components of the field **E** are functions of the coordinates, $E'_x(x', y', z')$ and $E_x(x, y, z) \equiv E_x(x'\cos\varphi - y'\sin\varphi, x'\sin\varphi + y'\cos\varphi, z')$. A *scalar* is a quantity that is invariant in rotations, such as the distance $\mathbf{r}^2$, the scalar product of two vectors **A.B** and the potential $V(x, y, z)$.

b) Invariance in reflections

To formulate physical laws, only right-handed reference frames $Oxyz$ are usually used. However, nothing forbids to use systematically left-handed frames $Ox'y'z'$. Typical transformations of a right-handed frame to a left-handed one are *reflections* such as the *total reflection* $\mathbf{r}' = -\mathbf{r}$ (i.e. $x' = -x$, $y' = -y$ and $z' = -z$) and the reflection with respect to the Oxy plane (i.e. $x' = x$, $y' = y$ and $z' = -z$ as in a mirror). The invariance of physical laws in reflections (that are discrete transformations) is not as evident as in translations and rotations (that are continuous transformations). However, the experiment shows that this invariance holds in mechanics, in electromagnetism and in the case of strong (nuclear) interactions. It is violated in the case of weak interactions (see section 1.9d).

Some vectors have components that transform in reflections exactly like the coordinates x_α; these are said to be *true vectors*. Similarly, some scalars do not change in reflections; these are *true scalars*. This is the case for the distance $d = \sqrt{x^2+y^2+z^2}$ and the scalar product of two true vectors **A.B**. On the other hand, the cross product of two true vectors $\mathbf{U} = \mathbf{A} \times \mathbf{B}$ transforms like **r** in rotations but, in reflections, the components U_α transform like x_α with an additional change of sign. For instance, in the total reflection ($x_\alpha \to -x_\alpha$), the components U_α remain unchanged ($U_\alpha \to U_\alpha$) and, in the reflection with respect to the Oxy plane ($x \to x$, $y \to y$ and $z \to -z$), the U_α transform according to the relations $U_x \to -U_x$,

$U_y \to -U_y$ and $U_z \to U_z$. We say that **U** is a *pseudo-vector*. This is also the case for the cross product of two pseudo-vectors, while the cross product of a true vector and a pseudo-vector is a true vector. The scalar product of a true vector and a pseudo-vector is a pseudo-scalar: it is invariant in rotations but it changes sign in reflections; this is the case of the triple scalar product of three true vectors $\mathbf{A}.(\mathbf{B} \times \mathbf{C})$.

A physical law, written as a mathematical relationship between physical quantities, can be valid only if it is covariant in the preceding transformations. Thus, we may add, subtract or write equalities of quantities of the same type. It is not valid to add a vector to a pseudo-vector or write the equality of one component of two vectors without having the other components equal. For instance, the fundamental law of mechanics $\mathbf{F} = m\, d^2\mathbf{r}/dt^2$ requires that **F** be a true vector (like **r**) and the definition of the potential energy by the relation $dU_P = -\mathbf{F}.d\mathbf{r}$ requires that U_P and the energy in general be true scalars.

1.8. Electric charges in nature

Although matter is neutral on the macroscopic scale, it is comprised of charged and neutral particles. The experiment shows that, on the microscopic scale, the electric charge takes only discrete values (0, $\pm e$, $\pm 2e$, $\pm 3e$, etc.) that are integer multiples of the *elementary charge*

$$e = 1.602\,189\,2 \times 10^{-19}\ \text{C}. \qquad [1.49]$$

This *quantization* was established for the first time in 1913 by Millikan's oil drop experiment (see Problem 14.3). The stable particles, which are the building blocks of matter, are the proton of charge $+e$, the electron of charge $-e$, and the neutron (which is neutral as its name indicates). The electrification by rubbing is simply a transfer of electrons from a body of low electronic affinity to another of higher affinity.

The equality of the charge of the proton and the charge of the electron in absolute values, i.e. the neutrality of the hydrogen atom, is verified by the absence of any deviation of this atom by electric or magnetic fields with a precision of 1 to 10^{20}. On the other hand, the electric charge of particles does not depend on their velocity or on physical conditions, such as temperature, pressure, etc., even in extreme conditions, as in the core of stars or in the early stage of the formation of the Universe. The electron and the proton are absolutely stable. It is not possible to eliminate one of them individually but an electron and a proton may interact and produce a neutron and a neutrino. Conversely, a neutron may decay into a proton, an electron and an antineutrino. More generally, physical, chemical or biological transformations may occur in an isolated system leading to the exchange of charged particles between the constituents of the system, the creation or the annihilation of pairs of oppositely charged particles, but the total charge of the system is conserved.

The quantization of electric charge and its numerical value as well as the equality of proton and electron charges in absolute values are not understood even today. On the macroscopic scale, the elementary charge is extremely small and often has no observable effect. For instance, a negative charge of 1 μC corresponds to 6×10^{12} electrons and a current of 1 A carries 3.2×10^{18} electrons per second! When speaking of a point charge, it may be an elementary particle or a macroscopic object of small size compared to the dimensions of the system. It is often a very good approximation to consider an extended macroscopic charge as a continuous charge distribution.

a) Macroscopic bodies and molecules

Molecules and atoms are constituted by charged or neutral particles (electrons, protons, and neutrons). Their electric interactions are responsible for the cohesion of matter and most of its physical and chemical properties. Materials may be classified as *conductors* if some of the electrons are more or less free to move, *insulators* if the electrons are strongly bound to the atoms, and *semiconductors* whose conduction is intermediary between conductors and insulators. In solids and liquids, the spacing between atoms is of the order of the atoms' diameter (i.e. a fraction of a nanometer = 10^{-9} m). In gases, the molecules are separated by much longer distances. They are normally neutral at normal and low temperatures but some may become ionized by collisions, which become more and more frequent and energetic at high temperature. A gas may also become ionized if an energetic particle or radiation passes through it. A gas that is totally or partially ionized is a *plasma*.

b) Atoms, electrons, protons, and neutrons

The late 19th Century experiments have shown that atoms contain negatively charged *electrons*. To be globally neutral, the atoms must also contain positively charged particles, *protons*. To explain the stability of atoms, Thomson assumed that positive charges as well as the negative charges are distributed within a sphere of radius of the order of 10^{-10} m. However, Rutherford's experiment in 1911 showed that the positive charge is concentrated in a nucleus with a radius of the order of 10^{-15} m (see section 14.3). To explain the stability of the atom, in 1913 Bohr proposed a model in which the electrons maintain circular or elliptical orbits with a radius of the order of 10^{-10} m around the nucleus, bound by electric force. This orbital motion is similar to that of the planets around the Sun via gravitational force. Later, quantum theory abandoned this simple model in favor of a negatively charged electronic cloud around a positively charged nucleus. The state of the electrons in the atom is governed by the laws of quantum mechanics and, in principle, the properties of macroscopic matter can be deduced, but this is a difficult procedure.

The number of charged particles in matter is enormous. The number of molecules or atoms in a mole[1] of substance is the Avogadro number $N_A \approx 6 \times 10^{23}\ \mathrm{mol}^{-1}$ and each atom of a given chemical element contains Z electrons and Z protons. The hydrogen atom ($Z = 1$), for instance, is formed by one proton and one electron. The helium atom ($Z = 2$) contains two electrons and two protons. If it contains no other particles, its atomic mass would be approximately twice that of hydrogen; experimentally it is four-times heavier. Thus the atomic nuclei must contain neutral particles. These particles, called *neutrons*, were observed by Chadwick in 1938 with a mass that is slightly higher than the mass of protons (i.e. about 1840 times the mass of the electron). The helium nucleus, called also *alpha particle*, is formed by two protons and two neutrons. The protons and neutrons, which constitute the atomic nuclei are referred to as *nucleons*. The atom is thus formed by Z electrons, Z protons, and N neutrons. Its mass is $M \cong Zm_e + Zm_p + Nm_n \cong (Z + N)\, m_H \equiv Am_H$, where we have neglected the mass difference between the proton and the neutron, the binding energy of the nucleons (responsible for the cohesion of the nucleus), and the binding energy of electrons to the nucleus (responsible for the cohesion of the atom). The chemical properties of elements are closely related to the *atomic number* Z, while the physical properties, in which mass plays an important part, are related to the *mass number* A.

c) Elementary particles and quarks

Particles are usually considered as *elementary* if they are the smallest part of matter that may be isolated. Besides electrons, protons and neutrons, which are the building blocks of ordinary matter, there are many additional particles, which are observed in cosmic rays or produced in collisions carried out in laboratories using accelerators. Particles are characterized by their *mass*, *charge*, *spin* (intrinsic angular momentum), *magnetic moment*, etc. A particle may be *stable* (the electron, the proton, the photon, and the neutrino) or *unstable* (the free neutron, for instance); in the latter case, they are characterized by their *average lifetime* ranging from 10^{-20} s to 898 s for the neutron. Some characteristics of stable particles are listed in Table 1.1.

It is well established that each particle has a corresponding *antiparticle* of the same mass but opposite charge and some other characteristics. For instance, the positron (e^+) is the antiparticle of the electron (e^-), the antiproton ($\bar{p}$) is the antiparticle of the proton (p), etc. The antiparticle and the particle may be identical as in the case of the photon; they are then necessarily neutral. However, the antiparticle of a neutral particle may be different from the particle. This is the case for the antineutron $\bar{n}$ whose gyromagnetic ratio is opposite to that of the neutron. A

---

1 A mole is the amount of substance that contains the same number of particles (molecules, atoms, ions, electrons as specified) as there are atoms in 12 g of pure carbon nuclide ^{12}C.

particle and its antiparticle may be produced simultaneously in a reaction. For instance, a photon (γ) of sufficiently high energy may be transformed into a pair ($e^+ + e^-$) if it collides with a nucleus or another particle N according to the reaction $\gamma + N \rightarrow N + e^- + e^+$. Conversely, if a positron encounters an electron, they may annihilate into two photons at least, according to the reaction $e^- + e^+ \rightarrow 2\gamma$. These two reactions are examples of the *conservation of the electric charge.* No nuclear reaction or particle reaction that violates this law of conservation has ever been observed until now. Thus, it is considered to be a fundamental law of nature. In a process, it is possible to have a transfer of charge, a creation or annihilation of particles of opposite charges, but the total charge of any electrically isolated system is conserved. The total charge of the Universe (which is an isolated system because there is nothing else) is a constant (and probably zero).

Particle and symbol	**Mass (kg)**	**Charge**	**Spin**	**Magnetic moment**	**Average lifetime**	**Decay mode**
Electron (e^-)	9.10953×10^{-31}	$-e$	$\hbar/2$	$-1.001145\ \mu_B$	Stable	
Proton (p)	1.67265×10^{-27}	$+e$	$\hbar/2$	$2.79275\ \mu_p$	Stable	
Neutron (n)	1.67495×10^{-27}	0	$\hbar/2$	$-1.91315\ \mu_p$	898 s	$n \rightarrow p + e^- + \bar{\nu}_e$
Photon (γ)	0	0	$\hbar$	0	Stable	
Neutrinos (ν)	Very small	0	$\hbar/2$	0	Stable	

Table 1.1. *Characteristic quantities of some particles.* $\mu_B = e\hbar/2m_e = 9.274\times10^{-24}$ A.m^2 *is Bohr's magneton and* $\mu_p = e\hbar/2m_p = 5.051\times10^{-27}$ A.m^2 *is the nuclear magneton.* $\hbar = h/2\pi = 1.054\ 589 \times 10^{-34}$ J.s *is Planck's reduced constant.* m_e *is the electron mass and* m_p *is the proton mass. The particles behave as small magnets with a magnetic moment parallel to their spin. There are several species of neutrinos*

Elementary particles are extremely small and the concept of size is ambiguous at this scale. The electron has an extremely small radius to be measured with present techniques; thus, for all purposes, it is considered to be a point particle. Protons and neutrons have radii of the order of 10^{-15} m. However, although neutrons are neutral, they have a magnetic moment. Protons are strictly stable, while neutrons may be stable inside the nucleus but, if free, a neutron decay into a proton, an electron, and an antineutrino (beta decay) with a mean lifetime of 898 s. A particle is considered as "stable" if its mean lifetime is long enough to be observed in a bubble chamber, for instance.

Besides the photon, fundamental particles can be classified into *leptons* and *hadrons*. The leptons (including the electron, the muon, the neutrinos, etc.) have electromagnetic and weak interactions but no strong interactions. They are actually considered as strictly elementary. *Hadrons* (counting about 300 types of particles including nucleons) have all types of interactions. They have a complex structure, so they are not considered elementary, but are comprised of more fundamental entities of charge $\pm e/3$ or $\pm 2e/3$, called *quarks*. However, until now, quarks have never been observed as a separate entity. Thus, the isolated charges are always integer multiples of the elementary charge e. Nucleons are formed by three quarks, but other hadrons are formed by two quarks.

1.9. Interactions in nature

Actually, we know four types of interactions. They can be distinguished by their strength and their range, i.e. the distance over which the forces are significant. They are also characterized by selection rules that we will not consider here.

a) Gravitational interactions

The interaction of two point masses m and M may be expressed by the law of universal attraction $\mathbf{F} = -GMm\mathbf{r}/r^3$, where $G = 6.67 \times 10^{-11}\ \text{m}^3/\text{kg.s}^2$. The corresponding interaction potential energy $U_G = GMm/r$ decreases slowly with the distance, like $1/r$. We say that this is a *long-range* force. As all bodies that have mass have gravitational interaction, this is the dominant force on the cosmic scale. It is responsible for the cohesion of celestial bodies, the binding of satellites to planets, of planets to stars, the stars to galaxies and the galaxies within the Universe.

b) Electromagnetic interactions

These interactions include the Coulomb force between electric charges $F_E = K_o q_1 q_2/r^2$ (where $K_o \approx 9 \times 10^9\ \text{N.m}^2/\text{C}^2$) and the magnetic forces between charges in motion, magnetic matter, and electric currents. These interactions are much more intense than gravitational forces. In the hydrogen atom, for instance, the electrons and the proton are separated by an average distance $r = 0.53 \times 10^{-10}$ m. Their electric attraction is $F_E = -K_o e^2/r^2 = -8.2 \times 10^{-8}$ N, while their gravitational attraction is only $F_G = -Gm_P m_e/r^2 = -3.6 \times 10^{-47}$ N, thus 10^{39} times weaker. However, the electric forces are rarely perceived on the macroscopic scale, as macroscopic bodies are usually neutral. The Coulomb interaction potential energy is $U_E = K_o q_1 q_2/r$, and decreases with distance like $1/r$; thus, electromagnetic interactions are long-range forces. The binding energy of particles by electromagnetic forces is of the order of the electron-volt (1 eV = 1.602189×10^{-19} J) and particles that decay by electromagnetic interactions have a mean lifetime of the order of 10^{-18} to 10^{-20} s.

c) Strong interactions

These interactions are responsible for the binding of nucleons within nuclei and the binding of quarks within hadrons. They are about 10^3 times more intense than electromagnetic forces. Their typical binding energy in nuclei is of the order of 8 MeV per nucleon. The particles, which decay by strong interactions (called *resonances*), have a mean lifetime of the order of 10^{-22} to 10^{-23} s. The strong interactions cannot be formulated as a classical law of force. However, we know that they have a very short range (of the order of the size of the nucleus, i.e. $\approx 10^{-15}$ m). For this reason, they play no part in atomic and molecular physics (where particles are separated by distances of the order of 10^{-10} m) in macroscopic physics and in chemistry.

d) Weak interactions

These interactions are responsible for beta decay of the neutron and atomic nuclei and the decay of most of the elementary particles. They are about 10^{12} times weaker than electromagnetic interactions, but much more intense than gravitational forces. They have an extremely short range. The particles, which decay by weak interactions, have a mean lifetime of the order of 10^{-8} to 10^{-10} s and sometimes much longer if the decay energy is small (for instance, the neutron has a mean lifetime of 898 s).

1.10. Problems

Scalar and vectors

P1.1 Designating the derivatives by primed quantities, show that

$$[\mathbf{A}(t).\mathbf{B}(t)]' = (\mathbf{A}'.\mathbf{B}) + (\mathbf{A}.\mathbf{B}') \qquad \text{and} \qquad |\mathbf{A}(t)|' = (\mathbf{A}.\mathbf{A}')/|\mathbf{A}|.$$

P1.2 a) Consider the rotation through an angle φ about Oz. Express the new basis $\mathbf{e}'_\alpha$ in terms of the basis $\mathbf{e}_\beta$. Write the transformation equations for the components of a vector field $\mathbf{A}$. Write this transformation in the matrix form $\mathbf{A}' = R\mathbf{A}$. What are the transposed matrix $\tilde{R}$ and the inverse matrix R^{-1}? Verify that R is orthogonal. **b)** Suppose that a magnetic field is given by $\mathbf{B} = (\mu_o I/2\pi r^2)(-y\ \mathbf{e}_x + x\ \mathbf{e}_y)$. Write its expression in the new frame. Considering this rotation, can the expression $\mathbf{B}' = (\mu_o I/2\pi r^2)(y\ \mathbf{e}_x + x\ \mathbf{e}_y)$ be a vector field?

P1.3 To handle complicated vector analysis, it is practical to introduce Levi-Civitta symbols of permutations $\varepsilon_{\alpha\beta\gamma}$. Any permutation ($\alpha$, β, γ) of (1, 2, 3) may be obtained by successive exchange of indices. We define $\varepsilon_{\alpha\beta\gamma}$ as equal to ± 1 depending on whether the number of exchanges is even or odd and $\varepsilon_{\alpha\beta\gamma} = 0$ if two indices are the same. Thus $\varepsilon_{\alpha\beta\gamma}$ is odd in the exchange of any two indices ($\varepsilon_{\alpha\beta\gamma} = -\varepsilon_{\beta\alpha\gamma} = \varepsilon_{\beta\gamma\alpha} = \varepsilon_{\gamma\alpha\beta}$). **a)** Verify that these symbols obey the relation $\Sigma_\alpha\, \varepsilon_{\alpha\beta\gamma}\, \varepsilon_{\alpha\mu\nu} = \delta_{\beta\mu}\,\delta_{\gamma\nu} - \delta_{\beta\nu}\,\delta_{\gamma\mu}$ and that the Kronecker symbols obey the contraction relations $\Sigma_\beta\ \delta_{\alpha\beta}\ \delta_{\beta\gamma} = \delta_{\alpha\gamma}$ and $\Sigma_{\alpha\beta}\ \delta_{\alpha\beta}\delta_{\beta\alpha} = 3$. Deduce that the symbols $\varepsilon_{\alpha\beta\gamma}$ verify the contraction relations $\sum_{\alpha\beta} \varepsilon_{\alpha\beta\gamma} \varepsilon_{\alpha\beta\mu} = 2\delta_{\gamma\mu}$ and $\sum_{\alpha\beta\gamma} \varepsilon_{\alpha\beta\gamma} \varepsilon_{\alpha\beta\gamma} = 6$. The second relation expresses simply that the number of different permutations of (1, 2, 3) is 3! = 6. **b)** Verify that the determinant of a matrix $M_{\alpha\beta}$ may be written as $\det(M_{\alpha\beta}) = \sum_{\alpha\beta\gamma} \varepsilon_{\alpha\beta\gamma} A_{1\alpha} A_{2\beta} A_{3\gamma}$ and, more generally, $\varepsilon_{\alpha\beta\gamma} \det(A_{\mu\nu}) = \sum_{\mu\nu\lambda} \varepsilon_{\mu\nu\lambda} A_{\mu\alpha} A_{\nu\beta} A_{\lambda\gamma}$. **c)** Verify the following relations of the cross product and the triple scalar product

$$\mathbf{e}_\alpha \times \mathbf{e}_\beta = \varepsilon_{\alpha\beta\gamma} \mathbf{e}_\gamma, \quad \mathbf{A} \times \mathbf{B} = \varepsilon_{\alpha\beta\gamma} A_\alpha B_\beta \mathbf{e}_\gamma, \quad (\mathbf{A} \times \mathbf{B}).\mathbf{C} = \varepsilon_{\alpha\beta\gamma} A_\alpha B_\beta C_\gamma.$$

d) Use these Levi-Civitta symbols to calculate the more complicated products:

$$(\mathbf{A} \times \mathbf{B})^2 = \mathbf{A}^2 \mathbf{B}^2 - (\mathbf{A}.\mathbf{B})^2,$$
$$\mathbf{A} \times (\mathbf{B} \times \mathbf{C}) = (\mathbf{A}.\mathbf{C})\, \mathbf{B} - (\mathbf{A}.\mathbf{B})\, \mathbf{C},$$
$$\nabla \times (f\mathbf{B}) = f(\nabla \times \mathbf{B}) + \nabla f \times \mathbf{B},$$
$$(\mathbf{A} \times \nabla) \times \mathbf{B} + (\mathbf{B} \times \nabla) \times \mathbf{A} = \nabla(\mathbf{A}.\mathbf{B}) - \mathbf{A}(\nabla.\mathbf{B}) - \mathbf{B}(\nabla.\mathbf{A}),$$
$$\mathbf{A} \times (\nabla \times \mathbf{A}) = \tfrac{1}{2} \nabla(\mathbf{A}^2) - (\mathbf{A}.\nabla)\, \mathbf{A},$$
$$\nabla \times (\nabla \times \mathbf{B}) = \nabla (\nabla.\mathbf{B}) - \nabla^2 \mathbf{B},$$
$$\nabla \times (\mathbf{A} \times \mathbf{B}) = \mathbf{A}(\nabla.\mathbf{B}) - \mathbf{B}(\nabla.\mathbf{A}) + (\mathbf{B}.\nabla)\, \mathbf{A} - (\mathbf{A}.\nabla)\, \mathbf{B}.$$

Integrals involving vectors

P1.4 Calculate the flux of the vector field $\mathbf{E} = f(r)\, \mathbf{e}_r$ through the sphere of center O and radius R. Calculate the divergence of $\mathbf{E}$ and its integral over the enclosed volume and verify Gauss-Ostrogradsky' theorem.

Gradient and curl, conservative field and scalar potential

P1.5 Verify that the differential operator ∇ is a vector operator. Deduce that the gradient ∇f is a vector, the divergence $\nabla.\mathbf{A}$ is a scalar and the curl $\nabla \times \mathbf{A}$ is a vector.

P1.6 a) Let V be a scalar potential. Show that $dV(\mathbf{r}) = \nabla V.d\mathbf{r}$. Deduce that the component of ∇V in the direction of the unit vector $\mathbf{e}$ is $\partial V/\partial u$, where du is the displacement in this direction. **b)** Show that $\mathbf{E} \equiv -\nabla V$ is orthogonal to the equipotential surface (V = constant) and it points in the direction of the higher rate of decrease of the potential. **c)** A scalar field $f(r)$ depends only on the distance r to the origin O. Calculate its gradient. Consider the special case $f = K/r$.

P1.7 Show that $\nabla \times \mathbf{r} = 0$ and that $\nabla \times (f\mathbf{B}) = f(\nabla \times \mathbf{B}) + \nabla f \times \mathbf{B}$. Deduce that the curl of the electrostatic field of a point charge $\mathbf{E} = Kq\mathbf{r}/r^3$ is equal to zero. As any electrostatic field is produced by point charges, the curl of any electrostatic field is equal to zero.

P1.8 a) The potential of an electric dipole moment $\mathbf{p}$ is $V = K(\mathbf{p}.\mathbf{r})/r^3$. Calculate the corresponding electric field $\mathbf{E} = -\nabla V$. Suppose that $\mathbf{p} = p\mathbf{e}_z$. Calculate V and $\mathbf{E}$ at the point $\mathbf{r}(0, 3, 4)$. **b)** The vector potential of a magnetic dipole moment $\mathcal{M}$ is

$\mathbf{A} = k(\boldsymbol{\mathcal{M}} \times \mathbf{r})/r^3$. Calculate the corresponding magnetic field $\mathbf{B} = \boldsymbol{\nabla} \times \mathbf{A}$. Suppose that $\boldsymbol{\mathcal{M}} = \mathcal{M}\,\mathbf{e}_z$. Calculate **A** and **B** at the point **r**(0, 3, 4).

P1.9 In a given frame of reference, a vector field **E** has the components $E_x = 6x - 5z$, $E_y = -8y$ and $E_z = -5x$. Is this field the gradient of a scalar field f? If yes, write the expression of f in this frame.

P1.10 Consider the uniform vector field $\mathbf{B} = B\mathbf{e}_z$. Show that there is a vector **A** such that $\mathbf{B} = \boldsymbol{\nabla} \times \mathbf{A}$. Write the expression of **A**. Show that **A** is not unique but it is always possible to impose the condition $\boldsymbol{\nabla}.\mathbf{A} = 0$.

P1.11 A surface $\mathcal{S}$ encloses a volume $\mathcal{V}$. Let **n** be the unit vector normal to $\mathcal{S}$. Show that, for any scalar field f and vector field **A**, we have

$$\iiint_{\mathcal{V}} d\mathcal{V}\, \boldsymbol{\nabla} f = \iint_{\mathcal{S}} d\mathcal{S}\, \mathbf{n} f\ , \qquad \iiint_{\mathcal{V}} d\mathcal{V}\, \boldsymbol{\nabla} \times \mathbf{A} = \iint_{\mathcal{S}} d\mathcal{S}\, (\mathbf{n} \times \mathbf{A}).$$

Divergence, conservative flux and vector potential

P1.12 a) Calculate the divergence of the vector fields $\mathbf{B} = k\mathbf{r}$ and $\mathbf{B} = \mathbf{r}f(\mathbf{r})$. **b)** Show that $\boldsymbol{\nabla}(f\mathbf{B}) = \mathbf{B}.\boldsymbol{\nabla} f + f(\boldsymbol{\nabla}.\mathbf{B})$.

Other properties of the vector differential operator

P1.13 Let f be a function of $r = |\mathbf{r}|$. Verify that $\Delta f(r) = d^2f/dr^2 + (2/r)\,(df/dr)$. Verify that $1/r$ is a solution of Laplace's equation $\Delta f = 0$.

P1.14 Let f and g be arbitrary scalar fields while **A** and **B** are vector fields. Show the following relations:

$$\boldsymbol{\nabla}(fg) = f\boldsymbol{\nabla} g + g\,\boldsymbol{\nabla} f, \qquad \boldsymbol{\nabla}\times\boldsymbol{\nabla} f = 0, \qquad \boldsymbol{\nabla}\times(f\boldsymbol{\nabla} g) = 0, \qquad \boldsymbol{\nabla}.(\boldsymbol{\nabla}\times\mathbf{A}) = 0,$$
$$\boldsymbol{\nabla}(f\mathbf{A}) = \boldsymbol{\nabla} f.\mathbf{A} + f(\boldsymbol{\nabla}.\mathbf{A}), \qquad \boldsymbol{\nabla}.(\mathbf{A}\times\mathbf{B}) = \mathbf{B}.(\boldsymbol{\nabla}\times\mathbf{A}) - \mathbf{A}.(\boldsymbol{\nabla}\times\mathbf{B}).$$

P1.15 a) Let Ψ and Φ be two scalar functions defined on a surface $\mathcal{S}$ and in the enclosed volume $\mathcal{V}$ and ∂_n be the differential operator with respect to the outgoing normal coordinate x_n on the surface $\mathcal{S}$. Show the following Green's theorems

$$\iint_{\mathcal{S}} d\mathcal{S}\,(\partial_n\Psi) = \iiint_{\mathcal{V}} d\mathcal{V}\,(\Phi\,\Delta\Psi + \boldsymbol{\nabla}\Phi.\boldsymbol{\nabla}\Psi),$$
$$\iint_{\mathcal{S}} d\mathcal{S}\,(\Phi\,\partial_n\Psi - \Psi\partial_n\Phi) = \iiint_{\mathcal{V}} d\mathcal{V}\,(\Phi\,\Delta\Psi - \Psi\,\Delta\Phi).$$

b) Show that any function Φ verifies the relations

$$\iint_{\mathcal{S}} d\mathcal{S}\,\mathbf{n}.(\Phi\,\boldsymbol{\nabla}\Phi) = \iiint_{\mathcal{V}} d\mathcal{V}\,[\Phi\,\Delta\Phi - (\Delta\Phi)^2], \qquad \iint_{\mathcal{S}} d\mathcal{S}\,\partial_n\Phi = \iiint_{\mathcal{V}} d\mathcal{V}\,\Delta\Phi).$$

c) Let Ψ and Φ be solutions of Laplace's equation (i.e. $\Delta\Phi = \Delta\Psi = 0$). Show that they verify the relation $\iint_{\mathcal{S}} d\mathcal{S}\,\Phi\,(\partial_n\Psi) = \iint_{\mathcal{S}} d\mathcal{S}\,\Psi\,(\partial_n\Phi)$. **e)** If Ψ, $\Delta\Psi$ and $\Delta\Phi$ are defined on a closed surface $\mathcal{S}$ and in the enclosed volume $\mathcal{V}$, show the Green representation

$$\iint_{\mathcal{S}} d\mathcal{S}\, \Phi\, (\partial_n \Psi) = \iiint_{\mathcal{V}} d\mathcal{V}\, \Psi\, (\Phi\, \Delta\Psi + \nabla\Phi . \nabla\Psi).$$

Invariances of physical laws

P1.16 The interaction potential energy of two particles located at $\mathbf{r}_1$ and $\mathbf{r}_2$ is a certain scalar function $U(\mathbf{r}_1, \mathbf{r}_2)$ if this interaction does not depend on any other physical quantity. **a)** Show that the homogeneity of space (i.e. the invariance under arbitrary translations of the reference frame) implies that U does not depend on $\mathbf{r}_1$ and $\mathbf{r}_2$ separately but on the relative position $\mathbf{r} = \mathbf{r}_1 - \mathbf{r}_2$. The force that the particle (1) exerts on particle (2) is $\mathbf{F}_{1\to2} = -\nabla_2 U(\mathbf{r})$, where ∇_2 is the gradient with respect to the coordinates of particle (2). Similarly, the force that particle (2) exerts on particle (1) is $\mathbf{F}_{2\to1} = -\nabla_1 U(\mathbf{r})$. Verify that $\mathbf{F}_{1\to2} = -\mathbf{F}_{2\to1}$ (principle of action and reaction). **b)** If the particles have no structure enabling the specification of special directions in space, show that space isotropy (i.e. the invariance in arbitrary rotations) implies that $U(\mathbf{r})$ does not depend on the direction of $\mathbf{r}$ but only on the distance r; thus we have $V = U(r)$. Calculate in this case $\mathbf{F}_{1\to2}$ and $\mathbf{F}_{2\to1}$ and verify that they are in the direction of $\mathbf{r}$ (central forces).

P1.17 a) The basis vectors $\mathbf{e}'_\beta$ and $\mathbf{e}_\alpha$ are related by the relation $\mathbf{e}_\alpha = \Sigma_\beta\, R_{\beta\alpha}\, \mathbf{e}'_\beta$. Show that the components of a vector $\mathbf{A}$ transform according to $A'_\beta = \Sigma_\alpha\, R_{\beta\alpha}\, A_\alpha$. **b)** Show that the transformation $R_{\beta\alpha}$ conserves the orthonormality of the basis if it is orthogonal, i.e. $\tilde{R}\, R = R\, \tilde{R} = I$. Show that, in this case, it conserves the scalar product of any two vectors $\mathbf{A}$ and $\mathbf{B}$, thus their angle. **c)** Show that these transformations must have a determinant $\det(M) = \pm 1$. Rotations are typical transformations such that $\det(M) = +1$ while reflections are typical transformations such that $\det(M) = -1$. Show that the cross product of two vectors $\mathbf{V} = \mathbf{A} \times \mathbf{B}$ transforms according to $V'_\alpha = \det(M)\, \Sigma_\beta\, M_{\alpha\beta} V_\beta$. Thus $\mathbf{V}$ transforms as a vector in rotations while, in reflections it acquires a supplementary change of sign. Verify that the triple scalar product $S = \mathbf{A}.(\mathbf{B} \times \mathbf{C})$ transforms according to $S' = \det(M)\, S$; thus it is a pseudo-scalar.

Electric charges and interactions in nature

P1.18 What is the number of electrons, protons, and neutrons in a piece of copper of mass 5 g ($Z = 29$, $N = 35$, and atomic mass 63.5)? How long it takes for a current of 10 A to carry the charge of these electrons? Assume that $1/10^9$ of these atoms lose one electron and that these electrons are transferred on an identical piece situated at 10 cm. What is the attraction force of these pieces? Compare this electric force to their gravitational attraction.

Chapter 2

Electrostatics in Vacuum

The interaction of electric charges, as expressed by Coulomb force, is formulated according to the Newtonian concept of *action-at-a-distance*: if a charge q' is produced at $\mathbf{r}'$ at a time t', a charge q located at $\mathbf{r}$ feels the action of q' instantaneously, whatever the distance $|\mathbf{r} - \mathbf{r}'|$ and the medium that separates the charges. The concept of *field* was developed by Faraday, Maxwell, Lorentz, Einstein, and many others. In modern physics, all interactions are conceived as *local*, i.e. involving quantities defined at the same point $\mathbf{r}$ and at the same time t. Fields are physical entities that are endowed with energy, momentum, etc., and they may propagate with some finite speed as *waves*. Furthermore, in quantum theory, the same objects (electrons for instance) have both particle and wave properties.

In this chapter, we introduce the concepts of electric field and potential, we derive the fundamental equations of electrostatics in vacuum, and we discuss some of their properties and the concept of electrostatic energy.

2.1. Electric forces and field

In a famous experiment, Coulomb used a torsion balance to measure the force of interaction of electric charges. He verified that a small charge q_1 acts on a small charge q_2 situated at a distance r with a force $F_E = K_o q_1 q_2 / r^2$ oriented along the line joining the charges. This force is repulsive between like charges and attractive between unlike charges. It has a similar mathematical form to Newton's *law of universal gravitation* $F_g = -G m_1 m_2 / r^2$. To specify both the direction and the magnitude, we write

$$\mathbf{F}_{1\to 2} = K_o q_1 q_2 \, \mathbf{R}_{12}/R_{12}^3, \qquad \text{where} \qquad \mathbf{R}_{12} = \mathbf{r}_2 - \mathbf{r}_1. \qquad [2.1]$$

Coulomb's force obeys the principle of action and reaction. K_o is a constant that depends on the adopted unit of charge. Using the *coulomb* (C) as the unit of charge and the *Heaviside* or *rationalized* system, we write

$$K_o = 1/4\pi\varepsilon_o = 8.987\,551\,79 \times 10^9\ \mathrm{N.m^2/C^2},$$
$$\text{where}\ \ \varepsilon_o = 8.854\,187\,82 \times 10^{-12}\ \mathrm{C^2/N.m^2}. \qquad [2.2]$$

ε_o is the *permittivity of vacuum*. The factor 4π is introduced to simplify the writing of equations. The electric force is much more intense than the gravitational force and the coulomb is an enormous charge on the human scale: electric sparks are produced by less than one microcoulomb ($1\ \mu\mathrm{C} = 10^{-6}$ C) and rubbing produces a charge of the order of the nanocoulomb per square centimeter ($1\ \mathrm{nC} = 10^{-9}$ C).

In accordance with the *superposition principle*, the total force that several charges q_i located at the points $\mathbf{r}_i$ exert on a charge q placed at $\mathbf{r}$ is the vector sum of the forces exerted by each charge q_i if it acts individually

$$\mathbf{F} = \Sigma_i\, K_o q q_i\, \mathbf{R}_i/R_i^3, \qquad \text{where}\ \ \mathbf{R}_i = \mathbf{r} - \mathbf{r}_i. \qquad [2.3]$$

In the following, the charge q on which the force acts is considered as a *test charge*, while the charges q_i that produce the force are considered as the *source charges*. If the source charges are continuously distributed in a volume $\mathcal{V}$, on a surface $\mathcal{S}$ or a curve $\mathcal{C}$, the source charge q_i must be replaced by $q_v(\mathbf{r}')\, d\mathcal{V}$, $q_s(\mathbf{r}')\, d\mathcal{S}$ or $q_L(\mathbf{r}')\, dL$, where q_v, q_s, and q_L are the charge densities, respectively, per unit volume, per unit area, and per unit length, and then integrate on the source charge distribution.

By analogy to the gravitational field represented by the acceleration $\mathbf{g}$ and the magnetic field near magnetized bodies, which exist independently of the test bodies, we define the *electric field* $\mathbf{E}$ such that the force acting on a test charge q is

$$\mathbf{F} = q\,\mathbf{E}(\mathbf{r}) \qquad [2.4]$$

without reference to the charges, which produce $\mathbf{E}$. The test charge q must be small in order that its action on the source charges and, consequently, on the field $\mathbf{E}$ itself be negligible.

From expression [2.3] of the force exerted by the point charges q_i at $\mathbf{r}_i$ on q at $\mathbf{r}$, we deduce the electric field produced by these charges $\mathbf{E} = \mathbf{F}/q$ and we may generalize it to continuous charge distributions; we get

$$\mathbf{E}(\mathbf{r}) = \Sigma_i\, K_o q_i \mathbf{R}_i/R_i^3, \qquad \text{where}\ \ \mathbf{R}_i = \mathbf{r} - \mathbf{r}_i\ \ \text{or}\ \ \mathbf{R} = \mathbf{r} - \mathbf{r}'.$$
$$\mathbf{E}(\mathbf{r}) = K_o \iiint_{\mathcal{V}} d\mathcal{V}'\, q_v(\mathbf{r}')\, \mathbf{R}/R^3, \quad K_o \iint_{\mathcal{S}} d\mathcal{S}'\, q_s(\mathbf{r}')\, \mathbf{R}/R^3, \quad \text{or}\ K_o\int_{\mathcal{C}} dL'\, q_L(\mathbf{r}')\, \mathbf{R}/R^3. \qquad [2.5]$$

We note that a distribution of point charges q_i at points $\mathbf{r}_i$ may be considered as a volume charge distribution of density $q_v(\mathbf{r}') = \Sigma_i\, q_i\, \delta^3(\mathbf{r}' - \mathbf{r}_i)$ where $\delta^3(\mathbf{r}' - \mathbf{r}_i)$ is the three-dimensional Dirac function centered at $\mathbf{r}_i$ (see section A.11 of Appendix A). Similarly, a surface charge density $q_s(\mathbf{r}')$ corresponds to $q_v(\mathbf{r}') = q_s(\mathbf{r}')\, \delta(z' - z_n)$, where z_n is a coordinate that is normal to the charged surface, and a linear charge density $q_L(\mathbf{r}')$ corresponds to $q_v(\mathbf{r}') = q_L(\mathbf{r}')\, \delta(x' - x_n)\, \delta(y' - y_n)$, where x_n and y_n are the coordinates that are normal to the charged line.

2.2. Electric energy and potential

The concept of energy is very important, especially in modern physics, because of its conservation in the case of isolated systems. Energy may have several forms. We are concerned here with the *electric potential energy* of the charges. The test charge q being subject to the conservative electric force $\mathbf{F}$, the analysis of section 1.4 allows us to define an electric potential energy U_E such that $\mathbf{F} = -\nabla U_E$. It may be shown that the electrostatic interaction of two charges q and q_i corresponds to a potential energy $U_E = K_o q q_i / R_i$, where $R_i = |\mathbf{r} - \mathbf{r}_i|$. In the case of a test charge $q = 1$ C, $\mathbf{F}$ becomes the electric field and the potential energy of the unit charge is the *electrostatic potential* V such that

$$E_x = -\partial_x V, \qquad E_y = -\partial_y V, \qquad E_z = -\partial_z V. \qquad [2.6]$$

The potentials produced by discrete or continuous charge distributions are given by

$$V(\mathbf{r}) = K_o\, \Sigma_i\, q_i / R_i,$$
$$V(\mathbf{r}) = K_o \iiint_{\mathcal{V}} d\mathcal{V}'\, q_v(\mathbf{r}')/R, \quad K_o \iint_{\mathcal{S}} d\mathcal{S}'\, q_s(\mathbf{r}')/R, \text{ or } \quad K_o \int_{\mathcal{C}} dL'\, q_L(\mathbf{r}')/R. \qquad [2.7]$$

The SI (*Système International*, the international system of units) unit of potential is the *joule per coulomb* (J/C) called the *volt* (V), and the unit of electric field is the (N/C), which may also be called *volt per meter* (V/m). In atomic, nuclear, and particle physics, the elementary charge e is frequently used. For this, it is convenient to use the *electron-volt* (eV) as the unit of energy; this is the energy that is gained or lost by an electron as it moves between two points with a difference of potential of 1 V; thus, 1 eV = 1.602 189 2 $\times 10^{-19}$ J. The keV = 10^3 eV, the MeV = 10^6 eV, the GeV = 10^9 eV, and the TeV = 10^{12} eV are also used.

The drop in potential is the work of the electric force on the positive unit charge. We may also interpret the increase of V as the work of an external agent in displacing the positive unit charge without varying its kinetic energy. This work is independent of the path because $\mathbf{E}$ is conservative. Particularly, if the potential is taken to be zero at infinity, as in the expressions [2.7], the potential $V(\mathbf{r})$ is the work that is required to bring the unit charge from infinity to the point $\mathbf{r}$ along any path.

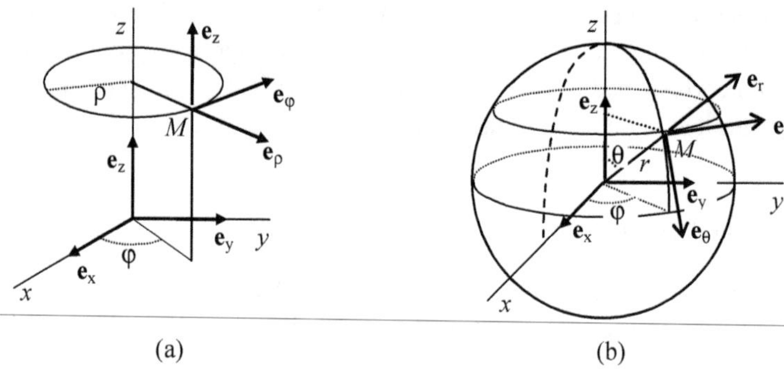

Figure 2.1. *a) Cylindrical coordinates, and b) spherical coordinates*

2.3. The two fundamental laws of electrostatics

a) Evaluation of the field from the potential and the potential from the field

In the case of time-independent phenomena, the electric field is conservative and we may introduce the electrostatic potential such that

$$\mathbf{E} = -\nabla V. \qquad [2.8]$$

Considering a displacement $\delta\mathbf{r} = \delta l\ \mathbf{e}$ along an arbitrary axis D of unit vector $\mathbf{e}$, the variation of the potential is $\delta V = \delta\mathbf{r}.\nabla V = -\ \delta l\ \mathbf{E}.\mathbf{e} \equiv -\ \delta l\ E_D$, where E_D is the component of $\mathbf{E}$ along D. We deduce that

$$E_D = -\,\delta V/\delta l\ |_D. \qquad [2.9]$$

This relation holds even if we know V only on the line D and it may be generalized to curvilinear coordinates. We obtain in the case of cylindrical coordinates and spherical coordinates (see Figure 2.1):

$$E_\rho = -\,\frac{\delta V}{\delta\rho} = -\,\partial_\rho V,\ E_\varphi = -\frac{\delta V}{\rho\,\delta\varphi} = -\frac{\partial_\varphi V}{\rho},\ \text{and}\ E_z = -\frac{\delta V}{\delta z} = -\,\partial_z V, \qquad [2.10]$$

$$E_r = -\frac{\delta V}{\delta r} = -\partial_r V,\ E_\theta = -\frac{\delta V}{r\,\delta\theta} = -\frac{\partial_\theta V}{r},\ \text{and}\ E_\varphi = -\frac{\delta V}{r\sin\theta\ \delta\varphi} = -\frac{\partial_\varphi V}{r\sin\theta}. \qquad [2.11]$$

The relation $\mathbf{E} = -\nabla V$ shows that, if V is constant, $\mathbf{E} = 0$ and conversely, if $\mathbf{E} = 0$, V is constant. The surface of equation $V(\mathbf{r}) = C$, where C is a constant, is an *equipotential surface* (see Figure 2.2). As the potential has a unique value at each point $\mathbf{r}$, two equipotential surfaces cannot intersect. The field $\mathbf{E}$ is orthogonal to the equipotential surface and it points toward the decreasing potential.

Conversely, the potential may be evaluated if the expression of the field $\mathbf{E}$ is known by using the equation

$$dV = dx\, \partial_x V + dy\, \partial_y V + dz\, \partial_z V = \nabla V.d\mathbf{r} = -\mathbf{E}.d\mathbf{r}. \qquad [2.12]$$

Sometimes, this equation may be directly integrated. For instance, the field of a charge q' is $\mathbf{E}(\mathbf{r}) = K_0 q'(\mathbf{r} - \mathbf{r}')/|\mathbf{r} - \mathbf{r}'|^3$. Setting $\mathbf{R} = \mathbf{r} - \mathbf{r}'$ and noting that $d\mathbf{R}^2 = 2\mathbf{R}.d\mathbf{R} = dR^2 = 2R\, dR$, equation [2.12] may be written as

$$dV = -K_0\, q' \frac{\mathbf{r}-\mathbf{r}'}{|\mathbf{r}-\mathbf{r}'|^3}.d\mathbf{r} = -K_0\, q' \frac{\mathbf{R}.d\mathbf{R}}{|\mathbf{R}|^3} = -K_0\, q' \frac{R\, dR}{R^3} = -K_0\, q' \frac{dR}{R^2}.$$

Integrating this equation, we find $V(\mathbf{r}) = K_0\, q'/R + C = K_0\, q'/|\mathbf{r} - \mathbf{r}'| + C$.

Generally, integrating equation [2.12] between $\mathbf{r}_0$ and $\mathbf{r}$, as $\mathbf{E}$ is conservative, we obtain the potential difference between $\mathbf{r}_0$ and $\mathbf{r}$ over any path

$$\int_{\mathbf{r}}^{\mathbf{r}_0} d\mathbf{r}'.\mathbf{E}(\mathbf{r}') = \int_{\mathbf{r}_0}^{\mathbf{r}} dV = V(\mathbf{r}) - V(\mathbf{r}_0) \qquad [2.13]$$

This equation determines the potential $V(\mathbf{r})$ up to an arbitrary additive constant. We may add to $V(\mathbf{r})$ the same constant everywhere without modifying the field or any physical law. It is possible to fix this constant by assigning a value V_0 at a particular point $\mathbf{r}_0$. This is possible if there is no electric charge or linear charge distribution at this point (as the potential is then infinite). A practical choice is to take $V = 0$ at infinity as in the expressions [2.7]. In this case, if $V(\infty) = 0$, we have

$$V(\mathbf{r}) = \int_{\mathbf{r}}^{\infty} d\mathbf{r}'.\, \mathbf{E}(\mathbf{r}'). \qquad [2.14]$$

The addition of an arbitrary constant V_0 to the potential ($V' = V + V_0$) is a *dynamic transformation*. According to Noether's theorem, the invariance of the laws of electrostatics in this continuous transformation is associated with a conservation law, which is the conservation of electric charge. To show this, let us consider a reaction $A + B \rightarrow C + D + \ldots$, where A, B, … are bodies of charges q_A, q_B, … If this reaction occurs in a region where the electric potential is V, the conservation law of the total energy may be written as $E_i + q_A V + q_B V = E_f + q_C V + q_D V + \ldots$, where E_i

and E_f are the non-electric initial energy and final energy, respectively. As this relation remains valid in the transformation $V' = V + V_o$ for any V_o, we must have $q_A V_o + q_B V_o = q_C V_o + q_D V_o + \ldots$ and, consequently, the conservation law of electric charge $q_A + q_B = q_C + q_D + \ldots$

b) The first fundamental equation of electrostatics: **E** *is conservative*

As the second derivatives do not depend on the order of differentiation, for instance $\partial_x\partial_y V = \partial_y\partial_x V$, we deduce from equations [2.6] that

$$\partial_y E_z - \partial_z E_y = 0, \qquad \partial_z E_x - \partial_x E_z = 0, \qquad \partial_x E_y - \partial_y E_x = 0. \qquad [2.15]$$

The left-hand sides of these equations are the components of $\nabla\times\mathbf{E}$. Thus, the fact that the electrostatic field **E** is conservative is equivalent to the equation

$$\nabla \times \mathbf{E} = 0. \qquad [2.16]$$

We may show this result in a different way: applying equation [2.13] to a closed path $\mathcal{C}$ ($\mathbf{r} \equiv \mathbf{r}_o$), we find

$$\oint_{\mathcal{C}} d\mathbf{r}.\mathbf{E} = 0. \qquad [2.17]$$

Using Stokes' theorem (see section A.8 of Appendix A), we may transform the integral into the flux of $\nabla\times\mathbf{E}$ through the surface $\mathcal{S}$ bounded by $\mathcal{C}$:

$$\int_{\mathcal{C}} d\mathbf{r}.\mathbf{E} = \iint_{\mathcal{S}} d\mathcal{S}\, \mathbf{n}.(\nabla \times \mathbf{E}) = 0. \qquad [2.18]$$

It vanishes for any $\mathcal{S}$ if $\nabla\times\mathbf{E} = 0$. This is the first fundamental equation of electrostatics.

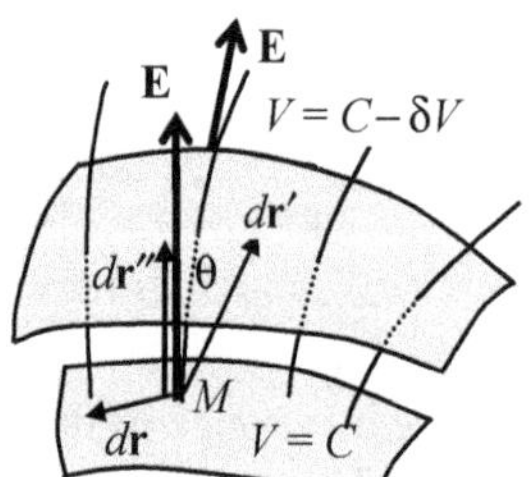

Figure 2.2. *Equipotential surfaces*

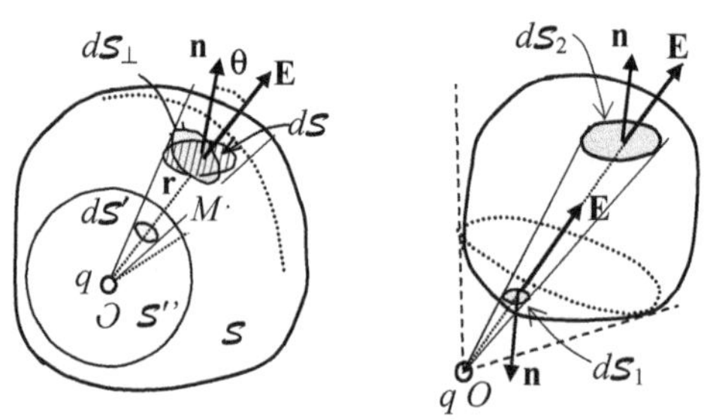

Figure 2.3. *Gauss's law*

c) The second fundamental equation of electrostatics: Gauss's law

Consider first the flux of the field of a point charge q taken at the origin O (Figure 2.3). Its field at M on the surface $\mathcal{S}$ is $\mathbf{E}(\mathbf{r}) = q\mathbf{r}/4\pi\varepsilon_o r^3$, where $\mathbf{r} \equiv \overrightarrow{OM}$. According to the analysis of section 1.3, the flux of $\mathbf{E}$ outgoing from a closed surface $\mathcal{S}$ is $q\Omega/4\pi\varepsilon_o$, where Ω is the solid angle of the cone whose apex is at O and which is subtended by the surface $\mathcal{S}$. If q is inside $\mathcal{S}$, the total solid angle Ω is 4π and the flux is $\Phi = q/\varepsilon_o$ and if q is outside $\mathcal{S}$, $\Omega = 0$ and so is Φ. Any field $\mathbf{E}$ is the resultant of the fields $\mathbf{E}_k$ produced by all the charges q_k. The flux of $\mathbf{E}$ through any closed surface $\mathcal{S}$ is the sum of the fluxes Φ_k of the fields $\mathbf{E}_k$, i.e. $\Phi_k = q_k/\varepsilon_o$ if q_k is inside $\mathcal{S}$ and $\Phi_k = 0$ if q_k is outside $\mathcal{S}$. Thus, the total flux of $\mathbf{E}$ is given by *Gauss's law*

$$\varepsilon_o\, \Phi_E = Q^{(in)}, \qquad [2.19]$$

where $Q^{(in)} = \Sigma_j\, q^{(in)}{}_j$ is the total charge inside $\mathcal{S}$. In the case of a volume charge distribution density $q_v(\mathbf{r})$ in the volume $\mathcal{V}$ enclosed by $\mathcal{S}$, we find *Gauss's law in the integrated form*

$$\varepsilon_o \iint_{\mathcal{S}} d\mathcal{S}\, \mathbf{n}.\mathbf{E} = \iiint_{\mathcal{V}} d\mathcal{V}\, q_v(\mathbf{r}). \qquad [2.20]$$

Using Gauss-Ostrogradsky's theorem (see section A.9 of Appendix A), we may transform the outgoing flux of $\mathbf{E}$ through $\mathcal{S}$ into the volume integral of the divergence of $\mathbf{E}$ over the enclosed volume $\mathcal{V}$. Thus, equation [2.20] takes the form $\varepsilon_o \iiint_{\mathcal{V}} d\mathcal{V}\, \nabla.\mathbf{E} = \iiint_{\mathcal{V}} d\mathcal{V}\, q_v(\mathbf{r})$ for any volume $\mathcal{V}$. Thus, we must have

$$\nabla.\mathbf{E}(\mathbf{r}) = q_v(\mathbf{r})/\varepsilon_o\,. \qquad [2.21]$$

This is the *local form of Gauss's law*. It is the second fundamental law of electrostatics. In the case of electrostatic phenomena, it is equivalent to Coulomb's law but, contrary to Coulomb's law in its simple form [2.5], we shall see that Gauss's law is valid even in the case of time-dependent phenomena.

2.4. Poisson's equation and its solutions

Substituting the expression $\mathbf{E} = -\nabla V$ in Gauss's equation $\nabla.\mathbf{E}(\mathbf{r}) = q_v(\mathbf{r})/\varepsilon_o$, we find that the potential obeys *Poisson's equation*

$$\Delta V(\mathbf{r}) = -\, q_v(\mathbf{r})/\varepsilon_o. \qquad [2.22]$$

Particularly if there is no charge ($q_v = 0$), this equation reduces to *Laplace's* equation

$$\Delta V(\mathbf{r}) = 0. \qquad [2.23]$$

The partial differential equations have many solutions that depend on arbitrary functions. It may be verified that Poisson's equation [2.22] has a solution

$$V(\mathbf{r}) = K_o \iiint d\mathcal{V}'\, q_v(\mathbf{r}')/|\mathbf{r} - \mathbf{r}'| \qquad [2.24]$$

with terms of the forms [2.7] in the case of surface, line and point charge distributions. These expressions are such $V \sim 1/r$ as $r \to \infty$. We may add to [2.24] any solution V_o of Laplace's equation [2.23] and obtain another solution

$$V(\mathbf{r}) = V_o(\mathbf{r}) + K_o \iiint d\mathcal{V}'\, q_v(\mathbf{r}')/|\mathbf{r} - \mathbf{r}'|. \qquad [2.25]$$

To determine the solution, we must know, besides q_v at each point $\mathbf{r}'$, the boundary conditions at infinity in the case of an infinite space or on the surfaces of the system if it is bounded. It is always possible to find V_o, such that V verifies these boundary conditions. Thus, there is always a solution V and this solution is unique. Once V is known, the equation $\mathbf{E} = -\boldsymbol{\nabla} V$ allows us to determine the field $\mathbf{E}$. It should be noted that only charges that contribute to $\mathbf{E}$ and V should appear on the right-hand side of equation [2.22] and its solution. For instance, if we study the action of a field $\mathbf{E}^{(ex)}$ on a body carrying charge q_i, the q_i should not appear in [2.22] and its solution even if the body is extended, as they do not contribute to the potential $V^{(ex)}$.

In some cases, some mathematical conditions have to be imposed (for instance, V has a unique determination and it is finite at points where there is no point charge or lines of charge). On the other hand, if the space is formed by different regions, Poisson's equation must be solved in each region and appropriate boundary conditions must be imposed on their interface. The problem may be further complicated if the charges are mobile and their positions depend on the field to be calculated (as in a conductor) or if the electric properties of the material depend on the field (as in the case of a dielectrics). On the other hand, often the solution is too complicated to be written in terms of known or simple functions. Approximation or numerical methods must be used in such cases.

To illustrate the use of Laplace and Poisson's equations, let us consider a ball of radius R and uniform charge density q_v. The potential V verifies Poisson's equation inside the ball and Laplace's equation outside the ball. Using the expression of the Laplacian in spherical coordinates and noting that V does not depend on θ and φ because of the spherical symmetry, we get the equations

$$\frac{1}{r^2}\frac{d}{dr}[r^2\frac{dV}{dr}] = -\frac{q_v}{\varepsilon_o} \text{ (for } r < R\text{)} \quad \text{and} \quad \frac{1}{r^2}\frac{d}{dr}[r^2\frac{dV}{dr}] = 0 \text{ (for } r > R\text{)}.$$

These differential equations have solutions that depend on two arbitrary constants:

$$V^{(\mathrm{in})} = -\,q_{\mathrm{v}}r^2/6\varepsilon_{\mathrm{o}} + B/r + C \ \ (\text{for } r < R) \quad \text{and} \quad V^{(\mathrm{ex})} = B'/r + C' \ \ (\text{for } r > R).$$

The condition that V is finite at the center of the ball (as there is no point charge or linear charge) imposes that $B = 0$ and, having no charge at infinity, we may take $V^{(\mathrm{ex})}(\infty) = 0$, thus $C' = 0$. Using equation [2.11] of the gradient, we obtain the field $\mathbf{E}^{(\mathrm{in})} = (q_{\mathrm{v}}r/3\varepsilon_{\mathrm{o}} + B/r^2)\,\mathbf{e}_{\mathrm{r}}$ (for $r < R$) and $\mathbf{E}^{(\mathrm{ex})} = (B'/r^2)\,\mathbf{e}_{\mathrm{r}}$ (for $r > R$). As the surface of the ball carries no point charge or linear and surface charge densities, the continuity of V and $\mathbf{E}$ at $r = R$ imposes the conditions $-q_{\mathrm{v}}R^2/6\varepsilon_{\mathrm{o}} + C = B'/R$ and $q_{\mathrm{v}}R/3\varepsilon_{\mathrm{o}} = B'/R^2$; thus, we find

$$V^{(\mathrm{in})} = K_{\mathrm{o}}\frac{Q}{2R^3}(3R^2 - r^2)\,, \quad \mathbf{E}^{(\mathrm{in})} = K_{\mathrm{o}}\frac{Qr}{R^3}\,\mathbf{e}_{\mathrm{r}}\,, \quad V^{(\mathrm{ex})} = K_{\mathrm{o}}\frac{Q}{r}\,, \text{ and } \mathbf{E}^{(\mathrm{ex})} = K_{\mathrm{o}}\frac{Q}{r^2}\,\mathbf{e}_{\mathrm{r}}.$$

2.5. Symmetries of the electric field and potential

The electric charge being the source of the electrostatic field $\mathbf{E}$ and potential V, the symmetries of the sources are reflected on the field and the potential. Although the vector analysis is simpler in the Cartesian coordinates, the determination of the field and the potential is greatly simplified if we use curvilinear coordinates that have some of the symmetries of the charge. The most commonly used are the cylindrical coordinates (ρ, φ and z) and spherical coordinates (r, θ and φ) illustrated in Figure 2.1. The basis vectors, the relations of the curvilinear components to the Cartesian components as well as the expressions of the vector differential operator are given in the section A.10 of Appendix A.

a) If the charge configuration has a translational symmetry in a direction D (Figure 2.4a), it is convenient to take one axis of coordinates, Oz for instance, in the direction of D and use the Cartesian coordinates or cylindrical coordinates around D. Then, V and the components of $\mathbf{E}$ do not depend on z and we may write:

$$V = V(x, y) \quad \text{and} \quad \mathbf{E} = -\nabla V = -\partial_{\mathrm{x}}V\,\mathbf{e}_{\mathrm{x}} - \partial_{\mathrm{y}}V\,\mathbf{e}_{\mathrm{y}}, \qquad [2.26]$$

$$V = V(\rho, \varphi) \quad \text{and} \quad \mathbf{E} = -\nabla V = -\,[\partial_{\rho}V\,\mathbf{e}_{\rho} + \rho^{-1}\,\partial_{\varphi}V\,\mathbf{e}_{\varphi}]. \qquad [2.27]$$

In this case, the field is orthogonal to the direction of translation D.

b) If the charge configuration has rotational symmetry about an axis Oz (Figure 2.4b), it is convenient to use cylindrical coordinates (or spherical coordinates) about Oz. Then, V and the cylindrical components E_{ρ}, E_{φ}, and E_{z} (or the spherical components E_{r}, E_{θ}, and E_{φ}) do not depend on the azimuthal angle φ about Oz:

$$V = V(\rho, z) \quad \text{and} \quad \mathbf{E}(\rho, z) = -\nabla V = -\,[\partial_{\rho}V\,\mathbf{e}_{\rho} + \partial_{z}V\,\mathbf{e}_{\mathrm{z}}], \qquad [2.28]$$

$$V = V(r, \theta) \quad \text{and} \quad \mathbf{E}(r, \theta) = -\nabla V = -[\partial_{\mathrm{r}}V\,\mathbf{e}_{\mathrm{r}} + r^{-1}\,\partial_{\theta}f\,\mathbf{e}_{\theta}]. \qquad [2.29]$$

As $E_\varphi = 0$, **E** is everywhere in the azimuthal plane Π_1 containing *Oz* and the point *M*. On *Oz* itself, if **E** is finite and it has a unique determination as a function of ρ (near ρ = 0), it must be collinear with *Oz*, thus $E_\rho(0, z) = -\partial_\rho V|_{\rho=0} = 0$. If *V* is finite, it must have a maximum or a minimum in ρ on *Oz* (ρ = 0).

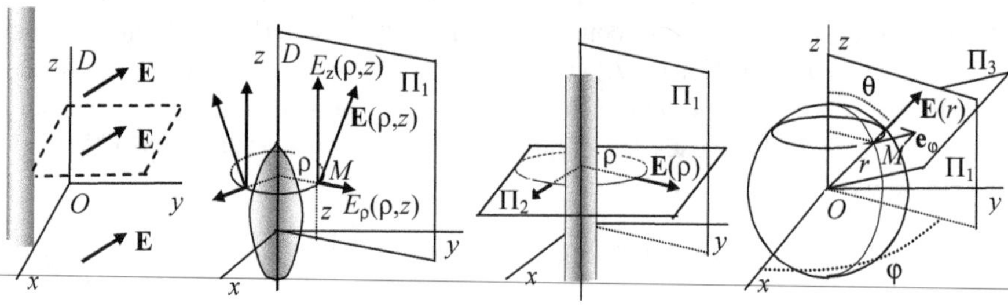

Figure 2.4. *Effects of charge symmetries on the electric field: a) translation in the direction of Oz, b) rotation about Oz, c) combined rotation about Oz and translation in the direction of Oz, and d) rotation around O*

c) In the case of a configuration of charge, which has both rotational symmetry about *Oz* and translational symmetry in the direction of *Oz* (Figure 2.4c), it is convenient to use cylindrical coordinates; then, *V* and **E** depend only on ρ:

$$V = V(\rho) \qquad \text{and} \qquad \mathbf{E}(\rho) = -\nabla V = -\partial_\rho V\, \mathbf{e}_\rho. \qquad [2.30]$$

In this case, **E** points in the radial direction $\mathbf{e}_\rho$. On *Oz* itself, if **E** is finite and it has a unique determination, it must vanish and, if *V* is finite, it verifies the condition $\partial_\rho V|_{\rho=0} = 0$. Thus, $V(\rho)$ must have a maximum or a minimum for ρ = 0.

d) If the charge configuration has a rotational symmetry around a point *O* (Figure 2.4d), it is convenient to use spherical coordinates *r*, θ and φ around *O*. Then, *V* and the components E_r, E_θ and E_φ do not depend on the angles θ and φ:

$$V = V(r) \qquad \text{and} \qquad \mathbf{E}(r) = -\nabla V = -\partial_r V\, \mathbf{e}_r. \qquad [2.31]$$

Thus **E** is radial in this case. This is also required by the rotational symmetry about the radial direction *OM* or the reflection symmetries with respect to the planes Π_1 and Π_3. At the center *O* itself, if **E** is finite and it has a unique determination, it must vanish and, if the potential *V* is finite, it must have a maximum or a minimum in *r*.

e) From the definition of the electric field, $\mathbf{F} = q\mathbf{E}$, and the definition of the potential by the relation $dV = -\mathbf{E}.d\mathbf{r}$, as *q* is a true scalar while **r** and **F** are true vectors, **E** must be a true vector and V a true scalar. In a reflection with respect to Oxy, for instance, the components of **E** transform like **r**, while V is unchanged:

$$E_x(x, y, z) = E'_x(x, y, -z), \ E_y(x, y, z) = E'_y(x, y, -z), \ E_z(x, y, z) = -E'_z(x, y, -z)$$
$$V(x, y, z) = V'(x, y, -z). \qquad [2.32]$$

– If the charge configuration is symmetric with respect to Oxy, we find

$$V(x,y,-z) = V(x,y,z), \quad \mathbf{E}_{//}(x,y,z) = \mathbf{E}_{//}(x,y,-z), \quad \text{and} \quad E_z(x,y,z) = -E_z(x,y,-z). \ [2.33]$$

From the third relation, we deduce that $E_z(x, y, 0) = 0$. The same symmetry or antisymmetry in z hold in cylindrical coordinates while, in spherical coordinates the symmetric points with respect to Oxy are $M(r, \theta, \varphi)$ and $M'(r, \pi-\theta, \varphi)$.

In the case of a charge distribution that is symmetric with respect to an arbitrary plane Π (Figure 2.5a), i.e. $q_v(M) = q_v(M')$ at any point M and M' symmetric with respect to Π, the potential and the field verify the *symmetry* conditions

$$V(M) = V(M'), \ \mathbf{E}_{//}(M) = \mathbf{E}_{//}(M'), \quad \text{and} \quad \mathbf{E}_{\perp}(M) = -\mathbf{E}_{\perp}(M'). \qquad [2.34]$$

Particularly, at the point M_o of Π, we must have $\mathbf{E}_{\perp}(M_o) = -\mathbf{E}_{\perp}(M_o)$. If **E** is finite at M_o, we must have $\mathbf{E}_{\perp}(M_o) = 0$. Thus the field at M_o lies in the plane Π.

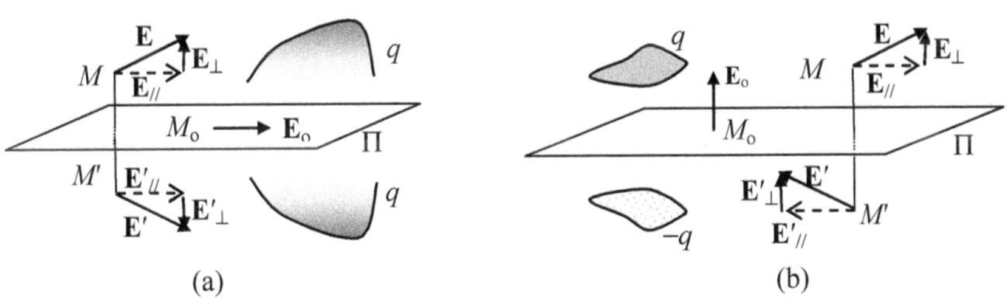

Figure 2.5. *a) Symmetry with respect to a plane* Π, *and b) antisymmetry with respect to* Π

– If the charge configuration is antisymmetric in the reflection with respect to Oxy, that is, $q_v(x, y, -z) = -q_v(x, y, z)$, **E** must verify the *antisymmetry* relations:

$$\mathbf{E}_{//}(x, y, z) = -\mathbf{E}_{//}(x, y, -z) \quad \text{and} \quad E_z(x, y, z) = E_z(x, y, -z). \qquad [2.35]$$

From the first relation, we deduce that $E_{//}(x, y, 0) = 0$. In the more general case of a charge distribution that is antisymmetric with respect to a plane Π (Figure 2.5b), that is, $q_v(M) = -q_v(M')$ at any points M and M' symmetric with respect to Π, **E** verifies the antisymmetry relations:

$$\mathbf{E}_{//}(M) = -\mathbf{E}_{//}(M') \quad \text{and} \quad \mathbf{E}_{\perp}(M) = \mathbf{E}_{\perp}(M'). \qquad [2.36]$$

Particularly, at points M_o of Π, we must have $\mathbf{E}_{//}(M_o) = -\mathbf{E}_{//}(M_o)$, thus $\mathbf{E}_{//}(M_o) = 0$, that is, **E** is normal to Π. The potential being defined up to a constant, it verifies the condition $V(M) = -V(M') + C$, where C is a constant. Particularly, at points M_o of Π, $V(M_o) = C/2$. The symmetry plane Π is thus equipotential.

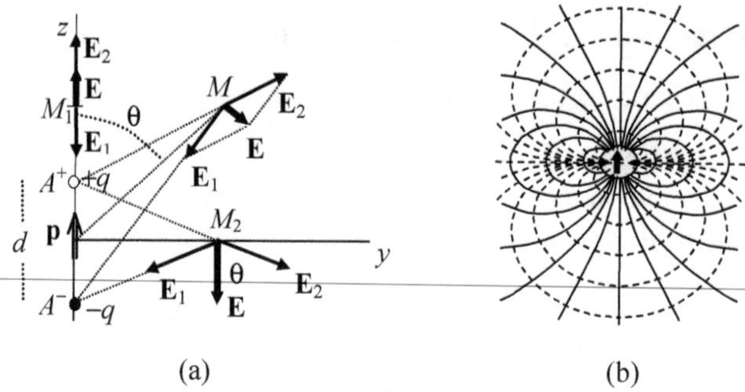

Figure 2.6. *a) Field of an electric dipole, and b) its lines of field (solid lines) and equipotential surfaces (dotted lines)*

2.6. Electric dipole

An *electric dipole* is a charge distribution that may be modeled as two charges $-q$ and $+q$ that we take at A^- and A^+ of coordinates $-d/2$ and $+d/2$ on the z axis (Figure 2.6a). The system having rotational symmetry about Oz, **E** and V do not depend on the azimuthal angle φ. Thus, we may evaluate **E** and V in the Oyz plane for instance. On the other hand, as the charge is symmetric in the reflection with respect to the Oyz plane, we have $E_y(0, y, z) = 0$ (see [2.34]) and, as the charge is antisymmetric in the reflection with respect to the Oxy plane, we have $E_z(x, y, z) = E_z(x, y, -z)$ (see [2.35]). Thus it is sufficient to calculate **E** and V at M with $z > 0$:

$$\mathbf{E}(\mathbf{r}) = K_o\,[q\frac{\overrightarrow{A^+M}}{A^+M^3} - q\frac{\overrightarrow{A^-M}}{A^-M^3}] \quad \text{and} \quad V(\mathbf{r}) = K_o\,[\frac{q}{A^+M} - \frac{q}{A^-M}]. \qquad [2.37]$$

As $\overrightarrow{A^\pm M} = \mathbf{r} \mp \mathbf{d}/2$, we find at large distances, to the first order in d/r,

$$\mathbf{E}(\mathbf{r}) \cong K_o\,[3(\mathbf{p.r})\frac{\mathbf{r}}{r^5} - \frac{\mathbf{p}}{r^3}] \quad \text{and} \quad V(\mathbf{r}) = K_o\,\frac{\mathbf{p.r}}{r^3}, \quad \text{where } \mathbf{p} = q\mathbf{d}. \qquad [2.38]$$

p is the *dipole moment.* Its field and potential decrease at large distances like $1/r^3$ and $1/r^2$, respectively, while those of a charge decrease like $1/r^2$ and $1/r$. Figure 2.6b illustrates the lines of field and the equipotential surfaces of a dipole.

More generally, consider the electric charges q_k occupying the positions $\mathbf{r}_k$ near the origin O. Their potential at a point $\mathbf{r}$ is $V(\mathbf{r}) = K_o\Sigma_k\, q_k/|\mathbf{r} - \mathbf{r}_k|$. At large distances ($r >> r_k$), expanding in a power series in $1/r$, we obtain to the first order in $\mathbf{r}_k/r$

$$|\mathbf{r} - \mathbf{r}_k|^{-1} = [r^2 - 2\mathbf{r}.\mathbf{r}_k + \mathbf{r}_k{}^2]^{-\frac{1}{2}} \cong r^{-1}[1 - 2(\mathbf{r}.\mathbf{r}_k)/r^2\,] \cong 1/r + (\mathbf{r}.\mathbf{r}_k)/r^3. \qquad [2.39]$$

The potential at large distances may be written as

$$V(\mathbf{r}) \cong K_o q/r + K_o(\mathbf{p}.\mathbf{r})/r^3, \qquad [2.40]$$

where we have set

$$q = \Sigma_k\, q_k, \qquad \mathbf{p} = \Sigma_k\, q_k\, \mathbf{r}_k. \qquad [2.41]$$

q is the *total charge* and $\mathbf{p}$ is the *electric dipole moment* of the charge distribution. If $q = 0$, $\mathbf{p}$ does not depend on the origin O in spite of the appearance of $\mathbf{r}_k$ in its expression. In the case of a continuous distribution of charge in a volume $\mathcal{V}$, the sums must be replaced by integrals and we obtain

$$q = \iiint_{\mathcal{V}} d\mathcal{V}'\, q_v(\mathbf{r}'), \qquad \mathbf{p} = \iiint_{\mathcal{V}} d\mathcal{V}' q_v(\mathbf{r}')\, \mathbf{r}'. \qquad [2.42]$$

At large distances, the potential is the sum of the potential of a charge q (which decreases like $1/r$) and that of an electric dipole moment $\mathbf{p}$ (which decreases like $1/r^2$). If $q = 0$ (as in the case of non-ionized atoms and molecules) the dominant term is that of the dipole. Some molecules (of water, for instance) are globally neutral but the barycenter A^+ of positive charges Q (the nuclei) is at a distance $\mathbf{d}$ from the barycenter A^- of negative charges $-Q$ (the electrons). Thus, the molecule has a permanent electric dipole moment $\mathbf{p} = Q\mathbf{d}$.

Let us consider now the action of an external electric field $\mathbf{E}$ on the electric dipole moment $\mathbf{p} = q\mathbf{d} = q\,\overrightarrow{A^-A^+}$ modeled by the charges $-q$ at A^- and $+q$ at A^+. The forces exerted by $\mathbf{E}$ on the charges $-q$ and $+q$ are $\mathbf{F}^- = -q\mathbf{E}(A^-)$ and $\mathbf{F}^+ = q\mathbf{E}(A^+)$ (Figure 2.7a). If the field is uniform $[\mathbf{E}(A^-) = \mathbf{E}(A^+)]$, the resultant of these forces is zero: the dipole undergoes no translational motion. However, the moment of these forces is

$$\mathbf{\Gamma} = \overrightarrow{OA^-} \times \mathbf{F}^- + \overrightarrow{OA^+} \times \mathbf{F}^+ = q\,(\overrightarrow{OA^+} - \overrightarrow{OA^-}\,) \times \mathbf{E} = q\,\mathbf{d} \times \mathbf{E} = \mathbf{p} \times \mathbf{E}. \qquad [2.43]$$

We take Ox in the direction of $\mathbf{E}$ and $\mathbf{p}$ in the Oxy plane and making an angle θ with $\mathbf{E}$. Then, $\mathbf{E} = E\mathbf{e}_x$, $\mathbf{p} = p\,(\cos\theta\, \mathbf{e}_x + \sin\theta\, \mathbf{e}_y)$ and

$$\mathbf{\Gamma} = -\,pE \sin\theta\, \mathbf{e}_z = \mathbf{p} \times \mathbf{E}. \qquad [2.44]$$

If $\theta > 0$, $\mathbf{\Gamma}$ is oriented in the opposite direction to Oz. Thus, the dipole rotates to align itself with $\mathbf{E}$ in a stable equilibrium position. Conversely, to maintain $\mathbf{p}$ at an angle θ' with $\mathbf{E}$, a moment $\mathbf{\Gamma}' = -\mathbf{\Gamma} = pE \sin\theta' \, \mathbf{e}_z$ must be exerted to counterbalance the electric moment of force. To rotate the dipole through an angle $d\theta'$, a work $dW' = \Gamma' d\theta' = pE \sin\theta' \, d\theta'$ is required. The total work required to rotate the dipole from the equilibrium position $\theta = 0$ to the position θ is

$$W'_{0\to\theta} = \int_0^\theta dW' = \int_0^\theta d\theta' \, pE \sin\theta' = -pE\cos\theta + pE. \qquad [2.45]$$

This is the *electric potential energy* of the dipole $\mathbf{p}$ if it makes an angle θ with $\mathbf{E}$. Dropping the constant term pE, we may write

$$U_E = -pE\cos\theta = -(\mathbf{p}.\mathbf{E}) = (\mathbf{p}.\boldsymbol{\nabla}V). \qquad [2.46]$$

In this expression, we do not take into account the binding energy of the dipole, since it is an internal and constant energy. Thus, U_E is the *interaction energy* of the dipole with the field $\mathbf{E}$.

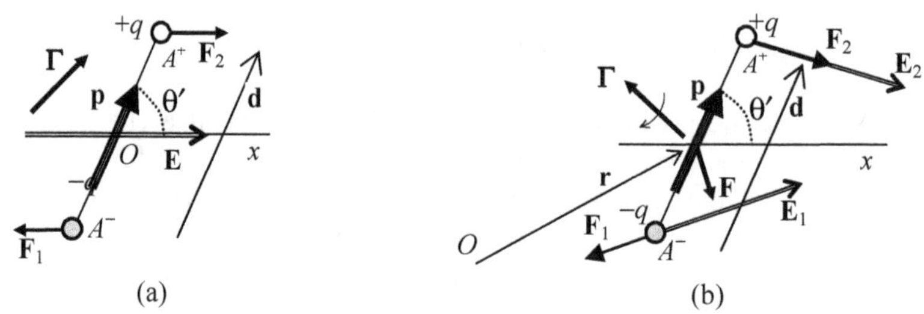

Figure 2.7. *a) Forces exerted by a uniform electric field on a dipole, and b) forces exerted by a non-uniform field on a dipole*

Consider now the case of a non-uniform electric field and the charges $-q$ and $+q$ located at $\mathbf{r}^+ = \mathbf{r} - \frac{1}{2}\mathbf{d}$ and $\mathbf{r}^- = \mathbf{r} + \frac{1}{2}\mathbf{d}/2$ (Figure 2.7b). If $\mathbf{E}$ varies slowly over the distance $\mathbf{d}$, we may write its components as power series of $\mathbf{d}$ up to the first order

$$\begin{aligned} E_\alpha(\mathbf{r} \pm \tfrac{1}{2}\mathbf{d}) &= E_\alpha(x \pm \tfrac{1}{2}d_x, y \pm \tfrac{1}{2}d_y, z \pm \tfrac{1}{2}d_z) \\ &= E_\alpha(x, y, z) \pm \tfrac{1}{2}d_x(\partial_x E_\alpha) \pm \tfrac{1}{2}d_y(\partial_y E_\alpha) \pm \tfrac{1}{2}d_z(\partial_z E_\alpha) = qE_\alpha(\mathbf{r}) \pm \tfrac{1}{2}(\mathbf{d}.\boldsymbol{\nabla})E_\alpha . \end{aligned}$$

The resultant of the electric forces acting on the dipole has the components

$$F_\alpha = q[E_\alpha(\mathbf{r}+\mathbf{d}/2) - E_\alpha(\mathbf{r}-\mathbf{d}/2)] = q(\mathbf{d}.\boldsymbol{\nabla})\, E_\alpha(\mathbf{r}) = (\mathbf{p}.\boldsymbol{\nabla})\, E_\alpha(\mathbf{r}). \qquad [2.47]$$

Using the expression of $\mathbf{E}$ in terms of the potential V and the expression of the potential energy of the dipole $U_E = -\Sigma_\beta p_\beta E_\beta$, we may write

$$F_\alpha = \Sigma_\beta p_\beta (\partial_\beta E_\alpha) = -\Sigma_\beta p_\beta (\partial^2_{\beta\alpha} V) = -\partial_\alpha (\Sigma_\beta p_\beta \partial_\beta V) = \partial_\alpha \Sigma_\beta (p_\beta E_\beta)$$
$$= -\partial_\alpha(-\mathbf{p}.\mathbf{E} + pE) = -\partial_\alpha U_E. \quad [2.48]$$

Thus we find the general expression

$$\mathbf{F} = -\nabla U_E. \quad [2.49]$$

Here U_E is a function of the coordinates $\mathbf{r}(x, y, z)$ of the center O of the dipole and ∇ is the vector differential operator with respect to these coordinates.

For instance, if $\mathbf{E}$ points in the direction of Ox ($\mathbf{E} = E\mathbf{e}_x$), it acts on the dipole $\mathbf{p}$ to orient it in the direction Ox (thus, $p_x > 0$ and $p_y = p_z = 0$). If $\mathbf{E}$ is non-uniform, the relation [2.48] gives the components of the resultant force $F_\alpha = p_x \partial_\alpha E$. For instance, if E is an increasing function of z ($\partial_x E = \partial_y E = 0$ and $\partial_z E > 0$), the resultant force has one component $F_z = p_x \partial_z E > 0$. Thus the force $\mathbf{F}$ points toward the increasing field.

We have shown that the field and the potential, which are produced at large distances by the distribution of charges q_k located at points $\mathbf{r}_k$ are the superposition of those of a single charge $q = \Sigma_k q_k$ and an electric dipole $\mathbf{p} = \Sigma_k q_k \mathbf{r}_k$. Let us now consider the action of an external field $\mathbf{E} = -\nabla V$ on these charges. If the distances r_k of the charges from the origin O are small and V varies slowly in the region that is occupied by these charges, we may make a power series expansion of V in the coordinates up to the first order

$$V(x_\alpha) = V(0) + \Sigma_\alpha \partial_\alpha V|_o x_\alpha + \ldots \cong V(0) - \Sigma_\alpha E_\alpha(0)\, x_\alpha = V(0) - \mathbf{r}.\mathbf{E}(0), \quad [2.50]$$

where the derivatives $\partial_\alpha V$ are evaluated at $x_\alpha = 0$. The potential energy of the charges q_k in the field is

$$U_E = \Sigma_k q_k V(\mathbf{r}_k) \approx \Sigma_k q_k [V(0) - \mathbf{r}.\mathbf{E}(0)] \approx q\, V(0) - \mathbf{p}.\mathbf{E}(0). \quad [2.51]$$

The first term is the potential energy of the total charge $q = \Sigma_k q_k$ and the second term is the potential energy of the electric dipole moment $\mathbf{p} = \Sigma_k q_k \mathbf{r}_k$, both located at O. Knowing U_E, we may show that the resultant force exerted by $\mathbf{E}$ on the charges is the force exerted on the total charge q and the resultant moment of the electric force is the moment $\boldsymbol{\Gamma} = \mathbf{p} \times \mathbf{E}$ exerted on $\mathbf{p}$. Particularly, if the total charge is zero (as in the case of a non-ionized atoms and molecules), the action of $\mathbf{E}$ on the charge distribution reduces to its action on the dipole moment $\mathbf{p}$.

2.7. Electric field and potential of simple charge configurations

The potential and the field of any charge distribution may be evaluated by using the integrals [2.5] and [2.7]. If the charge has some symmetry, it is sometimes possible to find a closed surface $\mathcal{S}$, on which **E** is normal with a uniform magnitude. The electric flux through this surface is $\Phi_E = E\mathcal{S}$. On a part of $\mathcal{S}$, the flux of **E** may be zero either because **E** is zero or **E** is tangent to $\mathcal{S}$. If $Q^{(in)}$ is the charge within $\mathcal{S}$, Gauss's law gives $E = Q^{(in)}/\varepsilon_o\mathcal{S}$. In this section we give the expressions of the field and the potential of some simple charge configurations.

– Field and potential on the axis of a uniformly charged ring of radius R and charge q at a distance z from its center:

$$\mathbf{E}(z) = \frac{K_o q z}{(R^2+z^2)^{3/2}}\,\mathbf{e}_z, \qquad V(z) = \frac{K_o q}{\sqrt{R^2+z^2}}. \qquad [2.52]$$

– Field and potential on the axis of a disk of radius R and uniform charge density q_s at a distance z from its center:

$$\mathbf{E}(z) = \frac{z q_s}{2\varepsilon_o}\left[\frac{1}{|z|} - \frac{1}{\sqrt{R^2+z^2}}\right]\mathbf{e}_z, \qquad V(z) = \frac{q_s}{2\varepsilon_o}\left[\sqrt{R^2+z^2} - |z|\right]. \qquad [2.53]$$

– Field and potential of a plane of uniform charge density q_s:

$$\mathbf{E}(z) = \pm\,(q_s/2\varepsilon_o)\,\mathbf{e}_z, \qquad V(z) = \mp\,(q_s z/2\varepsilon_o) + (q_s R/2\varepsilon_o). \qquad [2.54]$$

– Field and potential of an infinite thin rod of uniform charge density q_L at a distance ρ from the rod (ρ_o is a reference distance where the potential is taken equal to zero):

$$\mathbf{E}(\rho) = (2K_o q_L/\rho)\mathbf{e}_\rho, \qquad V(\rho) \to (2K_o q_L/\rho)\,\ln(\rho_o/\rho). \qquad [2.55]$$

– Field and potential of a uniformly charged infinite cylinder of radius R and charge density q_L per unit length at a distance ρ from its axis:

$$\mathbf{E}^{(in)} = (2K_o q_L \rho/R^2)\mathbf{e}_\rho, \qquad V^{(in)} = K_o q_L[1 - \rho^2/R^2 + 2\,\ln(\rho_o/R)], \qquad [2.56]$$

$$\mathbf{E}^{(ex)} = (2K_o q_L/\rho)\,\mathbf{e}_\rho, \qquad V^{(ex)} = 2K_o q_L\,\ln(\rho_o/\rho). \qquad [2.57]$$

– Field and potential at a distance r from the center of a spherical shell of radius R and uniform charge Q:

$$\mathbf{E}^{(in)}(\mathbf{r}) = 0, \qquad V^{(in)} = K_o Q/R,$$

$$\mathbf{E}^{(ex)}(\mathbf{r}) = (K_o Q/r^2)\,\mathbf{e}_z, \qquad V^{(ex)} = K_o Q/r. \qquad [2.58]$$

– Field and potential at a distance r from the center of a ball of radius R and uniform charge Q:

$$\mathbf{E}^{(in)}(\mathbf{r}) = K_o Q\mathbf{r}/R^3 = (q_v/3\varepsilon_o)\mathbf{r}, \qquad V^{(in)} = (K_o Q/2R)(3 - r^2/R^2) = (q_v/6\varepsilon_o)(3R^2 - r^2),$$

$$\mathbf{E}^{(ex)}(\mathbf{r}) = (K_o Q/r^3)\,\mathbf{r} = (q_v R^3/3\varepsilon_o r^3)\mathbf{r}, \qquad V^{(ex)} = K_o Q/r = (q_v R^3/3\varepsilon_o r). \qquad [2.59]$$

2.8. Some general properties of the electric field and potential

If the potential V is constant in a region, the field $\mathbf{E} = -\boldsymbol{\nabla} V$ is zero and Gauss's law $\boldsymbol{\nabla}.\mathbf{E} = q_v/\varepsilon_o$ implies that there is no charge density in this region. The reciprocal is not always true: if the charge density is zero in a region, the field $\mathbf{E}$ is not necessarily zero but Gauss's law implies that $\boldsymbol{\nabla}.\mathbf{E} = 0$. The flux of $\mathbf{E}$ through any closed surface $\mathcal{S}$ entirely in this region is zero. The electric field lines that enter $\mathcal{S}$ on one side leave it on the other. For instance, this is the case of a uniform field (the field lines are then parallel as in Figure 2.8a) and in the case of the field lines of a ball (the field lines are then radial as in Figure 2.8b).

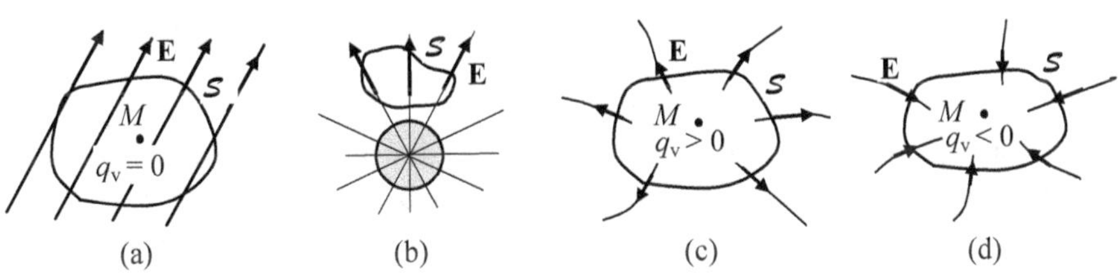

Figure 2.8. *Electric field lines in a region depending on the charge density*

If the potential has a maximum at M, as the field points toward the decreasing potential, it must diverge from M (Figure 2.8c). The outgoing flux from a surface $\mathcal{S}$ surrounding M is positive and Gauss's law implies that $\mathcal{S}$ contains a positive charge. As $\mathcal{S}$ may be taken arbitrarily small, we deduce that a positive charge must exist at M. By a similar argument, we show that, if the potential has a minimum at M, the field must converge toward M and a negative charge must exist at this point (Figure 2.8d). It is not necessary that the charge at M be a point charge. For instance, in the case of a charged sphere, the center is a maximum or a minimum of the potential without having a point charge.

Singularities and discontinuities of the field and the potential

The expressions of the field and the potential given in section 2.7 show that $\mathbf{E}$ and V are not always regular and continuous functions:

a) Near a point charge q, the field and the potential of this charge are much more important than those of the other charges. The field lines diverge in all directions from the position O of q (if q is positive) and converge toward O (if q is negative). Thus, the direction of **E** is not defined at O (Figure 2.9a). Furthermore, **E** and V are infinite like $\mathbf{E} \approx (K_o q/r^2)\mathbf{e}_r$ and $V \approx K_o q/r$ as $r \to \infty$ (Figure 2.9b).

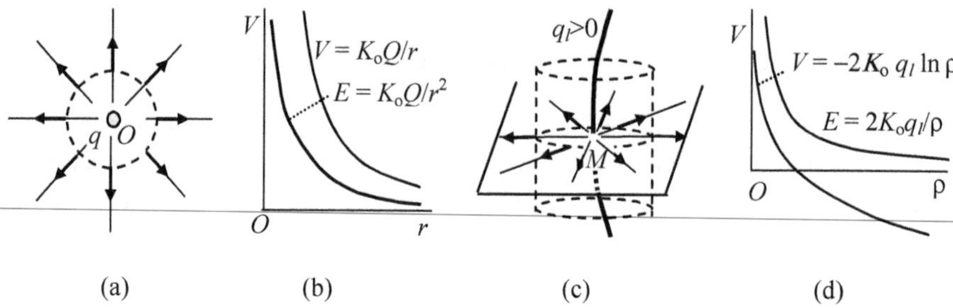

Figure 2.9. *a) At the position of a point charge, the direction of* **E** *is not defined, and b) E behaves like K_oq/r^2 and V like K_oq/r. c) At some point on a charged line, the direction of* **E** *is not defined, and c) E behaves like $2K_oq_l/\rho$ and V like $-2K_oq_l \ln \rho$*

b) Near a point M of a line of charge density q_l, the most important contribution to **E** and V are those of a small element of the line on both sides of M. The field lines diverge from M in all directions normally to the charged line (if q_l is positive) and converge toward M (if q_l is negative). Thus, the direction of **E** is not defined (Figure 2.9c). By considering a small Gaussian cylinder around an element of the charged line, we find that the dominant terms of **E** and V are $\mathbf{E} \approx (2K_o q_l/\rho)\mathbf{e}_\rho$ and $V \approx -2K_o q_l \ln \rho$. Thus **E** and V are infinite on the charged line (Figure 2.9d).

c) In the case of a surface $\mathcal{S}$ of charge density q_s, we consider a small cylindrical Gaussian surface, with a lateral surface that is very short and normal to $\mathcal{S}$ and its bases are situated on both sides of $\mathcal{S}$ (Figure 2.10a). Assuming that there is no point charge or line charge on $\mathcal{S}$, the total charge situated inside the cylinder is $q_s\, d\mathcal{S}$. Let $\mathbf{n}_{12}$ be the unit vector normal to $\mathcal{S}$ and oriented toward the medium (2). Gauss's law for the cylinder may be written as $d\mathcal{S}\,\mathbf{E}_2.\mathbf{n}_{12} - d\mathcal{S}\,\mathbf{E}_1.\mathbf{n}_{12} = d\mathcal{S}\, q_s/\varepsilon_o$. We deduce that the normal component of the field undergoes a discontinuity on the surface

$$\mathbf{E}_2.\mathbf{n}_{12} - \mathbf{E}_1.\mathbf{n}_{12} = q_s/\varepsilon_o, \quad \text{i.e. } E_{2\perp} - E_{1\perp} = q_s/\varepsilon_o. \tag{2.60}$$

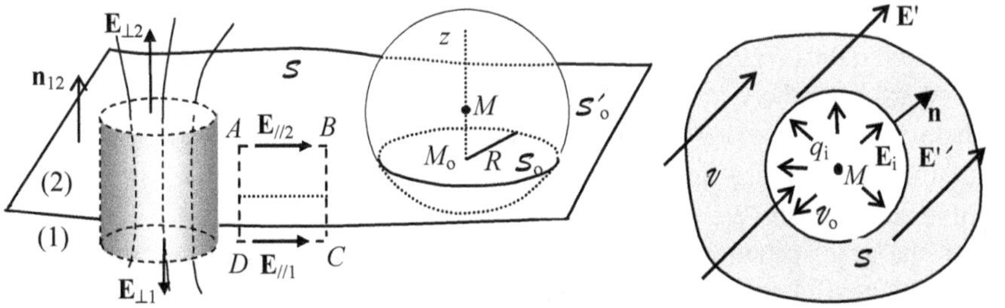

(a) (b)

Figure 2.10. *a) On a surface carrying a charge density q_s, the normal component of* **E** *is finite but it has a discontinuity q_s/ε_o, while the tangential component is finite and continuous. b) In the case of a volume charge distribution,* **E** *and V are finite and continuous*

Consider now a rectangular path $ABCD$ of sides AB and CD parallel to S and situated on one side S and the other, while BD and AC are very short. As **E** is conservative, its circulation on this path is equal to zero, thus, $\mathbf{E}_2.\overrightarrow{AB}+\mathbf{E}_1.\overrightarrow{CD}=0$, i.e. $E_{2//}\,AB - E_{1//}\,CD = 0$. As $AB = CD$, we deduce that

$$E_{2//} = E_{1//}. \tag{2.61}$$

This relation holds for any direction of AB and CD parallel to S. Thus the tangential component of **E** is continuous on the charged surface. In order to understand this result, we consider a point M near the charged surface and surround it by a small sphere that contains a small zone S_o of the charged surface (Figure 2.10a). Let S'_o be the remaining part of S that lies outside the sphere. The field and the potential at M may be written as

$$V(\mathbf{r}) = K_o \iint_{S_o} dS'\, q_s(\mathbf{r}')/|\,\mathbf{r}-\mathbf{r}'| + K_o \iint_{S'_o} dS'\; q_s(\mathbf{r}')/|\,\mathbf{r}-\mathbf{r}'|\,,$$

$$\mathbf{E}(\mathbf{r}) = K_o \iint_{S_o} dS'\, q_s(\mathbf{r}')(\mathbf{r}-\mathbf{r}')/|\,\mathbf{r}-\mathbf{r}'|^3 + K_o \iint_{S'_o} dS'\; q_s(\mathbf{r}')(\mathbf{r}-\mathbf{r}')/|\,\mathbf{r}-\mathbf{r}'|^3. \tag{2.62}$$

If M is close to S, the integrals over S'_o are continuous and finite as $|\mathbf{r}-\mathbf{r}'|$ does not vanish. The integrals over S_o are approximately the potential and the field of a disk of radius $(R^2 - z^2)^{1/2} \approx R$ and charge density q_s. According to [2.54], these are $V_{\text{disk}} \approx q_s R/2\varepsilon_o$ and $\mathbf{E}_{\text{disk}} \approx \pm(q_s/2\varepsilon_o)\mathbf{e}_z$. Thus, the potential of the disk vanishes in the limit $R\to 0$ and $z\to 0$ but $\mathbf{E}_{\text{disk}}$, which is normal to the disk, remains finite and has a discontinuity q_s/ε_o. We deduce that, on a charged surface, V and the tangential component of **E** are finite and continuous while the normal component of **E** is finite but it has a discontinuity q_s/ε_o.

d) In the case of a volume charge distribution, let us consider a small sphere $\mathcal{V}_o$ of center M and radius R (Figure 2.10b). According to [2.59], the field and the potential produced by the sphere at its center M are $E_o = 0$ and $V_o = q_V R^2/2\varepsilon_o$. Thus they vanish in the limit $R \to 0$ while $\mathbf{E}^{(ex)}$ and $V^{(ex)}$ that are produced by the charges outside the sphere $\mathcal{V}_o$ are finite and continuous, as the distance $|\mathbf{r} - \mathbf{r}'|$ in [2.5] and [2.7] does not vanish. We deduce that the field and the potential produced by any volume charge distribution are continuous and finite both inside and outside the charged volume if it contains no point charges, linear charge, or surface charge.

By comparing the linear charge to a cylinder and the point charge to a ball, we find that the singularities of $\mathbf{E}$ and V are due to the zero radius limit of the cylinder and the ball. Also, the discontinuity of $\mathbf{E}_\perp$ on a charged surface is due to the zero thickness of the charge distribution (see Problem 2.21). In fact, the point charge, the linear charge and the surface charge are mathematical idealizations. On the macroscopic scale, all bodies have always non-zero dimensions. On the microscopic scale, the elementary particles (such as electrons or protons) are considered as a point, but the concepts of position and radius lose their classical significations.

Because of the quantization of electricity as point-like particles, the superposition of their individual fields $\mathbf{E}_i$ and potentials V_i gives the *microscopic* field and potential, which undergo large fluctuations and they even become infinite at the positions of the particles. A macroscopic element of volume, area or length are assumed to be sufficiently large to contain a very large number of particles. The field $\mathbf{E}$ and the potential V, evaluated by using continuous charge densities, are said to be *macroscopic*. These are the averaged values of the microscopic field and potential over finite space element and time intervals.

Consider a small sphere $\mathcal{S}$ surrounding a particle (i) and containing no other charge (Figure 2.10b). The total average field $\langle\mathbf{E}\rangle$ in $\mathcal{S}$ is the vector sum $\langle\mathbf{E}_i\rangle + \langle\mathbf{E}'\rangle$ where $\mathbf{E}_i$ is the field of the particle (i) and $\mathbf{E}'$ is the field of the other particles located outside $\mathcal{S}$. As $\mathbf{E}_i$ is radial and it has a spherical symmetry, its average value in $\mathcal{S}$ is zero, while $\mathbf{E}'$ is regular inside $\mathcal{S}$. Thus the macroscopic field is regular. In the same way, we may show the regularity of the potential in the case of volume charge distribution and surface charge distribution.

2.9. Electrostatic energy of a system of charges

The energy U of a system of particles without intrinsic structure is the sum of their kinetic energies $U_{K,i} = \frac{1}{2} m_i v_i^2$ and their *interaction potential energy*. The electrostatic interaction potential energy U_E is the work required to bring the initially far away particles to their actual positions $\mathbf{r}_i$ without acquiring kinetic energy. As the electrostatic forces are conservative, U_E is a function of the relative positions $|\mathbf{r}_i - \mathbf{r}_j|$ of all the pairs of charged particles.

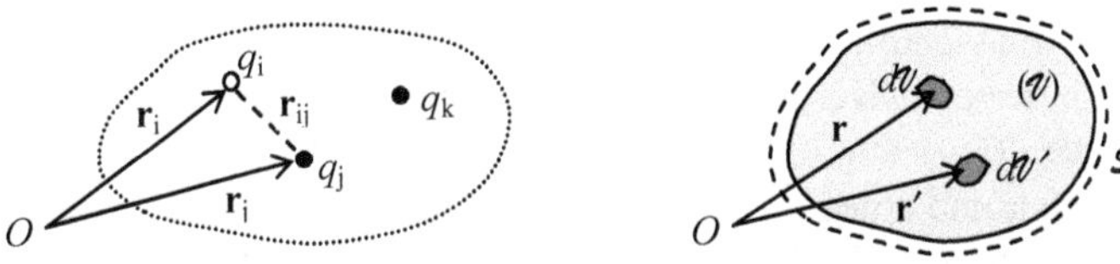

(a) (b)

Figure 2.11. *Interaction potential energy: a) for a system of discrete charges, and b) for a continuous charge distribution*

Consider, for instance three charged particles q_i (i = 1, 2, and 3) (Figure 2.11a). To bring q_1 from infinity to its position $\mathbf{r}_1$ in the absence of the other charges, no work is necessary. Then, to bring q_2 from infinity to its position $\mathbf{r}_2$ in the presence of q_1, a force $-\mathbf{F}_{12}$ must be exerted and a work $U_{12} \equiv K_o q_1 q_2 / r_{12}$ is required. Finally, to bring q_3 from infinity to its position $\mathbf{r}_3$, a force $-(\mathbf{F}_{13} + \mathbf{F}_{23})$ must be exerted and work $U_{13} + U_{23}$ is required. Thus, the total work necessary to assemble the three charges is

$$U_E = U_{12} + U_{13} + U_{23} = K_o (q_1 q_2 / r_{12} + q_1 q_3 / r_{13} + q_2 q_3 / r_{23}). \qquad [2.63]$$

This result may be easily generalized to systems of several charges q_i (i = 1, 2, N). Each pair of charges contributes a term U_{ij} to the total potential energy. Thus, we may write U_E as

$$U_E = \Sigma_{pairs}\, U_{ij} = \tfrac{1}{2} \Sigma_{i \neq j}\, U_{ij}, \quad \text{where} \quad U_{ij} = K_o q_i q_j / r_{ij}. \qquad [2.64]$$

In the expression $U_E = \Sigma_{pairs}\, U_{ij}$, the summation is carried over all the $\tfrac{1}{2}N(N-1)$ pairs of particles and, in the expression $U_E = \tfrac{1}{2} \Sigma_{i \neq j}\, U_{ij}$, the summation is carried over all distinct particles (the factor ½ takes into account the fact that each pair is counted twice). Explicitly, we find

$$U_E = \tfrac{1}{2} \sum_{i=1}^{i=N} q_i \,[\, K_o \sum_{j \neq i} q_j / r_{ij} \,] = \tfrac{1}{2} \sum_{i=1}^{i=N} q_i \, V'(\mathbf{r}_i), \qquad [2.65]$$

where $V'(\mathbf{r}_i) \equiv K_o \Sigma_{j \neq i}\, q_j / r_{ij}$ is the potential produced at $\mathbf{r}_i$ by all the charges except q_i. Note that V' is finite and different from the potential V produced by all the particles (which is infinite at the positions $\mathbf{r}_i$ of the point charges q_i).

As the potential energy is defined up to the addition of a constant, we choose this constant such that $U_{ij} \to 0$ in the limit $r_{ij} \to \infty$. The potential energy U_E may be positive or negative. If $U_E > 0$, a positive work must be done against the repulsive forces in order to assemble the charges in their actual positions. If the system is then left to itself, these repulsive forces disperse the charged bodies; the potential energy U_E is then transformed into kinetic energy. But, if $U_E < 0$, the particles are attracted toward each other and a negative work must be done against the attractive forces in order to prevent the particles from acquiring kinetic energy. If the system is then left to itself, these attractive forces maintain the particles bound together, as the negative potential energy U_E cannot be transformed into kinetic energy (which is always positive). A work $W = -U_E$ is required to completely separate the charges; W is the *binding energy* of the system of bound charges.

In the case of a continuous charge distribution with a density $q_v(\mathbf{r})$ (Figure 2.11b), an infinitesimal volume $d\mathcal{V}_i$ near $\mathbf{r}_i$ contains the charge $dq_i = q_v(\mathbf{r}_i)\, d\mathcal{V}_i$. The interaction energy of $d\mathcal{V}_i$ and $d\mathcal{V}_j$ is $U_{ij} = ½K_o d\mathcal{V}_i\, d\mathcal{V}_j\, q_v(\mathbf{r}_i)q_v(\mathbf{r}_j)/|\mathbf{r}_i-\mathbf{r}_j|$ and the total interaction energy is

$$U_E = ½K_o\iiint_{\mathcal{V}} d\mathcal{V}' \iiint_{\mathcal{V}} d\mathcal{V}\, q_v(\mathbf{r}')q_v(\mathbf{r})/|\mathbf{r}-\mathbf{r}'| = ½\iiint_{\mathcal{V}} d\mathcal{V}\, q_v(\mathbf{r})\, V(\mathbf{r}). \qquad [2.66]$$

In the second expression, $V(\mathbf{r}) = K_o\iiint_{\mathcal{V}} d\mathcal{V}'\, q_v(\mathbf{r}')/|\mathbf{r}-\mathbf{r}'|$ is the potential produced at $\mathbf{r}$ by all the charge of $\mathcal{V}$. As in [2.65], in principle, we should consider different elements of volume $d\mathcal{V}_i$ and $d\mathcal{V}_j$ and use the potential $V'(\mathbf{r})$ of the whole volume $\mathcal{V}$ except $d\mathcal{V}$. However, the potential of $d\mathcal{V}$ tends to zero with the dimensions of $d\mathcal{V}$ (see, for instance, equation [2.59] in the case of a small sphere of radius R). Thus, we may use the total potential V instead of V'. Similar relations to [2.66] may be written in the case of a surface charge but not in the case of a linear charge as the potential of an element of length dL does not go to zero with dL (see section 2.7 and Problem 2.18).

Using Gauss's law, we may write the electrostatic energy [2.66] in a form that uses only the potential and the field:

$$U_E = ½\varepsilon_o \iiint_{\mathcal{V}} d\mathcal{V}\, (\boldsymbol{\nabla}.\mathbf{E})\, V(\mathrm{r}) = ½\varepsilon_o \iiint_{\mathcal{V}} d\mathcal{V}\, \boldsymbol{\nabla}.(V\mathbf{E}) + ½\varepsilon_o \iiint_{\mathcal{V}} d\mathcal{V}\, \mathbf{E}^2, \qquad [2.67]$$

where we have used the relation

$$(\boldsymbol{\nabla}.\mathbf{E})V = \Sigma_\alpha \nabla_\alpha E_\alpha.V = \Sigma_\alpha[\nabla_\alpha(E_\alpha V) - E_\alpha.\nabla_\alpha V] = \Sigma_\alpha V_\alpha(E_\alpha V) + E_\alpha E_\alpha = \boldsymbol{\nabla}.(V\mathbf{E}) + \mathbf{E}^2.$$

If we are only interested in the total electrostatic energy of a charge distribution that occupies a finite $\mathcal{V}$, the first term of equation [2.67] is the integral of the divergence of $(V\mathbf{E})$ over this volume. It may be transformed into the outgoing flux of $(V\mathbf{E})$ from a large surface $\mathcal{S}$ enclosing the system on which V and $\mathbf{E}$ go to zero rapidly enough for the flux to be zero[1]. Thus we may write

$$U_E = \tfrac{1}{2}\varepsilon_0 \iiint_{\mathcal{V}} d\mathcal{V}\, \mathbf{E}^2. \qquad [2.68]$$

This expression is equivalent to an electric energy of volume density

$$U_{E,v} = \tfrac{1}{2}\, \varepsilon_0 \mathbf{E}^2 \qquad [2.69]$$

at each point in an electric field $\mathbf{E}$. For instance, in the case of a parallel plate capacitor, the field has a magnitude $E = q_s/\varepsilon_0 = Q/\mathcal{S}\varepsilon_0$ and it is localized between the plates. Thus, the electric energy is stored with a density $U_{E,v} = \tfrac{1}{2}\varepsilon_0(Q/\mathcal{S}\varepsilon_0)^2$ in this region and the total energy is $U_E = \mathcal{V}U_{E,v} = Q^2/2\varepsilon_0$.

This result is not a simple mathematical equivalence. The electric energy is effectively localized where there is an electric field, exactly as calorific energy is localized at hot places in a medium. Equation [2.68] gives not only the interaction energy of the whole system, but the energy in each volume $\mathcal{V}$ of the system. Contrary to equation [2.69], equation [2.66] does not interpret $\tfrac{1}{2}q_v(\mathbf{r})V(\mathbf{r})$ as the electric energy density, since it incorrectly implies that the energy density vanishes at points where $q_v(\mathbf{r}) = 0$. On the other hand, the expression $\tfrac{1}{2}q_v(\mathbf{r})V(\mathbf{r})$ is inadequate for the energy density, as it may be arbitrarily modified by adding a constant to V.

The practical relevance of the localization of energy is that the conservation of energy is not only global in the Universe but local in each element of volume. In each volume $\mathcal{V}$, there is a certain stored energy, a certain dissipated energy, and a certain energy being exchanged with the exterior through the surface $\mathcal{S}$ that encloses the volume $\mathcal{V}$. The dissipation of energy and its propagation are important in theoretical physics and in the applications of physics.

If two systems interact (two atoms within a molecule for instance), the total interaction energy may be split into internal interaction energies $U_{E,1}$ and $U_{E,2}$ (between the pairs of each systems considered separately) and the interaction energy $U_{E,(1,2)}$ of the particles of one system with the particles of the other:

$$U_E = U_{E,1} + U_{E,2} + U_{E,(1,2)}. \qquad [2.70]$$

---

1 On a sphere of large radius r, for instance, the element of area $d\mathcal{S}$ increases like r^2. If V decreases like r^{-n}, E decreases like r^{-n-1} and the product $V\mathbf{E}$ decreases like r^{-2n-1}. The flux of $V\mathbf{E}$ through this sphere goes to zero if $-2n-1+2$ is negative, i.e. if $n > \tfrac{1}{2}$.

Adding the kinetic energies and the other forms of energy for each system, the total energy may be written as:

$$U = U_1 + U_2 + U_{E,(1,2)}, \quad \text{where } U_1 = U_{K,1} + U_{E,1} \text{ and } U_2 = U_{K,2} + U_{E,2}. \qquad [2.71]$$

The interaction energy $U_{E,(1,2)}$ may be expressed in terms of the potential V_1 or V_2 produced by the particles of one of the systems at the positions of the charges of the other system

$$U_{E(1,2)} = \sum_{i_1} q_{i_1} V_2(\mathbf{r}_{i_1}) = \sum_{i_2} q_{i_2} V_1(\mathbf{r}_{i_2}), \qquad [2.72]$$

where the summations are carried over the charges of one of the systems. The knowledge of the interaction energy $U_{E(1,2)}$ allows the force exerted by one system on the other to be determined. For instance, if system (2) is rigid and its position is specified by a coordinate x, the x component of the force acting on it is $F_x = -\partial_x U_{E,(12)}(x)$. Similarly, if its orientation is specified by an angle θ about the axis Oz, the moment of the electric forces has a component $\Gamma_z = -\partial_\theta U_{E,(12)}$ about Oz.

If system (2) consists of a single particle of charge q and position $\mathbf{r}$, its interaction energy with system (1) is, according to [2.72], $U_{E(1,2)} = q\ V_1(\mathbf{r})$ where V_1 is the potential produced by system (1). The force exerted on this particle is

$$\mathbf{F} = -\nabla U_{E(1,2)} = -q\,\nabla V_1(\mathbf{r}) = -q\,[\partial_x V_1 \mathbf{e}_x + \partial_y V_1 \mathbf{e}_y + \partial_z V_1 \mathbf{e}_z] = q\mathbf{E}_1. \qquad [2.73]$$

More generally, if the interaction energy U_E of system (1) with a set of particles is expressed as a function of their coordinates, the force exerted on the particle (i) is

$$\mathbf{F}_i = -\nabla_i U_E(\mathbf{r}_1, \mathbf{r}_2, \ldots \mathbf{r}_N), \qquad [2.74]$$

where ∇_i is the vector differential operator with respect to the coordinates x_i, y_i, z_i of the particle (i).

In the case of time-dependent phenomena, this analysis must undergo two important modifications: the appearance of magnetic forces that are non-conservative (as they depend on velocities) and the non-validity of the action-at-a-distance. Then, it is imperative to use the interaction through the fields, which propagate and carry their proper energy, momentum, and angular momentum.

A system is in a state of equilibrium if it remains indefinitely in this state without any modification. The equilibrium is stable if it corresponds to the minimum potential energy U_E. Indeed, to move spontaneously away from this position, the potential energy of the system must increase (from the minimum), thus its kinetic energy U_K must decrease and this is impossible as U_K cannot be negative. If the system is slightly displaced away from this position and left without initial velocity, it may only come back and oscillate near this position. The potential energy of a charge q in a potential V being $U_E = qV$, a minimum of V is a stable equilibrium position for a positive charge q and a maximum of V is a stable equilibrium position for a negative charge q. For instance, the point situated halfway between two charges $+q$ in Figure 2.12a is unstable for a charge $-q'$ in longitudinal motion and for a charge $+q'$ in transversal motion. Thus, whatever the charge, it will not be in a stable equilibrium position for the motion in at least some directions.

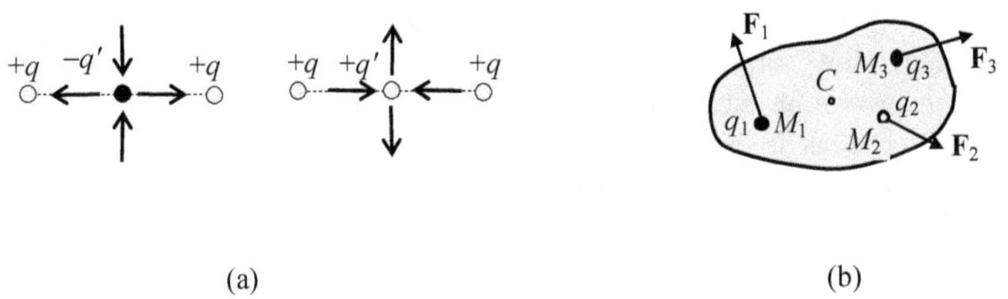

Figure 2.12. *a) A system of charges cannot be stable under the effect of electric forces alone. b) A body carrying bound charges q_i cannot be stable in an external field*

Consider now an extended, non-conducting, rigid body carrying bound charges q_i at points M_i (Figure 2.12b). Assume that this body is in equilibrium under the effect of mechanical forces and internal forces (including the electric interactions of the charges). If an external electric field acts on this body, new forces $q_j\mathbf{E}(\mathbf{r}_j)$ appear. Let $\mathbf{r}_C$ be the position of the center of mass of this body. The positions of the charges from the origin are $\mathbf{r}_j = \mathbf{r}_C + \overrightarrow{CM}_j$. The translational motion of the body is that of the center of mass, i.e. a point mass m equal to the total mass of the body and subject to the resultant of the electric forces $\mathbf{F}_E = \Sigma_i\, q_j\mathbf{E}(\mathbf{r}_j)$. If $\boldsymbol{\nabla}_C$ is the vector differential operator with respect to the coordinates of the center of mass and $\boldsymbol{\nabla}_j$ is the vector differential operator with respect to the components of $\mathbf{r}_j$, the force $\mathbf{F}_E$ verifies the equation

$$\boldsymbol{\nabla}_C.\mathbf{F}_E = \Sigma_j\, q_j\, \boldsymbol{\nabla}_C.\mathbf{E}(\mathbf{r}_j) = \Sigma_j\, q_j\, \boldsymbol{\nabla}_j.\mathbf{E}(\mathbf{r}_j). \qquad [2.75]$$

If there are no source charges of $\mathbf{E}$ at point $\mathbf{r}_j$, Poisson's equation may be written as $\boldsymbol{\nabla}_j.\mathbf{E}(\mathbf{r}_j) = 0$. Thus, we have $\boldsymbol{\nabla}_C.\mathbf{F}_E = 0$. This means that $\mathbf{F}_E$ cannot converge toward a given point. Thus, this point cannot be a stable equilibrium position for the motion of the center of mass. It may be that the body is in an indifferent equilibrium. For instance, an electric dipole placed in a uniform field $\mathbf{E}$ is in an indifferent equilibrium for translational motion as the resultant electric force is zero for any position of the dipole. It may also be shown that there is no stable configuration of charges on a rigid conductor placed in a non-uniform external field **E**, but an indifferent equilibrium is possible (in a uniform field **E** for instance).

2.10. Electrostatic binding energy of ionic crystals and atomic nuclei

In an ionic crystal, the ions are arranged in a periodic array. For instance, in the table salt (NaCl) crystal, the Cl^- and Na^+ ions form a face-centered cubic unit cell of sides d (Figure 2.13). The number of ions per mole N is of the order of Avogadro's number $N_A = 6.022 \times 10^{23}$. The cohesion of the crystal is due to the attractive electric forces. However, the binding energy is easier to analyze and to measure as it is related to several physical quantities. The relative configuration of the ions with respect to each other is the same everywhere in the crystal except at its surface. If the crystal is large, this surface effect may be neglected. The electrostatic energy of the crystal is then N times the interaction energy of one of the ions (the central ion in Figure 2.13, for instance) with the others. It is given by

$$U_E = \tfrac{1}{2} N K_o \sum_{n=2}^{N} q_1 q_j / r_{1j} \,. \qquad [2.76]$$

It is obviously unthinkable to carry out this summation, even by using the most powerful computer. The most important terms correspond to the shortest distances r_{1j}. If the ion (1) is Na^+, the closest neighbors are the six Cl^- ions situated at the centers of the faces of the cubic unit cell at a distance $r_{1j} = \tfrac{1}{2}d$; their contribution is $-12K_oe^2/d$. The next neighbors are the 12 Na^+ situated at the middle of the 12 edges at a distance $r_{1j} = d/2^{1/2}$; their contribution is $(12\times 2^{1/2})K_oe^2/d$. Then come the eight Cl^- ions at the vertexes at a distance $3^{1/2}d/2$; their contribution is $-(16/3^{1/2})K_oe^2/d$ and so on. Thus, the total energy is

$$U_E = (NK_oe^2/2d)\,(-12 + 12\sqrt{2} - 16/\sqrt{3} + \ldots) = -1.748\,(NK_oe^2/d). \qquad [2.77]$$

The final result is obtained using a computer. The negative value of U_E shows the dominance of the interaction with the closest neighbors and it explains the stability of the crystal. The energy that is required to extract one of the ions is

$$U_E(N) - U_E(N-1) = 1.748\,(K_oe^2/d). \qquad [2.78]$$

As it is positive, the crystal is stable against spontaneous disintegration.

Equation [2.77] shows that the energy of the crystal decreases (in algebraic value) if d decreases. You may think that the most stable configuration would be that with pairs of oppositely charged ions at the same point. This is not true because, at short distance, quantum effects become important and they are equivalent to a repulsive force between positive and negative ions. d is just the distance at which the repulsive quantum force counterbalances Coulomb attraction.

X-ray diffraction experiments give $d = 0.564$ nm for the NaCl crystal. Thus, $K_0e^2/d = 4.09 \times 10^{-19}$ J. A mole contains 6.022×10^{23} molecules, thus, $N = 1.204 \times 10^{24}$ ions, and the molar binding energy is $U_E = -8.61 \times 10^5$ J/mol. This is the energy required to completely disperse the ions. It may be compared to the evaporation energy of 7.64×10^5 J/mol. The agreement with experiment is of the order of 90%. To improve it, we must take into account other forms of energy such as the vibration energy of ions, surface effects, etc.

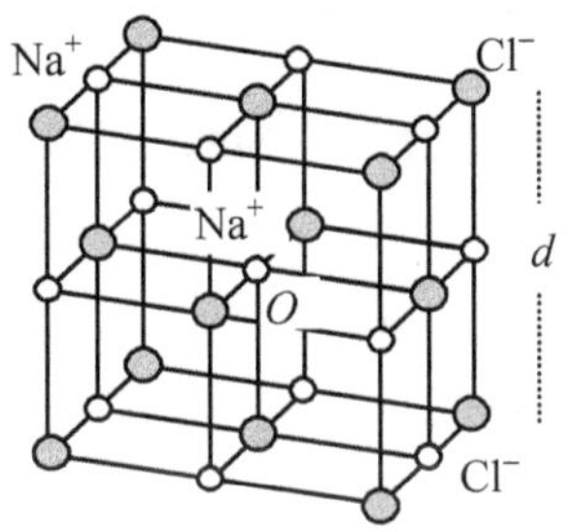

Figure 2.13. *Unit cell of* NaCl

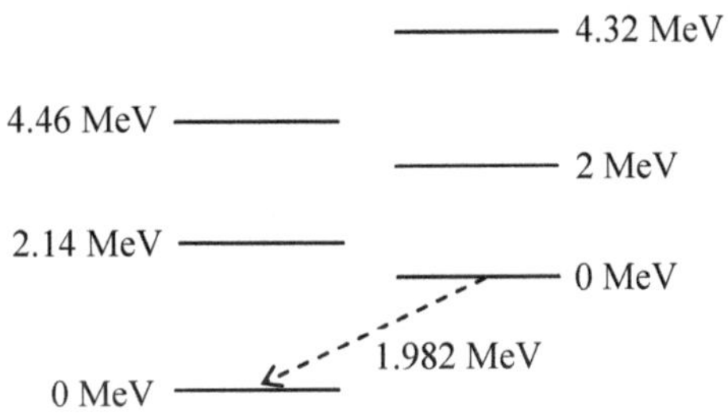

Figure 2.14. *Energy levels of* ^{11}B *and* ^{11}C

The dominant force in atomic nuclei is the nuclear force that binds the nucleons despite Coulomb repulsion between protons. It has a very short range (i.e. it is completely negligible at distances larger than about 10^{-15} m). This force also depends on the relative velocity and the orientation of the nucleons spins, but it does not depend on the electric charge of the nucleons (the nuclear p-p, n-n and n-p interactions are the same). The charge independence of nuclear forces is evident if we compare the spectrums of nuclei having the same mass number A (i.e. the same number of nucleons), but different atomic numbers Z (i.e. the same number of protons) such as ^{11}B ($Z = 5$) and the ^{11}C ($Z = 6$). The energy levels of these nuclei are shown in Figure 2.14. The levels of ^{11}B are at 2.14, 4.46, 5.03 MeV, etc., above the ground state, while those of ^{11}C are at 2,00, 4.32, 4.81 MeV, etc. If the interaction between the nucleons was only nuclear, the levels would be identical. In fact, those of ^{11}C seem to be 1.982 MeV higher. This difference may be explained by Coulomb repulsion, which is more important in the case of ^{11}C as it contains more protons. Taking into account the rest energy of the nucleons, the energy levels may be written as

$$E(\mathrm{B}) = 5m_pc^2 + 6m_nc^2 + U_N + U_E(^{11}\mathrm{B}), \quad E(\mathrm{C}) = 6m_pc^2 + 5m_nc^2 + U_N + U_E(^{11}\mathrm{C}),$$

where we have assumed the same nuclear energy U_N for both nuclei and U_E is the electrostatic energy. Thus, we find the difference of levels:

$$E(\mathrm{C}) - E(\mathrm{B}) = m_p c^2 - m_n c^2 + U_E(^{11}\mathrm{C}) - U_E(^{11}\mathrm{B}) = U_E(^{11}\mathrm{C}) - U_E(^{11}\mathrm{B}) - 1.2935 \text{ MeV}.$$

To evaluate the electrostatic energy U_E, let us assume that the nuclei are spherical of radius $R = R_o A^{1/3}$ with $R_o = 1.2 \times 10^{-15}$ m (which means that all nuclei have the same nuclear density). The radius of ^{11}B and ^{11}C is then $R = 2.67 \times 10^{-15}$ m. Using the expression $(3/5)K_o Q^2/R$ for the energy of a charge distribution in a ball (see Problem 2.31), we find $U_E = (3/5)(K_o Z^2 e^2/R)$. It is a better approximation to replace Z^2 by $Z(Z-1)$, which is twice the number of protons pairs, hence

$$U_E(Z) = (3/5)Z(Z-1)K_o e^2/R = 5.172 \times 10^{-14}\, Z(Z-1) \text{ J} = 0.3232\, Z(Z-1) \text{ MeV}. \quad [2.79]$$

We deduce that $U_E(^{11}\mathrm{C}) - U_E(^{11}\mathrm{B}) = 3.232$ MeV, thus

$$E(\mathrm{C}) - E(\mathrm{B}) = -1.2935 \text{ MeV} + 3.232 \text{ MeV} = 1.939 \text{ MeV}.$$

This result is in good agreement with the experimental values.

2.11. Interaction-at-a-distance and local interaction*

We have seen that it is possible to have two formulations of the electric interactions: the Coulombian formulation [2.1] in accordance with the Newtonian concept of the *instantaneous action-at-a-distance*, even if the charges are very far away, and the *local interaction* formulation [2.4] of a field emitted by a source charge q' and acting on a test charge q (Figure 2.15a). You may think that the two formulations are equivalent and that the concept of field is nothing but a convenient mathematical tool. This is true only in the case of static phenomena. If the system undergoes any modification (as in the case of moving particles, production or disappearance of charges), the instantaneous action-at-a-distance is no longer valid.

After the formulation of special relativity by Einstein in 1905, it became evident that the velocity of light in vacuum c is the upper limit of velocity for particles and the transfer of any physical quantity, interaction, or information. The existence of a higher speed would violate causality, as an observer moving at this speed may find that an effect precedes its cause. Thus, all interactions must be *local*, i.e. expressed in terms of quantities defined at the same space point **r** and time t.

The source charge q' emits a radiation (i.e. a field), which propagates in space with a certain speed v and modifies the space structure around q'. For instance, if q'

is created at time t_o, its field is not established instantaneously everywhere but it reaches at time t the points of the sphere of radius $R = v(t - t_o)$. At this time, no field exists outside this sphere (Figure 2.15a). An eventual test charge q at $\mathbf{r}$ will not detect the influence of q' before $t = t_o + |\mathbf{r} - \mathbf{r}'|/v$. The case of a charge source q' which moves on a curve $\mathcal{C}$ is much more complicated (Figure 2.15b). To evaluate the force that it exerts at time t on the test charge q located at $\mathbf{r}$, the position of q' must be taken at an earlier time $t' = t - |\mathbf{r} - \mathbf{r}'(t')|/v$. The determination of t' as the root of this equation and, consequently, the position $\mathbf{r}'(t')$ and the electric force may be a very complicated mathematical problem. It is possible that this equation has no roots (in that case $\mathbf{E} = 0$) or more than one root. It is even possible that the field exists while its source has already disappeared. Even in the static case, the charge q' may not act on a test charge q if they are separated by an obstacle in which the field does not propagate (a metallic plate, for instance).

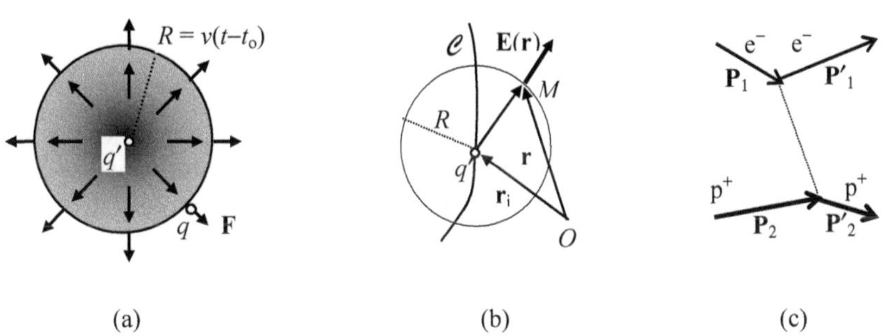

Figure 2.15. *a) The field produced by $q' > 0$ at $t = t_0$ propagates with a speed v and at time t reaches a sphere of radius $r = v(t - t_0)$. Then, it acts on the test charge $q > 0$. b) The field produced at* **r** *by a moving charge q'. c) Interaction by exchange of particles*

In the 19th Century, the Universe was considered to be composed of matter and radiation, two distinct entities but interacting. Both entities are characterized by measurable physical quantities (energy, momentum, etc.). In classical physics, a particle occupies a very small region of space, while radiation occupies an extended region. Particles obey second-order differential equations of motion, while radiation obeys partial differential equations of propagation. Initially, the concept of field was considered as a convenient representation (by lines of force) or a mathematical tool to study interactions. However, after the formulation of electromagnetism by Maxwell, it became clear that electromagnetic radiation (including light) are fields that propagate and carry energy and momentum. The interaction of two systems occurs via the exchange of radiation.

After the formulation of the special theory of relativity by Einstein, it became clear that the velocity of propagation never exceeds the velocity of light in vacuum and the interaction must be local (i.e. it involves quantities defined at the same space point and time). Thus, the concept of fields is necessary to formulate the interaction instead of the action-at-a-distance.

On the other hand, after the formulation of quantum theory, light waves and photons became two aspects of the same physical entity. This concept of particle-wave duality was extended by de Broglie to massive particles. The interaction between particles may be conceived as a process of emission and absorption of radiation or a process of exchange of particles. This exchange allows the transfer of energy, momentum, angular momentum, etc., which is the manifestation of forces. Figure 2.15c is a representation of the interaction between an electron and a proton by the exchange of a photon. It is called a *Feynman diagram.*

2.12. Problems

Electric forces and field

P2.1 a) A charge Q is distributed uniformly on a sphere of radius R. We use spherical coordinates around a diameter Oz. What is the force that the spherical band situated between the parallels θ and $\theta + d\theta$ exerts on a point charge q situated at the point of coordinate z on the z axis? What is the force exerted by the entire sphere? **b)** A charge Q is distributed uniformly in a ball of radius R. Decomposing this ball into successive spherical shells, calculate the force that the ball exerts on the charge q. What is the limit of this force if q is at the center of the ball? Interpret this result.

P2.2 Two charges equal to q are located at positions $-a$ and $+a$ on the x axis. What is the total force that they exert on a charge q' of coordinate y on the y axis? Deduce the work that is necessary to bring this charge from infinity to the origin along the axis Oy. Does this work depend on the path?

Electric energy and potential

P2.3 Consider a first set of charges q_i at the points $\mathbf{r}_i$. Calculate their potentials $V(\mathbf{r}'_k)$ at the points $\mathbf{r}'_k$. Consider a second set of charges q'_k at the points $\mathbf{r}'_k$ producing the potentials $V'(\mathbf{r}_i)$ at $\mathbf{r}_i$. Show Gauss identity $\Sigma_i\, q_i\, V'(\mathbf{r}_i) = \Sigma_k\, q'_k\, V(\mathbf{r}'_k)$.

P2.4 Two charges $q_1 = +10$ nC and $q_2 = -20$ nC are placed at $x = 0$ and $x = +5$ cm, respectively, on the x axis. **a)** Calculate the potential $V(x)$ at an arbitrary point of the x axis and plot V versus x. Deduce the expression of $\mathbf{E}(x)$ on the x axis. Using the expression of $\mathbf{E}(x)$ find $V(x)$. **b)** Calculate $V(x, y)$ at an arbitrary point of the Oxy plane. Deduce the expression of $\mathbf{E}$. Verify the expressions of question (a). **c)** What are the limits of V and $\mathbf{E}$ at large distance? Do you expect this result?

The two fundamental laws of electrostatics

P2.5 a) Show that the angle whose apex is at O and which is subtended by an element of length $d\mathbf{L}$ is $d\theta = dL\,(\mathbf{n.R})/R^2$ where $\mathbf{R}$ is the position of the middle of dL from O. Verify that the total angle for a closed plane curve $\mathcal{C}$ is 2π if O is inside $\mathcal{C}$ and 0 if O is outside $\mathcal{C}$. **b)** Show that the solid angle of the cone whose apex is at O and which is subtended by an element of area $d\mathcal{S}$ is $d\Omega = d\mathcal{S}\,(\mathbf{n.R})/R^3$ where $\mathbf{R}$ is the position of the center of $d\mathcal{S}$ from O. Verify that the solid angle for a closed surface $\mathcal{S}$ is 4π if O is inside $\mathcal{S}$ and 0 if O is outside $\mathcal{S}$. **c)** Deduce Gauss law $\varepsilon_o\Phi_E = \Sigma\, q^{(in)}$.

P2.6 Consider a uniform field $\mathbf{E}$ pointing in the direction Ox. Let $M(x, y)$ be a point of the Oxy plane and P its projection on Ox. Calculate the potential $V(x, y)$ by integrating the equation $\mathbf{E} = -\boldsymbol{\nabla} V$ and taking $V = V_o$ at the origin. Calculate $V(x, y)$ by integrating along the straight line OM, on the path $OP + PM$ formed by two rectilinear segments and on the half-circle OPM. What is the work required to displace an electron from O to M(2 cm, 1 cm) if $E = 2 \times 10^3$ V/m? Does this work depend on the path?

Poisson's equation and its solutions

P2.7 a) What are the charge densities that produce the potentials $V_1 = a(x^2 - y^2)$ and $V_2 = b(x^2 + y^2)$? Calculate the corresponding electrostatic fields and verify the local Gauss equation. **b)** Which one of the fields $\mathbf{E}_1 = (3x - y)\mathbf{e}_x + (3y + x)\,\mathbf{e}_y$ and $\mathbf{E}_2 = (3x - y)\mathbf{e}_x + (3y - x)\mathbf{e}_y$ may be effectively an electrostatic field? Calculate the corresponding potential and charge density. Verify the local Poisson's equation.

P2.8 A plate of thickness d has plane faces of very large dimensions, parallel to the Oyz plane at $x = -d/2$ and $x = +d/2$ respectively. It carries a uniform volume charge density q_v. **a)** Using the symmetries and Gauss law, determine the electric field $\mathbf{E}$ and deduce the potential V everywhere. Is it possible to take $V = 0$ at infinity in this case? **b)** Using Poisson's equation and the continuity conditions of V and $\mathbf{E}$, determine V and deduce $\mathbf{E}$. **c)** A particle of charge q, mass m, and velocity v is fired from far away perpendicularly to this plate. Depending on the sign of the charges, discuss whether the particle may reach the plate and cross it. The particle is assumed to have only electric interaction with the plate. It may help to plot the potential energy of the particle versus x by taking $V = 0$ at O. **d)** Assume now that the plate has surface charge densities $-q_s$ on the face $x = -d/2$ and $+q_s$ on the face $x = d/2$. Calculate the potential and the electric field inside and outside the plate and plot V and E versus x. Verify Laplace and Poisson's equations and the discontinuity of the electric field on the faces.

P2.9 Consider a model of the atom as a point-like nucleus of charge Ze surrounded by an electronic cloud of charge density $q_v = -q_o(1 - r^2/R^2)$. **a)** Interpret R and determine q_o if the total electronic charge is $-Ze$. **b)** What is the electronic charge that is enclosed inside the sphere of radius r? **c)** Using Gauss law, determine the field and deduce the potential. **d)** Using Poisson's equation, the symmetries and the boundary conditions, determine directly the potential $V(r)$ of the electronic cloud. What is the potential of the system nucleus+electronic cloud? Deduce the field.

P2.10 Using the relation $\Delta(1/|\mathbf{r}-\mathbf{r}'|) = -4\pi\,\delta^3(\mathbf{r}-\mathbf{r}')$ where $\delta^3(\mathbf{r}-\mathbf{r}')$ is the three-dimensional Dirac function (see section A11 of Appendix A), show that $V(\mathbf{r}) = K_o\iiint d\mathcal{V}\,q_v(\mathbf{r}')/|\mathbf{r}-\mathbf{r}'|$ is a solution of Poisson's equation and that $\boldsymbol{\nabla}.\mathbf{E} = q_v/\varepsilon_o$.

Electric dipole

P2.11 An *electric dipole* is modeled as two charges $-q$ and $+q$ situated at A^- and A^+ of coordinates $-d/2$ and $+d/2$ on the z axis. **a)** Analyze the symmetries of this system and their consequences on V and **E**. **b)** Evaluate directly V and **E** at points of Oz and Ox. **c)** Show that, at large distance, **E** and V are given by equation [2.38] and that, in spherical coordinates, $E_r \cong 2K_o p\cos\theta/r^3$ and $E_\theta \cong K_o p\sin\theta/r^3$. **d)** Using the expression $V = K_o(\mathbf{p}.\mathbf{r})/r^3$, show that $\mathbf{E}(\mathbf{r}) \cong K_o\,[3(\mathbf{p}.\mathbf{r})\mathbf{r}/r^5 - \mathbf{p}/r^3]$. Using the expression $V = K_o\,p\cos\theta/r^2$, show the expressions of E_r and E_θ.

P2.12 a) Show that the electric dipole moment **p** of a charge distribution given by equation [2.42] does not depend on the choice of the origin if the total charge is equal to 0. **b)** Calculate **p** in the case of a single charge q, in the case of the charges $\pm q$ of Figure 2.16a and in the case of two opposite dipole moments separated by a distance $2b$ (Figure 2.16b). **c)** A charge q is located at **r**. What is the mean value of its field in the sphere of center O and radius a? Generalize to the case of several charges.

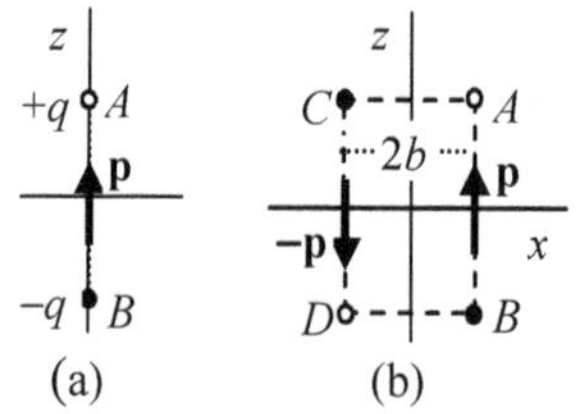

Figure 2.16. *Problem 2.12*

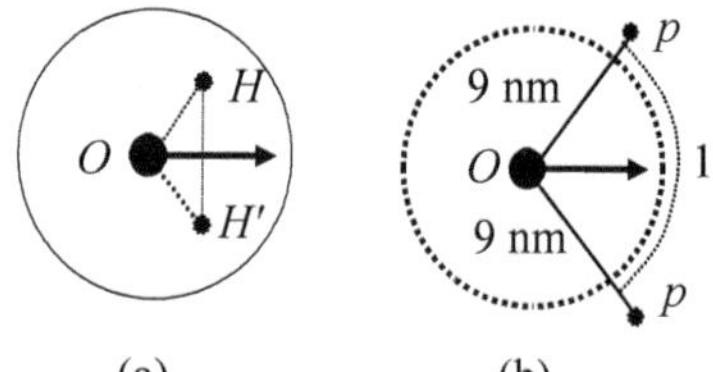

Figure 2.17. *Problem 2.13*

P2.13 In a water molecule, the protons of the hydrogen atoms are at a distance $r = 9$ nm from the oxygen nucleus O in directions making an angle $\theta = 104°$ (Figure 2.17). Assuming that the electron of each hydrogen atom has equal probability to be around its proton as around O, estimate the electric dipole moment of the molecule.

P2.14 a) Calculate the interaction energy of a dipole $\mathbf{p}$ at $\mathbf{r}$ and a charge q at $\mathbf{r}'$. Deduce the force and the moment of force acting on the dipole. **b)** A dipole $\mathbf{p}'$ is modeled as a charge $-q'$ at O and a charge $+q'$ at A such that $\overrightarrow{OA} = \mathbf{d}'$. Using the result of question (a), calculate the force and the moment of force exerted by $\mathbf{p}'$ on $\mathbf{p}$.

P2.15 Assume that a potential $V(\mathbf{r})$ is established in a region of space. Calculate the work required to successively bring charges $-q$ and $+q$ to points $\mathbf{r}_1 \equiv \mathbf{r} - \mathbf{d}/2$ and $\mathbf{r}_2 \equiv \mathbf{r} + \mathbf{d}/2$. Deduce the work required to bring a dipole $\mathbf{p}$ from infinity to a position making an angle θ with $\mathbf{E}$ by using a model of the dipole as two charges $-q$ and $+q$ separated by a distance $\mathbf{d}$.

Electric field and potential of simple charge configurations

P2.16 Evaluate the field and the potential of the simple charge configurations of section 2.7.

P2.17 a) A charge q is uniformly distributed on a circular ring of radius R. Analyze the symmetries of the field and the potential. Calculate the potential at $M(0, 0, z)$. Deduce the expression of $\mathbf{E}$ on the Oz axis. Analyze the variation of V and E as functions of z. Determine the points where the field is minimum and where it is maximum. **b)** An electron may move on Oz. What is the force that acts on this electron? What is the asymptotic limit of this force for large values of z? Justify this result. Is there any equilibrium position for the electron? Is it stable? What is the frequency of oscillations of the electron near this position? **c)** Would this analysis be different if the charge density was not uniform?

P2.18 A thin rod of length L and uniformly distributed charge Q lies on the z axis between A and B of coordinates $-L/2$ and $+L/2$. **a)** Discuss the symmetries and their consequences on V and $\mathbf{E}$ at a point $M(\rho, \varphi, z)$ in cylindrical coordinates. **b)** Let P be a running point of the rod and θ the angle that Oz forms with PM. Show that

$$\mathbf{E} = K_o(q_l/\rho)[(\cos\theta_1 - \cos\theta_2)\,\mathbf{e}_\rho + (\sin\theta_2 - \sin\theta_1)\,\mathbf{e}_z] \quad \text{and} \quad V = K_o q_l \ln(D^-/D^+)$$

where $D^{\pm} = [4\rho^2 + (L \pm 2z)^2]^{1/2} \mp L - 2z$, while θ_1 and θ_2 are the angles that Oz forms with AM and BM, respectively. Verify that V and $\mathbf{E}$ tend toward the potential and the field of a point charge Q at large distance. Verify that, in the limit of an infinite length L (with ρ finite) or in the limit $\rho \to 0$ (with L finite), we find $\mathbf{E} = 2K_o(q_l/\rho)\,\mathbf{e}_\rho$ and $V = 2K_o q_l \ln(L/\rho)$. As $V \to \infty$ if $L \to \infty$, we may take the potential to have the value V_o at $\rho = \rho_o$, thus $V(\rho, \varphi, z) = 2K_o q_l \ln(\rho_o/\rho) + V_o$.

P2.19 A disk has a uniform charge density q_s. Considering the element of area $d\mathcal{S} = \rho\, d\rho\, d\varphi$ in polar coordinates, calculate the field $\mathbf{E}$ and the potential V at points

M on its axis Oz. Using the expression of V, deduce the expression of **E**. Verify that, near the disk, V and **E** are the same as those of an infinite plane.

P2.20 Two parallel planes P and P' are separated by a distance d and carry the uniform charge densities q_s and q_s'. We take Oxy parallel to the planes and situated at equal distance from them. **a)** Using the expressions of the potential and the field of a uniformly charged plane, write the expressions of the potential and the field everywhere. **b)** Discuss the symmetries and deduce that V depends only on z. Deduce that **E** is parallel to Oz and uniform in the planes z = constant. **c)** Using the symmetries, establish directly these properties of **E**. Using Gauss's law, show that **E** is uniform in the three regions of this system and deduce the expressions of the field. Discuss the special cases $q_s = q'_s$ and $q_s = -q'_s$.

P2.21 Let the surface charge density q_s of a plane surface $\mathcal{S}$ be, in reality, a thin layer of thickness d and volume charge q_v. Show that $q_v = q_s/d$. Analyze the variation of V and **E** on both sides of this layer and inside it. Verify that the average field in the layer is ½($\mathbf{E}_1 + \mathbf{E}_2$) where $\mathbf{E}_1$ and $\mathbf{E}_2$ are the fields outside the layer. Verify that, in the limit $d \to 0$, one finds the field and the potential of a plane surface.

P2.22 **a)** A long cylindrical shell of radius R carries a uniform surface charge density q_s. Use Gauss's law to calculate the electric field inside and outside this shell. **b)** Find this result by using Coulomb law and decomposing the shell into infinitesimal elements of area in cylindrical coordinates. **c)** Use this result to calculate the field inside and outside a cylinder of infinite length and uniform volume charge density q_v. **d)** Using the relation $\mathbf{E} = -\nabla V$ in cylindrical coordinates, deduce the potential everywhere.

P2.23 **a)** Use Coulomb's law to calculate the field and the potential of a spherical ball of radius R and uniformly distributed charge Q in its volume. For this, decompose the ball into infinitesimal volume elements $d\mathcal{V} = r'^2 \sin\theta'\, dr'\, d\theta'\, d\varphi'$ of charge $dq' = q_v\, d\mathcal{V}'$ and integrate over the ball. **b)** Let us assume that there is an additional point charge q at the center of the charged ball. Determine the potential and the field of this charge configuration. Application: determine the potential and the field of an atom whose nucleus has a charge Ze and the Z electrons are uniformly distributed in a sphere of radius R. Plot $V(r)$ and $E(r)$.

P2.24 A ball of radius R_1 has a concentric cavity of radius R_2 and it carries a uniform volume charge density q_v between R_1 and R_2. **a)** Using Gauss's law, calculate the field everywhere. Deduce the potential. **b)** Assuming that the center of the cavity is on the z axis at a distance d from the center of the ball, calculate the potential at a point M of spherical coordinates r, θ and φ. Deduce the electric field.

Some general properties of the electric field and potential

P2.25 a) Consider a charge q at point $\mathbf{r}$ and a sphere $\mathcal{S}$ of center O and radius R. Let $<V>_{\mathcal{S}}$ be the average value of the potential of q on $\mathcal{S}$. Show that, if q is inside $\mathcal{S}$, $<V>_{\mathcal{S}}$ is equal to the potential produced at $\mathbf{r}$ by the sphere $\mathcal{S}$ carrying a uniformly distributed charge q. Show that, if q is outside $\mathcal{S}$, $<V>_{\mathcal{S}}$ is equal to the potential produced by q at the center of $\mathcal{S}$. Consider an arbitrary volume charge distribution. Show that $<V>_{\mathcal{S}} = V^{(ex)}(O) + K_o q^{(in)}/R$ where $V^{(ex)}(O)$ is the potential produced at O by all the charges that are outside $\mathcal{S}$ and $q^{(in)}$ is the total charge situated inside $\mathcal{S}$. Deduce that V is regular. **b)** Applying Gauss's law to a small sphere, show that $\mathbf{E}$ is regular and continuous even on the bounding surface of the charge distribution.

Electrostatic energy of a system of charge

P2.26 What is the SI unit of ε_o? Three charges $q_1 = +5$ μC, $q_2 = -10$ μC and $q_3 = +2$ μC are placed at points $x_1 = -4$ cm, $x_2 = 6$ cm and x, respectively, on the x axis (Figure 2.18). Calculate the interaction energy of this system as a function of x. Deduce the resultant force exerted by the charges q_1 and q_2 on q_3 if $x = 12$ cm.

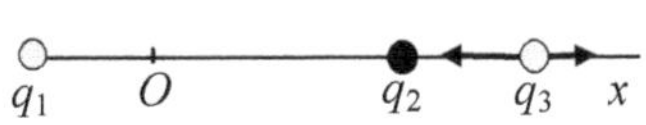

Figure 2.18. *Problem 2.26*

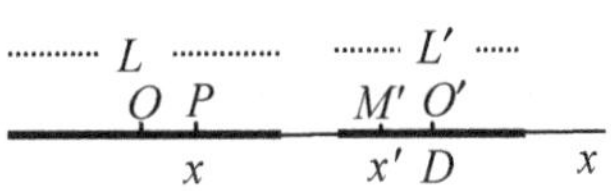

Figure 2.19. *Problem 2.27*

P2.27 A rod of length L and uniformly distributed charge q lies on the Ox axis with its middle at O (Figure 2.19). **a)** Calculate the force that it exerts on a point charge q' located at the point M' of coordinate x' on Ox. **b)** What is the force exerted by this rod on another rod of total charge q' uniformly distributed, of length L' and lying on the x axis with its middle at a distance D from O? Assume that the rods do not overlap. What is the value of this force for large D?

P2.28 Calculate the potential, which corresponds to the field of components $E_x = -3y$, $E_y = -3x + 10y - z$ and $E_z = -y$. Calculate directly the circulation of $\mathbf{E}$ over the line OM joining the origin O to the point $M(1, 2, 0)$ and verify that is equal to $V(M) - V(O)$.

P2.29 Consider a chain of $2N$ charges $q, -q, q, -q$, etc. Two consecutive charges are separated by a distance d. Calculate the electrostatic energy of this chain if N is very large. It helps to use the expansion of $\ln(1+x)$ as a power series.

P2.30 We consider a configuration of charges q_i at points $\mathbf{r}_i$ and we assume that the potential $V'(\mathbf{r}_k)$ produced at $\mathbf{r}_k$ by all the charges except q_k is finite. $V'(\mathbf{r}_k)$ is related to the charges by a linear relation $V'(\mathbf{r}_k) = \Sigma_{i\neq k} A_{ki}\, q_i$, where the coefficients A_{ki} depend on the relative positions of the charges. **a)** To calculate the electrostatic energy, we may assume that the charges are gradually increased from 0 to their final values q_i. Assume that, at a certain time, the charges are equal to αq_k, where α is increased from 0 to 1. Calculate the required energy to increase α to $\alpha + d\alpha$ (by bringing the charges $dq_k = q_k\, d\alpha$ from infinity to $\mathbf{r}_k$). Deduce that the energy of the charges may be written as $U_E = ½\Sigma_k V'(\mathbf{r}_k) q_k = ½\, \Sigma_k\, \Sigma_{i\neq k}\, A_{ki}\, q_k\, q_i$. **b)** Use this result to calculate the energy of a ball of radius R and uniform charge Q. **c)** To calculate the energy of the ball, let us assume that successive shells are brought from infinity. At a certain time, the radius is r. Calculate the work required to bring the charge of the shell of thickness dr and deduce the electrostatic energy of the ball.

P2.31 **a)** Using the expression $U_E = ½K_o\iint_S d\mathcal{S}' \iint_S d\mathcal{S}\; q_S(\mathbf{r}) q_S(\mathbf{r}')/|\mathbf{r} - \mathbf{r}'|$, show that the electrostatic energy of a sphere of radius R and charge Q that is uniformly distributed on the surface is $U_E = ½K_o Q/R$. Find the same result by using the energy density $U_{E,v} = ½\varepsilon_o \mathbf{E}^2$. **b)** Using the expression of the energy of a volume charge distribution $U_E = ½K_o\iiint_\mathcal{V} d\mathcal{V}' \iiint_\mathcal{V} d\mathcal{V}\; q_\mathcal{V}(\mathbf{r}) q_\mathcal{V}(\mathbf{r}')/|\mathbf{r} - \mathbf{r}'|$, show that the electrostatic energy of a ball of radius R and charge Q, which is uniformly distributed in the volume, is $U_E = (3/5)K_o Q/R$. Find the same result by using the energy density $U_{E,v} = ½\varepsilon_o \mathbf{E}^2$. **c)** In special relativity, a particle of mass m at rest has an energy mc^2, where c is the speed of light in vacuum. Deduce the radius of the electron if it is modeled as a ball of radius R.

P2.32 Two balls of radius R_1 and R_2 have their centers at points of coordinates $-a$ and $+a$ on the z axis and they have no common parts. Their charges Q_1 and Q_2 are uniformly distributed in their volumes. **a)** Calculate the potential and the electric field inside and outside these balls. **b)** Calculate the energy density. Deduce the proper energy of the balls, their interaction energy and their force of interaction. What is the work necessary to bring these balls from infinity to their actual positions? **c)** Calculate the force of interaction of these balls by using Coulomb interaction of their volume elements. **d)** Let us consider the particular case $Q_1 = -Q_2$, $R_1 = R_2$ and $a \ll R_1$. Show that the global field at large distance is the same as that of a sphere with a surface charge density proportional to $\cos\theta$ where θ is the angle with Oz as polar axis.

P2.33 Two electrons are separated by a distance d. The total field at any point is $\mathbf{E} = \mathbf{E}_1 + \mathbf{E}_2$ where $\mathbf{E}_1$ and $\mathbf{E}_2$ are their individual fields. The total energy density is $U_{E,v} = ½\varepsilon_o[\mathbf{E}_1^2 + \mathbf{E}_2^2 + 2\mathbf{E}_1.\mathbf{E}_2]$. The integrals of the first two terms over the whole space are the proper energies of the electrons (infinite in the limit of point particles). Show that the integral of the third term gives $U_E = e^2/4\pi\varepsilon_o d$ as expected.

P2.34 a) Calculate the electrostatic energy that is necessary to assemble the 92 protons of the uranium nucleus in a sphere of radius 7.4×10^{-15} m. Express your result in joules and in MeV. **b)** Using the nuclear radius $R = R_o A^{1/3}$, estimate the variation of the electrostatic energy of the uranium nucleus ($Z = 92$, $A = 236$), if it undergoes fission into two identical nuclei.

Chapter 3

Conductors and Currents

We have seen that materials may be classified as *insulators* or *conductors*, although the distinction is not clear cut, as semiconductors have a conductivity that is intermediary between conductors and insulators. Similar to heat conduction, some materials are better conductors than others. The historic Hall experiment has shown, even before the discovery of the electron, that conduction of metals is due to the motion of negative charges (see section 6.1). Actually, we know that the external electrons in some atoms are weakly bound; this makes them free to move from one atom to the other; these are the *free electrons* (or *conduction electrons*). In electrolytic solutions, the molecules are dissociated into two ions of opposite charge and both are free to move and contribute to electric conduction. In this chapter, we analyze the properties of solid conductors in equilibrium and study their conduction properties.

3.1. Conductors in equilibrium

In a body, each charge is subject to the electric field of the other bodies and that of the other charges of the body itself. If the body is a conductor, the free charges move very rapidly under the influence of these fields until they reach a stable electrostatic equilibrium configuration. Besides this orderly motion in a given direction, the particles have a random motion, called *thermal agitation*, which increases with temperature and which is equally probable in all directions if the temperature of the body is uniform. It does not correspond to a mean displacement of the particles and it persists even in the state of electrostatic equilibrium. In the following, we assume that the only forces that act on the charges are electric (thus, neglecting the weight, magnetic forces, etc.). We also assume that the external electric field is time-independent, as a variable electric field induces a magnetic field that exerts magnetic forces. Finally, we assume that the field inside the conductors obeys the same equations as in vacuum. This last assumption is justified by the agreement of its consequences with experiment.

A conductor may be either disconnected from other bodies (its total charge Q is then constant) or maintained at constant potential V (by generators). The analysis of its electrostatic equilibrium determines its voltage (or its charge), the distribution of the charge, and the electric field. These quantities depend on the electric influence of the nearby bodies. It may be shown that this electrostatic problem has a unique solution. The superposition principle is obviously valid and it may help to find this solution.

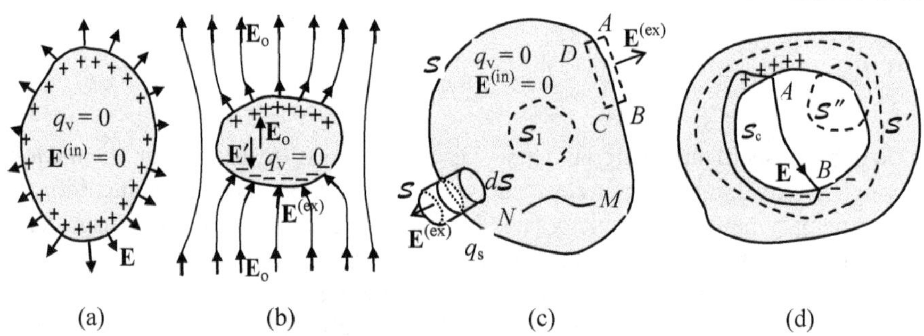

Figure 3.1. *Conductors in electrostatic equilibrium: a) the charges are distributed on the surface in such a way that* $\mathbf{E}^{(in)} = 0$ *and* $\mathbf{E}^{(ex)}_{//} = 0$. *b) Conductor placed in an external field. c) Inside the conductor,* $q_V = 0$*, on its external surface,* $\mathbf{E}^{(ex)}_{//} = 0$ *and* $\mathbf{E}^{(ex)}_{\perp} = q_s/\varepsilon_o$. *The conductor is equipotential and d) conductor with cavity*

Here are some properties of ideal conductors in electrostatic equilibrium:

a) *Inside a conductor in equilibrium, there is no electric field*[1]

$$\mathbf{E}^{(in)} = 0. \qquad [3.1]$$

Indeed, if there was a field $\mathbf{E}^{(in)}$, the free charges would move under the electric force and the conductor would not be in electrostatic equilibrium. If non-electric forces act on the conduction charges, these charges move to equilibrium positions and create in the conductor an electric field $\mathbf{E}^{(in)}$ such that the electric force $\mathbf{F} = q\mathbf{E}^{(in)}$

1 In this case the macroscopic field and charge density are averaged on volumes that are large enough to contain a large number of atoms. Because of the thermal agitation, the microscopic field and charge density undergo large variation in space and time and they may even be infinite at the positions of the charged particles.

counterbalances the non-electric forces. This is the case inside electric generators and a conductor that is immersed in a magnetic field (see Hall's experiment in section 6.1).

b) *Inside a conductor in electrostatic equilibrium, there is no net electric charge density* (Figure 3.1). Indeed, the electric field inside the conductor being zero, the local form of Gauss's law $\nabla.\mathbf{E}^{(\text{in})} = q_v/\varepsilon_o$ gives

$$q_v = 0. \qquad [3.2]$$

The absence of net charge inside the conductor may be easily explained by the mobility of the conduction electrons. If the net charge density was negative in a part of the conductor, due to an excess of electrons, these would repulse each other and would settle in equilibrium on the external surface of the conductor. Similarly, if there was a positive charge density in a part of the conductor, due to a lack of electrons, this positive charge would attract the free electrons to become neutralized and leave the surface of the conductor positively charged.

c) *A conductor in electrostatic equilibrium can carry charges only on its external surface.* Outside a conductor, just near its surface, the electric field is normal to the surface. As we have seen in section 2.8(b), the tangential component of **E** is continuous on boundary surfaces (since the circulation of **E** vanishes on any closed path), while the normal component is finite but it has a discontinuity $\mathbf{E}_2.\mathbf{n}_{12} - \mathbf{E}_1.\mathbf{n}_{12} = q_s/\varepsilon_o$ (given by Gauss law). Taking the conductor as medium (1) and the exterior as medium (2), $\mathbf{n}_{12}$ is the normal unit vector outgoing from the conductor. The field being zero inside the conductor, we must have

$$\mathbf{E}_{//2} = 0 \qquad \text{and} \qquad \mathbf{E}_2.\mathbf{n}_{12} \equiv E_{2\perp} = q_s/\varepsilon_o. \qquad [3.3]$$

Any charge of the conductor, whether it is due to the displacement of the conduction electrons, deposed on it or induced by the presence of other charged bodies nearby, can only be on its external surface. The surface charge density may be positive on certain parts of the surface and negative on others in a distribution such that the total field **E** in the conductor is zero (Figure 3.1b). In the usual conductors, the depth of the charged layer on the surface is of the order of the nanometer (i.e. atomic size).

d) *All points inside and on the surface of a conductor in electrostatic equilibrium are at the same potential*: indeed, as $\mathbf{E} = 0$ inside the conductor, its circulation inside the conductor along any path between any two points M and N is zero (Figure 3.1c). Using equation [2.13], we deduce that $V(M) = V(N)$. Thus, the surface of the conductor and its interior are equipotential and this agrees with the orthogonality of the field on this surface. This property may be established directly. Indeed, if $V(M)$

was different from $V(N)$, the conduction electron would move toward the higher potential to reduce the total potential energy and the equilibrium would be lost. To set a conductor at a positive potential V_0, it may be connected to the positive terminal of a generator of electromotive force V_0, and to set it to a negative potential $-V_0$, it may be connected to the negative terminal, the other terminal being grounded. In order to maintain a conductor to zero potential, it may be grounded.

3.2. Conductors with cavities, electric shielding

In the case of a conductor with cavities, the preceding results concerning **E** and V inside the conductor and on its external surface remain valid (Figure 3.1d). Let us consider a Gaussian surface $\mathcal{S}'$ entirely situated within the conductor and containing a cavity. The field in the conductor being zero, the flux of **E** through $\mathcal{S}'$ is zero and Gauss's law implies that the total charge inside $\mathcal{S}'$ is zero. Thus, the total charge carried by the surface of the cavity $\mathcal{S}_c$ is the opposite of the total charge that it contains.

If the cavity contains no charge, the total charge on $\mathcal{S}_c$ is zero. So, if it has some positive surface charge density at A and a negative charge density at B, the field points from the positive charge toward the negative charges. The circulation of **E** along a path going from A to B would be $\int_A^B d\mathbf{r}.\mathbf{E} = V_A - V_B > 0$ and this is impossible as the conductor is equipotential. Thus, if the cavity contains no charge, its surface charge density must be zero. We may reverse the argument and say that, points A and B of a conductor being equipotential and the cavity containing no charge, V has between A and B no maximum or minimum; it must be constant and the field $\mathbf{E} = -\nabla V$ must vanish in the cavity as well as in the conductor. Consider now a Gaussian surface $\mathcal{S}''$ having a part in the cavity and the other in the conductor. The flux through $\mathcal{S}''$ being zero, the total charge that it contains is zero. This is possible for any $\mathcal{S}''$ only if the cavity surface carries no charge.

It should be noted that, if we admit the property that $\mathbf{E}^{(\text{in})} = 0$ in a conductor, all the other properties result from Gauss's law, which is itself a consequence of the $1/r^2$ dependence of Coulomb force. The vanishing of the internal charge was verified experimentally by Cavendish in 1772. He introduced a charged metallic ball into a metallic box and let the ball touch the box; the ball became neutral. Thus, if the force law was $1/r^{2+\eta}$, he verified that $\eta < 0.02$. A modern version of this experiment was realized by Plimpton and Lawton. They used two metallic concentric enclosures and set the external one to a high alternating potential (of the order of 10^4 V) and low frequency (of the order of 2 Hz). Using an amplifier able of detect a difference of potential as low as 10^{-6} V, they did not find any difference of potential between the enclosures and deduced that $\eta \leq 2 \times 10^{-9}$.

More generally, the electrostatic phenomena in the cavity do not depend on the exterior. For instance, if the charge density in the cavity is q_v, the potential in the cavity is a solution of Poisson's equation $\Delta V = -q_v/\varepsilon_o$ and the field is $\mathbf{E} = -\nabla V$. Among these solutions, we must choose the one that verifies the conditions $V = V_o$ and $\mathbf{E.n} = q_s/\varepsilon_o$ on the surface $\mathcal{S}_c$ of the cavity. V_o is the potential of the enclosure, which is an unimportant constant, **n** is the unit vector normal to $\mathcal{S}_c$ and pointing toward the interior of the cavity and q_s is the charge density of $\mathcal{S}_c$. V must also be finite in the cavity if it contains no point charges or linear charges. It may be shown that this mathematical problem has a unique solution. We note that we consider here only static phenomena, as time-dependent electric fields induce magnetic fields. This constitutes an electromagnetic wave, which may propagate through the enclosure (if it is thin) and produce an electric and a magnetic field in the cavity. This property of the cavity results in *electric shielding*. Delicate electric devices or circuits are enclosed within a metallic, grounded, conducting casing to protect them against external electric disturbances. It is not necessary that the enclosure be hermetic. It may have small holes or it may be a simple grid allowing the evacuation of heat.

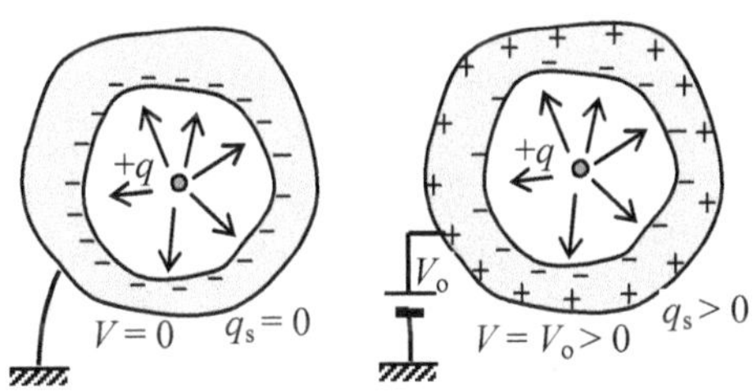

Figure 3.2. *a) Grounded conductor with cavity, b) the same conductor set to a potential V_o*

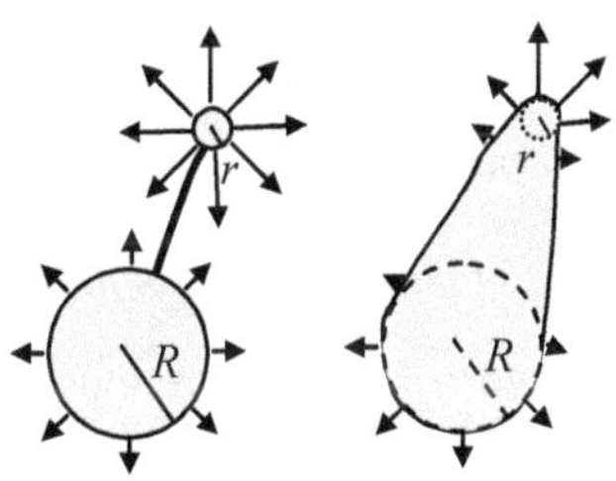

Figure 3.3. *Sharp point effect:* **E** *is intense at sharp points and edges*

The enclosure does not protect the exterior against electrostatic phenomena in the cavity. For instance, if a charge is introduced in the cavity, an equal charge appears on the external surface of the enclosure, this modifies its potential and produces an electric field outside it. The shielding acts in both directions if the enclosure is grounded (Figure 3.2a) or maintained at a fixed potential V_o (Figure 3.2b). The electric potentials $V^{(in)}$ in the cavity and $V^{(ex)}$ outside the enclosure are solutions of Poisson's equation in their respective regions with the boundary condition $V(\mathbf{r}) = V_o$ on the enclosure. The solution is unique in each region independently of the others.

Let us consider two distant conducting spheres of radii R and r, connected by a conducting wire (Figure 3.3a). They form a single conductor at a potential V. Their charges are $Q_R = 4\pi\varepsilon_o RV$ and $Q_r = 4\pi\varepsilon_o rV$ and the field near their surface is $E_R = V/R$ and $E_r = V/r$; thus, $E_r / E_R = q_{s,r}/q_{s,R} = R/r$. This example shows that the field and the charge density are very large at the sharp points or edges of a conducting body, as their radius of curvature is very small (Figure 3.3b). Lightning conductors are an application of this effect: a metallic rod ending by several needle-like conductors is placed on the top of buildings and connected to the ground. The Earth is not completely neutral but it has a charge density of about -10^{-9} C/m^2. Its field at ground level is about 100 V/m and it may be very intense near lightning conductors. This favors atmospheric discharge through the conductor.

The sharp point effect explains the so-called *corona discharge* brought on by the ionization of a fluid surrounding a sharp conductor at high potential. Air, for instance, is a poor conductor of electricity but it contains always some charged particles and ions that are produced by cosmic rays. The high electric field near the sharp points acts on these charged particles and ions and accelerates them strongly. In their displacement, they collide with the air atoms, producing other ions and emitting intense light. The motion of the charges also produces so-called *electric wind*, which may be detected by a flame placed near the sharp point.

3.3. Capacitors

a) Mutual influence of conductors, capacitance

If two conductors are close to each other, the field of each one acts on the free charges of the other and modifies the charge distribution and total charge if they are not isolated. We say that the conductors are in *mutual electric influence*. The field lines always start from the positive charges on one of them (or at infinity) and end at the negative charges on the other (or at infinity). As the Earth is a conductor, it must be considered as a part of the system.

We consider in this section the case of two conductors in mutual electric influence and in electrostatic equilibrium (Figure 3.4a). Let $\mathcal{S}'$ be a Gaussian surface formed by a thin tube of field lines and closed at the ends by two surfaces inside the conductors. The field being tangent to the lateral surface of the tube and equal to zero inside the conductors, its flux through $\mathcal{S}'$ is equal to zero. The surface elements of the conductors that it contains must have opposite charges dq_1 and dq_2. If all the field lines start on one of the conductors and end on the other, we say that the two conductors are in *total mutual influence*. Then, they carry opposite charges Q and $-Q$. This is the case, for instance, if one of the conductors is in a cavity within the other and also the case of two ideally isolated conductors carrying opposite charges. Two large plane parallel plates carrying opposite charge densities are almost in total mutual influence if edge effects are neglected (Figure 3.5a).

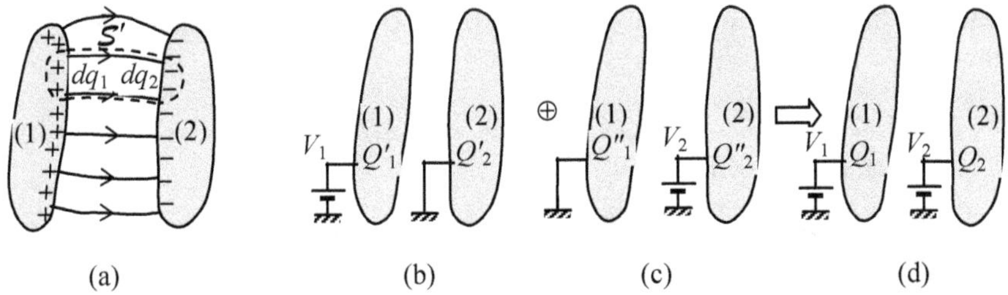

Figure 3.4. *a) Two conductors in total influence. The superposition of the states b) and c) is the state d)*

To derive the relation of the charges to the potentials, we consider a first configuration such that the first conductor is at potential V_1 and the second at zero potential (Figure 3.4b). The relation of the charges to the potentials being linear, the charges of the conductors are proportional to V_1, of the form $Q'_1 = C_{11}V_1$ and $Q'_2 = C_{21}V_1$. In a second configuration with the first conductor at zero potential and the second at potential V_2 (Figure 3.4c), the charges are proportional to V_2, of the form $Q''_1 = C_{12}V_2$ and $Q''_2 = C_{22}V_2$. It is evident that the configuration of the conductors at potentials V_1 and V_2 is a superposition of the preceding configurations (Figure 3.4d). We deduce that the charges Q_i are then the sums of the charges Q'_i and Q''_i:

$$Q_1 = C_{11}\,V_1 + C_{12}V_2, \qquad \text{and} \qquad Q_2 = C_{21}\,V_1 + C_{22}V_2. \qquad [3.4]$$

The C_{ik} are the *coefficients of electric influence*. They depend on the geometrical form of the bodies, their relative position, and also the non-conducting medium separating them. If the influence is total, we have $Q_2 = -Q_1$ for any potentials. This is possible if $C_{11} = -C_{21}$ and $C_{22} = -C_{12}$. We will show in section 3.4 the symmetry relation $C_{ik} = C_{ki}$. On the other hand, if, for instance, $V_2 = 0$ and V_1 is positive, Q_1 must be positive. Thus we have $C_{11} = C_{22} = -C_{12} = -C_{21} > 0$. This set-up of the conductors is then called a *capacitor* and the conductors are its *plates* (or *electrodes*). Designating the charge of the positive plate as Q and the common value of the mutual coefficients as C, called *capacitance*, we obtain the relation

$$Q = C\,V, \quad \text{where} \quad V = V_1 - V_2. \qquad [3.5]$$

The SI unit of capacitance is the coulomb per volt (C/V), called *farad* (F). Usual capacitances are of the order of the microfarad ($\mu F = 10^{-6}$ F) and picofarads (pF = 10^{-12} F). A single isolated conductor may be considered to be in total influence with the Earth. If it has a potential V, its charge is $Q = CV$. The coefficient C is also called capacitance of the body. For instance, the capacitance of a sphere is $C = 4\pi\varepsilon_o R$. We consider in this chapter only empty capacitors or, to a good approximation, air-filled capacitors.

b) Calculation of the capacitance

The simplest capacitor to analyze is the parallel plate capacitor, formed by two metallic plates of large area $\mathcal{S}$ and separated by a distance d (Figure 3.5a). The edge effects are negligible if the distance d is much smaller than the dimensions of the plates. For the calculation of V and $\mathbf{E}$, this assumes that the plates are infinite planes. The electric field is then the superposition of the fields $\pm q_s/2\varepsilon_o$ of two planes of uniform charge densities $\pm q_s$ (see section 2.7). The total field is zero outside the plates (where the fields of the plates are $+q_s/2\varepsilon_o$ and $-q_s/2\varepsilon_o$) and equal to q_s/ε_o between them (where both fields are equal to $q_s/2\varepsilon_o$). The difference of potential between the plates is then

$$V \equiv V(A) - V(B) = \int_A^B d\mathbf{r}.\mathbf{E} = Ed. \qquad [3.6]$$

The charge of the positive plate is

$$Q = \mathcal{S}q_s = \varepsilon_o E\mathcal{S}. \qquad [3.7]$$

Thus, the capacitance is

$$C = Q/V = \varepsilon_o \mathcal{S}/d. \qquad [3.8]$$

For instance, if $\mathcal{S} = 2\ \text{m}^2$ and $d = 1$ mm, we find $C = 17.7 \times 10^{-9}$ F $= 17.7$ nF. This example shows that the farad is an enormous capacitance.

A *cylindrical capacitor* consists of a cylindrical conductor of radius R_1 and length L surrounded by a coaxial cylindrical shell of internal radius R_2 (Figure 3.5b). Let Q be the charge of the cylinder and $-Q$ that of the shell. Applying Gauss's law to a coaxial cylindrical surface of radius r, we get $E = Q/2\pi\varepsilon_o Lr$ (see section 2.7). Thus, the difference of potential between the cylinder and the shell is

$$V \equiv V(A) - V(B) = \int_A^B d\mathbf{r}.\mathbf{E} = 2K_o(Q/L)\int_{R_1}^{R_2} dr/r = 2K_o(Q/L)\ln(R_2/R_1). \qquad [3.9]$$

The capacitance of this capacitor is

$$C = \frac{Q}{V} = \frac{2\pi\varepsilon_o L}{\ln(R_2/R_1)}. \qquad [3.10]$$

A Geiger counter uses ionization to detect particles. It is essentially a capacitor formed by a metallic cylinder surrounding a conducting wire whose potential is about 1 kV higher than the cylinder. This tube is filled with a gas at low pressure. If a particle enters though a small opening at its end, some gas atoms are ionized. The produced electrons are strongly attracted toward the wire and they ionize other atoms, producing an avalanche and a signal, which may be amplified and detected. This instrument may count particles but it cannot measure their energy.

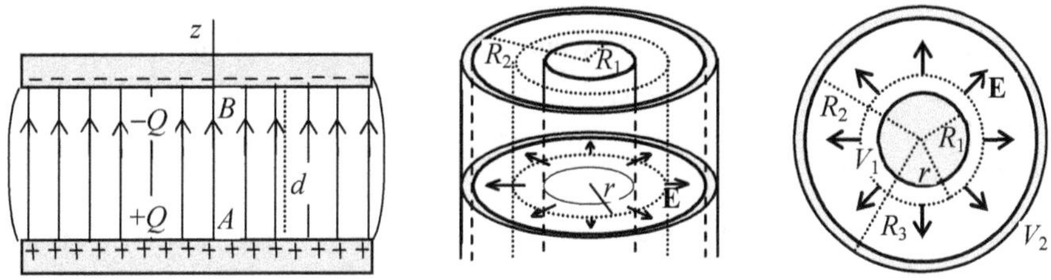

(a) (b) (c)

Figure 3.5. *a) Parallel plate capacitor, b) cylindrical capacitor, and c) spherical capacitor*

A *spherical capacitor* is a sphere of radius R_1 surrounded by a concentric spherical shell of internal radius R_2 (Figure 3.5c). Let Q be the charge of the sphere and $-Q$ that of the shell. Applying Gauss's law to a concentric spherical surface of radius r, we get $E = K_oQ/r^2$ (see section 2.7). Thus, the difference of potential between the sphere and the shell is

$$V \equiv V(A) - V(B) = \int_A^B d\mathbf{r}.\mathbf{E} = K_oQ \int_{R_1}^{R_2} dr/r^2 = K_oQ\,[1/R_1 - 1/R_2]. \qquad [3.11]$$

The capacitance of the capacitor is

$$C = Q/V = 4\pi\varepsilon_o\, R_1R_2/d, \qquad \text{where } d = R_2 - R_1. \qquad [3.12]$$

If $d << R_1$, the capacitance becomes $C = 4\pi\varepsilon_oR^2/d = \varepsilon_o\mathcal{S}/d$, i.e. the same as for a capacitor of thickness d and surface $\mathcal{S}$ of the sphere. On the other hand, if R_1 is finite and R_2 is infinite, we find $C_s = 4\pi\varepsilon_oR_1$. This is the capacitance of a sphere of radius R_1. The capacitance of Earth for instance is 710 μF.

c) Energy of capacitors

A capacitor stores electric energy. To evaluate it, assume that, at a certain time, the charges of the plates are $Q'_1 = uQ$ and $Q'_2 = -uQ$, where u is increased from 0 to 1. By linearity, the potentials of the plates are $V'_1 = uV_1$ and $V'_2 = uV_2$. To increase u

by du, a charge $dQ'_1 = du\, Q$ must be brought from infinity to the positive plate and $dQ'_2 = -du\, Q$ to the negative plate. This requires work

$$dW = dQ'_1\, V'_1 + dQ'_2\, V'_2 = du\, Q\, uV_1 - du\, Q\, uV_2 = du\, u\, QV.$$

The energy of the capacitor is the total work required to increase u from 0 to 1, i.e.

$$U_E = \int dW = QV \int_0^1 du\, u = \tfrac{1}{2}QV = \tfrac{1}{2}CV^2 = \tfrac{1}{2}Q^2/C. \qquad [3.13]$$

This is also the energy stored in the volume $\mathcal{V} = \mathcal{S}d$ with a uniform volume density $U_{E,v} = \tfrac{1}{2}\varepsilon_0 E^2$, where $E = V/d = Q/\varepsilon_0\mathcal{S}$.

d) Use of capacitors

Capacitors may be connected in an electric circuit in two ways with specific advantages and disadvantages for each one:

a) By connecting one of the terminals of all the capacitors to a point A and the other to a point B, we obtain capacitors in parallel (Figure 3.6a). If a voltage V is applied between A and B, the capacitors acquire the charge $Q_i = C_iV$. The total charge of the combination is $Q = \Sigma_i Q_i = V\Sigma_i C_i$. Thus, it is equivalent to a single capacitance

$$C = Q/V = \Sigma_i C_i. \qquad [3.14]$$

The stored energy is evidently the sum of the energies of the capacitors:

$$U_E = \Sigma_i U_{Ei} = \tfrac{1}{2}\Sigma_i C_i V^2 = \tfrac{1}{2}CV^2. \qquad [3.15]$$

It is the same as the energy of the equivalent capacitor.

This combination of capacitors increases the charge without increasing V. This is convenient if the individual capacitors cannot support high voltage. A practical way to achieve this is to pile metallic foils separated by insulating sheets and to connect the odd numbered foils to A and the even numbered foils to B. The number of capacitors in parallel is that of the insulating sheets.

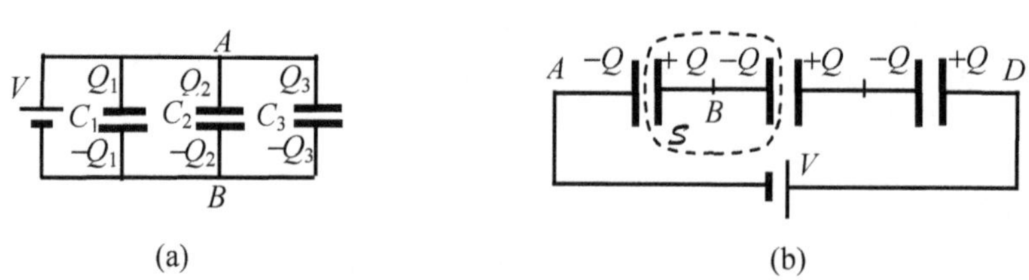

Figure 3.6. *a) Three capacitors in parallel and b) three capacitors in series*

b) By connecting the output plate of one capacitor to the input plate of another, we obtain capacitors in series (Figure 3.6b). The potential V is applied between the input A of the first capacitor and the output B of the last capacitor. If the capacitors were not initially charged, the charge inside the Gaussian surface $\mathcal{S}$, for instance, remains zero. Thus, the plates that it contains carry opposite charges $-Q$ and $+Q$ and the difference of potential between A and B may be written as

$$V = V(A) - V(B) = [V(A) - V(C)] + [V(C) - V(D)] + [V(D) - V(B)]$$
$$= Q/C_1 + Q/C_2 + Q/C_3 = Q(1/C_1 + 1/C_2 + 1/C_3).$$

Thus, this combination of capacitors has an equivalent capacitance C given by

$$1/C = V/Q = \Sigma_i\, 1/C_i. \qquad [3.16]$$

The energy stored in this set-up is

$$U_E = \Sigma_i\, U_{Ei} = \tfrac{1}{2}\, \Sigma_i\, Q^2/C_i = \tfrac{1}{2}\, Q^2\, \Sigma_i\, 1/C_i = \tfrac{1}{2}\, Q^2/C. \qquad [3.17]$$

It is the same as the energy of the equivalent capacitor.

The advantage of this combination of capacitors is to have a high voltage between the terminals, while the individual capacitors are at low voltage. A practical way to achieve this is to pile metallic foils separated by insulating sheets. The tension V is applied between the extreme metallic foils. The number of capacitors in series is that of the insulating sheets.

Capacitors are currently used as components in electronic circuits. They serve to regularize electric currents, to produce time delays, and transmit and detect electromagnetic signals in electronic equipment. Capacitors come in different forms. Small capacitors of the type used in electronic circuits are formed by two metallic foils (usually aluminum), separated by a sheet of dielectric (waxed paper, Mylar, etc.) and rolled up in a small cylinder. Large capacitors are obtained by immersing large metallic plates in insulating oil, and variable capacitors are obtained by rotating movable plates between a set of stationary plates. The effective surfaces of the capacitors are then only the parts facing each other. Capacitors of large capacitance, called *electrolytic capacitors*, are composed of a metallic foil (often in a spiral form), which acts as an electrode, immersed in an electrolytic solution in a metallic container, which acts as the second electrode. If a voltage is applied between the foil and the container, a layer of metal oxide is formed close to the foil. This acts as a very thin insulator while the solution remains conducting. This kind of capacitor has a specific polarity that must be maintained.

Capacitors may store a large amount of charge in a relatively small voltage and deliver them in a very short interval of time, avoiding the use of high voltages, which may cause electric breakdown. This discharge cannot be obtained by using batteries. It may be used to produce light flashes, accelerate particles, study nuclear fusion, etc. If the plates of a high-capacitance capacitor are short-circuited, its fast discharge may produce sparks and if one accidently touches both plates, the discharge current across the body may have grave consequences (especially if it passes the heart). Conversely, in the case of heart attack, a fast discharge of electrical energy through the heart may stop cardiac fibrillation (rapid and irregular heart beating).

3.4. Mutual electric influence of conductors

The results that we have established for two conductors may be generalized to several conductors in mutual influence (Figure 3.7). By the same argument using tubes of field, there is a correspondence between opposite charges on conductors in electric influence. Considering states of only one conductor (i) at the potential V_i and the others at zero potential and then making the superposition of these states, we get the relations:

$$Q_1 = C_{11} V_1 + C_{12}V_2 + C_{13}V_3 + \ldots = \Sigma_k C_{1k} V_k,$$
$$Q_2 = C_{21} V_1 + C_{22}V_2 + C_{23}V_3 + \ldots = \Sigma_k C_{2k} V_k,$$
$$\text{etc.,} \qquad \text{i.e. } Q_i = \Sigma_k C_{ik} V_k. \qquad [3.18]$$

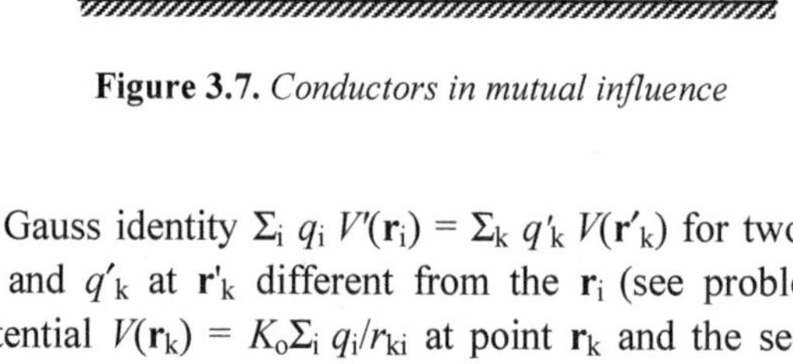

Figure 3.7. *Conductors in mutual influence*

Let us recall Gauss identity $\Sigma_i\, q_i\, V'(\mathbf{r}_i) = \Sigma_k\, q'_k\, V(\mathbf{r}'_k)$ for two configurations of charges q_i at $\mathbf{r}_i$ and q'_k at $\mathbf{r}'_k$ different from the $\mathbf{r}_i$ (see problem 2.3). The first produces the potential $V(\mathbf{r}_k) = K_o\Sigma_i\, q_i/r_{ki}$ at point $\mathbf{r}_k$ and the second produces the potential $V'(\mathbf{r}_i) = K_o\Sigma_{k\neq i}\, q'_k/r_{ki}$ at the point $\mathbf{r}_i$. Applying this identity to the configurations of the conductors (i) at potentials V_i and charges Q_i and the same conductors at potentials V'_k and charges Q'_k, we get the relation $\Sigma_i\, Q_i\, V'_i = \Sigma_k\, Q'_k\, V_k$. Using [3.18], this relation may be written also as $\Sigma_{i,k}\, C_{ik}V_kV'_i = \Sigma_{i,k}\, C_{ki}V'_iV_k$.

Considering the configurations such that all the potentials are zero except two of them, V'_i and V_k, we deduce the exchange symmetry relation

$$C_{ik} = C_{ki}. \quad [3.19]$$

Consider now a configuration of conductors i = 1, 2, etc., in total influence and at zero potential except the conductor (i) at potential $V_i = V > 0$. The field points from the conductor (i) toward the others. Thus, the charge of the conductor (i) is positive, while all the other conductors (k) have negative charge, hence

$$Q_i = \Sigma_k C_{ik} V_k = C_{ii} V > 0 \quad \text{and} \quad Q_{k\neq i} = \Sigma_m C_{km} V_m = C_{ki} V < 0 \ (k \neq j).$$

Thus, all the C_{ii} are positive and the C_{ik} (with $i \neq k$) are negative

$$C_{ii} > 0 \quad \text{and} \quad C_{ij} < 0 \ (i \neq j). \quad [3.20]$$

On the other hand, the potential is defined up to an additive constant and the charges are not modified if one adds the same quantity V_o to all the potentials. Thus, we must have $Q_i = \Sigma_k C_{ik} V_k = \Sigma_k C_{ik} (V_k + V_o)$ for any V_o, hence

$$\Sigma_k C_{ik} = C_{ii} + \Sigma_{k\neq i} C_{ik} = 0, \quad \text{i.e.} \quad C_{ii} = -\Sigma_{k\neq i} C_{ik}. \quad [3.21]$$

This means that $\Sigma_k Q_k = 0$, which is valid if the conductors are in total influence. If the influence is partial, we find the inequality $C_{ii} > -\Sigma_{k\neq i} C_{ik}$.

To deduce the energy, we may generalize the method of the preceding section. We may also directly use the expression of the interaction energy of the charges $q_{s,i}$ $d\mathcal{S}_i$, thus

$$U_E = ½ \Sigma_i \iint_{\mathcal{S}_i} d\mathcal{S}_i\, q_{s,i} V_i = ½\Sigma_i V_i \iint_{\mathcal{S}_i} d\mathcal{S}_i\, q_{s,i} = ½ \Sigma_i V_i Q_i = ½\Sigma_{ij} C_{ij} V_i V_j, \quad [3.22]$$

where we have used the expression [3.18] for the charge.

3.5. Electric forces between conductors

Let us consider the simple case of a parallel plate capacitor (Figure 3.5a). To calculate the force acting on one of the plates, we must consider it as a test body in the field of the other (not the field **E** of both plates). Taking the origin O on the positive plate and the axis Oz normal to the plates and oriented toward the negative plate, the field of the positive plate is $\mathbf{E}_1 = (q_s/2\varepsilon_o)\mathbf{e}_z$. An element of area $d\mathcal{S}$ of the negative plate has a charge $-q_s\, d\mathcal{S}$; thus, it is subject to a force exerted by the positive plate $d\mathbf{F} = -q_s\, d\mathcal{S}\, \mathbf{E}_1 = -(q_s^2/2\varepsilon_o)\, d\mathcal{S}\, \mathbf{e}_z$. The total force exerted on the negative plate is

$$\mathbf{F} = \iint_S d\mathbf{F} = -\frac{q_s^2}{2\varepsilon_o}\mathbf{e}_z \iint_{\mathcal{S}} dS = -\frac{q_s^2\,\mathcal{S}}{2\varepsilon_o}\mathbf{e}_z = -\frac{Q^2}{2\varepsilon_o\,\mathcal{S}}\mathbf{e}_z = -\frac{\varepsilon_o\,\mathcal{S}V^2}{2d^2}\mathbf{e}_z. \quad [3.23]$$

More generally, knowing the expression of the energy of a system of conductors U_E, we may use the method of virtual *displacements* to calculate the force **F** that is exerted on one of them; conductor (1), for instance. For this we may assume either that the conductors are disconnected from any other body or maintained at a constant potential.

a) If the conductors are disconnected from other bodies, their charges Q_i are fixed. If conductor (1), for instance, is subject to a force **F**, to displace it by δx, an external agent must exert a force $\mathbf{F}' = -\mathbf{F}$ and a work $\delta W = -F_x\,\delta x$. This work is transformed into stored electric energy $\delta U_E = -F_x\,\delta x$, hence (at constant charge)

$$F_x = -\,\delta U_E/\delta x = -\,\partial_x U_E\,|_Q. \quad [3.24]$$

b) If the conductors are maintained at constant potentials by using batteries, to displace conductor (1) by δx, the external agent must exert a force $\mathbf{F}' = -\mathbf{F}$ and a work $\delta W = -F_x\delta x$. However, the charges of the conductors vary by δQ_i and the generators must supply an electric energy $\delta U_g = \Sigma_i\, V_i\, \delta Q_i$. Thus the variation of the stored energy is $\delta U_E = -F_x\delta x + \Sigma_i\, V_i\,\delta Q_i$, hence (at constant potentials V_i)

$$F_x = \Sigma_i\, V_i\,\delta Q_i/\delta x - \partial_x U_E\,|_V. \quad [3.25]$$

In the case of the conductors being the plates of a parallel plate capacitor, using the virtual displacement at constant charge and writing $U_E = \frac{1}{2}Q^2/C = Q^2x/2\varepsilon_o\mathcal{S}$, equation [3.24] gives the force on the negative plate

$$F_x = \tfrac{1}{2}(Q^2/C^2)\,\partial_x C = -\,Q^2/2\varepsilon_o\mathcal{S}. \quad [3.26]$$

Using the virtual displacement at constant potential and the expressions $U_E = \frac{1}{2}CV^2 = \varepsilon_o\mathcal{S}V^2/2x$ and $Q_2 = -Q_1 = CV$, we find $\Sigma_i\, V_i\,\delta Q_i/\delta x = V^2\,\partial_x C = -\varepsilon_o\mathcal{S}V^2/x^2$. Thus, equation [3.25] gives

$$F_x = -\varepsilon_o\mathcal{S}V^2/x^2 + \varepsilon_o\mathcal{S}V^2/2x^2 = -\varepsilon_o\mathcal{S}V^2/2x^2. \quad [3.27]$$

A similar method may be used to calculate the moment $\mathbf{\Gamma}$ of the electric forces exerted on a conductor. To calculate the component Γ_x, we have only to consider a virtual rotation through an angle $\delta\theta$ about the axis Ox, thus

$$\Gamma_x = -\,\delta U_E/\delta\theta = -\,\partial_\theta U_E|_Q \qquad \text{(at constant charges } Q_i\text{)},$$
$$\Gamma_x = \Sigma_i\, V_i\,\delta Q_i/\delta\theta - \partial_\theta U_E\,|_V \qquad \text{(at constant potentials } V_i\text{)}. \quad [3.28]$$

For instance, if the plane plates of a capacitor form an angle θ, the capacitance depends on θ. The moment of the electric force on the negative plate is

$$\Gamma_x = -\partial_\theta U_E \,|_Q = (Q^2/2C^2)(\partial_\theta C), \qquad \Gamma_x = V^2(\partial_\theta C) - (\partial_\theta U_E)\,|_V = \tfrac{1}{2}V^2(\partial_\theta C). \quad [3.29]$$

In the case of several conductors, we may use the expression $U_E = \tfrac{1}{2} \sum_{ij} C_{ij} V_i V_j$ or $U_E = \tfrac{1}{2} \sum_{ij} C^{-1}_{ij} Q_i Q_j$. We find the relations

$$F_x = -\tfrac{1}{2} \sum_{ij} \partial_x C^{-1}_{ij} Q_i Q_j \quad \text{or} \quad F_x = \tfrac{1}{2} \sum_{ij} \partial_x C_{ij} V_i V_j, \quad [3.30]$$

$$\Gamma_x = -\tfrac{1}{2} \sum_{ij} \partial_\theta C^{-1}_{ij} Q_i Q_j \quad \text{or} \quad \Gamma_x = \tfrac{1}{2} \sum_{ij} \partial_\theta C_{ij} V_i V_j. \quad [3.31]$$

Application: Kelvin absolute electrometer

The Kelvin absolute electrometer, illustrated in Figure 3.8, uses the attraction force of the plates of a parallel plate capacitor to measure the potential difference of the plates. This force is balanced by a weight mg by using a precision balance. To eliminate edge effects, one of the capacitor plates is a circular disk of area $\mathcal{S}$ surrounded by a circular ring. The mechanical equilibrium is established if $mg = \varepsilon_o \mathcal{S} V^2/2d^2$, where we have used the relation [3.23], hence $V = d\sqrt{2mg/\varepsilon_o \mathcal{S}}$.

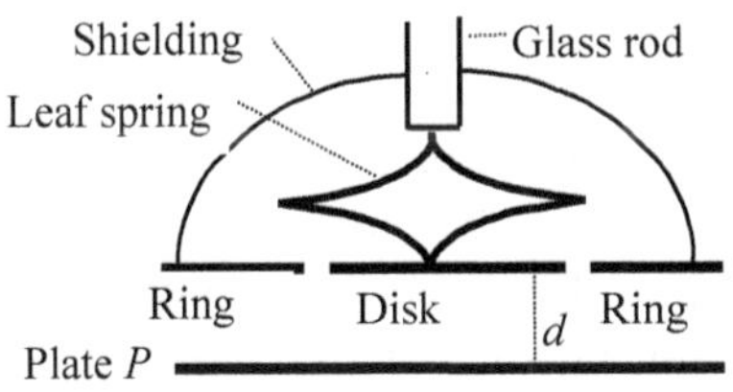

Figure 3.8. *Kelvin electrometer*

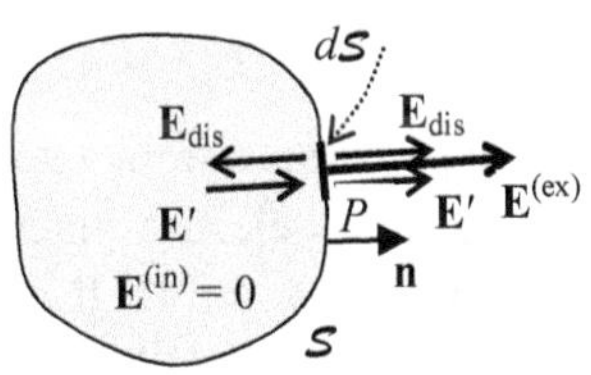

Figure 3.9. *Electrostatic pressure*

Electrostatic pressure

Consider a charged conductor and an element $d\mathcal{S}$ of its surface assimilated to a small disk (Figure 3.9). The electrostatic field $\mathbf{E}^{(ex)} = (q_s/\varepsilon_o)\,\mathbf{n}_{12}$ at point P just outside the conductor is the superposition of the field $\mathbf{E}_{disk} = (q_s/2\varepsilon_o)\mathbf{n}_{12}$ of the disk and the field $\mathbf{E}'$ of the other charges of the conductor and the other bodies. Inside the conductor, the total field is zero and it is the superposition of $\mathbf{E}'$ and the field of the disk $-(q_s/2\varepsilon_o)\mathbf{n}_{12}$. We deduce that $\mathbf{E}' = (q_s/2\varepsilon_o)\mathbf{n}_{12}$. The element of surface $d\mathcal{S}$, considered as a test charge, is subject to the field $\mathbf{E}'$, thus to a force $d\mathbf{F} = q_s\, d\mathcal{S}\, \mathbf{E}' =$

$d\mathcal{S}$ $(q_s^2/2\varepsilon_o)$ $\mathbf{n}_{12}$. This force is orthogonal to the surface of the conductor and it always points outward. It is equivalent to an *electrostatic pressure*

$$P_E = dF/d\mathcal{S} = (q_s^2/2\varepsilon_o) = \tfrac{1}{2}\,\varepsilon_o\mathbf{E}^2. \qquad [3.32]$$

This pressure tends to increase the volume of the conductor. It is numerically equal to the density of electrostatic energy just outside the surface.

3.6. Currents and current densities

An *electric current* is an ordered flow of charged particles. It may be of three types: a *beam of charged particles* moving in vacuum (electrons, protons, alpha particles, etc.), a *conduction current* in solid conductors and solutions, and a *convection current* produced by moving bodies carrying charges. In this chapter we consider mostly the conduction currents in solids. Electric currents have some physical and chemical effects and they transport energy and signals. The motion of charged particles is always impeded by friction forces, which dissipate energy and ultimately end the motion. Thus a sustained current can be established only if a generator acts on the charges and supplies them with energy.

If an average electric charge δQ traverses a surface $\mathcal{S}$ in the time interval δt (Figure 3.10a), we say that the *current intensity* (in amperes) through $\mathcal{S}$ is

$$I = \delta Q/\delta t. \qquad [3.33]$$

If a charge δq moves from a point A to a point B with a potential drop $V_{AB} \equiv V_A - V_B$, the charge loses electric potential energy $\delta q\ V_{AB}$. This energy is supplied to the circuit between A and B. It may be stored as electric energy in a capacitor or as magnetic energy in a coil, dissipated as heat in a resistor, transformed to mechanical energy or chemical energy, etc. If no generator exists between these points and the electric current flows from higher to lower potential, it supplies energy. The total energy in a circuit is obviously supplied by the generators. In the case of a continuous current I, the charge passing during δt is $\delta q = I\ \delta t$ and the energy supplied by the charge dq between A and B is $dW_{AB} = dq\ V_{AB} = I\ V_{AB}\ dt$. Thus the supplied power by the electric current is

$$P_{AB} = I\ V_{AB}. \qquad [3.34]$$

The motion of charged particles may be more significant at some places than at others. Thus, we define at each point $\mathbf{r}$ a *current density* $\mathbf{j}(\mathbf{r})$, such that the current intensity that traverses an element of area $d\mathcal{S}$ placed at $\mathbf{r}$ is

$$dI = (\mathbf{j.n})\ d\mathcal{S} = j \cos\theta\ d\mathcal{S}. \qquad [3.35]$$

Thus dI is the flux of $\mathbf{j}(\mathbf{r})$ through the element of area $d\mathcal{S}$. It is evident that dI vanishes if $\mathbf{j}$ is tangent to $d\mathcal{S}$ and dI is maximum if $\mathbf{j}$ is normal to $d\mathcal{S}$. The intensity that traverses a finite surface $\mathcal{S}$ is the flux of $\mathbf{j}$ through $\mathcal{S}$

$$I = \iint_{\mathcal{S}} dI = \iint_{\mathcal{S}} d\mathcal{S}\, (\mathbf{j}.\mathbf{n}). \qquad [3.36]$$

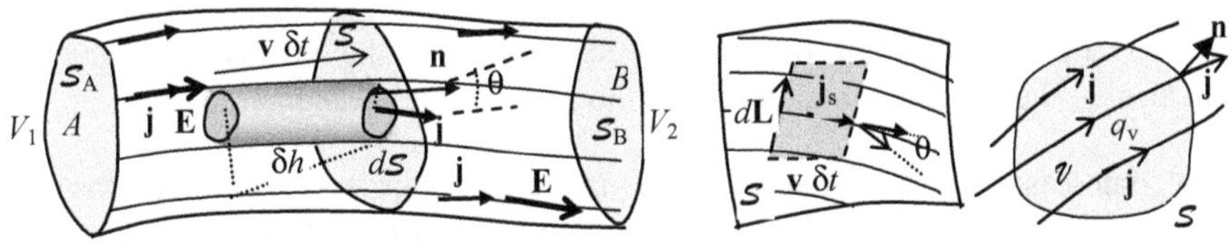

Figure 3.10. *a) Lines of current and volume current density, b) surface current density, and c) conservation of electric charge*

It is possible to relate the current density to the charge density $q_v(\mathbf{r})$ per unit volume and the mean velocity $\mathbf{v}(\mathbf{r})$ of the charge carriers. Indeed, the charge that traverses $d\mathcal{S}$ in the time interval δt is contained in the cylinder whose base is $d\mathcal{S}$ and length $\mathbf{v}\ \delta t$ (Figure 3.10a). The height of this cylinder is $\delta h = (\mathbf{n}.\mathbf{v}\ \delta t)$, its volume is $d\mathcal{S}\ \delta h = (\mathbf{n}.\mathbf{v})\ \delta t\ d\mathcal{S}$ and the charge that it contains is $d\delta q = q_v(\mathbf{r})\ (\mathbf{n}.\mathbf{v})\ \delta t\ d\mathcal{S}$. Thus, the current intensity that traverses $d\mathcal{S}$ is $dI = d\delta q/\delta t = q_v(\mathbf{r})\ (\mathbf{n}.\mathbf{v})\ d\mathcal{S}$. Comparing with the expression [3.35], we deduce that the current density may be written as

$$\mathbf{j}(\mathbf{r}) = q_v(\mathbf{r})\ \mathbf{v}(\mathbf{r}). \qquad [3.37]$$

Thus, $\mathbf{j}(\mathbf{r})$ points in the direction of the mean velocity of charge carriers if the charge carriers are positive and in the opposite direction if the charge carriers are negative. The lines of current are tangent to $\mathbf{j}$ at each point $\mathbf{r}$.

In a conductor at electrostatic equilibrium, we have $q_v = 0$. As we shall see in section 9.4, this property holds approximately in the quasi-static approximation (i.e. slowly varying phenomena). Thus, in metallic conductors, the negative charge density of the conduction electrons is counterbalanced by the positive charge density of the positively ionized atoms. As the ions are heavy, they do not move and do not contribute to the current. If the number of conduction electrons per unit volume is $n_e(\mathbf{r})$ and their average velocity is $\mathbf{v}_e(\mathbf{r})$, the current density may be written as

$$\mathbf{j}(\mathbf{r}) = -e\ n_e(\mathbf{r})\ \mathbf{v}_e(\mathbf{r}). \qquad [3.38]$$

If the current is due to the motion of several types (k) of particles of charges $q_{(k)}$, numbers $n_{(k)}(\mathbf{r})$ per unit volume and average velocity $\mathbf{v}_{(k)}(\mathbf{r})$, the current density may be written as $\mathbf{j}(\mathbf{r}) = \Sigma_{(k)}\, q_{(k)}\, n_{(k)}(\mathbf{r})\, \mathbf{v}_{(k)}(\mathbf{r})$. In ionic solutions, for instance, we have $q_{(k)}\, n_{(k)}^{+}(\mathbf{r}) + q_{(k)}^{-}\, n_{(k)}^{-}(\mathbf{r}) = 0$ and both positive and negative ions move, hence

$$\mathbf{j}(\mathbf{r}) = \Sigma_{(k)}\, q_{(k)}^{+}\, n_{(k)}^{+}(\mathbf{r})\, \mathbf{v}_{(k)}^{+}(\mathbf{r}) + \Sigma_{(k)}\, q_{(k)}^{-}\, n_{(k)}^{-}(\mathbf{r})\, \mathbf{v}_{(k)}^{-}(\mathbf{r}). \qquad [3.39]$$

The conventional direction of the current is that of the positive charges or, equivalently, the opposite direction of the motion of negative charges.

In some cases, the conduction charges move in a thin layer, forming a surface *current*. If the surface $\mathcal{S}$ contains a charge density q_s moving with a mean velocity $\mathbf{v}$, the surface $\mathcal{S}$ carries a *surface current density* $\mathbf{j}_s$ (Figure 3.10b). $\mathbf{j}_s$ is related to the intensity dI that traverses an infinitesimal segment $d\mathbf{L}$ by the relation $dI = d\mathbf{L}.\mathbf{j}_s$. The charge that traverses $d\mathbf{L}$ in the time interval δt is contained in the parallelogram of sides $d\mathbf{L}$ and $\mathbf{v}\ \delta t$. We deduce that

$$\mathbf{j}_s(\mathbf{r}) = q_s(\mathbf{r})\, \mathbf{v}(\mathbf{r}). \qquad [3.40]$$

Usual current intensities in electric circuit vary from a fraction of milliampere to few amperes. In electronic components, it varies from a picoampere (10^{-12} A) to a fraction of the ampere. It may attain hundreds of thousands of amperes in electrolysis solutions.

Consider a region of space in which the charge and current distributions are continuous and time-dependent (Figure 3.10c). The electric charge that leaves a closed surface $\mathcal{S}$ in the time interval δt is

$$\delta q_{\text{out}} = \delta t \iint_{\mathcal{S}} d\mathcal{S}\ (\mathbf{j}.\mathbf{n}) = \delta t \iiint_{\mathcal{V}} d\mathcal{V}\ (\nabla.\mathbf{j}\ ), \qquad [3.41]$$

where $\mathbf{n}$ is the unit vector normal to $\mathcal{S}$ and pointing outward $\mathcal{S}$. To write the last form, we have used Gauss-Ostrogradsky's theorem to transform the flux of $\mathbf{j}$ into the integral of $\nabla.\mathbf{j}$ over the volume $\mathcal{V}$ enclosed by $\mathcal{S}$. The total charge contained in $\mathcal{V}$ is $q^{(\text{in})} = \iiint_{\mathcal{V}} d\mathcal{V}\ q_v(\mathbf{r})$ and its decrease in the time interval δt is $-\delta q^{(\text{in})} = -\delta t\ (\partial_t q^{(\text{in})}) = -\ \delta t \iiint_{\mathcal{V}} d\mathcal{V}\ (\partial_t q_v)$. The conservation of charge requires that this decrease be equal to the charge [3.41] that leaves surface $\mathcal{S}$. Thus, we have

$$\iiint_{\mathcal{V}} d\mathcal{V}\ (\nabla.\mathbf{j}) = -\iiint_{\mathcal{V}} d\mathcal{V}\ (\partial_t q_v), \qquad [3.42]$$

which must be valid for any volume $\mathcal{V}$, hence the *continuity equation*

$$\nabla.\mathbf{j} + \partial_t q_v = 0. \qquad [3.43]$$

This is the local expression of the conservation of electric charge. Particularly, if the phenomena are time-independent ($\partial_t q_v = 0$), the continuity equation reduces to

$$\nabla.\mathbf{j} = 0. \qquad [3.44]$$

In this case, the charge [3.41] that flows out of a closed surface $\mathcal{S}$ vanishes. Thus, the total current intensity leaving $\mathcal{S}$ is equal to zero. This result holds in the case of stationary currents and approximately in the case of quasi-static currents. In the case of an electric circuit, if $\mathcal{S}$ contains no nodes but only one branch of current entering it and one branch leaving it, these branches carry the same current. If $\mathcal{S}$ encloses a node, the sum of the entering intensities is equal to the sum of the leaving intensities (*Kirchhoff node rule*).

3.7. Classical model of conduction, Ohm's law and the Joule effect

It is well known that in metals the current is carried by the conduction electrons (that is, the electrons that are weakly bound to atoms). Under the action of the electric field $\mathbf{F}_E = q\mathbf{E}$ a conduction electron is accelerated but, because of its collisions with other electrons and the immobile ions, it follows a zigzag path and its average velocity $\mathbf{v}$ is in the direction of $\mathbf{F}_E$. The effect of the collisions is equivalent to a resistance force $\mathbf{F}_f = -b\mathbf{v}$. Its velocity quickly attains a drift velocity $\mathbf{v}_d$ such that $\mathbf{F}_E + \mathbf{F}_f = 0$, that is, $q\mathbf{E} - b\mathbf{v}_d = 0$, thus $\mathbf{v}_d = (q/b)\mathbf{E}$. This velocity is always very small (of the order of the millimeter per second). The relation [3.37] may be written in the form, known as *Ohm's law*,

$$\mathbf{j} = \sigma\mathbf{E} \quad \text{or} \quad \mathbf{E} = \rho\,\mathbf{j}. \qquad [3.45]$$

$\sigma = n_e q^2/b$ is the *conductivity* of the material and $\rho = 1/\sigma = b/n_e q^2$ is its *resistivity*. If the medium is isotropic, σ is independent of the direction of the field, but if the medium is anisotropic, σ is a second rank tensor $\sigma_{\alpha\beta}$ and Ohm's law becomes

$$j_\alpha = \Sigma_\beta\, \sigma_{\alpha\beta}\, E_\beta. \qquad [3.46]$$

In this case, the current density is not necessarily in the direction of the field $\mathbf{E}$.

The friction constant b is a phenomenological parameter. In classical theory, b may be related to the collision time τ. For this, assume that between two collisions the electron experiences an electric force $\mathbf{F}_E = q\mathbf{E}$, thus an acceleration $\mathbf{a} = q\mathbf{E}/m$. Its velocity varies according to $\mathbf{v} = \mathbf{v}_o + (q\mathbf{E}/m)t$, where $\mathbf{v}_o$ is its velocity just after the last collision. $\mathbf{v}_o$ is randomly oriented in all directions for the conduction electrons. If we take the mean value for all the electrons, the average of $\mathbf{v}_o$ is zero and the equation reduces to $\mathbf{v} = (q\mathbf{E}/m)t$. In the average, the electrons drift with a velocity $\mathbf{v}_d = (1/\tau)\int_0^\tau dt\ \mathbf{v} = (q\tau/2m)\ \mathbf{E}$, where τ is the collision time, i.e. the average time between two collisions. Thus, the relation [3.38] gives $\mathbf{j} = (q^2 n_e \tau/2m)\mathbf{E}$ and we deduce *Drude's formula*

$$\mathbf{j} = \sigma\,\mathbf{E}, \qquad \text{with } \sigma = q^2 n_e \tau/2m. \qquad [3.47]$$

The drift velocity v_d is not to be mistaken for the average speed $\bar{v}$ of the electrons in the conductor, which may be as high as 10^6 m/s. Also, v_d should not be mistaken for the velocity of transmission of electric signals along the conductors used as a transmission line (telephone, electric current, etc.), which propagate with the speed of light as an electromagnetic wave in the medium outside the conductors. As the collision time is $\tau = l/\bar{v}$ where l is the *mean free path*, i.e. the average distance traveled by the electron between two collisions, the conductivity may be written as $\sigma = q^2 n_e l/2m\bar{v}$. The mean free path is of the order of the spacing between atoms ($\approx 10^{-10}$ m). The average speed $\bar{v}$ may be related to the thermal agitation by the statistical physics relation $\overline{U}_K = (3/2)k_B T = \frac{1}{2} m\overline{v}^2$, thus $\bar{v} = (3k_B T/m)^{1/2}$, where $k_B \cong 1.38 \times 10^{-23}$ is Boltzmann's constant and T is the absolute temperature. This classical model predicts a value of l about 10 times smaller than the real value and a resistivity ρ proportional to $\sqrt{T}$ instead of the experimental proportionality to T. This discrepancy is removed by quantum mechanical models.

Material	ρ (in Ω.m)	α_o (in K^{-1})
Aluminum	2.82×10^{-8}	3.9×10^{-3}
Constantan	$\approx 44 \times 10^{-8}$	2×10^{-6}
Copper	1.7×10^{-8}	3.9×10^{-3}
Iron	$\approx 10 \times 10^{-8}$	5×10^{-3}
Manganin	$\approx 44 \times 10^{-8}$	$\sim 10^{-7}$
Mercury	96×10^{-8}	8.9×10^{-4}
Nichrome	1×10^{-6}	4×10^{-4}
Platinum	11×10^{-8}	3.927×10^{-3}
Gold	2.44×10^{-8}	3.4×10^{-3}
Silver	1.59×10^{-8}	3.8×10^{-3}
Tungsten	5.6×10^{-8}	$4{,}5\times10^{-3}$
Carbon	3.5×10^{-5}	-5×10^{-4}
Lead	22×10^{-8}	3.9×10^{-3}

Material	ρ (Ω.m)	α_o (K^{-1})
Germanium	0.46	–0.048
Silicon	100-1000	–0.075
Glass	10^{10}-10^{14}	
Hard rubber	10^{13}	
Sulfur	10^{15}	
Fused quartz	75×10^{16}	
Polyethylene	10^{8} - 10^{9}	
Polystyrene	10^{7} - 10^{11}	
Porcelain	$10^{10} \times 10^{12}$	
Teflon	10^{14}	
Sodium chloride	0.044	–0.005
Blood	1.5	
Fat	25	

Table 3.1. *Resistivity* ρ *and temperature coefficient* α_0 *of some common materials. Constantan is* ~ 60 % Cu *and* ~ 40 % Ni, *manganin is* ~ 84 % Cu, ~ 12 % Mn, *and* ~ 4 % Ni, *and nichrome is* ~ 59 % Ni, ~ 23 % Cu, *and* ~ 16 % Cr. *The quoted values correspond to* 20°C

The electric power supplied by the electric field to each electron is $P_e = \mathbf{F}_E.\mathbf{v}_d = q\mathbf{E}.\mathbf{v}_d = b\mathbf{v}_d.\mathbf{v}_d = -\mathbf{F}_f.\mathbf{v}_d$ Thus, the supplied power is dissipated by the friction force as *Joule heat* with a density given by the local form of *Joule's law*

$$P_{J,v} = n_e P_e = \mathbf{E}.\mathbf{j} = \sigma\,\mathbf{E}^2 = \rho\,\mathbf{j}^2. \qquad [3.48]$$

As we have seen, there is no sharp distinction between conductors and insulators. Substances (such as metals) with a resistivity less than about 10^{-5} Ω.m are considered as *conductors*, while materials (such as glass, rubber, air, pure water, etc.) with a resistivity higher than about 10^5 Ω.m are considered as *insulators* (see Table 3.1). Materials whose resistivity lies between 10^{-5} and 10^5 Ω.m are called *semiconductors* (silicon, germanium, tellurium, etc.). Their resistivity depends strongly on the impurities they contain (small amounts of foreign atoms that are introduced into the semiconductor).

3.8. Resistance of conductors

Experimental studies starting with Ohm in 1826 showed that each segment of a conductor has a characteristic quantity, known as the *resistance R*, such that the drop in the potential between its ends is related to the current by *Ohm's law*

$$V = RI. \qquad [3.49]$$

The SI unit of R is the volt/ampere, called *ohm* (Ω). R depends on the material and the geometrical form of the conductor, on the impurities that it contains, and its temperature. Ohm's law applies to a variety of conductors said to be *Ohmic*. A plot of V versus I (called *characteristic*) for an Ohmic conductor is a straight line passing through the origin with a slope R (Figure 3.11a). In other words, for such materials, R (defined as V/I) is independent of I (or V) and, if the potential difference is reversed, the current through the conductor is reversed without change of intensity. Metals are almost Ohmic, but there are many conductors that are not (see Figure 3.11b and 3.11c). An ionized gas, a diode, etc., are non-Ohmic. Connection wires of reasonable length have usually a very small resistance and a conductor that serves to introduce a resistance in the circuit is called *resistor*. The dissipated power in the conductor of resistance R is

$$P_{AB} = IV_{AB} = R_{AB}I^2 = V_{AB}^2/R_{AB}. \qquad [3.50]$$

To see how the resistance depends on the shape of a conductor, let us consider a cylindrical wire of uniform cross section $\mathcal{S}$ and length L with the potential V applied between its ends. The conservation of electric charge implies that the current intensity is the same at any point of the wire and, if the section is uniform and the current density is uniform over the cross section, the current density $j = I/S$ is the same everywhere in the conductor. On the other hand, the translational symmetry implies that the electric field **E** has the same magnitude along the wire; thus, using the local form of Ohm's law $E = \rho j$, we find $V = EL = \rho jL = (\rho L/S)I$. This corresponds to a resistance $R = \rho L/S$. In more complicated shapes, the current density and the field are not uniform and the calculation is more complicated.

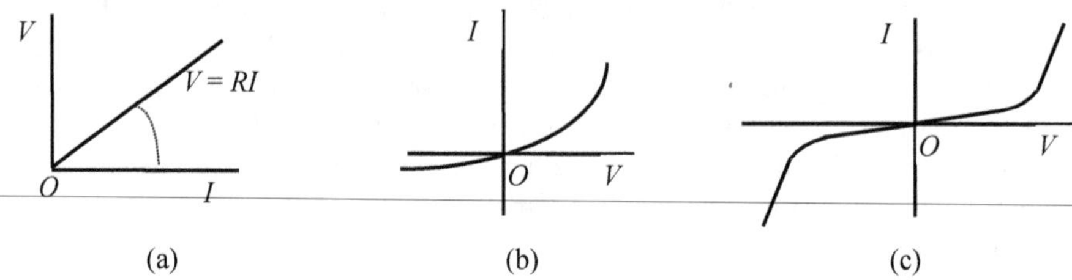

Figure 3.11. *a) The characteristic of an Ohmic conductor is a line of slope equal to R. b) The characteristic of a junction diode is not linear and the current does not take opposite values if V is reversed. c) Discharge current in a gas as a function of V*

3.9. Variation of resistivity with temperature, superconductivity

The experiment shows that the resistivity of a conductor increases with temperature. This can be explained by the increase of the thermal agitation and the scatterings of electrons with atoms and ions. Consequently, the electric resistance increases. For small variations of temperature, the variation of ρ is almost linear in T, of the form

$$\rho \approx \rho_o(1 + \alpha_o \Delta T), \qquad [3.51]$$

where ρ_o is the resistivity at some reference temperature T_o and $\Delta T = T - T_o$ is the difference in temperature from T_o. α_o is the *temperature coefficient of resistivity*. In fact, the relation [3.51] constitutes the first two terms of the power series of $\rho(T)$ about T_o. For large ΔT, higher terms may be important. For most pure metals in the temperature range 0 to 100°C, α_o varies between 3.2×10^{-3} and 6.2×10^{-3} K^{-1} (i.e. $\approx$ 1/273 K^{-1}). This means that, if the reistivity is extrapolated to very low temperatures it vanishes at absolute zero. In other words σ is proportional to T (Figure 3.12).

In 1911 Kamerlingh Onnes discovered that, at very low temperature, most metals depart radically from equation [3.51]. Their resistance falls to zero (Figure. 3.12); they become *superconductors*. Kamerlingh Onnes used liquid helium as a refrigerant down to about 1 K and found that the resistance of mercury slowly decreases with T, but at a *critical temperature* $T_c = 4.154$ K, it drops to an extremely small value. Kamerlingh Onnes concluded that mercury undergoes a phase transition to a *superconducting state*. The transition interval δT is of the order of only 10^{-3} K. Only 27 elements become superconducting under ordinary pressure, at critical temperatures below 4.2 K. They include aluminum (T_c = 1.175 K), lead (T_c = 7.23 K), niobium (T_c = 9.25 K), tin (T_c = 3.721 K), and tungsten (T_c = 0.0154 K). Good conductors, such as platinum, gold, silver, and copper do not become superconductors but thousands of alloys and compounds undergo this remarkable transition. The critical temperature depends on the presence of impurities and the internal stress in the sample. It depends also on the atomic weight of the isotope of the superconductor like $1/\sqrt{A}$.

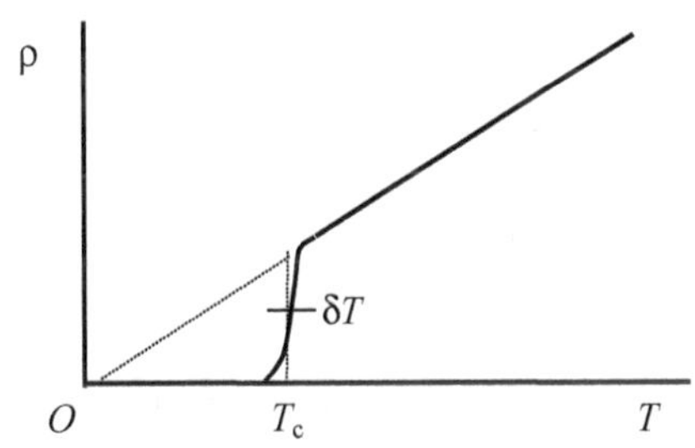

Figure 3.12. *Variation of resistivity as a function of temperature*

Contrary to substances in normal state, a superconductor has no resistivity (it is less than 4×10^{-25} Ω.m). An electric current, once established in the superconducting body, continues to circulate, perhaps indefinitely, without an applied electromotive force (emf) or an electric field. The so-called BCS theory (proposed in 1957 by Bardeen, Cooper and Schrieffer) explains the superconducting behavior as a quantum mechanical effect of pairing of electrons (because of a long range quantum mechanical attraction) unlike conductors in the normal state where electrons behave independently. The densely packed electrons are all linked together and act as a coherent unit and no single electron can be scattered to produce resistance.

In 1986, Berdnoz and Müller (1987 Nobel Prize) discovered that a ceramic compound (an oxide of barium, lanthanum, and copper) has a high critical temperature of 35 K. In 1987 another ceramic was found with T_c = 98 K and by 1988 another compound (Tl-Ba-Ca-Cu-O) was found with T_c = 125 K, and recently, a compound ($HgBa_2Ca_2Cu_3O_8$) was found with T_c = 134 K). Relatively high-temperature superconductors will certainly have very important technological applications.

3.10. Band theory of conduction, semiconductors*

The classical model of section 3.7 cannot explain some aspects of conduction, such as the free path in copper, which is ≈ 6 nm while the spacing of the atoms is ≈ 0.1 nm, and the variation of conductivity with temperature. On the other hand, classical physics cannot explain why some materials are insulators while others are conductors or semiconductors. Only quantum theory can answer these questions. As this theory is outside the scope of this book, the analysis in this section is only qualitative.

The basic idea of quantum mechanics is *wave-particle duality*. The electrons of momentum p also behave as a wave of wavelength given by the de Broglie formula

$$\lambda = h/p\,, \qquad [3.52]$$

where $h = 6.626\ 176 \times 10^{-34}$ J.s is Planck's constant. The square of the modulus of the *wave function* $\Psi(\mathbf{r},t)$ is interpreted as the probability of finding the particle at point $\mathbf{r}$ at time t. Ψ verifies Schrodinger's wave equation, which involves the potential energy of the electron in the medium. The electronic wave in the conductor being extended, the classical concepts of mean free path and of relaxation time become vague. In the case of the diffraction of a light wave by a diffraction grating, the wave is not diffracted by a particular slit of the grating but by all of them. Similarly, if the electronic wavelength is comparable to the distance between the atoms of the medium, we cannot say that the electron collides with a particular atom. The exact periodicity of the optical diffraction grating is essential: if a single slit is absent or displaced, the diffraction pattern is completely modified or even destroyed. The analogy is more striking with the diffraction of X-rays by a crystal (see section 11.12). If Bragg condition $2d \sin\theta = n\lambda$, which expresses that the waves reflected by the atomic planes are in phase, is satisfied, the wave is reflected with a high intensity, as if it traverses the crystal without collision with the atoms. In classical terms, the mean free path is then very large. This property is invalidated by impurities in the crystal: absence of atoms, the presence of different atoms, or simply a displacement of some atoms.

The wave behavior of electrons also explains the variation of conductivity with temperature. At lower temperature, there is less thermal agitation of the atoms and the crystal is more regular. Thus it is more transparent to the electronic wave and this results in higher conductivity. Similarly, if the crystal contains some impurities, its regularity is reduced as is its conductivity.

a) The free electrons model

To simplify, we consider a metallic parallelepiped of sides *a, b,* and *c* in the directions *Ox, Oy,* and *Oz*, respectively (Figure 3.13a). We treat the electrons as free particles in this box. Their wave function has the form $\Psi = A\ e^{i(\omega t - \mathbf{k.r})}$, where *A* is a constant, **k** is the wave vector, and ω is the angular frequency (see section 10.1), related to the momentum **p** and the energy *E* of the particle by the relations

$$k = 2\pi/\lambda = 2\pi p/h, \qquad \omega = 2\pi\,\tilde{\nu} = 2\pi E/h. \qquad [3.53]$$

The confinement of the electrons in the box means that $\Psi = 0$ on the faces $x = 0$, $x = a$, $y = 0$, $y = b$, $z = 0$, and $z = c$. Thus, the possible modes have the forms

$$\Psi = A\,\sin(xk_x)\,\sin(yk_y)\,\sin(zk_z)\,e^{i\omega t}, \quad k_x = \pi n_x/a, \quad k_y = \pi n_y/b, \quad k_z = \pi n_z/c. \qquad [3.54]$$

As only $|\Psi|^2$ has a direct physical meaning as a probability density, changing the signs of n_x, n_y, and n_z does not change the state. Thus we may take n_x, n_y, and n_z positive integers. The corresponding momentum **p** and energy of the electrons are

$$\mathbf{p} = \frac{hn_x}{2a}\mathbf{e}_x + \frac{hn_y}{2b}\mathbf{e}_y + \frac{hn_z}{2c}\mathbf{e}_z, \qquad E = \frac{\mathbf{p}^2}{2m} = \frac{h^2}{8m}\left[\frac{n_x^{\,2}}{a^2} + \frac{n_y^{\,2}}{b^2} + \frac{n_z^{\,2}}{c^2}\right]. \qquad [3.55]$$

Figure 3.13b illustrates *E* as a function of one component of the momentum. The representative points are on the parabola $E = p^2/2m$. The modes are represented by well-separated points.

On the other hand, electrons have a spin (i.e. an intrinsic angular momentum) with two possible spin states and *Pauli exclusion principle* forbids that two electrons occupy the same quantum state (determined by the numbers n_x, n_y, and n_z and the polarization state). Thus each state [3.54] may contain, at most, two electrons (one in each polarization state). At absolute zero, the electrons occupy the lowest energy levels up to the energy E_f, called *Fermi energy*; but, at finite temperature *T*, some electrons may be excited to occupy higher levels, leaving some unoccupied lower states, especially near E_f, as they are easier to excite.

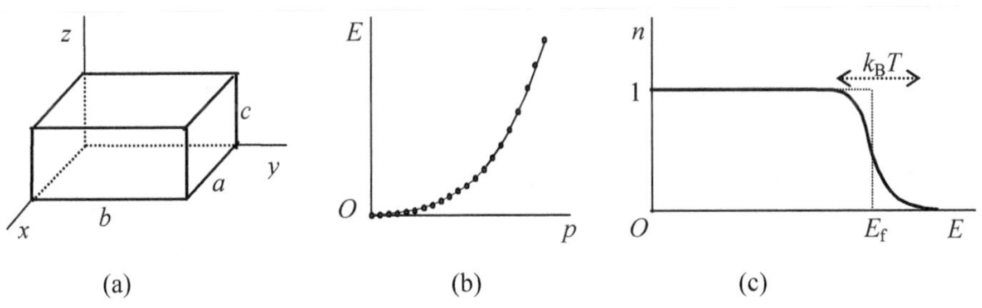

Figure 3.13. *Free electrons in a box: a) the box, b) possible states in the plane (p, E), and c) the distribution of electrons as a function of their energy at a given temperature*

If the medium contains n electrons per unit volume, quantum statistical physics gives the mean number of electrons that occupy a level E_i, called Fermi-Dirac distribution function, $<n_i> = 1/[C\, e^{E_i/k_BT} + 1\,]$, where C is determined by the condition that the total number of electrons is n. Figure 3.13c illustrates this distribution function at $T = 0$ (the rectangle up to the energy E_f) and at finite temperature. In the latter case the width of the transition band near E_f is about kT.

Taking the electric potential outside the conductor to be zero, the Fermi energy E_f is just the energy needed to extract an electron occupying this level from the conductor; it is called *work function*, usually written as eV where V is in volts. This quantum property of the metal is at the origin of the so-called *junction effect:* if two metals are in contact, a difference of potential $V_1 - V_2$ appears between them; it depends only on their chemical composition and their temperature.

If the metal is exposed to an external field, the electrons may be excited to higher levels; this gives them a momentum in the direction of the electric field. However, this free electron model cannot explain some properties of solids, particularly why certain materials are conductors, insulators, or semiconductors. Some of these properties are related to the interaction of the electrons with the individual atoms.

b) Bands theory of solids

Until now, we assumed that the electrons are free. In fact, they interact with the immobile ionized atoms. The fundamental property of a crystal is its periodicity in all directions. By analogy to the diffraction of X-rays by a crystal, whose atomic planes have a spacing d, let us assume that, if the electronic wave satisfies Bragg condition for $\theta = \pi/2$

$$2d = n\lambda = \frac{nh}{p}, \qquad \text{where } E = \frac{p^2}{2m} = \frac{n^2h^2}{8md^2}, \qquad [3.56]$$

it is almost entirely reflected by the atomic planes back and forth like a standing wave. Then, the wave cannot propagate in the metal and the electron cannot be at all considered as free. This means that the presence of ions forbids levels with a momentum near $p = nh/2d$. The same result is obtained if one solves Schrödinger's equation for electrons in a three-dimensional periodic potential. The states in the plane (p, E) are illustrated in Figure 3.14: the levels are grouped in *allowed bands*, each one containing a certain number of energy levels. The allowed bands are separated by *forbidden zones*. Let $i = 1, 2$, etc., label the allowed bands. For higher i, the forbidden zones become narrower and, at very large i, the spectrum become continuous.

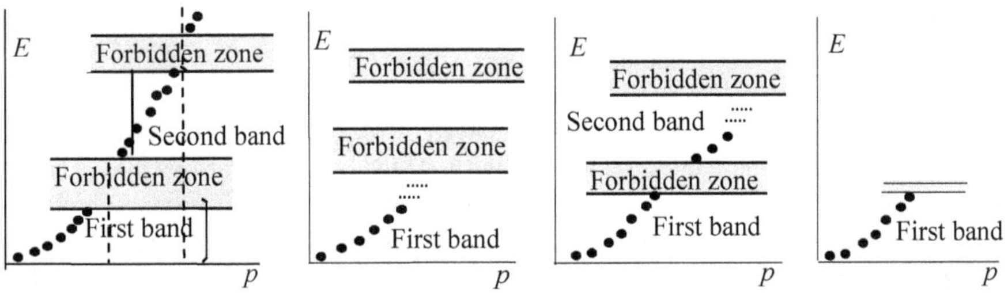

(a) (b) (c) (d)

Figure 3.14. *a) Band structure of solids. b) Case of an insulator with a filled first band and the second band empty. c) Case of a conductor whose first band is filled and the second band is partially filled. d) Case of a semiconductor whose first band is filled and the second band is empty, but with a very narrow forbidden zone*

Several electric properties of materials may be explained by their band structure. If the applied electric field is not very high, it cannot produce significant modifications of the energy of the electrons, because it acts only in a very short time interval, between two collisions. Thus, an electron cannot be excited from an allowed band to a higher one. However, if the last occupied allowed band contains non-occupied states, the electrons of this band may be excited by the electric field to a higher energy level in the same band; the material is then a conductor. Its last incomplete band is called *conduction band*, while the preceding filled band is the *valence band*. Conversely, if the last occupied allowed band is filled and the forbidden zone is large, the electrons cannot be excited from this last occupied band to a higher energy level in the next allowed band unless the electric field is very high. In this case the material is non-conductor. A narrowing of the forbidden zone or an overlap of allowed bands considerably increases the conductivity.

In a good conductor, such as copper, the atoms are singly ionized; this releases about one electron per atom. Its band structure is similar to that of Figure 3.14b with one allowed band having two unoccupied energy levels below the first forbidden zone. Even a weak electric field, may excite conduction electrons to one of these unoccupied levels; this makes copper a good conductor. The band structure of aluminum is similar to that of Figure 3.14c. Its first band is completely filled; thus it does not contribute to conduction. The electrons occupy some levels of the second allowed band and sometimes the third allowed band. The electrons of these partially occupied bands contribute to the electric conduction. The band structure of diamond is formed by a completely filled allowed band and a completely empty second allowed band and they are separated by a wide forbidden zone of 5 eV. As the thermic energy of electrons at normal temperatures is approximately 0.02 eV, the electrons of the first band cannot be excited to the second band. This makes diamond an almost perfect insulator.

If some impurities are added to the material, the periodicity of its crystalline structure is altered; this increases its resistivity. Thus, by using various alloys (iron-nickel, iron-nickel-chrome, etc.), it is possible to have materials to realize resistors of any resistance, for use in electronic circuits, rheostats, heaters, etc. It is also possible to have alloys with a temperature coefficient as small as 10^{-5}. The resistivity of some metals varies if they are exposed to a magnetic field. For instance, the resistivity of bismuth increases in a magnetic field perpendicular to the current; this enables the measurement of magnetic fields. Also, the resistance of certain alloys varies if they are under high pressure (because of the modification of their band structure) and the resistivity of some semiconductors, such as silicon, varies if they are exposed to radiation (because of the photoelectric effect).

c) Semiconductors

Figure 3.14d shows the band structure of a semiconductor, such as silicon or germanium. It is formed by a completely filled first band, but the forbidden zone is only 1.08 eV for silicon and 0.8 eV for germanium. Thus, the electrons may be easily excited from the first band to the second band, which becomes a conduction band. By doing so, the electrons leave the first band with vacancies, called *holes*. The conductivity is contributed to by the electrons of the formed conduction band and also from the electrons of the first band, which may move to occupy the holes. The latter are equivalent to positive charge carriers. This gives these materials conduction properties that are intermediate between insulators and conductors; they are called *intrinsic semiconductors*. The number of holes per unit volume is obviously equal to the number of free electrons n_e but the holes move with a velocity v_h in the direction of **E**, while the free electrons move with a velocity v_e in the opposite direction. If the field **E** is not very high, v_e and v_h are proportional to E. Thus, the global current density in the direction of **E** is

$$j = n_e(-e)(-v_e) + n_e(e)(v_e) = en_e(v_e + v_h) = \sigma E. \qquad [3.57]$$

σ increases rapidly with temperature, as the thermal energy of electrons $(3/2)kT$ increases, but it remains much lower than the conductivity of good metallic conductors. σ also increases if the body is exposed to a radiation, as the absorbed energy may excite an electron toward the conduction band and create a hole in the valence band. This formation of an electron-hole pair is called *photoconduction*.

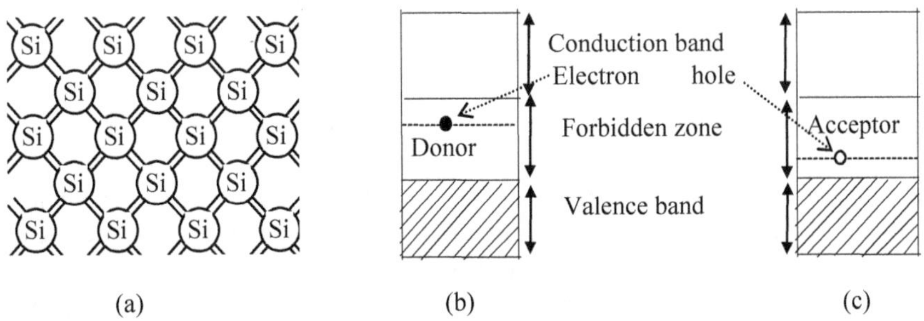

Figure 3.15. *a) Crystalline structure of silicon, b) impurity produced by a donor, and c) impurity produced by an acceptor*

The conductivity of a semiconductor is strongly affected by the presence of impurities, i.e. atoms of a different type replacing a small proportion (of the order of $1/10^6$) of the atoms of the crystalline lattice. Consider, for instance, silicon or germanium. These elements belong to the group 4 of the periodic table. They have the same crystalline structure as diamond; thus, each atom shares its four valence electrons with the neighboring atoms; two neighboring atoms share two valence electrons, one from each atom (Figure 3.15a). If an atom of silicon is replaced by an atom of the group 5 elements (As or Sb), which has five valence electrons, the additional valence electron remains in the vicinity of this atom in order to maintain its neutrality, but it is weakly bound to the atom. This atom is said to be a *donor*. This electron occupies an energy level situated in the forbidden zone (Figure 3.15b) and it may be easily excited to the conduction band. In this case, the *semiconductor* is said to be of the *type-n* (for negative). Conversely, if the impurity belongs to the group 3 elements (Al, Ga or In), one of the covalence electrons is missing; an electron of the neighboring atoms from the valence band may come to occupy this state. In this case, the impurity atom is said to be an *acceptor*. The energy level of this hole is in the forbidden zone (Figure 3.15c) and a valence electron may be excited to occupy it. Thus, the atoms become positive, one after the other; this is equivalent to the displacement of a positive charge (a hole), similar to the displacement of an air bubble in a liquid. In this case the semiconductor is said to be of the *type-p* (for positive).

In general, a semiconductor with P donor atoms, N acceptor atoms, n conduction electrons, and p conduction holes per unit volume has a total charge density

$$q_v = e\,(P - N + p - n). \quad [3.58]$$

The electric field is $\mathbf{E} = -\nabla V$ where the potential obeys Poisson's equation

$$\Delta V = -(e/\varepsilon)\,(P - N + p - n). \quad [3.59]$$

3.11. Electric circuits

The purpose of circuit analysis is to determine the electric current in branches of a circuit. Once the currents are determined, it is easy to determine the potentials. The currents are supplied by generators. A generator transforms a non-electric energy into electric energy to be supplied to the circuit. An ideal generator maintains a constant voltage $\mathcal{E}$ between its terminals A and B called the *emf.* The supplied current I_{AB} depends on the circuit. The total supplied power to the circuit is then $\mathcal{E}I_{AB}$. A real generator has some internal resistance r, which dissipates a power rI_{AB}^2. Thus, the supplied power to the external circuit is $\mathcal{E}I_{AB} - rI_{AB}^2$. If V_{AB} is the potential difference between the terminals, this power is also $V_{AB}I_{AB}$. Thus we must have $V_{AB} = \mathcal{E} - rI_{AB}$; this means that the generator is equivalent to an ideal generator of emf $\mathcal{E}$ and a resistance r in series.

To analyze an electric circuit, we may use Kirchhoff's rules, which are statements of the law of conservation of electric charge and the law of conservation of energy:

– the algebraic sum of the currents that meet at a node is zero;

– the algebraic sum of the potential drops and the emf around a closed path is zero.

These rules apply in the case of time-independent regimes and time-dependent regimes in quasi-permanent approximation, low-frequency alternating currents and slow transients, for instance (see section 9.4). In applying these rules, all voltages (including induced emf) and currents of various types must be taken into account.

These rules allow us to deduce the required equations. If the electric elements of the circuit are linear, the obtained equations are linear. Thus, it is possible to analyze the circuit with each generator separately and make the superposition of the solutions for all the generators. In the case of a sinusoidal emf $\mathcal{E} = \mathcal{E}_m \cos(\Omega t + \phi)$, the superposition principle allows us to consider the circuit with the exponential emf $\underline{\mathcal{E}} = \mathcal{E}_m\, e^{i(\Omega t+\phi)} = \underline{\mathcal{E}}_m\, e^{i\Omega t}$, to determine its solution and take its real part at the end of the calculation.

In the permanent regime and the quasi-permanent regime, the relations of the difference of potential at the terminals of any electric element to the current are valid for complex currents and potentials in the forms $\underline{V}_R = R\underline{I}$, $\underline{V}_C = \underline{Q}/C = \int dt\, \underline{I}/C$ and $\underline{V}_L = L\, d\underline{I}/dt$ at the terminals of a resistor, a capacitor and a self-inductance respectively. In the case of a sinusoidal current $\underline{I} = \underline{I}_m\, e^{i\Omega t}$, these relations become

$$\underline{V} = \underline{Z}\underline{I}, \qquad \text{where } \underline{Z}_R = R, \qquad \underline{Z}_C = 1/iC\Omega \qquad \text{and } \underline{Z}_L = i\Omega L. \qquad [3.60]$$

$\underline{Z}$ is the *complex impedance* of the circuit element; this is a generalization of the concept of resistance. It is useful because complex impedances in series are equivalent to a single impedance $\underline{Z} = \Sigma_i \underline{Z}_i$ and impedances in parallel are equivalent to a single impedance given by $1/\underline{Z} = \Sigma_i 1/\underline{Z}_i$.

As a simple example, consider a single loop circuit containing a resistance R, a capacitance C, and a self-inductance L connected in series to the terminals of a generator of emf $\underline{\mathcal{E}} = \mathcal{E}_m e^{i\Omega t}$ (Figure 3.16a). The impedance of these elements in series is $\underline{Z} = R + i(\Omega L - 1/C\Omega) \equiv Z e^{i\phi_z}$, where $Z = \sqrt{R^2 + (L\Omega - 1/C\Omega)^2}$ is the *real impedance* and $\phi_z = \arctan(L\Omega - 1/\Omega C)/R$ with $-\pi/2 < \phi_z < \pi/2$. The Kirchhoff loop rule gives $\underline{I}\,\underline{Z} = \underline{\mathcal{E}}$, hence

$$\underline{I} = \underline{\mathcal{E}}/\underline{Z} = (\mathcal{E}_m/Z)\, e^{i(\Omega t - \phi_z)}, \qquad I = I_m \cos(\Omega t - \phi_z) \text{ where } I_m = \mathcal{E}_m/Z. \qquad [3.61]$$

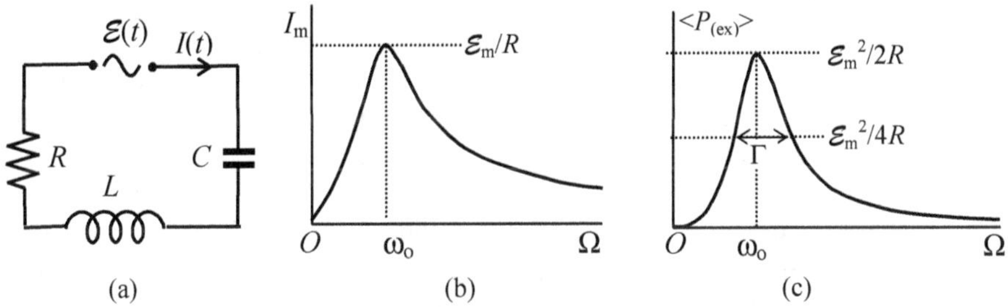

Figure 3.16. *a) Sustained LCR circuit, b) amplitude of the current versus Ω, and c) dissipated power versus Ω*

The variation of the amplitude of the current I_m as a function of Ω is illustrated in Figure 3.16b. It has a maximum equal to $\mathcal{E}_m/R$ for $\Omega = \omega_o = 1/\sqrt{LC}$. The instantaneous power supplied by the generator of the emf is

$$P_{(ex)} = I\mathcal{E} = I_m\mathcal{E}_m \cos(\Omega t)\cos(\Omega t - \phi_z) \qquad [3.62]$$

and its average value over a period is

$$< P_{(ex)} > = \tfrac{1}{2} I_m\mathcal{E}_m \cos\phi_z = \tfrac{1}{2}(RI_m/Z)\, I_m\mathcal{E}_m = \tfrac{1}{2}(R/Z^2)\,\mathcal{E}_m^2 = \tfrac{1}{2}RI_m^2. \qquad [3.63]$$

This is also the power that is dissipated in the resistor as Joule heat. The quantity $\cos\phi_z = R/Z$ is the power factor. We note that the instantaneous power is not a linear quantity in $\mathcal{E}$; thus, it must be calculated from real quantities. However, if we are only interested in averaged values, we may write $< P_{(ex)} > = \mathcal{R}e\,(\frac{1}{2}\underline{I}^*\underline{\mathcal{E}})$. The relation $< P_{(ex)} > = \frac{1}{2}RI_m^2$ shows that the average supplied power by an alternating current $I = I_m \cos(\omega t - \phi_z)$ in the circuit is the same as the power that is supplied by a direct current of intensity $I_{eff} = I_m/\sqrt{2}$ in the resistance R. I_{eff} is the *effective intensity* of the alternating current and $V_{eff} = V_m/\sqrt{2}$ is the effective voltage. The fact that I_m and $<P_{(ex)}>$ have sharp maximums for $\Omega = \omega_o$ is qualified as *resonance* and the frequency interval, which corresponds to $<P_{(ex)}>$ larger than half its maximum is the *resonance bandwidth* $\Gamma = R/2L$. Figure 3.16c illustrates the variation of $<P_{(ex)}>$ versus Ω.

3.12. Problems

Conductors in electrostatic equilibrium

P3.1 a) Assume that the charge on the surface of a conductor forms in fact a layer of thickness d and uniform volume charge density q_v. Express q_s in terms of d and q_v. Determine the electric field inside the conductor, in the layer, and just outside the conductor. **b)** Make the same analysis if the charge density varies with the depth x according to the relation $q_v = A \exp(-\delta x)$ where A and δ are two constants. Express q_s in terms of A and δ. **c)** Air may support a maximum electric field of 4×10^6 V/m without a risk of discharge. What should the maximum charge of a sphere of radius R be? What is the corresponding potential for spheres of radii 5 cm and 1 m? Assume that the charge in a silver sphere is due to one electron per atom. Calculate the thickness of charge on the surface of the sphere.

P3.2 A metallic ball of radius $R_1 = 10$ cm is surrounded by a concentric spherical shell of internal radius $R_2 = 25$ cm and external radius $R_3 = 30$ cm (see Figure 3.5c). These bodies were initially neutral. **a)** A charge $q = 5$ nC is placed on the ball. Determine the charge distributions, the field, and the potential everywhere. Does the ball act on a charge q' placed outside the shell? Does q' act on the ball? What can you say about the principle of action and reaction? **b)** Assume that the potential of the ball is $V_1 = 100$ V and that of the shell is $V_2 = 200$ V. Determine the charge distributions, the field and the potential everywhere. Can you say that there is an electric shielding of each region against the other?

Capacitors

P3.3 The Earth has a surface charge density $q_s = -0.9$ nC/m^2. **a)** What is its field **E** near the ground? What is the difference of potential between the ground and a point at a height 1.8 m (the top of your head)? Does this difference of potential produce any current in your body? **b)** A metallic plate of area 1 m^2 is initially neutral and placed horizontally at a height $h = 2$ m. Does it get any surface charge density? One connects this plate to the ground through a ballistic galvanometer. What charge does it indicate?

P3.4 A Geiger counter is formed by a conducting wire of radius 0.1 mm surrounded by a coaxial metallic cylindrical shell of internal radius 1 cm. The difference of potential between the wire and the shell being 1 kV, calculate the field E near the wire and near the cylinder. What is the charge of these conductors per unit length of the wire and the cylinder? What is the capacitance per unit length? The air may be ionized by a field $E > 10^6$ V/m. Up to which distance from the wire, this instrument may detect particles?

P3.5 Two capacitors of 5 and 10 μF are charged under 100 V. Calculate their charges and their energies. One disconnects them from the batteries and connects their plates of the same polarity. Calculate the new voltage and the new charges. Is there any loss of energy? Do the same if one connects plates of opposite polarities.

P3.6 a) Three metallic plates (1), (2) and (3) of area 3 m^2 are parallel. Plates (1) and (2) are 3 cm apart while (2) and (3) are 2 cm apart. Plate (2) is connected to the terminal A while (1) and (3) are connected to the terminal B. What is the capacitance of this set up between A and B? **b)** Assuming that plate (2) had a charge $Q = 50$ μC before placing the other plates. Determine the charge of the three plates.

Energy of capacitors

P3.7 a) Assume that the charge of a capacitor is increased gradually from 0 to Q. Show that the total required energy is $\frac{1}{2}Q^2/C$. **b)** Use the density of electrostatic energy to show this result in the cases of a parallel plate capacitor and a cylindrical capacitor. **c)** Consider the case of the cylindrical capacitor. What is the radius of a cylindrical surface that divides the energy into two equal parts?

P3.8 Two metallic spheres of radii R_1 and R_2 are at large distance from each other. The first has a charge Q, while the second is neutral. One connects them by a conducting wire. At a given moment, the charge of the first is q and that of the second is $Q - q$. What is the electric energy U_E of the system? Verify that U_E decreases to a minimum. Calculate the corresponding value of q and verify that the two sphere are then at the same potential. Is energy conserved in this process? Assuming that $R_1 = 1$ cm, $R_2 = 1$ m, and $Q = 5$ μC, calculate the charges, the potential of the spheres, the initial and the final energies.

Electric forces between conductors

P3.9 A parallel plate capacitor of thickness x and area S is charged under a potential V_0. **a)** The battery is disconnected and the thickness is varied by δx. What are the variations of the capacitance and the energy? Deduce the force of interaction of the plates. **b)** Assume now that the battery remains connected and the thickness is varied by δx. What is the corresponding variation of the energy? What is the energy supplied by the battery? Deduce the expression of the force.

P3.10 The discharge field in air is 4×10^6 V/m. This is the electric field strength that produces discharge. **a)** Two metallic plates have a difference of potential of 500 V. Up to which distance apart, they can be approached without a risk of discharge? What is then their force of attraction per unit area? **b)** A cloud of area 4 km^2 is at an altitude of 1 km and it carries a charge q. Neglecting edge effects, what is the capacitance of the capacitor that it forms with the ground? What is the field between the cloud and the ground? What should the charge q be to produce electric discharge? What then is the difference of potential between the Earth and the cloud? Assuming that the discharge is total, what is the liberated energy? **c)** A typical cardiac stimulator is essentially a capacitor of 100 μF. It is charged under 4.5 kV. What are the stored charge and energy? Assuming that the discharge is produced in 5 ms, what is the power of the instrument?

P3.11 a) How is the capacitance of a parallel plate capacitor of thickness d modified if a metallic sheet of thickness d' is introduced between the plates? Does the result depend on whether the sheet is parallel to the plates or not? **b)** Assuming that the plates and the sheet are squares of sides L and the sheet is introduced a distance x in the capacitor. Neglecting edge effects, write the capacitance as a function of x. What then is the energy of the capacitor if it has a charge Q or a potential V? Deduce the expression of the force exerted on the sheet.

P3.12 The plates of a capacitor are squares of sides L but they form a small angle θ. Show that the capacitance is $C \cong (\varepsilon_0 L^2/q)(1 - a\,\theta/2d)$. Calculate the moment of the forces $\mathbf{\Gamma}$ acting on one of the plates. To evaluate C, decompose it into thin bands.

Conduction and Joule effect

P3.13 Estimate v_d in copper, assuming one conduction electron per atom, a current of 2 A and a cross-sectional area of 1 mm^2.

P3.14 a) Calculate the resistance of a hollow cylindrical conductor between its internal and external cylindrical surfaces. **b)** Calculate the resistance of a hollow spherical conductor between its internal and external spherical surfaces. **c)** Assume that a difference of potential V is maintained between the end surfaces S_1 and S_2 of a conductor of arbitrary shape. Show that its resistance R and the capacitance of a capacitor having the same shape with $\mathcal{S}_1$ and $\mathcal{S}_2$ as electrodes are related by the relation $RC = \rho\varepsilon$.

P3.15 The electric energy of 1 MW produced by a power plant must be transported a distance of 100 km by using a line with a resistance of 0.20 Ω/km. Compare the lost energy by the Joule effect if the energy is transmitted at a voltage of 220 V, 22 kV, and at 220 kV?

P3.16 We consider the following model for a diode: electrons are emitted by thermionic effect by a cathode and collected by an anode. The electrodes are plane and parallel of area $\mathcal{S}$ and have a difference of potential V_o. Let N be the number of emitted electrons per second, $n(x)$ the electronic density at the distance x from the cathode and $v(x)$ their speed. **a)** Show that $N = \mathcal{S}\, n(x)v(x)$ and that the current intensity is $I = Ne$. **b)** Show that $n(x) = -(\varepsilon_o/e)\,\partial_x E(x)$. **c)** Assuming that the electrons are emitted without velocity, show that $I = (e/2m)^{1/2}\,(8\varepsilon_0\mathcal{S}/9d^2)\,V_o^{3/2}$.

P3.17 Consider an anisotropic parallelepiped of sides a, b and c in the directions Ox, Oy and Oz and of conductance σ_{pq}. **a)** A field **E** is applied in the Oz direction. Determine the current density? **b)** A current density **j** flows in the Oz direction. Determine the field **E** and the difference of potential between the opposite faces.

Figure 3.17. *Wheatstone bridge*

P3.18 A Wheatstone bridge, illustrated in Figure 3.17, allows the comparison of impedances. G is usually a sensitive galvanometer of impedance $\underline{z}$ or any detector of current. Calculate the current in the galvanometer. To measure $\underline{Z}_1$, for instance, one maintains two other impedances constant and varies the third until the galvanometer indicates no current. Show that this is the case if $\underline{Z}_2\,\underline{Z}_4 = \underline{Z}_1\underline{Z}_3$.

Chapter 4

Dielectrics

Insulators or *dielectrics* are mediums that contain no free charges. If a dielectric is placed in an electric field, it becomes *polarized.* We may consider the *electronic polarization* as due to the displacement of the electrons within the atoms and molecules and the *orientational polarization* due to the alignment of the polar molecules more or less in the direction of the electric field. The polarization of dielectrics explains some of their properties, particularly the propagation properties of electromagnetic waves (reflection, refraction, dispersion, etc.). Usually, the polarization disappears if the external field is removed, but some materials, called *electrets*, retain their polarization. These materials, (generally organic polymers, waxes, etc.) are the electrical analogs of permanent magnets. They are currently used in electrostatic microphones for modern phones. The purpose of this chapter is to study the polarization of dielectrics, the effects of dielectrics on the electric field, the field equations, and the energy.

4.1. Effects of dielectric on capacitors

In 1837, Faraday observed that, if a capacitor is maintained under a constant potential (by keeping it connected to a battery) and is filled with a dielectric, its charge is multiplied by a factor ε_r, which is a characteristic quantity of the dielectric called *relative electric permittivity*. Thus, the capacitance is multiplied by ε_r. On the other hand, if an empty capacitor is charged under a potential V_o (Figure 4.1a), the plates acquire charge densities $\pm q_s$ and Gauss's law gives the field $E_o = q_s/\varepsilon_o$ between the plates. If one disconnects the battery and fills the capacitor with the dielectric (Figure 4.1b), the charges of the isolated plates remains evidently the same, but the potential difference and hence the field are divided by ε_r to become

$$V = V_o/\varepsilon_r \quad \text{and} \quad E = V/d = V_o/\varepsilon_r d = E_o/\varepsilon_r = q_s/\varepsilon, \quad \text{where} \quad \varepsilon = \varepsilon_r\varepsilon_o. \qquad [4.1]$$

ε is the *absolute permittivity* (or the *dielectric constant*) of the dielectric. The capacitance becomes $C = \mathcal{S}q_s/V = \varepsilon_r C_o$ where C_o is the capacitance of the empty capacitor.

To increase the capacitance without increasing the area, it is possible to reduce the thickness d. However, d cannot be less than a certain limit determined by the electric discharge if the electric field attains the *electric strength* (or *breakdown field*) E_b. Thus, the breakdown voltage of a capacitor of thickness d is $V_b = E_b d$. The solution is to fill the capacitor with a dielectric of high permittivity. This allows the charge and stored energy to be increased without increasing the voltage. The values of ε_r and E_b for some common dielectrics are given in the Table 4.1.

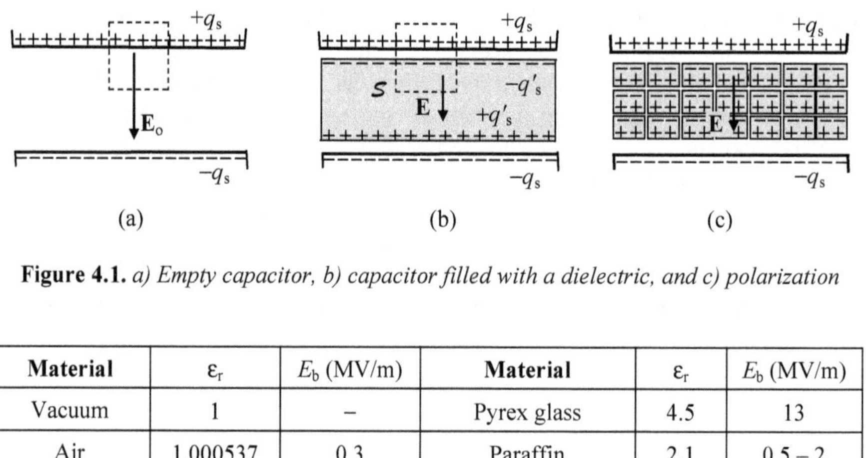

Figure 4.1. *a) Empty capacitor, b) capacitor filled with a dielectric, and c) polarization*

Material	ε_r	E_b (MV/m)	Material	ε_r	E_b (MV/m)
Vacuum	1	–	Pyrex glass	4.5	13
Air	1.000537	0.3	Paraffin	2.1	0.5 – 2
Water (20°)	80.36	–	Neoprene	6.9	12
Steam	1.0126	–	Mica	5.6	6 – 7
Paper	3.5	24	Glass	4 – 9	1 – 3
Porcelain	6.5	4	Strontium titanate	233	8
Quartz	3.8	13	Barium titanate	≈1500	8

Table 4.1. *Relative permittivity and dielectric strength E_b of some dielectrics under normal conditions (20°C and 1 atm). Those of steam correspond to 110° and 1 atm*

4.2. Polarization of dielectrics

The reduction in the electric field from E_o to $E = E_o/\varepsilon_r$ if a dielectric is introduced into a capacitor may be interpreted as due to uniform densities of *bound charges* $\mp q_s'$ on the dielectric faces, of opposite signs to the *free charges* $\pm q_s$ of the plates (Figure 4.1b). Thus, the surface charge densities of the plates become $\pm (q_s - q_s')$. The electric field is then $E = (q_s - q'_s)/\varepsilon_o$. As $E = E_o/\varepsilon_r = q_s/\varepsilon_o\varepsilon_r$, we deduce that

$$q_s' = q_s (1 - 1/\varepsilon_r). \qquad [4.2]$$

The densities of bound charges $\mp q_s'$ on the faces of a dielectric body do not depend on its dimensions. According to [4.2], they depend only on the relative permittivity ε_r and the densities of free charges $\pm q_s$ on the plates, i.e. the external electric field acting on the dielectric. We may always consider the dielectric body as a juxtaposition of small cubes of sides d (Figure 4.1c). If an electric field acts on the dielectric, bound charges of surface densities $\pm q_s'$ appear on the faces of the cubes, which are normal to **E**. The signs '±' correspond to a field leaving the cube or entering the cube, respectively. Thus, each cube has an electric dipole moment in the direction of **E**. We say that the dielectric becomes *polarized* and the bound charges of densities $\pm q_s'$ are also called *polarization charges*. If the dielectric is uniform (i.e. ε_r is the same at all points of the dielectric) and the field **E** is uniform, the faces of two cubes, which are in contact, have opposite polarization charge densities $+q_s'$ and $-q_s'$, which neutralize each other. Only the polarization charge densities $\pm q_s'$ on the external faces of the dielectric remain. The electric dipole moment of each cube in the direction of **E** is $p = (q'_s\mathcal{S})d = q_s'\mathcal{V}$, where $\mathcal{S} = d^2$ is the surface of the cube faces and $\mathcal{V} = d^3$ is its volume. The electric dipole moment being proportional to the volume $\mathcal{V}$, we may consider infinitesimal volume elements $d\mathcal{V}$ and define the *polarization density* (also called *polarization vector* or simply *polarization*) as the electric dipole moment per unit volume

$$\mathbf{P} = d\mathbf{p}/d\mathcal{V} = q_s'\ \mathbf{E}/E = (\varepsilon - \varepsilon_o)\ \mathbf{E}. \qquad [4.3]$$

Thus, **P** is proportional to the field **E** within the dielectric. We write

$$\mathbf{P} = \varepsilon_o\chi_E\mathbf{E}\ , \qquad \text{where } \chi_E = \varepsilon_r - 1. \qquad [4.4]$$

χ_E is the *electric susceptibility* of the dielectric. In the case of an electret, we may define the polarization **P** and the polarization charge, but the proportionality of **E** and **P** does not hold (as the permanent polarization **P** is independent of **E**).

4.3. Microscopic interpretation of polarization

In conductors at electrostatic equilibrium, the free electrons are distributed in such a way that the field **E** vanishes. In dielectrics, the electrons may move only within the molecules, this reduces the field **E** without making it equal to zero. The molecule is globally neutral and, in most cases, if no external electric field acts on it, the barycenter of the nuclei and that of the electrons coincide (Figure 4.2a). Thus, the molecule has no permanent electric dipole moment. If an external electric field acts on the molecule, it pushes the positive nuclei in its direction and it pulls the negative electrons in the opposite direction. The barycenter of the total negative charge $-q$ and that of the positive charge $+q$ are then separated by a distance $\mathbf{d}$ (Figure 4.2b) and the molecule gets an electric dipole moment $\mathbf{p}_e = q\mathbf{d}$. This polarization is called *electronic polarization* as it is essentially due to the deformation of the electronic cloud in the molecule.

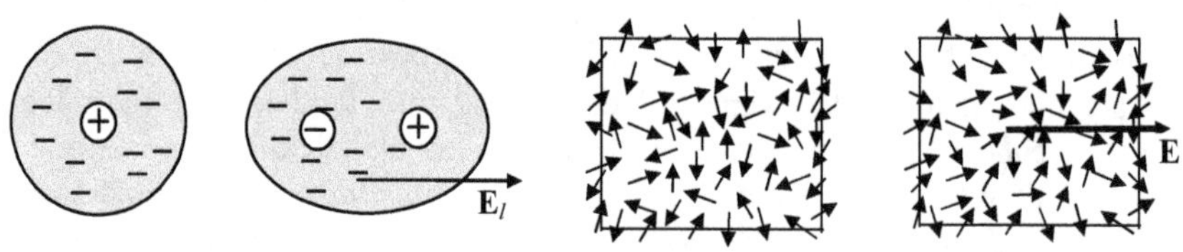

Figure 4.2. *a) Molecule non-submitted to an electric field, b) molecule submitted to a local field* $\mathbf{E}_l$, *c) polar dielectric non-submitted to an external field, and d) the same dielectric submitted to a macroscopic field* **E**

To study the polarization of a dielectric body (A), we should distinguish various electric fields. What we call *external electric field* $\mathbf{E}^{(ex)}$ is the field that exists before we place the body (A). The so-called *macroscopic electric field* **E** is the resultant of the field $\mathbf{E}^{(ex)}$ and the field $\mathbf{E}_p$ produced by the polarized body (A). $\mathbf{E}_p$ is the macroscopic field (that is, the averaged field) produced by all the molecules of (A). The so-called *local field* $\mathbf{E}_l$ that acts on a given molecule to polarize it, does not include the field of that molecule. Thus, we have $\mathbf{E}_l = \mathbf{E}^{(ex)} + \mathbf{E}'_p$ where $\mathbf{E}'_p$ is the averaged field of all the molecules of (A) except the considered molecule. The electric dipole moment of the molecule $\mathbf{p}_e$ is proportional to $\mathbf{E}_l$; it may be written as

$$\mathbf{p}_e = \alpha_e \mathbf{E}_l . \qquad [4.5]$$

α_e is the *polarizability* of the molecule, which depends on the nature of the molecule not on the physical conditions of the medium.

Some molecules are non-symmetric. The barycenter of positive charges and that of the negative charges do not coincide. Thus, the molecule has a *permanent electric dipole moment* $\mathbf{p}_m$ and it is said to be a *polar molecule*. This is the case of the molecules H_2O, SO_2, and NH_3. If no electric field acts on the body (Figure 4.2c), the molecules are randomly oriented in all directions because of thermal agitation and the frequent collisions of molecules if the medium is a gas. A macroscopic element of volume δV contains a large number of molecules. Thus, the total electric dipole moment of δV, which is equal to the vector sum of the dipole moments of all its molecules, is equal to zero. If an external field acts on the body, the local field $\mathbf{E}_l$ acts on the molecules with a moment of force $\mathbf{\Gamma}_m = \mathbf{p}_m \times \mathbf{E}_l$ to orient them in the direction of $\mathbf{E}_l$. However, the alignment cannot be complete because of the thermal agitation (Figure 4.2d). The mean electric dipole moment of the molecule is proportional to $\mathbf{E}_l$. It is called the *orientation dipole moment* and it may be written as

$$< \mathbf{p} > = \alpha_o \, \mathbf{E}_l. \qquad [4.6]$$

The constant α_o, called *orientation polarizability*, depends on the physical conditions, especially temperature (see section 4.13). The orientation polarization is always accompanied by the electronic polarization, but the latter is often less important. The total average electric dipole moment is then $\mathbf{p} = \mathbf{p}_e + <\mathbf{p}>$ and the *total polarizability* is $\alpha = \alpha_e + \alpha_o$.

A macroscopic element of volume $\delta\mathcal{V}$ of the dielectric contains $\delta N = N_v \, \delta\mathcal{V}$ molecules, where N_v is the number of molecules per unit volume. Thus, it has an electric dipole moment $\delta\mathbf{P} = \mathbf{p} \, N_v \, \delta\mathcal{V}$. In other words, the *polarization* $\mathbf{P} = \delta\mathbf{P}/\delta\mathcal{V}$ is

$$\mathbf{P} = N_v \, \mathbf{p} = N_v \, \alpha \, \mathbf{E}_l. \qquad [4.7]$$

To have an estimate of the order of magnitudes, let us consider the hydrogen atom. To simplify, we assume that the electronic cloud of radius $r_o \approx 10^{-10}$ m is not put out of shape but simply displaced with respect to the nucleus by a distance d in the opposite direction to $\mathbf{E}_l$. The nucleus is then subject to the electric field of this cloud $\mathbf{E}_e = -K_o e\mathbf{d}/r_o^3$ and to the local field $\mathbf{E}_l$. It is in equilibrium if $e\mathbf{E}_e + e\mathbf{E}_l = 0$, i.e. $K_o e\mathbf{d}/r_o^3 = \mathbf{E}_l$. The electric dipole moment of the atom is then $\mathbf{p} = e\mathbf{d} = (r_o^3/K_o)\mathbf{E}_l$. It has the form $\mathbf{p} = \alpha_e \mathbf{E}_l$ with $\alpha_e = r_o^3/K_o \approx 10^{-40}$ Cm^2/V. In some cases, α_e may be much higher (3×10^{-39} Cm^2/V for sodium, for instance). Even if an electric field of 10^6 V/m acts on the hydrogen atom, the induced dipole moment is only 10^{-34} C.m. It is much smaller than the permanent electric dipole moments (6.10×10^{-30} C.m for a water molecule, for instance). Obviously, if the medium has a permanent polarization, it does not depend on the electric field and it must be added to the induced polarization.

4.4. Polarization charges in dielectric

To simplify, we model the dipole moment **p** of the molecule as two charges $-q$ and $+q$ separated by a distance **d** such that $\mathbf{p} = q\mathbf{d}$. In the absence of an external electric field, the polar molecules are randomly oriented (Figure 4.2c). On the surface of a dielectric, there is equal probability to find bound charges $-q$ as bound charges $+q$. Thus, the surface charge density due to polarization is zero for any orientation of the dielectric surface. This is also true in the case of non-polar molecules since the charges $-q$ and $+q$ coincide. Now, if an electric field acts on a parallelepiped of dielectric *ABED* normally to the faces *AB* and *CD* (Figure 4.3a), the molecules acquire an average dipole moment **p** in the direction of **E**. Everything is the same as if all the charges $+q$ of the molecules are displaced by **d**/2 in the direction of **E** and all the charges $-q$ displaced by $-\mathbf{d}/2$ in the opposite direction. If $\mathcal{S}$ is the area of the face of the parallelepiped, the displaced positive charge near *CD* is $q' = N_V q \mathcal{S} d$; it corresponds to a surface charge density $q'_s = N_V qd = N_V p = P$. Similarly, the negative charge near *AB* is $-q'$ and this corresponds to a surface polarization charge density $q'_s = -P$. The faces *AD* and *BC*, which are parallel to the electric field, acquire no surface polarization charge density.

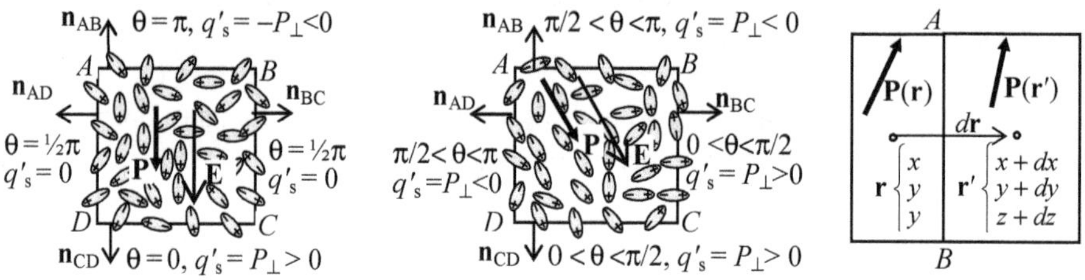

(a) (b) (c)

Figure 4.3. *Polarization charges in a dielectric: a) surface charge on a face normal to* **E**, *b) surface charge on an oblique face, and c) volume charge*

If **P** makes an angle θ with the outward normal **n** to the face *CD*, for instance, (Figure 4.3b), the positively charged layer at this face has a thickness $d \cos\theta$ and it contains a charge $N_V \mathcal{S} qd \cos\theta$. This corresponds to a surface polarization charge

$$q_s' = Nqd\cos\theta = Np\cos\theta = P\cos\theta = \mathbf{P.n}. \quad [4.8]$$

This relation holds for any form of the face with **n** always pointing outward from the dielectric. On the faces BC and DC, **P** and **n** form acute angles (**P** leaving the dielectric); thus, q_s' is positive. On the contrary, on the faces AB and AD, **P** and **n** form obtuse angles (**P** entering to the dielectric); thus, q_s' is negative. Within the dielectric, the volume polarization charge density is equal to zero if the polarization **P** is uniform.

Consider now the case of a non-uniform polarization and two parallelepipeds centered at points $\mathbf{r}(x, y, z)$ and $\mathbf{r}'(x+dx, y, z)$ with a common face AB of area $\mathcal{S}$ parallel to the Oyz plane (Figure 4.3c). Inside each one of these parallelepipeds, **P** has approximately the uniform values $\mathbf{P}(x, y, z)$ and $\mathbf{P}(x+dx, y, z)$. The normal to the face AB outgoing from the first parallelepiped is $\mathbf{e}_x$; thus, it has a surface charge density $q_s' = \mathbf{P}(x, y, z).\mathbf{e}_x$. The normal to the face AB outgoing from the second parallelepiped is $-\mathbf{e}_x$; thus, it has a surface charge density $q_s' = -\mathbf{P}(x+dx, y, z).\mathbf{e}_x$. The total polarization charge density on the face AB is $\mathcal{S}[P_x(x, y, z) - P_x(x+dx, y, z)] = -\mathcal{S}\, dx\, (\partial P_x/\partial x)$. It is equivalent to a volume polarization charge density $q'_v = -\partial_x P_x$. If the parallelepipeds are centered at the points $\mathbf{r}(x, y, z)$ and $\mathbf{r}'(x+dx, y+dy, z+dz)$, we find

$$q_v' = -\partial_x P_x - \partial_y P_y - \partial_z P_z = -\nabla.\mathbf{P}. \qquad [4.9]$$

We conclude that in a dielectric there are *polarization charges* of *surface density* $q'_s = \mathbf{P}.\mathbf{n}$ on the faces of the dielectric and a *volume density* $q'_v = -\nabla.\mathbf{P}$ within the dielectric. On the contrary to the conduction charges, the polarization charges are completely bound to the molecules of the dielectric.

4.5. Potential and field of polarized dielectrics

Let us consider a dielectric occupying a volume $\mathcal{V}$ and bounded by a surface $\mathcal{S}$. An element of volume $d\mathcal{V}'$ of dielectric near the point $\mathbf{r}'$ is equivalent to an electric dipole moment $d\mathbf{p}' = \mathbf{P}(\mathbf{r}')\, d\mathcal{V}'$ where $\mathbf{P}(\mathbf{r}')$ is the polarization. Using the results of section 2.6, the potential produced by this element of volume at a point **r** outside the dielectric (thus at large distance from the dipoles) is

$$dV_p(\mathbf{r}) = K_o\, d\mathcal{V}'\mathbf{R}.\mathbf{P}(\mathbf{r}')/R^3, \qquad \text{where } \mathbf{R} = \mathbf{r} - \mathbf{r}'. \qquad [4.10]$$

The total potential due to the polarization is obtained by integration over $\mathcal{V}$, hence

$$V_p(\mathbf{r}) = K_o \iiint_{\mathcal{V}} d\mathcal{V}'\mathbf{P}(\mathbf{r}').\mathbf{R}/R^3 = K_o \iiint_{\mathcal{V}} d\mathcal{V}'\mathbf{P}(\mathbf{r}').\nabla'(1/R), \qquad [4.11]$$

where we have used the relation $\mathbf{R}/R^3 = \boldsymbol{\nabla}'(1/R)$ with $\boldsymbol{\nabla}'$ designating the vector differential operator with respect to the coordinates (x', y', z'). Then, using the relation $\boldsymbol{\nabla}'[\mathbf{P}(\mathbf{r}')/R] = (1/R)\,\boldsymbol{\nabla}'.\mathbf{P}(\mathbf{r}') + \mathbf{P}(\mathbf{r}').\boldsymbol{\nabla}'(1/R)$, we may write

$$V_p(\mathbf{r}) = K_o \iiint_{\mathcal{V}} d\mathcal{V}'\boldsymbol{\nabla}'[\mathbf{P}(\mathbf{r}')/R] - K_o \iiint_{\mathcal{V}} d\mathcal{V}'(1/R)\nabla'.\mathbf{P}(\mathbf{r}'). \qquad [4.12]$$

Using Gauss-Ostrogradsky's theorem, we may transform the first integral into the flux of $\mathbf{P}(\mathbf{r}')/R$ over the surface $\mathcal{S}$ of the dielectric, hence

$$V_p(\mathbf{r}) = K_o \iint_{\mathcal{S}} d\mathcal{S}'\mathbf{n}'.\mathbf{P}(\mathbf{r}')/R - K_o \iiint_{\mathcal{V}} d\mathcal{V}'(1/R)\,\boldsymbol{\nabla}'.\mathbf{P}(\mathbf{r}'), \qquad [4.13]$$

where $\mathbf{n}'$ is the outward unit vector normal to the surface $\mathcal{S}$. This is the same potential as that of a polarization charge of surface density $q_s'(\mathbf{r}') = \mathbf{n}'.\mathbf{P}(\mathbf{r}')$ and volume density $q_v'(\mathbf{r}') = -\boldsymbol{\nabla}'.\mathbf{P}(\mathbf{r}')$. Thus, we may write the potential and the electric fields by using these polarization charge densities:

$$V_p(\mathbf{r}) = K_o \iint_{\mathcal{S}} d\mathcal{S}'q'_s(\mathbf{r}')/R + K_o \iiint_{\mathcal{V}} d\mathcal{V}'q'_v(\mathbf{r}')/R\,, \qquad [4.14]$$
$$\mathbf{E}_p(\mathbf{r}) = K_o\iint_{\mathcal{S}} d\mathcal{S}'q_s'(\mathbf{r}')\,\mathbf{R}/R^3 + K_o \iiint_{\mathcal{V}} d\mathcal{V}'q_v'(\mathbf{r}')\,\mathbf{R}/R^3\,. \qquad [4.15]$$

It is not obvious that these expressions hold for the potential and the electric field at points M inside the dielectric, as the expression [4.10] used to derive them is not valid for the potential near the dipole moment and it becomes infinite if $R \to 0$. Let us surround the point M by a small sphere of radius R_1 (Figure 4.4a). It may be verified that, for some geometrical configurations of molecules, the potential of this sphere of dielectric at its center M vanishes in the limit $R_1 \to 0$. To evaluate the potential of the dielectric occupying the volume $\mathcal{V}_2$ outside the sphere, we may use the expression [4.14] as M is outside $\mathcal{V}_2$. However, for this calculation, we must include the potential of the polarization charges on the surface $\mathcal{S}_1$ of the sphere as on the external surfaces $\mathcal{S}$ of the dielectric. The normal unit vector outgoing from $\mathcal{S}_1$ is $-\mathbf{n}_1 = -\,\mathbf{R}/R$ pointing toward M, hence

$$V_p(\mathbf{r}) = K_o \iiint_{\mathcal{V}_2} d\mathcal{V}'\, q'_v(\mathbf{r}')/R + K_o\iint_{\mathcal{S}} d\mathcal{S}'q'_s(\mathbf{r}')/R + K_o \iint_{\mathcal{S}_1} d\mathcal{S}'\; q'_s(\mathbf{r}')/R. \quad [4.16]$$

We may extend the integral on $\mathcal{V}_2$ to the whole volume $\mathcal{V}$ of the dielectric. Indeed, by doing so, we add the integral over $\mathcal{V}_1$, i.e. a term

$$\delta_1 V_p(\mathbf{r}) = K_o \iiint_{\mathcal{V}_1} d\mathcal{V}'\, q'_v(\mathbf{r}')/R \approx K_o\, q'_v \int_0^{R_1} dR\, R^2 \int_0^{\pi} d\theta' \sin\theta' \int_0^{2\pi} d\varphi'\,/R \approx 2\pi K_o\, q'_v R_1^{\,2},$$

which vanishes in the limit $R_1 \to 0$. Also, the third term in [4.16] may be written as

$$\delta_2 V_p(\mathbf{r}) = K_o \iint_{\mathcal{S}_1} d\mathcal{S}'\; q'_s(\mathbf{r}')/R = -\,K_o \iint_{\mathcal{S}_1} d\mathcal{S}'\; \mathbf{P}(\mathbf{r}').\mathbf{n}_1/R$$
$$\approx -\,K_o R_1 P \int_0^{\pi} d\theta' \sin\theta' \cos\theta' \int_0^{2\pi} d\varphi' = -\,\pi K_o R_1 P \sin^2\theta'\Big|_{\theta'=0}^{\theta'=\pi} = 0.$$

We deduce that the expression [4.14] for the potential and hence the expression [4.15] for the field are valid both inside and outside of the dielectric.

In the particular case of a uniform polarization ($\nabla.\mathbf{P} = 0$), the effect of the polarization reduces to that of the surface charge density $q_s' = \mathbf{n}.\mathbf{P}$. This is effectively the case if a plate of dielectric is introduced in a parallel plate capacitor (Figure 4.1b). **P** is then oriented from the positive plate to the negative plate. On the dielectric face that is close to the positive plate, **P** points toward the dielectric and $q_s' = \mathbf{P}.\mathbf{n}$ is negative. On the contrary, on the dielectric face that is close to the negative plate, **P** points outward the dielectric and $q_s' = \mathbf{P}.\mathbf{n}$ is positive. Thus, the field of the polarization $\mathbf{E}_p$ is in the opposite direction to that of the capacitor plates. We say the $\mathbf{E}_p$ is a *depolarizing field.*

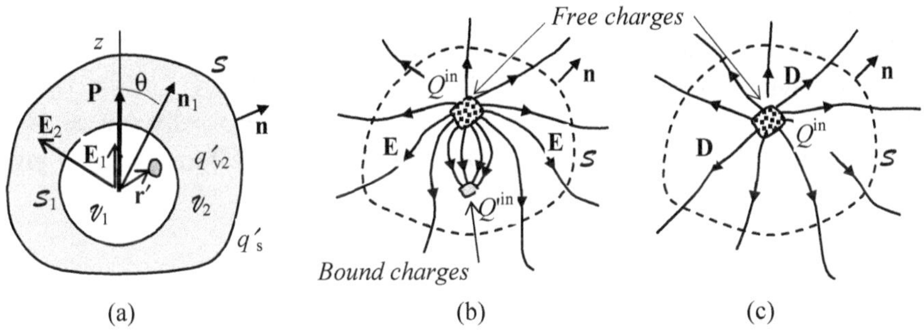

Figure 4.4. *a) Evaluation of **E** in a dielectric. b) Gauss's law $\Phi_E = (Q^{(in)}+Q'^{(in)})/\varepsilon_o$ uses both free charges $Q^{(in)}$ and bound charges $Q'^{(in)}$ inside $\mathcal{S}$. c) It may be written also as $\Phi_D = Q^{(in)}/\varepsilon_o$*

4.6. Gauss's law in the case of dielectrics, electric displacement

To write Gauss's law for **E**, we must use both the polarization charges and the free charges (Figure 4.4b). Thus, the integral form of Gauss's law may be written as

$$\Phi_E \equiv \iint_{\mathcal{S}} d\mathcal{S}\ \mathbf{n}.\mathbf{E} = (Q^{(in)} + Q'^{(in)})/\varepsilon_o,$$
$$\text{where } Q^{(in)} = \iiint_{\mathcal{V}} d\mathcal{V}\ q_v \text{ and } Q'^{(in)} = -\iiint_{\mathcal{V}} d\mathcal{V}\ \nabla.\mathbf{P}. \quad [4.17]$$

$\mathcal{S}$ is a Gaussian surface assumed not to have point charges, linear charge, or surface charge and $\mathcal{V}$ is the enclosed volume by $\mathcal{S}$. Using Gauss-Ostrogradsky's theorem, we may transform the integral of $\nabla.\mathbf{P}$ into the flux of **P** outgoing from $\mathcal{S}$ and write

$$\iint_{\mathcal{S}} d\mathcal{S}\ \mathbf{n}.(\mathbf{E} + \mathbf{P}/\varepsilon_o) = Q^{(in)}/\varepsilon_o, \quad \text{i.e.,} \quad \Phi_D \equiv \iint_{\mathcal{S}} d\mathcal{S}\ \mathbf{n}.\mathbf{D} = Q^{(in)}. \quad [4.18]$$

D is the *electric displacement field* (or *electric induction*)

$$\mathbf{D} = \varepsilon_o \mathbf{E} + \mathbf{P}. \qquad [4.19]$$

Gauss's law, written in the form [4.18], expresses that the flux of **D** outgoing from a closed surface $\mathcal{S}$ is equal to the free charge that it contains (Figure 4.4c). If the dielectric is linear and isotropic, we have seen that the polarization is $\mathbf{P} = (\varepsilon - \varepsilon_o)\mathbf{E}$ (see equation [4.3]). Thus, the electric displacement may be written as

$$\mathbf{D} = \varepsilon \mathbf{E}. \qquad [4.20]$$

For instance, let us consider a plate of linear and isotropic dielectric introduced in a parallel plate capacitor. Let $\mathcal{S}$ be a cylindrical Gaussian surface having a base of area $\mathcal{A}$ in the dielectric and the other within the positive plate (Figure 4.1b). Because of the symmetries, **E** and **D** are uniform and perpendicular to the plates. Their fluxes outgoing from $\mathcal{S}$ are $E\mathcal{A}$ and $D\mathcal{A}$. The free charge inside $\mathcal{S}$ is $Q^{(in)} = q_s\mathcal{A}$ and the polarization charge is $Q'^{(in)} = -q_s'\mathcal{A} = q_s\mathcal{A}(\varepsilon_o/\varepsilon - 1)$, where we have used [4.2]. Gauss's law in the form [4.17] gives $E\mathcal{A} = q_s\mathcal{A}/\varepsilon$ and, in the form [4.18], it gives $D\mathcal{A} = q_s\mathcal{A}$. We deduce that $E = q_s/\varepsilon$ and $D = q_s = \varepsilon E$.

4.7. Electrostatic equations in dielectrics

The electrostatic phenomena in the presence of dielectrics are specified by two fields: the electric field **E** and the electric displacement **D** (or **E** and the polarization **P**). Two fundamental laws govern these phenomena:

a) The *electric field* **E** is *conservative:* its circulation between any two points A and B depends on these points but not on the path

$$\int_A^B d\mathbf{r}.\mathbf{E} = V_A - V_B. \qquad [4.21]$$

Particularly, the circulation of **E** over a closed path $\mathcal{C}$ ($A \equiv B$) is zero:

$$\oint_{\mathcal{C}} d\mathbf{r}.\mathbf{E} = 0. \qquad [4.22]$$

As we have seen in section 2.3b, this is equivalent to the local equation

$$\nabla \times \mathbf{E} = 0, \qquad [4.23]$$

which is obviously satisfied if

$$\mathbf{E} = -\nabla V. \qquad [4.24]$$

b) The *electric field verifies Gauss's law*: this law may be written in the integral form [4.17] that uses **E** and **P**. Transforming the flux into a volume integral of the divergence by using Gauss-Ostrogradsky's theorem, we find

$$\iiint_{\mathcal{V}} d\mathcal{V}\, \nabla.[\varepsilon_o \mathbf{E}(\mathbf{r}) + \mathbf{P}(\mathbf{r})] = \iiint_{\mathcal{V}} d\mathcal{V}\, q_v(\mathbf{r})\,. \quad [4.25]$$

As this equation is valid for any volume $\mathcal{V}$, we must have

$$\varepsilon_o\, \nabla.\mathbf{E}(\mathbf{r}) - \nabla.\mathbf{P}(\mathbf{r}) = q_v(\mathbf{r}), \qquad \text{i.e.,} \quad \varepsilon_o \nabla.\mathbf{E}(\mathbf{r}) = q_v(\mathbf{r}) + q'_v(\mathbf{r})\,. \quad [4.26]$$

If, instead of **P**, we use the *electric displacement* $\mathbf{D} = \varepsilon_o\mathbf{E} + \mathbf{P}$, the integral form of Gauss's law may be written as

$$\iint_{\mathcal{S}} d\mathcal{S}\, (\mathbf{n}.\mathbf{D}) = \iiint_{\mathcal{V}} d\mathcal{V}\, q_v(\mathbf{r}), \quad [4.27]$$

while equation [4.26] becomes

$$\nabla.\mathbf{D}(\mathbf{r}) = q_v(\mathbf{r}). \quad [4.28]$$

Although the solution of the electrostatic problem exists and is unique, the field equations $\nabla \times \mathbf{E} = 0$ and $\varepsilon_o\nabla.\mathbf{E} = q_v + q'_v$ are not sufficient to determine **E**. Even in the absence of dielectrics ($q'_v = 0$), the solution of these equations is not unique. On the other hand, in the presence of dielectrics, the polarization charge density $q'_v = -\nabla.\mathbf{P}$ depends on **P** hence on **E** that we have to determine. The problem is even more complicated if the exact position of the free charges is not completely known as in the case of charges on conductors and if the region in which the field has to be calculated is confined by surfaces.

The analysis is slightly simplified if the dielectric is linear and isotropic. Then, the polarization **P** and **D** are proportional to **E**

$$\mathbf{P} = \varepsilon_o\, \chi_E\, \mathbf{E}\,, \qquad \text{and} \qquad \mathbf{D} = \varepsilon\mathbf{E}\,, \qquad \text{with } \varepsilon = \varepsilon_o(1 + \chi_E) \quad [4.29]$$

and the two fundamental equations [4.23] and [4.28] take the form

$$\nabla \times \mathbf{E} = 0, \qquad \text{and} \qquad \nabla.\, \mathbf{E} = q_v/\varepsilon\,. \quad [4.30]$$

Knowing **E**, we may determine the potential V and the fields **D** and **P** by using the relation [4.29] and the polarization charge density $q'_v = (\varepsilon_o/\varepsilon - 1)\, q_v$.

It is often more convenient to determine the potential first and then deduce the field **E**. Indeed, substituting the expression $\mathbf{E} = -\nabla V$ in Gauss's equation $\nabla.\mathbf{E} = q_v/\varepsilon$, we find the Poisson equation in the dielectric

$$\Delta V = -\, q_v/\varepsilon. \quad [4.31]$$

Assume that we look for a solution in a region $\mathcal{V}$, which contains a dielectric of permittivity ε (which may be non-uniform). The region $\mathcal{V}$ may be infinite or bounded by some surfaces $\hat{\mathcal{S}}_i$ (of conductors or any other material) with known boundary conditions. If we know the free charges at each point of $\mathcal{V}$ (excluding the charges on the surfaces $\hat{\mathcal{S}}_i$, which are taken into account by the boundary conditions), the Poisson equation [4.31] has the general solution

$$V(\mathbf{r}) = V_0(\mathbf{r}) + (1/4\pi) \iiint_{\mathcal{V}} d\mathcal{V}' q_v(\mathbf{r}')/\varepsilon R + (1/4\pi)\iint_{\mathcal{S}} d\mathcal{S}' q_s(\mathbf{r}')/\varepsilon R$$
$$+ (1/4\pi) \int_{\mathcal{C}} dl' \, q_l(\mathbf{r}')/\varepsilon R + (1/4\pi) \Sigma_k \, q_k/\varepsilon R_k . \qquad [4.32]$$

$V_0(\mathbf{r})$ is any solution of Laplace equation ($\Delta V = 0$) and the four other terms correspond to volume charges, surface charges, linear charges and point charges with $\mathbf{R} = \mathbf{r} - \mathbf{r}'$ and $\mathbf{R}_k = \mathbf{r} - \mathbf{r}_k$. In these terms, the permittivity ε may depend on the position $\mathbf{r}'$ or $\mathbf{r}_k$ of the source charges. The corresponding field $\mathbf{E}$ is

$$\mathbf{E}(\mathbf{r}) = \mathbf{E}_0(\mathbf{r}) + (1/4\pi) \iiint_{\mathcal{V}} d\mathcal{V}' q_v(\mathbf{r}')\mathbf{R}/\varepsilon R^3 + (1/4\pi)\iint_{\mathcal{S}} d\mathcal{S}' q_s(\mathbf{r}')\mathbf{R}/\varepsilon R^3$$
$$+ (1/4\pi) \int_{\mathcal{C}} dl' \, q_l(\mathbf{r}')\mathbf{R}/\varepsilon R^3 + (1/4\pi) \Sigma_k \, q_k \, \mathbf{R}_k /\varepsilon R_k^3 \qquad [4.33]$$

where $\mathbf{E}_o = -\nabla V_0$. It may be shown that it is always possible to choose $V_0(\mathbf{r})$ in order to respect the boundary conditions on the surfaces $\hat{\mathcal{S}}_i$. Thus, the physical problem has a unique solution. In the particular case of a uniform and isotropic dielectric filling all the volume $\mathcal{V}$, the solution is the same as without dielectric but divided by the relative permittivity ε_r.

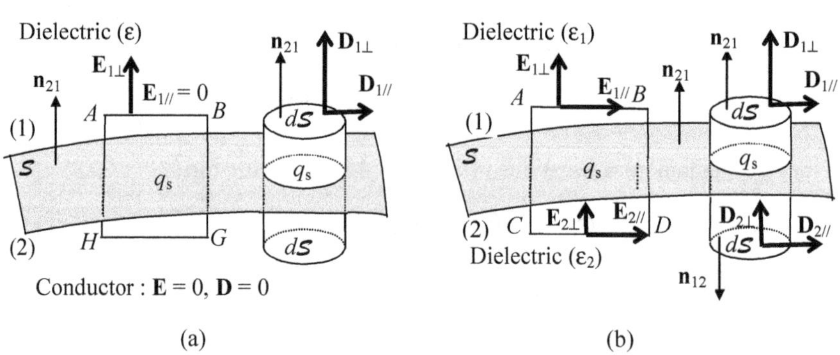

Figure 4.5. *a) The field at the interface of a conductor and a dielectric. b) The field at the interface of two dielectrics*

Let $\mathcal{S}$ be the interface of a dielectric (1) and a medium (2) (which may be a dielectric or a conductor) and q_s the surface density of free charge on $\mathcal{S}$ (Figure 4.5). Because of the discontinuity on $\mathcal{S}$, the fields **E** and **D** may have discontinuities and we cannot use the two fundamental laws of electrostatics in the local forms $\nabla\times\mathbf{E} = 0$ and $\nabla.\mathbf{D} = q_v$, but in the integral forms [4.22] and [4.27]. Repeating the analysis of section 2.8 by considering a narrow rectangular path *ABCD* and a cylindrical thin Gaussian surface on both sides of $\mathcal{S}$, we find that the tangential component of **E** is continuous, while the normal component of **D** has a discontinuity equal to q_s

$$\mathbf{E}_{1//} = \mathbf{E}_{2//} \quad \text{and} \quad \mathbf{n}_{21}.\mathbf{D}_1 - \mathbf{n}_{21}.\mathbf{D}_2 = q_s. \tag{4.34}$$

– If the medium (1) is a linear and isotropic dielectric of permittivity ε_1 and the medium (2) is a conductor (Figure 4.5a), we have $\mathbf{E}_2 = 0$ and $\mathbf{D}_2 = 0$. Equations [4.34] and the relation $\mathbf{D}_1 = \varepsilon_1\mathbf{E}_1$ give $\mathbf{E}_{1//} = 0$ and $\mathbf{n}_{21}.\mathbf{E}_1 = q_s/\varepsilon$. We deduce that

$$\mathbf{E}_1 = (q_s/\varepsilon_1)\mathbf{n}_{21}, \quad \mathbf{D}_1 = \varepsilon_1\mathbf{E}_1 = q_s\mathbf{n}_{21}, \quad \text{and} \quad \mathbf{P}_1 = \mathbf{D}_1 - \varepsilon_o\mathbf{E}_1 = q_s(1-\varepsilon_o/\varepsilon_1)\mathbf{n}_{21}. \tag{4.35}$$

The polarization charge density on $\mathcal{S}$ is

$$q_s' = \mathbf{P}_1.\mathbf{n}_{12} = -(1 - \varepsilon_o/\varepsilon_1)\, q_s. \tag{4.36}$$

– If $\mathcal{S}$ separates two dielectrics (1) and (2) and carries no free charge (Figure 4.5b), equations [4.34] and the relations $\mathbf{D}_1 = \varepsilon_1\mathbf{E}_1$ and $\mathbf{D}_2 = \varepsilon_2\mathbf{E}_2$ give

$$\begin{aligned} &\mathbf{E}_{1//} = \mathbf{E}_{2//}, \quad \mathbf{n}_{21}.(\varepsilon_1\mathbf{E}_1 - \varepsilon_2\mathbf{E}_2) = 0, \\ &\mathbf{P}_1 = (\varepsilon_1 - \varepsilon_o)\mathbf{E}_1 \quad \text{and} \quad \mathbf{P}_2 = (\varepsilon_2 - \varepsilon_o)\mathbf{E}_2. \end{aligned} \tag{4.37}$$

The polarization charge density on $\mathcal{S}$ is

$$q_s' = \mathbf{P}_1.\mathbf{n}_{12} + \mathbf{P}_2.\mathbf{n}_{21} = \varepsilon_o\, \mathbf{n}_{21}.(\mathbf{E}_1 - \mathbf{E}_2)\,. \tag{4.38}$$

Using the relation [4.21] of the potential to the field **E**, we deduce that V cannot have a discontinuity on the interface of a dielectric and a conductor or another dielectric. However, V is infinite at points where there are point charges or linear charges.

4.8. Field and potential of permanent dielectrics

We consider in this section a dielectric of given permanent uniform polarization **P** and we neglect the induced polarization. In this case, the relations $\mathbf{D} = \varepsilon_o\mathbf{E} + \mathbf{P}$,

$\nabla \times \mathbf{E} = 0$ and $\nabla.\mathbf{D} = q_v$ are valid but the proportionality relations of **P**, **D**, and **E** are not. The field is the same as that of a surface polarization charge density $q'_s = \mathbf{P.n}$.

a) Field of a uniformly polarized cylinder in the direction of its axis

The field of a cylinder, which is polarized in the direction of its axis, is the same as that of two disks, which carry the surface charge densities $\pm q'_s = \pm P$ and coincide with the bases of the cylinder (Figure 4.6). The field at points, which are off the axis, cannot be expressed in terms of simple functions.

If the cylinder is very thin (polarized plate as in Figure 4.6a), the field is uniform like that of a parallel plate capacitor. Applying Gauss's law to the surfaces S_1 and S_2, we get

$$\mathbf{D}^{(ex)} = 0, \qquad \mathbf{D}^{(in)} = 0, \qquad \text{thus} \qquad \mathbf{E}^{(ex)} = 0, \quad \mathbf{E}^{(in)} = -\mathbf{P}/\varepsilon_o \qquad [4.39]$$

where we have used the relation $\mathbf{D} = \varepsilon_o\mathbf{E} + \mathbf{P}$. The field **E** may also be evaluated directly by using the polarization charge densities $\pm q'_s = \pm P$ of the bases assimilated to infinite planes.

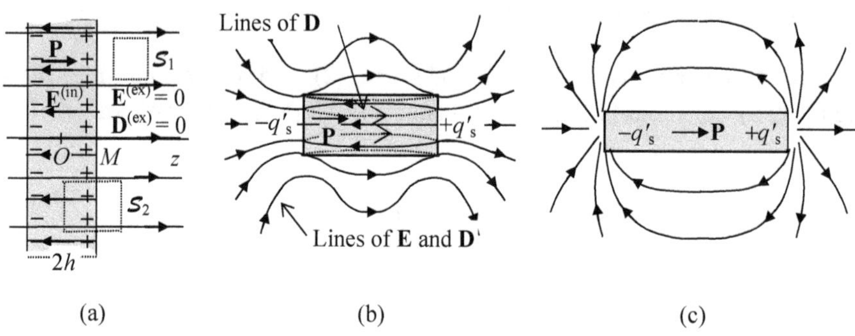

Figure 4.6. *Field of a cylinder that is uniformly polarized in the direction of its axis in the cases a) of a thin cylinder, b) of a thick cylinder, and c) of a long rod*

If the length of the cylinder $2h$ is not very small, compared to R (Figure 4.6b), using the expression [2.53] for the field of a disk, we may write the expressions of the field on the axis of the cylinder outside and inside the cylinder

$$\mathbf{E}^{(ex)}(z) = (\mathbf{P}/2\varepsilon_o)\left[(z+h)/\sqrt{R^2+(z+h)^2} - (z-h)/\sqrt{R^2+(z-h)^2}\right],$$

$$\mathbf{E}^{(in)}(z) = -(\mathbf{P}/2\varepsilon_o)\left[2 - (z+h)/\sqrt{R^2+(z+h)^2} + (z-h)/\sqrt{R^2+(z-h)^2}\right]. \quad [4.40]$$

Particularly, the field at the center is $\mathbf{E}^{(in)}(0) = -(\mathbf{P}/\varepsilon_o)[1 - h/\sqrt{R^2+h^2}\,]$. If the length of the cylinder $2h$ is much larger than R (a polarized rod, Figure 4.6c), the field at the center is $\mathbf{E}^{(in)}(0) \cong -(R^2/2\varepsilon_o h^2)\mathbf{P}$. Outside the dielectric, the lines of the fields **D** and **E** coincide, as $\mathbf{D} = \varepsilon_o\mathbf{E}$, but they are distinct within the dielectric.

b) Field of a uniformly polarized ball

We use spherical coordinates with the origin O at the center of the ball and Oz pointing in the direction of **P** (Figure 4.7a). The surface polarization charge density is $q_s' = \mathbf{P.n} = P\cos\theta$. The potential at a point $M(r, \theta, \varphi)$ is independent of φ, because of the symmetry about Oz; thus, we may calculate it for $\varphi = 0$. An element of area $d\mathcal{S}' = R^2 \sin\theta'\, d\theta'\, d\varphi'$ at the point $P(R, \theta', \varphi')$ on the surface of the ball, has the charge $dQ' = q'_s\, d\mathcal{S}'$; the potential that it produces at M is

$$dV(\mathbf{r}) = K_o \frac{q'_s\, d\mathcal{S}}{|\mathbf{r}-\mathbf{r}'|} = K_o P \frac{R^2\cos\theta'\sin\theta'\, d\theta'\, d\varphi'}{\sqrt{R^2+r^2-2Rr\cos\theta\cos\theta' - 2Rr\sin\theta\sin\theta'\cos\varphi'}}.$$

Instead of making a complicated integration, we use an approximation method based on a simple physical idea. Let us assume that the polarization of each molecule is due to the displacement of a positive charge q by $\mathbf{d}/2$ in the direction of **P** and a displacement of a negative charge $-q$ by $-\mathbf{d}/2$. The electric dipole moment of a molecule is then $\mathbf{p} = q\mathbf{d}$ and the polarization of the medium is $\mathbf{P} = N_v q\mathbf{d}$ where N_v is the number of molecules per unit volume. The polarized ball is thus equivalent to two balls of charge densities $N_v q$ and $-N_v q$ and of centers O_1 and O_2 separated by a distance $\overrightarrow{O_1O_2} = \mathbf{d}$ (Figure 4.7b). The total charges of these balls are $\pm Q = \pm(4/3)\pi R^3 N_v q$ and the total potential that they produce at a point M outside them is the same as that of two point charges $\pm Q$ located at their centers O_1 and O_2, i.e.

$$V^{(ex)}(\mathbf{r}) = K_o Q(1/r_1 - 1/r_2) = K_o Q[1/|\mathbf{r}-\mathbf{d}/2| - 1/|\mathbf{r}+\mathbf{d}/2|] \approx K_o Q\,(\mathbf{r.d})/r^3. \quad [4.41]$$

As $Qd = (4/3)\pi R^3 N_v qd = P\mathcal{V}$, where $\mathcal{V} = (4/3)\,\pi R^3$ is the volume of the ball, we may express the electric potential and field outside the ball in terms of **P** as

$$V^{(ex)}(\mathbf{r}) = K_o\mathcal{V}\,(\mathbf{r.P})/r^3 = (R^3/3\varepsilon_o r^3)\,(\mathbf{r.P}), \quad [4.42]$$

$$\mathbf{E}^{(ex)}(\mathbf{r}) = -\nabla V^{(ex)}(\mathbf{r}) = K_o\mathcal{V}\,[3(\mathbf{r.P})\frac{\mathbf{r}}{r^5} - \frac{\mathbf{P}}{r^3}] = \frac{R^3}{3\varepsilon_o}[3(\mathbf{r.P})\frac{\mathbf{r}}{r^5} - \frac{\mathbf{P}}{r^3}]. \quad [4.43]$$

Particularly, on the surface of the ball

$$V^{(ex)}(\mathbf{r}) = \mathbf{R.P}/3\varepsilon_o, \qquad \mathbf{E}^{(ex)}(\mathbf{r}) = (1/3\varepsilon_o R^2)\,[3(\mathbf{R.P})\,\mathbf{R} - R^2\,\mathbf{P}]. \quad [4.44]$$

Using the expression [2.59] for the potential inside a charged ball, we may write the total potential inside the dielectric ball and hence the electric field

$$V^{(\text{in})}(\mathbf{r}) = K_0 \frac{Q}{R^3}\,[(3R^2 - r_1^2) - (3R^2 - r_2^2)] = K_0 \frac{Q}{R^3}\,(\mathbf{d}.\mathbf{r}) = \frac{\mathbf{P}.\mathbf{R}}{3\varepsilon_0} = \frac{Pz}{3\varepsilon_0}. \quad [4.45]$$

$$\mathbf{E}^{(\text{in})}(\mathbf{r}) = -\nabla V^{(\text{in})}(\mathbf{r}) = -\mathbf{P}/3\varepsilon_0. \quad [4.46]$$

The electric field inside the ball is uniform and oriented in the opposite direction to **P**. The electric displacement $\mathbf{D} = \mathbf{P} + \varepsilon_0\mathbf{E}$ is then given by the expressions

$$\mathbf{D}^{(\text{ex})}(\mathbf{r}) = \varepsilon_0\,\mathbf{E}^{(\text{ex})}(\mathbf{r}) = (R^3/3)\,[3(\mathbf{r}.\mathbf{P})\,\mathbf{r}/r^5 - \mathbf{P}/r^3], \qquad \mathbf{D}^{(\text{in})}(\mathbf{r}) = (2/3)\,\mathbf{P}. \quad [4.47]$$

The fields **D** and **E** verify the conditions [4.34] on the surface of the dielectric. The lines of **E** and **D** are represented in Figures 4.7c and 4.7d, respectively.

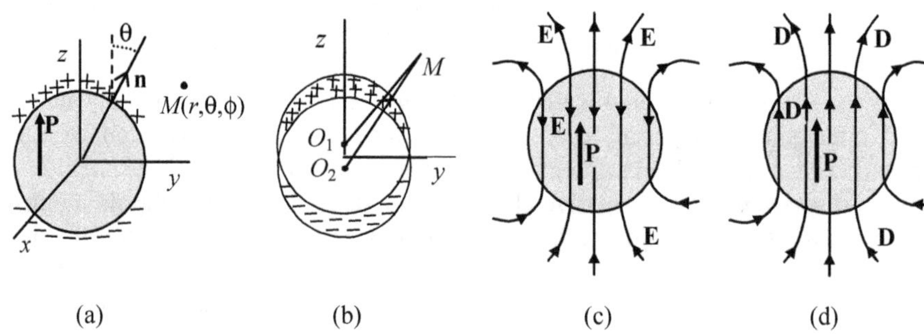

Figure 4.7. *Polarized ball: a) the surface polarization charge density, b) its equivalence to two balls of opposite charges, c) lines of the field* **E**, *and d) lines of the field* **D**

The expressions [4.39], [4.40] and [4.46] show that the electric field due to the polarized dielectric is in the opposite direction to the polarization **P**. This is a general property for any form of the dielectric body. We say that **E** is a *depolarizing field.*

c) Field in a cavity within a dielectric

Consider a dielectric body (*B*) of polarization **P** containing a cavity with no free charges in it. Let $\mathbf{E}^{(\text{C})}$ be the field at a point *M* in the cavity (Figure 4.8) and $\hat{\mathbf{E}}$ the field that is produced at the same point *M* by a dielectric body that may fill the cavity exactly. It is evident that, if the dielectric body had no cavity, the field at *M* would be $\mathbf{E} = \mathbf{E}^{(\text{C})} + \hat{\mathbf{E}}$. This relation holds even if there are free charges and other dielectric bodies than the body (B) by including their fields in $\mathbf{E}$ and $\mathbf{E}^{(C)}$. Thus, we have

$$\mathbf{E}^{(C)} = \mathbf{E} - \hat{\mathbf{E}}\,. \qquad [4.48]$$

– If the cavity is spherical (Figure 4.8a), $\hat{\mathbf{E}} = -\mathbf{P}/3\varepsilon_o$, thus $\mathbf{E}^{(C)} = \mathbf{E} + \mathbf{P}/3\varepsilon_o$.

– If the cavity is cylindrical and thin in the direction of $\mathbf{P}$ (Figure 4.8b), we have $\hat{\mathbf{E}} = -\mathbf{P}/\varepsilon_o$, thus $\mathbf{E}^{(C)} = \mathbf{E} + \mathbf{P}/\varepsilon_o$. We also get this result by applying Gauss law to the cylinder $\mathcal{S}$.

– If the cavity is cylindrical and long in the direction of $\mathbf{P}$ (Figure 4.8c), we have $\hat{\mathbf{E}} = 0$, thus $\mathbf{E}^{(C)} = \mathbf{E}$. We get also this result by writing that the circulation of $\mathbf{E}$ over the closed path $\mathcal{C}$ is equal to zero.

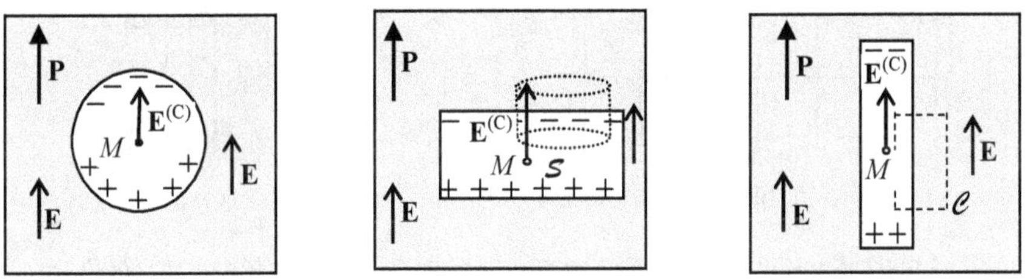

Figure 4.8. *Field in a cavity within a dielectric: a) spherical cavity, b) cylindrical cavity that is thin in the direction of* **P**, *and c) cylindrical cavity that is long in the direction of* **P**

4.9. Polarization of a dielectric in an external field

If a dielectric is placed in an external electric field $\mathbf{E}_o$, it becomes polarized and it produces its own field $\mathbf{E}_p$, which superposes to the external field $\mathbf{E}_o$ so that the total field is $\mathbf{E} = \mathbf{E}_o + \mathbf{E}_p$. We assume that $\mathbf{E}_o$ is uniform and it is produced by a system, which is not influenced by the dielectric. At large distances from the dielectric, the total field $\mathbf{E}$ approaches $\mathbf{E}_o$ asymptotically but, at small distances and inside the dielectric, $\mathbf{E}$ depends on the shape of the dielectric body and its polarization $\mathbf{P}$, which is unknown (as it depends on $\mathbf{E}$ that we have to calculate). In the general case, the problem is complicated and it may be solved only by approximations or numerically. In this section we consider two cases, where the simple geometry facilitates the solution.

a) Dielectric plate in a uniform and normal field

Consider an infinite plate of dielectric of thickness d placed normally to a uniform field $\mathbf{E}_o$. We take Oz in the direction of $\mathbf{E}_o$; thus, $\mathbf{E}_o = E_o\,\mathbf{e}_z$ (Figure 4.9a). Because of the obvious symmetries, the polarization $\mathbf{P}$ is in the direction of Oz and, if the plate is thin, we may admit that $\mathbf{P}$ is uniform. The polarization $\mathbf{P}$ produces polarization charge densities $\pm P$ on the faces of the plate similar to those of a parallel plate capacitor. Their field is equal to q'_s/ε_o, i.e. $\mathbf{E}_p^{(in)} = -(q'_s/\varepsilon_o)\mathbf{e}_z = -\mathbf{P}/\varepsilon_o$ inside the plate and $\mathbf{E}_p^{(ex)} = 0$ outside the plate. Thus, the total field is

$$\mathbf{E}^{(ex)} = \mathbf{E}_o + \mathbf{E}_p^{(ex)} = \mathbf{E}_o\,, \qquad \mathbf{E}^{(in)} = \mathbf{E}_o + \mathbf{E}_p^{(in)} = \mathbf{E}_o - \mathbf{P}/\varepsilon_o. \qquad [4.49]$$

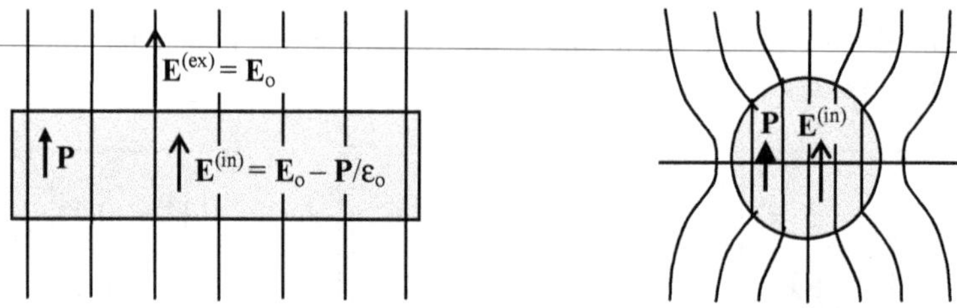

Figure 4.9. *a) Polarization of an infinite plate and b) polarization of a ball*

To determine $\mathbf{P}$, we must use the properties of the dielectric. If the medium is linear, isotropic with a susceptibility χ_E, we find $\mathbf{P} = \varepsilon_o\chi_E\,\mathbf{E}^{(in)} = \varepsilon_o\chi_E\,(\mathbf{E}_o - \mathbf{P}/\varepsilon_o)$, hence

$$\mathbf{P} = \frac{\varepsilon_o\chi_E}{1+\chi_E}\,\mathbf{E}_o, \qquad \mathbf{E}^{(in)} = \frac{\mathbf{E}_o}{1+\chi_E} = \frac{\mathbf{E}_o}{\varepsilon_r}. \qquad [4.50]$$

The electric displacement is

$$\mathbf{D}^{(ex)} = \varepsilon_o\,\mathbf{E}^{(ex)} = \varepsilon_o\mathbf{E}_o\,, \qquad \mathbf{D}^{(in)} = \varepsilon_o\mathbf{E}^{(in)} + \mathbf{P} = \varepsilon_o\mathbf{E}_o\,. \qquad [4.51]$$

We may use Gauss's law in the integral form to show that $\mathbf{E}$ is uniform outside the plate and that $\mathbf{D}^{(in)} = \mathbf{D}^{(ex)} = \varepsilon_o\mathbf{E}_o$.

b) Dielectric ball in a uniform field

Let us assume also in this case that the polarization $\mathbf{P}$ of the ball is uniform (Figure 4.9b). According to the section 4.8b, the field of the ball inside it is $-\mathbf{P}/3\varepsilon_o$. Thus, the total field inside the ball is

$$\mathbf{E}^{(in)} = \mathbf{E}_o - \mathbf{P}/3\varepsilon_o\,. \qquad [4.52]$$

If the dielectric is linear and isotropic with a susceptibility χ_E, the polarization is $\mathbf{P} = \varepsilon_o\chi_E\,\mathbf{E}^{(in)} = \varepsilon_o\chi_E\,(\mathbf{E}_o - \mathbf{P}/3\varepsilon_o)$, hence

$$\mathbf{P} = \frac{3\varepsilon_o\chi_E}{3+\chi_E}\,\mathbf{E}_o, \qquad \mathbf{E}^{(in)} = \frac{3}{3+\chi_E}\,\mathbf{E}_o\,. \qquad [4.53]$$

Two facts concerning these examples are to be noted: first, the field $\mathbf{E}_p$ due to the polarization has the opposite sign to $\mathbf{P}$, thus to $\mathbf{E}_o$. The effect of the polarization is thus to reduce the electric field inside the dielectric ($\mathbf{E}_p$ is a depolarizing field)[1]. The second is that our starting assumption of a uniform polarization without justification leads to a uniform field $\mathbf{E}$ and, consequently, to a uniform polarization. The analysis is thus coherent. The solution, that we have found, verifies all the required conditions. As the solution is unique, we are sure that it is the solution and that there is no other one.

4.10. Energy and force in dielectrics

We have seen in section 2.9 that the electrostatic energy of a system of particles of charges q_i is $U_E = \frac{1}{2}\sum_i q_i\,V'(\mathbf{r}_i)$, where $V'(\mathbf{r}_i)$ is the potential produced at the point $\mathbf{r}_i$ by all the charges except q_i itself (whose potential at $\mathbf{r}_i$ is infinite). The interpretation of U_E as the work required to assemble the charges in their actual configuration allows us to admit this expression in the case of dielectrics provided that the q_i include only the free charges (as the bound charges may not be displaced on the macroscopic scale), while the potential is obviously produced by all the charges (free and bound). In the case of a continuous distribution of free charges with a density $q_v(\mathbf{r})$, we have shown that V' may be replaced by the total potential V; thus, U_E may be written as

$$U_E = \tfrac{1}{2}\iiint d\mathcal{V}'\,q_v(\mathbf{r}')\,V(\mathbf{r}'). \qquad [4.54]$$

Using Gauss's law in the local form $\nabla.\mathbf{D} = q_v$, we may also write

$$U_E = \tfrac{1}{2}\iiint d\mathcal{V}'[\nabla'.\mathbf{D}(\mathbf{r}')]\,V(\mathbf{r}') = \tfrac{1}{2}\iiint d\mathcal{V}'\,\nabla'.[\mathbf{D}(\mathbf{r}')\,V(\mathbf{r}')] - \tfrac{1}{2}\iiint d\mathcal{V}'\mathbf{D}(\mathbf{r}').[\nabla'.V(\mathbf{r}')].$$

The first term may be transformed into the flux of the vector $V\mathbf{D}$ through a surface $\mathcal{S}$, which contains the system. $\mathcal{S}$ may be a sphere of large radius on which $V(\mathbf{r}')$ and $\mathbf{D}(\mathbf{r}')$ tend to 0. Thus, the first term gives no contribution to U_E. Using the relation $\mathbf{E} = -\nabla V$ in the second term, we may write

$$U_E = \tfrac{1}{2}\iiint d\mathcal{V}'\mathbf{D}(\mathbf{r}').\mathbf{E}(\mathbf{r}'). \qquad [4.55]$$

1 The field of the dielectric $\mathbf{E}_p$ cannot be in the direction of $\mathbf{E}$ as this would lead to spontaneous polarization of the dielectric if it is subject to any small electric disturbance.

This relation allows us to express the energy in terms of the fields **E** and **D** without reference to their sources. This means that the energy is localized in the region of the fields with a volume density

$$U_{E,v} = \tfrac{1}{2}\, \mathbf{D}.\mathbf{E}. \qquad [4.56]$$

Particularly, if the medium is linear, homogeneous and isotropic with a permittivity ε, the energy density may be written as

$$U_{E,v} = \tfrac{1}{2}\, \varepsilon\, \mathbf{E}^2. \qquad [4.57]$$

For instance, in the case of a capacitor, the energy is localized in the dielectric (or the vacuum) which fills it. We note that, contrary to [4.56], $\tfrac{1}{2} q_v V$ in [4.54] cannot be interpreted as the electric energy density, as it vanishes in a region where there is no electric charge and V is defined only up to an additive constant. However, if the potential is taken equal to zero at infinity, the total energy of the whole system has the same value if we use either expression [4.54] or [4.55].

The electric forces may be calculated by using Coulomb's law or the fields. Sometimes, it is practical to use the energy and the method of virtual displacements, as we have done in the case of the conductors in vacuum. We have seen in section 4.7 that the potential and the field of charges in a dielectric are ε_r times less intense than in vacuum. This is true also for the energy of point charges or charged conductors. Equation [4.57] may also be written in the form $U_{E,v} = \mathbf{D}^2/2\varepsilon$. As **D** depends only on the free charges, if particles or conductors with given charges are immersed in a dielectric, the energy and the forces are reduced by the factor ε_r. On the other hand, if conductors with given potentials are immersed in a dielectric, the energy and the forces are multiplied by ε_r.

4.11. Action of an electric field on a polarized medium

We have shown in section 2.6 that a uniform electric field **E** acts on an electric dipole **p** with a moment of force $\mathbf{\Gamma} = \mathbf{p} \times \mathbf{E}$ and that a non-uniform field acts on the dipole with a force $\mathbf{F} = (\mathbf{p}.\nabla)\, \mathbf{E}$. If the dielectric contains N_v dipoles per unit volume (producing its polarization $\mathbf{P} = N_v\mathbf{p}$), it is subject to a *moment of force per unit volume*

$$\mathbf{\Gamma}_v = N_v\, \mathbf{p} \times \mathbf{E} = \mathbf{P} \times \mathbf{E} \qquad [4.58]$$

and a force per unit volume

$$\mathbf{F}_v = N_v\, (\mathbf{p}.\nabla)\, \mathbf{E} = (\mathbf{P}.\nabla)\, \mathbf{E}. \qquad [4.59]$$

If $\mathbf{E}$ is uniform, $\mathbf{F}_v$ vanishes, but if $\mathbf{E}$ is non-uniform, its component (k) is given by

$$(F_v)_\alpha = \Sigma_\beta P_\beta \partial_\beta E_\alpha = \Sigma \partial_\beta(P_\beta E_\alpha) - \Sigma_\beta (\partial_\beta P_\beta) E_\alpha.$$

The total force that acts on the dielectric is obtained by integrating $\mathbf{F}_v$ over the volume $\mathcal{V}$ of the dielectric. The integral of the first term, which is the divergence of the vector ($E_\alpha \mathbf{P}$), may be transformed into the flux of this vector outgoing from the surface $\mathcal{S}$ of the dielectric and the force that acts on the dielectric may be written as

$$F_\alpha = \iiint_{\mathcal{V}} d\mathcal{V}\, (F_v)_\alpha = \iint_{\mathcal{S}} d\mathcal{S}\, \Sigma_\beta n_\beta (P_\beta E_\alpha) + \iiint_{\mathcal{V}} d\mathcal{V}\, (-\Sigma_\beta \partial_\beta P_\beta)\, E_\alpha$$
$$= \iint_{\mathcal{S}} d\mathcal{S}\, (\mathbf{n}.\mathbf{P}) E_\alpha + \iiint_{\mathcal{V}} d\mathcal{V}\, (-\boldsymbol{\nabla}.\mathbf{P}) E_\alpha.$$

Using the polarization charge densities $q'_s = (\mathbf{n}.\mathbf{P})$ and $q'_v = -(\boldsymbol{\nabla}.\mathbf{P})$, we may write the force in the vector form

$$\mathbf{F} = \iint_{\mathcal{S}} d\mathcal{S}\, q'_s\, \mathbf{E} + \iiint_{\mathcal{V}} d\mathcal{V}\, q'_v\, \mathbf{E}. \qquad [4.60]$$

This expression means that the force exerted by the field $\mathbf{E}$ on a dielectric body is the resultant of the forces that it exerts on the surface and volume polarization charges.

If the dielectric is linear and isotropic, the force density $\mathbf{F}_v$ may be related to the electrostatic energy density. Indeed, if ε is the permittivity, the force density [4.59] may be written as

$$\mathbf{F}_v = (\varepsilon - \varepsilon_o)\, (\mathbf{E}.\boldsymbol{\nabla})\, \mathbf{E}. \qquad [4.61]$$

Writing explicitly the x component, we find

$$(F_v)_x = (\varepsilon - \varepsilon_o)(E_x \partial_x + E_y \partial_y + E_z \partial_z)\, E_x = (\varepsilon - \varepsilon_o)(E_x \partial_x E_x + E_y \partial_x E_y + E_z \partial_x E_z)$$
$$= \tfrac{1}{2}\, (\varepsilon - \varepsilon_o)\, \partial_x\, \mathbf{E}^2$$

where, to write the second form, we have used the field equation $\boldsymbol{\nabla}\times\mathbf{E} = 0$, which gives $\partial_x E_y = \partial_y E_x$ and $\partial_x E_z = \partial_z E_x$. Similar relations may be written for the other components of $\mathbf{F}_v$. Thus, we may write the vector relation

$$\mathbf{F}_v = \tfrac{1}{2}\, (\varepsilon - \varepsilon_o)\, \boldsymbol{\nabla}(\mathbf{E}^2) = (1 - \varepsilon_o/\varepsilon)\, \boldsymbol{\nabla} U_{E,v}\,, \qquad [4.62]$$

where we have used equation [4.57] for the energy density. This expression shows that the force is independent of the direction of $\mathbf{E}$; it points in the direction of increasing $\mathbf{E}^2$, i.e. of the increasing energy density. The force that acts on an element of volume $d\mathcal{V}$ is $d\mathbf{F} = \mathbf{F}_v\, d\mathcal{V}$. To displace it by $\delta\mathbf{r}$, a force $d\mathbf{F}' = -d\mathbf{F}$ must be exerted; its work is

$$\delta\, dW' = \delta\mathbf{r}.d\mathbf{F}' = -\,\delta\mathbf{r}.d\mathbf{F} = -\delta\mathbf{r}.\,\mathbf{F}_v\, d\mathcal{V} = -(1-\varepsilon_o/\varepsilon)\,\delta\mathbf{r}.\boldsymbol{\nabla}\, U_{E,v}\, d\mathcal{V}$$
$$= -(1-\varepsilon_o/\varepsilon)\,\delta U_{E,v}\, d\mathcal{V} \qquad [4.63]$$

where $\delta U_{E,v} = \delta\mathbf{r}.\boldsymbol{\nabla} U_{E,v}$ is the variation of the energy density in the displacement $\delta\mathbf{r}$.

4.12. Electric susceptibility and permittivity

The polarization $\mathbf{P}$ is related to the mean electric dipole moment $\mathbf{p}$ of the molecules, which is itself proportional to the local field $\mathbf{E}_l$ according to the relation $\mathbf{p} = \alpha\mathbf{E}_l$. Thus, to derive the relation of $\mathbf{P}$ to the macroscopic field $\mathbf{E}$, we need the relations between $\mathbf{E}$, $\mathbf{P}$, and $\mathbf{E}_l$. The difference between $\mathbf{E}$ and $\mathbf{E}_l$ comes from the field of the molecule, whose polarization is analyzed. As the polarized molecule has a field in the opposite direction to $\mathbf{E}_l$, we expect that E_l is less than E.

Consider a dielectric of susceptibility χ_E. By definition, its polarization density is related to the macroscopic field $\mathbf{E}$ by the relation

$$\mathbf{P} = \varepsilon_o\chi_E\mathbf{E}. \qquad [4.64]$$

The local field $\mathbf{E}_l$, which acts on a molecule at M, is produced by all bodies except the considered molecule. To determine $\mathbf{E}_l$, we surround this molecule by a small sphere that contains no free charge but a large number of polarized molecules (Figure 4.8a). We may write $\mathbf{E}_l = \mathbf{E}^{(ex)} + \mathbf{E}'^{(sph)}$, where $\mathbf{E}^{(ex)}$ is the field of other bodies except the dielectric situated inside the sphere and $\mathbf{E}'^{(sph)}$ is the field of the molecules of the sphere except the molecule at M. The molecules inside the sphere, being at short distance from M, the field $\mathbf{E}'^{(sph)}$ cannot be evaluated as the field of dipoles (thus, the same as the field of the polarization charges q'_v and q'_s). Its value depends on the configuration of the molecules within the sphere. The calculation shows that it vanishes for a cubic crystal lattice. We admit that it is equal to zero for any crystal and for amorphous dielectrics. On the other hand, $\mathbf{E}^{(ex)}$ is equal to the field $\mathbf{E}^{(cav)}(0)$ at the center of the sphere if it is empty. According to [4.48], it is given by $\mathbf{E}^{(cav)}(0) = \mathbf{E} + \mathbf{P}/3\varepsilon_o$, hence

$$\mathbf{E}_l = \mathbf{E} + \mathbf{P}/3\varepsilon_o. \qquad [4.65]$$

If the medium contains N_v molecules per unit volume with an average dipole moment $\mathbf{p} = \alpha\mathbf{E}_l$, the polarization density may be written as

$$\mathbf{P} = N_v\mathbf{p} = N_v\alpha\mathbf{E}_l = N_v\alpha[\mathbf{E} + \mathbf{P}/3\varepsilon_o], \quad \text{hence } \mathbf{P} = N_v\alpha\mathbf{E}/(1 - N_v\alpha/3\varepsilon_o). \quad [4.66]$$

As $\mathbf{P} = \varepsilon_o\chi_E\mathbf{E}$, the electric susceptibility and the relative permittivity are given by

$$\chi_E = \frac{N_v\alpha/\varepsilon_o}{1 - N_v\alpha/3\varepsilon_o}, \qquad \text{and} \qquad \varepsilon_r = 1 + \chi_E = \frac{1 + 2N_v\alpha/3\varepsilon_o}{1 - N_v\alpha/3\varepsilon_o}. \qquad [4.67]$$

Conversely, these equations allow the microscopic coefficient α to be determined in terms of the macroscopic quantity χ_E or ε_r. The index of refraction of a non-magnetic dielectric is related to ε_r by the relation $n^2 = \varepsilon_r$. Thus, we find the so-called *Clausius-Mossotti* equation for non-polar substances

$$\frac{N_v\alpha}{3\varepsilon_o} = \frac{\varepsilon_r - 1}{\varepsilon_r + 2} = \frac{n^2 - 1}{n^2 + 2}. \qquad [4.68]$$

The polarizability α depends only on the nature of the molecule, not on the physical conditions (temperature, pressure, etc.) but the number of molecules N_v per unit volume depends on them according to the relation $N_v = m_v N_A/m_M$, where m_v is the mass density, m_M is the molar mass, and N_A is Avogadro's number. Thus, the relation [4.68] may be written as

$$\alpha_{molar} \equiv \frac{\alpha N_A}{3\varepsilon_o} = \frac{m_M}{m_v}\frac{\varepsilon_r - 1}{\varepsilon_r + 2} = \frac{m_M}{m_v}\frac{\chi_E}{\chi_E + 3} \quad \text{or} \quad \chi_E = \frac{3\alpha_M m_v}{m_M - \alpha_M m_v}. \qquad [4.69]$$

α_{molar} is called *molar polarization* (although it is not really a polarization). For a given substance, α_{molar} does not depend on physical conditions. Thus, the ratio $(n^2 - 1)/(n^2 + 2)$ is proportional to the mass density.

The expression [4.69] may be easily compared to experimental values. Consider oxygen O_2, for instance. As a gas under standard conditions, its mass density is $m_v = 32 \times 10^{-3}/2.241 \times 10^{-2} = 1.428$ kg/m^3. The experiment shows that its susceptibility is 5.23×10^{-4}. We deduce that $\alpha_{molar} = 3.91 \times 10^{-6}$ m^3/mole. In the liquid state, its mass density is 1190 kg/m^3. Equation [4.69] gives $\chi_E = 0.509$ compared with the experimental value of 0.507.

If the molecules are well separated, the term $N_v\alpha/\varepsilon_o$ is much smaller than 1 and the electric susceptibility becomes $\chi_E = \alpha N_v/\varepsilon_o$. This approximation is equivalent to neglecting the interaction of the molecules. It is a good approximation in the case of a gas but certainly poor in the case of solids and liquids, particularly if the substance is polar. The large interaction between molecules increases the electric susceptibility. On the other hand, if the polarizability and N_v are high enough to make $N_v\alpha/3\varepsilon_o$ comparable to 1, the electric susceptibility is large and so is the polarization, even if the field **E** is weak. We shall see that this is effectively the case if the temperature is close to a critical value T_c. Under these conditions, the linear approximation $\mathbf{P} = \varepsilon_0\chi_E\mathbf{E}$ is not valid and the preceding analysis does not hold.

4.13. Variation of polarization with temperature

Statistical physics enables the analysis of the variation of the polarization of a polar substance with temperature. The dielectric body being formed by a large number of molecules, the probability for the energy of a molecule to have a value u is proportional to e^{-u/k_BT}, where $k_B = 1.380\ 658 \times 10^{-23}$ J/K is Boltzmann's constant and T is the absolute temperature. We take the Oz axis in the direction of the local field $\mathbf{E}_l$ and specify the orientation of the electric dipole moment $\mathbf{p}_o$ by its angles θ and φ of spherical coordinates around Oz (Figure 4.10a). The energy of the dipole $\mathbf{p}_o$ in the local electric field $\mathbf{E}_l$ is $u = p_oE_l - \mathbf{p}_o\cdot\mathbf{E}_l = p_oE_l(1 - \cos\theta)$. Thus, the probability that $\mathbf{p}_o$ points within the solid angle $d\Omega$ close to the direction (θ, φ) is $d\Pi = \eta\, e^{-u/k_BT} d\Omega$, where η is a constant to be determined by the normalization of the probability to 1:

$$\int d\Pi = \eta \int_0^{\pi} d\theta \sin\theta \int_0^{2\pi} d\varphi\, e^{-u/k_BT} = 1. \qquad [4.70]$$

Using the expression of u, we find

$$\eta = (1/4\pi)\, x\, e^{x}/\mathrm{sh}(x), \qquad \text{where } x = p_oE_l/k_BT. \qquad [4.71]$$

Because of the rotational symmetry about Oz, the average value of $\mathbf{p}$ points necessarily in the direction of Oz. Thus we have to calculate only the average value of $p_z = p_o \cos\theta$; we find

$$\langle p_z \rangle = \int d\Pi\, p_o \cos\theta = \eta p_o \int d\Omega \cos\theta\, e^{-u/k_BT} = 4\pi\eta\, p_o(e^{-x}/x^2)[x\, \mathrm{ch}(x) - \mathrm{sh}(x)].$$

Using the expression [4.71] of η, we may write

$$\langle p_z \rangle = p_o L(x), \qquad \text{where } L(x) = \coth x - 1/x. \qquad [4.72]$$

$L(x)$ is called *Langevin's function*. Thus, the polarization density is

$$P = N \langle p_z \rangle = N p_o L(x). \qquad [4.73]$$

Its variation as a function of x is illustrated in Figures 4.10b and 4.10c.

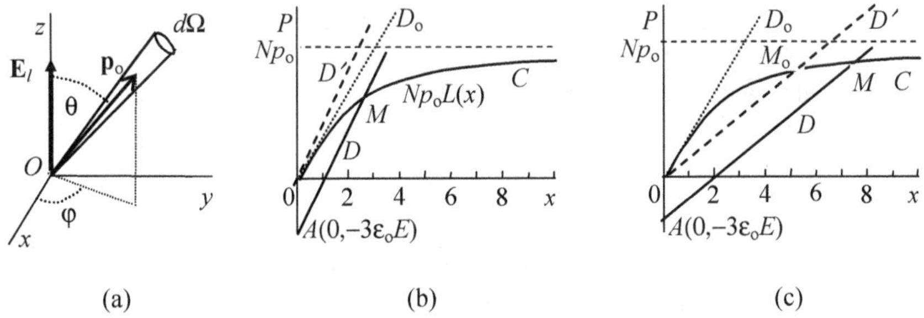

Figure 4.10. *Polarization of a polar substance: a) orientation of the dipole moment* $\mathbf{p}_o$, *b) polarization for* $T > T_c$, *and c) polarization for* $T < T_c$

– If x is very large ($x >> 1$), i.e. in the case of a strong field or very low temperature, $L(x)$ increases asymptotically to 1 and the polarization density tends toward a *saturation* value N_vp_o. At this limit, all the dipoles point in the direction of the local field $\mathbf{E}_l = \mathbf{E} + \mathbf{P}/3\varepsilon_o = (E + N_vp_o/3\varepsilon_o)\mathbf{e}_z$. Under standard conditions, $k_BT \approx 4 \times 10^{-21}$J. Noting that p_o is of the order of 10^{-30} C.m for polar molecules and N_v is of the order of 10^{25} molecules/m^3, the condition $x >> 1$ corresponds to an exceptional field $E >> 4 \times 10^9$ V/m. Thus, to have $x >> 1$, the temperature must be very low.

– If x is very small ($x << 1$), i.e. in the case of a weak field or high temperature, $L(x) \approx x/3$, the curve $L(x)$ may be replaced by the tangent D_o at the origin. Using [4.73], we may write

$$P \approx N_vp_ox/3 = N_vp_o^2E_l/3k_BT = (3\varepsilon_oT_c/T)E_l. \quad [4.74]$$

$T_c = N_vp_o^2/9\varepsilon_ok_B$ is the *critical temperature* of the dielectric. Then, the polarizability α, the susceptibility and the local field $\mathbf{E}_l$ are given by

$$\alpha_o = \frac{3\varepsilon_o}{N_v}\frac{T_c}{T}, \qquad \chi_E = \frac{3T_c}{T - T_c} \qquad \text{and} \quad \mathbf{E}_l = \mathbf{E}\,\frac{T}{T - T_c}, \quad [4.75]$$

where we have used the equations [4.67] and [4.65]. If T approaches T_c, the polarizability tends toward $3\varepsilon_o/N_v$ and the electric susceptibility becomes very high (it may be as high as 10^5). In the case of a gas ($N_v \approx 10^{25}$ molecules/m^3), T_c is of the order of 10^{-2} K, but, in the case of liquids and solids ($N_v \approx 10^{27}$ molecules/m^3), T_c lies between nearly 1 to 100 K. If the temperature is much higher than T_c, the susceptibility is given by Debye-Langevin equation

$$\chi_E \approx 3\frac{T_c}{T} = \frac{N_vp_o^2}{3\varepsilon_ok_BT} \qquad (T >> T_c). \quad [4.76]$$

In order to verify the $1/T$ dependence experimentally, the number N_v of molecules per unit volume must be kept constant by varying the pressure simultaneously with the temperature. Otherwise, if the pressure is maintained constant, the variation of temperature leads to a variation of N_v like $1/T$ and the susceptibility varies effectively like $1/T^2$ with temperature.

For intermediary values of x, it is not possible to write the expression of the polarization P as a function of E. The relation $E_l = k_B Tx/p_o = E + P/3\varepsilon_o$ (see [4.65]) allows the expression of P in terms of x and E. Thus, we have the two equations

$$P = N_v\, p_o L(x)\,, \qquad P = 3\varepsilon_o k_B Tx/p_o - 3\varepsilon_o E\,, \qquad [4.77]$$

which may be solved numerically. The two expressions of P versus x may also be plotted on the same graph. The first is the Langevin curve (C) and the second is a straight line (D) that intercepts the P axis at the point $A(0, -3\varepsilon_o E)$. The coordinates of the intersection M of (C) and (D) determine P and x (thus E) at the temperature T. If the macroscopic field E is reduced, the line D moves parallel to itself toward the origin. If its slope $3\varepsilon_o kT/p_o$ is larger than the slope $Np_o/3$ of the curve (C) at the origin, the points M and A tend toward the origin (Figure 4.10b) and this corresponds to $P = 0$ and $E = 0$. This occurs effectively if T is higher than the critical temperature $T_c = N_v p_o^2/9\varepsilon_o k_B$. On the other hand, if $T < T_c$ (Figure 4.10c) the point A approaches O but M approaches a point M_o. This case corresponds to $E = 0$ but a non-zero polarization. Thus, we obtain a permanent polarization (electret).

4.14. Nonlinear dielectrics and non-isotropic dielectrics

Similar to the deformation of a rigid body caused by mechanical forces, the polarization **P** is a response of a dielectric body to electric excitation **E**. We may write the components of **P** in terms of the components of **E** as $P_\alpha = f_\alpha(E_1, E_2, E_3)$. If the field **E** is not strong, we may expand the P_α as power series of the components of the E_β in the form

$$P_\alpha(\mathbf{E}) = P_\alpha(0) + \Sigma_\beta\, (\partial P_\alpha/\partial E_\beta)|_o\, E_\beta + \tfrac{1}{2}\, \Sigma_{\beta,\gamma}\, (\partial^2 P_\alpha/\partial E_\beta\, \partial E_\gamma)|_o\, E_\beta E_\gamma + \ldots \qquad [4.78]$$

In the case on a non-permanent dielectric, the polarization vanishes if $\mathbf{E} = 0$; thus, we must have $P_\alpha(0) = 0$. If the quadratic terms in E_α are negligible and $P_\alpha(0) = 0$, the dielectric is said to be *linear*. In this case, the components P_α may be written as

$$P_x = \varepsilon_o\, [\chi_{xx}\, E_x + \chi_{xy}\, E_y + \chi_{xz}\, E_z\,],$$
$$P_y = \varepsilon_o\, [\chi_{yx}\, E_x + \chi_{yy}\, E_y + \chi_{yz}\, E_z\,],$$
$$P_z = \varepsilon_o\, [\chi_{zx}\, E_x + \chi_{zy}\, E_y + \chi_{zz}\, E_z\,]. \qquad [4.79]$$

The nine constants $\chi_{\alpha\beta}$ are the components of the *electric susceptibility tensor* of the dielectric medium. In general, **P** and **E** are not parallel and the polarization **P** is more important if the field **E** is applied in some directions rather than others. The dielectric is then *non-isotropic*. It is isotropic if the tensor $\chi_{\alpha\beta}$ is diagonal and the diagonal elements are equal

$$\chi_{\alpha\beta} = 0 \quad \text{if} \quad \alpha \neq \beta \qquad \text{and} \qquad \chi_{11} = \chi_{22} = \chi_{33} \equiv \chi. \tag{4.80}$$

Then, we may write the vector relation

$$\mathbf{P} = \varepsilon_o \chi \mathbf{E}. \tag{4.81}$$

In this case, the medium is said to be *linear and isotropic*. The non-isotropy of some crystalline materials underlies some electric and optical properties.

If a dielectric medium is isotropic but the second order terms (or higher order terms) in [4.78] are important, the medium is said to be *nonlinear*. In this case, **D** points in the direction of **E** but its magnitude has the form

$$D = \varepsilon E + \gamma E^2 + \ldots \tag{4.82}$$

If a wave of frequency $\tilde{\nu}$ is incident on this medium, the analysis of the propagation shows that waves of frequencies $2\tilde{\nu}$, $3\tilde{\nu}$, etc., are generated in the medium. Effectively a crystal of barium niobate or sodium niobate, for instance, transforms an infrared laser beam of wavelength 1060 nm into visible green light of wavelength 530 nm. Nonlinear optics is used in modern communication systems.

In the case of a field **E** that varies rapidly in time, the dielectric is not polarized instantaneously in response to the electric excitation and the polarization does not disappear instantaneously if **E** is turned off. In this case, the polarization at time t depends on the field at earlier moments t'. Mathematically, this is expressed by a relation of the form

$$\mathbf{P}(t) = \varepsilon_o \int_{-\infty}^{t} dt' \, \chi_E(t-t') \, \mathbf{E}(t'). \tag{4.83}$$

It is more appropriate in this case to analyze the field **E** and the response **P** in terms of their Fourier transforms. Then, it may be shown that, if the field has a frequency $\tilde{\nu}$, the polarization has the same frequency with a relation of the form

$$\mathbf{P}(\tilde{\nu}) = \varepsilon_o \chi_E(\tilde{\nu}) \, \mathbf{E}(\tilde{\nu}), \tag{4.84}$$

where the susceptibility $\chi_E(\tilde{\nu})$ is the Fourier transform of $\chi_E(t - t')$. The variation of χ_E as a function of $\tilde{\nu}$ is at the origin of the *dispersion* of electromagnetic waves (particularly light) in dielectrics.

4.15. Problems

Effects of dielectrics in capacitors

P4.1 a) The plane and parallel plates of a capacitor of area S are maintained with a difference of potential V and they are separated by a distance d. A dielectric plate of thickness b ($b < d$) and permittivity ε is introduced in the capacitor. Calculate the electric field outside and inside the plate, the polarization charge density, and the capacitance. **b)** An empty capacitor with $S = 0.1\ \text{m}^2$, $d = 1$ mm is charged at a potential $V = 350$ V. Calculate its capacitance C_o and its charge Q_o. If this capacitor is disconnected and filled with a dielectric, it is found that the potential becomes 280 V. Calculate the new capacitance C, the relative dielectric constant ε_r and the polarization charge density on the faces of the plates.

Polarization of dielectrics

P4.2 Applying Gauss-Ostrogradsky's theorem, show that the total polarization charge induced on the surface and the volume of a dielectric is equal to zero. Explain this result by considering the polarization as due to the displacement of charges in the molecules.

Potential and field in dielectrics, electric displacement

P4.3 A capacitor is formed by a metallic cylinder of radius R_1 and length L surrounded by a coaxial cylindrical shell of internal radius R_2. A dielectric of relative permittivity ε_r fills the cylindrical region between R_3 and R_4 (such that $R_1 < R_3 < R_4 < R_2$). Assuming a potential V, calculate the vectors **E**, **D**, and **P** as well as the surface polarization charge densities and volume polarization charge density.

Equations of the field in dielectrics

P4.4 A charge q is placed at the center of a ball of radius R of a dielectric of permittivity ε. Calculate **E** and **D** inside and outside the ball. Determine the vector polarization and the polarization charge densities. Discuss the conservation of charge.

Field of permanent dielectrics

P4.5 A sphere of radius R is polarized with a uniform polarization density **P**. **a)** Calculate the polarization charge density on its surface. What is the total positive polarization charge $+q'$? Where is its barycenter? What is the total negative polarization charge $-q'$? Where is its barycenter? What is the electric dipole moment of the sphere calculated by using the charges $\pm q'$ at their respective barycenters? Compare with the electric dipole moment calculated by using **P** and the volume of the sphere. **b)** Using the polarization density, calculate the potential inside and outside the sphere. Deduce the field **E** and the electric displacement **D**. Verify the continuity of the tangential component of **E** and the normal component of **D**.

P4.6 A cylinder of radius R and length h has a polarization **P** in the direction of its axis. Calculate the fields **E** and **D** at a point of its axis inside and outside the cylinder. Find the limits of these fields outside the cylinder if $h \ll R$ and $|z| \gg R$ and inside it if $h \ll R$. Verify that, at large distance ($|z| \gg R$ and $|z| \gg h$), the field is the same as that of an electric dipole moment $\mathbf{p} = \mathcal{V}\mathbf{P}$, where $\mathcal{V}$ is the volume of the cylinder.

Polarization of a dielectric in an external field

P4.7 Assume that a sphere of a linear dielectric is placed in an initially uniform field $\mathbf{E}_o$. If the susceptibility is low, we may write the polarization and the field as the power series: $\mathbf{P} = \mathbf{p}_o + \mathbf{p}_1 \chi_E + \mathbf{p}_2 \chi_E^2 + \ldots$ and $\mathbf{E} = \mathbf{e}_o + \mathbf{e}_1\chi_E + \mathbf{e}_2\chi_E^2 + \ldots$ Determine the $\mathbf{p}_i$ and the $\mathbf{e}_i$ by using the polarization field $\mathbf{E}_p = -\mathbf{P}/3\varepsilon_o$ of a polarized sphere.

P4.8 Show that the field of a sphere of a crystal of cubic lattice is equal to zero at its center.

Energy and force in dielectrics

P4.9 After charging a capacitor at a potential V_o, it is disconnected and a dielectric plate of permittivity ε and nearly the same thickness as the capacitor is introduced. Calculate the stored energy before and after the introduction of the plate. Interpret the variation of the energy in terms of the work required to introduce the plate without variation of its kinetic energy.

P4.10 A parallel plate capacitor is formed by two square plates of sides L and spacing d. It is maintained at a voltage V. A dielectric plate of permittivity ε, width L and thickness d' parallel to the plates is introduced in the capacitor. Let x be the length of the plate already introduced in the capacitor. What is the electric energy in this position? What is the force F of attraction of the dielectric plate by the capacitor plates?

Chapter 5

Special Techniques and Approximation Methods

A fundamental problem in the application of electrostatics is to determine the potential V, knowing the charge distribution and the dielectric properties of the medium. V obeys Poisson's equation $\Delta V = -q_v/\varepsilon$, whose solution [2.25] contains an arbitrary term $V_o(\mathbf{r})$ that verifies Laplace's equation $\Delta V_o = 0$. In fact, the expression [2.25] is not always useful, because we do not know the positions of all the charges of the Universe and, even if we know some of them, the solution is often too complicated. On the other hand, the positions of charges on the surface of conductors and the polarization of dielectrics depend on the electric field that we have to determine. Finally, the region, in which we have to determine the potential, is often bounded by surfaces whose potential is given or whose total charge is given. This imposes boundary conditions on the field and the potential. The linearity of electrostatic equations (relating the sources, the field, and the potential) may bring some helpful simplifications to the problem. If a first configuration of charges $q^{(1)}_i$ produces the field $\mathbf{E}^{(1)}$ and the potential $V^{(1)}$ and a second configuration $q^{(2)}_j$ produces the field $\mathbf{E}^{(2)}$ and the potential $V^{(2)}$, the configuration formed by the superposition of charges $\alpha q^{(1)}_i + \beta q^{(2)}_i$ produces the field $\alpha\mathbf{E}^{(1)} + \beta\mathbf{E}^{(2)}$ and the potential $\alpha V^{(1)} + \beta V^{(2)}$.

In this chapter, we analyze some mathematical techniques and approximation methods that are frequently used in the study of these problems. They include the method of images, the solution of Laplace's equation in Cartesian, spherical and cylindrical coordinates and the multipole expansion.

5.1. Unicity of the solution

The potential V that we have to determine is a solution of Poisson's equation

$$\Delta V(\mathbf{r}) = -q_v(\mathbf{r})/\varepsilon_o \qquad \text{where} \quad \mathbf{r} \in \mathcal{V}. \tag{5.1}$$

This is a partial differential equation. It has an infinity of solutions that depend on arbitrary functions. We show in this section that, if the boundary conditions are imposed, the solution V is completely determined ant it is unique. Knowing the potential, the field and the other physical quantities may be determined. In general, it is only possible to write an exact analytic solution for some simple geometrical configurations. Approximation methods must be used in the other situations.

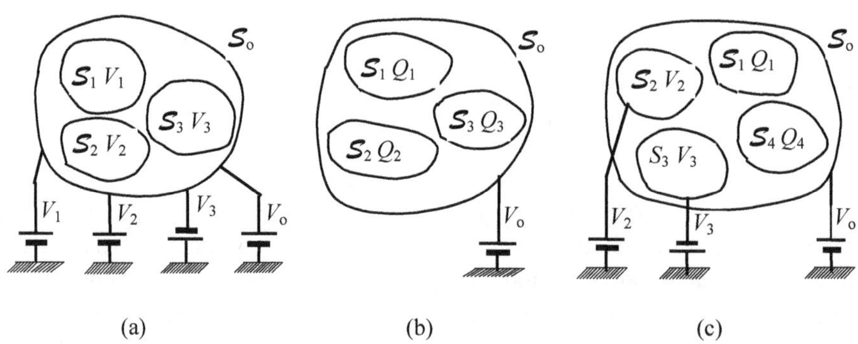

Figure 5.1. *a) A set of conductors of given potentials (Dirichlet boundary conditions), b) a set of conductors of given total charges (Neumann boundary conditions), and c) mixed boundary conditions. The whole set is surrounded by a metallic enclosure S_o whose potential is V_o. If S_o does not really exist, it may be considered as a sphere of infinite radius and potential $V_o = 0$*

Often, the region $\mathcal{V}$, in which we have to determine the potential and the field, is bounded by conductors with unknown charge distributions, but known total charge or potential. Thus, we have two types of boundary conditions.

a) If a boundary conductor (i) is maintained at a given potential, V must verify on its surface the *Dirichlet condition*

$$V(\mathbf{r}) \to V_i \qquad \text{if } \mathbf{r} \in S_i. \tag{5.2}$$

In particular, the whole system may be bounded by a metallic enclosure S_o at a given potential V_o. An unbounded space is equivalent to a system in a spherical enclosure of infinite radius at zero potential.

b) If a boundary conductor (i) was charged previously and then disconnected from the battery, the potential on its surface $\mathcal{S}_i$ is unknown but its total charge Q_i is known. The field just near $\mathcal{S}_i$ is normal to the surface with a component $E_n = -\partial V/\partial x_n$, where dx_n is the displacement normal to $\mathcal{S}_i$. The surface charge density is $\varepsilon_o E_n$ and we must have the *Neuman boundary condition*

$$Q_i = \iint_{\mathcal{S}_i} d\mathcal{S}'\, \varepsilon_o E_n = -\varepsilon_o \iint_{\mathcal{S}_i} d\mathcal{S}'\, \partial V/\partial x_n = -\varepsilon_o \iint_{\mathcal{S}_i} d\mathcal{S}'\, \mathbf{n}.\nabla V(\mathbf{r}') . \qquad [5.3]$$

In general, we may have Dirichlet conditions on all the surfaces (Figure 5.1a), Neumann conditions on all the surfaces (Figure 5.1b), or mixed boundary conditions (Figure 5.1c). If we have to determine V in different regions separated by interfaces $\mathcal{S}'$, we determine the solution of Poisson's equation in each region with the appropriate boundary conditions on $\mathcal{S}'$.

To show the unicity of the solution, let us assume that two solutions V and V' exist with the same boundary conditions. The function $\delta V = V - V'$ is obviously a solution of Poisson's equation with $q_v = 0$ (i.e. Laplace's equation $\Delta\delta V = 0$) with the boundary conditions $\delta V_i = 0$ on the Dirichlet-type surfaces and $\delta Q_i = 0$ on the Neumann-type surfaces. Thus, δV is the potential in a space that is empty of charges and bounded by conductors of zero potential or zero charge. However, in the absence of charges, δV can have neither a maximum nor a minimum; thus, δV must be constant and the value of this constant is irrelevant. As for the existence of this solution, we shall not study this problem from the mathematical point of view; but we know that the physical problem always has a solution. In principle, these considerations hold for both free and bound charges. Thus, the electrostatic problem always has a unique solution even in the presence of dielectrics (see problem 5.1).

As an application, consider an enclosure at the potential V_o containing no charge. The problem consists of finding the solution to Laplace's equation $\Delta V = 0$ with the boundary condition $V \rightarrow V_o$ on the internal face of the enclosure. It is evident that the constant potential $V = V_o$ is a solution to this problem and there is no other solution. It corresponds to a field $\mathbf{E} = 0$ in the cavity and a surface charge density on the internal face of the enclosure $q_s = E_n/\varepsilon = 0$. Similarly, the potential in the region situated outside an enclosure is a solution of Poisson's equation involving only the external charges with the boundary condition $V = V_o$ on the external face of the enclosure. The existence and the unicity of the solution imply that V and $\mathbf{E}$ outside the enclosure are completely independent of the charges inside the enclosure. Thus, the enclosure at the potential V_o completely separates the interior from the exterior.

5.2. Method of images

Consider the problem of determining V and **E** in a region of space $\mathcal{V}$ containing some charge distribution and bounded by a surface $\mathcal{S}$ with some boundary conditions on it. The unicity of the solution implies that this problem has exactly the same solution in $\mathcal{V}$ as any other problem having the same charge in $\mathcal{V}$ and the same boundary conditions on $\mathcal{S}$. In some cases, it is possible to replace the system that is behind $\mathcal{S}$ by some fictive charges (called *image charges*) whose values and positions are chosen in such a way that they produce the same boundary conditions on $\mathcal{S}$. The total potential produced by the real charges in $\mathcal{V}$ and these image charges is the solution to our problem in $\mathcal{V}$. The image charges are always outside $\mathcal{V}$ (thus, they do not produce infinite V or **E** in $\mathcal{V}$) and they are chosen by analogy to other known problems. In this section we study three examples.

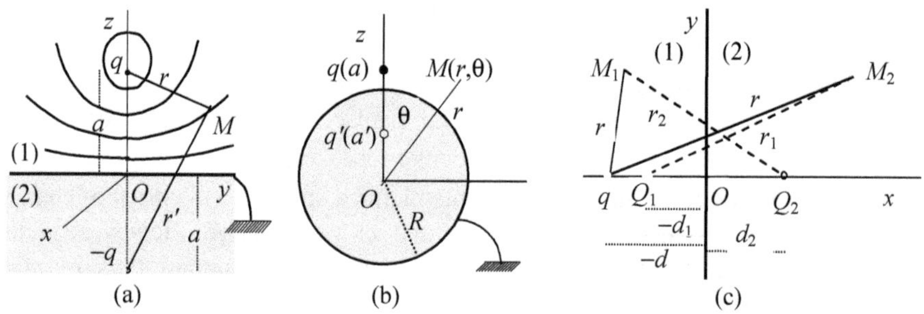

Figure 5.2. *Method of images: a) a charge q near the plane face of a conductor at $V = 0$, b) charge q near a metallic sphere, and c) charge q near the interface of two dielectrics*

A) Point charge near the plane face of a conductor at zero potential

Consider a point charge q placed on the axis Oz at a distance a from the face Oxy of a conductor maintained at zero potential (Figure 5.2a). To study this problem we recall that two charges $+q$ and $-q$ produce a potential $V = 0$ on their median plane. Thus, the solution of both problems is the same in front of the conductor. The charge $-q$ is the *image* of the real charge q. The potential at $M(x, y, z)$ is

$$V(\mathbf{r}) = K_o q\left(\frac{1}{r} - \frac{1}{r'}\right) = K_o q\left\{\frac{1}{\sqrt{x^2 + y^2 + (z-a)^2}} - \frac{1}{\sqrt{x^2 + y^2 + (z+a)^2}}\right\}. \quad [5.4]$$

This potential verifies the boundary condition $V(x, y, 0) = 0$ on the surface of the conductor; thus, it is the solution of our problem. The corresponding electric field is

$$\mathbf{E}(\mathbf{r}) = -\nabla V = K_o q\left\{\left[\frac{x}{r^3} - \frac{x}{r'^3}\right]\mathbf{e}_x + \left[\frac{y}{r^3} - \frac{y}{r'^3}\right]\mathbf{e}_y + \left[\frac{z-a}{r^3} - \frac{z+a}{r'^3}\right]\mathbf{e}_z\right\}. \quad [5.5]$$

Particularly, on the surface of the conductor, $z = 0$ and $r = r'$, thus

$$\mathbf{E}(x, y, 0) = -2K_o \frac{qa}{r^3} \mathbf{e}_z = -2K_o \frac{qa}{(x^2 + y^2 + a^2)^{3/2}} \mathbf{e}_z. \quad [5.6]$$

This field is normal to the surface of the conductor, as it should be. It corresponds to a surface charge density

$$q_s = \varepsilon_o(\mathbf{E.n})|_{z=0} = \varepsilon_o(\mathbf{E.e}_y)|_{z=0} = -\frac{qa}{2\pi(x^2+y^2+a^2)^{3/2}}. \quad [5.7]$$

Integrating this charge density over the plane Oxy, we find $-q$. The electric field $\mathbf{E}_c$ produced by the charged conductor is the same as that of the image charge $-q$. In particular, the force exerted by the conductor on the charge q is the same as the force exerted by the image charge $-q$, i.e. $\mathbf{F} = -(K_o q^2/4a^2)\, \mathbf{e}_z$. We note that the potential and the field in the region (2) inside the conductor (which are equal to zero) are not the same as those of the charge q and the charge $-q$.

B) Point charge near a metallic sphere

Consider a point charge q placed on the axis Oz at a distance a from the center O of a metallic sphere of radius R and zero potential (Figure 5.2b). To determine the potential V at any point M outside the sphere, we try to replace the sphere by a charge q', such that the potential of the sphere is zero. By symmetry, q' must be on the axis Oz. Let a' be the unknown coordinate of q'. Using spherical coordinates, the potential of the charges q and q' is

$$V(\mathbf{r}) = K_o\left\{ \frac{q}{\sqrt{r^2 + a^2 - 2ar\cos\theta}} + \frac{q'}{\sqrt{r^2 + a'^2 - 2a'r\cos\theta}} \right\}.$$

$V = 0$ on the sphere for any θ if $q(R^2 + a'^2 - 2a'R\cos\theta)^{1/2} = -q'(R^2 + a^2 - 2aR\cos\theta)^{1/2}$. This equation shows that q' must have an opposite sign to q. Squaring this equation and identifying the constant term and the term proportional to $\cos\theta$, we obtain the equations $q^2a' = q'^2a$ and $q^2(R^2 + a'^2) = q'^2(R^2 + a^2)$. This gives $q' = -qR/a$ and $a' = R^2/a$. Thus, the potential and the electric field may be written as

$$V(r,\theta) = K_o q\left\{ \frac{1}{\sqrt{r^2+a^2-2ar\cos\theta}} - \frac{R}{\sqrt{r^2a^2+R^4-2arR^2\cos\theta}} \right\}, \quad [5.8]$$

$$\mathbf{E}(\mathbf{r}) = -\nabla V = -\frac{\partial V}{\partial r}\,\mathbf{e}_r - \frac{1}{r}\frac{\partial V}{\partial \theta}\,\mathbf{e}_\theta - \frac{\partial V}{\partial \varphi}\,\mathbf{e}_\varphi$$

$$= K_o q\,\{\frac{r - a\cos\theta}{(r^2 + a^2 - 2ar\cos\theta)^{3/2}} - Ra\frac{ar - R^2\cos\theta}{(r^2a^2 + R^4 - 2arR^2\cos\theta)^{3/2}}\}\,\mathbf{e}_r$$

$$+ K_o qa\sin\theta\{\frac{1}{(r^2+a^2-2ar\cos\theta)^{3/2}} - \frac{R^3}{(r^2a^2+R^4-2arR^2\cos\theta)^{3/2}}\}\mathbf{e}_\theta. \quad [5.9]$$

In particular, we find on the surface of the sphere ($r = R$)

$$\mathbf{E}(\mathbf{R}) = K_o\frac{q}{R}\,\frac{R^2 - a^2}{(R^2 + a^2 - 2aR\cos\theta)^{3/2}}\,\mathbf{e}_r. \quad [5.10]$$

$\mathbf{E}(\mathbf{R})$ is normal to the sphere, as it should be. The charge density on the sphere is

$$q_s = \varepsilon_o(\mathbf{E}.\mathbf{n}) = \varepsilon_o(\mathbf{E}.\mathbf{e}_r) = \frac{q}{4\pi R}\,\frac{R^2 - a^2}{(R^2 + a^2 - 2aR\cos\theta)^{3/2}}. \quad [5.11]$$

The total charge that is induced on the sphere is obtained by integration

$$Q = \iint_S dS\, q_s = \frac{q}{4\pi R}(R^2 - a^2)\int_0^\pi d\theta\, R^2\sin\theta\,\frac{1}{(R^2+a^2-2aR\cos\theta)^{3/2}}\int_0^{2\pi} d\phi = -q\frac{R}{a} = q'.$$

The force exerted by the sphere on the charge q is the same as the force exerted by the image charge, that is

$$\mathbf{F} = K_o\frac{qq'}{(a - a')^2}\,\mathbf{e}_z = -K_o\frac{q^2 Ra}{(a^2 - R^2)}\,\mathbf{e}_z. \quad [5.12]$$

We note that the potential and the field inside the sphere are not the same as those of the charges q and the charge image q'.

If the sphere is at a potential V_o, we may use the principle of superposition. Consider a first state with the sphere at the potential V_o in the absence of the charge q. The charge of the sphere is then $Q_1 = 4\pi\varepsilon_o RV_o$, uniformly distributed. The potential that it produces at M outside the sphere is $V_1(\mathbf{r}) = Q_1/4\pi\varepsilon_o r = V_oR/r$. Consider also a second state with the sphere at zero potential in the presence of the charge q; the corresponding potential V_2 is [5.8]. The superposition of these two states corresponds to a sphere at the potential V_o in the presence of the charge q. Thus, the potential at point $M(r,\theta,\phi)$ is

$$V = \frac{R}{r}V_o + K_o q\{\frac{1}{\sqrt{r^2 + a^2 - 2ar\cos\theta}} - \frac{R}{\sqrt{r^2a^2 + R^4 - 2arR^2\cos\theta}}\}. \quad [5.13]$$

This is the same as the potential produced by the charge q, the fictive charge $q' = -Rq/a$ at the distance $a' = R^2/a$ and the fictive charge $q'' = Q_1 = 4\pi\varepsilon_o RV_o$ placed at the center of the sphere.

Let us assume that a charge q is brought near an isolated sphere initially carrying a charge Q_o. Consider the state (1) of the charge q near the sphere at zero potential. The corresponding solution V_1 is [5.8] and the sphere has the total charge $q' = -Rq/a$. Consider also the state (2) with the sphere carrying the charge $Q_o - q'$ in the absence of the charge q. The superposition of these two states corresponds to the sphere carrying a charge Q_o in the presence of the charge q. The potential is then

$$V = \frac{K_o}{r}(Q_o + q\frac{R}{a}) + K_o q\{\frac{1}{\sqrt{r^2+a^2-2ar\cos\theta}} - \frac{R}{\sqrt{r^2a^2+R^4-2arR^2\cos\theta}}\}. \quad [5.14]$$

C) Point charge near the plane interface of two dielectrics

Let us consider two dielectrics separated by a plane surface Oyz and a charge q at $x = -d$ in the medium (1) (Figure 5.2c). The dielectrics become polarized, each under the influence of the field of the charge q and also the field of the other polarized dielectric. To calculate the field in the medium (2) let us try to replace the effect of the dielectric (1) by a point charge Q_1 at $x = -d_1$. Similarly, to calculate the field in the medium (1), we try to replace the effect of the dielectric (2) by a point charge Q_2 at $x = d_2$. The system being symmetric about Ox, the charges Q_1 and Q_2 should be on this axis and they should not be in the region in which we calculate their field in order not to have singular fields. We calculate the fields at the point M of the Oxy plane. The fields $\mathbf{E}_1$ in medium (1) and $\mathbf{E}_2$ in medium (2) are then

$$\mathbf{E}_1(\mathbf{r}) = K_o q\frac{\mathbf{r}}{r^3} + K_o Q_2\frac{\mathbf{r}_2}{r_2^3}, \qquad \mathbf{E}_2(\mathbf{r}) = K_o q\frac{\mathbf{r}}{r^3} + K_o Q_1\frac{\mathbf{r}_1}{r_1^3}$$

and the corresponding electric displacements are $\mathbf{D}_1 = \varepsilon_1\mathbf{E}_1$ and $\mathbf{D}_2 = \varepsilon_2\mathbf{E}_2$. We now impose the boundary conditions on the interface: the tangential component of $\mathbf{E}$ (i.e. E_y and E_z) and the normal component of $\mathbf{D}$ (i.e. D_x) are continuous for $x = 0$. We find the equations

$$Q_2(y^2 + d_2^2)^{-3/2} = Q_1(y^2 + d_1^2)^{-3/2}$$
$$\varepsilon_1 qd(y^2 + d^2)^{-3/2} - \varepsilon_1 Q_2 d_2(y^2 + d_2^2)^{-3/2} = \varepsilon_2 qd(y^2 + d^2)^{-3/2} + \varepsilon_2 Q_1 d_1(y^2 + d_1^2)^{-3/2}.$$

These equations are satisfied at any point of the interface (i.e. for any y and z) if $d_1 = d_2 = d$ and $Q_1 = Q_2 = q(\varepsilon_1 - \varepsilon_2)/(\varepsilon_1 + \varepsilon_2)$. Thus, the electric fields are given by

$$\mathbf{E}_1(\mathbf{r}) = K_o q\{\frac{\mathbf{r}}{r^3} + \frac{\mathbf{r}_2}{r_2^3}\frac{\varepsilon_1 - \varepsilon_2}{\varepsilon_1 + \varepsilon_2}\}, \qquad \mathbf{E}_2(\mathbf{r}) = K_o q\frac{2\varepsilon_1}{\varepsilon_1 + \varepsilon_2}\frac{\mathbf{r}}{r^3}. \quad [5.15]$$

The corresponding electric displacements are $\mathbf{D}_1 = \varepsilon_1\mathbf{E}_1$ and $\mathbf{D}_2 = \varepsilon_2\mathbf{E}_2$, while the polarizations are $\mathbf{P}_1 = \mathbf{D}_1 - \varepsilon_o\mathbf{E}_1 = (\varepsilon_2 - \varepsilon_o)\mathbf{E}_1$ and $\mathbf{P}_2 = \mathbf{D}_2 - \varepsilon_o\mathbf{E}_2 = (\varepsilon_2 - \varepsilon_o)\,\mathbf{E}_2$. The polarization charge densities on the interface is

$$q'_s = P_{1x} - P_{2x} = (\varepsilon_1 - \varepsilon_o)E_{1x} - (\varepsilon_2 - \varepsilon_o)\,E_{2x} = \frac{q}{2\pi}\,\frac{\varepsilon_1 - \varepsilon_2}{\varepsilon_1 + \varepsilon_2}\,\frac{d}{(y^2 + d^2)^{3/2}}\,. \qquad [5.16]$$

The volume polarization charge density is $q'_v = -\,\nabla.\mathbf{P} = (\varepsilon - \varepsilon_o)\,\nabla.\mathbf{E} = 0$ in both mediums.

5.3. Method of analytic functions

Let $f(z)$ be a function of the complex variable $z = x + iy$. We may consider f as a function of the variables x and y. Separating its real part from its imaginary part, we may write

$$f(z) \equiv f(x,y) = U(x,y) + \mathrm{i}V(x,y). \qquad [5.17]$$

If f is a differentiable function of z and if x varies by dx, the variation of f is

$$df = \partial_x U\,dx + \mathrm{i}\,\partial_x V\,dx = (df/dz)\,(\partial z/\partial x)\,dx = (df/dz)\,dx.$$

Similarly, if y varies by dy, the variation of f is

$$df = \partial_y U\,dy + \mathrm{i}\,\partial_y V\,dy = (df/dz)\,(\partial z/\partial y)\,dy = \mathrm{i}\,(df/dz)\,dy.$$

Comparing the two expressions, we deduce that $\partial_x U + \mathrm{i}\,\partial_x V = -\mathrm{i}\,\partial_y U + \partial_y V$, hence

$$\partial_x U = \partial_y V \qquad \text{and} \qquad \partial_y U = -\,\partial_x V. \qquad [5.18]$$

Thus, U and V satisfy the partial differential equations

$$\partial^2_{xx} U + \partial^2_{yy} U = 0 \qquad \text{and} \qquad \partial^2_{xx} V + \partial^2_{yy} V = 0. \qquad [5.19]$$

Let us consider an electrostatic problem with translational symmetry in the direction Oz. This is the case, for instance, for cylindrical conductors parallel to Oz. Thus, the potential V does not depend on z and Laplace's equation reduces to [5.19]. Considering $V(x, y)$ as the real part or the imaginary part of an analytic function $f(z)$, the normal to the equipotential surface $V(x, y) =$ Constant is in the direction of

$$\mathbf{E} = -\,\nabla V = -\partial_x V\,\mathbf{e}_x - \partial_y V\,\mathbf{e}_y = \partial_y U\,\mathbf{e}_x - \partial_x U\,\mathbf{e}_y. \qquad [5.20]$$

The last expression shows that **E** is normal to the vector $\nabla U = \partial_x U\, \mathbf{e}_x + \partial_y U\, \mathbf{e}_y$, which is itself normal to the lines U = Constant. Thus, U = Constant represents the lines of field and V = Constant represents the equipotentials (Figure 5.3a).

Let us consider, for instance, the analytic function $f = z^2$, whose real part and imaginary part are $U = x^2 - y^2$ and $V = 2xy$. The lines $x^2 - y^2 = C_i$ (where C_i are constants), which asymptotically approach the bisectors of the axes, may represent the equipotential lines, while the hyperbolas $2xy = C'_j$ may represent the lines of field and vice versa (Figure 5.3b). The components of the electric field are in this case $E_x = -2x$ and $E_y = -2y$. The field is weaker near the Ox and Oy axes. This system, called a quadrupole lens, is used to focalize a beam of charged particles.

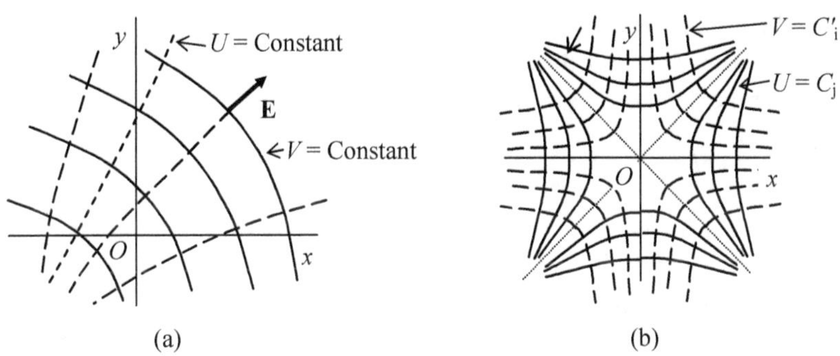

Figure 5.3. *Method of analytic functions: a) interpretation of the curves U = Constant and V = Constant as lines of field and equipotential lines, and b) the example $f = z^2$.*

5.4. Method of separation of variables

The electrostatic problem in linear and isotropic mediums consists in finding the solution of Poisson's equation $\Delta V = -q_v/\varepsilon$. We have seen that the solution is

$$V(\mathbf{r}) = V_o(\mathbf{r}) + \frac{1}{4\pi} \iiint d\mathcal{V}' \frac{q_v(\mathbf{r}')}{\varepsilon|\mathbf{r} - \mathbf{r}'|}, \qquad [5.21]$$

where V_o is a solution of Laplace's equation $\Delta V_o = 0$. It may always be chosen to have V satisfy the imposed boundary conditions.

Contrarily to ordinary differential equations that always have a finite number of independent solutions, Laplace's equation, which is a second-order partial differential equation, has an infinity of independent solutions. Let $f_n(\mathbf{r})$ be a set of solutions. We say that these functions are *complete* if any function $V(\mathbf{r})$ may be written as a linear superposition of these functions:

$$V(\mathbf{r}) = \Sigma_n a_n f_n(\mathbf{r}), \qquad [5.22]$$

where the a_n are constant coefficients. The functions $f_n(\mathbf{r})$ are orthogonal; they may be normalized to verify the *orthonormalization relations*

$$\iiint_D d\mathcal{V}\, p(\mathbf{r})\, f_m(\mathbf{r})\, f_n(\mathbf{r}) = \delta_{m\,n}. \qquad [5.23]$$

D is an appropriate domain of integration, $p(\mathbf{r})$ is a characteristic *weight function* and $\delta_{m\,n}$ are *Kronecker symbols* such that $\delta_{m\,n} = 1$ if $m = n$ and $\delta_{m\,n} = 0$ if $m \neq n$.

Using the orthonormalization relation, we may determine the coefficients a_m for any given function $V(\mathbf{r})$. For this, we multiply both sides of [5.22] by $p(\mathbf{r})\, f_m(\mathbf{r})$ and integrate over the domain D; we find

$$\iiint_D d\mathcal{V}\, p(\mathbf{r})\, f_m(\mathbf{r})\; V(\mathbf{r}) = \Sigma_n a_n \iiint_D d\mathcal{V}\, p(\mathbf{r})\, f_m(\mathbf{r})\, f_n(\mathbf{r}) = \Sigma_n a_n \,\delta_{m\,n} = a_m. \qquad [5.24]$$

The functions $f_n(\mathbf{r})$ play in the "function space" a part similar to that of the orthonormalized basis in a vector space. The choice of the functions $f_n(\mathbf{r})$ is not unique. In particular, each system of space coordinates corresponds to an adapted set of functions $f_n(\mathbf{r})$.

To determine the function $V_o(\mathbf{r})$, which allows the boundary conditions to be imposed, we write $V_o(\mathbf{r})$ in the form [5.22] and we determine the coefficients a_n. The functions $f_m(\mathbf{r})$ that are adapted to a given problem are those that have the same geometrical symmetries as the studied system (or most of them). In some cases, the series [5.22] may contain a finite number of terms. In others, the coefficients a_n become negligible for high values of n. In some other cases, the series is very slowly convergent; the method is then inappropriate.

5.5. Laplace's equation in Cartesian coordinates

In Cartesian coordinates, Laplace's equation takes the form

$$\partial^2_{xx} V + \partial^2_{yy} V + \partial^2_{zz} V = 0. \qquad [5.25]$$

If we try solutions of the form $V = X(x)\, Y(y)\, Z(z)$, equation [5.25] takes the form $X''YZ + XY''Z + XYZ'' = 0$. Dividing by XYZ, we obtain

$$X''/X + Y''/Y + Z''/Z = 0. \qquad [5.26]$$

The terms of this equation are functions of x, y, and z, respectively. The equation may be identically satisfied only if each one of these terms is constant:

$$X''/X = C_1, \qquad Y''/Y = C_2, \qquad Z''/Z = C_3 \qquad \text{with } C_1 + C_2 + C_3 = 0. \qquad [5.27]$$

Consider one of these equations, $X'' = C_1 X$, for instance. The form of its solution depends on the sign of C_1:

– if $C_1 > 0$, we set $C_1 = \kappa^2$. The general solution has the exponential form

$$X = A_1 e^{-\kappa x} + B_1 e^{\kappa x}; \qquad [5.28]$$

– if $C_1 < 0$, we set $C_1 = -k^2$. The general solution is simple harmonic of the form

$$X = A_1 e^{-ikx} + B_1 e^{ikx}, \qquad \text{or } A_1 \cos(kx) + B_1 \sin(kx); \qquad [5.29]$$

– if $C_1 = 0$, the general solution is algebraic of the form

$$X = A_1 + B_1 x. \qquad [5.30]$$

The values of the constants C_i, A_i, and B_i are determined by the boundary conditions.

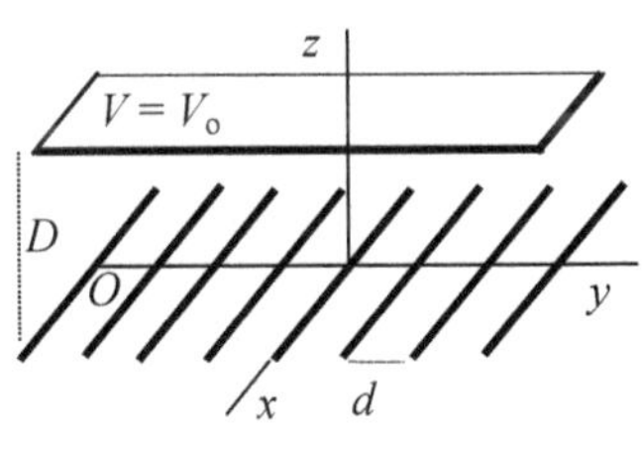

Figure 5.4. *A grid parallel to a conducting plate*

As an application, let us consider a grid formed by thin metallic wires lying in the Oxy plane, parallel to the x-axis, and separated by a distance d. A plane metallic plate P is parallel to the grid at a distance D and it has a potential V_0, while the grid has zero potential (Figure 5.4). We assume that the grid and the plate are infinite and we determine the potential $V(\mathbf{r})$ everywhere. The potential is independent of x (because of the translational symmetry in the direction Ox) and it is a periodic function of y with a period d. Thus, it may be written as a linear combination of $\cos(n\pi y/d)$ and $\sin(n\pi y/d)$ with coefficients that may depend on z. If the origin is taken on one of the wires, the system has a reflection symmetry ($y \rightarrow -y$); thus, the potential is an even function of y. This excludes the terms $\sin(n\pi y/d)$ and we write

$V(x, y, z) = \Sigma_{n\geq 0} F_n(z) \cos(n\pi y/d)$. This expression verifies Laplace's equation $\Delta V = 0$ if $F_n'' - (n\pi/d)^2 F_n = 0$, whose general solution is

$$F_o = A_o + B_o z, \qquad \text{and} \qquad F_n = A_n \exp(n\pi z/d) + B_n \exp(-n\pi z/d) \quad \text{for } n \neq 0$$

The coefficients A_n and B_n are determined from the boundary conditions:

– on the plate ($z = D$), we have

$$V(x,y,D) = A_o + B_o D + \Sigma_{n>0} [A_n \exp(n\pi D/d) + B_n \exp(-n\pi D/d)] \cos(n\pi y/d) = V_o.$$

This condition is satisfied for any y if

$$V_o = A_o + B_o D, \qquad \text{and} \qquad A_n \exp(n\pi D/d) + B_n \exp(-n\pi D/d) = 0.$$

– The potential of the wires ($z = 0$ and $y = md$) is zero if

$$0 = A_o + \Sigma_{n>0} (A_n + B_n) \cos(nm\pi) = A_o + \Sigma_{n>1} (-1)^{nm} (A_n + B_n).$$

This condition is satisfied for any m if $\Sigma_{n>0} (A_n + B_n) = \Sigma_{n>0} (-1)^n (A_n + B_n) = -A_o$. All these conditions are satisfied if $A_o = 0$, $B_o = V_o/D$ and $A_n = B_n = 0$ for $n \neq 0$. Thus, the potential is $V = V_o z/D$; it is the same as if the grid was replaced by a continuous plate.

5.6. Laplace's equation in spherical coordinates

In spherical coordinates, Laplace's equation may be written as

$$\Delta V \equiv \frac{1}{r^2} \left[\frac{\partial}{\partial r}\left(r^2 \frac{\partial V}{\partial r}\right) + \frac{1}{\sin\theta}\left(\frac{\partial}{\partial\theta}\right)\left(\sin\theta \frac{\partial V}{\partial\theta}\right) + \frac{1}{\sin^2\theta} \frac{\partial^2 V}{\partial\varphi^2}\right] = 0. \qquad [5.31]$$

Let us find solutions of the form $V(r, \theta, \varphi) = F(r)\, G(\theta)\, H(\varphi)$. Substituting this expression into the equation, and multiplying by $r^2 \sin^2\theta / FGH$, we obtain

$$\frac{\sin^2\theta}{F} \frac{d}{dr}\left(r^2 \frac{dF}{dr}\right) + \frac{\sin\theta}{G} \frac{d}{d\theta}(\sin\theta\, G') = -\frac{H''}{H}. \qquad [5.32]$$

The left-hand side of this equation is a function of r and θ, while its right-hand side is a function of φ. It is identically satisfied only if both sides are equal to a constant:

$$\frac{\sin^2\theta}{F} \frac{d}{dr}\left(r^2 \frac{dF}{dr}\right) + \frac{\sin\theta}{G} \frac{d}{d\theta}(\sin\theta\, G') = C, \quad \frac{H''}{H} = -C. \qquad [5.33]$$

As φ is defined up to 2π, the function $H(\varphi)$ has a single determination only if it is a periodic function with period 2π. This requires that C be positive (as, if C were negative, H would be exponential). Setting $C = m^2$, the general solution for H is

$$H_{\mathrm{m}}(\varphi) = A\ e^{im\varphi} + B\,e^{-im\varphi}. \qquad [5.34]$$

Effectively, it has a period 2π if m is an integer that we may take as positive or zero (a negative value of m is equivalent to exchange the constants A and B). In particular, the solution with $m = 0$ is symmetric about Oz. Replacing C by m^2, the equation of F and G may be written as

$$(1/F)\partial_r(r^2\,\partial_r F) = -\,(1/G\sin\theta)\;\partial_\theta(\sin\theta\;\partial_\theta G) + m^2/\sin^2\theta. \qquad [5.35]$$

The left-hand side of this equation being a function of r and the right-hand side a function of θ, the equation is identically satisfied if both sides are equal to a constant k. Thus, we find the equations

$$(1/G\sin\theta)\;\partial_\theta(\sin\theta\;\partial_\theta G) - m^2/\sin^2\theta + k = 0, \quad (1/F)\partial_r(r^2\,\partial_r F) = k. \qquad [5.36]$$

Setting $u = \cos\theta$, the equation of G becomes

$$(1-u^2)\;\partial^2_{\mathrm{uu}}G(u) - 2u\;\partial_{\mathrm{u}}G(u) + [k - \frac{m^2}{1-u^2}]G(u) = 0. \qquad [5.37]$$

In particular, for $m = 0$, we find the simpler equation

$$(1-u^2)\;\partial^2_{\mathrm{uu}}G(u) - 2u\;\partial_{\mathrm{u}}G(u) + k\,G(u) = 0, \qquad [5.38]$$

called the *Legendre equation,* while [5.37] is the *associate Legendre equation.* The solution of the Legendre equation is singular at the points $u = \pm 1$ (and this is unacceptable as these points are the limits of the physical domain), unless

$$k = l(\,l+1) \qquad \text{with} \quad l = 0, 1, 2, 3, \ldots \qquad [5.39]$$

For each l, the solution is a polynomial of degree l, called the *Legendre polynomial,* given by *Rodrigues' formula*

$$P_l(u) = \frac{1}{2^l\,l!}\;\frac{d^l}{du^l}\;(u^2-1)^l, \qquad [5.40]$$

where the normalization factor $1/2^l l!$ is chosen so that $P_l(1) = 1$ by convention. On the other hand, the Legendre polynomials verify the *orthogonality relation*

$$\int_{-1}^{1} du\;\; P_l(u)\; P_k(u) = 2\delta_{l,k}/(2l+1), \qquad [5.41]$$

where the $\delta_{l,k}$ are Kronecker symbols. The first Legendre polynomials are

$$P_o = 1, \qquad P_1 = u, \qquad P_2 = ½(3u^2 - 1), \qquad P_3 = ½(5u^3 - 3u). \qquad [5.42]$$

These polynomials are even or odd according to whether l is even or odd

$$P_l(-u) = (-1)^l P_l(u). \qquad [5.43]$$

The Legendre polynomials form a complete set for the functions of u. Thus, any function of u may be written as a linear combination of these polynomials:

$$f(u) = \Sigma_n \, a_n P_n(u). \qquad [5.44]$$

In the general case $m \neq 0$, the associate Legendre equation has the solutions

$$P_l^{|m|}(u) = (-1)^m (1 - u^2)^{|m|/2} \frac{d^{|m|}}{du^{|m|}} P_l(u), \qquad [5.45]$$

called *associate Legendre functions*. They vanish if $m > l$. The first functions are

$$P_1^1 = -(1-u^2)^{½} = -\sin\theta, \qquad P_2^2 = 3(1-u^2) = 3\sin^2\theta,$$
$$P_2^1 = -3u(1-u^2)^{½} = -3\sin\theta\cos\theta, \qquad P_3^3 = -15(1-u^2)(1-u^2)^{½} = -15\sin^3\theta,$$
$$P_3^2 = 15(1-u^2)u = 15\sin^2\theta\cos\theta,$$
$$P_3^1 = -(3/2)(5u^2-1)(1-u^2)^{½} = -(3/2)\sin\theta\,(5\cos^2\theta - 1). \qquad [5.46]$$

They verify the relations of symmetry, orthogonality, and differentiation

$$P_l^m(-u) = (-1)^{l+m} P_l^m(u), \qquad [5.47]$$

$$\int_{-1}^{1} du \; P_l^m(u)\, P_k^m(u) = \frac{2}{2l+1}\frac{(l-m)!}{(l+m)!}\delta_{l,k}, \qquad [5.48]$$

$$(1 - u^2)\,\partial_u P_l^m(u) = (l+m)\, P_{l-1}^m(u) - lu\, P_l^m(u). \qquad [5.49]$$

We consider now the radial equation in [5.36] with $k = l(l + 1)$; it takes the form

$$r^2 \partial^2_{rr} F(r) + 2r\,\partial_r F(r) - l(l+1)\,F = 0. \qquad [5.50]$$

This equation has solutions of the form $F = r^p$ if $p(p + 1) = l(l + 1)$, i.e. $p = l$ or $p = -l - 1$. Thus, the general solution of [5.50] is

$$F(r) = Ar^l + B\,r^{-l-1}, \qquad [5.51]$$

where A and B are arbitrary constants.

We conclude that any solution of Laplace's equation in spherical coordinates is a superposition of the solutions of the form

$$V(r, \theta, \varphi) = \Sigma_{l\geq 0}\ \Sigma_{0\leq m\leq l}\ [A_{l,m}e^{im\varphi} + B_{l,m}e^{-im\varphi}][A_{l,m}r^l + B_{l,m}r^{-l-1}]\ P_l^m(\cos\theta). \quad [5.52]$$

We note that we may also write

$$V(r, \theta, \varphi) = \Sigma_{l\geq 0}\ \Sigma_{-l\leq m\leq l}\ [\ A_{l,m}\ r^l + B_{l,m}\ r^{-l-1}]\ e^{im\varphi}\ P_l^m(\cos\theta). \quad [5.53]$$

In particular, if the system is symmetric about Oz, V is independent of φ; thus, only the $m = 0$ terms contribute to V, hence

$$V(r, \theta) = \Sigma_{l\geq 0}\ [A_l r^l + B_l r^{-l-1}]P_l(\cos\theta). \quad [5.54]$$

Sometimes, it helps to use the *spherical harmonics*

$$Y_l^m(\theta,\varphi) = \sqrt{\frac{2l+1}{4\pi}\frac{(l-|m|)!}{(l+|m|)!}}\ e^{im\varphi}\ P_l^{|m|}(\cos\theta), \quad \text{where } -l < m < l. \quad [5.55]$$

They verify the relations of complex conjugation, symmetry, and orthogonality

$$Y_l^m(\theta,\varphi)^* = (-1)^m\, Y_l^{-m}(\theta,-\varphi) = Y_l^m(\theta,-\varphi), \quad [5.56]$$

$$Y_l^m(\theta+\pi,\varphi+2\pi) = (-1)^l\, Y_l^m(\theta,\varphi), \quad [5.57]$$

$$\int_0^{2\pi} d\varphi \int_0^{\pi} d\theta\ \sin\theta\ Y_l^m(\theta,\varphi)^*\ Y_{l'}^{m'}(\theta,\varphi) = \delta_{l,l'}\,\delta_{m,m'}. \quad [5.58]$$

They form a complete set of functions of θ and φ: any function of θ and φ may be written as $f(\theta, \varphi) = \Sigma_{l\geq 0}\ \Sigma_{-l\leq m\leq l}\ a_l^m\, Y_l^m(\theta,\varphi)$. Thus, the general solution of Laplace's equation in spherical coordinates may be written as

$$V(r, \theta, \varphi) = \Sigma_{l\geq 0}\ \Sigma_{-l\leq m\leq l}\ (A_{l,m}r^l + B_{l,m}\ r^{-l-1})\ Y_l^m(\theta,\varphi). \quad [5.59]$$

The coefficients $A_{l,m}$ and $B_{l,m}$ may be chosen so that the potential verifies the boundary conditions of any electrostatic problem. If V is symmetric about Oz, only the $m = 0$ terms contribute to the series.

As an application, we consider a metallic sphere of radius R maintained at zero potential and placed in an initially uniform field $\mathbf{E}_0$. We determine the potential, the field, and the surface charge density on the sphere (Figure 5.5). We take the origin at the center of the sphere and Oz in the direction of the field. Oz is an axis of symmetry for the system. Writing the solution in the form [5.54], the condition $V = 0$ on the sphere ($r = R$) is satisfied if $V(R, \theta) = \Sigma_{l\geq 0} [A_l R^l + B_l R^{-l-1}] P_l(\cos\theta) = 0$, which must be verified for any θ. The Legendre polynomials being linearly independent, we must have $A_l R^l + B_l R^{-l-1} = 0$. This allows us to write the potential only with the coefficients A_l: $V(r, \theta) = \Sigma_{l\geq 0} A_l [r^l - R^{2l+1} r^{-l-1}] P_l(\cos\theta)$. At large distance r from O, the field reduces to the uniform field $\mathbf{E}_o$, whose potential is $V = -E_o z = -E_o r\cos\theta$. Thus, we must have in the limit $r >> R$

$$V(r, \theta) = \Sigma_{l\geq 0} A_l [r^l - R^{2l+1} r^{-l-1}] P_l(\cos\theta) \to -Er\cos\theta.$$

$P_l(\cos\theta)$ being a polynomial of degree l in $\cos\theta$, only the polynomials P_o and P_1 contribute to V, as the polynomials P_2, P_3, etc. give an asymptotic form which depends on $\cos^2\theta$, $\cos^3\theta$, etc. Thus, using the expressions [5.42], we find

$$V = A_o[1 - R/r]P_o(\cos\theta) + A_1[r - R^3/r^2]P_1(\cos\theta) = A_o[1 - R/r] + A_1[r - R^3/r^2]\cos\theta,$$

whose limit at large distance is $A_o + A_1 r\cos\theta$. Comparing with the asymptotic form $-E_o r\cos\theta$, we deduce that $A_o = 0$ and $A_1 = -E_o$. Thus, the potential is given by

$$V(r, \theta, \varphi) = -E_o[r - R^3/r^2]\cos\theta. \qquad [5.60]$$

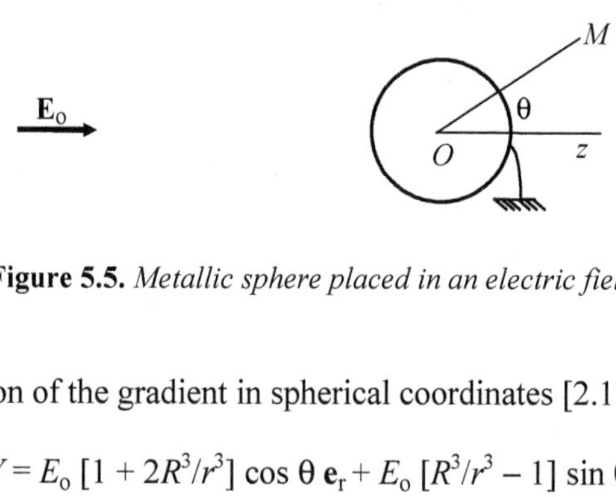

Figure 5.5. *Metallic sphere placed in an electric field*

Using the expression of the gradient in spherical coordinates [2.11], we may write

$$\mathbf{E}(\mathbf{r}) = -\nabla V = E_o [1 + 2R^3/r^3]\cos\theta\,\mathbf{e}_r + E_o [R^3/r^3 - 1]\sin\theta\,\mathbf{e}_\theta. \qquad [5.61]$$

In particular, we find on the sphere $\mathbf{E}(\mathbf{r}) = 3E_o\cos\theta\,\mathbf{e}_r$, which is normal to the sphere. The surface charge density on the sphere is

$$q_s = \varepsilon_o(\mathbf{E}.\mathbf{n}) = \varepsilon_o(\mathbf{E}.\mathbf{e}_r) = 3\varepsilon_o E_o\cos\theta. \qquad [5.62]$$

5.7. Laplace's equation in cylindrical coordinates

In cylindrical coordinates, Laplace's equation takes the form

$$\partial^2_{\rho\rho}V + \frac{1}{\rho}\partial_\rho V + \frac{1}{\rho^2}\partial^2_{\varphi\varphi}V + \partial^2_{zz}V = 0. \qquad [5.63]$$

Let us try solutions of the form $V(\rho, \varphi, z) = F(\rho)\ H(\varphi)\ Z(z)$. Substituting this expression in the equation and dividing by FHZ, we find the equation

$$F''/F + F'/\rho F + H''/\rho^2 H = -Z''/Z.$$

The left-hand side is a function of ρ and φ, while the right-hand side is a function of z. The equation may be identically satisfied only if both sides are equal to a constant D. Thus, we have

$$Z''/Z = D, \qquad F''/F + F'/\rho F + H''/\rho^2 H = -D.$$

Depending on the physical situation, the solution of the equation $Z''/Z = D$ is simple harmonic if D is negative, exponential if D is positive or algebraic if $D = 0$.

Separating the variables in the equation of F and H, we find

$$\rho^2 F''/F + \rho F'/F + D\rho^2 = -\ H''/H.$$

The left-hand side is a function of ρ, while the right-hand side is a function of φ. The equation is identically satisfied if both sides are equal to a constant C. Thus, we have

$$H''/H = -C\,, \qquad \rho^2 F''/F + \rho F'/F + D\rho^2 = C.$$

By the same argument that we used in the previous section, we must have $C = m^2$ where m is an integer and the solution for H is of the form

$$H_m(\varphi) = A\ e^{im\varphi} + B\ e^{-im\varphi}.$$

Replacing C by m^2, we find that $F(\rho)$ is a solution of *Bessel equation*

$$F'' + \frac{1}{\rho}F' + (D - \frac{m^2}{\rho^2})F = 0. \qquad [5.64]$$

a) The case $D > 0$:

If D is positive, setting $k = \sqrt{D}$ and $u = k\rho$, equation [5.64] takes the form

$$u^2 F'' + u\,F' + (u^2 - m^2)\,F = 0. \qquad [5.65]$$

It has a solution, called *Bessel function of the first kind*, which is regular at the origin and may be expressed as a power series about $u = 0$

$$J_m(u) = \Sigma_{p\geq 0} \frac{(-1)^p}{p!(m+p)!} \left(\frac{u}{2}\right)^{m+2p}. \quad [5.66]$$

If m is not equal to an integer, a second solution is obtained by changing m into $-m$. The general solution of [5.65] is then $F(u) = A\ J_m(u) + B\ J_{-m}(u)$. But, if m is an integer, $J_{-m}(u)$ is not independent of $J_m(u)$, as $J_{-m}(u) = (-1)^m J_m(u)$ for $m = 0, 1, 2\ldots$ However, equation [5.65] also has a solution called the *Bessel function of the second kind* (or *Neumann function*), which is singular at the origin and given by

$$N_m(u) = \lim_{\mu\to m} \frac{J_\mu(u)\cos(\mu\pi) - J_{-\mu}(u)}{\sin(\mu\pi)}. \quad [5.67]$$

Thus, the general solution of [5.65] is

$$F(u) = A_m\ J_m(u) + B_m\ N_m(u). \quad [5.68]$$

Figure 5.6a illustrates the first three Bessel and Neumann functions.

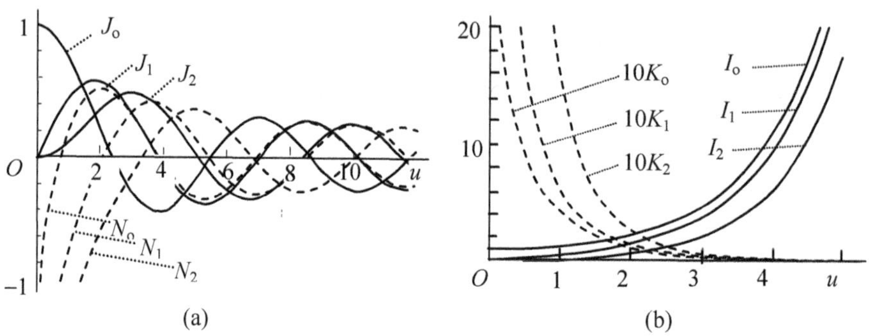

Figure 5.6. *Bessel functions*

Here are some useful properties of Bessel functions (X_m stands for J_m or N_m):

$$-X_{m-1}(u) + X_{m+1}(u) = (2m/u)\,X_m(u), \quad [5.69]$$

$$-dX_m/du = ½\,X_{m-1}(u) - ½\,X_{m+1}(u) = X_{m-1}(u) - (m/u)\,X_m(u), \quad [5.70]$$

$$-(d/du)[u^m X_m(u)] = u^m X_{m-1}(u) = -u^{-m}\ X_{m+1}(u), \quad [5.71]$$

$$-J_m(u) = (i^{-m}/2\pi)\int_{-\pi}^{\pi} d\phi\ e^{i(u\cos\phi + m\phi)} = (-1)^m/2\pi \int_0^{2\pi} d\phi\ e^{i(u\sin\phi + m\phi)}. \quad [5.72]$$

The asymptotic expressions of functions J_m and N_m are

$$J_m(u) \underset{u\to 0}{\to} \frac{1}{m!}\left(\frac{u}{2}\right)^m, \qquad J_m(u) \underset{u\to\infty}{\to} \sqrt{\frac{2}{\pi u}}\cos(u - m\pi/2 - \pi/4),$$

$$N_{m\neq 0}(u) \underset{u\to 0}{\to} -\frac{(m-1)!}{\pi}\left(\frac{2}{u}\right)^m, \qquad N_{m\neq 0}(u) \underset{u\to\infty}{\to} \sqrt{\frac{2}{\pi u}}\sin(u - m\pi/2 - \pi/4),$$

$$N_0(u) \underset{u\to 0}{\to} \frac{2}{\pi}\left[\ln\left(\frac{u}{2}\right) + 0.5772...\right], \qquad N_0(u) \underset{u\to\infty}{\to} \sqrt{\frac{2}{\pi u}}\sin(u - \pi/4). \quad [5.73]$$

The general solution of equation [5.64] is then

$$F(\rho) = A_m J_m(k\rho) + B_m N_m(k\rho). \quad [5.74]$$

We note that, if V is regular at the origin, the functions of the second kind N_m are excluded. The solution for the function $Z(z)$ is then $Z(z) = M e^{kz} + N e^{-kz}$ and the general solution of Poisson's equation in cylindrical coordinates is a superposition of the modes specified by the integer m and the constant k:

$$V(\rho, \varphi, z) = \Sigma_m \Sigma_k J_m(k\rho)[M_m e^{kz} + N_m e^{-kz}]\,[A_m e^{im\varphi} + B_m e^{-im\varphi}]. \quad [5.75]$$

The electric field is given by

$$\mathbf{E} = -\nabla V = -\frac{\partial V}{\partial \rho}\mathbf{e}_\rho - \frac{1}{\rho}\frac{\partial V}{\partial \varphi}\mathbf{e}_\varphi - \frac{\partial V}{\partial z}\mathbf{e}_z. \quad [5.76]$$

To evaluate it, we use equation [5.70] for the derivative of the Bessel functions.

b) The case $D < 0$:

If D is negative, setting $\kappa = \sqrt{-D}$ and $u = \kappa\rho$, equation [5.64] takes the form

$$u^2 F'' + u F' - (u^2 + n^2) F = 0. \quad [5.77]$$

This is the so-called *modified Bessel equation.* It has the independent solutions

$$I_m(u) = \Sigma_{p\geq 0} \frac{1}{p!(m+p)!}\left(\frac{u}{2}\right)^{m+2p}, \quad K_m(u) = \lim_{\mu\to m} \frac{\pi}{2}\frac{I_{-\mu}(u) - J_\mu(u)}{\sin(\mu\pi)}. \quad [5.78]$$

Figure 5.6b illustrates the first three functions I_n and K_n. The asymptotic expressions of the functions I_n and K_n at small and large values of u are:

$$I_m(u) \underset{u\to 0}{\to} \frac{1}{m!}\left(\frac{u}{2}\right)^m, \qquad I_m(u) \underset{u\to \infty}{\to} \frac{1}{\sqrt{2\pi u}}\, e^u,$$

$$K_o(u) \underset{u\to 0}{\to} -\ln(u) + \ldots, \qquad K_o(u) \underset{u\to \infty}{\to} \sqrt{\frac{\pi}{2u}}\, e^{-u},$$

$$K_m(u) \underset{u\to 0}{\to} \frac{1}{2}(m-1)!\left(\frac{2}{u}\right)^m, \qquad K_m(u) \underset{u\to \infty}{\to} \sqrt{\frac{\pi}{2u}}\, e^{-u} \quad (m \neq 0). \qquad [5.79]$$

Thus, the general solution of equation [5.64] is

$$F(\rho) = A_m I_m(\kappa\rho) + B_m K_m(\kappa\rho). \qquad [5.80]$$

If V is regular at the origin, the functions K_m are excluded. The solution for the function $Z(z)$ is then $Z(z) = M\cos(\kappa z) + N\sin(\kappa z)$ and the general solution of Poisson's equation is a superposition of the modes specified by the integer m and the constant κ:

$$V(\rho,\varphi,z) = \Sigma_m\Sigma_\kappa A_m I_m(\kappa\rho)[M\cos(\kappa z) + N\sin(\kappa z)]\,[A_m e^{in\varphi} + B_m e^{-in\varphi}]. \quad [5.81]$$

The electric field is evaluated by using equation [5.76] and the derivative of the function I_m given by

$$\frac{dI_m(u)}{du} = I_{m-1}(u) - \frac{m}{u} I_m(u) = I_{m+1}(u) + \frac{m}{u} I_m(u). \qquad [5.82]$$

5.8. Multipole expansion

Let us consider the expression [5.21] of the potential of a charge distribution $q_v(\mathbf{r})$. We write $|\mathbf{r} - \mathbf{r}'| = (r^2 + r'^2 - 2rr'\cos\theta)^{1/2}$ where θ is the angle of $\mathbf{r}$ and $\mathbf{r}'$. As the Legendre polynomials form a complete set for the functions of θ, we may write

$$1/|\mathbf{r} - \mathbf{r}'| = (r'^2 - 2rr'u + r^2)^{-1/2} = \Sigma_{l\geq 0}\,(r'^l/r^{l+1})\,P_l(u) \quad (\text{if } r' < r). \qquad [5.83]$$

If the charge distribution is localized in a volume $\mathcal{V}$, its potential at large distance may be written as

$$\begin{aligned} V(\mathbf{r}) &= K_o\iiint_{\mathcal{V}} d\mathcal{V}'\, q_v(\mathbf{r}')/|\mathbf{r} - \mathbf{r}'| = K_o\iiint_{\mathcal{V}} d\mathcal{V}'\, q_v(\mathbf{r}')\Sigma_{l\geq 0}\,(r'^l/r^{l+1})\,P_l(u) \\ &= K_o \iiint_{\mathcal{V}} d\mathcal{V}'\, q_v(\mathbf{r}')\{(1/r)P_o(u) + (r'/r^2)\,P_1(u) + (r'^2/r^3)\,P_2(u) + \ldots\} \\ &= K_o \iiint_{\mathcal{V}} d\mathcal{V}'\, q_v(\mathbf{r}')\{1/r + (r'/r^2)\,u + (r'^2/2r^3)(3u^2 - 1) + \ldots\}. \end{aligned} \qquad [5.84]$$

We may replace $u = \cos\theta$ par $(\mathbf{r}.\mathbf{r}')/rr'$ and obtain

$$V(\mathbf{r}) = K_o \iiint d\mathcal{V}' q_v(\mathbf{r}')\{\frac{1}{r} + \frac{\mathbf{r}.\mathbf{r}'}{r^3} + \frac{1}{2r^5}\Sigma_{\alpha\beta}(3x'_\alpha x'_\beta - r'^2\delta_{\alpha\beta})\, r_\alpha r_\beta\} + \ldots\}, \quad [5.85]$$

where α, β = 1, 2 and 3. We may also write

$$V(\mathbf{r}) = K_o q/r + K_o\mathbf{r}.\mathbf{p}/r^3 + (K_o/2r^5)\,\Sigma_{\alpha\beta}\, Q_{\alpha\beta}\, r_\alpha\, r_\beta + \ldots, \quad [5.86]$$

where we have introduced the total charge q, the electric dipole moment $\mathbf{p}$, and the electric quadrupole moment defined by

$$q = \iiint d\mathcal{V}' q_v(\mathbf{r}'), \quad \mathbf{p} = \iiint d\mathcal{V}' q_v(\mathbf{r}')\mathbf{r}' \quad \text{and } Q_{\alpha\beta} = \iiint_{\mathcal{V}} d\mathcal{V}' q_v(\mathbf{r}')[3x'_\alpha x'_\beta - r'^2\delta_{\alpha\beta}]. \quad [5.87]$$

The electric field is then

$$\mathbf{E} = -\nabla V = K_o\Sigma_{l\geq 0}\,(1/r^{l+3})\iiint d\mathcal{V}' q_v(\mathbf{r}')\, r'^{l-1}\{(l+1)P_l(u)r'\mathbf{r} + P'_l(u)[-r\mathbf{r}' + \mathbf{r}(\mathbf{r}.\mathbf{r}')/r]\},$$

$$E_\beta = \frac{K_o q}{r^3}\, x_\beta + \frac{K_o}{r^5}[3\, x_\beta(\mathbf{r}.\mathbf{p}) - r^2 p_\beta] + \frac{K_o}{r^7}[(5/2)\mathbf{r}\; Q_{\alpha\beta}\, x_\alpha x_\beta - r^2\, Q_{\alpha\beta}\, x_\alpha]. \quad [5.88]$$

5.9. Other methods

To determine the potential obeying given boundary conditions, some other methods may be used. We mention the variational method, which is used in several branches of physics. It is based on the property that the distribution of charge in electrostatic equilibrium is the one that makes the electrostatic energy

$$U_e = ½ \iiint d\mathcal{V}\, \varepsilon\, \mathbf{E}^2 = ½ \iiint d\mathcal{V}\, (\mathbf{E}.\mathbf{D}) \quad [5.89]$$

minimal. Indeed, if it is not, a certain amount of energy is available to displace the charge and supply them with kinetic energy. Thus, we may look for the solution of Poisson or Laplace's equations, which makes the energy minimal. For instance, we may choose a superposition of the modes with the coefficients as variational parameters. We calculate the total energy as a function of these parameters and we determine their values in order to have the minimum energy.

At present, it is possible to use computers to numerically solve physical problems by using easily programmable methods. The procedure consists in making all of the physical quantities discrete. In particular, a continuous domain of the variation of a coordinate x is divided into intervals of width δ_x. A continuous volume $\mathcal{V}$ is divided into discrete cells of sides δ_x, δ_y, and δ_z in the directions Ox, Oy, and Oz and the continuous points $M(x, y, z)$ of $\mathcal{V}$ are replaced by the discrete nodes $M_{i,j,k}$ of a lattice. To integrate a function $f(x)$ over the interval $a < x < b$, we divide this interval into N parts by the points $x_o \equiv a, x_1, x_2, \ldots, x_{N-1}, x_N \equiv b$ and we replace the continuous integral by a discrete sum

$$\int_a^b dx\, f(x) = \frac{b-a}{N}\,[\tfrac{1}{2} f(a) + \textstyle\sum_{n=1}^{n=N-1} f(x_n) + \tfrac{1}{2} f(b)]. \qquad [5.90]$$

Very often, the problems encountered in the applications of electromagnetism have no simple analytic solution. Numerical methods can be used to solve the electrostatic problems, that is, to determine the potential obeying Poisson's equation $\Delta V = -q_v/\varepsilon$ and verifying some boundary conditions. The numerical method replaces the partial differential equation by a set of algebraic equations relating the discrete potentials $V_{i,j,k}$ to the nodes of the lattice.

The method of *finite differences* allows the expression of the derivative in terms of the difference of the function at the nodes. Let us consider a function of a single variable $f(x)$ taking the discrete values $f_i \equiv f(x_i)$ and a function of two variables $V(x, y)$ taking the discrete values $V_{i,j} \equiv V(x_i, y_j)$. We define the successive derivatives by the symmetric expressions

$$\begin{aligned}
&\partial f/\partial x\,|_i = (f_{i+1} - f_{i-1})/2\delta_x, && \partial^2 f/\partial x^2\,|_i = (f_{i+1} - 2f_i + f_{i-1})/\delta_x^2,\\
&\partial V/\partial x\,|_{i,j} = (V_{i+1,j} - V_{i-1,j})/2\delta_x, && \partial V/\partial y\,|_{i,j} = (V_{i,j+1} - V_{i,j-1})/2\delta_y,\\
&\partial^2 V/\partial x^2\,|_{i,j} = (V_{i+1,j} - 2V_{i,j} + V_{i-1,j})/\delta_x^2, && \partial^2 V/\partial y^2\,|_{i,j} = (V_{i,j+1} - 2V_{i,j} + V_{i,j-1})/\delta_y^2,\\
&\partial^2 V/\partial x\partial y\,|_{i,j} = (V_{i+1,j+1} + V_{i-1,j-1} - V_{i+1,j-1} - V_{i-1,j+1})/4\delta_y\,\delta_y, \text{ etc.}
\end{aligned} \qquad [5.91]$$

The two-dimensional Poisson's equation $\partial^2_{xx} V + \partial^2_{yy} V = \eta \equiv -q_v/\varepsilon$ becomes

$$(V_{i+1,j} - 2V_{i,j} + V_{i-1,j})/\delta_x^2 + (V_{i,j+1} - 2V_{i,j} + V_{i,j-1})/\delta_y^2 = \eta_{ij}. \qquad [5.92]$$

In particular, if $\delta_x = \delta_y \equiv \delta$, we find a set of linear algebraic equations

$$V_{i+1,j} + V_{i-1,j} + V_{i,j+1} + V_{i,j-1} - 4V_{i,j} = \eta_{ij}\,\delta^2. \qquad [5.93]$$

If a node $M_{i,j,k}$ is situated on the boundary surface $\mathcal{S}_o$ whose potential is V_o, the Dirichlet boundary condition is imposed simply by setting $V_{i,j,k} = V_o$. The Neumann's condition $E_\perp = -\partial V/\partial x_\perp = q_s/\varepsilon$ on $\mathcal{S}_o$ is more complicated. To simplify, we consider the one-dimensional case with a node x_1 on $\mathcal{S}_o$. To calculate the derivative of V at x_1 by using equations [5.91], we must assume that a fictive node x' exists on the other side of $\mathcal{S}_o$ with a potential V' given by $(V' - V_2) = (q_s/\varepsilon)_1$ (Figure 5.7). Knowing V', we may calculate the second derivative at x_1 and write the discrete Poisson's equation at the points situated on the boundaries.

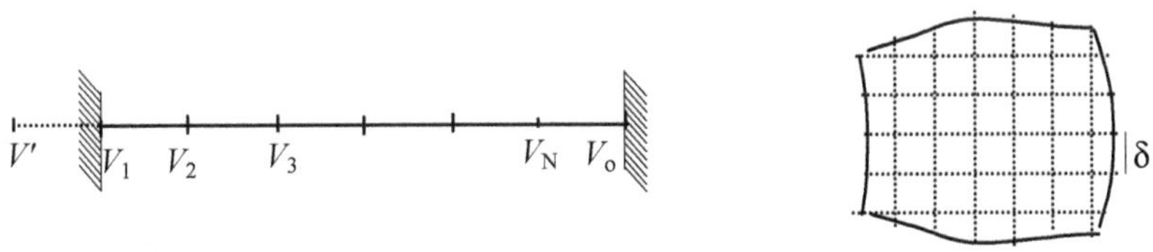

Figure 5.7. *Making V discrete on a segment and on a surface*

The linear equations [5.92] enable the determination of the potential $V_{i,j,k}$ at all the nodes. We may solve them by using appropriate computer programs. It is also possible to use the so-called *relaxation method.* It consists in writing equation [5.93] as

$$V_{i,j} = \overline{V}_{ij} - \tfrac{1}{4}\eta_{ij}\,\delta^2, \qquad \text{where} \quad \overline{V}_{ij} = \tfrac{1}{4}(V_{i+1,j} + V_{i-1,j} + V_{i,j+1} + V_{i,j-1}). \quad [5.94]$$

We start by taking some reasonable values $V_{i,j}^{(1)}$ (that we call first approximation), use [5.94] to calculate the second approximation $V_{i,j}^{(2)} = \overline{V}_{ij}^{(1)} - \tfrac{1}{4}\eta_{ij}\,\delta^2$, then use these values and equation [5.94] to calculate the third approximation and so on, obtaining, by iteration, $V_{i,j}^{(k+1)} = \overline{V}_{ij}^{(k)} - \tfrac{1}{4}\eta_{ij}$ until the relative difference between successive iterations $[V_{i,j}^{(k+1)} - \overline{V}_{ij}^{(k)}]/V_{i,j}^{(k)}$ becomes less than a certain value, for instance, 10^{-3}.

5.10. Problems

P5.1 Consider a volume $\mathcal{V}$ containing free charges and dielectrics (assumed linear and isotropic for simplicity) and bounded by surfaces $\mathcal{S}_i$ and an enclosure $\mathcal{S}_o$ (eventually at infinity). To show the unicity of the solution, assume that there are two solutions V and V'. We set $\delta V = V - V'$, $\delta\mathbf{E} = \mathbf{E} - \mathbf{E}'$ and $\delta\mathbf{D} = \mathbf{D} - \mathbf{D}'$ the corresponding electric field and displacement. **a)** What are the boundary conditions of δV on the conductors whose potential is given and on the conductors whose total charge is given? **b)** Consider the vector identity $\boldsymbol{\nabla}(\delta V\,\delta\mathbf{D}) = \delta V\,(\boldsymbol{\nabla}.\delta\mathbf{D}) + (\delta\mathbf{D}.\boldsymbol{\nabla}\delta V)$. Show Gauss's equation $\boldsymbol{\nabla}.\delta\mathbf{D} = 0$, thus $\boldsymbol{\nabla}(\delta V\,\delta\mathbf{D}) = (\delta\mathbf{D}.\boldsymbol{\nabla}\delta V)$. Consider the integral of both sides of the identity on $\mathcal{V}$. Using Gauss-Ostrogradsky's theorem, show that

$$\iiint_{\mathcal{V}} d\mathcal{V}\, \boldsymbol{\nabla}(\delta V\,\delta\mathbf{D}) = \iint_{\mathcal{S}_i} d\mathcal{S}'\ \mathbf{n}.\delta\mathbf{D}(\mathbf{r}')\,\delta V(\mathbf{r}') = 0,$$

where the $\mathcal{S}_i$ stands for all boundary surfaces including the enclosure. Deduce that

$$\iiint_{\mathcal{V}} d\mathcal{V}\ (\delta\mathbf{D}.\boldsymbol{\nabla}\delta V) = -\iiint_{\mathcal{V}} d\mathcal{V}\,\varepsilon\,\delta\mathbf{E}^2 = 0,$$

which is impossible unless $\delta\mathbf{E} = 0$, thus $\mathbf{E} = \mathbf{E}'$, $\mathbf{D} = \mathbf{D}'$ and $V = V'$.

Method of images

P5.2 A horizontal line is formed by a long cylindrical conductor of radius R and it lies at height h from the ground ($h >> R$). Calculate the difference in potential between the ground and this conductor. Deduce the capacitance per unit length of this line. What is the capacitance of a line formed par two conductors separated by a distance d if they lie in a horizontal plane and if they lie in a vertical plane?

Method of analytic functions

P5.3 Consider the logarithmic function $f(z) = a \ln z + b$. Using the exponential representation $z = \rho e^{i\varphi}$, show that $f(z) = V + iU = (a \ln \rho + b) + i\varphi$. If the lines $V = C_i$ are the equipotential lines, the curves $\varphi = C'_j$ are the lines of field (where C_i and C'_j are constants). This is the case of a symmetric field about Oz. Consider a cylindrical conductor of radius r and zero potential surrounded by a cylindrical shell of radius R and potential V_o. Calculate the potential and the field.

Laplace's equation in Cartesian coordinates

P5.4 Use Laplace's equation and the method of separation of variables to analyze the potential and the field of a parallel plate capacitor if one of its plates has zero potential and the other a potential V_o.

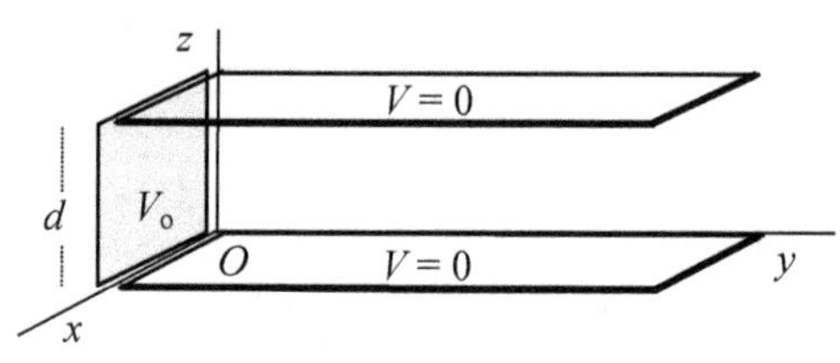

Figure 5.8. *Problem 5.5*

P5.5 Determine the potential and the field in a region $\mathcal{V}$ bounded by two plane plates parallel to Oxy and an electrode in the Oxz plane (Figure 5.8). The plates have large dimensions and are separated by a distance d. They have zero potential and the electrode has a potential V_o. Is this set up possible?

Laplace's equation in spherical coordinates

P5.6 Consider a linear charge of uniform density q_L between the points of coordinates $-a$ and $+a$ of the z axis. Using Legendre polynomials, show that the potential at large distance may be written in the form

$$V = K_o q_L (2a/r)\,[P_o(\cos\theta) + (a^2/3r^2)\,P_2(\cos\theta) + (a^4/5r^4)\,P_4(\cos\theta) + \ldots$$

P5.7 A dielectric ball of radius R and permittivity ε is placed in an initially uniform field $\mathbf{E}_o = E_o\mathbf{e}_z$. Using the method of separation of variables, write the general solution of Laplace's equation. Impose the continuity conditions on the ball and deduce that the external and the internal potentials are given by

$$V^{(ex)} = A_o - E_o z[1 - (\varepsilon - \varepsilon_o)R^3/(\varepsilon + 2\varepsilon_o)r^3\,] \quad \text{and} \quad V^{(in)} = A_o - E_o z[3\varepsilon_o/(\varepsilon + 2\varepsilon_o].$$

Deduce that the field is uniform inside the ball and that the polarization is $\mathbf{P} = [3\varepsilon_o(\varepsilon - \varepsilon_o)/(\varepsilon + 2\varepsilon_o)]\mathbf{E}_o$, while the external field is the superposition of the field $\mathbf{E}_o$ and that of an electric dipole $\mathbf{P}\mathcal{V}$ where $\mathcal{V}$ is the volume of the sphere.

Chapter 6

Magnetic Field in Vacuum

The Earth's magnetic field and the magnetism of some natural ores or iron rods that have been stroked by a magnet, have been known in the Middle East and China since antiquity. In 1821 Oersted discovered that an electric current produces a magnetic field. This effect was studied by Ampère, Biot, Savart, and others. Ampère assumed that permanent magnetism is due to microscopic currents in matter; this idea is retained in modern physics. Conversely, Faraday discovered in 1831 that a variable magnetic field induces an electric current in circuits. In 1888, Maxwell unified electricity and magnetism in a single theory, called *electromagnetism*. Currently, magnetism has many technological applications: magnets and electromagnets are used in generators and motors, instruments, computers, telecommunications, etc.

In this chapter, we introduce the concept of magnetic field and we study its action on magnetic currents. Then we study the creation of magnetic fields by moving charges and currents, magnetic energy and the interactions of circuits.

6.1. Force exerted by a magnetic field on a moving charge

The magnetic field is defined by its action on a charged particle in motion (Figure 6.1a in the case of positive charge). The experiment shows that this force is given by:

$$\mathbf{F}_\mathrm{M} = q\,\mathbf{v} \times \mathbf{B}. \qquad [6.1]$$

B is the *magnetic field* or, more precisely, the *magnetic induction field.* The magnetic force $\mathbf{F}_M$ vanishes if the particle is at rest or if its velocity is oriented in the direction of the field **B**. The SI unit of magnetic field is the kg/s^2. A called *tesla* (T).

The magnetic force always being orthogonal to the charged particle velocity, the work of this force $dW = \mathbf{F}_M.d\mathbf{r} = \mathbf{F}_M.\mathbf{v}\,dt$ is equal to zero. If the particle is subject to no other forces, its kinetic energy remains constant. Thus, its speed remains constant but the direction of its velocity changes. Conversely, to displace a charged particle in a field **B** without modification of its kinetic energy, an external agent must exert a force $\mathbf{F}' = -\mathbf{F}_M = -q(\mathbf{v} \times \mathbf{B})$, but no work is required for this displacement. Thus, it is not possible to define a potential energy of the particle in the field **B**. In other words, this force is not conservative, contrarily to the electric force.

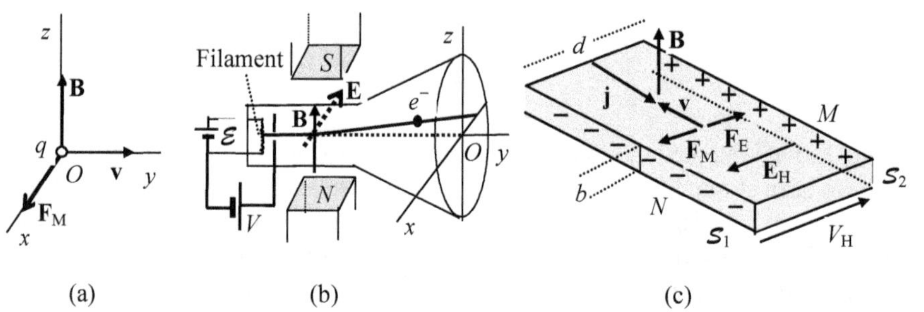

Figure 6.1. *a) Action of a magnetic field on a positive charge. b) Thomson experiment. c) Hall effect in the case of negative charge carrier*

If a particle of charge q moves with a velocity **v** in both an electric field **E** and a magnetic field **B**, it is subject to the *Lorentz force*,

$$\mathbf{F} = q\,(\mathbf{E} + \mathbf{v} \times \mathbf{B}). \qquad [6.2]$$

In a famous experiment in 1897, Thomson observed the action of a known magnetic field **B** and an adjustable orthogonal electric field **E** on a focalized cathode ray (Figure 6.1b). As the beam was deviated by either field acting separately, it is formed by charged particles. Turning on both fields, the beam suffers no deviation if $\mathbf{E} + \mathbf{v} \times \mathbf{B} = 0$, thus $v = E/B = \sqrt{2eV/m}$, where V is the accelerating potential of the particles. This experiment allowed Thomson to determine the ratio e/m for electrons.

The Lorentz force manifests itself in the Hall effect. In a famous experiment in 1879, Hall showed (before the discovery of the electron) that the charge carriers in metallic conductors are negatively charged particles. Figure 6.1c illustrates the motion of charges in a conducting strip of width d and thickness b, carrying a current I and placed in a magnetic field B orthogonal to the strip. If the charges are negative, the magnetic force pushes the charges and makes them accumulate on the lateral face $\mathcal{S}_1$ of the strip, leaving the other face $\mathcal{S}_2$ positively charged. These surface charges produce a Hall electric field $\mathbf{E}_H$ and the flow of charge becomes stationary if $\mathbf{E}_H + \mathbf{v} \times \mathbf{B} = 0$. The surface $\mathcal{S}_2$ is then at a higher potential than $\mathcal{S}_1$. This is what Hall observed. In the case of positive charge carriers, the directions of $\mathbf{F}_E$, $\mathbf{F}_B$ and $\mathbf{E}_H$ would be reversed and $\mathcal{S}_2$ would be at a lower potential than $\mathcal{S}_1$. Actually, we know that the charge carriers in metals are the free electrons of charge $-e$. The Hall field $\mathbf{E}_H = -\mathbf{v} \times \mathbf{B}$ produces a measurable Hall potential $V_H = d|\mathbf{v}{\times}\mathbf{B}| = vBd$. The current density is $j = I/bd = eN_v v$, where N_v is the number of free electrons per unit volume. A measurement of V_H enables us to determine v, and hence, the number of conduction electrons per unit volume $N_v = IB/beV_H$. For instance, in the case of a strip of silver with $d = 1$ cm, $b = 0.1$ mm, carrying an intensity $I = 10$ A in a field $B = 1$ T, the Hall potential is of the order of 10 μV. This corresponds to $N_v = 6 \times 10^{22}$ electrons/cm^3 (about 1 free electron per atom). In the case of polyvalent metals and magnetic metals (iron, nickel, etc.) and in the case of semiconductors, it is not possible to give a simple interpretation of the Hall effect by using a simple classical model with electrons as charge carriers. Quantum models are required to give reasonable agreement with experiment.

The Hall effect is an important means of investigation of the properties of solid conductors and semiconductors. In the case of semiconductors, we have a superposition of the Hall effect of electrons and that of positive *holes*. The number of charge carriers is much smaller than in metals; thus, the Hall effect is more important (although it is somehow attenuated by the weakness of the current density). The Hall potential V_H being proportional to the field B, the Hall effect in a strip of semiconductor may be used to measure the magnetic field.

6.2. Force exerted by a magnetic field on a current, Laplace's force

The force that a field $\mathbf{B}$ exerts on the conduction charges $\mathbf{F}_M = \Sigma_i\, q_i(\mathbf{v}_i \times \mathbf{B})$ is transmitted to the conductor if it is rigid. To simplify, we assume that the current is due to the displacement of N_v particles per unit volume, of charge q and average velocity $\mathbf{v}$. An element of volume $d\mathcal{V}$ of the conductor contains $N_v d\mathcal{V}$ particles. Thus, it is subject to a force $d\mathbf{F}_M = q\, N_v d\mathcal{V}\, \mathbf{v} \times \mathbf{B} = (\mathbf{j} \times \mathbf{B})\, d\mathcal{V}$, where $\mathbf{j} = qN_v\mathbf{v}$ is the current density (see section 3.6). The force that acts on the unit volume of the conductor is

$$\mathbf{F}_{M,v} = \mathbf{j} \times \mathbf{B}. \qquad [6.3]$$

Consider an element $d\mathbf{L}$ of a thin conductor of section $\mathcal{S}$ and volume $d\mathcal{V} = \mathcal{S}\, dL$ (Figure 6.2a). The force exerted by the field $\mathbf{B}$ on $d\mathbf{L}$ is $d\mathbf{F}_M = d\mathcal{V}\, \mathbf{F}_{M,v} = dL\mathcal{S}(\mathbf{j} \times \mathbf{B})$.

As **j** is oriented in the direction of $d\mathbf{L}$, we may write $dL\ \mathbf{j} = j\ d\mathbf{L}$. The intensity being $I = \mathcal{S}j$, the force may be written as

$$d\mathbf{F}_M = I\ d\mathbf{L} \times \mathbf{B}. \quad [6.4]$$

This is *Laplace's law* for the force on $d\mathbf{L}$. This force is orthogonal to $d\mathbf{L}$ and **B**.

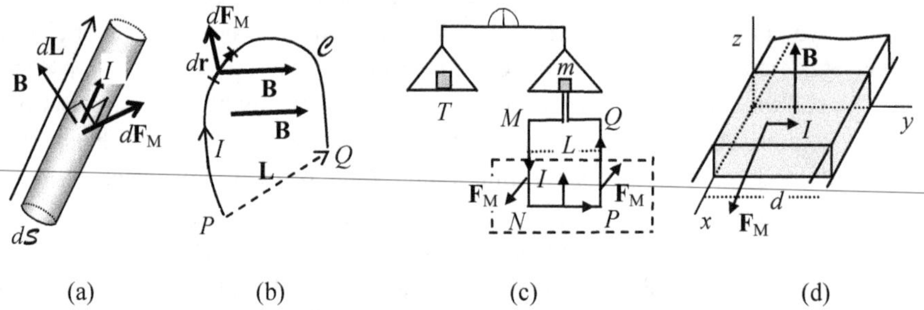

Figure 6.2. *Force exerted by a field* **B***: a) on an element* $d\mathbf{L}$ *of a circuit, and b) on a finite circuit. c) Absolute measurement of* **B** *using a Cotton's balance. d) Electromagnetic pump*

The resultant force exerted by a field **B** on a finite circuit $\mathcal{C}$ is the integral of Laplace's force over all the elements $d\mathbf{r}$ of the circuit, we find:

$$\mathbf{F}_M = \int_{\mathcal{C}} d\mathbf{F}_M = I \int_{\mathcal{C}} d\mathbf{r} \times \mathbf{B}, \quad [6.5]$$

where **B** is the field acting on $d\mathbf{r}$ at each point of $\mathcal{C}$. If **B** is uniform, we may write

$$\mathbf{F}_M = I\ \{\int_{\mathcal{C}} d\mathbf{r}\} \times \mathbf{B} = I\,(\mathbf{L} \times \mathbf{B}), \quad [6.6]$$

where $\mathbf{L} = \int_{\mathcal{C}} d\mathbf{r} = \lim \Sigma_i\, d\mathbf{r}_i$ is the space vector that joins the origin P of $\mathcal{C}$ to its end Q (Figure 6.2b). This relation shows that $\mathbf{F}_M$ does not depend on the shape of the circuit between P and Q; it is the same as the force acting on a rectilinear circuit joining P to Q. If a circuit $\mathcal{C}$ is closed and placed in a uniform field, $\int_{\mathcal{C}} d\mathbf{r} = \lim \Sigma_i\, d\mathbf{r}_i = 0$ and $\mathbf{F}_M = 0$. Thus, a uniform **B** field exerts no resultant force on a closed circuit.

Cotton's balance (Figure 6.2c) provides an absolute measurement of a magnetic field **B** by measuring the force that it exerts on a circuit $MNPQ$ of n turns transporting a known current I. This force is measured by using a balance. The forces exerted on the vertical parts MN and PQ cancel and the force acting on the part NP, equal to $nLIB$, is counterbalanced by the weight mg. We deduce that $B = mg/nLI$.

The electromagnetic pump (Figure 6.2d) is another application of the magnetic force on currents. Consider a conducting liquid flowing in a rectangular pipe whose lateral faces are two electrodes, between them an electric current is established in the liquid in the direction *Oy* over a distance *d*. A magnetic field **B** pointing in the direction *Oz* acts on the liquid with a force *IBd*. This system may be used in a nuclear reactor to pump liquid sodium that is used to transfer the heat generated in the core of the reactor.

6.3. Magnetic flux and vector potential

The flux of **B** through a surface $\mathcal{S}$ is

$$\Phi = \iint_{\mathcal{S}} d\mathcal{S}\, \mathbf{n}.\mathbf{B}. \qquad [6.7]$$

The magnetic flux plays an important part in the analysis of magnetic forces and energy and in the phenomena of induction. One of the important properties of the magnetic flux is that it is *conservative*, i.e. the flux through a closed surface is zero

$$\iint_{\mathcal{S}} d\mathcal{S}\, \mathbf{n}.\mathbf{B} = 0. \qquad [6.8]$$

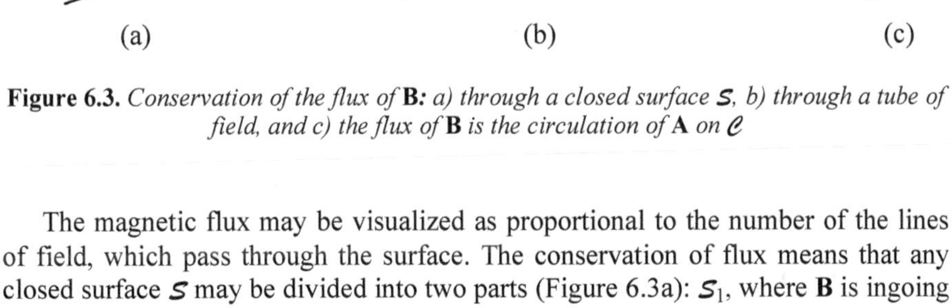

Figure 6.3. *Conservation of the flux of* **B***: a) through a closed surface $\mathcal{S}$, b) through a tube of field, and c) the flux of* **B** *is the circulation of* **A** *on $\mathcal{C}$*

The magnetic flux may be visualized as proportional to the number of the lines of field, which pass through the surface. The conservation of flux means that any closed surface $\mathcal{S}$ may be divided into two parts (Figure 6.3a): $\mathcal{S}_1$, where **B** is ingoing (thus $\Phi_1 < 0$) and $\mathcal{S}_2$, where **B** is outgoing (thus $\Phi_2 > 0$). We must have $\Phi_1 + \Phi_2 = 0$. In other words, each field line entering $\mathcal{S}$, leaves it. Contrary to the electric field, the magnetic field cannot diverge from "positive magnetic charges" or converge toward "negative magnetic charges". In other words, there are no magnetic charges. If we consider a field tube, i.e. having its lateral surface tangent to the field, and ending by two sections $\mathcal{S}_1$ and $\mathcal{S}_2$ normal to **B** (Figure 6.3b). The magnetic fluxes Φ_1 and Φ_2

through these sections are positive and the flux through the lateral surface is equal to zero. As Φ_1 is inward while Φ_2 is outward, the total outward flux through the tube is $\Phi = \Phi_2 - \Phi_1$ and this must be zero. We deduce that $B_1\mathcal{S}_1 = B_2\mathcal{S}$; thus, if $\mathcal{S}_2 < \mathcal{S}_1$, we must have $B_2 > B_1$.

Using Gauss-Ostrogradsky's theorem, we may transform the flux through a closed surface into the integral of the divergence of **B** over the enclosed volume $\mathcal{V}$, thus $\iiint_{\mathcal{V}} d\mathcal{V}\, \boldsymbol{\nabla}.\mathbf{B} = 0$ for any $\mathcal{V}$ and we have the equation:

$$\boldsymbol{\nabla}.\mathbf{B} = 0. \qquad [6.9]$$

Similar to Gauss's law, this is a fundamental equation of electromagnetism, which remains valid even in the case of time-dependent phenomena. The analogy with Gauss's law indicates that there are no magnetic charges. This fact is confirmed experimentally.

The SI unit of magnetic flux is the kg.m^2/s^2.A or N.m/A called *weber* (Wb) and the SI unit of magnetic field is the Wb/m^2, called also *tesla*. The gauss (1 G = 10^{-4} T) is another unit of magnetic field that is frequently used. The Earth's field is about 0.5 G, the fields produced by electric circuits are of the order of the gauss and it may reach 10 to 20 kG near the poles of an iron-filled electromagnet and 100 kG for a superconducting magnet.

Equation $\boldsymbol{\nabla}.\mathbf{B} = 0$ implies that $\mathbf{B}(\mathbf{r})$ may be written as the curl of a *vector potential* $\mathbf{A}(\mathbf{r})$ (see section A.7 of the appendix A)

$$\mathbf{B} = \boldsymbol{\nabla} \times \mathbf{A}, \qquad [6.10]$$

It should be noted that the gradient of an arbitrary function f may be added to **A**

$$\mathbf{A}(\mathbf{r}) \rightarrow \mathbf{A}'(\mathbf{r}) = \mathbf{A}(\mathbf{r}) + \boldsymbol{\nabla} f(\mathbf{r}) \qquad [6.11]$$

without changing the field **B**, because of the identity $\boldsymbol{\nabla}\times(\boldsymbol{\nabla} f) = 0$. The transformation [6.11] is called a *gauge transformation*. It is always possible to find a gauge function f such that **A** has zero divergence in the case of static phenomena

$$\boldsymbol{\nabla}.\mathbf{A} = 0. \qquad [6.12]$$

Using Stokes' theorem, it is possible to express the flux of the field **B** through an open surface $\mathcal{S}$ as the circulation of **A** on the contour $\mathcal{C}$ bounding $\mathcal{S}$ (Figure 6.3c)

$$\Phi = \iint_{\mathcal{S}} d\mathcal{S}\,(\mathbf{n}.\mathbf{B}) = \iint_{\mathcal{S}} d\mathcal{S}\,\mathbf{n}.(\boldsymbol{\nabla} \times \mathbf{A}) = \int_{\mathcal{C}} d\mathbf{r}.\mathbf{A}. \qquad [6.13]$$

This result explains why the flux of **B** depends only on $\mathcal{C}$ and not on the surface $\mathcal{S}$.

6.4. Magnetic field of particles and currents, Biot-Savart's law

A particle of charge q, position $\mathbf{r}'$, and velocity $\mathbf{v}$ (Figure 6.4a) produces at each point $\mathbf{r}$ a magnetic field

$$\mathbf{B}(\mathbf{r}) = \frac{\mu_o}{4\pi}\frac{q(\mathbf{v}\times\mathbf{R})}{R^3}, \qquad \text{where } \mathbf{R} = \mathbf{r} - \mathbf{r}'. \qquad [6.14]$$

$\mathbf{R}$ is the relative position of the point, where $\mathbf{B}$ is evaluated, measured from the position of the particle. It may be shown that the corresponding vector potential is

$$\mathbf{A}(\mathbf{r}) = \frac{\mu_o}{4\pi}\frac{q\mathbf{v}}{R}. \qquad [6.15]$$

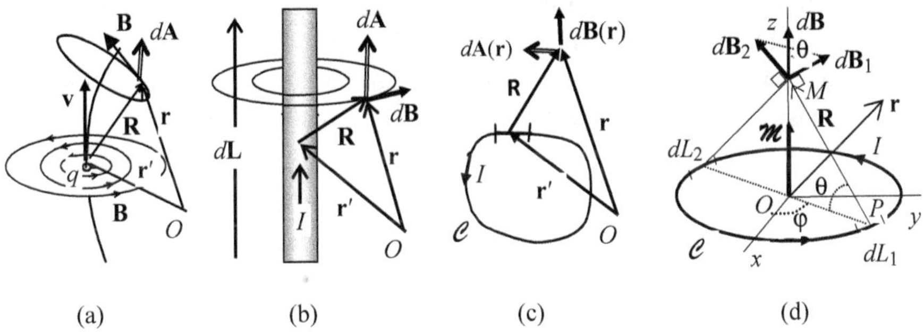

Figure 6.4. *Magnetic field and vector potential a) of a charged particle of velocity* **v**, *b) of a circuit element d***L** *(Biot-Savart's law), c) of a finite thin circuit, and d) of a circular loop*

The expression [6.14], which is assumed here without proof, is verified by all its consequences. We note that it is postulated for a particle of constant velocity. This is the case for charge carriers in a conductor if the current is constant but not for a free particle (as it emits radiation and hence energy, momentum, etc.). The constant μ_o is the *magnetic permeability of vacuum*. Its numerical SI value is:

$$\mu_o = 4\pi \times 10^{-7}\,\text{kg.m.A}^{-2}.\text{s}^{-2}. \qquad [6.16]$$

As we shall see later, electromagnetic theory predicts that the permittivity of vacuum ε_o and its magnetic permeability μ_o are related to the speed of light in vacuum c by the relation $\varepsilon_o\mu_o = 1/c^2$. The SI unit of intensity (the *ampere*) is defined in relation to the magnetic interaction of two conductors (see section 6.11C), which is proportional to μ_o and the ampere is chosen so that μ_o is given by [6.16]. As the meter is actually defined so that c is exactly 299 792 458 m/s, the result is that $\varepsilon_o = 1/c^2\mu_o = 8.854\ 187\ 82 \times 10^{-12}\ \text{A}^2.\text{s}^4/\text{m}^3\,\text{kg}$.

The magnetic field obeys the superposition principle: the field and the vector potential produced by several systems (1), (2) … are the vector sums $\Sigma_i \mathbf{B}_i$ and $\Sigma_i \mathbf{A}_i$ of the fields and the vector potentials of the individual systems. For instance, the field and vector potential produced at **r** by the particles of charges q_i at points $\mathbf{r}_i$ are

$$\mathbf{B}(\mathbf{r}) = \frac{\mu_o}{4\pi}\Sigma_i \frac{q_i(\mathbf{v}_i \times \mathbf{R}_i)}{R_i^3}, \qquad \mathbf{A}(\mathbf{r}) = \frac{\mu_o}{4\pi}\Sigma_i \frac{q_i \mathbf{v}_i}{R_i^3}, \text{ where } \mathbf{R}_i = \mathbf{r} - \mathbf{r}_i. \qquad [6.17]$$

Consider now an element of a thin circuit of length $d\mathbf{L}$ and section $\mathcal{S}$ carrying a current I (Figure 6.4b). To simplify, we assume that the charge carriers have equal charge q and velocity **v**. The volume of the element being $d\mathcal{V} = \mathcal{S}\, dL$, the number of conduction charges that it contains is $dN = N_v\, \mathcal{S}\, dL$, where N_v is the number of charges per unit volume. If dL is small compared to the distance R to the point **r**, where the field is evaluated, the distances $\mathbf{R}_i$ from the charges to the point **r** and hence their fields $\mathbf{B}_i(\mathbf{r})$ are equal. Thus, the field produced by the element dL at **r** is $d\mathbf{B}(\mathbf{r}) = (\mu_o/4\pi)\, qN_v\mathcal{S}\, dL\, (\mathbf{v} \times \mathbf{R})/R^3$. Noting that the current density is $\mathbf{j} = N_v q\mathbf{v}$, the current intensity is $I = j\mathcal{S}$ and $\mathbf{j}\, dL = j\, d\mathbf{L}$ (as **j** is the direction of $d\mathbf{L}$), we get

$$d\mathbf{B}(\mathbf{r}) = \frac{\mu_o}{4\pi} I \frac{d\mathbf{L} \times \mathbf{R}}{R^3}, \qquad \mathbf{R} \equiv \mathbf{r} - \mathbf{r}'. \qquad [6.18]$$

This result is known as *Biot-Savart's* law. It expresses the field produced at **r** by the element $d\mathbf{L}$ in terms of macroscopic quantities (making no reference to the conduction particles). This elementary field decreases as $1/R^2$. It may be used to evaluate the field of a finite circuit $\mathcal{C}$ by integration. As I is the same at any point of the circuit, we find the field and the vector potential (Figure 6.4c).

$$\mathbf{B}(\mathbf{r}) = \int_{\mathcal{C}} d\mathbf{B} = \frac{\mu_o}{4\pi} I \int_{\mathcal{C}} \frac{d\mathbf{r}' \times \mathbf{R}}{R^3}, \qquad \mathbf{A}(\mathbf{r}) = \frac{\mu_o I}{4\pi} \int_{\mathcal{C}} \frac{d\mathbf{r}'}{R} \qquad \mathbf{R} \equiv \mathbf{r} - \mathbf{r}'. \qquad [6.19]$$

In the case of currents distributed in a volume $\mathcal{V}$ with a volume current density $\mathbf{j}(\mathbf{r}')$ or on a surface $\mathcal{S}$ with a current density $\mathbf{j}_s(\mathbf{r}')$, we decompose $\mathcal{V}$ and $\mathcal{S}$ into elements of volume $d\mathcal{V}'$ or of area $d\mathcal{S}'$ and we get by integration

$$\mathbf{B}(\mathbf{r}) = \frac{\mu_o}{4\pi} \iiint_{\mathcal{V}} d\mathcal{V}' \frac{\mathbf{j}(\mathbf{r}') \times \mathbf{R}}{R^3}, \qquad \mathbf{A}(\mathbf{r}) = \frac{\mu_o}{4\pi} \iiint_{\mathcal{V}} d\mathcal{V}' \frac{\mathbf{j}(\mathbf{r}')}{R},$$

$$\mathbf{B}(\mathbf{r}) = \frac{\mu_o}{4\pi} \iint_{\mathcal{S}} d\mathcal{S}' \frac{\mathbf{j}_s(\mathbf{r}') \times \mathbf{R}}{R^3}, \qquad \mathbf{A}(\mathbf{r}) = \frac{\mu_o}{4\pi} \iint_{\mathcal{S}} d\mathcal{S} \frac{\mathbf{j}_s(\mathbf{r}')}{R}. \qquad [6.20]$$

We note that a surface current density $\mathbf{j}_s(\mathbf{r}')$ may be considered as a volume current density $\mathbf{j}(\mathbf{r}') = \mathbf{j}_s(\mathbf{r}')\ \delta(z' - z_n)$, where z_n is the normal coordinate to the surface, and a line current I parallel to Oz corresponds to $\mathbf{j}_v(\mathbf{r}') = I\,\delta(x'-x_n)\delta(y'-y_n)\mathbf{e}_z$, where x_n and y_n are the x and y coordinates to the current carrying line.

As an application, consider a circular loop $\mathcal{C}$ of radius a and current I, which we take in the plane Oxy with its center at O (Figure 6.4d). We evaluate the field $\mathbf{B}$ at a point M on the axis Oz at a distance $OM = z$. Using Biot-Savart's law [6.18], an element $d\mathbf{L}$ at P produces at M a field $d\mathbf{B}$ that is orthogonal to $d\mathbf{L}$ and $\mathbf{R} \equiv \overrightarrow{PM}$. Thus $d\mathbf{B}$ is in the azimuthal plane OMP and its magnitude is $dB = (\mu_o I/4\pi)(dL/R^2)$. The elements $d\mathbf{L}$ of the loop being at the same distance $R = (a^2 + z^2)^{½}$ to M, two elements $d\mathbf{L}_1$ and $d\mathbf{L}_2$ symmetric with respect to O produce two fields $d\mathbf{B}_1$ and $d\mathbf{B}_2$ that are symmetric with respect to Oz. Their components perpendicular to Oz cancel out. Adding their components along Oz, we get

$$\mathbf{B} = \mathbf{e}_z \int_{\mathcal{C}} dB_z = \mathbf{e}_z \int_{\mathcal{C}} dB \cos\theta = \mathbf{e}_z \int_{\mathcal{C}} dB\,\frac{a}{R} = \frac{\mu_o I a}{4\pi R^3}\mathbf{e}_z \int_{\mathcal{C}} dL = \frac{\mu_o I a^2}{2R^3}\,\mathbf{e}_z. \quad [6.21]$$

6.5. Magnetic moment

A) Moment of the magnetic forces on a circuit

A magnetic field $\mathbf{B}$ may exert a moment of force on a closed circuit carrying a current I. This moment may provoke a rotation of the circuit. Consider for instance a rectangular circuit $MNPQ$ free to rotate about the axis Oz that joins the mid points of MP and NQ (Figure 6.5a). The forces $\mathbf{F}_1$ and $\mathbf{F}_2$ that a uniform field $\mathbf{B}$ parallel to Ox exerts on the sides MQ and PN are opposite and oriented in the direction of Oz. Thus, they produce no moment with respect to O. The sides MN and PQ of length L are orthogonal to $\mathbf{B}$. They are subject to opposite forces $\mathbf{F}_3 = -ILB\,\mathbf{e}_y$ and $\mathbf{F}_4 = ILB\,\mathbf{e}_y$. Let $\mathbf{n}$ be the unit vector normal to the circuit and oriented according to the right-hand rule and let θ be the angle that $\mathbf{n}$ forms with $\mathbf{B}$ measured algebraically about Oz. The total moment of the magnetic forces with respect to O is the vector sum of the moments of $\mathbf{F}_3$ and $\mathbf{F}_4$:

$$\boldsymbol{\Gamma}_M = -\,LL'IB\sin\theta\,\mathbf{e}_z = -\,\mathcal{S}IB\sin\theta\,\mathbf{e}_z, \qquad \text{(uniform field)} \qquad [6.22]$$

where $\mathcal{S} = LL'$ is the area of the circuit. We define the *magnetic moment* of the circuit as the vector

$$\mathcal{M} = I\mathcal{S}\,\mathbf{n}. \qquad [6.23]$$

It has a magnitude $I\mathcal{S}$ and it is normal to the circuit and oriented according to the right-hand rule with the circuit oriented in the direction of the current. Thus, the moment of the magnetic forces may be written as

$$\boldsymbol{\Gamma}_{\mathrm{M}} = \boldsymbol{\mathcal{M}} \times \mathbf{B} \qquad \text{(uniform field).} \qquad [6.24]$$

This expression is similar to the moment of the electric forces $\boldsymbol{\Gamma}_{\mathrm{E}} = \mathbf{p} \times \mathbf{E}$ exerted by an electric field on an electric dipole moment $\mathbf{p}$. It is valid if the dimensions of the circuit are small enough for the field $\mathbf{B}$ to be considered as constant.

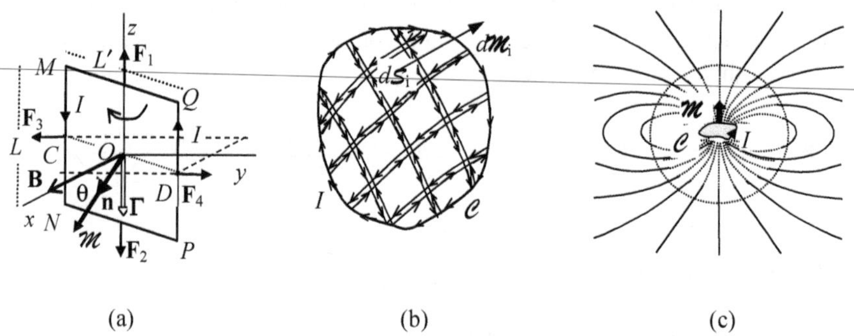

Figure 6.5. *a) Moment of the magnetic forces exerted by a uniform field* **B** *on a rectangular circuit, and b) magnetic moment of a circuit, and c) field lines of a magnetic dipole*

A large circuit $\mathcal{C}$ may be considered as a juxtaposition of infinitesimal small circuits $\mathcal{C}_i$ as in Figure 6.5b. The coinciding sides carry opposite currents and the magnetic force on them cancel; thus, we are left with the magnetic force on the circuit $\mathcal{C}$. Also, the magnetic moment $\boldsymbol{\Gamma}_{\mathrm{M}}$ on $\mathcal{C}$ is the vector sum of the magnetic moments $\Sigma_i\ I\ d\mathcal{S}_i\ \mathbf{n}_i{\times}\mathbf{B}_i$ on these infinitesimal circuits and, in the limit of infinitesimal $d\mathcal{S}_i$, it may be written as the integral

$$\boldsymbol{\Gamma}_{\mathrm{M}} = I \iint_{\mathcal{S}} d\mathcal{S}'\mathbf{n}(\mathbf{r}') \times \mathbf{B}(\mathbf{r}'). \qquad [6.25]$$

$\mathcal{S}$ is a surface bounded by the circuit $\mathcal{C}$, $\mathbf{n}(\mathbf{r}')$ is the unit vector normal to $\mathcal{S}$ and $\mathbf{B}(\mathbf{r}')$ is the field at the running point $\mathbf{r}'$ of $\mathcal{S}$.

It is only in the case of a uniform magnetic field $\mathbf{B}$ over the surface $\mathcal{S}$ that the expression [6.25] may be written as

$$\boldsymbol{\Gamma}_{\mathrm{M}} = \boldsymbol{\mathcal{M}} \times \mathbf{B}, \qquad \text{where } \boldsymbol{\mathcal{M}} = I \iint_{\mathcal{S}} d\mathcal{S}'\mathbf{n}(\mathbf{r}') \qquad \text{(uniform field).} \qquad [6.26]$$

$\mathcal{M}$ is the magnetic moment of the circuit $\mathcal{C}$. We note that the integral may be evaluated over any surface $\mathcal{S}$ bounded by $\mathcal{C}$. Particularly, if the circuit is planar, we may take for $\mathcal{S}$ the plane surface; then, $\mathbf{n}$ is the same at all the points of $\mathcal{S}$ and

$$\mathcal{M} = \mathcal{S}I\,\mathbf{n}. \qquad [6.27]$$

We note also that, if the circuit is formed by N turns, $\mathcal{M}$ must be multiplied by N. The expressions [6.26] may be generalized to a distribution of current with density $\mathbf{j}$ in a volume $\mathcal{V}$; its magnetic moment is

$$\mathcal{M} = \tfrac{1}{2} \iiint_{\mathcal{V}} d\mathcal{V}'\,\mathbf{r}' \times \mathbf{j}(\mathbf{r}'). \qquad [6.28]$$

Electric motors use the moment of the magnetic forces acting on a coil but in a radial magnetic field. *Galvanometers* also use the moment of the magnetic forces acting on a coil, which is proportional to the current intensity. The magnetic moment is $\mathcal{M} = N\mathcal{S}I\,\mathbf{n}$ and the torque exerted by a radial field $\mathbf{B}$ is $N\mathcal{S}IB$. If, in addition, the coil is subject to the restoring torque $\Gamma' = -C\theta$ of a spiral spring, the equilibrium condition is $N\mathcal{S}IB = C\theta$. Thus, the current intensity is proportional to the rotation angle of the coil.

B) Field of a small circuit at large distance

The field of the circular loop of Figure 6.4d at a large distance z is

$$\mathbf{B} = (\mu_0 \mathcal{M}/2\pi z^3), \qquad \text{where } \mathcal{M} = I\mathcal{S}\,\mathbf{e}_z = \pi a^2 I\,\mathbf{e}_z\,. \qquad [6.29]$$

To calculate the field off the axis Oz, as the system has rotation symmetry about Oz, we evaluate $\mathbf{B}$ at points M in the plane Oyz for instance. We specify M by its spherical coordinates $(r, \theta, \pi/2)$ and the point P of the loop by its polar coordinates (a, φ) in the Oxy plane. The element $d\mathbf{L}$ corresponds to a variation $d\varphi$, hence

$$\overrightarrow{OP} = a\cos\varphi\,\mathbf{e}_x + a\sin\varphi\,\mathbf{e}_y, \quad \overrightarrow{OM} = r\sin\theta\,\mathbf{e}_y + r\cos\theta\,\mathbf{e}_z,$$
$$\mathbf{R} = \overrightarrow{PM} = -a\cos\varphi\,\mathbf{e}_x + (r\sin\theta - a\sin\varphi)\,\mathbf{e}_y + r\cos\theta\,\mathbf{e}_z,$$
$$R = PM = [\,a^2 + r^2 - 2\,ar\sin\theta\sin\varphi]^{1/2}, \quad d\mathbf{L} = a(-\sin\varphi\,\mathbf{e}_x + \cos\varphi\,\mathbf{e}_y)\,d\varphi.$$

Thus, using Biot-Savart's law, the field at M may be written as

$$\mathbf{B} = \frac{\mu_0 I}{4\pi}\int_{\mathcal{C}} \frac{d\mathbf{L}\times\mathbf{R}}{R^3} = \frac{\mu_0}{4\pi}\,Ia\int_0^{2\pi}\frac{d\varphi}{R^3}\,[r\cos\theta\,(\cos\varphi\,\mathbf{e}_x + \sin\varphi\,\mathbf{e}_y) + (a - r\sin\theta\sin\varphi)\,\mathbf{e}_z].$$

This integral cannot be expressed with simple functions. If we are interested only in the field at large distance ($r >> a$), we may write to first order in a/R

$$R \cong r\,[\,1 - (a/r)\sin\theta\sin\varphi\,], \qquad 1/R^3 = (1/r^3)\,[1 + 3\,(a/r)\sin\theta\sin\varphi\,],$$

$$\mathbf{B} \cong (\mu_o I a^2/4r^3)\,[\,(3\sin\theta\cos\theta)\,\mathbf{e}_y + (2 - 3\sin^2\theta)\,\mathbf{e}_z\,]\,. \qquad [6.30]$$

We may also write **B** in the vector form

$$\mathbf{B} \cong \frac{\mu_o}{4\pi r^5}[3(\mathcal{M}.\mathbf{r})\,\mathbf{r} - r^2\mathcal{M}], \qquad \text{where } \mathcal{M} = I\mathcal{S}\,\mathbf{e}_z. \qquad [6.31]$$

At the same approximation, the expression [6.19] gives the vector potential at **r**

$$\begin{aligned}\mathbf{A}(\mathbf{r}) &= (\mu_o I/4\pi)\textstyle\int_{\mathcal{C}} d\mathbf{r}'/R \\ &= (\mu_o I a/4\pi r)\int_0^{2\pi} d\varphi'\,[-\sin\varphi'\,\mathbf{e}_x + \cos\varphi'\,\mathbf{e}_y][1 + (a/r)\cos\varphi\cos\varphi' + (a/r)\sin\varphi\sin\varphi'] \\ &= (\mu_o I a^2/4r^2)\,\mathbf{e}_\varphi = (\mu_o\mathcal{M}/4\pi r^2)\,\mathbf{e}_\varphi = (\mu_o/4\pi)\,(\mathcal{M}\times\mathbf{r})/r^3. \qquad [6.32]\end{aligned}$$

The expression [6.31] is similar to the expression [2.38] of the electric field **E** of an electric dipole **p**. We say that, at large distances, the loop is equivalent to a magnetic moment $\mathcal{M}$. This result is valid for any circuit $\mathcal{C}$: its field **B** and vector potential **A** at large distances are given by [6.31] and [6.32] with a magnetic moment

$$\mathcal{M} = I\iint_{\mathcal{S}} d\mathcal{S}\,\mathbf{n}. \qquad [6.33]$$

The field lines of a magnetic dipole at large distances are illustrated in Figure 6.5c.

C) Earth's magnetic field

Like many celestial bodies, the Earth has its own magnetic field. The field outside the Earth is almost the same as that of a large magnet *NS* placed at the center of the Earth. The south pole of this magnet is in the northern hemisphere while its north pole is in the southern hemisphere. The geographic poles are the points where the rotation axis of the Earth intercepts its surface. The magnetic axis makes an angle of approximately 11.5° with the rotation axis (Figure 6.6). For this reason, a compass will not align itself exactly due north but toward a point situated at about 1600 km from the geographic North Pole and, in fact, this is the *magnetic south pole* of the Earth. Aside from extremely small daily and annual variations, this angle undergoes important variations with a period of 960 years and has even reversed many times in the Earth's geological history. It is widely assumed that the Earth's magnetic field is generated by the motion of the liquid metallic core of the Earth.

The Earth's magnetic field is not horizontal except at the equator; it makes an angle, α, called *inclination*, with the horizontal plane. This angle and the magnitude of the field depend on the geographic location. At a latitude of 45° north, $B \cong 5.8 \times 10^{-5}$ T and $\alpha \cong 73°$. Thus, the vertical component is 5.5×10^{-5} T downward and the horizontal component is 1.7×10^{-5} T, and it makes an angle called *declination*, which is approximately 15° to the west, with the geographical meridian. The Earth's magnetic field extends thousands of kilometers in altitude. Charged cosmic rays (mostly from the Sun) are trapped by this magnetic field (see section 14.4), they spiral around the field and form Van Allen radiation belts surrounding the planet. When these particles collide with air molecules, the latter emit light in a wonderful display of colors called the aurora borealis and aurora australis at more than 60° latitude north and south.

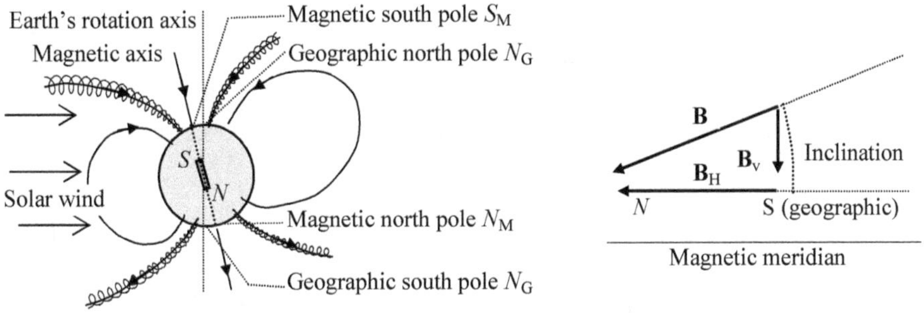

Figure 6.6. *Earth's magnetic field and the equivalent magnet. The magnetic south pole and north pole are close to the geographic North and South Poles respectively. The field extends to thousands of kilometers in altitude and is almost symmetrical about the Earth's magnetic axis*

6.6. Symmetries of the magnetic field

As the source of the magnetic field is the electric current, a symmetry of the distribution of currents implies the same symmetry for the field **B** and the vector potential **A**. To analyze **B** and **A**, it is practical to use coordinates that are convenient to impose this symmetry. As in the case of the electric field (section 2.5), if the configuration of currents has a translational symmetry in a direction *D*, it is convenient to have one of the axes of coordinates, *Oz* for instance, parallel to *D* and to use Cartesian or cylindrical coordinates about *Oz*. The components of **B** and **A** will not depend on *z*. If the currents have a rotational symmetry about an axis, it is convenient to use cylindrical or spherical coordinates about this axis taken as *Oz*. The components B_ρ, B_φ, and B_z (or B_r, B_θ, and B_φ) will not depend on the angle φ.

Consider now the reflections. The Lorentz force $\mathbf{F} = q(\mathbf{E} + \mathbf{v} \times \mathbf{B})$ being a true vector and q a true scalar, the cross product $\mathbf{v} \times \mathbf{B}$ is a true vector. As **v** is a true vector, **B** must be a *pseudo-vector* (see section 1.7b) and, as the vector operator ∇ is a true vector, the relation $\mathbf{B} = \nabla \times \mathbf{A}$ shows that **A** is a true vector, like **j**. This means that, in reflections, **A** transforms like **r** or **j**, while **B** transforms like **r** or **j** with an additional change of sign. As **A** is defined up to a gauge transformation [6.11], the gauge function $f(\mathbf{r}, t)$ must be a true scalar function.

– If the configuration of currents is symmetric in the reflection with respect to the plane *Oxy* for instance, it is convenient to use Cartesian or cylindrical coordinates about *Oz*. Then, **B** and **A** transform according to:

$$\mathbf{B}_{//}(x, y, z) = -\mathbf{B}_{//}(x, y, -z) \quad \text{and} \quad B_z(x, y, z) = B_z(x, y, -z),$$
$$\mathbf{A}_{//}(x, y, z) = \mathbf{A}_{//}(x, y, -z) \quad \text{and} \quad A_z(x, y, z) = -A_z(x, y, -z). \qquad [6.34]$$

Particularly, at the points M_o of *Oxy*, we find $\mathbf{B}_{//}(x, y, 0) = 0$ and $A_z(x, y, 0) = 0$.

More generally, if the distribution of currents is symmetric with respect to a plane Π (Figure 6.7a), i.e. $\mathbf{j}_{//}(M) = \mathbf{j}_{//}(M')$ and $\mathbf{j}_{\perp}(M) = -\mathbf{j}_{\perp}(M')$ at points M and M' symmetric with respect to Π, the field **B** is antisymmetric and **A** is symmetric, hence

$$\mathbf{B}_{//}(M) = -\mathbf{B}_{//}(M'), \quad \text{and} \quad \mathbf{B}_{\perp}(M) = \mathbf{B}_{\perp}(M'),$$
$$\mathbf{A}_{//}(M) = \mathbf{A}_{//}(M'), \quad \text{and} \quad \mathbf{A}_{\perp}(M) = -\mathbf{A}_{\perp}(M'). \qquad [6.35]$$

Particularly, at the points M_o of Π, we must have $\mathbf{B}_{//}(M_o) = -\mathbf{B}_{//}(M_o)$ and $\mathbf{A}_{\perp}(M_o) = -\mathbf{A}_{\perp}(M_o)$, thus $\mathbf{B}_{//}(M_o) = 0$ and $\mathbf{A}_{\perp}(M_o) = 0$.

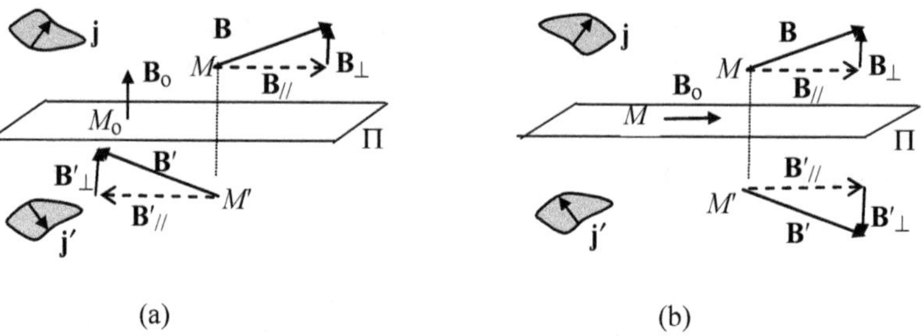

Figure 6.7. *Field of a distribution of currents: a) symmetric with respect to a plane Π, and b) antisymmetric with respect to Π*

– If the configuration of currents is antisymmetric in the reflection with respect to the plane *Oxy* for instance, **B** and **A** transform according to:

$$\mathbf{B}_{//}(x, y, z) = \mathbf{B}_{//}(x, y, -z) \quad \text{and} \quad B_z(x, y, z) = -B_z(x, y, -z),$$
$$\mathbf{A}_{//}(x, y, z) = -\mathbf{A}_{//}(x, y, -z) \quad \text{and} \quad A_z(x, y, z) = A_z(x, y, -z). \qquad [6.36]$$

We deduce that $B_z(x, y, 0) = 0$ and we may take $\mathbf{A}_{//}(x, y, z) = 0$ in some gauges. More generally, if the distribution of current is antisymmetric with respect to a plane Π (Figure 6.7b), that is $\mathbf{j}_{//}(M) = -\mathbf{j}_{//}(M')$ and $\mathbf{j}_{\perp}(M) = \mathbf{j}_{\perp}(M')$ at points M and M' symmetric with respect to Π, the field **B** is symmetric and **A** is antisymmetric, hence

$$\mathbf{B}_{//}(M) = \mathbf{B}_{//}(M') \quad \text{and} \quad \mathbf{B}_{\perp}(M) = -\,\mathbf{B}_{\perp}(M'),$$
$$\mathbf{A}_{//}(M) = -\mathbf{A}_{//}(M') \quad \text{and} \quad \mathbf{A}_{\perp}(M) = \mathbf{A}_{\perp}(M')\;. \qquad [6.37]$$

Particularly, at the points M_o of Π, we have $\mathbf{B}_{\perp}(M_o) = -\mathbf{B}_{\perp}(M_o)$, hence $\mathbf{B}_{\perp}(M_o) = 0$ and $\mathbf{A}_{//}(M_o) = -\mathbf{A}_{//}(M_o)$, and we may take $\mathbf{A}_{//}(M_o) = 0$. At M_o, the field $\mathbf{B}$ is in the plane Π and $\mathbf{A}$ is perpendicular to Π.

In the case of currents such $\mathbf{j}(\mathbf{r}) = -\,\mathbf{j}(-\mathbf{r})$, we must have $\mathbf{B}(\mathbf{r}) = \mathbf{B}(-\mathbf{r})$ and $\mathbf{A}(\mathbf{r}) = -\mathbf{A}(-\mathbf{r})$ and in the case of currents such that $\mathbf{j}(\mathbf{r}) = \mathbf{j}(-\mathbf{r})$, we must have $\mathbf{B}(\mathbf{r}) = -\mathbf{B}(-\mathbf{r})$ and $\mathbf{A}(\mathbf{r}) = \mathbf{A}(-\mathbf{r})$.

6.7. Ampère's law in the integral form

Contrary to the electric field, the magnetic field is not conservative. Consider a circuit $\mathcal{C}$ carrying a current I and a closed and oriented path $\mathcal{A}$, which we designate as an *Ampérian contour*. Let $\mathcal{S}$ be a surface bounded by $\mathcal{A}$. The field $\mathbf{B}(\mathbf{r}_A)$ produced by $\mathcal{C}$ at a running point $\mathbf{r}_A$ of $\mathcal{A}$ is given by [6.19] and the circulation of $\mathbf{B}$ over $\mathcal{A}$ is

$$\int_{\mathcal{A}} d\mathbf{r}_A.\mathbf{B}(\mathbf{r}_A) = (\mu_o I/4\pi)\int_{\mathcal{A}} d\mathbf{r}_A.\int_{\mathcal{C}} d\mathbf{r}_C \times \mathbf{R}/R^3 = (\mu_o I/4\pi)\int_{\mathcal{A}}\int_{\mathcal{C}} (d\mathbf{r}_A \times d\mathbf{r}_C).\mathbf{R}/R^3, \qquad [6.38]$$

where $\mathbf{r}_C$ is the position of the running point on $\mathcal{C}$ and $\mathbf{R} \equiv \mathbf{r}_A - \mathbf{r}_C$. We may show that the double integral over $\mathcal{C}$ and $\mathcal{A}$ is zero if $\mathcal{C}$ does not pass within $\mathcal{A}$ (Figure 6.8a) and it is equal to $\pm\, 4\pi$ if $\mathcal{C}$ passes within $\mathcal{A}$. The sign is ($\pm$) according to the right-hand rule (see section 6.9B). Thus, the right-hand side of [6.38] is $\pm\,\mu_o I$. In the case of several current-carrying circuits $\mathcal{C}_i$, the total field $\mathbf{B}$ is the superposition of the fields of the various circuits taken individually. The circuits that do not pass within $\mathcal{A}$ produce fields whose circulations are equal to zero, and those that pass within $\mathcal{A}$ produce fields whose circulations are $\pm\mu_o I_i$. For instance, in the case of the contour $\mathcal{A}$ of Figure 6.8a, we find $\int_{\mathcal{A}} d\mathbf{r}.\mathbf{B} = \mu_o(-I_2 + I_3 - 2I_4)$. Designating by $I^{(\text{in})}$ the total intensity that passes within $\mathcal{A}$ (i.e. crosses $\mathcal{S}$), we may write Ampère's law in the integral form:

$$\int_{\mathcal{A}} d\mathbf{r}.\mathbf{B} = \mu_o\, I^{(\text{in})}, \qquad \text{where } I^{(\text{in})} = \Sigma_i\, I_i\,. \qquad [6.39]$$

It should be noted that $I^{(\text{in})}$ includes all types of electric current (conduction currents, beams of charged particles, convection currents, etc.). On the other hand, only the currents that contribute to the field $\mathbf{B}$ in the integral $\int_{\mathcal{A}} d\mathbf{r}.\mathbf{B}$ must be included. For instance, if we analyze the field $\mathbf{B}_1$ produced by a circuit $\mathcal{C}_1$ and acting on a circuit $\mathcal{C}_2$, the current of $\mathcal{C}_2$ should not be included in $I^{(\text{in})}$, because it does not contribute to $\mathbf{B}_1$. It is to be noted that this expression of Ampère's law holds only in time-independent phenomena. It does not hold if the current varies (as in the case of alternating current) or if charged particles pass inside the Ampèrian contour $\mathcal{A}$.

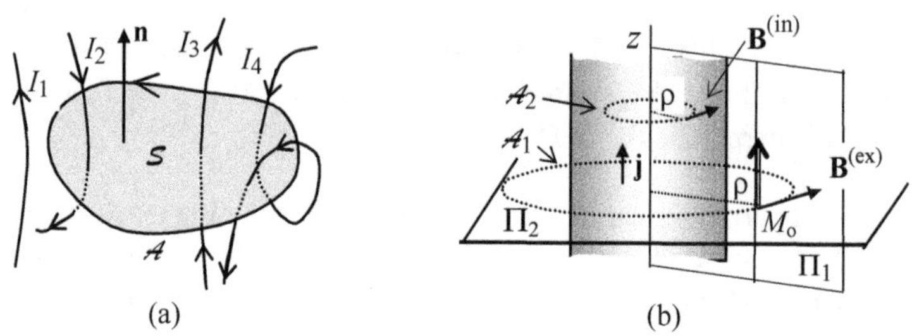

Figure 6.8. *a) Ampère's law, and b) evaluation of* **B** *and* **A** *of a cylinder*

In some simple symmetrical configurations of currents, it is possible to use Ampère's law to calculate **B** at points M. If we can find an Ampèrian contour $\mathcal{A}$, passing by M and such that **B** has a uniform magnitude and is tangent to $\mathcal{A}$, the circulation of **B** along $\mathcal{A}$ is simply LB, where L is the length of $\mathcal{A}$, hence $B = \mu_o I^{(in)}/L$. The contour $\mathcal{A}$ may include a part, where **B** is equal to zero or normal to $\mathcal{A}$. This part will not be included in evaluating L.

To illustrate this method, we consider a very long cylinder of radius R carrying a volume current density $\mathbf{j}(\rho) = j(\rho)\,\mathbf{e}_z$ symmetric about the axis Oz (Figure 6.8b). We analyze the symmetries of the distribution of current and their consequences first.

– The current density $\mathbf{j}(\rho)$ having a translational symmetry in the direction of Oz, the cylindrical components B_ρ, B_φ, B_z, A_ρ, A_φ and A_z do not depend on z.

– $\mathbf{j}(\rho)$ having a rotational symmetry about Oz, the components B_ρ, B_φ, B_z, A_ρ, A_φ and A_z do not depend on φ.

– $\mathbf{j}(\rho)$ has a reflection symmetry with respect to the azimuthal plane Π_1 containing the point M_o and Oz, i.e. $\mathbf{j}_{//}(M') = \mathbf{j}_{//}(M)$ and $\mathbf{j}_\perp(M') = -\mathbf{j}_\perp(M)$ at M and M' symmetric with respect to Π_1. The equation $\mathbf{B}_{//}(M) = -\mathbf{B}_{//}(M')$ and $\mathbf{A}_\perp(M) = -\mathbf{A}_\perp(M')$ in [6.35] imply that $B_\rho = 0$, $B_z = 0$ and $A_\varphi = 0$. The current density also has a reflection antisymmetry with respect to the plane Π_2 containing M_o and normal to Oz, i.e. $\mathbf{j}_{//}(M') = -\mathbf{j}_{//}(M)$ and $\mathbf{j}_\perp(M') = \mathbf{j}_\perp(M)$. The equation $B_z(x, y, z) = -B_z(x, y, -z)$ and $\mathbf{A}_{//}(x, y, z) = -\mathbf{A}_{//}(x, y, -z)$ in [6.36] imply that $B_z = 0$, $A_\rho = 0$ and $A_\varphi = 0$. Thus, **B** is tangent to the circle $\mathcal{A}$ of axis Oz and passing by M_o and its magnitude is uniform on this circle, while **A** has one component $A_z(\rho)$. The circulation of **B** on $\mathcal{A}$ is $2\pi\rho B$ and Ampère's law gives $\mathbf{B} = (\mu_o I^{(in)}/2\pi\rho)\,\mathbf{e}_\varphi$. If M_o is inside the cylinder (ρ <

R), the intensity that passes inside $\mathcal{A}$ is $I^{(in)} = \iint_{\mathcal{S}} d\mathcal{S}\, \mathbf{n}.\mathbf{j} = 2\pi \int_0^{\rho} d\rho' \rho' j(\rho')$ but, if M_o is outside the cylinder ($\rho > R$), $I^{(in)} = I = 2\pi \int_0^{R} d\rho' \rho' j(\rho')$. Thus, the field may be written as

$$\mathbf{B}^{(ex)} = (\mu_o I/2\pi\rho)\, \mathbf{e}_{\varphi}, \qquad \mathbf{B}^{(in)} = (\mu_o/\rho)\, \mathbf{e}_{\varphi} \int_0^{\rho} d\rho' \; \rho' \; j(\rho'). \tag{6.40}$$

Particularly, if the current density is uniform, we find $j = I/\pi R^2$, hence

$$\mathbf{B}^{(ex)} = (\mu_o I/2\pi\rho)\, \mathbf{e}_{\varphi}, \quad \mathbf{B}^{(in)} = (\mu_o I\rho/2\pi R^2)\, \mathbf{e}_{\varphi} \tag{6.41}$$

The vector potential $\mathbf{A}$ may be obtained by integrating the equation $\nabla\times\mathbf{A} = \mathbf{B}$, which reduces in cylindrical coordinates to the differential equation $dA_z/d\rho = -B_{\varphi}$.

Note that the field is finite and continuous everywhere. It increases linearly from 0 on the axis to a maximum $B = (\mu_o I/2\pi R)$ on the surface of the cylinder and then decreases like $1/\rho$. The field outside the cylinder is independent of its radius and it remains valid in the case of a wire carrying the current I.

6.8. Field and potential of some simple circuits

A) Field and potential of a thin rectilinear conductor

Consider a thin and straight rod P_1P_2 of length $2L$ carrying a current I (Figure 6.9a). Taking Oz in the direction of the rod and O at its middle, the system has a rotational symmetry about Oz. Using cylindrical coordinates, an element of length dz' situated at the point $K(0, 0, z')$ produces at $M(\rho, \varphi, z)$ a field given by Biot-Savart's law $d\mathbf{B}(\mathbf{r}) = (\mu_o I/4\pi)\, dz'\, \mathbf{e}_z \times \mathbf{R}/R^3$, where $\mathbf{R} = \rho\, \mathbf{e}_{\rho} + (z - z')\, \mathbf{e}_z$ and $R = \sqrt{\rho^2 + (z - z')^2}$. To integrate over P_1P_2, it is convenient to use instead of z' the angle θ' that $\mathbf{R}$ makes with Oz. We have $\cos\theta' = (z - z')/R$ and $\sin\theta' = \rho/R$, $R = \rho/\sin\theta'$ and $dz' = \rho\, d\theta'/\sin^2\theta'$. Thus, the field may be written as

$$\mathbf{B}(\mathbf{r}) = \int_A^B d\mathbf{B}' = \frac{\mu_o I}{4\pi\rho}\, \mathbf{e}_{\varphi} \int_{\theta_1}^{\theta_2} d\theta' \; \sin\theta' = \frac{\mu_o I}{4\pi\rho} (\cos\theta_1 - \cos\theta_2)\, \mathbf{e}_{\varphi}, \tag{6.42}$$

where θ_1 and θ_2 are the extreme values of θ'. Setting $R_{\pm} = [\rho^2 + (z \pm L)^2]^{1/2}$, we find

$$\mathbf{B}(\mathbf{r}) = \frac{\mu_o I}{4\pi\rho} \left[\frac{L+z}{R_+} + \frac{L-z}{R_-}\right] \mathbf{e}_{\varphi}. \tag{6.43}$$

A similar analysis gives for the vector potential

$$\mathbf{A}(\mathbf{r}) = \frac{\mu_o I}{4\pi}\int_{\mathscr{C}}\frac{d\mathbf{r}'}{R} = \frac{\mu_o I}{4\pi}\,\mathbf{e}_z\int_{-L}^{L}\frac{dz'}{R} = \frac{\mu_o I}{4\pi}\ln\frac{z-L-R_-}{z+L-R_+}\,\mathbf{e}_z. \quad [6.44]$$

If the circuit is very long or, equivalently, if the point M is very close to O ($z \ll L$ and $\rho \ll L$), $\theta_1 \to 0$ and $\theta_2 \to \pi$, we find

$$\mathbf{B}(\mathbf{r}) \to (\mu_o I/2\pi\rho)\,\mathbf{e}_\varphi \quad \text{and} \quad \mathbf{A}(\mathbf{r}) \to -(\mu_o I/2\pi)\ln\rho\,\mathbf{e}_z. \quad [6.45]$$

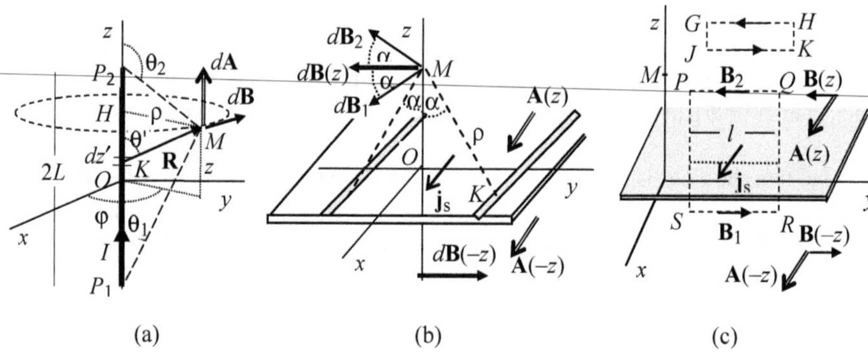

Figure 6.9. *Evaluation of the magnetic field: a) in the case of a thin rod by using Biot-Savart's law, b) in the case of a sheet of width L carrying a current density* $\mathbf{j}_s$ *by using Biot-Savart's law, and c) in the case of a wide sheet by using Ampère's law*

B) Field and potential of a sheet carrying a current density j_s

Consider a sheet lying in the plane Oxy between $y = -L/2$ and $y = L/2$, very long in the direction of Ox and carrying a surface current density $\mathbf{j}_s = j_s\mathbf{e}_x$ (Figure 6.9b). To calculate its field and vector potential at a point M of the normal axis Oz, we consider a narrow strip of infinite length in the direction Ox and situated between y and $y + dy$ and the symmetric strip with respect to Oxz. They carry the intensities $dI = j_s\,dy$ and they produce the fields $d\mathbf{B}_1$ and $d\mathbf{B}_2$ of magnitude $\mu_o\,dI/2\pi\rho$ at M, where $\rho = (y^2 + z^2)^{1/2}$ is the distance of M to the strips. These fields form the same angle α with Oy as MK with Oz; thus $\cos\alpha = z/\rho$. The resultant of these fields is

$$d\mathbf{B}(z) = d\mathbf{B}_1 + d\mathbf{B}_2 = 2\,dB_1\cos\alpha\,\mathbf{e}_y = -(\mu_o j_s\,z\,dy/\pi\rho^2)\,\mathbf{e}_y. \quad [6.46]$$

The total field is obtained by integration on y from 0 to $L/2$:

$$\mathbf{B}(z) = \int_0^{L/2} d\mathbf{B}(z) = -(\mu_o j_s z/\pi)\,\mathbf{e}_y\int_0^{L/2} dy\,/(y^2+z^2) = -(\mu_o j_s/\pi)\tan^{-1}(L/2z)\mathbf{e}_y. \quad [6.47]$$

Setting D as the length of the sheet, the vector potential is

$$\mathbf{A}(z) = (\mu_o/4\pi)\iint_S dS'\,\mathbf{j}_s/R = (\mu_o\mathbf{j}_s/4\pi)\int_{-D/2}^{D/2} dx' \int_{-L/2}^{L/2} dy'/(x'^2 + y'^2 + z^2)^{1/2}$$
$$= (\mu_o\mathbf{j}_s/4\pi)\,[-L\ln(L^2 + 4z^2) - 4z\tan^{-1}(L/2z) + C], \qquad [6.48]$$

where C is a constant, which diverges like $2L \ln D$ as $D \to\infty$, but this is irrelevant because $\mathbf{A}$ is defined up to an arbitrary ∇f. At a point near the sheet ($z << L$), we find

$$\mathbf{B}(z) \to -\tfrac{1}{2}\mu_o j_s \,\text{sign}(z)\,\mathbf{e}_y \quad \text{and} \quad \mathbf{A}(z) = -\tfrac{1}{2}\mu_o\mathbf{j}_s|z| + \text{Constant}. \qquad [6.49]$$

The magnetic field of the infinite sheet may be easily determined using Ampère's law (Figure 6.9c). The translational symmetry in any direction parallel to the plane Oxy of the sheet implies that $\mathbf{B}$ and $\mathbf{A}$ depend only on z. The reflection symmetry with respect to Oxz implies that $B_x = 0$, $B_z = 0$ and $A_y = 0$. The antisymmetry of $\mathbf{j}$ in the reflection with respect to the plane Oyz implies that $A_y = A_z = 0$. Thus, we have $\mathbf{B} = B_y\,\mathbf{e}_y$ and $\mathbf{A} = A_x\,\mathbf{e}_x$. The reflection symmetry with respect to Oxy implies that $B_y(-z) = -B_y(z)$ and $A_x(-z) = A_x(z)$. Consider the rectangular contour $GHKJ$ situated on one side of the sheet. As no current crosses it, Ampère's law gives $lB_{GH} - lB_{KJ} = 0$. Thus, B_y does not depend on z in each one of the regions $z > 0$ and $z < 0$. Consider now the contour $PQRS$ whose side PQ is on the side $z > 0$ and the side RS is in the side $z < 0$. As the current that crosses it is $j_s l$, Ampère's law gives $lB_{RS} - lB_{PQ} = \mu_o j_s l$. As $B_{PQ} = -B_{RS}$, we find $B_y(z) = -B_y(-z) = \tfrac{1}{2}\mu_o j_s$. Knowing the field, the relation $\mathbf{B} = \nabla \times \mathbf{A}$ reduces to the equation $B_y = \partial_z A(z)$, hence $A(z) = -\tfrac{1}{2}\mu_o j_s|z| + C$. Thus, we find again the result [6.49].

C) Field and potential of a solenoid

Consider a solenoid of radius a and length L constituted by n turns per unit length and carrying a current I (Figure 6.10a). We take the origin O at the center of the solenoid and Oz along its axis. An element of the solenoid situated between z' and $z' + dz'$ near the point P contains $n\,dz'$ turns; thus, it may be assimilated to a circular loop carrying a current $dI = nI\,dz'$. According to equation [6.21], its field at a point $M(z)$ on the axis is $d\mathbf{B}(z) = (\mu_o\,nIa^2/2R^3)\,dz'\mathbf{e}_z$, where $R = \sqrt{a^2 + (z - z')^2}$ and the total field is obtained by integrating over z' from $-L/2$ to $+L/2$; we get

$$\mathbf{B}(z) = \tfrac{1}{2}\mu_o nIa^2\,\mathbf{e}_z \int_{-L/2}^{+L/2} dz'/R^3 = \tfrac{1}{2}\mu_o nI\,[(L/2 - z)/R_- + (L/2 + z)/R_+]\,\mathbf{e}_z, \qquad [6.50]$$

where the extreme values of R are $R_\pm = [a^2 + (z \pm L/2)^2]^{1/2}$. The integral may easily be evaluated using the angle θ that Oz forms with MP. We find $\sin\theta = a/R$ and $\cos\theta = (z' - z)/R$, hence $dz' = -a\,d\theta/\sin^2\theta$ and $R = a/\sin\theta$; thus,

$$\mathbf{B}(z) = \tfrac{1}{2}\,\mu_o nI\,\mathbf{e}_z \int_{\theta_1}^{\theta_2} d\theta\,\sin\theta = \tfrac{1}{2}\,\mu_o nI\,(\cos\theta_1 - \cos\theta_2)\,\mathbf{e}_z\,. \qquad [6.51]$$

Note that $\theta_1 < \pi/2$ and $\theta_2 > \pi/2$ for M situated inside the solenoid, while for M situated outside the solenoid, $\theta_1 > \pi/2$ and $\theta_2 > \pi/2$ if $z > L/2$ and $\theta_1 < \pi/2$ and $\theta_2 < \pi/2$ if $z \cong \pm L/2$. The lines of the field **B** are illustrated in Figure 6.10b. The field has its maximum value at the center (Figure 6.10c)

$$B_{max} = B(0) = \tfrac{1}{2}\,\mu_o nIL/\sqrt{4a^2+L^2}\,. \qquad [6.52]$$

B is almost uniform in a long solenoid and it decreases quickly at the ends (where $B \approx \tfrac{1}{2}B_{max}$ for $z < \pm L/2$). Particularly, if the solenoid is very long ($L >> a$ and $L >> z$), setting $j_s = nI$ for the surface current density, the field is given by

$$B = \mu_o nI\,\mathbf{e}_z = \mu_o j_s\,\mathbf{e}_z. \qquad [6.53]$$

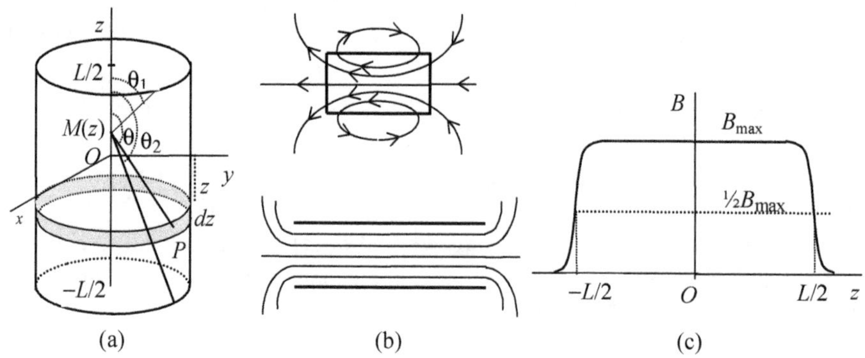

Figure 6.10. *a) Field of a solenoid, b) field lines in the case of a solenoid of finite length L, and c) variation of the field as a function of z*

If the solenoid is ideally infinite, we may use Ampère's law to determine the field **B**. Let us first analyze the symmetries (see Figure 6.11a). Because of the translational symmetry in the direction of Oz, the field **B** does not depend on z. In the approximation of the solenoid made of circular current loops, the field **B** at point $M(\rho, \varphi, z)$ does not depend on the azimuthal angle φ because of the rotational symmetry about Oz. Thus, **B** is a function only of the distance ρ of M from the axis Oz. By the same approximation, the current density is antisymmetric in the reflection with respect to the azimuthal plane Π_1 containing M and the axis Oz. As per equations [6.37], the field has no normal component B_φ. The current density has a reflection symmetry with respect to the plane Π_2 containing M and normal to the axis Oz. As per equation [6.35], the field has no parallel components B_ρ and B_φ. These combined symmetries imply that $\mathbf{B} = B(\rho)\mathbf{e}_z$ and we may assume that M is in the plane Ozy and use Ampère's law:

– Consider a rectangular path $CDPQ$ situated inside the solenoid with the sides CD and PQ of length l parallel to Oz. As the field is independent of z and oriented in the direction Oz and no current passes inside this path, Ampère's law gives $lB^{(in)}_{CD} - lB^{(in)}_{PQ} = 0$. This means that the field is uniform inside the solenoid.

– Consider a rectangular path $C'D'P'Q'$ completely situated outside the solenoid. As no current passes inside it, Ampère's law gives $B^{(ex)}_{C'D'} = B^{(ex)}_{P'Q'}$. Thus, the field **B** is uniform outside the solenoid. As it is obviously equal to zero at large distance, it must be equal to zero everywhere outside the solenoid.

– Finally, consider the rectangular path $C''D''P''Q''$ such that $C''D''$ is inside the solenoid and $P''Q''$ is outside it. The intensity that passes inside it is lj_s; thus, Ampère's law may be written as $lB^{(in)} - lB^{(ex)} = \mu_o lj_s$. As $B^{(ex)} = 0$, we deduce that

$$B^{(in)} = \mu_o j_s = \mu_o nI . \quad [6.54]$$

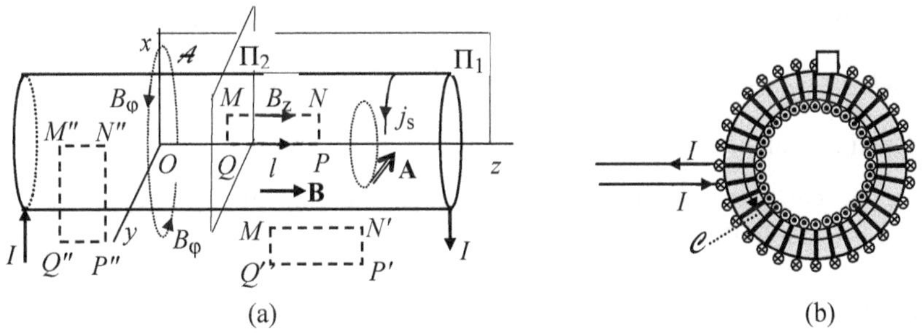

Figure 6.11. *a) Field of an infinitely long solenoid, and b) toroidal coil*

Our analysis of the solenoid as formed by circular loops is approximately valid if they are thin and almost in contact; then, the translational and rotational symmetries about Oz are almost exact. In reality, the solenoid being almost helical carries a current I in the direction of the axis. If we consider an Ampèrian circular path $\mathcal{A}$ of radius $\rho < R$ and axis Oz, no current passes inside it and Ampère's law gives $2\pi\rho B^{(in)}_{\varphi} = 0$. Thus the field inside the solenoid has no component $B^{(in)}_{\varphi}$. On the contrary, if $\mathcal{A}$ is external ($\rho > R$), a current I passes inside it and Ampère's law gives $2\pi\rho B^{(ex)}_{\varphi} = \mu_o I$, hence, $B^{(ex)}_{\varphi} = \mu_o I/2\pi\rho$. This is the same field as that of an infinite thin conductor carrying a current I along the axis. It is possible to eliminate this component of the field by coiling the solenoid an even number of layers, in such a way that the current enters and leaves the solenoid at the same end; in that case, no net current flows in the direction Oz.

In the case of an infinitely long solenoid, the translational symmetry in the direction of Oz and the rotational symmetry about this axis imply that **A** depends only on ρ. The reflection symmetry about any sectional plane, implies that $A_z = 0$. Thus, the vector potential has the form $\mathbf{A} = A(\rho)\,\mathbf{e}_\varphi$. Applying equation [6.13] to a circular contour of radius ρ, we find $\pi\rho^2 B = 2\pi\rho A$, thus $A(\rho) = \frac{1}{2}\rho B(\rho) = \frac{1}{2}\,\mu_o n\,I\rho$. We obtain also this result by writing $\mathbf{B} = \nabla\times\mathbf{A}$, which gives the differential equation $\partial_\rho(\rho A_\varphi) = \mu_o\, nI\,\rho$, whose solution is $A_\varphi = \frac{1}{2}\mu_o nI\rho + C/\rho$, where C is an integration constant. As A_φ is regular on the axis, we must have $C = 0$.

To avoid the leakage of the field, the N turns of the solenoid may be uniformly distributed round a closed circular ring (Figure 6.11b). The field in this *toroidal coil* is not uniform but it has a rotational symmetry about the axis. Writing Ampère's law over a circle of radius r, we find $2\pi r B = \mu_o NI$. We deduce that $B = \mu_o NI/2\pi r$.

6.9. Equations of time-independent magnetism in vacuum, singularities of B

A) Basic equations of time-independent magnetism in vacuum

The time-independent magnetic field in vacuum obeys two basic laws:

1. The conservation of the magnetic flux: the flux of **B** through any closed surface is equal to 0:

$$\iint_{\mathcal{S}} d\mathcal{S}\,(\mathbf{n}.\mathbf{B}) = 0. \qquad [6.55]$$

Using Gauss-Ostrogradsky's theorem, we may transform the flux into the integral of $\nabla.\mathbf{B}$ over the enclosed volume, thus $\iiint_{\mathcal{V}} d\mathcal{V}\,\nabla.\mathbf{B} = 0$ for any volume $\mathcal{V}$, hence

$$\nabla.\mathbf{B} = 0. \qquad [6.56]$$

This equation implies the existence of a vector potential such that

$$\mathbf{B} = \nabla\times\mathbf{A}, \qquad [6.57]$$

where **A** is determined up to the addition of the gradient of any function, $\mathbf{A}' = \mathbf{A}+\nabla f$.

2. *Ampère's law*: the magnetic field is related to the electric current by Ampère's law in the integral form

$$\int_{\mathcal{A}} d\mathbf{r}.\mathbf{B} = \mu_o I^{(\text{in})}, \qquad \text{where } I^{(\text{in})} = \Sigma_i I_i = \iint_{\mathcal{S}} d\mathcal{S}\,\mathbf{j}.\mathbf{n}. \qquad [6.58]$$

I_i are the currents that pass through $\mathcal{S}$. Using Stokes' theorem, the circulation of **B** on the closed path $\mathcal{A}$ may be transformed into the flux of $\nabla\times\mathbf{B}$, thus $\iint_{\mathcal{S}} d\mathcal{S}\, \mathbf{n}.(\nabla\times\mathbf{B}) = \mu_o \iint_{\mathcal{S}} d\mathcal{S}\, \mathbf{j}.\mathbf{n}$ for any $\mathcal{S}$, hence Ampère's law in the local form

$$\nabla\times\mathbf{B} = \mu_o\,\mathbf{j}. \qquad [6.59]$$

Expressing **B** in terms of **A**, using the identity $\nabla\times(\nabla\times\mathbf{A}) = \nabla(\nabla.\mathbf{A}) - \Delta\mathbf{A}$ and imposing the condition $\nabla.\mathbf{A} = 0$ (by eventually making a suitable gauge transformation $\mathbf{A}' = \mathbf{A} + \nabla f$), Ampère's equation [6.59] becomes

$$\Delta\mathbf{A} = -\,\mu_o\,\mathbf{j}. \qquad [6.60]$$

All these equations are verified by the expressions of the field and vector potential of charges in motion and current densities [6.17], [6.19] and [6.20].

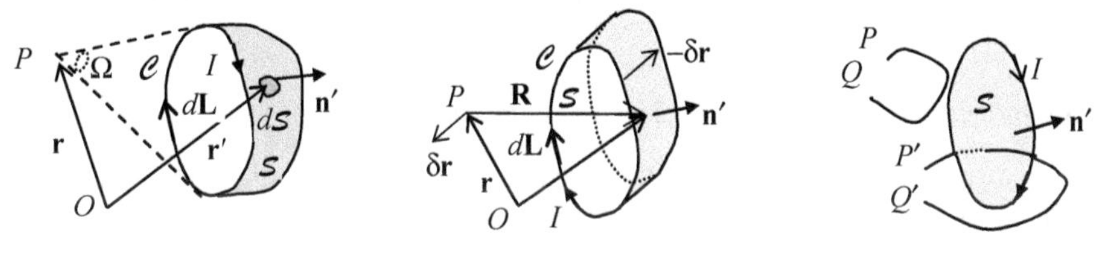

Figure 6.12. *Scalar potential of* **B** *in a region, where* **j** = 0

B) Concept of scalar magnetic potential

In a region where there is no electric current density, Ampère's equation [6.59] reduces to $\nabla\times\mathbf{B} = 0$. This means that it is possible to define a scalar potential V_M such that $\mathbf{B} = -\nabla V_M$. Then, the circulation of **B** on any closed path in this region vanishes. Let us consider a circuit $\mathcal{C}$ carrying a current I and a surface $\mathcal{S}$ bounded by $\mathcal{C}$. This surface is subtended from a point $P(\mathbf{r})$ by a cone of solid angle:

$$\Omega = \iint_{\mathcal{S}} d\mathcal{S}'\, \mathbf{n}'.\mathbf{R}/R^3, \qquad \text{where } \mathbf{R} = \mathbf{r}' - \mathbf{r}. \qquad [6.61]$$

$\mathbf{n}'$ is the unit vector normal to $\mathcal{S}$ at $\mathbf{r}'$ and oriented with respect to the circuit carrying the current I according to the right-hand rule (Figure 6.12a). If the point P moves by $\delta\mathbf{r}$, the variation of the solid angle $d\Omega = \Omega' - \Omega$ is the same as if P was fixed and the circuit displaced by $-\delta\mathbf{r}$ (Figure 6.12b). As the solid angle, which subtends the whole closed cylinder of bases $\mathcal{S}$ and $\mathcal{S}'$, is equal to zero, $d\Omega$ is the opposite of the solid angle $\delta\Omega$ of the strip $\delta\mathcal{S}$ of width $-\delta\mathbf{r}$ along the circuit $\mathcal{C}$, hence

$$d\Omega = -\delta\Omega = -\iint_{\delta\mathcal{S}} d\mathcal{S}'\mathbf{n}'.\mathbf{R}/R^3 = \int_{\mathcal{C}} (d\mathbf{L}\times\delta\mathbf{r}).\mathbf{R}/R^3 = -\,\delta\mathbf{r}.\int_{\mathcal{C}} (d\mathbf{L}\times\mathbf{R})/R^3.$$

The field of $\mathcal{C}$ at P being $\mathbf{B}(\mathbf{r}) = (\mu_o I/4\pi)\int_{\mathcal{C}} d\mathbf{L}\times(\mathbf{r}-\mathbf{r}')/|\mathbf{r}-\mathbf{r}'|^3 = -(\mu_o I/4\pi)\int_{\mathcal{C}} d\mathbf{L}\times\mathbf{R}/R^3$, we may write $\mathbf{B}.\delta\mathbf{r} = (\mu_o I/4\pi)\,\delta\Omega$, thus $\mathbf{B} = -\nabla V_M$, where $V_M = -(\mu_o I/4\pi)\,\Omega$.

The circulation of $\mathbf{B}$ along a path PQ, which does not cross $\mathcal{S}$ (Figure 6.12c), is $\int_{PQ} d\mathbf{r}.\mathbf{B} = (\mu_o I/4\pi)(\Omega_P - \Omega_Q)$. It vanishes if the path is closed ($P \equiv Q$), as Ω is a continuous function of the position. On the other hand, the circulation along a path $P'Q'$ that crosses $\mathcal{S}$ is $\int_{P'Q'} d\mathbf{r}.\mathbf{B} = (\mu_o I/4\pi)(\Omega_P - \Omega_Q) = \pm\mu_o I$, as, by crossing $\mathcal{S}$, Ω changes from 2π to -2π, or conversely, depending on whether $P'Q'$ crosses $\mathcal{S}$ in the direction of $\mathbf{n}$ or in the opposite direction. This result is equivalent to Ampère's law.

Contrary to the electric potential V, which is the potential energy of the unit charge, the magnetic potential V_M is not an energy. This is evident because $\mathbf{B}$ being a pseudo vector, the function V_M (such that $\mathbf{B} = -\nabla V_M$) is a pseudo-scalar, while the energy is a true scalar quantity.

C) Singularities and discontinuities of the magnetic field

The examples of evaluation of the magnetic field that we have considered show that $\mathbf{B}$ and $\mathbf{A}$ are not always regular and continuous:

a) Near a point M of a thin conductor carrying a current I, the dominant part of $\mathbf{B}$ is the field of a small element of the conductor at M. Its field lines are circular around the conductor (Figure 6.13a); thus, $\mathbf{B}$ has no well-defined direction. On the other hand, Ampère's law applied to a small circle of radius ρ around the conductor shows that B is infinite like $B = \mu_o I/2\pi\rho$. The vector potential also becomes very large like $-(\mu_o I/2\pi)\ln\rho$ in the direction of the current.

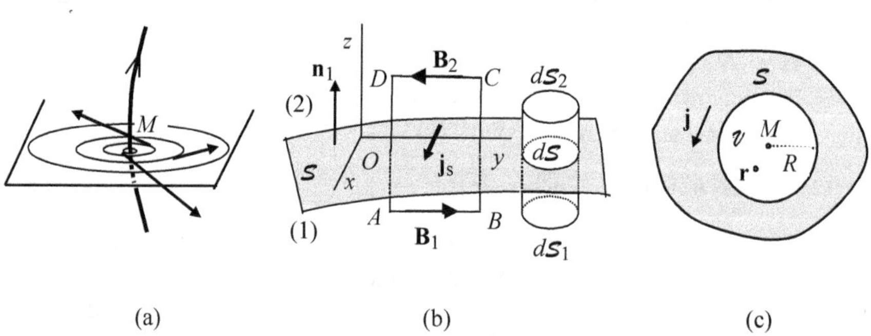

Figure 6.13. *a) Singularity of* **B** *on a line of current, b) discontinuity of* **B** *on a sheet of current, and c) continuity of* **B** *in a volume distribution of current*

b) Consider a sheet of current $\mathcal{S}$ with a surface density $\mathbf{j}_s$ (Figure 6.13b). Let us apply the law of conservation of magnetic flux to a very short cylinder of bases $d\mathcal{S}_1$ and $d\mathcal{S}_2$ situated on both sides of the sheet. We designate as $\mathbf{n}_{12}$ the unit vector that is normal to $\mathcal{S}$ and oriented from side (1) toward side (2). The flux of $\mathbf{B}$ through the very short lateral surface may be neglected and the outgoing flux from $d\mathcal{S}_1$ and $d\mathcal{S}_2$ are $d\mathcal{S}\ \mathbf{n}_{12}.\mathbf{B}_2$ and $-d\mathcal{S}\ \mathbf{n}_{12}.\mathbf{B}_1$, respectively. The conservation of the flux of $\mathbf{B}$ implies that $\mathbf{n}_{12}.\mathbf{B}_2 = \mathbf{n}_{12}.\mathbf{B}_1$. In other words, the normal component of $\mathbf{B}$ is continuous on the sheet of current. We take the axes of coordinates, such that Oz is in the direction of $\mathbf{n}_{12}$ and $\mathbf{j}_s$ in the direction of Ox. Consider a rectangular path $PQCD$ such that PQ and CD have a length l and they are parallel to Oy, while PD and QC are very short and parallel to Oz. The circulation of $\mathbf{B}$ on this contour is $l(B_{2y} - B_{1y})$ and it is equal to $\mu_o j_s l$ according to Ampère's law. Thus, the y component of $\mathbf{B}$ has a discontinuity $B_{2y} - B_{1y} = \mu_o j_s$. If the sides PQ and CD are parallel to Ox, no current crosses the rectangle and we find $B_{2x} = B_{1x}$. These two relations may be written in a single vector equation

$$\mathbf{B}_1 - \mathbf{B}_2 = \mu_o\ \mathbf{n}_{12} \times \mathbf{j}_s. \qquad [6.62]$$

Thus, on a sheet of current $\mathcal{S}$, the tangential component of $\mathbf{B}$, which is normal to $\mathbf{j}_s$, undergoes a discontinuity equal to $\mu_o j_s$, while the component of $\mathbf{B}$ that is parallel to $\mathbf{j}_s$ and the normal component to $\mathcal{S}$ are continuous.

c) In the case of a volume distribution of current, the field $\mathbf{B}$ is finite and continuous. Indeed, an eventual singularity at a point M may be produced by the current density that is very close to M (making $|\mathbf{r} - \mathbf{r}'| \to 0$ in the denominator of the expression of the field). Let us surround M by a small sphere of radius R enclosing a volume $\mathcal{V}$ (Figure 6.13c). According to the theorem of the mean, the integral of a function $f(\mathbf{r})$ over $\mathcal{V}$ is equal to the product $\mathcal{V} f(\mathbf{r}_o)$, where $\mathbf{r}_o$ is a certain point of $\mathcal{V}$. Thus the field of the sphere, evaluated by using [6.20], may be written as $\mathbf{B}(\mathbf{r}) = (\mu_o/4\pi)\mathcal{V}\ \mathbf{j}(\mathbf{r}_o)\times(\mathbf{r} - \mathbf{r}_o)/|\mathbf{r} - \mathbf{r}_o|^3$. Designating by θ the angle that $\mathbf{j}(\mathbf{r}_o)$ makes with $(\mathbf{r} - \mathbf{r}_o)$, we find for the magnitude of $\mathbf{B}$

$$B(\mathbf{r}) < (\mu_o/4\pi)\mathcal{V} j(\mathbf{r}_o) \sin\theta/|\mathbf{r}-\mathbf{r}_o|^2 < (\mu_o R^3/3)\, j(\mathbf{r}_o)/R^2 = (\mu_o R/3)\, j(\mathbf{r}_o).$$

Thus, the field of the sphere tends to 0 like R as $R \to 0$. This shows that the field of a volume current distribution is finite and continuous.

The singularities and the discontinuities of $\mathbf{B}$ are due to the zero limit of the diameter of the line of current and the thickness of the sheet of current. In reality, the macroscopic bodies always have finite dimensions. At the microscopic level, we must distinguish between the *microscopic field*, which undergoes fluctuations and become infinite at the position of point charges, and the *macroscopic field*, which is regular and continuous (see the section 2.8).

6.10. Magnetic energy of a circuit in a field B

We have seen that a uniform field **B** exerts *a* moment of force $\mathbf{\Gamma}_M = \mathcal{M} \times \mathbf{B}$ on a rigid circuit carrying a current I. Thus, $|\mathbf{\Gamma}_M| = \mathcal{M} B \sin\theta$, where θ is the angle that $\mathcal{M}$ makes with **B**. If the circuit is free to rotate around a point O, it may rotate about an axis whose unit vector **u** is parallel to $\mathbf{\Gamma}_M$ (thus, orthogonal to $\mathcal{M}$ and **B**). The circuit is in equilibrium if $\mathbf{\Gamma}_M = 0$, thus $\theta = 0$ ($\mathcal{M}$ pointing in the direction of **B**) or $\theta = \pi$ ($\mathcal{M}$ pointing in the opposite direction to **B**). We may verify that the position $\theta = 0$ is stable, while the position $\theta = \pi$ is not. The work of the magnetic forces in the rotation of the circuit about **u** from its equilibrium position $\theta = 0$ to an arbitrary position θ is

$$W_{0\to\theta} = \int_0^\theta d\theta' \, \mathbf{\Gamma}_M.\mathbf{u} = -\mathcal{M}B \int_0^\theta d\theta' \sin\theta' = \mathcal{M}B(\cos\theta - 1) = (\mathcal{M}.\mathbf{B}) - \mathcal{M}B. \quad [6.63]$$

Conversely, to rotate the circuit from its stable equilibrium position $\theta = 0$ to the position θ, a moment of force $\mathbf{\Gamma}' = -\mathbf{\Gamma}_M$ must be exerted; the required work is

$$W'_{0\to\theta} = -W = \mathcal{M}B - (\mathcal{M}.\mathbf{B}). \quad [6.64]$$

This work remains stored as *magnetic potential energy* of the circuit in the external magnetic field.

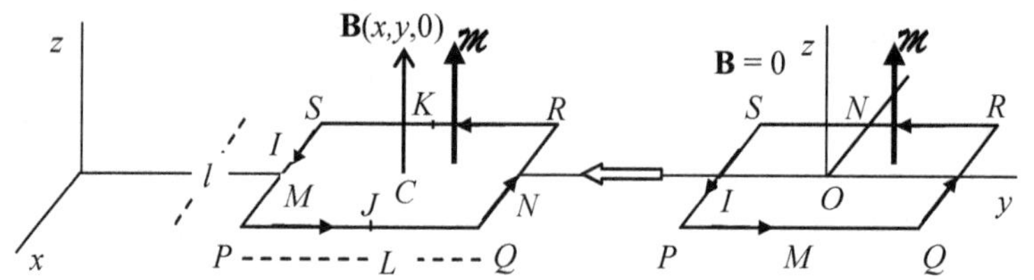

Figure 6.14. *Evaluation of the interaction energy of an electric circuit in a field* **B**

Consider now the case of a non-uniform field **B**. Let us evaluate the work in bringing a circuit from a position where $\mathbf{B} = 0$ to a position where the field is **B** while maintaining the current intensity I constant. Let us consider first the simple case of a rectangular circuit $PQRS$ lying in the plane Oxy (Figure 6.14). We assume that it is small enough to neglect all terms of the second order in its dimensions L and l. We evaluate the magnetic force that acts on each side by taking the field at its middle; this is a good approximation if the field is slowly varying over the circuit. We have for instance $\mathbf{B}(M) = \Sigma_\alpha B_\alpha(x, y - \frac{1}{2}L, 0)\mathbf{e}_\alpha \cong \Sigma_\alpha [B_\alpha(x, y, 0) - \frac{1}{2}L\,\partial_y B_\alpha]\,\mathbf{e}_\alpha$. Thus, the total force that acts on the circuit may be written as

$$\begin{aligned}\mathbf{F}_M &= I\overrightarrow{SP} \times \mathbf{B}(M) + I\overrightarrow{PQ} \times \mathbf{B}(J) + I\overrightarrow{QR} \times \mathbf{B}(N) + I\overrightarrow{RS} \times \mathbf{B}(K)\\ &= Il\,\mathbf{e}_x \times \Sigma_\alpha [B_\alpha(x, y, 0) - \tfrac{1}{2}L\partial_y B_\alpha]\,\mathbf{e}_\alpha + IL\,\mathbf{e}_y \times \Sigma_\alpha [B_\alpha(x, y, 0) + \tfrac{1}{2}l\partial_x B_\alpha]\,\mathbf{e}_\alpha\\ &\quad - Il\,\mathbf{e}_x \times \Sigma_\alpha [B_\alpha(x, y, 0) + \tfrac{1}{2}L\,\partial_y B_\alpha]\,\mathbf{e}_\alpha - IL\,\mathbf{e}_y \times \Sigma_\alpha [B_\alpha(x, y, 0) - \tfrac{1}{2}l\,\partial_x B_\alpha]\,\mathbf{e}_\alpha\\ &= I\mathcal{S}\,(-\partial_y B_y\,\mathbf{e}_z + \partial_y B_z\,\mathbf{e}_y - \partial_x B_x\,\mathbf{e}_z + \partial_x B_z\,\mathbf{e}_x).\end{aligned}$$

By using equation [6.9], which implies that $\partial_x B_x + \partial_y B_y = -\partial_z B_z$, we find:

$$\mathbf{F}_M = I\mathcal{S}\,(\partial_x B_z\,\mathbf{e}_x + \partial_y B_z\,\mathbf{e}_y + \partial_z B_z\,\mathbf{e}_z), \quad \text{i.e.,} \quad \mathbf{F}_{M,\alpha} = \partial_\alpha(\boldsymbol{\mathcal{M}}.\mathbf{B}). \qquad [6.65]$$

To displace the circuit from y_o (where $\mathbf{B}_o = 0$) to the actual position y (where $\mathbf{B} \neq 0$) while I is maintained constant, a force $\mathbf{F}' = -\mathbf{F}_M$ and a work must be exerted

$$W'_{0\to\mathbf{B}} = \int_{y_o}^{y} d\mathbf{r}.\,\mathbf{F}'_1 = -\int_{y_o}^{y} d\mathbf{r}.\,\mathbf{F}_1 = -\int_{y_o}^{y} dy\; I\mathcal{S}(\partial_y B_3) = I\mathcal{S}[B_3(y_o) - B_3(y)] = -\boldsymbol{\mathcal{M}}.\mathbf{B}.$$

This is also the variation of the magnetic energy in the displacement of the circuit from the region where $\mathbf{B} = 0$ to the region where the field is $\mathbf{B}$. It differs from the expression [6.64] only by a constant term $\mathcal{M}B$, which has no physical importance. Thus, the magnetic force exerted by a magnetic field on a circuit carrying a constant current corresponds to a potential energy $U_M = -\boldsymbol{\mathcal{M}}.\mathbf{B} = -IS\,\mathbf{n}.\mathbf{B}$. We may also write:

$$U_M = -I\Phi, \qquad [6.66]$$

where Φ is the magnetic flux. A circuit $\mathcal{C}$ of arbitrary shape may be considered as a juxtaposition of small rectangular circuits $\mathcal{C}_i$ (Figure 6.5b). As I is the same for all these circuits and the sum of the Φ_i is the total flux Φ through the circuit $\mathcal{C}$, equation [6.66] is valid for circuits of any shape $\mathcal{C}$ immersed in a field $\mathbf{B}$, even if the field is non-uniform. We note that U_M is only the interaction energy of the circuit with the external field $\mathbf{B}$. It is not the total magnetic energy of the circuit, which must include its proper magnetic energy, that is the energy necessary to establish the current in the circuit. Although the force exerted by the field $\mathbf{B}$ on a charge in motion or an element of a length of circuit is not conservative, we have defined here the interaction energy of the closed circuit in the field $\mathbf{B}$ and this energy depends only on the position of the circuit. In other words, the force of interaction of a closed circuit with a magnetic field is conservative.

The interaction energy $U_M = -I\Phi$ determines the motion or the deformation of an electric circuit in a magnetic field if the intensity I is maintained constant. By evolving from a state (1) to a state (2), the work of the magnetic forces is equal to the decrease of the interaction energy, $U_{M,1} - U_{M,2} = I(\Phi_2 - \Phi_1)$ and this work cannot be negative. Thus, Φ_2 must be higher than Φ_1: the circuit moves or deforms in such a way that U_M decreases, i.e. Φ increases. This is *Maxwell's maximum flux rule*. The equilibrium is reached if U_M is minimum, i.e. if Φ is maximum.

In the case of a rigid circuit, the ratio $\mathcal{M}/I$ is constant. If **B** is uniform, the flux $\Phi = (\boldsymbol{\mathcal{M}}.\mathbf{B})/I = (\mathcal{M}/I)\, B \cos\theta$ is maximum for $\theta = 0$. Thus, the circuit rotates until its magnetic moment points in the direction of **B**. This position corresponds to a maximum of Φ, i.e. a minimum of U_M; thus, it is stable. The moment of force vanishes also for $\theta = \pi$. However, this position corresponds to a minimum of Φ, i.e. a maximum of U_M; thus, it is unstable. If a rigid circuit is free to move in a non-uniform field, the maximum flux rule implies that it moves toward the region of stronger field. Finally, if the circuit is deformable, it deforms until its magnetic moment $\mathcal{M}$ is maximum (thus, its surface is maximum). In general, it may undergo rotation, translation, and deformation.

6.11. Magnetic forces

A) Magnetic force on a circuit in terms of the flux

Let us consider a circuit $\mathcal{C}$ carrying a constant current I (maintained by a generator). The other systems act on $\mathcal{C}$ only by the intermediary of the magnetic field **B**. The magnetic force acting on the circuit may be evaluated by using Laplace's force $d\mathbf{F}_M = I\, d\mathbf{L} \times \mathbf{B}$. It may be calculated also by using the interaction energy $U_M = -I\Phi$ of the circuit in the field **B**. Indeed, if the element $d\mathbf{L}$ undergoes a translation $\delta\mathbf{r}$ (Figure 6.15a), the work of the magnetic force is

$$\delta dW_M = d\mathbf{F}_M.\delta\mathbf{r} = I\,(d\mathbf{L} \times \mathbf{B}).\,\delta\mathbf{r} = I\,\mathbf{B}.(\delta\mathbf{r} \times d\mathbf{L}) = I\, d\delta\mathcal{S}\,(\mathbf{B}.\mathbf{n}) = I\, d\delta\Phi,$$

where **n** is the unit vector orthogonal to the parallelogram of sides $\delta\mathbf{r}$ and $d\mathbf{L}$ and $d\delta\mathcal{S} = |\delta\mathbf{r} \times d\mathbf{L}|$ is the swept area; $d\delta\Phi$ is then the magnetic flux through this area. In the case of a finite part $\mathcal{C}$ of an electric circuit (Figure 6.15b) or a closed circuit $\mathcal{C}$ (Figure 6.15c), the work of the magnetic forces is obtained by integration over the circuit; we find $\delta W_M = I\,\delta\Phi$, where $\delta\Phi$ is the flux that is swept by $\mathcal{C}$ in the course of its motion.

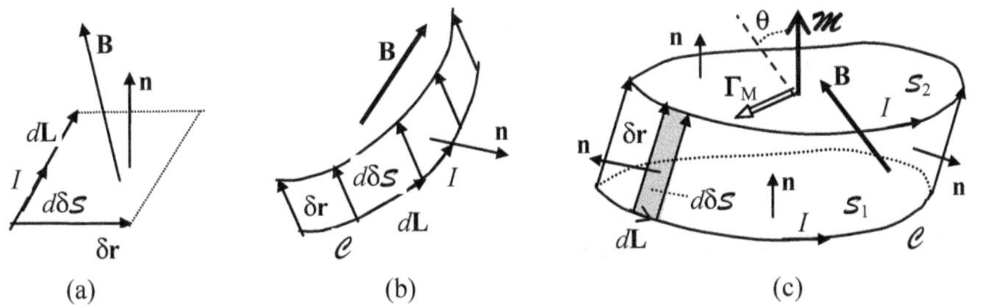

Figure 6.15. *a) Work of the magnetic forces exerted on an element* **dL** *of circuit, b) work on a finite circuit $\mathcal{C}$, and c) work on a closed circuit and moment of the magnetic forces*

If a rigid circuit $\mathcal{C}$ undergoes a translation, we define its position by the position **r** of one of its points. In a virtual translation $\delta\mathbf{r}$ of the circuit, the work of $\mathbf{F}_M$ is equal to the decrease of the interaction energy of the circuit in the field **B**, hence $\delta W_M = \mathbf{F}_M.\,\delta\mathbf{r} = -\,\delta U_M = I\,\delta\Phi$, where Φ is the flux through $\mathcal{C}$. As U_M and Φ are functions of **r**, their variations are $\delta U_M = \delta\mathbf{r}.\nabla U_M$ and $\delta\Phi = \delta\mathbf{r}.\nabla\Phi$. Thus, the force may be written as

$$\mathbf{F}_M = -\,\nabla U_M = I\,\nabla\Phi. \qquad [6.67]$$

A similar analysis may be undertaken to determine the moment of the magnetic forces $\mathbf{\Gamma}_M$ exerted by a field **B** on a circuit $\mathcal{C}$ with respect to a point O. For this, let us assume that the circuit undergoes a virtual rotation through an infinitesimal angle $\delta\theta$ about an axis at O of unit vector **u**. The work of the magnetic forces is $\delta W_M = \mathbf{\Gamma}_M.\mathbf{u}\;\delta\theta = -\delta U_M = I\;\delta\Phi$. We deduce that the component of $\mathbf{\Gamma}_M$ in the direction of **u** is

$$\mathbf{\Gamma}_M.\mathbf{u} = -\,\partial U_M/\partial\theta = I\,(\partial\Phi/\partial\theta). \qquad [6.68]$$

The interaction energy of a circuit $\mathcal{C}$ of magnetic moment $\mathcal{M}$ with a magnetic field **B** is $U_M = -\,\mathcal{M}.\mathbf{B} = -\Sigma_\beta\,\mathcal{M}_\beta B_\beta$. Thus, the components of the magnetic force exerted on $\mathcal{C}$ are $F_\alpha = -\partial_\alpha U_M = \partial_\alpha(\mathcal{M}.\mathbf{B}) = \Sigma_\beta\,\mathcal{M}_\beta\,(\partial_\alpha B_\beta)$. As **B** is produced by permanent magnets or other circuits than $\mathcal{C}$, Ampère's law [6.59] implies that $\nabla\times\mathbf{B} = 0$ at the points on the circuit $\mathcal{C}$, hence $\partial_\alpha B_\beta = \partial_\beta B_\alpha$. Thus, we may write

$$F_{M,\alpha} = \Sigma_\beta\,\mathcal{M}_\beta\,(\partial_\beta B_\alpha), \qquad \text{i.e. } \mathbf{F}_M = (\mathcal{M}.\nabla)\,\mathbf{B}. \qquad [6.69]$$

Particularly, if the field is uniform, any translation does not modify the flux, and equation [6.67] implies that the resultant of the magnetic forces on the circuit is equal to zero and the equation $\mathbf{F}_M = (\mathcal{M}.\nabla)\,\mathbf{B}$ gives the same result. Similarly, if the magnetic moment of the circuit makes an angle θ with **B**, the interaction energy is $U_M = -\mathcal{M}B\cos\theta$; thus, $\mathbf{\Gamma}_M.\mathbf{u} = -\mathcal{M}B\sin\theta$ and this is equivalent to $\mathbf{\Gamma}_M = \boldsymbol{\mathcal{M}} \times \mathbf{B}$.

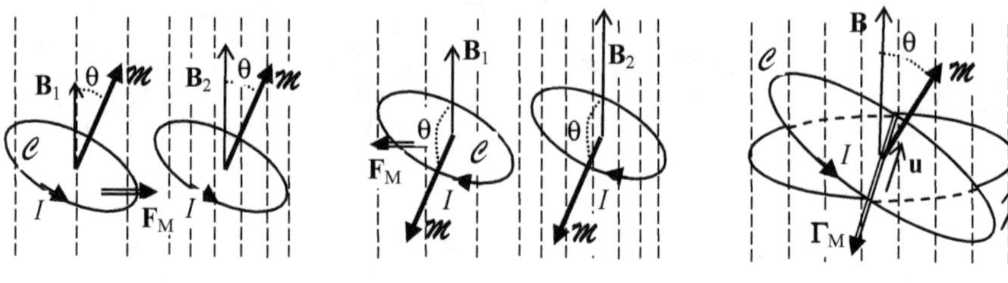

Figure 6.16. *Maxwell's maximum flux rule: a) the magnetic force on $\mathcal{C}$ moves it toward a stronger field (if θ < π/2) or b) toward a weaker field (if θ > π/2). c) The moment of magnetic forces on $\mathcal{C}$ rotates it toward smaller θ*

Figure 6.16 illustrates the use of the maximum flux rule to determine the direction of the magnetic force and the moment of force. In the case of Figure 6.16a, $\boldsymbol{\mathcal{M}}$ forms an acute angle with $\mathbf{B}_1$. According to the right-hand rule, the flux Φ is positive. By moving the circuit in a direction such that $B_2 > B_1$, Φ increases. According to [6.67], the component of the magnetic force in this direction is positive. On the contrary, in the case of Figure 6.16b, $\boldsymbol{\mathcal{M}}$ forms an obtuse angle with $\mathbf{B}_1$ and Φ is negative. By moving the circuit in a direction such that $B_2 > B_1$, Φ decreases (in algebraic values). According to [6.67], the component of $\mathbf{F}_M$ in this direction is negative. In the case of Figure 6.16c, $\boldsymbol{\mathcal{M}}$ forms an acute angle θ with **B** and Φ is positive. If θ is increased by rotating the circuit about **u** (according to the right-hand rule), the flux Φ decreases. According to [6.68], the component of $\mathbf{\Gamma}_M$ in the direction of **u** is negative, $\mathbf{\Gamma}_M$ is thus in the opposite direction to **u**, in agreement with the relation $\mathbf{\Gamma}_M = \boldsymbol{\mathcal{M}} \times \mathbf{B}$.

If the circuit is rigid and it carries a constant current *I*, the forces exerted by its own field are counterbalanced by the internal mechanical forces that keep it rigid. Thus, if the circuit is subject to no external forces, it undergoes no translation, rotation, or deformation. Thus, the equation $\delta W_M = I\,\delta\Phi$ holds with δΦ representing only the flux of the external field **B** that is swept by the circuit. In the case of a closed circuit (Figure 6.16c), δΦ is the flux of **B** through the lateral surface. As the flux of **B** leaving any closed surface is equal to zero, δΦ may be written as $(\Phi_2 - \Phi_1)$, where Φ_1 and Φ_2 are the fluxes of **B** through the circuit in the initial and the final positions, respectively. Thus, we may write $\delta W = I\,(\Phi_2 - \Phi_1)$.

B) Magnetic interactions of charges and circuits

Besides the electric interaction, two charges q at $\mathbf{r}$ and q' at $\mathbf{r}'$ have a magnetic interaction if they are in motion. Indeed, the charge q' of velocity $\mathbf{v}'$ produces a magnetic field $\mathbf{B}'(\mathbf{r}) = (\mu_o/4\pi)\, q'(\mathbf{v}' \times \mathbf{R})/R^3$ at $\mathbf{r}$ where $\mathbf{R} = \mathbf{r} - \mathbf{r}'$. This field acts on the charge q of velocity $\mathbf{v}$ with a force

$$\mathbf{F}_{q'\to q} = q(\mathbf{v}\times\mathbf{B}') = \frac{\mu_o qq'}{4\pi R^3}[\mathbf{v}\times(\mathbf{v}'\times\mathbf{R})] = \frac{\mu_o qq'}{4\pi R^3}[\,\mathbf{v}'(\mathbf{v}.\mathbf{R}) - \mathbf{R}\,(\mathbf{v}.\mathbf{v}')]. \quad [6.70]$$

Conversely, charge q produces a field $\mathbf{B}(\mathbf{r}')$ at $\mathbf{r}'$, which acts on q' with a force

$$\mathbf{F}_{q\to q'} = q'(\mathbf{v}'\times\mathbf{B}) = \frac{\mu_o qq'}{4\pi R^3}\{\mathbf{v}'\times[\mathbf{v}\times(-\mathbf{R})\} = \frac{\mu_o qq'}{4\pi R^3}[-\mathbf{v}(\mathbf{v}'.\mathbf{R}) + \mathbf{R}(\mathbf{v}.\mathbf{v}')]. \quad [6.71]$$

We find that the forces $\mathbf{F}_{q'\to q}$ and $\mathbf{F}_{q\to q'}$ are not opposite and they are not along the axis of $\mathbf{R}$, which joins the two charges (Figure 6.17a). Thus, the magnetic interaction of charges does not obey the principle of action and reaction of classical mechanics and it does not conserve angular momentum. On the other hand, it is not conservative; thus, it is not possible to associate a potential energy with it.

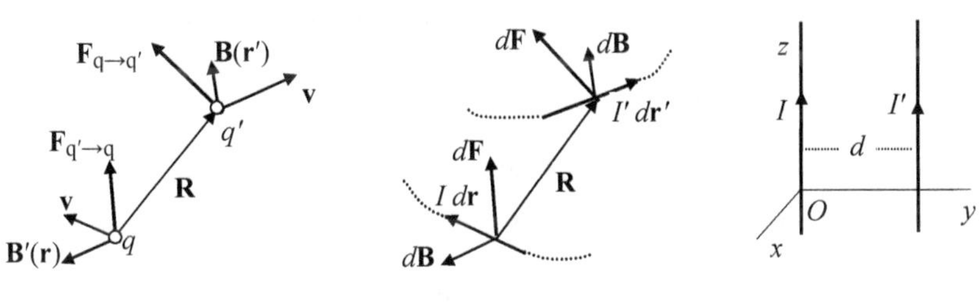

Figure 6.17. *Magnetic interaction: a) of two charges, b) of two elements of circuits, and c) of two parallel thin rods carrying currents I and I′*

C) Magnetic interaction of two circuits

The field $\mathbf{B}(\mathbf{r}')$ produced by $\mathcal{C}$ at a point $\mathbf{r}'$ of $\mathcal{C}'$ is

$$\mathbf{B}(\mathbf{r}') = (\mu_o I/4\pi)\int_{\mathcal{C}} d\mathbf{r}\times\mathbf{R}/\,R^3, \qquad \text{where } \mathbf{R} = \mathbf{r}' = -\,\mathbf{r}. \quad [6.72]$$

This field acts on an element $d\mathbf{r}'$ of $\mathcal{C}'$ with a force $d\mathbf{F}' = I'\, d\mathbf{r}' \times \mathbf{B}(\mathbf{r}')$ (Figure 6.17b). The force $\mathbf{F}_{\mathcal{C}\to\mathcal{C}'}$ exerted by $\mathcal{C}$ on $\mathcal{C}'$ is obtained by integration over $\mathcal{C}'$, hence

$$\begin{aligned}\mathbf{F}_{\mathcal{C}\to\mathcal{C}'} &= \int_{\mathcal{C}'} d\mathbf{F}' = I'\int_{\mathcal{C}'} d\mathbf{r}' \times \mathbf{B}(\mathbf{r}') = (\mu_o II'/4\pi)\int_{\mathcal{C}'} d\mathbf{r}' \times \int_{\mathcal{C}} d\mathbf{r} \times \mathbf{R}/R^3 \\ &= (\mu_o II'/4\pi)\int_{\mathcal{C}}\int_{\mathcal{C}'}[d\mathbf{r}(d\mathbf{r}'.\mathbf{R}) - \mathbf{R}(d\mathbf{r}.d\mathbf{r}')]/R^3.\end{aligned} \qquad [6.73]$$

Let us consider two linear, thin and parallel conductors of length L carrying currents I and I' (Figure 6.17c). Taking $\mathcal{C}$ along Oz and $\mathcal{C}'$ in the Oyz plane at a distance d from Oz, we find

$$\begin{aligned}\mathbf{F}_{\mathcal{C}\to\mathcal{C}'} &= (\mu_o II'/4\pi)\int_{-L/2}^{L/2} dz \int_{-L/2}^{L/2} dz'\,[\mathbf{e}_z(z'-z) - \mathbf{R}]/R^3 \\ &= -\frac{\mu_o}{4\pi} II'\,\mathbf{e}_y \int_{-L/2}^{L/2} dz \int_{-L/2}^{L/2} \frac{dz'}{[d^2+(z'-z)^2]^{3/2}} = \frac{\mu_o}{2\pi} II'[1-\sqrt{L^2/d^2+1}\,]\mathbf{e}_y.\end{aligned} \qquad [6.74]$$

Particularly, if $L >> d$, we find

$$\mathbf{F}_{\mathcal{C}\to\mathcal{C}'} = -\frac{\mu_o}{2\pi}\frac{II'L}{d}\,\mathbf{e}_y. \qquad [6.75]$$

This force lies in the plane of the conductors, it is attractive if the currents are in the same direction and repulsive if they are in opposite directions. It is proportional to the length of the conductors and inversely proportional to their spacing d. This force is used to define the unit intensity with the choice $\mu_o = 4\pi \times 10^{-7}$ T.m.A^{-1}.

In the case of closed circuits $\mathcal{C}$ and $\mathcal{C}'$, the first term of [6.73] provides no contribution. Indeed, by making the change of variable $\mathbf{r}' = \mathbf{R} + \mathbf{r}$, we get

$$\int_{\mathcal{C}} d\mathbf{r} \int_{\mathcal{C}'} (d\mathbf{r}'.\mathbf{R})/R^3 = \int_{\mathcal{C}} d\mathbf{r} \int_{\mathcal{C}'} (d\mathbf{R}.\mathbf{R})/R^3 = \tfrac{1}{2}\int_{\mathcal{C}} d\mathbf{r}\int_{\mathcal{C}'} d\mathbf{R}^2/R^3 = \int_{\mathcal{C}} d\mathbf{r}\int_{\mathcal{C}'} dR/R^2 = \int_{\mathcal{C}} d\mathbf{r}\,(1/R)\,|_{\mathcal{C}'},$$

where R is the distance from a running point $\mathbf{r}$ of $\mathcal{C}$ to the running points $\mathbf{r}'$ of $\mathcal{C}'$. The integral over $\mathcal{C}'$ is $(1/R)\,|_{\mathcal{C}'}$, i.e. the variation of $1/R$ over the closed path $\mathcal{C}'$ and this variation is equal to zero. Thus, the force exerted by $\mathcal{C}$ on $\mathcal{C}'$ may be written as

$$\mathbf{F}_{\mathcal{C}\to\mathcal{C}'} = -(\mu_o II'/4\pi)\int_{\mathcal{C}}\int_{\mathcal{C}'} (d\mathbf{r}.d\mathbf{r}')\,\mathbf{R}/R^3. \qquad [6.76]$$

The force that $\mathcal{C}'$ exerts on $\mathcal{C}$ is obtained simply by exchanging $\mathcal{C}$ and $\mathcal{C}'$ (thus $\mathbf{R}$ to $-\mathbf{R}$). Contrarily to the magnetic interaction of two charged particles or two elements of circuits $d\mathbf{r}$ and $d\mathbf{r}'$, we find that $\mathbf{F}_{\mathcal{C}'\to\mathcal{C}}$ is the opposite of $\mathbf{F}_{\mathcal{C}\to\mathcal{C}'}$ in the case of two closed circuits or two open but rectilinear and parallel circuits, as indicated by equation [6.74]. The reason for this apparent violation of the principle of action and reaction is that the used fields are instantaneous (i.e. evaluated at the same time t as the positions and velocities of the particles or the currents). The right formulation of the interaction must take into account the propagation of the field with a finite speed. On the other hand, the radiated electric and magnetic fields carry energy, momentum, and angular momentum, which we have to take into account to write the conservation laws of these quantities. These effects appear in the case of non-stationary phenomena, as in the case of moving particles or open circuits (which cannot carry permanent currents).

Magnetic forces exist also between different parts of the same circuit. An infinitesimal element $d\mathbf{r}$ of the circuit is subject to a force $d\mathbf{F} = I\, d\mathbf{r} \times \mathbf{B}'$, where $\mathbf{B}'$ is the field produced by all the parts of the circuit except $d\mathbf{r}$. This internal magnetic force has no effect on the circuit if it is rigid, as it is counterbalanced by the internal mechanical forces. However, if the circuit is deformable, the effect of the internal magnetic forces increases its area so that the flux of the proper magnetic field $\mathbf{B}$ increases, in agreement with Maxwell's maximal flux rule.

Two points should be noted. The first is that the magnetic interaction is different from Coulomb's interaction. Indeed, the magnetic force vanishes if the charges are at rest, contrarily to Coulomb's force. On the other hand, the elements of the circuits have no net charge; thus, they have a magnetic interaction but no Coulomb interaction. However, as we shall see in section 8.4 and Chapter 13, the electric and magnetic forces are not completely independent, especially if we consider electromagnetic phenomena in different referential frames. The second point is that, in all our analysis of the fields of electric circuits and their interactions, we have assumed that the electric current is permanent; thus, it is the same at all points of the circuit. This holds approximately in the case of quasi-permanent phenomena (i.e. slowly variable in time) but not in the case of high-frequency currents, for instance.

D) Magnetic interaction of two volume distributions of current

Let us consider two distributions of current in two distinct volumes $\mathcal{V}$ and $\mathcal{V}'$. The field $\mathbf{B}(\mathbf{r}')$ produced by $\mathcal{V}$ at a point $\mathbf{r}'$ of $\mathcal{V}'$ is

$$\mathbf{B}(\mathbf{r}') = (\mu_o/4\pi) \iiint_{\mathcal{V}} d\mathcal{V}\, \mathbf{j}(\mathbf{r}) \times \mathbf{R} / R^3, \qquad \text{where } \mathbf{R} = \mathbf{r}' - \mathbf{r}. \qquad [6.77]$$

This field acts on an element of volume $d\mathcal{V}'$ with a force $d\mathbf{F}' = d\mathcal{V}'\, \mathbf{j}'(\mathbf{r}') \times \mathbf{B}(\mathbf{r}')$. The force $\mathbf{F}_{\mathcal{V}\to\mathcal{V}'}$ exerted by $\mathcal{V}$ on $\mathcal{V}'$ is obtained by integration over $\mathcal{V}'$

$$\mathbf{F}_{\mathcal{V}\to\mathcal{V}'} = \iiint_{\mathcal{V}'} d\mathbf{F}' = (\mu_o/4\pi) \iiint_{\mathcal{V}'} d\mathcal{V}'\, \mathbf{j}'(\mathbf{r}') \times \iiint_{\mathcal{V}} d\mathcal{V}\, \mathbf{j}(\mathbf{r}) \times \mathbf{R}/ R^3 \quad \text{where } \mathbf{R} = \mathbf{r}' - \mathbf{r}$$
$$= (\mu_o/4\pi) \iiint_{\mathcal{V}'} d\mathcal{V}' \iiint_{\mathcal{V}} d\mathcal{V}\, \{\, \mathbf{j}(\mathbf{r})[\mathbf{R}.\mathbf{j}'(\mathbf{r}')] - \mathbf{R}\, [\mathbf{j}(\mathbf{r}).\mathbf{j}'(\mathbf{r}')]\}/ R^3.$$

Here too, it may be shown that the first term offers no contribution. Indeed, it may be written as

$$-(\mu_o/4\pi) \iiint_{\mathcal{V}} d\mathcal{V}\, \mathbf{j}(\mathbf{r}) \iiint_{\mathcal{V}'} d\mathcal{V}' \mathbf{j}'(\mathbf{r}').\, \boldsymbol{\nabla}'(1/R)$$
$$= -(\mu_o/4\pi) \iiint_{\mathcal{V}} d\mathcal{V}\, \mathbf{j}(\mathbf{r}) \iiint_{\mathcal{V}'} d\mathcal{V}' \boldsymbol{\nabla}'.[\mathbf{j}'(\mathbf{r}')/R] + (\mu_o/4\pi) \iiint_{\mathcal{V}} d\mathcal{V}\, \mathbf{j}(\mathbf{r}) \iiint_{\mathcal{V}'} d\mathcal{V}' \boldsymbol{\nabla}'.\mathbf{j}'(\mathbf{r}')/R.$$

The integral over $\mathcal{V}'$ in the first term is that of a divergence; it may be transformed into the flux of $\mathbf{j}'/R$ through a surface $\mathcal{S}'$,which contains $\mathcal{V}'$, and we may chose it such that $\mathbf{j}' = 0$ at each point. The second term is equal to zero, as $\boldsymbol{\nabla}'.\mathbf{j}' = 0$ because of the conservation of charge in the case of time-independent phenomena. Thus, the force may be written as:

$$\mathbf{F}_{\mathcal{V}\to\mathcal{V}'} = \frac{\mu_o}{4\pi} \iiint_{\mathcal{V}} d\mathcal{V} \iiint_{\mathcal{V}'} d\mathcal{V}' [\mathbf{j}(\mathbf{r}).\, \mathbf{j}'(\mathbf{r}')] \frac{\mathbf{r}-\mathbf{r}'}{|\mathbf{r}-\mathbf{r}'|^3}. \qquad [6.78]$$

This force verifies the principle of action and reaction. On the other hand, $\mathbf{F}_{\mathcal{V}\to\mathcal{V}'}$ reduces to [6.76] if the current densities $\mathbf{j}(\mathbf{r})$ and $\mathbf{j}'(\mathbf{r}')$ are restricted to the thin volumes of the closed circuits $\mathcal{C}$ and $\mathcal{C}'$. We note that [6.78] does not hold if $\mathcal{V}$ or $\mathcal{V}'$ is a part of current distribution (i.e. if the current density does not vanish on its boundaries). This is the case of a finite circuit as in [6.73], for instance, as the relation $\boldsymbol{\nabla}'.\mathbf{j}' = 0$, which we used to derive [6.78], does not hold at the ends of the circuit.

6.12. Question of magnetic monopoles*

We have seen that the conservation of the magnetic flux and the equation $\boldsymbol{\nabla}.\mathbf{B} = 0$, compared to Gauss'a law in electrostatics ($\boldsymbol{\nabla}.\mathbf{E} = q_v/\varepsilon_o$), indicate the absence of magnetic charges. Indeed, all experiments up to now have failed to detect magnetic monopoles, although their existence is required by some modern theories. In 1931, Dirac showed that the existence of magnetic monopoles explains the quantization of electric charge. By analogy to electric dipole moments, we may consider electric circuits and magnets (which have dipole moments) as two opposite monopoles. A solenoid is thus equivalent to two monopoles of opposite signs located at its ends. Dirac conceived a monopole as the end of a very long and very thin solenoid if no experiment can detect the solenoid (Figure 6.18a).

In quantum theory, a particle is represented by a complex wave function. As in the case of a light wave, we can imagine that the solenoid can be detected by producing an interference or diffraction pattern with a beam of electrons or other charged particles. If a particle of charge q moves along a path PQ, its wave function is multiplied by a phase factor $\exp(i\alpha_{PQ})$ where the phase shift α_{PQ} is related to the circulation of the vector potential $\mathbf{A}$ on the path PQ by the relation $\alpha_{PQ} = iq \int_P^Q d\mathbf{r}.\mathbf{A}$.

In particular, if the phase shift $\alpha_{\mathcal{C}}$ on any closed contour $\mathcal{C}$ around the solenoid is an integer multiple of 2π, no experiment is able to detect this phase shift, exactly as if the solenoid does not exist.

Figure 6.18. *a) Dirac monopole and b) vector potential of a magnetic monopole*

By analogy to the electric field of a charged particle, let us assume that a magnetic monopole produces a magnetic field $\mathbf{B}(\mathbf{r}) = \kappa\mathbf{r}/r^3$. We may show that the vector potential of this field in spherical coordinates is

$$\mathbf{A}(\mathbf{r}) = \frac{\kappa}{r}\left[\frac{1-\cos\theta}{\sin\theta}\,\mathbf{e}_\varphi + \nabla U(r, \theta, \varphi)\right], \tag{6.79}$$

where U is an arbitrary scalar function (as $\mathbf{A}$ is specified up to a gradient). The phase shift $\alpha_{\mathcal{C}}$ associated with this closed contour $\mathcal{C}$ is equal to $2n\pi$ if

$$q = n/2\kappa, \qquad \text{where } n = \text{integer}. \tag{6.80}$$

This relation explains the quantization of the electric charge q. Considering the elementary charge e of the electron, the quantity $\kappa = 1/2e$ plays the part of an elementary magnetic charge (*Dirac monopole*).

For $\theta = \pi$, the expression [6.79] shows that $\mathbf{A}$ is infinite. The existence of this singularity of $\mathbf{A}$ is not physically acceptable and, although the solenoid is not observable, it is preferable to formulate the theory of magnetic monopoles without the concept of solenoid. To study the vector potential on the sphere, we divide it into two hemispheres (Figure 6.18b). The solutions on the upper and the lower hemispheres are, respectively,

$$\mathbf{A}_U = \frac{\kappa}{r}\frac{1-\cos\theta}{\sin\theta}\,\mathbf{e}_\varphi \qquad (\text{for } 0 < \theta < \pi/2),$$

$$\mathbf{A}_L = -\frac{\kappa}{r}\frac{1+\cos\theta}{\sin\theta}\,\mathbf{e}_\varphi \qquad (\text{for } \pi/2 < \theta < \pi). \tag{6.81}$$

Together, $\mathbf{A}_U$ and $\mathbf{A}_L$ constitute a non-singular solution in their respective regions and both correspond to $\mathbf{B} = \kappa\mathbf{r}/r^3$. On the common equator ($\theta = \pi/2$), $\mathbf{A}_U$ and $\mathbf{A}_L$ do not coincide, but $\mathbf{A}_U - \mathbf{A}_L = (2\kappa/r)\mathbf{e}_\varphi$. However, both expressions represent the same physical situation, as they are related by a gauge transformation $\mathbf{A}_U - \mathbf{A}_L = \nabla f$, with $f = 2\kappa\varphi$. The phase shift received by the wave function is the same for both solutions if Dirac condition is satisfied.

We note that, if the magnetic monopoles exist, they show up by their radial magnetic field, which decreases like $1/r^2$ instead of $1/r^3$ for magnetic dipole moments, by a magnetic force exerted by a uniform field **B** in the direction of **B,** and finally, by a violation of the conservation of the magnetic flux.

More recently, in the so-called *Grand Unification Theory* (which attempts to unify all the interactions of Nature), magnetic monopoles are assumed to have been produced during the "Big Bang", which formed the Universe. They must be very heavy (probably 10^{16} times the mass of the proton) and probably with a lifetime so short that very few of them still survive.

6.13. Problems

Force exerted by a magnetic field on moving charges and currents, Laplace's force

P6.1 A strip of copper of width 10 cm and thickness 4 mm carries a current of 100 A and is immersed in an orthogonal magnetic field $B = 2$ T. Assuming 1 conduction electron per atom, calculate the drift velocity of the electrons and the magnetic force exerted on these electrons. What is the electric field whose force counterbalances this magnetic force? What is then the Hall voltage? Calculate the charge density on the narrow faces of the strip. Verify that the ratio of the Hall field to the electric field that produces the current is $E_H/E = B/\rho N_v e$. What would be the Hall voltage if the conductor has the same number of positive and negative charge carriers? The copper density is 8920 kg/m^3 and its atomic mass is 63.6.

P6.2 Using Laplace's law, calculate the force exerted by a magnetic field **B** on a wire having the shape of a half-circle and carrying a current *I*. Assume that **B** is orthogonal to the plane of the wire. Calculate also the force exerted on a wire carrying the current *I* along the diameter.

P6.3 Barlow wheel is a simple electric motor (Figure 6.19). This is a metallic disk of radius *R* that may rotate about its axis *O* in a magnetic field **B** parallel to the axis. An electric current arrives by the axis and leaves from a point *A* of the periphery in contact with mercury. Let $\mathbf{j}_s(r, \varphi)$ be the surface current density in polar coordinates in the disk. Calculate the force that acts on the element of area $dS = r\, dr\, d\varphi$ of the disk and show that its moment with respect to O is $d\mathbf{\Gamma}_s = -\, dS\ \mathbf{B}\ (\mathbf{r}.\mathbf{j}_s)$. Show that $\int_0^\pi d\varphi\ (\mathbf{r}.\mathbf{j}_s)$ does not depend on r and it is equal to the current intensity. Deduce that the moment of force is $\Gamma_M = -\frac{1}{2} IR^2\mathbf{B}$, as if the current was rectilinear along OA.

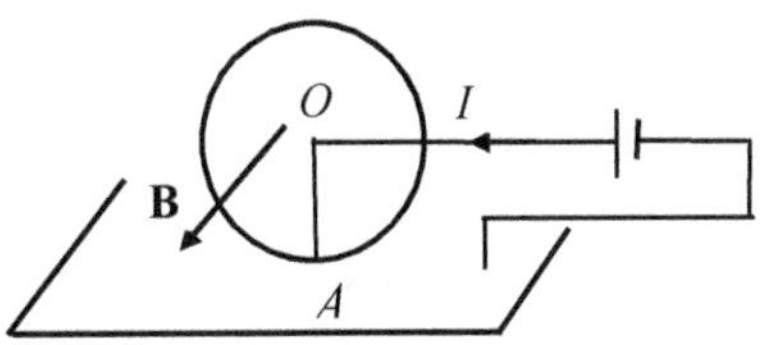

Figure 6.19. *Problem 6.3*

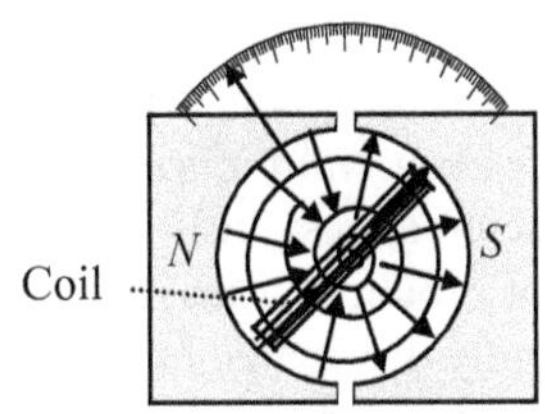

Figure 6.20. *Problem 6.3*

Magnetic fields of particles and currents, Biot-Savart's law

P6.4 a) Show the identity $\nabla\times(f\ \mathbf{V}) = \nabla f\times\mathbf{V} + f\ \nabla\times\mathbf{V})$. Show that $\mathbf{A} = \mu_0 q\mathbf{v}/4\pi R$ corresponds to the field $\mathbf{B} = \mu_0 q\ (\mathbf{v}\times\mathbf{R})/4\pi R^3$ and that $\mathbf{A}(\mathbf{r}) = (\mu_0/4\pi)\iiint_{\mathcal{V}} d\mathcal{V}'\ \mathbf{j}(\mathbf{r}')/R$ corresponds to $\mathbf{B}(\mathbf{r}) = (\mu_0/4\pi)\iiint_{\mathcal{V}} d\mathcal{V}'\ \mathbf{j}(\mathbf{r}')\times\mathbf{R}/R^3$. Verify that $\nabla.\mathbf{B} = 0$. **b)** Determine the vector potential of the uniform field $\mathbf{B} = B\ \mathbf{e}_Z$. Verify that the flux of $\mathbf{B}$ through a circle of axis Oz is equal to the circulation of $\mathbf{A}$ along this circle.

P6.5 a) A narrow coil of N turns and radius a is supplied with a current I. Analyze its field $\mathbf{B}$ on its axis. What is the limit expression of $\mathbf{B}$ if $z >> R$? Compare it with the field of a magnetic moment. At which distance do these fields differ by less than 1 %? Calculate the circulation of $\mathbf{B}$ along the axis and show that Ampère's law is verified although the path is not closed. Explain why. **b)** Two narrow coils (called *Helmholtz coils*) of radius a and N turns are put parallel to one another a distance D apart. They are supplied with the same current I. Study the variation of $\mathbf{B}$ on their common axis Oz. Verify that $\mathbf{B}$ is almost constant near the point O situated at equal distance from the coils. Show that, if $D = a$, both the first and the second derivatives of $\mathbf{B}$ are equal to 0 at O. Thus, the field $\mathbf{B}$ is almost uniform near O.

Magnetic moment

P6.6 A d'Arsonval galvanometer consists of a narrow rectangular coil of N turns placed in a radial magnetic field B of a permanent magnet (Figure 6.20). **a)** Show that it is subject to a magnetic torque $NI\mathcal{S}B$, where I is the current and $\mathcal{S}$ is the area of one rectangular run. **b)** The coil is subject also to the restoring moment $\Gamma' = -C\theta$ of a spring. Show that the intensity is related to the deviation angle θ by the relation $I = C\theta/NB\mathcal{S}$.

P6.7 a) Show that the magnetic moment $\mathcal{M} = I \iint_{\mathcal{S}} d\mathcal{S}' \, \mathbf{n}(\mathbf{r}')$ of a circuit $\mathcal{C}$ is independent of the surface $\mathcal{S}$ bounded by $\mathcal{C}$ and used to calculate $\mathcal{M}$. **b)** This circuit is free to rotate about a point O in a uniform external magnetic field **B**. Show that the moment of the magnetic forces is independent of O.

P6.8 The current due to the motion of charges that are carried by a moving body is called *convection current* to distinguish it from the *conduction current* in a conductor at rest. Rowland's experiment in 1876 has shown that the convection current produces a magnetic field exactly as conduction currents. **a)** A charge q is uniformly distributed on a non-conducting disk of radius a. The disk is rotated about its axis at N turns per second. By analyzing the symmetries, determine the direction of **B** at a point of the axis situated at a distance z from the center O of the disk. Calculate **B** and show that, at large distance ($z >> a$), the disk is equivalent to a magnetic moment $\mathcal{M} = \frac{1}{2}\pi q N a^2$. **b)** Make a similar analysis in the case of a charge q uniformly distributed on a sphere of radius R. **c)** Verify that we may write in both cases $\mathcal{M} = q\mathbf{L}/2m$ where **L** is the angular momentum of the body. This result holds for a body of any shape if the charge and the mass were uniformly distributed.

Field and potential of some simple circuits

P6.9 A long cylindrical conductor of radius R and axis Oz has a cylindrical cavity of radius a with its axis displaced a distance b from Oz and it carries a current I uniformly distributed on its section. Show that the field **B** inside the cavity is uniform with a magnitude $B = \mu_0 bI/2\pi(R^2 - a^2)$.

P6.10 a) A thin rod of length $2L$ carries a current I. Calculate the field and the vector potential at a distance ρ in its median plane. Use this result to calculate the field and the vector potential of a square circuit of sides $2L$ at a distance z from its center on its normal axis. What are the limit expressions of **B** and **A** if $z >> L$? **b)** A very long rod and radius a carries a current I uniformly distributed on the section. Determine **B** and **A** at a distance ρ from the axis both inside and outside the rod. What are the limit expressions of **B** and **A** if $a \to 0$? **c)** A plate of very large dimensions but a small thickness d carries a uniform current density **j**. Calculate **B** and **A** outside and inside this plate. What are the limits of **B** and **A** if $d \to 0$.

P6.11 A coaxial cable is formed by a cylinder of radius r_1 surrounded by a cylindrical shell of internal radius r_2 and external radius r_3. It carries in one direction and the other a current I uniformly distributed on the section of the conductors. Calculate **B** as a function of the distance r to the axis.

P6.12 A long copper cylindrical conductor of radius a = 1 cm carries a current I = 20 A uniformly distributed on the section. **a)** Calculate **B** inside the conductor at a distance ρ from the axis. Compare B on the surface of the conductor to the Earth's magnetic field of 0.4 G. **b)** Assuming that copper has 1 conduction electron per atom, calculate the drift velocity of the electrons. **c)** Calculate the magnetic force exerted on a conduction electron situated at a distance ρ from the axis. What is the direction of this force? **d)** Calculate the radial electric field, which counterbalances this force. What is the charge density, which produces this electric field? What is the number of electrons in excess that are able to produce this charge density? Is it reasonable to assume that the current density is uniform?

P6.13 a) Consider a cylindrical beam of particles of charge q, density N_v particles per unit volume, and moving with the velocity **v**. The Coulomb repulsion tends to increase the beam radius, while the magnetic force tends to decrease it. Calculate the electric and the magnetic fields produced by this beam at a distance ρ from its axis and compare the electric force and the magnetic force. Show that a charge at the periphery is submitted to a total force $\mathbf{F} = CI/\rho$, where I is the intensity of the beam and $C = (q/2\pi\varepsilon_o v)(1-v^2/c^2)$. **b)** Show that the divergence of the beam is $d\rho/dt = = 2(CI/m)\ \ln(\rho/\rho_o)^{1/2}$, where ρ_o is the distance to the axis of the points, where the radial velocity is equal to zero.

P6.14 a) Use the symmetries to show that the field of a very long solenoid is independent of z and φ in cylindrical coordinates. **b)** Use the equation $\nabla.\mathbf{B} = 0$ with the condition of regularity on the axis (or that the flux of **B** through an appropriate closed surface is equal to zero) to show that $B_\rho = 0$ everywhere. **c)** Use Ampère's law $\nabla \times \mathbf{B} = 0$ to show that, outside the conductor, B_z does not depend on the distance ρ from the axis. **d)** Let n be the number of turns per unit length. Show that, outside the solenoid, $B_z = 0$ and $B_\varphi = \mu_o I/2\pi\rho$ and that, inside the solenoid, $B_\varphi = 0$ and $B_z = \mu_o nI$, **e)** Using a simple argument in the case of a solenoid of finite length, show that the field B_z just at the end of the solenoid is $\frac{1}{2}\mu_o nI$.

P6.15 a) Using the symmetries, show that the vector potential inside a solenoid is oriented in the direction of $\mathbf{e}_\varphi$ with a magnitude that depends only on ρ. **b)** Express the circulation of **A** along an arbitrary closed path $\mathcal{C}$ in terms of the flux of **B**. Choosing appropriate paths, establish the expression of the curl in cylindrical coordinates. **c)** Using the expression $\mathbf{B}^{(in)} = \mu_o nI\ \mathbf{e}_z$ of the field inside a solenoid (where n is the number of turns per unit length) and $\mathbf{B}^{(ex)} = (\mu_o I/2\pi r)\mathbf{e}_\varphi$ outside it, determine the vector potential **A** inside and outside the solenoid.

P6.16 A toroidal coil of N turns carries a current I. Use the symmetry and Ampère's law to show that $B_\varphi = \mu_o NI/2\pi\rho$ inside the coil and $B_\varphi = 0$ outside it. Using an Ampèrien contour $\mathcal{A}$ around the coil, show that the field **B** outside the coil has a magnitude comparable to the field of a single loop carrying an intensity I.

Equations of time-independent magnetism in vacuum, singularities of B and A

P6.17 a) Show that $\mathbf{A}(\mathbf{r}) = (\mu_o/4\pi)\iiint d\mathcal{V}'\mathbf{j}(\mathbf{r}')/R$, where $\mathbf{R} = \mathbf{r} - \mathbf{r}'$, is a solution of the equation $\Delta\mathbf{A} = -\mu_o\mathbf{j}$. For this, use the identity $\Delta(1/|\mathbf{r} - \mathbf{r}'|) = -4\pi\,\delta(\mathbf{r} - \mathbf{r}')$, where $\delta(\mathbf{r} - \mathbf{r}')$ is the three-dimensional Dirac function (see section A.11 of the appendix A). Show that $\mathbf{A}$ verifies the condition $\nabla.\mathbf{A} = 0$. **b)** Deduce that the field may be written as $\mathbf{B}(\mathbf{r}) = (\mu_o/4\pi)\iiint_{\mathcal{V}} d\mathcal{V}'\,[\nabla'\times\mathbf{j}(\mathbf{r}')]/R$, where ∇' is the vector differential operator with respect to the coordinates (x', y', z'). Using the equation $\mathbf{B} = \nabla\times\mathbf{A}$, show that $\nabla\times\mathbf{B} = \mu_o\mathbf{j}$ and $\nabla.\mathbf{B} = 0$. Show that the above expression of $\mathbf{B}$ verifies these equations. **c)** A particle of charge q, which is at the position $\mathbf{r}_o$ with a velocity $\mathbf{v}$, is equivalent to a current density $\mathbf{j}(\mathbf{r}) = q\mathbf{v}\,\delta(\mathbf{r} - \mathbf{r}_o)$. Show that its vector potential at the point $\mathbf{r}$ is $\mathbf{A}(\mathbf{r}) = (\mu_o q/4\pi)\,\mathbf{v}/R$ and that its field is $\mathbf{B}(\mathbf{r}) = (\mu_o q/4\pi)\,\mathbf{v}\times\mathbf{R}/R^3$, where $\mathbf{R} = \mathbf{r} - \mathbf{r}_o$

P6.18 a) Let us assume that the field $\mathbf{B}$ is symmetric about Oz. Using the cylindrical components of the field, show that the conservation of the magnetic flux through a small cylinder of axis Oz implies that $B_\rho = -½\rho\,\partial_z B_z$ near the axis. **b)** Consider the field of a charged particle situated at the origin with a velocity $\mathbf{v}$ in the direction of Oz. Verify that the flux of its magnetic field through a sphere centered at O is zero. Calculate the cylindrical components of $\mathbf{B}$ and verify that $B_\rho = -½\rho\,\partial_z B_z$.

P6.19 A very long cylinder of radius R carries a current I uniformly distributed on its section. Using cylindrical coordinates, integrate the equation $\Delta\mathbf{A} = -\mu_o\mathbf{j}$ inside and outside the cylinder. Use the continuity conditions of $\mathbf{A}$ and $\mathbf{B}$ and that $\mathbf{B}$ is finite everywhere to write the expressions of $\mathbf{A}$ and $\mathbf{B}$ inside and outside the cylinder.

P6.20 A long solenoid of n turns per unit length carries a current I. The winding forms a cylindrical shell of internal radius a and external radius b. Using the symmetries, show that the equation $\nabla\times\mathbf{A} = -\mu_o\mathbf{j}$ reduces in cylindrical coordinates to the equation $(d/d\rho)[(1/\rho)(d/d\rho)(\rho A_\varphi)] = -\mu_o j_\varphi$. Determine A_φ as a function of ρ. Impose the continuity of the solutions on the surfaces $\rho = a$ and $\rho = b$ and verify that $\mathbf{B}^{(in)} - \mathbf{B}^{(ex)} = \mu_o j_s\,\mathbf{e}_z$.

P6.21 a) To analyze the field and the vector potential near a point O of a thin wire $\mathcal{C}$, we consider a segment $\mathcal{C}_1$ situated between $-L$ and $+L$ and the remaining $\mathcal{C}_2$. Obviously, the contribution of $\mathcal{C}_2$ is regular. Show that, in the limit $L \to 0$, the contribution of $\mathcal{C}_1$ to the field is $(\mu_o IL/2\pi\rho)\mathbf{e}_\varphi$ and its contribution to the vector potential is $(\mu_o I/2\pi)\ln(2L/\rho)\mathbf{e}_z$. Thus, the field and the vector potential are infinite on the wire. **b)** To show that the field is regular at points M of a surface $\mathcal{S}$ carrying a current density $\mathbf{j}_s$, we consider a small disk $\mathcal{S}_1$ of radius r_1 around M and the remaining of the surface $\mathcal{S}_2$. The contribution of $\mathcal{S}_2$ is regular. Show that, in the limit $r_1 \to 0$, the contribution of $\mathcal{S}_1$ to the field is $\pm\frac{1}{2}\mu_o \mathbf{j}_s \times \mathbf{e}_z$ near the surface and its contribution to the vector potential tends to 0 proportionally to r_1. **c)** To show that the field and the vector potential are regular at points M of a volume $\mathcal{V}$ carrying a current density **j**, we consider a small sphere $\mathcal{V}_1$ of radius r_1 and center M and the remaining $\mathcal{V}_2$. The contribution of $\mathcal{V}_2$ is regular. Show that, in the limit $r_1 \to 0$, the contribution of $\mathcal{V}_1$ tends to zero.

Magnetic energy and forces

P6.22 Show that the resultant of the forces exerted by a field **B** on the sides of a small rectangular circuit may be written as $d\mathbf{F} = (d\boldsymbol{\mathcal{M}}.\,\nabla)\,\mathbf{B}$.

P6.23 a) A circular loop of radius R is parallel to the Oxy plane and its center is at $z = h$ on the z axis. It is immersed in a field $\mathbf{B}_1 = B_o(1 + px)\mathbf{e}_z$ where B_o and p are constant. Calculate the resultant of the magnetic forces exerted on this loop by using Laplace's law and by using the potential energy of the circuit in the magnetic field. **b)** Do the same if the field was $\mathbf{B}_2 = B_o(1 + px)\mathbf{e}_x$. Verify that, in this case, the two methods do not give the same result. Show that the expression $\mathbf{B}_2$ cannot be really that of a magnetic field.

P6.24. A very long rectilinear and thin conducting wire carries a current of 5 A. What is the force that it exerts on an electron situated at a distance $d = 10$ cm from the wire and moving with a velocity $v = 3 \times 10^5$ m/s parallel to the wire? What is the field **B** produced by the electron at points of the wire? What is the force that the electron exerts on the wire? Is the principle of action and reaction verified?

P6.25 Two conducting rectilinear wires are parallel and separated by a distance $d = 10$ cm. They carry the intensities $I_1 = 20$ A and $I_2 = 30$ A in the same direction. Calculate the magnetic field at a distance $D = 30$ cm from both wires. Calculate the field of the first wire on the second. What is the force exerted per unit length on the second wire?

P6.26 a) A magnetic dipole $\mathcal{M}$ is placed at the origin O and oriented in the direction of Oz. Calculate its field at a point M of the axis Oz such that $OM = z$. Another dipole $\boldsymbol{\mathcal{M}}$ is placed at M and it makes an angle θ with Oz. Calculate the moment of force exerted on the second dipole. What is its equilibrium position? **b)** Calculate the interaction energy of these dipoles. What is the required energy to reverse the direction of the second dipole from its equilibrium position if $\mathcal{M} = 2 \times 10^{-23}$ A/m^2 and $z = 10^{-10}$ m ? Using the expression of this energy, derive the expression of the moment of force and the expression of the force of interaction of these dipoles if they point in the direction of Oz. **c)** According to the Bohr model for the hydrogen atom, the electron may move on a circular or elliptical orbit around the proton. We consider the case of a circular orbit of radius r. Calculate the period of this orbit, the equivalent current intensity and the magnetic moment. What is its field B at the center? The proton has an intrinsic magnetic moment $\mathcal{M}_p$. What is the magnetic interaction energy of the electron with the proton?

P6.27 Assume that a surface carries a surface current density $\mathbf{j}_s$ and that the magnetic field is zero on the side (1) of this surface. What is the field on the other side? An element of area $d\mathcal{S}$ of this surface is subject to a magnetic force exerted by the field $\mathbf{B}'$ of the other parts of the surface. Show that $\mathbf{B}'$ is continuous on the surface and it is equal to ½$\mathbf{B}$. Deduce that the element of area $d\mathcal{S}$ is subject to a force $d\mathbf{f}_M = \frac{1}{2}\, d\mathcal{S}\, |\mathbf{j}_s \times \mathbf{B}|$.

Chapter 7

Magnetism in Matter

Before the discovery of the magnetic effects of electric current and charges, the understanding of magnetism pertained to permanent magnets. Even today, some of the magnetic properties of matter remain little understood and other properties remain to be explored. This does not prevent magnetism from underlying many applications, ranging from the magnetic compass to measurement instruments, electric generators and motors, magnetic tapes for sound and video recording and for computer data storage, magnetic levitation, etc. The purpose of this chapter is to introduce some basic elements of magnetism in matter.

7.1. Types of magnetism

Some materials, said to be *ferromagnetic*, become magnetized if they are exposed to a magnetic field and they remain permanently magnetized if the magnetic field is removed. A magnetized body is equivalent to a magnetic moment $\boldsymbol{\mathcal{M}}$ in a characteristic direction *SN*. An external magnetic field acts on this body and orients it in such a way that the field **B** enters the body at *S* and leaves it at *N* (Figure 7.1a). Particularly, in the Earth's magnetic field, *N* points approximately toward the geographic North and *S* toward the South (Figure 7.1b). However, contrary to the electric charges, which constitute an electric dipole, the "magnetic poles" cannot be separated and the concept of magnetic pole is simply an analogy with electric charges. Similar to dielectrics, which polarize if they are placed in an electric field, all materials become *magnetized* to some extent if they are submitted to the magnetic field **B** of an electric current or another magnetized body. Some materials (such as aluminum, chrome, platinum, etc.) acquire a magnetic moment in the direction of **B**, they are said to be *paramagnetic*. Other materials (such as silver, gold, copper, mercury, lead, etc.) acquire a magnetic moment in the opposite direction to **B**, they are said to be *diamagnetic*. A magnetized body produces its own magnetic field, which leaves the body near N and enters the body near S (Figure 7.1c). It acts on nearby magnets in such a way that like poles repulse each other while unlike poles attract each other.

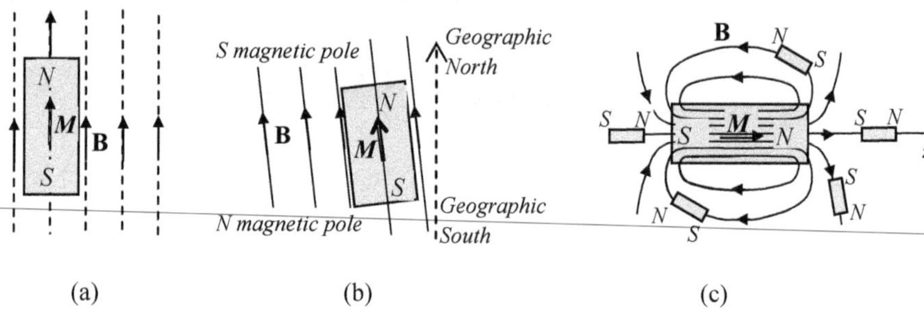

Figure 7.1. *a) Magnetization of a rod in an external field* **B**, *b) a bar magnet aligns itself in the direction of the Earth's magnetic field in such a way that the pole N of the magnet points approximately toward the geographic North (S magnetic pole of the Earth), and c) field lines of a bar magnet. A nearby magnet orients itself such that unlike poles attract each other*

The macroscopic magnetic properties of materials have their origin in their atomic structure. An electron in orbital motion in the atom is equivalent to a microscopic electric circuit, which is subject to the magnetic field of other systems and which produces a magnetic field exactly like a magnetic moment $\mathcal{M}_o = -e\mathbf{L}_e/2m_e$, where $\mathbf{L}_e$ is the orbital angular momentum of the electron. The magnetic moments of the various electrons of the atom add up vectorially to form the magnetic moment of the atom $\mathcal{M}_a = -e\mathbf{L}_a/2m_e$, where $\mathbf{L}_a$ is the total angular momentum of the atom. We must add also the intrinsic magnetic moments of the electrons and the nuclear magnetic moment. The magnetic properties of materials may be explained only by using quantum mechanics. In this theory, the three components of angular momentum $\hat{\mathbf{L}}$ cannot be determined simultaneously. It is possible to determine only the squared magnitude and one component of $\hat{\mathbf{L}}$, in the z direction, for instance. $\hat{\mathbf{L}}^2$ takes the values $\hbar^2 l(l+1)$, where $\hbar = h/2\pi$ is the reduced Planck's constant and the quantum number l takes the values 0, 1, 2, etc. For a given l, L_z takes the values $\hbar m_l$, where m_l may take one of the values $-l$, $-l+1$, … $l-1$, l. It is convenient to express the angular momentum in unit of $\hbar$ and write the orbital magnetic moment of the electron as $\mathcal{M}_o = -\mu_B \hat{\mathbf{L}}_e$, where $\mu_B = (e\hbar/2m_e)$ is *Bohr magneton*. On the other hand, the electron also has an intrinsic angular momentum or spin $s = ½$, thus two states of polarization $m_s = -½$ and $m_s = +½$. The spin corresponds to an intrinsic magnetic moment $\mathcal{M}_s = g\mu_B \hat{\mathbf{s}}_e$ where g is the *gyromagnetic ratio* of the electron that is very approximately equal to –2. Similarly, the proton has an orbital magnetic moment $\mathcal{M}_o = \mu_N \hat{\mathbf{L}}_p$ and an intrinsic magnetic moment $\mathcal{M}_s = g_p\mu_N \hat{\mathbf{s}}_p$, where $g_p = 2.793$ and $\mu_N = e\hbar/2m_p$ (about 1839 times smaller than μ_B). Although the neutron is neutral, it has an intrinsic magnetic moment $\mathcal{M}_s = g_n\mu_N \hat{\mathbf{s}}_n$, where $g_n = -1.913$.

Although electrons have individual magnetic moments, they are often paired in such a way that the magnetic moment of the atom is zero. On the other hand, in solids, the magnetic moment of the tightly packed atoms is not the same as that of the free atoms. The most important contribution to magnetism comes from the electron spin.

7.2. Diamagnetism and paramagnetism

Diamagnetism may be explained by the action of an external magnetic field on the electron orbits in the atom. To simplify, let us consider an electron in a circular orbit of radius r_o about the nucleus under the effect of Coulomb's force. Thus, we have $mv_o^2/r_o = e^2/4\pi\varepsilon_o r_o^2$. This rotation of period $2\pi r_o/v_o$ is equivalent to a current $I_o = -ev_o/2\pi r_o$. The corresponding orbital magnetic moment is $\mathcal{M}_o = \pi r_o^2 I_o \mathbf{n} = -\frac{1}{2} e r_o v_o \mathbf{n} = -e\mathbf{L}_e/2m_e$, where $\mathbf{n}$ is the unit vector normal to the orbit and oriented in such a way that the electron circulates according to the right-hand rule (Figure 7.2a).

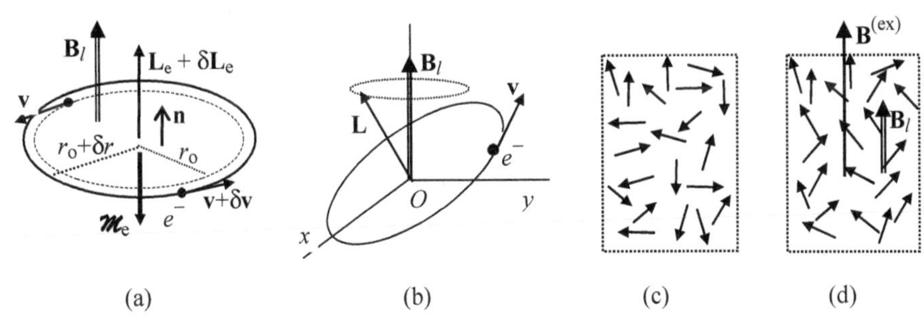

Figure 7.2. *Action of a field* **B**$_l$*: a) on an electronic orbit perpendicular to* **B**$_l$*, and b) on an oblique orbit, c) random orientation of the magnetic moments in the absence of* **B**$_l$*, and d) their orientation in an external field* **B**$^{(ex)}$

If a sample is placed in an external field $\mathbf{B}^{(ex)}$, each atom is subject to a local magnetic field $\mathbf{B}_l$, which results from the superposition of $\mathbf{B}^{(ex)}$ and the fields of the other atoms of the sample, excluding the field of the considered atom. Consider first the case of a field $\mathbf{B}_l$ that is weak and parallel to $\mathbf{L}_e$ (Figure 7.2a), let us assume that the radius of the orbit changes slightly to become $r = r_o + \delta r$ and the speed of the electron changes to becomes $v = v_o + \delta v$. The electron is now subject to an additional force $- e(\mathbf{v}\times\mathbf{B}_l) = \mp evB_l\, \mathbf{e}_r$ depending on whether it circulates in the right-hand or the left-hand directions about $\mathbf{B}_l$. The new condition of stationary motion is $m_e v^2/r = e^2/4\pi\varepsilon_o r^2 \pm evB_l$. Then, the magnetic energy of the atom in the magnetic field is $-I\Phi_M = (\pm ev/2\pi r)\pi r^2 B_l = \pm \frac{1}{2}evrB_l$. The law of conservation of energy implies that $\frac{1}{2}m_e v^2 - e^2/4\pi\varepsilon_o r = \frac{1}{2}m_e v_o^2 - e^2/4\pi\varepsilon_o r_o \pm \frac{1}{2}evrB_l$. These equations give $\delta r = 0$ and $\delta v = \pm(e/2m_e)r_o B_l$ to the first order in B_l. The angular momentum of the electron varies by $\delta\mathbf{L}_e = \pm m_e r\, \delta v\, \mathbf{n} = \frac{1}{2}er_o^2\mathbf{B}_l$ and its magnetic moment varies by

$$\delta\mathcal{M}_o = -(e^2/4m_e)\, r_o^2\mathbf{B}_l. \quad [7.1]$$

This additional magnetic moment points in the opposite direction to $\mathbf{B}_l$.

If $\mathbf{B}_l$ is not perpendicular to the orbit (Figure 7.2b), it acts on the magnetic moment with a moment of force $\mathbf{\Gamma}_M = \mathcal{M}_o \times \mathbf{B}_l = -(e/2m_e)\, \mathbf{L}_e \times \mathbf{B}_l$. Thus, the angular momentum of the electron $\mathbf{L}_e$ varies according to the equation $d\mathbf{L}_e/dt = \mathbf{\Gamma}_M$, i.e. $d\mathbf{L}_e/dt = -(e/2m_e)(\mathbf{L}_e\times\mathbf{B}_l)$. Taking Oz parallel to $\mathbf{B}_l$, this equation is equivalent to the three equations:

$$dL_x/dt = -\,\omega_L L_y, \quad dL_y/dt = \omega_L L_x, \quad dL_z/dt = 0 \quad \text{where } \omega_L = eB_l/2m. \quad [7.2]$$

From these equations, we deduce that $d\mathbf{L}_e^2/dt = 2\mathbf{L}_e . d\mathbf{L}_e/dt = 0$. Thus, the magnitude of $\mathbf{L}_e$ is not modified by the magnetic field $\mathbf{B}_l$. On the other hand, the equation $dL_z/dt = 0$ implies that the component of $\mathbf{L}_e$ in the direction of the field remains constant. Thus, the tip of the vector $\mathbf{L}_e$ moves on a circle about $\mathbf{B}_l$. Explicitly, we may solve the equations [7.2] and obtain

$$L_x = A\cos(\omega_L t + \varphi), \quad L_y = A\sin(\omega_L t + \varphi) \quad \text{and } L_z = \text{Constant}. \quad [7.3]$$

The vector $\mathbf{L}_e$ undergoes a precession about $\mathbf{B}_l$ with an angular frequency $\omega_L = eB_l/2m_e$, called the *Larmor frequency*. This precession may be generally shown by using Larmor's theorem: the motion of a charged particle in a weak magnetic field $\mathbf{B}_l$ is the superposition of its motion in the absence of $\mathbf{B}_l$ and a rotation ω_L about $\mathbf{B}_l$. This precession gives the electron an additional velocity $\delta\mathbf{v} = \boldsymbol{\omega}_L \times \mathbf{r}$ and an additional angular momentum

$$\delta\mathbf{L}_e = m_e\, \mathbf{r} \times \delta\mathbf{v} = \tfrac{1}{2}\, e\mathbf{r} \times (\mathbf{B}_l \times \mathbf{r}) = \tfrac{1}{2}\, e\, [\mathbf{B}_l\, r^2 - \mathbf{r}\, (\mathbf{B}_l . \mathbf{r})]. \quad [7.4]$$

The corresponding additional magnetic moment is

$$\delta\mathcal{M}_e = -(e/2m)\delta\mathbf{L}_e = -(e^2/4m_e)[\mathbf{B}_l r^2 - \mathbf{r}(\mathbf{B}_l.\mathbf{r})] = \tfrac{1}{2}\omega_L[xz\,\mathbf{e}_x + yz\,\mathbf{e}_y - (x^2+y^2)\mathbf{e}_z].$$

Because of the rotational symmetry about Oz, the average values over a period are $\langle xz\rangle = \langle yz\rangle = 0$ and $\langle x^2\rangle = \langle y^2\rangle = \langle z^2\rangle = \langle r^2\rangle/3$ where $\langle r^2\rangle$ is the average value of the distance squared of the electron from the nucleus. Thus, the mean value of the additional magnetic moment is $\langle \delta\mathcal{M}_e \rangle = -(e^2/6m_e)\langle r^2\rangle \mathbf{B}_l$ and the additional magnetic moment of the atom is

$$\langle \delta\mathcal{M}_a \rangle = \Sigma_k\, \delta\mathcal{M}_k = \alpha_M \mathbf{B}_l, \qquad \text{where } \alpha_M = -(e^2/6m_e)\,\Sigma_k \langle r_k^2\rangle. \qquad [7.5]$$

α_M is the *magnetic polarizability*. This is a characteristic constant of the atom. It does not depend on physical conditions such as temperature and pressure. It is amazing that quantum theory does not modify the expression [7.5], although it abandons the concept of the electron orbit. Thus, magnetization allows us to have some information about $\Sigma_k\langle r_k^2\rangle$, i.e. the atomic structure of the material. For an estimate of the order of magnitude, let us take r_k as the radius of the atom 0.5×10^{-10} m, hence $\Sigma_k \langle r_k^2\rangle \approx 0.25 \times 10^{-20} Z$, where Z is the atomic number. We find a polarizability $\alpha_M \approx -1.17 \times 10^{-29} Z$. All materials are diamagnetic and, among materials of high diamagnetism, we find metallic bismuth and some organic substances such as benzene.

The atoms of some substances have permanent magnetic moments with a weak interaction between these moments. These substances are said to be *paramagnetic*. This may occur if the number of atomic electrons is odd; thus, they cannot form pairs of zero magnetic moment. For instance, this is the case for the NO molecule, which has 15 electrons. Even if the number of electrons is even, it may happen that the total magnetic moment is non-zero because of some particular electronic structure. This is effectively the case of the O_2 molecule, the transition elements (chromium, manganese, iron, cobalt, nickel, and copper), and the rare earth elements, i.e. the lanthanides going from $Z = 58$ (cerium) to 71 (thecerium) and the actinides going from $Z = 90$ (thorium) to $Z = 103$ (lawrencium). We note also that the conduction electrons in metals may contribute to paramagnetism. The atoms or the molecules may become paramagnetic or lose their paramagnetism if they are ionized.

If a paramagnetic body is not exposed to an external field, because of the thermal agitation, the magnetic moments are randomly oriented in all directions with the same probability. The mean magnetic moment is thus equal to zero (Figure 7.2c). If the body is exposed to an external field $\mathbf{B}^{(ex)}$, the magnetic moment $\mathcal{M}_a$ of an atom or molecule is subject to a moment of force $\mathbf{\Gamma}_M = \mathcal{M}_a \times \mathbf{B}_l$, where $\mathbf{B}_l$ is the local field.

$\mathbf{\Gamma}_M$ tends to orient $\mathcal{M}_a$ in the direction of $\mathbf{B}_l$. However, the alignment cannot be complete because of the thermal agitation (Figure 7.2d). If the medium is isotropic, the transverse component of $\mathcal{M}_a$ (i.e. perpendicular to $\mathbf{B}_l$), has the same probability to point in opposite directions; thus, the average magnetic moment $<\mathcal{M}_a>$ is parallel to $\mathbf{B}_l$. We have seen that the induced mean magnetic moment in diamagnetic materials is also in the direction of $\mathbf{B}_l$. Thus, if the medium is linear and isotropic and the external field is not very strong, we may assume that the mean magnetic moment is proportional to $\mathbf{B}_l$ (the terms of the order $\mathbf{B}_l^2$ or higher being negligible):

$$<\mathcal{M}_a> = \alpha_M \mathbf{B}_l. \quad [7.6]$$

The atoms of all substances may acquire a small diamagnetic magnetization. The corresponding polarizability α_M is negative and independent of the physical conditions. On the other hand, in the case of paramagnetic and ferromagnetic substances, the polarizability α_M is positive, much higher than the diamagnetic polarizability, and it depends on physical conditions (especially temperature). Thus, diamagnetism is a general property of matter, while paramagnetism and ferromagnetism are properties of some specific materials that hide diamagnetism.

If the correlation of the magnetic moments is weak, an element of volume $d\mathcal{V}$ acquires a magnetic moment $d\mathcal{V}\, N_v<\mathcal{M}_a>$ where N_v is the number of atoms or molecules per unit volume. Thus, the medium becomes magnetized with an *intensity of magnetization* (or simply a *magnetization*)

$$\mathbf{M} = N_v <\mathcal{M}_a> = N_v\, \alpha_M \mathbf{B}_l. \quad [7.7]$$

We expect that, in the case of paramagnetism, $\mathbf{M}$ increases with $\mathbf{B}_l$ and increases if the temperature decreases (as the thermal agitation decreases). However, $\mathbf{M}$ reaches a *saturation value* $\mathbf{M}_{max} = N_v\mathcal{M}_a$ if all the magnetic moments point in the same direction; this occurs if $\mathbf{B}_l$ is very strong or if the temperature is very low.

The magnetization of a sample is studied by placing it in the uniform field of a solenoid carrying a current I (Figure 7.3a). In order to have the field almost uniform, the conducting wire is wound around a torus of the substance. This set-up is called *Rowland's ring* (Figure 7.3b). If the radius of the section is small, compared to the radius of the ring, the field is almost uniform in the sample. The conducting wire, wound in many turns and carrying the current I, is the *primary circuit*. The field outside matter may be measured from its action on an electric current or by using the Hall effect. Inside a substance, it is measured indirectly from the induction that it produces in a secondary circuit (a coil of few turns wound around the ring and connected to a ballistic galvanometer). If the magnetic flux $\mathbf{\Phi}$ through a circuit varies, an electromagnetic force $\mathcal{E} = -d\Phi/dt$ is induced. It produces a current $I = \mathcal{E}/R$

in a secondary of resistance R (see section 8.1). The induced charge is $Q = \int|I|\, dt = (1/R)\int dt\, |d\Phi/dt| = \Phi/R$. As the current is set up in the primary, the field increases from zero to **B**. The flux through the N_s turns of the secondary increases from 0 to $\Phi = N_s SB$ and induces a charge $Q = \Phi/R = N_s\, SB/R$. A measurement of Q using the ballistic galvanometer enables the average field B within the substance to be determined. For this reason, **B** is called *magnetic induction field.*

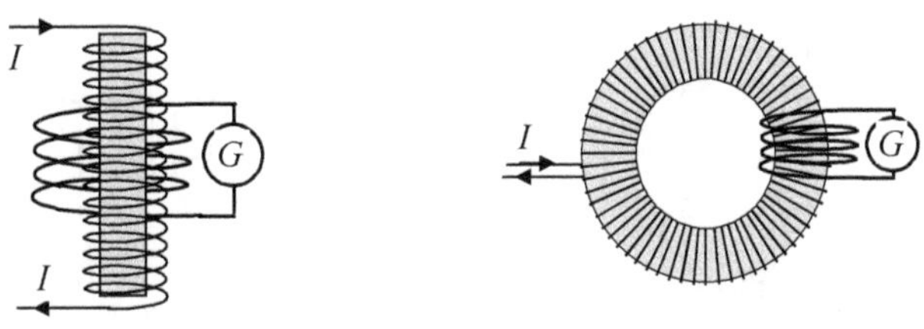

Figure 7.3. *Magnetization of a sample placed a) in a solenoid, and b) in a Rowland ring*

7.3. Magnetization current

Let us consider a long solenoid of n turns per unit length carrying a current I (Figure 7.4a). If the solenoid is empty, the field inside it is uniform and given by $B_o = \mu_o j_s$, where $j_s = nI$ is the surface conduction current density. If it is filled with a magnetic substance, experiment shows that the magnetic induction field is multiplied by a factor μ_r, called the *relative magnetic permeability*, to become

$$B = \mu\, j_s, \qquad \text{where } \mu = \mu_r\, \mu_o. \qquad [7.8]$$

μ is the *magnetic permeability* of the substance. In the case of a paramagnetic substance, μ depends on the physical conditions (temperature and pressure in the case of a gas and temperature in the case of a dense medium). As μ does not depend on the size of the sample, we may interpret the modification of the value of B as due to a *magnetization current* of density j'_s on the surface of the magnetic substance. j'_s is superposed to the conduction current density j_s, such that

$$B = \mu_o(j_s + j'_s). \qquad [7.9]$$

Comparing with the expression [7.8], we deduce that

$$j'_s = (\mu_r - 1)\, j_s = (\mu_r - 1)\,(B/\mu). \qquad [7.10]$$

j'_s does not depend on the dimensions of the sample but on **B** and the permeability μ. A cylinder of the substance of radius r and infinitesimal length dz in the direction of

B carries a magnetization current of intensity $dI' = dz\, j'_s$ on its lateral surface (Figure 7.4b). The magnetic moment of this cylinder is $d\mathcal{M} = \pi r^2\, dI' = \pi r^2\, dz\, j'_s = d\mathcal{V}\, j'_s$, where $d\mathcal{V}$ is the volume of the cylinder. Thus, we may define the *intensity of magnetization*

$$M = d\mathcal{M}/d\mathcal{V} = j'_s = (\mu_r - 1)\, B/\mu. \qquad [7.11]$$

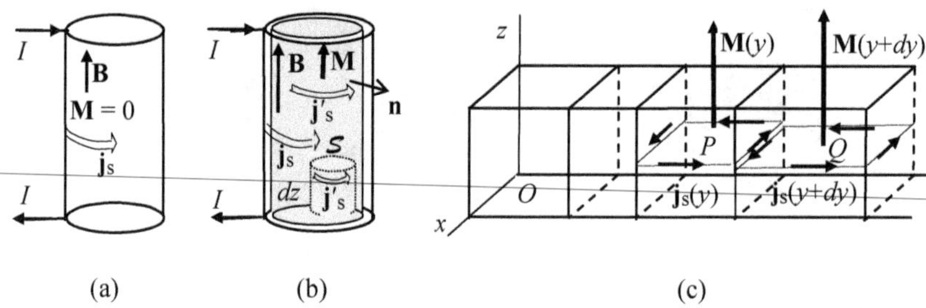

Figure 7.4. *a) Surface conduction current in the case of an empty solenoid. b) Magnetization current in a solenoid filled with a magnetic substance. c) Volume magnetization current*

In fact, the intensity of magnetization is a vector. If we define a *surface magnetization current density* $\mathbf{j}'_s$, the magnetization **M** is normal to $\mathbf{j}'_s$ and oriented in a direction such that $\mathbf{j}'_s$ circulates about **M** according to the right-hand rule, hence

$$\mathbf{j}'_s = \mathbf{M} \times \mathbf{n}, \qquad [7.12]$$

where **n** is the unit vector normal to the surface of the magnetized body and pointing outward. Thus, if a body has a uniform magnetization **M**, it produces a magnetic field exactly as if the surface of the body carries a surface current density $\mathbf{j}'_s = \mathbf{M}\times\mathbf{n}$.

If the local field $\mathbf{B}_l$ is non-uniform, it will be so for the magnetization **M**. For instance, let us consider a medium whose magnetization is parallel to Oz and depends only on y (Figure 7.4c). We may consider the medium as a juxtaposition of rectangular parallelepipeds of sides dx, dy, and dz, which are small enough for the magnetization to be approximately uniform in each one of them. The magnetic field produced by the parallelepiped centered at $P(x, y, z)$ is the same as that of the surface current density $j'_s = M_z(y)$ on the faces that are parallel to Oz and in the illustrated direction. This holds also for the parallelepiped centered at $Q(x, y + dy, z)$, but with a surface current density $j'_s(y+dy) = M_z(y+dy)$. The surface current density on the common face is $M_z(x, y+dy, z) - M_z(x, y, z) = dy\, (\partial_y M_z)$ and the current intensity carried by this face is $I'_x = (\partial_y M_z)\, dx\, dy\, dz = (\partial_y M_z)\, d\mathcal{V}$ where $d\mathcal{V}$ is the volume of the parallelepiped. This is equivalent to a *volume magnetization current density*

$\mathbf{j}' = \partial_y M_z \, \mathbf{e}_x$. If M_z depends on x, y and z, the volume magnetization current density is $\mathbf{j}' = (\partial_y M_z)\, \mathbf{e}_x - (\partial_x M_z)\, \mathbf{e}_y$. In the general case of **M** non-uniform and pointing in an arbitrary direction, it may be shown that the volume magnetization current density is

$$\mathbf{j}' = \nabla \times \mathbf{M}. \qquad [7.13]$$

Similar to the polarization charges in dielectrics, the magnetization current of surface density $\mathbf{j}'_s = \mathbf{M} \times \mathbf{n}$ and volume density $\mathbf{j}' = \nabla \times \mathbf{M}$ are bound in the medium. They are not only a mathematical trick to calculate the magnetic field of the magnetized body; they can be interpreted as currents due to the motion of charges within the atoms and the molecules. Contrary to conduction currents, once established, the magnetization currents produce no energy dissipation by Joule or ether effects, as they correspond to the motion of charged particles in vacuum.

7.4. Magnetic field and vector potential in the presence of magnetic matter

If a medium is magnetized, an element of volume $d\mathcal{V}'$ near a point $\mathbf{r}'$ has a magnetic moment $d\mathcal{M} = \mathbf{M}(\mathbf{r}')d\mathcal{V}'$. It produces at point **r** a vector potential

$$d\mathbf{A}(\mathbf{r}) = \frac{\mu_o}{4\pi R^3} d\mathcal{V}'[\,\mathbf{M}(\mathbf{r}') \times \mathbf{R}\,], \quad \text{where } \mathbf{R} \equiv \mathbf{r} - \mathbf{r}'. \qquad [7.14]$$

The total vector potential produced at **r** by the magnetized body is obtained by integration on the volume $\mathcal{V}$ of the body

$$\mathbf{A}_m(\mathbf{r}) = (\mu_o/4\pi)\iiint_{\mathcal{V}} d\mathcal{V}'\mathbf{M}(\mathbf{r}') \times \mathbf{R}\,/R^3 = (\mu_o/4\pi)\iiint_{\mathcal{V}} d\mathcal{V}'\mathbf{M}(\mathbf{r}') \times \nabla'\,(R^{-1}), \; [7.15]$$

where ∇' is the vector differential operator with respect to the coordinates (x', y', z'). Using the identity $\nabla'[f(\mathbf{r}')\,\mathbf{M}(\mathbf{r}')] = \nabla' f(\mathbf{r}') \times \mathbf{M}(\mathbf{r}') + f(\mathbf{r}')\,\nabla' \times \mathbf{M}(\mathbf{r}')$, where f is any scalar function, we may write:

$$\mathbf{A}_m(\mathbf{r}) = (\mu_o/4\pi)\iiint_{\mathcal{V}} d\mathcal{V}'[\nabla' \times \mathbf{M}(\mathbf{r}')]/R - (\mu_o/4\pi)\iiint_{\mathcal{V}} d\mathcal{V}'\nabla' \times [\mathbf{M}(\mathbf{r}')/R]. \; [7.16]$$

The second integral may be transformed into a surface integral by using the identity $\iiint_{\mathcal{V}} d\mathcal{V}\, \nabla \times \mathbf{V} = \iint_{\mathcal{S}} d\mathcal{S}\, \mathbf{n} \times \mathbf{V}$ (see problem 7.2). Thus, we may write

$$\mathbf{A}_m(\mathbf{r}) = (\mu_o/4\pi)\iiint_{\mathcal{V}} d\mathcal{V}'[\nabla' \times \mathbf{M}(\mathbf{r}')]/R + (\mu_o/4\pi)\iint_{\mathcal{S}} d\mathcal{S}'[\mathbf{M}(\mathbf{r}') \times \mathbf{n}]/R, \qquad [7.17]$$

where **n** is the normal unit vector outgoing from the surface $\mathcal{S}$. Comparing with the expressions [6.20] of the vector potential produced by a volume current density $\mathbf{j}(\mathbf{r}')$

and a surface current density $\mathbf{j}_s(\mathbf{r}')$, the expression [7.17] means that a magnetized body produces the same vector potential as the volume current density $\mathbf{j}' = \nabla\times\mathbf{M}$ and the surface current density $\mathbf{j}'_s = \mathbf{M}\times\mathbf{n}$. These are the magnetization current densities.

The magnetic field produced at $\mathbf{r}$ by the element of volume $d\mathcal{V}'$ at the point $\mathbf{r}'$ is

$$d\mathbf{B}_m(\mathbf{r}) = \frac{\mu_o}{4\pi R^5}\, d\mathcal{V}'\{3[\mathbf{M}(\mathbf{r}').\mathbf{R}]\mathbf{R} - R^2\,\mathbf{M}(\mathbf{r}')\}, \quad \text{where } \mathbf{R} \equiv \mathbf{r} - \mathbf{r}'. \qquad [7.18]$$

The field $\mathbf{B}_m$ of the magnetized body is obtained by integration over $\mathcal{V}$. We may also obtain it by using the relation $\mathbf{B}_m = \nabla\times\mathbf{A}_m$ and the expression [7.17], hence

$$\mathbf{B}_m(\mathbf{r}) = (\mu_o/4\pi)\iiint_{\mathcal{V}} d\mathcal{V}'[\mathbf{j}'(\mathbf{r}') \times \mathbf{R}]/R^3 + (\mu_o/4\pi)\iint_{\mathcal{S}} d\mathcal{S}'[\mathbf{j}'_s(\mathbf{r}') \times \mathbf{R}]/R^3. \qquad [7.19]$$

To have the total vector potential and the total magnetic field, we must add to $\mathbf{A}_m$ and $\mathbf{B}_m$ the vector potential and field produced by the free currents.

It is not evident that the expressions [7.17] and [7.19] hold for the vector potential $\mathbf{A}_m$ and the field $\mathbf{B}_m$ evaluated inside the magnetized body or outside it near its surface, because they are consequences of the expression [7.14] of the vector potential of a magnetic moment at large distance. As in the case of the potential V_p and the field $\mathbf{E}_p$ of a polarized dielectric, it may be shown that the expressions of $\mathbf{A}_m$ and $\mathbf{B}_m$ are valid everywhere. We note that $\mathbf{A}_m$ and $\mathbf{B}_m$ are the *macroscopic vector potential and field*, i.e. the average values of the microscopic vector potential and field produced by the magnetic moments of atoms. Contrary to microscopic quantities, which undergo important variations in space and time and even become infinite at the position of atoms, macroscopic quantities are slowly varying in space and time and are time-independent in the case of stationary phenomena.

7.5. Ampère's law in the integral form in the presence of magnetic matter

The experiment shows that the field in a paramagnetic or diamagnetic cylinder placed in a very long solenoid is $\mathbf{B} = \mu_r\mathbf{B}_o$, where μ_r is the relative permeability and $\mathbf{B}_o$ is the field in the empty solenoid (Figure 7.5). If $\mathbf{j}_s$ is the conduction current surface charge density in the winding, the circulation of $\mathbf{B}$ over the path $\mathcal{C}$ is

$$\int_{\mathcal{C}} d\mathbf{r}.\mathbf{B} = Bl = \mu_r B_o l = \mu_r\,\mu_o j_s l = \mu\, I^{(\text{in})}, \qquad [7.20]$$

where we have used the relation $B_o = \mu_o j_s$, and $\mu = \mu_r\mu_o$ is the permeability. $I^{(\text{in})}$ is the conduction current, which passes inside $\mathcal{C}$. Using [7.10], we may also write

$$\int_{\mathcal{C}} d\mathbf{r}.\mathbf{B} = Bl = \mu_o l\,(j_s + j'_s) = \mu_o(I^{(\text{in})} + I'^{(\text{in})}). \qquad [7.21]$$

$I'^{(in)}$ is the total intensity of the magnetization currents, which pass inside $\mathcal{C}$. Equations [7.20] and [7.21], which were derived in a particular case, hold in all cases. Thus, we may formulate Ampère's law in two ways: by using only the conduction currents

$$\int_{\mathcal{C}} \mathbf{dr}\,.(\mathbf{B}/\mu) = I^{(in)}, \qquad [7.22]$$

or by using both the conduction currents and the magnetization currents,

$$\int_{\mathcal{C}} d\mathbf{r}.\mathbf{B} = \mu_o(I^{(in)} + I'^{(in)}). \qquad [7.23]$$

It should be noted that, if the magnetic medium is non-homogeneous, the permeability μ in [7.22] varies along the path $\mathcal{C}$.

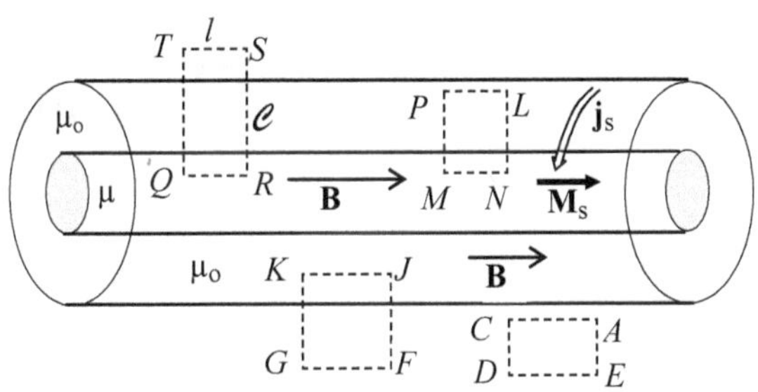

Figure 7.5. *Using Ampère's law to determine the field of a solenoid containing a magnetic material*

Using the expression $\mathbf{j}' = \nabla\times\mathbf{M}$ for the magnetization current density, equation [7.23] may be written as

$$\int_{\mathcal{C}} d\mathbf{r}'.\mathbf{B}(\mathbf{r}')/\mu_o = I^{(in)} + \iint_{\mathcal{S}} d\mathcal{S}'\ \nabla\times\mathbf{M}(\mathbf{r}') = I^{(in)} + \int_{\mathcal{C}} d\mathbf{r}'.\mathbf{M}(\mathbf{r}').$$

We may also write this equation in the form

$$\int_{\mathcal{C}} d\mathbf{r}'.\mathbf{H}(\mathbf{r}') = I^{(in)}, \qquad \text{where} \quad \mathbf{H} = \mathbf{B}/\mu_o - \mathbf{M}. \qquad [7.24]$$

H is the *magnetic field* (sometimes it is called *magnetic excitation*).

7.6. Equations of time-independent magnetism in the presence of matter

In the presence of matter, magnetic phenomena are specified by the magnetic induction field **B** and the intensity of magnetization **M**. We may also use the fields **B** and **H**. Time-independent magnetic phenomena obey two fundamental laws: the conservation of the flux of **B** and Ampère's law.

A) Law of conservation of magnetic flux

The expression of the conservation of the flux of **B** and its consequences are the same as in vacuum. The absence of magnetic charges implies that the flux of the magnetic induction field **B** through any closed surface is equal to zero

$$\iint_{\mathcal{S}} d\mathcal{S}\, \mathbf{n}.\mathbf{B} = 0. \quad [7.25]$$

Transforming the flux of **B** through $\mathcal{S}$ into the volume integral of $\boldsymbol{\nabla}.\mathbf{B}$ by using Gauss-Ostrogradsky's theorem, this law may be written in the local form

$$\boldsymbol{\nabla}.\mathbf{B} = 0, \quad [7.26]$$

which is identically satisfied if **B** is the curl of a vector potential **A**:

$$\mathbf{B} = \boldsymbol{\nabla} \times \mathbf{A}. \quad [7.27]$$

Using Stokes' theorem, the flux of **B** through any open surface $\mathcal{S}$ bounded by a closed curve $\mathcal{C}$ may be written as the circulation of **A** over $\mathcal{C}$:

$$\iint_{S} d\mathcal{S}\, \mathbf{n}.\mathbf{B} = \int_{\mathcal{C}} d\mathbf{r}.\mathbf{A}. \quad [7.28]$$

Thus, the flux through all surfaces $\mathcal{S}$ bounded by the same curve $\mathcal{C}$ is the same.

B) Ampère's law in the local form

Let us write Ampère's law in the form [7.23], transform the circulation of **B** over $\mathcal{C}$ into the flux of $\boldsymbol{\nabla} \times \mathbf{B}$ through $\mathcal{S}$, and express the conduction and the magnetization currents as the fluxes of **j** and **j**' through $\mathcal{S}$. We obtain the equation

$$\iint_{\mathcal{S}} d\mathcal{S}\, \mathbf{n}.(\boldsymbol{\nabla} \times \mathbf{B}) = \mu_o \iint_{\mathcal{S}} d\mathcal{S}\, \mathbf{n}.(\mathbf{j} + \mathbf{j}'). \quad [7.29]$$

This equation must be satisfied for any surface $\mathcal{S}$, hence

$$\boldsymbol{\nabla} \times \mathbf{B} = \mu_o\, \mathbf{j} + \mu_o\, \boldsymbol{\nabla} \times \mathbf{M}, \quad [7.30]$$

where we have used the equation $\mathbf{j}' = \nabla \times \mathbf{M}$. This is the local form of Ampère's law. Instead of **M**, we may use the magnetic field **H** defined by

$$\mathbf{H} = \mathbf{B}/\mu_o - \mathbf{M} \quad \text{or} \quad \mathbf{B} = \mu_o(\mathbf{H} + \mathbf{M}). \qquad [7.31]$$

The local form of Ampère's law [7.30], becomes

$$\nabla \times \mathbf{H} = \mathbf{j}. \qquad [7.32]$$

If the magnetic medium is linear and isotropic (thus, excluding ferromagnetic mediums), fields **B**, **M,** and **H** are proportional to each other:

$$\mathbf{M} = \chi_M \mathbf{H}, \quad \mathbf{B} = \mu\mathbf{H}, \qquad \text{where } \chi_M = \mu_r - 1 = \mu/\mu_o - 1. \qquad [7.33]$$

χ_M is the *magnetic susceptibility* of the medium. In this case, Ampère's law takes the integral form [7.22] or the local form

$$\nabla \times (\mathbf{B}/\mu) = \mathbf{j}. \qquad [7.34]$$

The magnetic susceptibility of some materials is given in the Table 7.1.

In the case of permanent magnetic materials of known **M**, knowing the conduction current density **j**, equation [7.32] and the boundary conditions, we may determine **H**. Then, equation [7.31] determines **B**. Also, knowing **j** and **M**, we may use equation [7.30] and the boundary conditions to determine directly **B** and, then, use [7.31] to determine **H**. In the case of a paramagnetic or diamagnetic material of known μ, **B** is sufficient to specify the magnetostatic field.

Diamagnetic material	**bismuth (solid)**	**Carbon (diamond)**	**Sodium**	**Copper**	**Water**	**Nitrogen**
$10^6 \chi_M$ SI	–166	–22	–2.4	–9.2	–9.0	–0.005

Paramagnetic material	**Tungsten**	**Cesium**	**Aluminum**	**Magnesium**	**Oxygen**
$10^6 \chi_M$ SI	68	0.505	23	12	1.9

Table 7.1. *Magnetic susceptibility of some materials at* 20°C

C) Vector potential

In the following, we consider the case of a linear and isotropic medium of permeability μ. As $\mathbf{B} = \mu\mathbf{H}$, the field **B** is sufficient to specify the magnetic field.

The fundamental equation $\nabla.\mathbf{B} = 0$ is identically satisfied if $\mathbf{B} = \nabla \times \mathbf{A}$. Then, we may write Ampère's equation $\nabla \times \mathbf{B} = \mu\mathbf{j}$ in the form

$$\Delta\mathbf{A} - \nabla(\nabla.\mathbf{A}) = -\mu\mathbf{j}. \qquad [7.35]$$

This equation enables **A** to be determined if **j** is known everywhere. Knowing **A**, we may calculate $\mathbf{B} = \nabla \times \mathbf{A}$, then $\mathbf{H} = \mathbf{B}/\mu$ and $\mathbf{M} = \mathbf{B}/\mu_o - \mathbf{H}$.

We note that the vector potential **A** is not unique. A *gauge transformation*, $\mathbf{A} \rightarrow \mathbf{A}' = \mathbf{A} + \nabla f$ does not change the field **B**. We may always choose the function f to make **A** verify the condition

$$\nabla.\mathbf{A} = 0. \qquad [7.36]$$

This enables equation [7.35] to be written in a simpler form

$$\Delta\mathbf{A} = -\mu\,\mathbf{j}. \qquad [7.37]$$

Taking the Cartesian components, this vector equation is equivalent to three equations (for the three components of **A**) similar to Poisson's equation in electrostatics. Thus, it has the solution

$$\mathbf{A}(\mathbf{r}) = \mathbf{A}_o(\mathbf{r}) + (\mu/4\pi) \iiint_{\mathcal{V}} d\mathcal{V}'\,\mathbf{j}(\mathbf{r}')/R, \qquad [7.38]$$

where $\mathbf{A}_o(\mathbf{r})$ is a solution of the homogeneous equation $\Delta\mathbf{A}_o = 0$. We may always choose $\mathbf{A}_o(\mathbf{r})$ to make **A** satisfy any imposed boundary conditions.

In the case of a point charge q_i of velocity $\mathbf{v}_i$ and located at the point $\mathbf{r}_i$, the current density at a point $\mathbf{r}'$ is equal to zero if $\mathbf{r}'$ is different from $\mathbf{r}_i$ and infinite if $\mathbf{r}' = \mathbf{r}_i$. This is a three-dimensional Dirac function $q_i\mathbf{v}_i\,\delta(\mathbf{r}' - \mathbf{r}_i)$ (see section A.11 of Appendix A). In the case of several charges, the current density is $\Sigma_i\, q_i\mathbf{v}_i\,\delta(\mathbf{r}' - \mathbf{r}_i)$. Substituting this expression in [7.38] and integrating by using the δ-function, we get the expression of the vector potential and the field produced by a system of charges:

$$\mathbf{A}(\mathbf{r}) = \mathbf{A}_o(\mathbf{r}) + (\mu/4\pi)\, q_i\, \mathbf{v}_i/R_i, \qquad \text{where } \mathbf{R}_i = \mathbf{r} - \mathbf{r}_i, \qquad [7.39]$$

$$\mathbf{B}(\mathbf{r}) = \nabla \times \mathbf{A}_o(\mathbf{r}) + (\mu/4\pi)\Sigma_i\, q_i\mathbf{v}_i \times \mathbf{R}_i\,/R_i^{\,3}. \qquad [7.40]$$

Similarly, the vector potential and the magnetic field produced at **r** by a magnetic moment $\boldsymbol{\mathcal{M}}$ at $\mathbf{r}'$ are

$$\mathbf{A}(\mathbf{r}) = (\mu/4\pi R^3)\,(\boldsymbol{\mathcal{M}} \times \mathbf{R}), \qquad \text{where } \mathbf{R} = \mathbf{r} - \mathbf{r}', \qquad [7.41]$$

$$\mathbf{B}(\mathbf{r}) = (\mu/4\pi R^5)\,[3\,(\boldsymbol{\mathcal{M}}.\mathbf{R})\,\mathbf{R} - R^2\,\boldsymbol{\mathcal{M}}]. \qquad [7.42]$$

D) Comparison of the laws of electrostatics and time-independent magnetism

– The field **B** may be compared to the field **E**, while the field **H** may be compared to the electric displacement **D**. Indeed, **D** and **H** are calculated by using the free electric charges and free currents, while **E** and **B** are calculated by using both free and bound electric charges and currents. Thus, **E** and **B** depend on the properties of matter.

– The equation $\nabla\times\mathbf{E} = 0$, which expresses that **E** is conservative (thus $\mathbf{E} = -\nabla V$) is similar to the equation $\nabla.\mathbf{B} = 0$, which expresses that the flux of **B** is conservative (thus $\mathbf{B} = \nabla\times\mathbf{A}$). On the other hand, Gauss equation $\nabla.\mathbf{D} = q_v$ is similar to Ampère's equation $\nabla\times\mathbf{H} = \mathbf{j}$. In the general case, **D** is not conservative and **H** does not have a conservative flux.

– In the case of time-dependent phenomena, the equations $\nabla.\mathbf{D} = q_v$ and $\nabla.\mathbf{B} = 0$ remain valid, while $\nabla \times \mathbf{E} = 0$ and $\nabla \times \mathbf{H} = \mathbf{j}$ must be modified.

– As for the invariance properties, the fields **E** and **D**, the polarization **P**, the dipole moment **p**, the current density **j**, and the vector potential **A** are true vectors, the potential V and the charge are true scalars. The magnetic fields **B** and **H**, the intensity of magnetization **M**, and the magnetic moment $\boldsymbol{\mathcal{M}}$ are pseudovectors.

7.7. Discontinuities of the magnetic field

As we have seen in section 6.9C, the magnetic field has singularities at points where there are charges and linear currents and discontinuities at a surface carrying a current or the interface $\mathcal{S}$ of two mediums. Thus, the two fundamental laws of time-independent magnetism cannot be used on $\mathcal{S}$ in the local forms $\nabla\times\mathbf{H} = \mathbf{j}$ and $\nabla.\mathbf{B} = 0$ but in the integral forms [7.24] and [7.25].

Let $\mathcal{S}$ be the interface of two mediums (1) and (2), in which the magnetization are $\mathbf{M}_1$ and $\mathbf{M}_2$, respectively (Figure 7.6a). If (at least) one of the mediums is a conductor, $\mathcal{S}$ may carry a conduction surface current density $\mathbf{j}_s$ and the conductor may have a conduction volume current density. We use coordinates such that the plane Oxy is tangent to $\mathcal{S}$, Ox parallel to $\mathbf{j}_s$ and Oz oriented in the direction of the normal unit vector $\mathbf{n}_{12}$ (pointing from medium 1 toward medium 2). Let $EFGH$ be a rectangular path whose sides EF and GH are parallel to $\mathbf{j}_s$ and situated on one side of $\mathcal{S}$ and the other, while the sides GH and HF are very short. The flux of the volume currents $\mathbf{j}_1$ and $\mathbf{j}_2$ through this path is negligible and no conduction surface current density passes through it. Thus, Ampère's law may be written as

$$\overrightarrow{EF}.\mathbf{H}_2 + \overrightarrow{GH}.\mathbf{H}_1 = 0, \qquad \text{hence} \quad H_{1x} - H_{2x} = 0. \qquad [7.43]$$

Next, consider the rectangular path *MNPQ* oriented according to the right-hand rule about $\mathbf{j}_s$ and whose sides *MN* and *PQ* of length l are parallel to *Oy*. The conduction current intensity that passes through it is lj_s. Thus, Ampère's law gives

$$\overrightarrow{MN}.\mathbf{H}_2 + \overrightarrow{PQ}.\mathbf{H}_1 = l\,j_s, \qquad \text{hence } -H_{1y} + H_{2y} = j_s. \qquad [7.44]$$

Designating by $\mathbf{H}_{1//}$ and $\mathbf{H}_{2//}$ the components of **H**, which are parallel to $\mathcal{S}$, equations [7.43] and [7.44] may be written in a single vector relation

$$\mathbf{H}_{1//} - \mathbf{H}_{2//} = \mathbf{n}_{12} \times \mathbf{j}_s. \qquad [7.45]$$

Thus, if the interface $\mathcal{S}$ carries a conduction surface current density $\mathbf{j}_s$, the component of **H**, which is parallel to $\mathbf{j}_s$ is continuous, while the component, which is parallel to $\mathcal{S}$ but normal to $\mathbf{j}_s$, has a discontinuity equal to j_s according to [7.45].

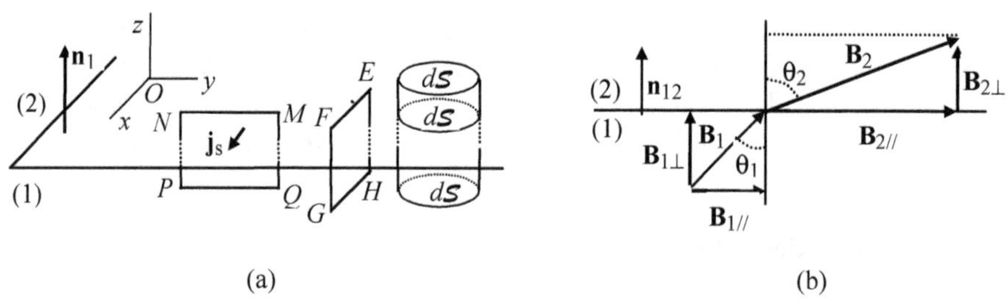

Figure 7.6. *a) Discontinuities of* **H** *and* **B** *at the interface of two magnetic mediums, and b) the field lines* **B** *are broken at the interface of two linear mediums*

Let us now apply the conservation law of the flux of **B** through a very short cylinder of section $d\mathcal{S}$, and situated on both sides of $\mathcal{S}$ (Figure 7.6a). We find

$$d\mathcal{S}\,\mathbf{B}_2.\,\mathbf{n}_{12} - d\mathcal{S}\,\mathbf{B}_1.\mathbf{n}_{12} = 0, \qquad \text{hence } \mathbf{B}_{\perp 1} = \mathbf{B}_{\perp 2}. \qquad [7.46]$$

This relation expresses the continuity of the normal component of **B**. Using the intensity of magnetization **M** instead **H** and equation [7.31], equation [7.45] gives the discontinuity of the component of **B** parallel to the interface:

$$\mathbf{B}_{1//} - \mathbf{B}_{2//} = \mu_o\,\mathbf{n}_{12} \times \mathbf{j}_s + \mu_o\,(\mathbf{M}_{1//} - \mathbf{M}_{2//}). \qquad [7.47]$$

If the mediums are linear and isotropic, and $\mathcal{S}$ carries no surface conduction current, the component $\mathbf{H}_{//}$ parallel to $\mathcal{S}$ and the component $\mathbf{B}_{\perp}$ normal to $\mathcal{S}$ are continuous. As $\mathbf{B}_1 = \mu_1\mathbf{H}_1$ and $\mathbf{B}_2 = \mu_2\mathbf{H}_2$, we find:

$$B_{1//}/\mu_1 = B_{2//}/\mu_2, \qquad B_{1\perp} = B_{2\perp}. \qquad [7.48]$$

The lines of the field **B** are broken on $\mathcal{S}$. If θ_1 and θ_2 are the angles of $\mathbf{B}_1$ and $\mathbf{B}_2$ with the normal $\mathbf{n}_{12}$ (Figure 7.6b), we find:

$$\frac{\tan\theta_1}{\tan\theta_2} = \frac{B_{//1}/B_{\perp 1}}{B_{//2}/B_{\perp 2}} = \frac{\mu_1}{\mu_2}. \qquad [7.49]$$

The angle of **B** with $\mathbf{n}_{12}$ is larger in the medium with the higher permeability μ.

7.8. Examples of calculation of the field of permanent magnets

In this section we analyze the magnetic field produced by a permanent magnet, of given intensity of magnetization **M** in the absence of any conduction current density and external field. The fields **B** and **H** obey the relations $\nabla \times \mathbf{H} = 0$, $\nabla.\mathbf{B} = 0$ and $\mathbf{B} = \varepsilon_o(\mathbf{H} + \mathbf{M})$, without having the ratio B/H necessarily constant. We assume that the magnetization is uniform, the volume magnetization current density $\mathbf{j}' = \nabla \times \mathbf{M}$ is then equal to zero and the field and vector potential are the same as those of a surface current density $\mathbf{j}'_s = \mathbf{M} \times \mathbf{n}$. Sometimes we may determine **A** as a solution of the equation $\Delta\mathbf{A} = 0$ with the boundary conditions and deduce **B**.

A) Field of a magnetized cylinder in the direction of its axis

Consider a cylinder of axis Oz, length $2h$, radius R and density of magnetization **M**, which is uniform and parallel to Oz (Figure 7.7a). The surface magnetization current density is $\mathbf{j}'_s = \mathbf{M} \times \mathbf{n} = M\,\mathbf{e}_\varphi$ only on the lateral surface. The magnetized cylinder is thus equivalent to a solenoid of finite length $2h$ and radius R (see section 6.8C). Its field at a point P of coordinate z on the axis is

$$\mathbf{B} = \tfrac{1}{2}\,\mu_o(\cos\theta_1 - \cos\theta_2)\,\mathbf{M}, \qquad [7.50]$$

where θ_1 and θ_2 are the half-angles of the cone whose vertex is at P and which subtends the bases at $z = \pm\, h$. They are given by

$$\cos\theta_1 = \frac{h-z}{\sqrt{R^2 + (h-z)^2}}, \qquad \cos\theta_2 = \frac{-h-z}{\sqrt{R^2 + (z+h)^2}}. \qquad [7.51]$$

Using equation [7.31], **H** inside and outside the cylinder may be written as

$$\mathbf{H}^{(in)} = \tfrac{1}{2}[\cos\theta_1 - \cos\theta_2 - 2]\mathbf{M}, \qquad \mathbf{H}^{(ex)} = \tfrac{1}{2}[\cos\theta_1 - \cos\theta_2]\,\mathbf{M}. \qquad [7.52]$$

In the case of a thin disk ($h \ll R$), we find inside it

$$\mathbf{B}^{(in)} = \mu_o\,(h/R)\,\mathbf{M} \qquad \text{and} \qquad \mathbf{H}^{(in)} = -\,[1 - h/R]\,\mathbf{M}. \qquad [7.53]$$

In the case of a thin rod, we find at its middle-point

$$\mathbf{B}^{(in)}(0) \approx \mu_o\mathbf{M}, \qquad \mathbf{H}^{(in)}(0) \approx 0. \qquad [7.54]$$

Figure 7.7a illustrates the lines of the fields **B** and **H** in the case of a magnetized cylinder.

B) Field of a uniformly magnetized ball

Consider a sphere of radius R and uniform intensity of magnetization **M** in the direction of Oz (Figure 7.7b). Outside the sphere, where $\mathbf{M} = 0$ and $\mathbf{j} = 0$, the field $\mathbf{B}^{(ex)}$ obeys the equations $\nabla.\mathbf{B}^{(ex)} = 0$ and $\nabla\times\mathbf{B}^{(ex)} = 0$. Thus, it is conservative and we may use a scalar magnetic potential $\mathbf{B}^{(ex)} = -\nabla V_M$. The scalar potential of a magnetic moment $\mathcal{M}$ has the same expression as the electric potential of an electric dipole **p** with the replacement of $\mathbf{p}/\varepsilon_o$ by $\mu_o\mathcal{M}$. Thus, the scalar magnetic potential of a sphere of magnetization **M** has the same expression as the electric potential of a sphere of dielectric of polarization **P**, but with $\mathbf{P}/\varepsilon_o$ replaced by $\mu_o\mathbf{M}$ (see section 4.8B), hence

$$V_M^{(ex)}(\mathbf{r}) = \mu_o (R^3/3r^3)(\mathbf{M.r}),$$
$$\mathbf{B}^{(ex)}(\mathbf{r}) = -\nabla V_M = \mu_o(R^3/3r^5)[3(\mathbf{M.r})\mathbf{r} - r^2\mathbf{M}] \quad \text{and} \quad \mathbf{H}^{(ex)} = \mathbf{B}^{(ex)}/\mu_o. \qquad [7.55]$$

This is the field of a magnetic moment $\mathcal{M} = (4/3)(\pi R^3)\mathbf{M}$ placed at the sphere center.

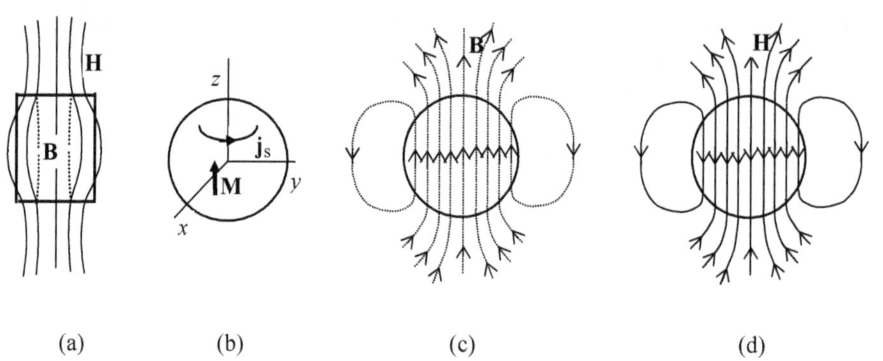

Figure 7.7. *a) Field of a magnetized cylinder. b) Uniformly magnetized sphere and the corresponding surface magnetization current. c) Lines of* **B** *and d) lines of* **H**

We note that the field **H** also verifies the equations $\nabla.\mathbf{H} = 0$ and $\nabla\times\mathbf{H} = 0$ outside the sphere. Thus, it is conservative and we may write $\mathbf{H} = -\nabla V'_M$ with $V'_M = V_M/\mu_o$. Particularly, on the surface of the sphere ($r = R$), we find

$$V_M^{(ex)}(\mathbf{R}) = \mu_o V'_M = (\mu_o/3)(\mathbf{M.R}) = (\mu_o/3) M z,$$
$$\mathbf{B}^{(ex)}(\mathbf{R}) = \mu_o\mathbf{H}^{(ex)} = (\mu_o/3)M[\cos\theta\, \mathbf{e}_r - \mathbf{e}_z]. \qquad [7.56]$$

Inside the magnetized sphere, the field $\mathbf{H}^{(\mathrm{in})} = \mathbf{B}^{(\mathrm{in})}/\mu_o - \mathbf{M}$ verifies Ampère's law $\nabla \times \mathbf{H}^{(\mathrm{in})} = 0$ since $\mathbf{j} = 0$. We deduce that $\mathbf{H}^{(\mathrm{in})} = -\nabla V'^{(\mathrm{in})}_{\mathrm{M}}$. It also verifies the equation $\nabla.\mathbf{H}^{(\mathrm{in})} = 0$ as $\nabla.\mathbf{B}^{(\mathrm{in})} = 0$ and $\nabla.\mathbf{M} = 0$ (in the case of uniform $\mathbf{M}$). On the other hand, $\mathbf{H}$ is determined only by the conduction currents (which are equal to zero in this case). Thus, there is no reason for $V'^{(\mathrm{in})}_{\mathrm{M}}$ to be discontinuous on the surface of the sphere. We may guess that $V'^{(\mathrm{in})}_{\mathrm{M}} = (1/3)Mz$ is a solution of Ampère's equation with the right boundary conditions [7.56]. As the solution of the field equations (which verifies the boundary conditions) is unique, we deduce that this is the solution of our problem, hence

$$V'^{(\mathrm{in})}_{\mathrm{M}} = (1/3)\, Mz; \qquad \mathbf{H}^{(\mathrm{in})} = -\nabla V'^{(\mathrm{in})}_{\mathrm{M}} = -(1/3)\mathbf{M},$$
$$\mathbf{B}^{(\mathrm{in})} = \mu_0(\mathbf{H}^{(\mathrm{in})} + \mathbf{M}) = (2/3)\,\mu_0\mathbf{M}. \qquad [7.57]$$

We note that $\mathbf{B}^{(\mathrm{in})} = \mu_0\mathbf{H} + \mathbf{M}$ verifies the equation $\nabla\times\mathbf{B}^{(\mathrm{in})} = 0$. Thus, we may write $\mathbf{B}^{(\mathrm{in})} = -\nabla V^{(\mathrm{in})}_{\mathrm{M}}$, where $V^{(\mathrm{in})}_{\mathrm{M}} = -(2/3)\mu_0 Mz + C$. This potential does not match $V^{(\mathrm{ex})}_{\mathrm{M}}$ given by [7.56] for any value of the constant C. Thus, V_{M} is not continuous.

The fields $\mathbf{B}^{(\mathrm{ex})}$ and $\mathbf{B}^{(\mathrm{in})}$ correspond to the vector potentials

$$\mathbf{A}^{(\mathrm{ex})}(\mathbf{r}) = (\mu_o/4\pi r^3)\,(\boldsymbol{\mathcal{M}}\times\mathbf{r}) = (\mu_o R^3/3r^2)\, M\,\mathbf{e}_\varphi,$$
$$\mathbf{A}^{(\mathrm{in})}(\mathbf{r}) = \tfrac{1}{2}\, Br\,\mathbf{e}_\varphi = (\mu_o/3)Mr\,\mathbf{e}_\varphi. \qquad [7.58]$$

$\mathbf{A}^{(\mathrm{ex})}$ is the vector potential of a magnetic moment $\boldsymbol{\mathcal{M}} = (4/3)\pi R^3\mathbf{M}$. The expressions of $\mathbf{A}^{(\mathrm{in})}$ and $\mathbf{A}^{(\mathrm{ex})}$ coincide on the sphere; this shows the continuity of $\mathbf{A}$.

Inside a magnetized cylinder, sphere, or a body of any shape, the field $\mathbf{B}$ is parallel to $\mathbf{M}$ and in the same direction, while the field $\mathbf{H}$ points in the opposite direction to $\mathbf{M}$. We say that $\mathbf{H}^{(\mathrm{in})}$ is a *demagnetizing field.* The calculated fields verify the continuity conditions of the normal component of $\mathbf{B}$ and the tangential component of $\mathbf{H}$ on the surface of a body if it carries no conduction current.

C) Magnetic field in cavities

Using the superposition principle and the preceding results, we may calculate the field in a cavity excavated in a medium in which the magnetization $\mathbf{M}$ and the field $\mathbf{B}$ are uniform. Let us assume that the fields are $\mathbf{B}^{(\mathrm{cav})}(\mathbf{r})$ and $\mathbf{H}^{(\mathrm{cav})}(\mathbf{r})$ at the points $\mathbf{r}$ in the cavity and that the fields are $\mathbf{B}^{(\mathrm{in})}(\mathbf{r})$ and $\mathbf{H}^{(\mathrm{in})}(\mathbf{r}) = \mathbf{B}^{(\mathrm{in})}/\mu_o - \mathbf{M}$ at the same point $\mathbf{r}$ inside a body, which has the same magnetization $\mathbf{M}$ and may fill the cavity. If the medium had no cavity, the fields would be obviously $\mathbf{B}$ and $\mathbf{H} = \mathbf{B}/\mu_o - \mathbf{M}$. The superposition principle enables us to write

$$\mathbf{B}^{(\mathrm{cav})}(\mathbf{r}) = \mathbf{B} - \mathbf{B}^{(\mathrm{in})}(\mathbf{r}) \quad \text{and} \quad \mathbf{H}^{(\mathrm{cav})}(\mathbf{r}) = \mathbf{H} - \mathbf{H}^{(\mathrm{in})}(\mathbf{r}) = \mathbf{B}/\mu_o - \mathbf{B}^{(\mathrm{in})}/\mu_o. \quad [7.59]$$

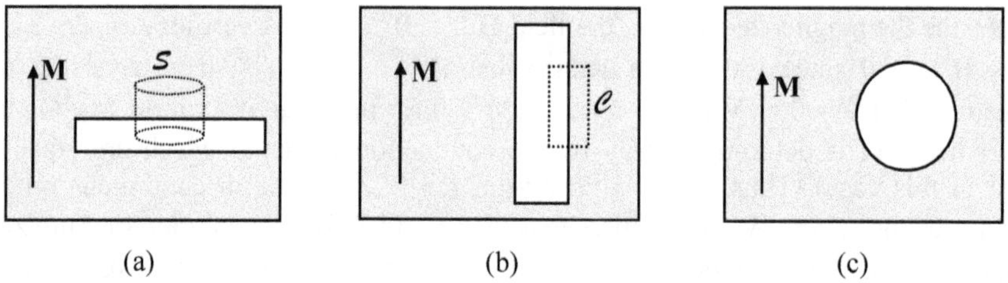

Figure 7.8. *Magnetic field: a) in a thin cylindrical cavity, whose axis is in the direction of* **M**, *b) a long cylindrical cavity in the direction of* **M**, *and c) in a spherical cavity*

– In the case of a cylindrical cavity that is very thin in the direction of **M** (Figure 7.8a), we find $\mathbf{B}^{(in)}(0) = \mu_o(h/R)\,\mathbf{M}$ and $\mathbf{H}^{(in)} = -(1 - h/R)\,\mathbf{M}$ at its center; hence

$$\mathbf{B}^{(cav)}(0) = \mu_o\mathbf{H}^{(cav)}(0) = \mathbf{B} - \mu_o(h/R)\,\mathbf{M}. \qquad [7.60]$$

– In the case of a cylindrical cavity that is very long in the direction of **M** (Figure 7.8b), we find $\mathbf{B}^{(in)}(0) \approx \mu_o\mathbf{M}$ and $\mathbf{H}^{(in)}(0) \approx 0$ at its center; hence

$$\mathbf{B}^{(cav)}(0) = \mu_o\mathbf{H}^{(cav)}(0) = \mathbf{B} - \mu_o\mathbf{M}. \qquad [7.61]$$

– In the case of a spherical cavity (Figure 7.8c), $\mathbf{B}^{(in)}(\mathbf{r}) = (2/3)\,\mu_o\mathbf{M}$, hence

$$\mathbf{B}^{(cav)}(\mathbf{r}) = \mu_o\mathbf{H}^{(cav)}(\mathbf{r}) = \mathbf{B} - (2/3)\,\mu_o\mathbf{M}. \qquad [7.62]$$

7.9. Magnetization of a body in an external field

In the absence of magnetic matter, the field **B** is sufficient to analyze magnetic phenomena as $\mathbf{B} = \mu_o\mathbf{H}$. The field **B** may be calculated using Biot-Savart's law or Ampère's law. In the presence of magnetic matter, the problem is more complicated because **B** is the superposition of the field $\mathbf{B}_I$ produced by the conduction currents and the field $\mathbf{B}_m$ produced by the magnetic matter. We may first calculate **H**, which depends only on the conduction currents and deduce the field $\mathbf{B} = \mu_o(\mathbf{H} + \mathbf{M})$. However, **M** is not known as it depends on **B**, which we have to calculate. Thus, the problem is complicated and generally, it may not be solved analytically. In this section, we consider some simple cases and we assume that the medium is linear and isotropic with a known relative permeability μ_r and that it is exposed to a uniform field $\mathbf{B}_o$ (in the absence of the sample).

A) Magnetization and field of a thin rod parallel to a uniform field

Consider a thin rod that we place parallel to the field $\mathbf{B}_o$ (Figure 7.9a). By symmetry, the intensity of magnetization is parallel to the axis of the rod. The conservation of the flux of $\mathbf{B}$ through the sections of the rod implies that $\mathbf{B}$ is uniform and so is $\mathbf{M}$. Thus the field due to the magnetization is the same as that of a solenoid carrying a surface current of density $j_s = M$. Its field is $\mathbf{B}_m{}^{(in)} = \mu_o\mathbf{M}$ inside the rod. Thus, the total field in the central region of the rod is

$$\mathbf{B}^{(in)} = \mathbf{B}_o + \mathbf{B}_m{}^{(in)} = \mathbf{B}_o + \mu_o\mathbf{M}, \quad \mathbf{B}^{(ex)} = \mathbf{B}_o,$$
$$\mathbf{H}^{(in)} = \mathbf{H}^{(ex)} = \mathbf{B}^{(in)}/\mu_o - \mathbf{M} = \mathbf{B}_o/\mu_o. \qquad [7.63]$$

The permeability being μ_r, we must have $\mathbf{B}^{(in)} = \mu_r\mu_o\mathbf{H}^{(in)}$, hence:

$$\mathbf{M} = (\mu_r - 1)\mathbf{B}_o/\mu_o, \quad \text{and} \quad \mathbf{B}^{(in)} = \mu_r\mathbf{B}_o. \qquad [7.64]$$

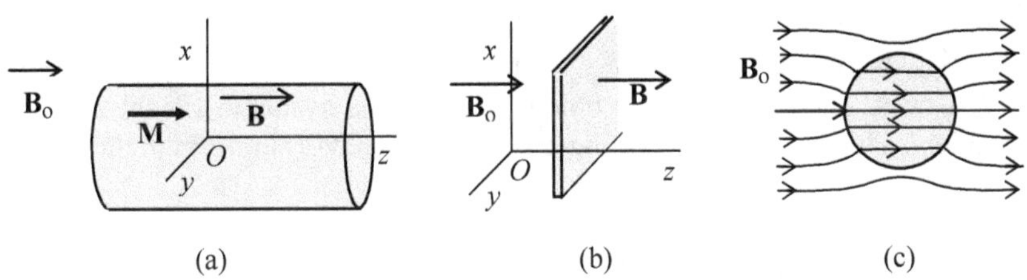

Figure 7.9. *a) Magnetization of a rod, b) of a thin plate, and c) of a sphere*

B) Magnetization and field of a plate perpendicular to a uniform field

If a thin plate is placed perpendicular to a field $\mathbf{B}_o$, because of the symmetries, the magnetization $\mathbf{M}$ and the field $\mathbf{B}$ are perpendicular to the plate and uniform in any plane parallel to the faces. The conservation of the flux of $\mathbf{B}$ through these planes implies that $\mathbf{B}$ is uniform in the plate and so is $\mathbf{M}$ (Figure 7.9b). The volume magnetization current density is $\mathbf{j}' = \nabla \times \mathbf{M} = 0$ and the surface magnetization current density on the faces of the plate $\mathbf{j}'_s = \mathbf{M} \times \mathbf{n}$ is equal to zero. On the other hand, if the plate is large, its thin lateral faces are far away, their surface magnetization current produces a negligible field in the central region. Thus, the field $\mathbf{B}$ is not modified by the plate, hence $\mathbf{B}^{(in)} = \mathbf{B}^{(ex)} = \mathbf{B}_o$ and the magnetization is

$$\mathbf{M} = (\mu_r - 1)\mathbf{H}^{(in)} = (\mu_r - 1)\,\mathbf{B}_o/\mu_o\mu_r. \qquad [7.65]$$

C) Magnetization and field of a ball in a uniform field

If a ball of radius R is placed in an initially uniform field $\mathbf{B}_o$, it becomes magnetized and it produces its own field (Figure 7.9c). Let us assume that it acquires a uniform magnetization $\mathbf{M}$. Adding the field of the magnetized sphere to the external field, we get the total field outside and inside the ball:

$$\mathbf{B}^{(ex)} = \mathbf{B}_o + (\mu_o R^3/3r^5)[3(\mathbf{M.r})\mathbf{r} - r^2\mathbf{M}], \qquad \mathbf{B}^{(in)} = \mathbf{B}_o + (2/3)\mu_o\mathbf{M}.$$
$$\mathbf{H}^{(ex)} = \mathbf{B}_o/\mu_o + (R^3/3r^5)[3(\mathbf{M.r})\mathbf{r} - r^2\mathbf{M}], \qquad \mathbf{H}^{(in)} = \mathbf{B}_o/\mu_o - \mathbf{M}/3. \qquad [7.66]$$

If the sphere is diamagnetic or paramagnetic of relative permeability μ_r, we must have $\mathbf{B}^{(in)} = \mu_r\mu_o\mathbf{H}^{(in)}$, hence

$$\mathbf{M} = \frac{3(\mu_r - 1)}{\mu_o(2+\mu_r)}\mathbf{B}_o, \qquad \mathbf{B}^{(in)} = \frac{3\mu_r}{2+\mu_r}\mathbf{B}_o \qquad \text{and} \qquad \mathbf{H}^{(in)} = \frac{3}{\mu_o(2+\mu_r)}\mathbf{B}_o,$$
$$\mathbf{B}^{(ex)} = \mu_o\mathbf{H}^{(ex)} = \mathbf{B}_o + \frac{\mu_r - 1}{2+\mu_r}\frac{R^3}{r^5}[3(\mathbf{B}_o.\mathbf{r})\mathbf{r} - r^2\mathbf{B}_o]. \qquad [7.67]$$

This solution verifies the equations of the field $\nabla.\mathbf{B} = 0$, $\nabla\times\mathbf{H} = 0$ inside and outside the ball, the relation $\mathbf{B}^{(in)} = \mu_o[\mathbf{H}^{(in)} + \mathbf{M}]$ inside the ball, the continuity relations $\mathbf{B}^{(ex)}.\mathbf{n} = \mathbf{B}^{(in)}.\mathbf{n}$ (with $\mathbf{n} = \mathbf{r}/r$) and $\mathbf{H}^{(ex)}_{//} = \mathbf{H}^{(in)}_{//}$ on the ball and the limit $\mathbf{B}^{(ex)} \to \mathbf{B}_o$ at large distances. The solution being unique, we may assert that this is the solution to our problem. This justifies our starting assumption that $\mathbf{M}$ is uniform in the ball.

7.10. Magnetic susceptibility, nonlinear mediums and non-isotropic mediums

Diamagnetism and paramagnetism are due to the mean magnetic moment acquired by the atoms under the influence of a magnetic field acting on the sample. To analyze magnetization, we must distinguish between the *external field* $\mathbf{B}_o$ (produced by all currents and magnetic bodies except the studied sample) and the *macroscopic field* $\mathbf{B}$, which is the superposition of $\mathbf{B}_o$ and the field $\mathbf{B}_m$ produced by the magnetized sample. The field, which acts on the atom to magnetize it, is the *local field* $\mathbf{B}_l$, which excludes the field of the considered atom. The mean magnetic moment acquired by the atom is thus $\mathcal{M} = \alpha_M\mathbf{B}_l$. To evaluate $\mathbf{B}_l$, let us imagine that the atom is at the center O of a sphere of radius R_1 and volume $\mathcal{V}_1$. The total field at O is obviously $\mathbf{B} = \mathbf{B}_o + \mathbf{B}_1 + \mathbf{B}_2$, where $\mathbf{B}_1$ is the field produced by the atoms of $\mathcal{V}_1$ and $\mathbf{B}_2$ is that of the atoms of the sample situated outside the sphere. On the other hand, the local field is $\mathbf{B}_l = \mathbf{B}_o + \mathbf{B}'_1 + \mathbf{B}_2$ where $\mathbf{B}'_1$ is the field of the atoms of $\mathcal{V}_1$ except the atom at O. As in our study of polarization, $\mathbf{B}'_1$ vanishes in the case of a cubic lattice. Thus, we have $\mathbf{B}_l = \mathbf{B}_o + \mathbf{B}_2$ and this is the field at the center O of a spherical cavity, which is $\mathbf{B}^{(cav)} = \mathbf{B} - (2/3)\,\mu_o\mathbf{M}$ according to [7.62]. We assume the validity of this expression for any magnetized medium. Thus, we have

$$\mathbf{B}_l = \mathbf{B} - (2/3)\,\mu_o\mathbf{M}. \qquad [7.68]$$

Using equation [7.7], the intensity of magnetization may be written as

$$\mathbf{M} = N_v\alpha_M\mathbf{B}_l = N_v\alpha_M[\mathbf{B} - \frac{2}{3}\mu_o\mathbf{M}], \quad \text{where } \mathbf{M} = \frac{3N_v\alpha_M}{3 + 2\mu_o N_v\alpha_M}\mathbf{B}. \qquad [7.69]$$

The magnetic susceptibility is defined by the relation $\mathbf{M} = \chi_M\mathbf{H} = \chi_M(\mathbf{B}/\mu_o - \mathbf{M})$, hence $\mathbf{M} = \chi_M\,\mathbf{B}/\mu_o(1+\chi_M)$. On comparison with [7.69], we deduce that the magnetic susceptibility and the relative magnetic permeability are given by

$$\chi_M = \frac{3N_v\alpha_M\mu_o}{3 - N_v\alpha_M\mu_o}, \qquad \mu_r = \frac{\mu}{\mu_o} = 1 + \chi_M = \frac{3 + 2N_v\alpha_M\mu_o}{3 - N_v\alpha_M\mu_o}. \qquad [7.70]$$

Particularly, if the magnetization is weak ($N_v\alpha_M\mu_o << 1$), we find $\mu_o M << B$ and

$$\chi_M \approx N_v\alpha_M\,\mu_o, \qquad \mu_r \approx 1 + N_v\alpha_M\mu_o \ \text{ and } \ \mathbf{B}_l \approx \langle\mathbf{B}\rangle. \qquad [7.71]$$

This is a good approximation in the case of a diamagnetic medium. If the substance has a mass density m_v and molar mass m_M, the number of molecules per unit volume is $N_v = (m_v/m_M)N_A$, where N_A is Avogadro's number. Thus, the magnetic susceptibility may be written as

$$\chi_M \approx \mu_o\,\alpha_M(m_v/m_M)N_A = (m_v/m_M)\,\chi_{mol}, \qquad [7.72]$$

where $\chi_{mol} = \mu_o\alpha_M N_A$ is the *molar susceptibility*, i.e. the susceptibility of a sample which counts 1 mole per unit volume (thus $m_v = m_M$). The analysis of section 7.2 gives $\alpha_M \approx -1.17 \times 10^{-29}\,Z$. We deduce that the magnetic susceptibility is of the order of $\chi_M \approx -8.85 \times 10^{-9}\,m_v Z/A$, where A is the mass number. The molar susceptibility of diamagnetic solids and liquids is of the order of -10^{-10} to -10^{-9} and that of paramagnetic substances is of the order of $+10^{-10}$.

The magnetization **M** is a response of the medium to the magnetic excitation **H**. If **H** is not very strong, we may write the components of **M** as power series of the components of **H** in the form

$$M_\alpha(\mathbf{H}) = M_\alpha(0) + \Sigma_\beta\,(\partial M_\alpha/\partial H_\beta)\,|_o\,H_\beta + \tfrac{1}{2}\Sigma_{\beta,\gamma}\,(\partial^2 M_\alpha/\partial H_\beta\,\partial H_\gamma)\,|_o\,H_\beta\,H_\gamma + \ldots \qquad [7.73]$$

If the medium has no permanent magnetization ($\mathbf{M} = 0$ if $\mathbf{H} = 0$), the term $M_\alpha(0)$ is equal to zero and if, in addition, the quadratic term is negligible, we say that the medium is *linear*. In this case, we may write $M_\alpha = \Sigma_\beta\,\chi_{\alpha\beta}\,H_\beta$ or explicitly

$$M_1 = \chi_{11}H_1 + \chi_{12}H_2 + \chi_{13}H_3,$$
$$M_2 = \chi_{21}H_1 + \chi_{22}H_2 + \chi_{23}H_3,$$
$$M_3 = \chi_{31}H_1 + \chi_{32}H_2 + \chi_{33}H_3. \qquad [7.74]$$

The nine coefficients $\chi_{\alpha\beta}$ are the components of the *magnetic susceptibility tensor,* which is characteristic of the magnetic medium. In the general case, **M** and **H** are not parallel. In order to be parallel, the tensor $\chi_{\alpha\beta}$ must be diagonal with equal diagonal elements: $\chi_{\alpha\beta} = 0$ if $\alpha \neq \beta$ and $\chi_{11} = \chi_{22} = \chi_{33} \equiv \chi_M$. In this case, the medium is *isotropic* and we may write the vector relation

$$\mathbf{M} = \chi_M \mathbf{H}. \qquad [7.75]$$

Even if a magnetic medium is isotropic, the terms of the second order in [7.73] may become important in the case of a strong magnetic field. The approximation $M = \chi_M H$ is no longer valid. There may be a quadratic or higher-order terms in H. In this case, the medium is said to be *nonlinear.* This is effectively the case of ferromagnetic substances. If the susceptibility χ_M is the same at all the points of the medium, we say that it is *homogeneous*. A magnetic medium is said to be *perfect* if it is isotropic and homogeneous.

7.11. Action of a magnetic field on a magnetic body

If a sample of magnetic material is placed in an external magnetic field $\mathbf{B}_o(\mathbf{r})$, its field $\mathbf{B}_m(\mathbf{r})$ superposes to $\mathbf{B}_o$ to produce the total field $\mathbf{B}(\mathbf{r}) = \mathbf{B}_o(\mathbf{r}) + \mathbf{B}_m(\mathbf{r})$. The field that acts on an element of volume $d\mathcal{V}$ at the point $\mathbf{r}$ of the sample is $\mathbf{B}' = \mathbf{B} - \mathbf{B}_{d\mathcal{V}}$ where $\mathbf{B}_{d\mathcal{V}}$ is the field of the element of volume $d\mathcal{V}$. This element of volume has a magnetic moment $d\mathcal{M} = d\mathcal{V}\, \mathbf{M}(\mathbf{r})$, where the intensity of magnetization **M** is the vector sum of an eventual permanent magnetization and the induced magnetization $N_v\alpha_M\mathbf{B}'$. According to [6.69], the magnetic force acting on $d\mathcal{V}$ has the components

$$dF_{M,\alpha} = \Sigma_\beta\, d\mathcal{M}_\beta\, \partial_\alpha B'_\beta = d\mathcal{V}\, \Sigma_\beta\, M_\beta\, \partial_\alpha B'_\beta. \qquad [7.76]$$

In the general case, the different parts of the sample are subject to different force densities and the body is under stress forces; thus, it may be deformed.

A) Case of a diamagnetic or a paramagnetic linear and isotropic medium

The field $\mathbf{B}_{d\mathcal{V}}$ of the element of volume $d\mathcal{V}$ is proportional to the magnetization $\mathbf{M} = \chi_M\mathbf{B}/\mu$ (see section 7.6B). In the case of a diamagnetic or a paramagnetic medium, χ_M is small and $B_{d\mathcal{V}} << B$. Thus, we may replace $\mathbf{B}'$ by $\mathbf{B}$ in [7.76], hence

$$dF_{M,\alpha} \cong d\mathcal{V}\, \Sigma_\beta\, M_\beta\, \partial_\alpha B_\beta = (\chi_M/2\mu)\, d\mathcal{V}\, \partial_\alpha \mathbf{B}^2. \qquad [7.77]$$

The field $\mathbf{B}_m$ of a paramagnetic or diamagnetic sample usually being weak, we may replace $\mathbf{B}$ by the external field $\mathbf{B}_o$ acting on the sample. Thus, the sample is subject to the total magnetic force

$$\mathbf{F}_M = \iiint_{\mathcal{V}} d\mathbf{F}_M \cong \iiint_{\mathcal{V}} d\mathcal{V}\, (\chi_M/2\mu)\, \nabla(\mathbf{B}_o^2). \qquad [7.78]$$

If $\mathbf{B}_o$ is uniform, $d\mathbf{F}_M \cong (\chi_M/2\mu)\, \nabla(\mathbf{B}_o^2)\, d\mathcal{V} = 0$ and the sample is not subject to a total magnetic force or a magnetic stress. On the contrary, if $\mathbf{B}_o$ is non-uniform but the dimensions of the sample are small, so that $\mathbf{B}_o$ and its derivatives vary very little over the sample, we may write

$$F_{M,\alpha} \cong \Sigma_k\, \partial_\alpha B_{o\,\beta} \iiint_{\mathcal{V}} d\mathcal{V}\, M_\beta = \mathcal{M}_\beta\, \partial_\alpha B_{o\,\beta} = (\chi_M/\mu)\mathcal{V} B_{o\beta} \partial_\alpha B_{o\,\beta} = (\chi_M/2\mu)\mathcal{V}\, \partial_\alpha \mathbf{B}_o^2, \quad [7.79]$$

where $\mathcal{M} = \iiint_{\mathcal{V}} d\mathcal{V}\, \mathbf{M}$ is the magnetic moment of the sample. Thus, the force acting on the sample is proportional to its volume, to its magnetic susceptibility, to the field $\mathbf{B}_o$, and to its gradient. For instance, if $\mathbf{B}_o$ depends only on z and it points in the direction Oz, the force may be written as $\mathbf{F}_M = (\chi_M/2\mu)\mathcal{V}(\partial_z B_{oz}^2)\mathbf{e}_z$. For instance, if a small paramagnetic sample ($\chi_M > 0$) is placed near the N pole of a permanent magnet, it becomes weakly magnetized in the direction of the field, as in Figure 7.1c. In this case, B_{oz}^2 decreases if z increases, thus $\mathbf{F}_M$ is attractive. It is of the order of a few newtons per kilogram and of the order of 10^2 N/kg in the exceptional case of liquid oxygen. Conversely, in the case of a diamagnetic sample, χ_M is negative and $\mathbf{F}_M$ is repulsive but very small.

B) Case of a permanent magnet

In the case of a permanent magnet, the magnetization $\mathbf{M}$ is usually much higher than the induced magnetization and the field $\mathbf{B}_{d\mathcal{V}}$ in [7.76], which is proportional to $\mu_o\mathbf{M}$, is large. The problem may be simplified if the magnetization is saturated, so that $\mathbf{B}'$ differs from $\mathbf{B}$ by a constant term. The gradient of this term is then equal to zero and we may replace $\mathbf{B}'$ by $\mathbf{B}$ in [7.76] to become $dF_{M,\alpha} \cong d\mathcal{V}\, \Sigma_\beta\, M_\beta\, \partial_\alpha B_\beta$. If the sample is small, so that the gradient of $\mathbf{B}$ varies very little over the sample, the total force acting on the sample may be written as

$$F_{M,\alpha} \cong \Sigma_\beta\, \mathcal{M}_\beta\, \partial_\alpha B_\beta = \partial_\alpha(\mathcal{M}.\mathbf{B}), \qquad \text{i.e. } \mathbf{F}_{M,} \cong \nabla(\mathcal{M}.\mathbf{B}), \qquad [7.80]$$

where $\mathcal{M} = \iiint_{\mathcal{V}} d\mathcal{V}\, \mathbf{M}$ is the magnetic moment of the sample. Thus, in the case of a saturated permanent magnet, the force depends only on the gradient of $\mathbf{B}$.

The magnetic field also exerts on the element of volume $d\mathcal{V}$ a moment of force

$$d\boldsymbol{\Gamma} = d\mathcal{M} \times \mathbf{B} = d\mathcal{V}\,(\mathbf{M} \times \mathbf{B}). \quad [7.81]$$

The total moment of force acting on the permanent magnet is

$$\boldsymbol{\Gamma} = \iiint_{\mathcal{V}} d\mathcal{V}\,(\mathbf{M} \times \mathbf{B}). \quad [7.82]$$

Particularly, if **B** varies little over the sample, we find:

$$\boldsymbol{\Gamma} = \mathcal{M} \times \mathbf{B}. \quad [7.83]$$

7.12. Magnetic energy in matter

Consider a long solenoid of length l and n turns per unit length (Figure 7.4b). The field **H** inside the solenoid is given by Ampère's law, $H = nI$. Let us assume that the solenoid is filled with a magnetic material. As the current and the field **B** are established, an induced electromotive $\mathcal{E} = -d\Phi/dt$ appears in the solenoid (see Chapter 8). It is equivalent to a back-electromotive force $d\Phi/dt$, storing in the interval of time dt a magnetic energy

$$dU_{\mathrm{M}} = \mathcal{E}'I\,dt = (d\Phi/dt)\,I\,dt = d\Phi\,I. \quad [7.84]$$

The total number of turns being nl, the total flux is $\Phi = nl\mathcal{S}B = n\mathcal{V}B$, hence $dU_{\mathrm{M}} = n\mathcal{V}.dB.H/n = \mathcal{V}H.dB$, where $\mathcal{V} = l\mathcal{S}$ is the volume of the solenoid. Thus, the variation of the magnetic energy density is

$$dU_{\mathrm{M,v}} = H\,dB. \quad [7.85]$$

This expression, shown here for a solenoid, may be generalized to any medium in a solenoid or not, even if it is nonlinear and anisotropic, in the form

$$dU_{\mathrm{M,v}} = \mathbf{H}.d\mathbf{B}. \quad [7.86]$$

The magnetic energy per unit volume that is necessary to set up the field, by increasing it from 0 to the final value **B** is

$$U_{\mathrm{M,v}} = \int_0^{\mathbf{B}} d\mathbf{B}\,.\mathbf{H}. \quad [7.87]$$

To evaluate this integral, we need the relation of **H** to **B**. In the case of a linear and isotropic medium of permeability μ, $\mathbf{B} = \mu\mathbf{H}$, hence

$$U_{\mathrm{M,v}} = (1/\mu)\int_0^{\mathbf{B}} d\mathbf{B}\,.\mathbf{B} = (1/2\mu)\int_0^{\mathbf{B}} d\mathbf{B}^2 = \mathbf{B}^2/2\mu. \quad [7.88]$$

In the case of a permanent magnet of magnetization **M**, we have $\mathbf{H} = \mathbf{B}/\mu_o - \mathbf{M}$. Thus, the magnetic energy density may be written as

$$U_{M,v} = \int_0^{\mathbf{B}} d\mathbf{B}.(\mathbf{B}/\mu_o - \mathbf{M}) = \mathbf{B}^2/2\mu_o - \mathbf{M.B}. \qquad [7.89]$$

Very often, we have $\mathbf{M.B} >> \mathbf{B}^2/2\mu_o$, hence $U_{M,v} \cong -\mathbf{M.B}$. If **B** is uniform, the energy of a permanent magnetic moment $\mathcal{M} = \mathcal{V}\mathbf{M}$ is then $U_M = -\mathcal{M}.\mathbf{B}$.

In the case of a linear medium, using the vector potential, the magnetic energy may be written also in the form

$$U_M = \iiint_{\mathcal{V}} d\mathcal{V}\, U_{M,v} = \tfrac{1}{2}\iiint_{\mathcal{V}} d\mathcal{V}\, \mathbf{H}.(\nabla\times\mathbf{A}) = \tfrac{1}{2}\iiint_{\mathcal{V}} d\mathcal{V}\, [\, \nabla.(\mathbf{A}\times\mathbf{H}) + \mathbf{A}.(\nabla\times\mathbf{H})]. \qquad [7.90]$$

The first term is the volume integral of the divergence of $(\mathbf{A} \times \mathbf{H})$; it may be transformed into the flux of this vector through a surface $\mathcal{S}$, which contains the system; it may be taken eventually at infinity so that this flux is equal to zero. Using the local form of Ampère's law, the second term may be written as

$$U_M = \tfrac{1}{2} \iiint_{\mathcal{V}} d\mathcal{V}\, (\mathbf{A.j}), \qquad [7.91]$$

where **j** is the conduction current density. We note that $\tfrac{1}{2}(\mathbf{j.A})$ cannot be interpreted as the energy density, as it vanishes in a region, where there is no current density and **A** is defined only up to an arbitrary gradient contrarily to $\mathbf{B}^2/2\mu$. However, the total energy of the whole system has the same value if we use either [7.90] or [7.91]. We also note the similarity of the expressions [7.88] and [7.91] for the magnetic energy with the expressions $U_{E,v} = \tfrac{1}{2}\varepsilon\mathbf{E}^2$ and $U_E = \tfrac{1}{2}\iiint_{\mathcal{V}} d\mathcal{V}\, q_v V$, respectively, for the electrostatic energy.

The magnetic energy is often very small, compared to the electric energy. Thus, magnetic phenomena require a small amount of energy and they often have very little influence on physical, chemical and biological processes.

7.13. Variation of magnetization with temperature

Contrarily to diamagnetism, paramagnetism and ferromagnetism depend on temperature. As in the case of the polarization of dielectrics, we use statistical physics to analyze this dependence (see section 4.13). Let $\mathbf{B}_l$ be the local field that acts on the atom. We take Oz in the direction of $\mathbf{B}_l$ and specify the orientation of the atom magnetic moment $\mathcal{M}_a$ by its angles θ and φ about Oz (Figure 7.10a). The probability that $\mathcal{M}_a$ is oriented in the solid angle dΩ is $d\Pi = \eta\, e^{-U_m/k_B T}\, d\Omega$, where

$U_M = \mathcal{M}_a B_l(1 - \cos\theta)$ is the magnetic energy of the atom in the magnetic field $\mathbf{B}_l$ and $k_B = 1.381\times10^{-23}$ J/K is Boltzmann's constant.

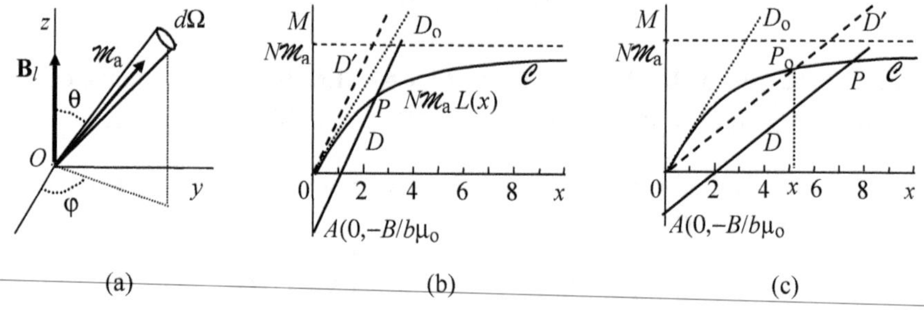

Figure 7.10. *a) Orientation of the magnetic moment, b) magnetization if $T > T_W$, and c) magnetization if $T < T_W$*

Because of the rotational symmetry about Oz, the mean value of $\mathcal{M}_a$ is parallel to $\mathbf{B}_l$. Thus, we have only to calculate the mean value of $\mathcal{M}_z = \mathcal{M}_a \cos\theta$. We find

$$<\mathcal{M}_z> = \int d\Pi\, \mathcal{M}_a \cos\theta = \eta \mathcal{M}_a \int d\Omega \cos\theta\, e^{-U_m/k_B T} = 4\pi\eta\, \mathcal{M}_a(\, e^{-x}/x^2)[x\, \mathrm{ch}(x) - \mathrm{sh}(x)].$$

where $x = \mathcal{M}_a B_l / k_B T$. Using the relation $\eta = (1/4\pi)\, x\, e^x / \mathrm{sh}(x)$, we may write:

$$<\mathcal{M}_z> = \mathcal{M}_a L(x), \qquad L(x) = \coth x - 1/x. \qquad [7.92]$$

$L(x)$ is the *Langevin function*. Thus, the intensity of magnetization is

$$M = N\mathcal{M}_a L(x). \qquad [7.93]$$

The variation of the magnetization as a function of x is illustrated in Figure 7.10b. If x is very small ($x \ll 1$, i.e. B_l weak or T high), we may write $L(x) \approx x/3$. This replaces the curve $L(x)$ by the tangent D_o at the origin. Thus, we find

$$\alpha_M = N\mathcal{M}_a^2/3k_B T \qquad \text{and} \qquad M = N\mathcal{M}_a^2 B_l/3k_B T. \qquad [7.94]$$

Instead of [7.68], Weiss proposed a more general form $\mathbf{B}_l = \mathbf{B} + b\mu_o\mathbf{M}$ (called the *molecular field*). Using the expression [7.94] for M, we deduce that

$$M = CB/(T - T_W) \qquad \text{where} \qquad C = N_v\mathcal{M}_a^2/3k_B \text{ and } T_W = Cb\mu_o. \qquad [7.95]$$

T_W is called the *Weiss temperature*. As $M = \chi_M H = \chi_M B/\mu_o(1+\chi_M)$, we obtain the *Curie-Weiss law*

$$\chi_M = \frac{C\mu_o}{T - T_c} \quad \text{and} \quad \mu = \mu_o(1+\chi_M) = \mu_o \frac{T - T_w}{T - T_c}, \quad \text{where } T_c = T_w(1+\frac{1}{b}). \quad [7.96]$$

C is the *Curie constant* and T_c is the *Curie temperature* (or *critical temperature*). At temperatures higher than T_c, the body loses its ferromagnetic properties; it becomes simply paramagnetic. These relations are effectively verified by several substances. If T is close to T_c, the magnetic susceptibility and the magnetic permeability become very high. We say that the sample undergoes a *phase transition* at T_c. If T is much higher than T_c, we find

$$\chi_M \to T_w/bT \quad \text{and} \quad \mu \to \mu_o. \quad [7.97]$$

The term $b\mu_o\mathbf{M}$ in the expression $\mathbf{B}_l = \mathbf{B} + b\mu_o\mathbf{M}$ takes into account the influence of the nearby magnetic moments. If the magnetization is weak, we may neglect it. In this case, equation [7.95] reduces to the *Curie law*

$$M \approx CB/T \quad \text{and} \quad \chi_M \approx C\mu_o/T, \quad [7.98]$$

which was derived experimentally by Pierre Curie in 1895; it holds for $B/T \approx 0.1$ T/K.

If x is very large ($x >> 1$, i.e. B_l very strong or T very low), $L(x)$ tends asymptotically to 1 and the magnetization tends to the saturation value $N\mathcal{M}_a$. In this limit, all the atoms have their magnetic moment pointing in the direction of $\mathbf{B}_l$.

To write the expression of M as a function of B for intermediary values of x, we have to eliminate B_l between the equations $B_l = B + b\mu_o M$ and $x = \mathcal{M}_a B_l/k_B T$, that is,

$$M = N\mathcal{M}_a L(x) \quad \text{and} \quad M = (k_B T/\mathcal{M}_a b\mu_o)\, x - B/b\mu_o. \quad [7.99]$$

This can be done by using numerical methods. We may also use a graphical method by plotting these two expressions of M as functions of x on the same graph. The first equation is the curve C representing $N\mathcal{M}_a L(x)$ and the second is a straight line D, which meets the M-axis at the point $A(0, -B/b\mu_o)$. The coordinates of the crossing point P of the curves [7.99] determine M and x (hence, B) at temperature T. If the macroscopic field B is reduced, the line D moves parallel to itself toward the origin. If its slope $k_B T/\mathcal{M}_a b\mu_o$ is larger than the slope $N\mathcal{M}_a/3$ of the curve C at the origin, that is $T > T_W$, where $T_W = b\mu_o N\mathcal{M}_a^2/3k_B$, the points A and P approach the origin (Figure 7.10b); this corresponds to $M = 0$ and $B = 0$ (i.e. no permanent magnetization). On the contrary, if $T < T_W$, the point A tends toward O but the crossing point tends toward a point P_o, which corresponds to a field B equal to 0 but a non-zero magnetization (Figure 7.10c). In this case, we have a permanent magnetization M_o, which is a solution of the equation

$$M_o = N\mathcal{M}_a L(b\mu_o\mathcal{M}_a M_o/k_B T). \tag{7.100}$$

This classical theory of paramagnetism was proposed by Langevin in 1905. The quantum formulation was derived by Brillouin and Debye in 1927, the only difference being that the direction of the atomic magnetic moment $\mathcal{M} = -(ge/2m_e)\mathbf{L}$ is quantized. The energy of $\mathcal{M}$ in the field $\mathbf{B}_l$ is

$$U_M = \mathcal{M}B_l - \boldsymbol{\mathcal{M}}.\mathbf{B}_l = -(ge/2m_e)B_l(L - L_z), \tag{7.101}$$

where L_z may take only the values $-L$, $-L + 1$, ... $L - 1$, L. This changes $\mathcal{M}^2 = (ge/2m_e)^2L^2$ into $\mathcal{M}^2 = (ge/2m_e)^2L(L + 1)$. We find the same $1/T$ Curie law, but with a different coefficient. The two theories neglect the interaction of the magnetic moments. The energy of the system is then the sum of the energies of the magnetic moments in the field. This is effectively the case of gases and the rare earth salts.

Under normal conditions, the permeability μ of paramagnetic substances exceeds μ_o by less than 1% but these magnetic effects may be measured with a very high precision.

We have seen that many materials become superconductors at very low temperature, below a superconductivity critical temperature T_{sc}. Such substances have remarkable magnetic properties. If a magnetic field B is applied to the sample, the critical temperature is lowered to a value $T_B \approx T_{sc}(1 - B/B_{sc})^{1/2}$, where B_{sc} is a characteristic critical field. If B is weaker than the critical field, it does not enter into the superconductor. It decreases with the depth x proportionally to $e^{-x/\delta}$, where δ is the *penetration depth*. Thus, the superconductor behaves as an ideal diamagnetic material with a magnetic susceptibility $\chi_M = -1$. The applied magnetic field induces electric currents that can be damped by no resistance in the sample. The magnetic field of these currents exactly counterbalances the external magnetic field. This property of superconductors may be used for magnetic levitation.

7.14. Ferromagnetism

Some materials, such as iron, nickel, cobalt, gadolinium, and dysprosium, and some of their alloys and chemical compounds are not normally magnetized but, if they are exposed to a magnetic field, they become strongly magnetized and they remain magnetized after the removal of the field; they are said to be *ferromagnetic*. The analysis of the preceding section shows that this is possible if the temperature is less than a critical temperature T_c characteristic of the material. T_c is proportional to Weiss parameter b, which increases with the interaction of nearby magnetic moments. Thus, if a ferromagnetic substance is heated, its permanent magnetization decreases and disappears completely at T_c; the substance becomes paramagnetic. The Curie temperature is about 1043 K for iron and 631 K for nickel.

The susceptibility of ferromagnetic substances is of the order of 10^3 for steel and it may be as high as 10^5 for some iron-nickel alloys. The force that a magnetic field exerts on a ferromagnetic substance may be as intense as thousands of newtons per kilogram. Contrary to the force exerted on a paramagnetic or diamagnetic body, which depends on **B** and its gradient, the force exerted on a ferromagnetic body depends only on the gradient of **B** and this may be explained by the saturation of the magnetization (which is then independent of **B**; see section 7.11b).

In 1907, Weiss introduced the concept of *molecular field* and *ferromagnetic domains*, which were understood only after the formulation of quantum theory and the works of Heisenberg in 1928. According to this theory, ferromagnetism is due to the intrinsic magnetic moment of electrons. A quantum effect, called *exchange interaction*, strongly couples the spins of the electrons of nearby atoms and favors their alignment in the same direction, as they then have less potential energy. Thus, the magnetic moments of a large number of atoms become held up in a given direction over small regions of the sample, called *Weiss domains*, which are strongly magnetized in this direction (Figure 7.11a). This exchange interaction is about 10^3 times stronger than the ordinary magnetic interaction between magnetic moments. The Weiss domains have dimensions that vary between 1 μm and 1 mm and each one contains billions of atoms. Two nearby domains are separated by a transition region, called a *Bloch wall*, having a thickness about 100 times the inter-atomic distance. In these walls, the magnetic moment varies continuously from one domain to another. Despite thermal agitation, each domain has a certain magnetic moment. Naturally, in the absence of an external field, the sample is formed by a very large number of domains, whose magnetic moments are randomly oriented in all directions. The macroscopic magnetization of the sample is then zero.

The spins on both sides of a Weiss wall do not align themselves in the same direction to form a single domain for the whole sample because a single domain has a very strong magnetic field, thus a large energy density $B^2/2\mu_0$. The equilibrium configuration is that of the minimum of the total energy of the system, and this corresponds effectively to a sample formed by domains of dimensions of the order of 1 μm.

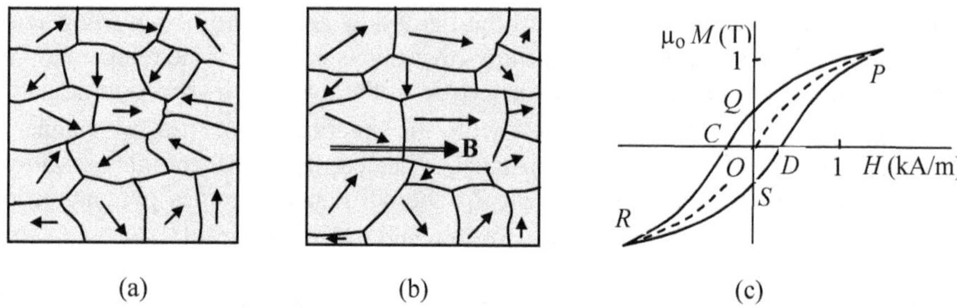

Figure 7.11. *a) Magnetic domains in the absence of an external field, b) deformation of the domains under the influence of a field* **B**, *and c) hysteresis loop*

If the sample is immersed in an external field $\mathbf{B}_o$, the domains with magnetization in the direction of $\mathbf{B}_o$ are favored energetically (Figure 7.11b). They expand by acting on the magnetic moments of the neighboring domains. This produces a global magnetization parallel to $\mathbf{B}_o$. This magnetization increases with the intensity of the field $\mathbf{B}_o$ and remains after the removal of $\mathbf{B}_o$. Then, the sample becomes a permanent magnet. If the external field is very high and the temperature is very low, the sample becomes a single domain and the magnetization becomes saturated. In addition, the magnetization of a sample (i.e. its magnetic susceptibility) depends on its previous history; this is the hysteresis phenomenon.

The magnetization of a substance is studied using a Rowland's ring (see section 7.2). If the ferromagnetic sample had never been magnetized, M increases with H along the dotted curve OP of Figure 7.11c (called curve of *first magnetization*). For an intense field H, the magnetization tends to a constant value corresponding to saturation. If, from a point P, H is reduced to 0, the magnetization follows the curve PQ. The point Q corresponds to $H = 0$ and a magnetization M_r, called *remanent magnetization*, which is of the order of 10^6 A/m and corresponds to a *remanent field* B_r of the order of 1 T. The sample becomes a permanent magnet. If the field **H** is exerted in the opposite direction, M vanishes at point C corresponding to a certain value $-H_c$, called the *coercive field.* Afterwards, M becomes negative, increases in absolute value and tends to saturation in the opposite direction. If, at the point R that is symmetrical to P, the intensity of **H** is decreased, the magnetization decreases and, as H reaches 0, there will be remanent magnetization $-M_r$. If the direction of **H** is reversed again and gradually increased, the magnetization follows the curve SDP, describing an *hysteresis loop*. Thus, the magnetic properties of a ferromagnetic substance depend on its previous treatment. This enables these substances to store information as magnetic tapes and disks.

The evolution of a magnetic sample is accompanied by a variation $H\,dB$ of its energy. Thus, to accomplish the hysteresis loop, the variation of energy is proportional to the area under the curve $H(B)$, therefore to the area of the cycle. This can only be an energy loss as heat. If the remanent magnetization is weak, the area of the loop is small and the material is said to be *soft ferromagnetic*. This is the case of iron; it becomes easily magnetized and loses its magnetization quickly with little loss of energy. It is used for devices using alternating currents (e.g. electromagnets, motors, transformers, etc.). On the contrary, if the remanent magnetization is intense, the area of the loop is large and the material is said to be *hard ferromagnetic*. This is the case for cobalt and nickel, which are difficult to magnetize but they keep their magnetization for a long time. They are used as permanent magnets.

Some substances, said to be *antiferromagnetic*, have antiparallel instead of parallel neighboring spins. In this case, the very small distance between the electrons makes the exchange interaction favor their antiparallel coupling. Their domains may be considered as two superposed lattices with opposite magnetic moments; their resultant magnetization is then equal to zero. However, above the *Neil temperature*, the substance becomes paramagnetic. Some other substances, said to be *ferrimagnetic*, also have antiparallel neighboring spins, but unlike antiferromagnetic substances, the lattices have different magnetic moments. Thus, the superposition of their magnetizations produces a global magnetization of the sample. Contrary to ferromagnetic materials, the ferrimagnetic materials are electric insulators; so they have some interesting applications.

7.15. Magnetic circuits

A) Magnetic circuits without permanent magnets

The field lines of a coil carrying a current I are illustrated in Figure 7.12a. They are almost parallel inside the coil and they disperse outside it. If pieces of soft iron are juxtaposed to form a closed circuit as in Figure 7.12b and N turns carry a current I around a part of this circuit, the magnetic field is very intense in the iron and it almost vanishes outside the circuit as if the field lines are canalized in the circuit. The intense field in iron may be explained by its magnetization, which increases the field **B** considerably. This set-up is called a *magnetic circuit* with the same magnetic flux at all its sections. It is analogous to an electric circuit with the same electric current at all its points. It is not necessary for the magnetic circuit to be closed as the field may exist in vacuum as well as in matter (Figure 7.12c).

Consider a magnetic circuit formed by pieces (i) of length l_i, section $\boldsymbol{S}_i$ and permeability μ_i. The field is produced by a coil of N turns carrying a current I

(Figure 7.12b). Let H_i be the magnetic field in the piece (i). The conservation of magnetic flux implies that $\Phi = \mathcal{S}_i B_i = \mu_i \mathcal{S}_i H_i$, hence $H_i = \Phi/\mu_i\mathcal{S}_i$. On the other hand, applying Ampère's law to a path in the magnetic circuit, we get

$$NI = \int_{\mathcal{C}} d\mathbf{r}.\mathbf{H} = \Sigma_i\, l_i\, H_i = \Phi\, \Sigma_i\, l_i /\mu_i\, \mathcal{S}_i. \qquad [7.102]$$

This equation is similar to the relation $\mathcal{E} = I\, \Sigma_i\, R_i$ for electric circuits. By analogy to the electromotive force, $\mathcal{E}_M = NI$ is called *magnetomotive force*; it supplies the magnetic flux Φ. Similarly, $\mathcal{R}_i = l_i /\mu_i\, \mathcal{S}_i$ is the *reluctance* of the piece of iron. The expression of reluctance is similar to that of resistance $R = l/\sigma\mathcal{S}$ with conductivity σ replaced by permeability μ. Similar to resistances, reluctances in series add up; thus, the total reluctance of the circuit is $\mathcal{R} = \Sigma_i\, \mathcal{R}_i$. The reciprocal of the reluctance is the *permeance* $1/\mathcal{R}$. The law of magnetic circuits may be written as

$$\mathcal{E}_M = \mathcal{R}\Phi, \qquad \text{where } \mathcal{R} = \Sigma_i\, \mathcal{R}_i \text{ and } \mathcal{R}_i = l_i/\mu_i\, \mathcal{S}_i. \qquad [7.103]$$

The practical utility of magnetic circuits is to have an air-gap, i.e. a region in which an intense field may be used (Figure 7.12c). Its permeability is usually μ_o. It is possible to cut the poles in such a way to have a concentrated field. An air gap of section $\mathcal{S}_a$ and length l_a has a reluctance $\mathcal{R}_a = l_a/\mu_o\mathcal{S}_a$. The flux is $\Phi = \mathcal{E}_M/(\mathcal{R}_{iron}+\mathcal{R}_a)$, where $\mathcal{R}_{iron}$ is the reluctance of the iron part. Thus, the field in the air-gap is

$$B_a = \Phi/\mathcal{S}_a = \mathcal{E}_M/\mathcal{S}_a(\mathcal{R}_{iron} + \mathcal{R}_a) \text{ and } H_a = B_a/\mu_o = \mathcal{E}_M/\mu_o\mathcal{S}_a(\mathcal{R}_{iron} + \mathcal{R}_a). \qquad [7.104]$$

Very often, $\mathcal{R}_{iron}$ is much smaller than $\mathcal{R}_a$, hence

$$B_a \cong \mathcal{E}_M/\mathcal{S}_a\mathcal{R}_a = \mu_o\mathcal{E}_M/l_a \qquad \text{and} \qquad H_a \cong \mathcal{E}_M/l_a. \qquad [7.105]$$

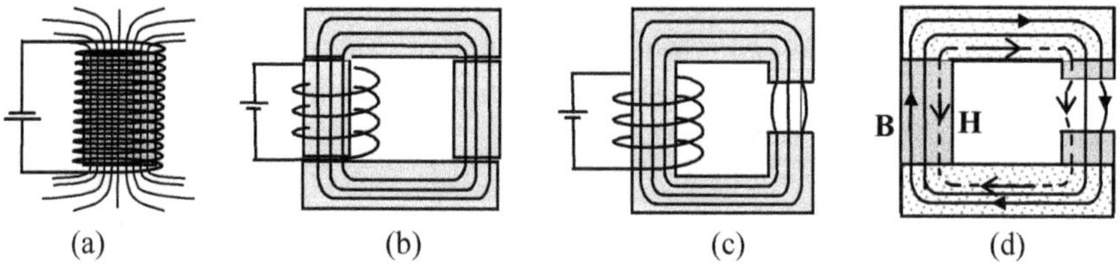

Figure 7.12. *a) Field lines of a coil, b) canalization of the field lines in a magnetic circuit, c) magnetic circuit with an air-gap, and d) magnetic circuit with a permanent magnet*

B) Magnetic circuits including permanent magnets

A magnetic circuit may be formed of a permanent magnet and pieces of soft iron instead of a coil (Figure 7.12d). In this case, **B** and **H** have the same direction in iron (and in the air-gap), but opposite directions in the permanent magnets. Let H_{per} be the field in the permanent magnet and l_{per} its length. Ampère's law gives

$$\int_{\mathcal{C}} d\mathbf{r}.\mathbf{H} = \Sigma_i\, l_i\, H_i + l_a\, H_a - l_{\text{per}}\, H_{\text{per}} = 0. \tag{7.106}$$

The $\mathcal{R}_i$ are usually negligible, compared to $\mathcal{R}_a$, the terms l_iH_i are negligible, compared to l_aH_a, and we may write

$$l_a\, H_a = l_{\text{per}}\, H_{\text{per}}. \tag{7.107}$$

In this case, the calculation is complicated because of the nonlinearity of the relation between **B** and **H** in the permanent magnet. The magnetic energy in the air-gap, whose volume is $\mathcal{S}_a l_a$, is

$$U_M = \tfrac{1}{2}\, B_a H_a(\mathcal{S}_a l_a) = \tfrac{1}{2}\, H_{\text{per}}\, l_{\text{per}}\, \mathcal{S}_{\text{per}} B_{\text{per}}. \tag{7.108}$$

The product $\mathcal{S}_{\text{per}}\, l_{\text{per}}$ is the volume of the permanent magnet. The field is intense in the air-gap if the magnetic energy U_M is large. Thus, the permanent magnet must operate at a point of the hysteresis loop such that the product $H_{\text{per}}B_{\text{per}}$ is maximal. This condition determines H_{per} and B_{per} and equation [7.105] allows H_a and, consequently, B_a to be determined.

7.16. Problems

Magnetic moment of the electron

P7.1 a) Assume that the charge of the electron $-e$ is uniformly distributed on its spherical surface of radius R and that it spins with an angular velocity ω. Show that its magnetic moment is $\mathcal{M}_e = -eR^2\omega/3$. **b)** Assume now that the charge is uniformly distributed in a ball of radius R. Show that the magnetic moment is $\mathcal{M}_e = -eR^2\omega/5$. Verify that, in both models, $\mathcal{M}_e = -(e/2m_e)\mathbf{S}$ where **S** is the intrinsic angular momentum. The experiment shows that the spin is 0.53×10^{-34} J.s and the magnetic moment is 9.3×10^{-24} A.m^2. Do these models agree with experiment?

Equations of the time-independent magnetism

P7.2 Consider a volume $\mathcal{V}$ enclosed in a surface $\mathcal{S}$. Let **n** be the outgoing unit vector normal to $\mathcal{S}$. **a)** Show the identity $\iiint_{\mathcal{V}} d\mathcal{V}\, \partial_z f = \iint_{\mathcal{S}} d\mathcal{S}\, n_z f$ and similarly for the x and y components, hence $\iiint_{\mathcal{V}} d\mathcal{V}\, \partial_i f = \iint_{\mathcal{S}} d\mathcal{S}\, n_i f$, i.e. $\iiint_{\mathcal{V}} d\mathcal{V}\, \nabla f = \iint_{\mathcal{S}} d\mathcal{S}\, \mathbf{n} f$. **b)** Deduce from this relation Gauss-Ostragradsky's theorem $\iiint_{\mathcal{V}} d\mathcal{V}\, \nabla\mathbf{V} = \iint_{\mathcal{S}} d\mathcal{S}\, \mathbf{n}.\mathbf{V}$ and the identity $\iiint_{\mathcal{V}} d\mathcal{V}\, \nabla\times\mathbf{V} = \iint_{\mathcal{S}} d\mathcal{S}\, \mathbf{n}\times\mathbf{V}$.

P7.3 Consider a region in which $\mathbf{j} = 0$ and assume that the field $\mathbf{B}$ is time-independent, parallel to the plane Oxy and independent of z. Thus, it obeys the equations $\nabla.\mathbf{B} = 0$ and $\nabla \times \mathbf{B} = 0$ similar to the equations $\nabla.\mathbf{E} = 0$ and $\nabla\times\mathbf{E} = 0$ of the electrostatic field in vacuum. **a)** Show that it is possible to choose the vector potential $\mathbf{A}$ parallel to Oz. Then, we have $B_x = \partial A/\partial y$ and $B_y = \partial A/\partial y$, where A is the magnitude of $\mathbf{A}$. **b)** Show that the lines A = Constant are the lines of the field $\mathbf{B}$. **c)** Show that $\mathbf{B} = -\nabla U_M$ and that the lines U_M = Constant are perpendicular to the lines A = Constant. Show that $\Delta U_M = 0$.

P7.4 Show that, on the interface of two mediums (1) and (2), the vector potential obeys the boundary conditions

$$\mathbf{n}_{12}.[\nabla \times (\mathbf{A}_1 - \mathbf{A}_2)] = 0 \qquad \text{and} \qquad \mathbf{n}_{12} \times [\nabla \times (\mathbf{A}_1/\mu_1) - \nabla \times (\mathbf{A}_2/\mu_2)] = \mathbf{j}_S,$$

where $\mathbf{n}_{12}$ is the unit vector normal to the interface and oriented from the medium (1) toward the medium (2). Show that the magnetization current density on the interface of two materials of magnetizations $\mathbf{M}_1$ and $\mathbf{M}_2$ is $(\mathbf{M}_2 - \mathbf{M}_1) \times \mathbf{n}_{12}$.

Examples of the calculation of the field of a magnetized body

P7.5 A long cylindrical shell of internal radius R_1 and external radius R_2 is uniformly magnetized parallel to its axis. Calculate $\mathbf{B}$ on the axis but far away from its ends.

P7.6 A sphere of radius R has a uniform magnetization $\mathbf{M}$ parallel to Oz. **a)** Calculate the field $\mathbf{B}$ at a point P of Oz by using the magnetization currents. **b)** Discuss the symmetries and deduce that the vector potential has a single component A_φ that obeys the partial differential equation

$$2r \sin^2\theta\, \partial_r A_\varphi + r^2 \sin^2\theta\, \partial^2_{rr} A_\varphi - A_\varphi + \sin\theta \cos\theta\, \partial_\theta A_\varphi + \sin^2\theta\, \partial^2_{\theta\theta} A_\varphi = 0.$$

Try to solve this equation by setting $A_\varphi = g(r) + f(r) \sin\theta$. Verify that we must have $g = 0$ and $f = ar + b/r^2$. Imposing the regularity of the solution at $r = 0$ and $r \to \infty$, verify that $b^{(in)} = 0$ and $a^{(ex)} = 0$. Express $\mathbf{B}$ and $\mathbf{H}$ in terms of $a^{(in)}$ and $b^{(ex)}$. Imposing the continuity conditions on the sphere and setting $\mathcal{M} = (4/3)\, \pi R^3 \mathbf{M}$, deduce that

$$\mathbf{A}^{(in)} = (1/3)\, \mu_o M r \sin\theta\, \mathbf{e}_\varphi, \qquad \mathbf{B}^{(in)} = (2/3)\, \mu_o \mathbf{M},$$
$$\mathbf{A}^{(ex)} = (\mu_o/4\pi r^3)\, \mathcal{M} \times \mathbf{r} \quad \text{and} \quad \mathbf{B}^{(ex)} = (\mu_o/4\pi\, r^5)\, [3(\mathcal{M}.\mathbf{r})\mathbf{r} - r^2\mathcal{M}].$$

P7.7 An atom of magnetic moment $\mathcal{M}_a$ is located at a point $\mathbf{r}'$ outside a sphere of radius ρ. Show that the average of its field in the sphere is equal to its field at the center O. Show that, if the atom is inside the sphere, its average field in the sphere is $\langle\mathbf{B}\rangle^{(in)} = (\mu_o/4\pi\rho^3 r'^2)[3(\mathbf{r}'.\mathcal{M})\mathbf{r}' - r'^2\mathcal{M}]$. Assuming that many atoms are distributed at random in the sphere, show that the average value of their field is equal to 0.

Magnetization of a sample placed in an external field

P7.8 A non-magnetic long wire of radius R_1 carries a current I. It is surrounded by a cylindrical shell of a medium of permeability μ, internal radius R_2 and external radius R_3. **a)** Calculate the fields **H** and **B** everywhere. **b)** Determine the magnetization current density in the shell and on its faces.

P7.9 Estimate the magnetic susceptibility of helium and air under normal conditions.

Action of a magnetic field on a magnetized medium

P7.10 An iron disk of density 7800 kg/m^3, thickness e = 2 mm and radius R = 5 cm has a magnetization $M = 2 \times 10^5$ A.m^{-2} parallel to its axis. **a)** Calculate **H** and **B** at its center O and at the point P of coordinate z = 3 cm on the axis. At what distance z, the field differs from that of a magnetic moment $\mathcal{M} = M\mathcal{V}$ by less than 5%? **b)** At what distance an identical disk must be placed above the first one in order to remain in equilibrium? Treat the disks as magnetic moments.

P7.11 a) A bar magnet of length l and section $\mathcal{S}$ has a magnetization M. What is its magnetic moment? Two small bar magnets of magnetic moments $\mathcal{M}_1$ and $\mathcal{M}_2$ have their centers at O and **r**. Calculate their force of interaction. **b)** By analogy to an electric dipole, suppose that the magnetic moment is modeled as two "magnetic charges" $\pm q_M$ placed at the ends of the bar magnet and that the interaction of two "magnetic charges" is $\mathbf{F}_{1\to 2} = K q_{M1} q_{M2} \mathbf{R}_{12}/R_{12}^3$ where $\mathbf{R}_{12} = \mathbf{r}_2 - \mathbf{r}_1$. Calculate the force that a bar magnet at O exerts on another one at **r**. Verify that it is the same as the interaction of two magnetic moments if $K = \mu_o/4\pi$, $\boldsymbol{\mathcal{M}}_{1} = q_{M1}\mathbf{l}_1$ and $\boldsymbol{\mathcal{M}}_{2} = q_{M2}\mathbf{l}_2$.

Variation of the magnetization with temperature

P7.12 To see how quantum mechanics modifies Langevin's theory of paramagnetism, we consider atoms of spin $s = ½$. Contrarily to the classical theory, the projection S_z of the spin on Oz takes only discrete values. Thus, in the case of a particle of spin ½, S_z takes only the values $-½\hbar$ and $½\hbar$. To the spin **S**, corresponds an intrinsic magnetic moment $\boldsymbol{\mathcal{M}} = -(ge/2m_e)\mathbf{S}$, where g is the gyromagnetic ratio. The energy of the atom placed in a magnetic field **B** is $U_M = -\boldsymbol{\mathcal{M}}.\mathbf{B}$. According to statistical mechanics, the probability of a state of energy U_M is proportional to $\exp(-U_M/k_BT)$. Deduce that the number of atoms per unit volume whose spin points in the direction of **B** and in the opposite direction are, respectively, $N_+ = \eta\, e^x$ and $N_- = \eta\, e^{-x}$, where $x = geB/4m_e k_B T$ and η is given by the condition of normalization $N_+ + N_- = N_v$, hence $\eta = N_v/2\ \mathrm{ch}(x)$. Show that the magnetization density is $M = \mathcal{M}N_+ - \mathcal{M}N_- = N_v\, \mathcal{M}\ \mathrm{tgh}(x)$. Deduce that the saturation magnetization is $M_s = N_v\mathcal{M}$ while, at low temperature, $M \approx N_v B\mathcal{M}^2/kT$. Note the absence of the classical factor 1/3. On the other hand, according to quantum mechanics, $\mathcal{M}^2 = (ge\hbar/2m_e)^2\mathbf{S}^2 = (ge\hbar/2m_e)^2 s(s+1)$. Thus, the density of magnetization is $M = N_v g^2\, s(s+1)\, \mu_B{}^2 B/3kT$ with $\mu_B = e\hbar/2m_e$.

Ferromagnetism and magnetic circuits

P7.13 The iron mass density is 7860 kg/m^3 and the magnetic moment of its atom is 1.8×10^{-23} A.m^2. **a)** What should its magnetic susceptibility at 300 K be if iron was paramagnetic and the field is weak? What should be the magnetic moment of a rod of iron of cross-sectional area 1 cm^2 and length 5 cm be if it is placed in a field of 0.5 T? **b)** In fact, because of the spin coupling, iron is ferromagnetic. To simplify, assume that the rod is a single domain with all the atomic magnetic moments pointing in the direction of **B**. What then is the magnetic moment of the rod? What is the moment of the magnetic forces acting on this rod if it is placed in a field $B = 0.01$ T perpendicular to the axis of the rod?

P7.14 a) Iron has a mass density $m_v = 7.8 \times 10^3$ kg/m^3, a mass number $A = 56$ and an atomic number $Z = 26$. Determine the number of atoms and the number of electrons per m^3 of iron. The magnetic moment of the electron is $\mathcal{M} = 0.93 \times 10^{-23}$ J/T. What should the magnetization of iron be if all the electrons have their spins pointing in the same direction? In fact, the saturation magnetization of iron is $M_s = 2 \times 10^6$ A/m. Deduce the number of electrons whose spins point in the same direction and the number of paired electrons per atom. **b)** What is the magnetic moment of 1 kg of iron and what is the force that acts on 1 kg of iron if the gradient of B is 20 T/m?

P7.15 a) Consider a coil of $N = 500$ turns around a torus of iron whose relative permeability is $\mu_r = 10^3$, its average radius is $R = 40$ cm and its cross-sectional area is $\mathcal{S} = 10$ cm^2. Calculate the flux of B through a section of this torus if $I = 1$ A in the coil. **b)** Let us assume that the torus is not complete but it has an air-gap of length 4 cm. Calculate the field B_{ir} in the iron and B_a in the air-gap.

Chapter 8

Induction

After the discovery in 1819 by Oersted that an electric current produces a magnetic field, scientists turned their attention to search for the inverse effect, that is, the production of an electric current by a magnetic field. In 1831, Faraday in England and Henry in the United States showed that a varying magnetic field or more generally a varying magnetic flux in a circuit induces an electric current. Faraday was the first to publish his results and the discovery of this *induction law* was attributed to him. Henry later discovered *self-induction*, i.e. the induction of an electromotive force (emf) in a circuit if its own magnetic flux varies. The discovery of induction had many important applications. One of them was the large-scale production of electricity, which led to a new technological era.

The induction phenomenon is quite complicated, because there are really two kinds: *Neumann's induction*, which appears even in vacuum if the magnetic field varies in time, and *Lorentz induction*, which appears in moving conductors in a constant magnetic field. In this chapter we will study induction and some of its applications.

8.1. Induction due to the variation of the flux, Faraday's and Lenz's laws

Faraday's historic experiment showed that a current is induced in a circuit (without generators) if the magnetic flux through this circuit varies. The flux may be varied by displacing a magnet or by varying the field of a nearby electromagnet by varying its current. The induced current lasts as long as the magnetic flux varies. Faraday formulated his results by stating that *the induced emf in a circuit is equal to the rate of variation of the magnetic flux through this circuit*. In 1834, Lenz formulated the law that *the direction of the induced current is such that it opposes the cause that produces it*. These results may be expressed by the relation

$$\mathcal{E} = -\dot{\Phi}, \qquad \text{where} \qquad \Phi = \iint_{\mathcal{S}} d\mathcal{S}\,\mathbf{n}.\mathbf{B}. \tag{8.1}$$

The variation of Φ may be due either to a variation of $\mathbf{B}$ in a circuit $\mathcal{C}$ at rest (Neumann's induction), to the deformation or the displacement of $\mathcal{C}$ in a time-independent field (Lorentz induction) or to both causes.

If the circuit has a resistance R, the induced emf produces a current of intensity $I = \mathcal{E}/R = -\dot{\Phi}/R$. The total induced charge between the times t_1 and t_2 may be written as $Q = \int_{t_1}^{t_2} dt\, I = (\Phi_1 - \Phi_2)/R$. This induced charge is thus proportional to the variation of the magnetic flux in the interval of time. It may be measured with a ballistic galvanometer, for instance, and this allows measuring the flux variation.

The induced emf $\mathcal{E}$, which produces a current I in a circuit $\mathcal{C}$, supplies a power $P = \mathcal{E}I$, which may be used or dissipated as Joule heat. It is evidently supplied by the external agent, which produces the variation of the flux or due to the variation of the stored electromagnetic energy in nearby circuits $\mathcal{C}'$ or in the circuit $\mathcal{C}$ itself.

To interpret the induced emf, we recall that, in a battery for instance, the emf is equivalent to a non-electric generating force $\mathbf{f}_g$ pushing thc positive charges toward the positive terminal P and the negative charges toward the negative terminal N. Thus, $\mathbf{f}_g$ is equivalent to a non-electric generating field $\mathbf{E}_g = \mathbf{f}_g/q$. If no current is drawn, the accumulated charges produce between the terminals a potential $V_{PN} \equiv V_p - V_N$ equal to the emf $\mathcal{E}$. Then, the generating field counterbalances the electric field inside the battery ($\mathbf{E}_g = -\mathbf{E}^{in}$). If the battery is connected to a circuit $\mathcal{C}$, it acts like a pump, exerting a force $\mathbf{f}_g = q\mathbf{E}_g$ on the conduction charges and supplying an energy $q\mathcal{E}$. A part qV_{PN} of this energy is consumed in the external circuit and the remaining qrI is dissipated as Joule heat in the battery itself.

In the case of induction in a circuit without generators, the emf is not localized in a particular element of the circuit but along all the circuit $\mathcal{C}$. The generating force, which sets the charges in motion, can only be the Lorentz force $\mathbf{f}_g = q(\mathbf{E} + \mathbf{v}\times\mathbf{B})$ due to an "induced electric field" $\mathbf{E}$ and the magnetic field $\mathbf{B}$ produced by set-ups that are external to the circuit $\mathcal{C}$ (magnets and other circuits) and by the circuit itself. Here, $\mathbf{v}$ is the total velocity of the conduction charges, that is the vector sum of their drift velocity $\mathbf{v}_d$ with respect to the conductor ($\mathbf{v}_d$ produces the electric current in conductors at rest) and an eventual drag velocity $\mathbf{v}_o$ if the conductor is moving. We may say also that there is an induced generating field $\mathbf{E}_g = \mathbf{E} + \mathbf{v} \times \mathbf{B}$. The induced emf is the circulation of $\mathbf{E}_g$ along the circuit $\mathcal{C}$ in the chosen direction, i.e.

$$\mathcal{E} = \int_{\mathcal{C}} d\mathbf{r}.\,(\mathbf{E} + \mathbf{v} \times \mathbf{B}) = \int_{\mathcal{C}} d\mathbf{r}.\mathbf{E} + \int_{\mathcal{C}} d\mathbf{r}.(\mathbf{v}_\mathrm{o} \times \mathbf{B}) + \int_{\mathcal{C}} d\mathbf{r}.(\mathbf{v}_\mathrm{d} \times \mathbf{B}). \qquad [8.2]$$

As we shall see in the next section, the first term on the right-hand side is Neumann's induction (due to the variation of **B** in time). The second term is Lorentz induction (due to the displacement of the conductor). In the third term, if the circuit is thin, the drift velocity $\mathbf{v}_\mathrm{d}$ points in the direction of the element $d\mathbf{r}$ of the circuit, the triple product $d\mathbf{r}.(\mathbf{v}_\mathrm{d} \times \mathbf{B})$ is then equal to zero. In the case of a conductor of any shape, $\mathbf{v}_\mathrm{d}$ is often very small and we may neglect this term.

8.2. Neumann's induction

If the circuit $\mathcal{C}$ is at rest ($\mathbf{v}_\mathrm{o} = 0$), the induced emf reduces to the first term of [8.2] called *Neumann's induction*, for which we have to specify the field **E**. It is well known that the field **E** produced by static charges is conservative. Thus, its circulation along the closed circuit $\mathcal{C}$ is equal to zero. On the contrary, moving electric charges, magnets, or circuits and variable electric currents produce a time-dependent field **B**. According to Faraday's law [8.1], this variable **B** induces an emf

$$\mathcal{E}_\mathrm{N} = -\dot{\Phi} = -\iint_{\mathcal{S}} d\mathcal{S}\, \mathbf{n}.\partial_t \mathbf{B} \qquad \text{(Neumann's induction).} \qquad [8.3]$$

Here $\mathcal{S}$ is any surface bounded by $\mathcal{C}$. Identifying $\mathcal{E}_\mathrm{N}$ to the first term of [8.2] written in the form $\int_{\mathcal{C}} d\mathbf{r}.\,\mathbf{E} = \iint_{\mathcal{S}} d\mathcal{S}\, \mathbf{n}.\nabla \times \mathbf{E}$ (by using Stokes' theorem), we deduce that

$$\nabla \times \mathbf{E} = -\partial_t \mathbf{B}. \qquad [8.4]$$

This is the local form of Faraday's law in the case of Neumann's induction. **B** is the total magnetic field (including that of the circuit itself) and **E** is the induced electric field (due to the variation of **B** produced by magnets, moving charges, and variable currents). The electric field of static electric charges being conservative, its circulation along any closed path $\mathcal{C}$ vanishes and its curl vanishes. Thus, it may be added to the induced field without modifying equations [8.3] and [8.4], and we may consider, in these equations, **E** and **B** as the total fields.

The magnetic field, even if it is time-dependent has a conservative flux, hence

$$\iint_{\mathcal{S}} d\mathcal{S}\, (\mathbf{n}.\mathbf{B}) = 0 \qquad \text{and} \qquad \nabla.\mathbf{B} = 0, \qquad [8.5]$$

where $\mathcal{S}$ is any closed surface and **n** is the outgoing normal unit vector. We deduce that there is a *vector potential* **A**, such that

$$\mathbf{B} = \nabla \times \mathbf{A}. \qquad [8.6]$$

Using this expression of **B**, equation [8.4] may be written as $\nabla\times(\mathbf{E} + \partial_t\mathbf{A}) = 0$. So, the vector $(\mathbf{E} + \partial_t\mathbf{A})$ is conservative; it may be written as $-\nabla V$, hence

$$\mathbf{E} = -\nabla V - \partial_t\mathbf{A}. \qquad [8.7]$$

Evaluating the circulation of both sides of this relation along the circuit $\mathcal{C}$, the circulation of ∇V vanishes. Thus, the Neumann's induced emf may be written as

$$\mathcal{E}_N = \int_{\mathcal{C}} d\mathbf{r}.\mathbf{E} = -\int_{\mathcal{C}} d\mathbf{r}.\partial_t\mathbf{A}. \qquad [8.8]$$

This result allows $\mathcal{E}_N$ to be considered as the circulation of $-\partial_t\mathbf{A}$ along the circuit $\mathcal{C}$ in accordance with the concept of local interaction. To clarify this point, consider a circuit $\mathcal{C}$ that encircles a thin cylindrical region of a variable magnetic field, a thin solenoid carrying a variable current, for instance. The magnetic flux through $\mathcal{C}$ is $\mathcal{S}B$, the same as through a section $\mathcal{S}$ of the solenoid and the induced emf is $\mathcal{E}_N = -\iint_{\mathcal{S}} d\mathcal{S}\,(\mathbf{n}.\,\partial_t\mathbf{B})$. Although this relation is correct, it is hard to understand what force acts on the conduction charges of $\mathcal{C}$ to set them in motion (as there is no significant fields acting on them) and how the field $\mathbf{B}(t)$, which is restricted inside the solenoid, can induce instantaneously an emf in a circuit $\mathcal{C}$ that is totally outside the solenoid even if it is at large distance. Equation [8.8], considers this effect as the result of the local action of $-\partial_t\mathbf{A}$ on each point of the circuit. This equation shows that the potentials are not only a convenient mathematical trick to calculate the fields. In some cases, they are more fundamental than the fields. Some other effects (such as the quantum mechanical Aharonov-Bohm effects) confirm this conclusion.

8.3. Lorentz induction

If a body moves with a velocity $\mathbf{v}_o$ in a field **B**, the second term of [8.2] suggests that an emf $\mathcal{E} = \int_{\mathcal{C}} d\mathbf{r}.\,(\mathbf{v}_o \times \mathbf{B})$ is induced in this body. This is true for any field **B**, uniform or not and time-dependent or not and for any $\mathcal{C}$, closed or open. This phenomenon is called *Lorentz induction.*

To analyze this induction, let us assume that a uniform field **B** is established in a region of space, where there is no electric field (Figure 8.1a). We take Oz in the direction of **B** and assume that a metallic rod of length D parallel to Oy is moved with a velocity $\mathbf{v}_o = v_o\mathbf{e}_x$ in the direction Ox. The magnetic field exerts on the conduction electrons of the rod a generating force $\mathbf{f}_g = -e\mathbf{v}_o \times \mathbf{B} = ev_oB\,\mathbf{e}_y$. It pushes them toward end N, making it negatively charged and leaving end M positively charged. These charges produce an electric field $\mathbf{E}'$ pointing toward the positive y.

The displacement of the electrons stops when the generating force $\mathbf{f}_g$ counterbalances the electric forces $-e\mathbf{E}'$, hence

$$\mathbf{E}' = -\mathbf{v}_o \times \mathbf{B} = v_o B\, \mathbf{e}_y. \quad [8.9]$$

This field, being produced by stationary charges, is conservative in the proper frame of the rod. The induced emf is the corresponding difference of potential

$$\mathcal{E} \equiv V_M - V_N = \int_M^N d\mathbf{r}\,.\mathbf{E}' = v_o BD. \quad [8.10]$$

If the rod slides on two metallic rails connected by a resistance R, an intensity

$$I = \mathcal{E}/R = v_o\, \frac{DB}{R} \quad [8.11]$$

is induced in the direction NM. The magnetic force that the field $\mathbf{B}$ exerts on the rod carrying this current is

$$\mathbf{F}_M = I\,(\overrightarrow{NM} \times \mathbf{B}) = -\,IDB\,\mathbf{e}_x = -\,\frac{v_o D^2 B^2}{R}\,\mathbf{e}_x. \quad [8.12]$$

It points in the opposite direction to the motion (in agreement with Lenz's law) and it is proportional to the velocity. To displace the rod, it must be pulled with a force $\mathbf{F}_{op} = -\mathbf{F}_M$. The work of $\mathbf{F}_{op}$ in the interval of time dt is $\mathbf{F}_{op}.\mathbf{v}_o\, dt = -\mathbf{F}_M.\mathbf{v}_o\, dt = (v_o^2 D^2 B^2/R)\, dt$. It is equal to the dissipated energy as Joule heat, $dU_J = I^2 R\, dt$. We note that our analysis neglects the magnetic field produced by the induced current (i.e. self-induction).

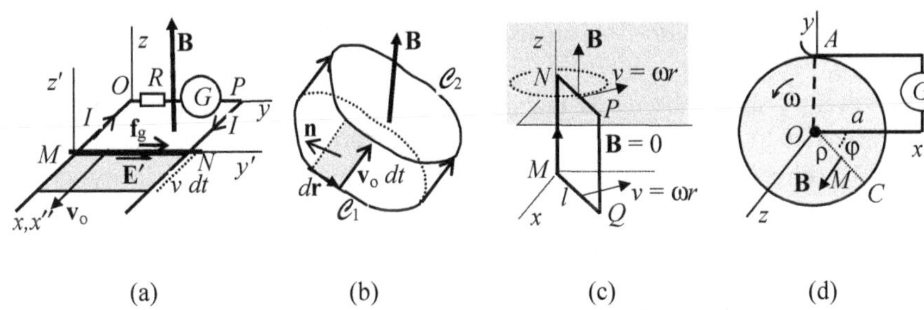

Figure 8.1. *Lorentz induction: a) in a moving rod, and b) in a circuit in motion, c) and d) induction without variation of flux*

The induced emf [8.10] agrees with Faraday's law [8.1] with the variation of the flux due to the variation of the circuit area. Indeed, let us orient the circuit $\mathcal{C}$ formed by the rod and rails in the direction *OMNP* (according to the right-hand rule about *Oz*). The unit normal vector **n** is then oriented in the direction of *Oz* and the magnetic flux through the circuit in the direction of **n** is positive and equal to *BDx*, where *x* is the coordinate of *M*. Its rate of variation is $\dot{\Phi} = BD\,\partial_t x = BDv_o$. Thus, we have an induced emf $\mathcal{E} = -\dot{\Phi} = -\,BDv_o$ and an induced current $-BDv_o/R$. The negative sign means that it is effectively in the opposite direction (i.e. in the direction *OPNM*). Another way to apply Lenz's law is to say that if v_o is positive, the flux in the direction of **n** increases. In order to oppose this increase, the induced current must produce a flux in the opposite direction. Thus, the current must be in the direction *OPNM*.

This result ($\mathcal{E} = -\dot{\Phi}$) holds for any open or closed circuit $\mathcal{C}$ moving or deformed in a field **B** (Figure 8.1b). Indeed, let us assume that $\mathcal{C}$ has an infinitesimal translational motion $\mathbf{v}_o dt$ from the position $\mathcal{C}_1$ to the position $\mathcal{C}_2$ generating a cylindrical surface. We orient $\mathcal{C}$ and we consider an elements $d\mathbf{r}$, over which **B** is approximately uniform. The induced emf over $d\mathbf{r}$ is $d\mathcal{E} = d\mathbf{r}.(\mathbf{v}_o \times \mathbf{B}) = \mathbf{B}.(d\mathbf{r} \times \mathbf{v}_o)$. We may write $d\mathbf{r} \times \mathbf{v}_o\, dt = d\mathcal{S}\, \mathbf{n}$, where $d\mathcal{S}$ is the lateral surface that is swept by the element $d\mathbf{r}$ in the interval of time dt and **n** is the normal unit vector outgoing from the lateral surface. We deduce that $\mathbf{B}.(d\mathbf{r} \times \mathbf{v}_o) = (d\mathcal{S}/dt)(\mathbf{n}.\mathbf{B}) = d\Phi_l/dt$, where $d\Phi_l$ is the lateral magnetic flux that is cut by $d\mathbf{r}$. Thus the induced emf in $\mathcal{C}$ is

$$\mathcal{E} = \int_{\mathcal{C}} d\mathbf{r}.(\mathbf{v}_o \times \mathbf{B}) = \int_{\mathcal{C}} d\Phi_l/dt = \Delta\Phi_l/dt. \qquad [8.13]$$

$\Delta\Phi_l$ is the total flux outgoing from the lateral surface. The conservation of magnetic flux implies that $\Delta\Phi_l + \Phi_2 - \Phi_1 = 0$. Thus, the induced emf in the circuit may be written as $\mathcal{E} = (\Phi_1 - \Phi_2)/dt = -\,\dot{\Phi}$ in accordance with Faraday's law [8.1].

However, it should be noted that it is not necessary that the flux through the circuit $\mathcal{C}$ varies to have a Lorentz-induced emf. Figure 8.1c illustrates the case of a rectangular circuit *MNPQ* rotating about *MN* with an angular velocity ω in a magnetic field **B**. This field acts only on the upper part, i.e. *NP* and a part of *PQ* and *MN*. In this case, **B** has no flux through the circuit but it acts on the electrons of *MN*. These electrons move with a velocity ωr and they are pushed toward *N*. This induces an emf $\mathcal{E} = \frac{1}{2}\omega B l^2$. Figure 8.1d shows a metallic disk, which rotates in a field **B**. Its center *O* and a point *A* of the periphery are connected to a galvanometer *G*. An emf is induced in the circuit, even though **B** is not variable and its flux through the circuit *OAG* does not vary. The explanation of this induction is that, during the rotation of the disk, the conduction electrons along the radius *OA* move with respect to the observer with a speed ωr in the field **B**; thus, they are subject to a generating force $-e\mathbf{v}_o \times \mathbf{B}$ that pushes them toward O or A depending on the direction of rotation and the direction of **B**. This induces an emf between O and A (see problem 8.5). A correct analysis of the Lorentz induction in a circuit must always be based on the generating field $\mathbf{E}_g$ that acts on the conduction charges in the various parts of the conductor.

8.4. Lorentz induction and the Galilean transformation of fields

Although the arguments that we have used in the preceding sections use the conduction electrons, the expression of the emf does not depend on the properties of the conductor. This means that induction is not due to a direct action of the magnetic field on the charges of a conductor. It does not require that the body be a conductor or even that there is a body at all! We show in this section that Lorentz induction may be understood as the result of the transformation law of the fields **E** and **B** from one frame of reference to another. This may be shown also for Neumann's induction but in the framework of special relativity (see Chapter 13).

Consider a particle of charge q moving with a velocity **v** in an electric field **E** and a magnetic field **B** measured with respect to a Galilean frame (S). Thus, the particle is subject to a Lorentz force

$$\mathbf{F} = q(\mathbf{E} + \mathbf{v} \times \mathbf{B}). \quad [8.14]$$

In another Galilean frame (S'), which moves with a velocity $\mathbf{v}_o$ with respect to (S), the velocity of the charged particle is $\mathbf{v}' = \mathbf{v} - \mathbf{v}_o$. Let us assume that the fields measured in (S') are $\mathbf{E}'$ and $\mathbf{B}'$. Thus, the force that acts on the particle in (S') is

$$\mathbf{F}' = q\,(\mathbf{E}' + \mathbf{v}' \times \mathbf{B}'). \quad [8.15]$$

In the framework of classical mechanics, the forces do not depend on the Galilean frame ($\mathbf{F}' = \mathbf{F}$). Substituting the expression $\mathbf{v}' = \mathbf{v} - \mathbf{v}_o$ in [8.15] and comparing with [8.14], we find the relation $\mathbf{E} + \mathbf{v}\times\mathbf{B} = \mathbf{E}' - \mathbf{v}_o\times\mathbf{B}' + \mathbf{v}\times\mathbf{B}'$. The law of transformation of the fields is obviously independent of the eventual existence of a charged test particle. Thus, the relation $\mathbf{E} + \mathbf{v}\times\mathbf{B} = \mathbf{E}' - \mathbf{v}_o\times\mathbf{B}' + \mathbf{v}\times\mathbf{B}'$ must be identically verified for any **v**. This gives the Galilean law of transformation of the fields

$$\mathbf{E} = \mathbf{E}' - \mathbf{v}_o \times \mathbf{B}', \qquad \mathbf{B} = \mathbf{B}'. \quad [8.16]$$

The inverse relation is obviously obtained by changing $\mathbf{v}_o$ into $-\mathbf{v}_o$, i.e.

$$\mathbf{E}' = \mathbf{E} + \mathbf{v}_o \times \mathbf{B}, \qquad \mathbf{B}' = \mathbf{B}. \quad [8.17]$$

These equations show that **E** and **B** cannot be considered as two distinct physical objects. If the field is pure magnetic for an observer in (*S*) it may be both magnetic and electric for another observer in (*S*').

The emf is defined as the circulation of the generating force $\mathbf{F}_g$ acting on the unit charge along the circuit. As both $\mathbf{F}_g$ and the circuit element are invariant in the Galilean transformation, we deduce that the induced emf is invariant in the transformation from (*S*) to (*S*') (i.e. $\mathcal{E} = \mathcal{E}'$). For instance, in the case of the rod moving normally to a field **B**, we have in the frame of the observer (*S*) only a magnetic field. According to [8.17], we have in the proper frame of the rod (*S*′) both a magnetic field $\mathbf{B}' = \mathbf{B}$ and an electric field $\mathbf{E}' = \mathbf{v}_o \times \mathbf{B} = -v_o B \mathbf{e}_y$. The velocity of the rod $\mathbf{v}'$ in (*S*′) being equal to zero, the induced emf is identical to [8.10]:

$$\mathcal{E} = \mathcal{E}' = \int_M^N d\mathbf{r}'.(\mathbf{E}'+\mathbf{v}'\times\mathbf{B}') = \int_M^N d\mathbf{r}'.\mathbf{E}' = \int_0^D dy\, v_o B = v_o BD,$$

8.5. Mutual inductance and self-inductance

Consider two circuits $\mathcal{C}_1$ and $\mathcal{C}_2$ orientated and near one another (Figure 8.2a). A current I_1 in $\mathcal{C}_1$ produces a field $\mathbf{B}_1$, whose flux through $\mathcal{C}_2$ is Φ_{12}. As the field $\mathbf{B}_1$ is proportional to the current I_1 that produces it, Φ_{12} is proportional to I_1 in the form

$$\Phi_{12} = M_{12} I_1. \qquad [8.18]$$

M_{12} is the *mutual inductance* of $\mathcal{C}_1$ in $\mathcal{C}_2$. It depends only on the geometrical configuration of the two circuits. If I_1 varies, the field $\mathbf{B}_1$ and its flux Φ_{12} vary. In the following, we assume that this variation is sufficiently slow to calculate the fields as in the case of time-independent intensities. As we shall see in section 9.4, this so-called *quasi-permanent approximation* is valid if the characteristic time τ of the current in $\mathcal{C}_1$ is much longer than the time of propagation D/c, where D is the distance separating the points of the circuits. If the circuits are rigid and stationary with respect to each other, M_{12} is time-independent and the variation of I_1 in $\mathcal{C}_1$ leads to a variation of Φ_{12} and, consequently, the induction of an emf in $\mathcal{C}_2$ given by

$$\mathcal{E}_2 = -\dot{\Phi}_{12} = -M_{12}\dot{I}_1. \qquad [8.19]$$

If R_2 is the resistance of the circuit $\mathcal{C}_2$, the induced intensity in $\mathcal{C}_2$ is

$$I_2 = \mathcal{E}_2/R_2 = -M_{12}\dot{I}_1/R_2. \qquad [8.20]$$

Conversely, a current I_2 in $\mathcal{C}_2$ produces through $\mathcal{C}_1$ a flux

$$\Phi_{21} = M_{21}\, I_2, \quad [8.21]$$

where M_{21} is the mutual inductance of $\mathcal{C}_2$ in $\mathcal{C}_1$. If I_2 varies, it induces in $\mathcal{C}_1$ an emf

$$\mathcal{E}_1 = -\, \dot{\Phi}_{21} = -M_{21}\, \dot{I}_2\,. \quad [8.22]$$

If R_1 is the resistance of $\mathcal{C}_1$, the induced intensity in $\mathcal{C}_1$ is

$$I_1 = \mathcal{E}_1/R_1 = -\, M_{21}\, \dot{I}_2\,/R_1. \quad [8.23]$$

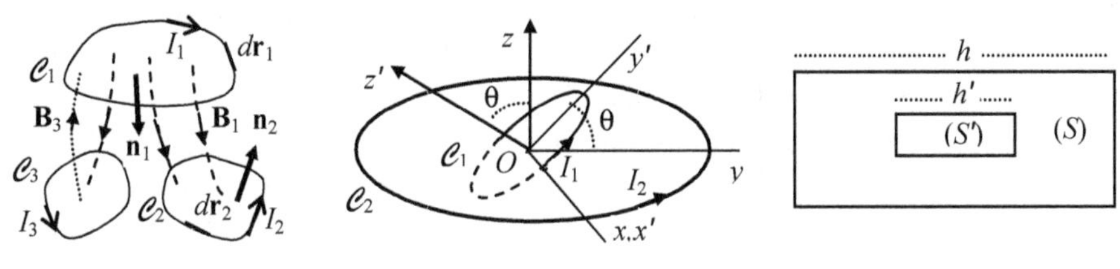

Figure 8.2. *Circuits in mutual influence: a) three arbitrary circuits, b) two concentric circular loops making an angle θ, and c) two coaxial and concentric solenoids*

We note that M_{12} and M_{21} are positive or negative depending on the orientation of the circuits and their relative position. The intensities are positive or negative depending on whether or not the current circulates effectively in the chosen direction. The orientation of the circuit determines also the direction of the unit vector normal to any surface bounded by the circuit and the sign of the flux. To calculate the mutual inductance, we write the expression of the magnetic flux through $\mathcal{C}_2$ as

$$\Phi_{12} = \iint_{\mathcal{S}_2} d\mathcal{S}'_2\, [\mathbf{n}'_2.\mathbf{B}_1(\mathbf{r}'_2)] = \int_{\mathcal{C}_2} d\mathbf{r}'_2\, .\mathbf{A}_1(\mathbf{r}'_2), \quad [8.24]$$

where $\mathbf{A}_1(\mathbf{r}'_2)$ is the vector potential produced by $\mathcal{C}_1$ at the points of $\mathcal{C}_2$ and given by

$$\mathbf{A}_1(\mathbf{r}'_2) = \frac{\mu}{4\pi} I_1 \int_{\mathcal{C}_1} \frac{d\mathbf{r}'_1}{|\mathbf{r}'_1 - \mathbf{r}'_2|}\,. \quad [8.25]$$

Thus we can write

$$\Phi_{12} = \frac{\mu}{4\pi} I_1 \int_{\mathcal{C}_1} \int_{\mathcal{C}_2} \frac{d\mathbf{r}'_1 . d\mathbf{r}'_2}{|\mathbf{r}'_1 - \mathbf{r}'_2|}\,. \quad [8.26]$$

As $\Phi_{12} = M_{12} I_1$, we get *Neumann's formula:*

$$M_{12} = \frac{\mu}{4\pi} \int_{\mathcal{C}_1} \int_{\mathcal{C}_2} \frac{d\mathbf{r}'_1 . d\mathbf{r}'_2}{|\mathbf{r}'_1 - \mathbf{r}'_2|} . \tag{8.27}$$

This symmetric relation in the exchange of $\mathcal{C}_1$ and $\mathcal{C}_2$ implies the reciprocity relation

$$M_{12} = M_{21}, \tag{8.28}$$

which may also be derived by energy considerations.

The use of [8.27] is often too complicated; a direct use of the definition $\Phi_{12} = M_{12} I_1$ with an approximate evaluation of the flux may be more simple. Consider for instance two loops of radii R_1 and R_2 (with $R_2 << R_1$), having the same center O and lying in two planes making an angle θ (Figure 8.2b). As the field of the large loop $\mathcal{C}_1$, $\mathbf{B}_1 = (\mu I_1/2R_1)\mathbf{n}_1$, varies little over the small loop, it is easier to calculate the flux of $\mathcal{C}_1$ through $\mathcal{C}_2$. It is $\Phi_{12} \cong \mathbf{B}_1.\mathbf{n}_2\, \mathcal{S}_2 \cong \pi\mu I_1(R_2^2/2R_1) \cos\theta$. As, by definition $\Phi_{12} \equiv M_{12} I_1$, we deduce that $M_{12} \cong \pi\mu\, (R_2^2/2R_1) \cos\theta$. As a second example, consider two coaxial solenoids (S) and (S') of lengths h and h' ($h > h'$), radii R and R' ($R > R'$) and formed by n and n' turns, respectively, per unit length (Figure 8.2c). To reduce the effect of the finite length, we calculate the flux of the longer solenoid S through the second S'. The field of S is almost uniform near the center and given by $B = \mu nIh/(4R^2 + h^2)^{½}$. Its flux through the n' turns of S' is $\Phi = \pi\mu R'^2 n' nIh/(4R^2 + h^2)^{½}$. We deduce that $M = \mu\pi R'^2 n' nh/(4R^2 + h^2)^{½}$.

A) Concept of inductance

The experiment shows that an emf is induced in a circuit $\mathcal{C}$ if it's current varies. This *self-induction* is due to the variation of the flux of its own magnetic field through the circuit itself. This effect is important in the case of an alternating current and in the case of a direct current that is turned on or turned off.

The field **B** produced by a circuit $\mathcal{C}$ at each point **r** is proportional to its current and so is the flux Φ of this field through the circuit itself, hence

$$\Phi = LI. \tag{8.29}$$

It may be easily verified that L is a positive quantity called *self-inductance* of the circuit (or just *inductance,* for short). It depends only on the geometrical form of the circuit. If the intensity I of the circuit varies, Φ varies, and this induces an e.m.f.

$$\mathcal{E} = -\dot{\Phi} = -L\dot{I} . \tag{8.30}$$

This induced emf is superposed to the other emf $\mathcal{E}_0$ and, according to Lenz's law, it opposes the variation of Φ (thus, of I): if I increases ($\dot{I} > 0$), $\mathcal{E}$ is negative; thus, it produces a current in the opposite direction. On the contrary, if I decreases ($\dot{I} < 0$), $\mathcal{E}$ is positive; thus, it produces a current in the same direction. For instance, if a direct current is switched on, the self-induction opposes its instantaneous set up; so, I increases gradually from 0 to its final value. Similarly, if the current is turned off, the self-induction opposes its instantaneous vanishing; so I decreases gradually to 0 (see section 8.6).

B) Calculation of the inductance

We may consider the inductance L as the mutual inductance of two circuits that coincide (Figure 8.3a). The expression [8.27] becomes

$$L = \frac{\mu}{4\pi} \int_{\mathcal{C}} \int_{\mathcal{C}} \frac{d\mathbf{r}'_1 . d\mathbf{r}'_2}{|\mathbf{r}'_1 - \mathbf{r}'_2|} \ . \qquad [8.31]$$

However, this expression is not easy to use and it may diverge in the case of a thin circuit, as the denominator $|\mathbf{r}'_1 - \mathbf{r}'_2|$ vanishes. This is due to the singularity of the field and the vector potential at the points of linear current.

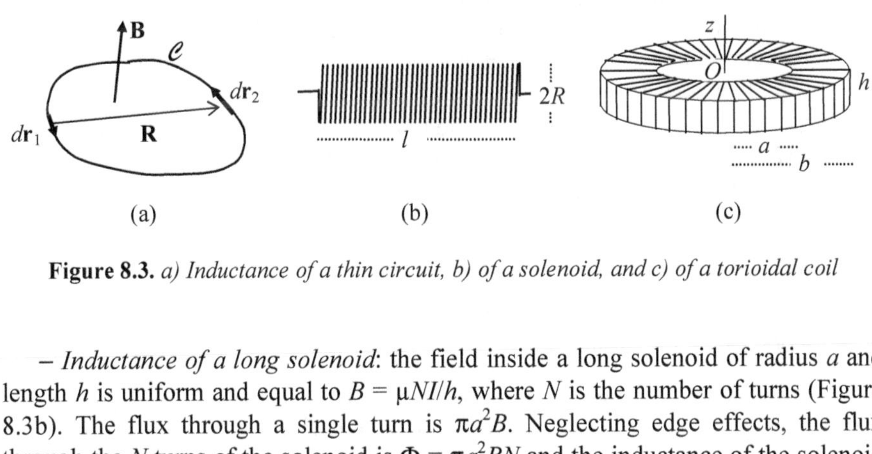

Figure 8.3. *a) Inductance of a thin circuit, b) of a solenoid, and c) of a torioidal coil*

– *Inductance of a long solenoid*: the field inside a long solenoid of radius a and length h is uniform and equal to $B = \mu NI/h$, where N is the number of turns (Figure 8.3b). The flux through a single turn is $\pi a^2 B$. Neglecting edge effects, the flux through the N turns of the solenoid is $\Phi = \pi a^2 BN$ and the inductance of the solenoid is approximately

$$L = \Phi/I \cong \pi\mu N^2 a^2/h. \qquad [8.32]$$

For instance, the inductance of a solenoid of length $h = 25$ cm, radius $a = 5$ cm and 1,000 turns is $L = 0.04$ H. If it is filled with a magnetic material of relative permeability μ_r, the field and the flux are multiplied by μ_r and so is L.

– *Inductance of a narrow coil:* the inductance of a circular narrow coil of N turns is difficult to evaluate. If we approximate the field of a circular loop by its value at the center $\mathbf{B}(O) = \frac{1}{2}\mu(I/a)\mathbf{e}_z$, the field of the coil is $NB(O)$ and the flux of this field through the N loops is $\Phi \cong \pi a^2 N^2 B(O) \cong \frac{1}{2}\pi\mu a N^2 I$. Thus, the inductance is approximately $L \cong \frac{1}{2}\,\pi\mu N^2 a$.

– *Inductance of a toroidal coil:* Let us consider a toroidal coil formed by N rectangular turns of height h, whose parallel sides to the axis Oz are situated at the distances a and b, respectively, from Oz (Figure 8.3c). Applying Ampère's law to a circle of radius r about Oz, we find a field $B = \mu NI/2\pi r$. The flux through a rectangular turn is

$$\Phi_1 = \int_a^b dr\, hB = \frac{\mu}{2\pi} NIh \int_a^b \frac{dr}{r} = \frac{\mu}{2\pi} NIh \ln\frac{b}{a}.$$

Thus, the flux through the N turns is $\Phi = N\Phi_1$ and the inductance is approximately

$$L = \frac{\Phi}{I} \cong \frac{\mu}{2\pi} N^2 h \ln\frac{b}{a}. \qquad [8.33]$$

The mutual inductance M and the self-inductance L have the dimensions of Φ/I. In the SI, they are expressed in units of weber per ampere, also called *henry* (H). Every circuit has a more or less important self-inductance. If all the dimensions of a circuit are multiplied by a factor k, the quantities $d\mathbf{r}_1$, $d\mathbf{r}_2$, $|\mathbf{r}_1 - \mathbf{r}_2|$ and, consequently, L are multiplied by k. In practice, instead of increasing the dimensions of the circuit to increase L, electronic components, called *inductors*, are connected in series in the circuit. An inductor consists of a coil of wire wound on a hollow cylinder that may contain air or a ferromagnetic core.

8.6. LR circuit

A) Case of a single LR circuit

A self-inductance L connected between two points A and B of a circuit and carrying a current I_{AB} is equivalent to an induction emf $\mathcal{E}^{(in)} = -L\,\partial_t I_{AB}$ or a receiver of back-emf $\mathcal{E}'^{(in)} = L\,\partial_t I_{AB}$. Here I_{AB} is algebraic, that is, positive if the current circulates effectively from A to B. Let us consider a circuit formed by a self-inductance L, a total resistance R, and a generator of emf $\mathcal{E}$ connected in series (Figure 8.4a). The potential drops along the circuit are $V_{AC} = L\,\partial_t I$, $V_{CB} = IR$ and $V_{BA} = -\mathcal{E}$, hence the equation of the circuit

$$L\,\partial_t I + RI = \mathcal{E}. \qquad [8.34]$$

This equation has a particular solution $I_0 = \mathcal{E}/R$. Setting $I = I_0 + f(t)$, we find that $f(t)$ verifies the equation $L\,\partial_t f + Rf = 0$, that is, the equation of the circuit without emf This equation may be written as $df/f = -dt/\tau$, where we have introduced the characteristic time $\tau = L/R$, called also *relaxation time* of the circuit. Integrating this equation, we find $f = B\,e^{-t/\tau}$. Thus, the general solution of [8.34] is

$$I(t) = \mathcal{E}/R + B\,e^{-t/\tau}. \tag{8.35}$$

The arbitrary constant B is determined from the initial conditions.

a) Switching on the current: let us assume that the circuit is closed at $t = 0$. The initial condition $I(0) = 0$ implies that $B = -\mathcal{E}/R$, hence

$$I(t) = \frac{\mathcal{E}}{R}\,[1 - e^{-t/\tau}]. \tag{8.36}$$

Thus, the intensity increases exponentially from 0 to its final value $\mathcal{E}/R$, that it attains, in principle, after an infinite time. Practically after $t \approx 10\tau$, I differs from its limit value by a few millionths, and for usual electronic circuits, τ is very short. For instance, for $L = 1$ mH and $R = 1$ kΩ, we find $\tau = 1$ µs. Figure 8.4b illustrates the variation of I versus t.

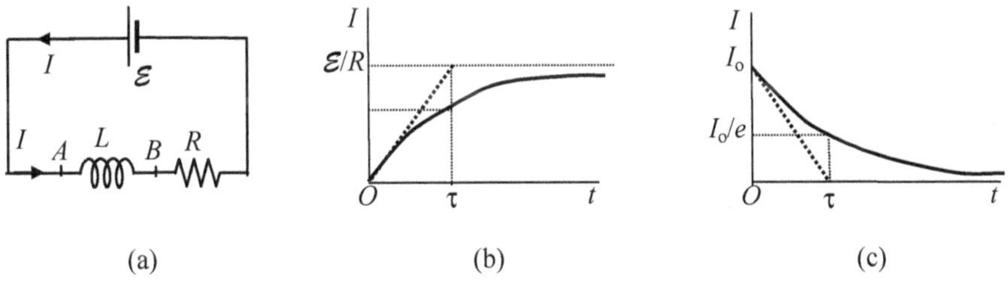

Figure 8.4. *a) LR circuit, b) switching on the current, and c) turning off the current*

b) Turning off the current: let us assume that, after having the generator turned on for a certain time, the generator is disconnected but the circuit is maintained closed. Then, we have an *LR* circuit without a generator. Its equation is

$$L\partial_t I + RI = 0. \tag{8.37}$$

Its general solution is $I(t) = B\,e^{-t/\tau}$. The initial condition $I = I_0$ at $t = 0$ is satisfied if $B = I_0$. Thus, the solution is

$$I(t) = I_0\,e^{-t/\tau}. \tag{8.38}$$

The intensity decreases exponentially from its initial value to 0 with the characteristic time τ. Figure 8.4c illustrates the variation of I versus t.

B) Case of several circuits in magnetic influence

In the case of n circuits $\mathcal{C}_k$ of resistances R_k, self-inductances L_k and mutual inductances M_{kj}, the flux through the circuit $\mathcal{C}_k$ is

$$\Phi_k = L_k I_k + \Sigma_{j\neq k} M_{kj} I_j. \qquad [8.39]$$

If the circuits are stationary and rigid, the L_k and M_{kj} are time-independent. The induced emf in the circuit (k) is then $\mathcal{E}'_k = L_k \dot{I}_k + \Sigma_{j\neq k} M_{kj} \dot{I}_j$. If the circuits contain generators of emf $\mathcal{E}_k$, their equations take the form

$$L_k \dot{I}_k + \Sigma_{j\neq k} M_{kj} \dot{I}_j + R_k I_k = \mathcal{E}_k, \qquad \text{where } k = 1, 2, \ldots n. \qquad [8.40]$$

This is a system of n coupled differential equations. They allow the n intensities I_k to be determined. They have particular solutions $I^{(o)}_k = \mathcal{E}_k/R_k$ if the $\mathcal{E}_k$ are constant. Setting $I_k = I^{(o)}_k + f_k(t)$ and substituting in equations [8.40], we find that the f_k must verify the system of homogeneous equations

$$L_k \dot{f}_k + \Sigma_{j\neq k} M_{kj} \dot{f}_j + R_k f_k = 0, \qquad \text{where } k = 1, 2, \ldots n. \qquad [8.41]$$

To solve this system of differential equations, we try the solutions $f_k = A_k e^{-\alpha_k t}$, where A_k and α_k are constants. These functions are solutions if the α_k are all equal to a certain value α. Then, the system [8.41] reduces to a system of linear and homogeneous algebraic equations to determine the coefficients A_k

$$(\alpha L_k - R_k) A_k + \alpha \Sigma_{j\neq k} M_{kj} A_j = 0, \qquad \text{where } k = 1, 2, \ldots n. \qquad [8.42]$$

This system of equations has a non-trivial solution (i.e. not all the A_j equal to zero) if the determinant of the coefficients is equal to zero. This gives an algebraic equation of degree n in α. It has, in general, n roots $\alpha_{(i)}$. For each root, the system [8.42] allows $(n–1)$ coefficients A_k to be determined in terms of the nth. Thus, we have a solution, called *mode*, which depends of an arbitrary parameter. The general solution of [8.40] is a superposition of these n modes and the particular solution $I^{(o)}_k$. It depends on n integration constants, which may be determined from the initial conditions.

8.7. Magnetic energy

A) Proper magnetic energy of a single circuit

Let us assume that an LR circuit is connected to a generator of constant emf $\mathcal{E}$. In the interval of time dt during the build-up of the current, the generator supplies an energy $dU_g = dq\, \mathcal{E} = I\mathcal{E}\, dt$. A part of it, $dU_J = I^2R\, dt$, is dissipated as Joule heat and the remaining dU_M is stored as magnetic energy in the inductor. Taking into account the circuit equation [8.34], we may write

$$dU_M = dU_g - dU_J = I(\mathcal{E} - IR)\, dt = LI\,(dI/dt)\, dt = LI\, dI. \qquad [8.43]$$

By increasing the intensity from 0 to I, the inductor accumulates a magnetic energy

$$U_M = \int_0^I dU_M = \int_0^I dI\ LI = \tfrac{1}{2}LI^2 = \tfrac{1}{2}\,\Phi I, \qquad [8.44]$$

where Φ is the magnetic flux through the circuit.

To show that this energy is effectively stored, let us look to what happens if the generator is disconnected at $t = 0$, but the circuit is kept closed. The intensity decreases gradually from I at $t = 0$ to the final value $I = 0$ at $t = \infty$. In the interval of time dt, the magnetic energy decreases by $dU_M = d(\tfrac{1}{2}LI^2) = LI\,(dI/dt)\, dt = -RI^2\, dt$ where we have used the equation of the circuit [8.37]. Thus, the decrease of the magnetic energy $\tfrac{1}{2}LI^2$ is just the dissipated energy as Joule heat in the resistor.

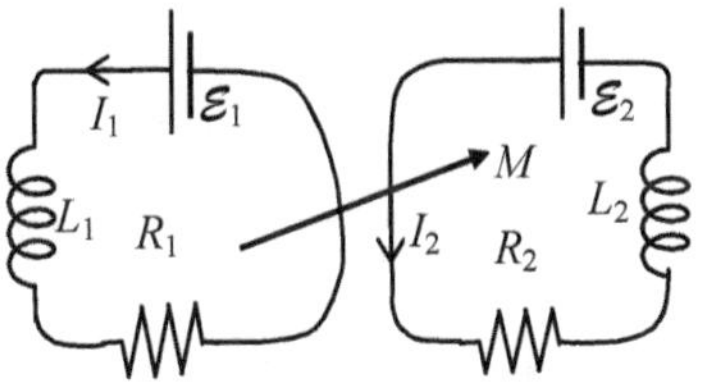

Figure 8.5. *Circuits in mutual influence*

B) Magnetic energy of circuits in mutual influence

Let us consider two circuits in mutual influence (Figure 8.5). The generalization to several circuits is straightforward. Let R_k, L_k, and $\mathcal{E}_k$ ($k = 1, 2$) be the resistances, the inductances and the emf of the generators and $M = M_{12} = M_{21}$ the mutual inductance of these circuits. Their equations may be written as

$$L_1 \dot{I}_1 + M\dot{I}_2 + R_1 I_1 = \mathcal{E}_1, \qquad L_2 \dot{I}_2 + M\dot{I}_1 + R_2 I_2 = \mathcal{E}_2. \qquad [8.45]$$

The energy that the generators supply in the interval of time dt is $dU_E = \mathcal{E}_1 I_1 dt + \mathcal{E}_2 I_2 dt$ and the total energy dissipated as Joule heat is $dU_J = R_1 (I_1)^2\, dt + R_2 (I_2)^2\, dt$. Thus, the variation of the magnetic energy that is stored in the circuits is

$$\begin{aligned} \mathrm{d}U_M &= dU_E - dU_J = I_1\, dt\, (E_1 - I_1 R_1) + I_2\, dt\, (E_2 - I_2 R_2) \\ &= L_1 I_1\, dI_1 + L_2 I_2\, dI_2 + M(I_1\, dI_2 + I_2\, dI_1), \end{aligned} \qquad [8.46]$$

where we have used equations [8.45]. Integrating between the initial state with the intensities equal to zero (consequently, the energy is equal to zero by definition) and the final state with the intensities equal to I_1 and I_2, we find

$$U_M = \int_0^{I_f} dU_M = \tfrac{1}{2} L_1 I_1^{\,2} + \tfrac{1}{2} L_2 I_2^{\,2} + M I_1 I_2 \equiv U_1 + U_2 + U_{12}. \qquad [8.47]$$

The terms $U_i = \tfrac{1}{2} L_i I_i^2$ represent the magnetic energies of the individual circuits carrying the currents I_i and isolated. The term $U_{12} = M I_1 I_2$ represents the *magnetic interaction energy*. This last quantity is the energy that is required to bring first the circuit $\mathcal{C}_1$ then the circuit $\mathcal{C}_2$ from infinity to their actual position while their intensities are maintained constant. We note that the magnetic interaction energy may be positive negative, or zero depending on the values of M and the intensities. As $\Phi_{21} = M I_2$ and $\Phi_{12} = M I_2$, the magnetic interaction energy may also be written as the product of the current of one of the circuit by the magnetic flux of the other through it

$$U_{12} = I_1\, \Phi_{21} = I_2\, \Phi_{12}. \qquad [8.48]$$

The generalization to the case of several circuits is

$$U_M = \textstyle\sum_j U_j + \sum_{j<k} U_{jk}, \qquad U_j = \tfrac{1}{2} L_j I_j^2, \qquad U_{jk} = M_{jk} I_j I_k = I_j\, \Phi_{kj} = I_k\, \Phi_{jk}. \qquad [8.49]$$

In order to have the total magnetic interaction energy, we must sum the U_{ij} for all the pairs (or all i and j such that $i < j$ or $j < i$ so that each pair is counted once).

The magnetic energy of two circuits $U_M = \tfrac{1}{2} L_1 I_1^{\,2} + \tfrac{1}{2} L_2 I_2^{\,2} + M I_1 I_2$ cannot be negative for any values of the intensities; otherwise currents would be established in the circuits without generators! This implies that the quadratic form U_M has a negative discriminant, i.e.:

$$M^2 \leq L_1 L_2. \qquad [8.50]$$

This relation holds for any geometrical configuration of the circuits. The ratio $\kappa = |M|/\sqrt{L_1 L_2} < 1$ is the *coupling coefficient* of the circuits. The coupling is said to be *perfect* if $\kappa = 1$. In this case, all the lines of the field of one of the circuit pass within the other. This is the case of two coinciding circuits; then, we have $M = L_1 = L_2$.

Consider, for instance, a long solenoid of length h and N turns carrying a current I. According to [8.32], its self-inductance is $L = \pi\mu N^2R^2/h$. The field inside it is $B = \mu NI/h$ and it is equal to zero outside it. The magnetic energy stored in this solenoid is $U_M = \frac{1}{2}LI^2 = \pi\mu N^2R^2I^2/2h = \frac{1}{2}(B^2/\mu)\,\mathcal{V}$, where $\mathcal{V} = \pi R^2h$ is the volume of the solenoid. Thus, we may consider that the density of magnetic energy is

$$U_{M,v} = B^2/2\mu. \qquad [8.51]$$

This result, derived in a particular case, holds for any configuration of currents: In a region of space where there is a magnetic field, there is magnetic energy with a volume density $B^2/2\mu$ (see section 7.12). The energy stored in a finite volume $\mathcal{V}$ is

$$U_M = \iiint_{\mathcal{V}} d\mathcal{V}\, U_{M,v} = (1/2\mu) \iiint_{\mathcal{V}} d\mathcal{V}\, \mathbf{B}^2. \qquad [8.52]$$

We may express the total magnetic energy in terms of the current density $\mathbf{j}$ and the vector potential $\mathbf{A}$ as

$$U_M = \tfrac{1}{2} \iiint_{\mathcal{V}} d\mathcal{V}\, (\mathbf{A}.\mathbf{j}), \qquad [8.53]$$

However $\frac{1}{2}(\mathbf{A}.\mathbf{j})$ cannot be interpreted as the magnetic energy density as we have seen in section 7.12.

8.8. Magnetic forces acting on circuits

If a circuit carries an electric current, its magnetic field acts on the circuit itself and on the neighboring circuits. The circuits may undergo a deformation or a displacement. The forces and the moments of force may be evaluated by using Laplace's law in the quasi-permanent approximation. However, it is often easier to evaluate them by using the magnetic energy and the method of *virtual work*.

A) Force exerted by a deformable circuit on itself

If a circuit of inductance L carries a current I, its magnetic energy is $U_M = \frac{1}{2}LI^2 = \Phi^2/2L$. If it undergoes a virtual deformation, its inductance varies by δL. To calculate the magnetic force, which produces this deformation, we may assume that it occurs while maintaining constant either the intensity I or the magnetic flux Φ.

a) If the intensity is kept constant, the magnetic flux varies by $\delta\Phi = I\,\delta L$, inducing an emf $\mathcal{E}' = -\dot{\Phi}$. To maintain I constant, we must assume that the emf of the generator in the circuit undergoes an opposite variation $\mathcal{E}'' = \dot{\Phi}$, thus supplying an additional energy

$$\delta U_g|_{I=\text{constant}} = \mathcal{E}''\,I\,\delta t = I\,\dot{\Phi}\,\delta t = I\,\delta\Phi = I^2\,\delta L. \qquad [8.54]$$

A part of it, $\delta U_M = \frac{1}{2}I^2\delta L$, is stored as magnetic energy and the remaining is used as a mechanical work δW of the force F, which produces the deformation, hence

$$\delta W = \delta U_g|_{I=\text{constant}} - \delta U_M|_{I=\text{constant}} = \tfrac{1}{2}I^2\,\delta L. \qquad [8.55]$$

For instance, if the circuit has a part that may move in the x direction, we have $\delta W = F_x\,\delta x$, thus

$$F_x = \delta W/\delta x = \tfrac{1}{2}\,I^2\,(\delta L/\delta x) = \tfrac{1}{2}\,I^2\,\partial_x L. \qquad [8.56]$$

b) Let us assume now that the flux is kept constant. Using the expression $U_M = \frac{1}{2}\Phi^2/L$, we find $\delta U_M|_{\Phi=\text{constant}} = -\frac{1}{2}(\Phi^2/L^2)\,\delta L$. In this case, no emf is induced in the circuit and the generator needs to supply no energy, hence

$$\delta W = -\,\delta U_M|_{\Phi=\text{constant}} = \tfrac{1}{2}\,(\Phi^2/L^2)\,\delta L = \tfrac{1}{2}\,I^2\,\delta L. \qquad [8.57]$$

We find the same result as [8.56].

B) Interaction of two circuits

Let us consider two rigid circuits $\mathcal{C}_1$ and $\mathcal{C}_2$ in mutual influence (Figure 8.5). We define the positions of $\mathcal{C}_1$ and $\mathcal{C}_2$ as those of two particular points $\mathbf{r}_1$ and $\mathbf{r}_2$ of them. The mutual inductance M is then a function of their relative position $\mathbf{r}_2 - \mathbf{r}_1$ and their relative orientation. To evaluate the force $\mathbf{F}_{12}$ exerted by $\mathcal{C}_1$ on $\mathcal{C}_2$, we assume that $\mathcal{C}_1$ is fixed and $\mathcal{C}_2$ undergoes a virtual translational motion $\delta\mathbf{r}_2$. We may also assume in this case that either the intensities or the fluxes are maintained constant.

a) Let us assume that the intensities are kept constant. The circuits being rigid, the self-inductances L_i do not change while the mutual inductance M varies. The fluxes through the circuits vary by $\delta\Phi_1 = I_2\,\delta M$ and $\delta\Phi_2 = I_1\,\delta M$. This induces the emf $\mathcal{E}'_1 = -\dot{\Phi}_1$ and $\mathcal{E}'_2 = -\dot{\Phi}_2$. To maintain constant intensities, the emf of the generators must vary by $\mathcal{E}''_1 = -\mathcal{E}'_1$ and $\mathcal{E}''_2 = -\mathcal{E}'_2$ and supply the energy

$$\delta U_g|_{I=\text{constant}} = \mathcal{E}''_1\,I_1\,\delta t + \mathcal{E}''_2\,I_2\,\delta t = 2\,I_1\,I_2\,\delta M. \qquad [8.58]$$

The variation of the magnetic energy [8.47] is then $\delta U_M = I_1 I_2\, \delta M$. The difference between the energy that is supplied by the generators and the variation of the stored energy is the mechanical work

$$\delta W = \mathbf{F}_{12} .\delta \mathbf{r}_2 = \delta U_G|_{I = \text{constant}} - \delta U_M|_{I = \text{constant}} = I_1\, I_2\, \delta M = I_1\, I_2\, \boldsymbol{\nabla}_2 M.\delta \mathbf{r}_2. \quad [8.59]$$

We deduce that

$$\mathbf{F}_{12} = \boldsymbol{\nabla}_2 U_M = I_1\, I_2\, \boldsymbol{\nabla}_2 M = \boldsymbol{\nabla}_2 U_M. \quad [8.60]$$

This expression shows that the magnetic force $\mathbf{F}_{12}$ is conservative.

A similar analysis may be performed to evaluate the moment of force $\boldsymbol{\Gamma}_{12}$ exerted by $\mathcal{C}_1$ on $\mathcal{C}_2$ with respect to a point O. Let us consider an axis of rotation of unit vector $\mathbf{u}$ passing by O. The projection of $\boldsymbol{\Gamma}_{12}$ on $\mathbf{u}$ is related to the mechanical work in a rotation through an angle $\delta\theta$ about $\mathbf{u}$ by the relation $\delta W = \boldsymbol{\Gamma}_{12}.\mathbf{u}\ \delta\theta$. The variation of U_M in this rotation is $\delta U_M = I_1 I_2\, \partial_\theta M\, \delta\theta$, hence the relation

$$\boldsymbol{\Gamma}_{12}.\mathbf{u} = I_1 I_2\, \partial_\theta M = \partial_\theta U_M. \quad [8.61]$$

b) If we assume that the fluxes are kept constant in the displacement, there will be no induction of emf in the circuits and the generators supply no additional energy. Expressing U_M in terms of the fluxes, we may write

$$U_M = [½(L_1 L_2 - M^2)][L_2 \Phi_1^2 + L_1 \Phi_2^2 - 2M\, \Phi_1 \Phi_2]. \quad [8.62]$$

In this virtual displacement, L_1, L_2, Φ_1 and Φ_2 remain constant. Calculating $\delta U_M|_{\Phi = \text{constant}}$, which corresponds to δM, and then expressing it in terms of the intensities, we find $\delta U_M|_{\Phi = \text{constant}} = -I_1 I_2\, \delta M$. The conservation of energy implies that $\delta W|_{\Phi = \text{constant}} = -\delta U_M$. Thus, we find again the expressions [8.60] in the case of a translational motion and [8.61] in the case of a rotation about the unit vector $\mathbf{u}$.

In the case of several circuits in interaction, the superposition principle allows us to write the force and the moment of force acting on one of them, $\mathcal{C}_i$, as the vector sums of the forces and the moments of force exerted on it by the other circuits acting individually. We may also write the expression of the total magnetic energy U_M and use the generalizations of [8.60] and [8.61]:

$$\mathbf{F}_i = \boldsymbol{\nabla}_i U_M \qquad \text{and} \qquad \boldsymbol{\Gamma}_i.\mathbf{u} = \partial_\theta U_M. \quad [8.63]$$

8.9. Some applications of induction

The induction due to the displacement of a conductor in a magnetic field enables electromechanical coupling to transform mechanical energy into electric energy.

A) Simple devices using induction

– *Circuit breaker*. This is a simple application of induction, which protects people and electrical equipment in the case of a short-circuit. The current, which supplies the equipment, passes within a soft iron ring (Figure 8.6a). A coil encircles the ring and is connected to the circuit breaker. The current, which circulates in both directions, produces two opposite magnetic fields and no current is normally induced in the coil. If a short-circuit is provoked in the equipment, there will be no return of the current and only one of the alternating magnetic fields remains. It induces a current in the coil, which releases the circuit breaker.

– *Microphones.* Microphones transform sounds into electric signals. There are many types. The moving coil microphone consists of a membrane that vibrates with an amplitude proportional to that of the sound wave and with the same frequency (Figure 8.6b). The membrane acts on a small coil, which moves in a magnetic field. The induced current in the coil may be recorded or amplified to feed a loudspeaker, for instance. In an electric guitar, a magnetized string vibrates in front of a soft iron rod with a coil encircling it. The vibration of the string produces a variation of the magnetic flux in the coil. The induced current is amplified to feed a loudspeaker.

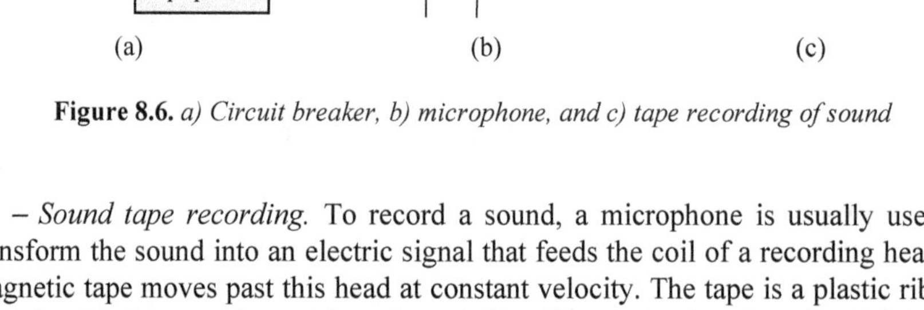

Figure 8.6. *a) Circuit breaker, b) microphone, and c) tape recording of sound*

– *Sound tape recording.* To record a sound, a microphone is usually used to transform the sound into an electric signal that feeds the coil of a recording head. A magnetic tape moves past this head at constant velocity. The tape is a plastic ribbon coated with iron oxide or chromium oxide (Figure 8.6c). The electric signals produce a magnetic field near the recording head and a permanent magnetization of small parts of the tape in patterns that reproduce the amplitude and the frequency of the signal. To reproduce the sound, the tape is passed at the same speed in front of the head operating in playback mode. The magnetized tape induces electric signals in the head coil. These signals are amplified and sent to a speaker.

B) Electric generators

The production of electricity is the most important application of induction. Electric generators convert mechanical energy to electric energy. The mechanical energy is supplied by a hydraulic turbine or a steam turbine (the steam comes either from burning fossil fuels or from nuclear reactors). The rod of Figure 8.1a, sliding in a magnetic field **B** on two rails connected by a resistance R, is an example of a simple generator. The disk of Figure 8.1d that is rotated in a field **B** is another example of a generator. The dynamo is a generator that produces electric energy more efficiently. It comprises a coil rotated in a magnetic field **B** (Figure 8.7a). If it is rotated with an angular velocity ω, the axis of the coil makes an angle $\theta = \omega t$ with **B**. The magnetic flux through the coil is $\Phi = N\mathcal{S}B\cos(\omega t)$, where N is the number of turns of area $\mathcal{S}$. Thus, the induced emf is

$$\mathcal{E} = -\dot{\Phi} = NB\mathcal{S}\omega\sin(\omega t). \qquad [8.64]$$

To derive this result, we have neglected the self-induction of the coil. To estimate this effect, let us assume that the coil is connected to a circuit of total resistance R. The induced intensity is $I = (NB\mathcal{S}\omega/R)\sin(\omega t)$ and the self-induced emf (due to the variation of the current I) is $\mathcal{E}' = -L(dI/dt) = -LNB\mathcal{S}\omega^2\cos(\omega t)$. The ratio of its amplitude to that of $\mathcal{E}$ is $\mathcal{E}'_m/\mathcal{E}_m = L\omega \approx \frac{1}{2}\pi\mu\omega N^2 R$. It is often very small.

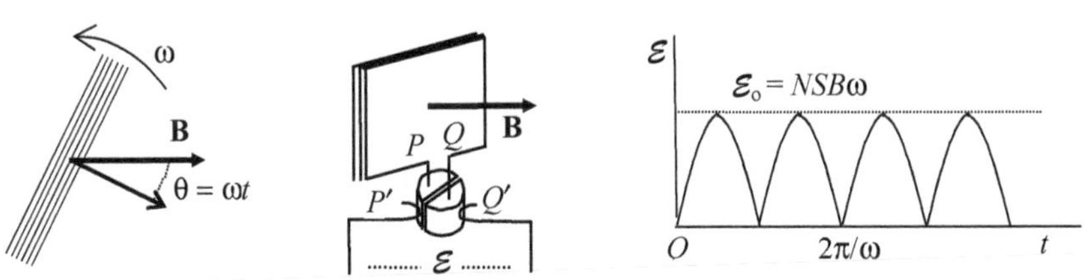

(a) (b) (c)

Figure 8.7. *Generators: a) rotating coil producing an alternating induced emf, b) commutator used to rectify the current, and c) the rectified emf versus t*

It is possible to obtain a unidirectional induced voltage from an alternating supply by suppressing the parts of the wave that have the opposite polarity or by reversing the polarity of those parts (by using diodes for instance). Rectification can also be achieved using a split-ring commutator: the terminals of the rotating coil change from one slit to the other, once in each rotation (Figure 8.7b). Figure 8.7c illustrates the rectified output emf as a function of t.

C) Measurement of B by using a ballistic galvanometer

A ballistic galvanometer is essentially a rectangular coil having N turns of area $\mathcal{S}$, which may rotate in a radial field **B** (Figure 8.8). It has a small damping and a large moment of inertia J, and it is submitted to a restoring moment of force $\Gamma' = -K\varphi$, where φ is the rotation angle. If a current I passes through the coil, it is subject to a moment of force $\Gamma_M = N\mathcal{S}BI$. Let us assume that the current passes between $t = 0$ and $t = \tau$. Integrating the equation of motion $J\ddot{\varphi} = N\mathcal{S}BI$ with respect to time between 0 and τ, we find $J\dot{\varphi} = N\mathcal{S}Bq$ where q is the total charge that has passed through the coil. If the time τ is sufficiently short, as the coil has a large moment of inertia, it does not move noticeably during this time. The effect of this charge is to provoke an initial angular velocity $\dot{\varphi}_o = N\mathcal{S}Bq/J$. This corresponds to an initial kinetic energy $\frac{1}{2}J(N\mathcal{S}Bq)^2$, which makes the coil oscillate with an amplitude $\varphi_m = N\mathcal{S}Bq(K/J)^{1/2}$ (reached when the initial kinetic energy changes totally to potential energy $\frac{1}{2}K\varphi_m^2$). Thus, φ_m is proportional to the charge q to be measured.

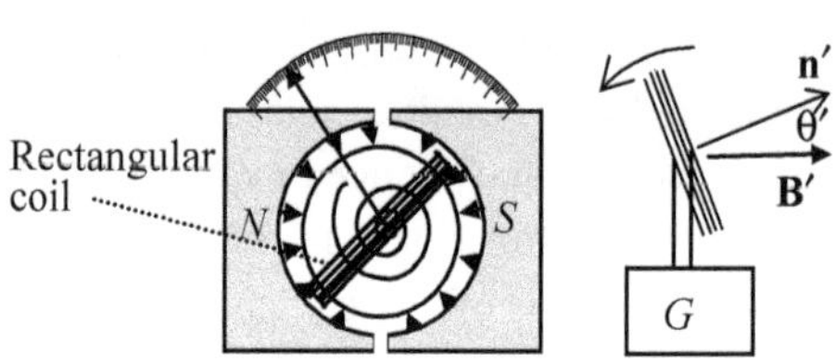

Figure 8.8. *Galvanometer and measurement of B*

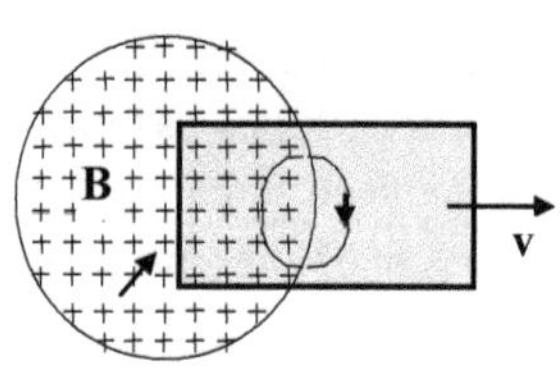

Figure 8.9. *Eddy current*

The ballistic galvanometer may be used to measure the magnetic flux and, consequently, the magnetic field. For this, we use a coil of N' turns and area $\mathcal{S}'$ connected to the terminals of the ballistic galvanometer. If this coil is placed normal to a field **B′**, the magnetic flux through it is $\Phi' = N'\mathcal{S}'B'$. By withdrawing it quickly from the field, a charge $q = \Phi'/R$ is induced. A measurement of q using the ballistic galvanometer determines Φ', thus B'.

D) Eddy currents

Eddy currents (also known as *Foucault currents*) are those that are induced in the mass of a conductor if it is moving in a magnetic field or exposed to a variable magnetic field. If the extended body moves in a non-uniform field or if different parts of the body have different velocities (as in the case of rotation), the induced electric currents circulate in closed swirls (Figure 8.9). According to Lenz's law, these currents circulate in directions such that they oppose the variation of **B** or such that the Laplace's force exerted by **B** on them opposes the motion. This effect may be observed by letting a copper disk oscillate near a magnet. The induced Eddy currents strongly slow down the disk by dissipating its mechanical energy as Joule heat; hence, the use of this effect in the conception of electromagnetic brakes and for induction heating with high-frequency alternating magnetic fields. Another illustration of this effect is magnetic levitation: if a magnet is placed above a superconducting plate, it may stay in equilibrium, as if it starts to fall, Eddy currents oppose its fall. To reduce the undesirable large-scale circulation of Eddy current effects in motors, transformers, etc., the metallic pieces are made of insulated sheets.

In some cases, if a metallic body subject to a force field moves in a magnetic field, as its velocity increases, the Eddy force increases and may counterbalance the force field. Thus, the body attains a limit velocity that is proportional to the force field. This effect is used in some instruments, such as *wattmeters*. These are essentially an aluminum disk rotated by a small motor exerting a torque proportional to the current intensity (thus, proportional to the consumed power under a given voltage). The rotation velocity increases until the Eddy torque counterbalances the exerted torque; thus, it is proportional to power consumed, and the number of turns is proportional to the consumed energy.

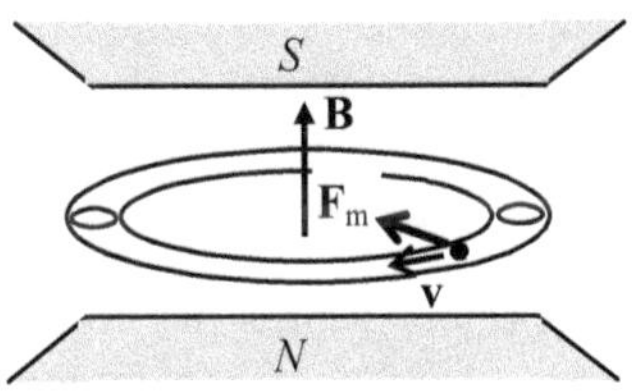

Figure 8.10. *The betatron*

E) The betatron

The betatron (invented by Kerst in 1941) is an electron accelerator that uses induction. Electrons travel on circular orbits of radius about 1 m in a toroidal tube of elliptical section between the poles of a powerful electromagnet (Figure 8.10). The alternating current in the electromagnets produces a magnetic field oscillating at the same frequency. The electrons are emitted and pre-accelerated by a voltage of about 50 kV. They are injected tangentially into the tube and they circulate under the influence of the magnetic field of value B_{orb} along the orbit. The orbit is effectively circular of radius r if the magnitude of the magnetic force is equal to the centripetal inertial force ($evB_{orb} = mv^2/r$), thus $erB_{orb} = mv$. The magnetic flux through the orbit is $\Phi = \pi r^2 B_m$, where B_m is the average field within the orbit. If the field of the electromagnet is increased, an electric field E is induced such that $2\pi rE = \dot{\Phi}$. We deduce that $E = \frac{1}{2} r \dot{B}_m$. This electric field acts on the electrons with a force of magnitude eE. Thus, the velocity of the electrons increases at a rate $\dot{v} = eE/m = (er/2m)\dot{B}_m$. The orbit is stable if the equation $erB_{orb} = mv$ remains verified in spite of the variation of the velocity. This is possible if the geometry of the electromagnet is such that $B_{orb} = \frac{1}{2}B_m$. In this case, the magnetic field guides the electrons in their motion and the variation of B accelerates them. In this way, it is possible to attain energies of several hundred mega-electronvolts for electrons. This higher limit is hindered by the difficulty of creating an intense magnetic field over a large area (this sets an upper limit on the radius r of the orbit). On the other hand, at high energy, the equations of classical mechanics that we have used should be replaced by relativistic equations. These factors limit the possibilities of the betatron.

8.10. Problems

Faraday's law and Lenz's law

P8.1 A circuit of surface $\mathcal{S}$ and resistance R is placed normal to a uniform magnetic field **B**, whose magnitude depends on time according to the relation $B = B_o + kt$. Calculate the induced emf, the intensity and the total charge in the circuit between $t = 0$ and t. Calculate the dissipated energy as Joule heat. Discuss the direction of the current depending on the values of the constants B_o and k.

Neumann's induction and Lorentz induction

P8.2 A long solenoid of radius a and n turns per unit length carries a current I. It is surrounded by a circular circuit of radius a' coaxial with the solenoid. Calculate the vector potential **A** everywhere. Assuming that I depends on time, calculate the induced electric field everywhere. Calculate the circulation of **E** over the circuit. Find this result again using Farasay's law.

P8.3 A solenoid of length h, radius r, and n turns per unit length carries a current $I(t)$. **a)** Using Farasay's equation, calculate the induced electric field in this solenoid. What is the induced emf? **b)** Calculate the corresponding vector potential **A**.

Lorentz induction

P8.4 A rod of mass m is placed normally on two horizontal rails separated by a distance D in a vertical field **B** (Figure 8.1a). The rails are connected by a wire of resistance R. **a)** What is the induced emf in the rod if it is displaced with a constant velocity $v_o\mathbf{e}_x$? What is the induced intensity? Determine the exerted force and the power needed to displace this rod. Numerical applications: $D = 0.5$ m, $B = 0.2$ T, $v_o = 2$ m/s and $R = 4\ \Omega$. **b)** Assume that a generator G, that is connected in series with the resistance R, produces a constant current I. Determine the position $x(t)$ of the rod if it was initially at rest. What should the emf of the generator be in order to maintain I constant? Discuss the conservation of energy. **c)** G is replaced by a battery of constant emf $\mathcal{E}$. Determine $x(t)$ and show that the velocity of the rod attains a limit value. What then is the intensity I? Discuss the conservation of energy. **d)** The generator is removed, but the resistance R is maintained and the rod is launched on the rails with an initial velocity v_0. Study its motion and discuss the conservation of energy.

P8.5 A uniform magnetic field **B** points in the direction Oz. **a)** A rod of length L has one end at the origin O and it rotates in the Oxy plane with an angular velocity ω about Oz. Calculate the induced emf between its ends. Determine the polarity of this emf. **b)** A copper disk of radius R rotates at N revolutions per second about its axis Oz. Calculate the induced difference of potential between its center and its periphery. Numerical application: $B = 3$ T, $R = 10$ cm and $N = 10$ rev/s.

P8.6 A rectangular circuit $MQPN$, of sides a and b parallel to Ox and Oy respectively, moves with a constant velocity **v** parallel to Oy in a field $\mathbf{B} = B\mathbf{e}_z$ (Figure 8.11). **a)** Assuming that **B** is uniform and using the proper frame of the circuit, verify that QP becomes positively charged, while MN becomes negatively charged. Show that the induced emf is equal to zero. **b)** Assuming that **B** is non-uniform (field of an electromagnet, for instance) and that the circuit is small, calculate the force **f** that acts on a conduction electron and show that the circulation of **f** over the circuit is $-eva(B_{PN} - B_{MQ})$, where B_{PN} and B_{MQ} are the average fields on the sides PN and MQ, respectively. Deduce that the induced emf is $va(B_{MQ} - B_{PN})$. **c)** Considering two positions of the circuit separated by an interval of time dt, show that the variation of the flux is $d\Phi = (B_{MQ} - B_{PN})va\, dt$. Deduce that the induced emf is $-\partial_t\Phi$. **d)** Let us consider a circuit $\mathcal{C}$, of any shape, which moves in a field **B**. Show that the variation of the flux during dt is

$$\Phi(t+dt) - \Phi(t) = \iint_{\mathcal{S}} d\mathcal{S}\,(\mathbf{B.n}) = \int_{\mathcal{C}} \mathbf{B}.(\mathbf{v}\,dt \times d\mathbf{r}).$$

Deduce that the induced emf is $\mathcal{E}' = \int_{\mathcal{C}} (\mathbf{v} \times \mathbf{B}).d\mathbf{r}$.

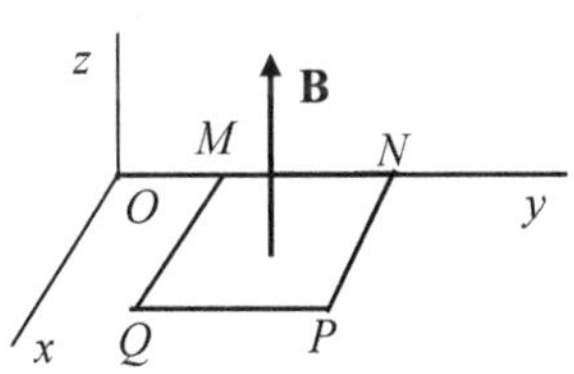

Figure 8.11. *Problem 8.6*

Mutual inductance and self-inductance

P8.7 Two circular loops of radii R and r are coaxial and their centers are on the axis Oz at a distance h apart (Figure 8.12). **a)** Determine the direction of the induced current in the small loop if the intensity I in the large loop increases and if it decreases. **b)** I is maintained constant and the small loop is displaced. Determine the direction of the induced current if it is moved toward the large loop. **c)** Calculate the mutual inductance M in the cases $r << R$ and $r = R$.

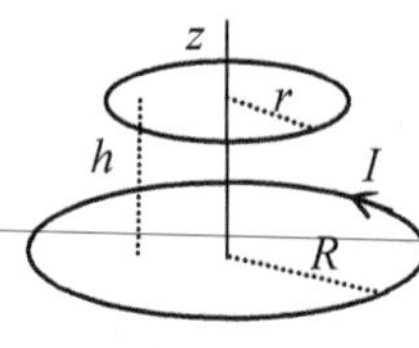

Figure 8.12. *Problem 8.7*

P8.8 a) Calculate the inductance L of a finite solenoid of n turns per unit length, of radius a and length $2h$ by taking, for each loop, the value of **B** at the center of that loop. Calculate L if $h = 1$ m, $a = 10$ cm and $n = 500$ turns/m. Compare L with that of a solenoid of the same characteristics but treated as very long. **b)** A toroidal coil of average radius R has a transversal circular section of radius a and it is formed by N turns. Show that its inductance is $L = \mu_0 N^2[R - (R^2 - a^2)^{1/2}]$. Compare L with the inductance of a very long solenoid with the same characteristics.

LR and LCR circuits

P8.9 The intensity in an electromagnet of inductance 5 H and resistance 10 Ω decreases linearly from 10 A to 2 A in 0.05 s. Determine the induced emf, the variation of the magnetic energy and the energy that is dissipated as Joule heat. Calculate the energy supplied by the generator in the same interval of time.

P8.10 A capacitor of capacitance $C = 10$ μF is charged under 100 V and then connected at time $t = 0$ to the terminals of an inductance $L = 20$ H in series with a resistance $R = 2$ kΩ. **a)** Write the differential equation for the charge Q. What is the solution that corresponds to the given initial conditions? **b)** Write the expressions of the current intensity, the electric energy, the magnetic energy, and the total energy. **c)** Calculate the energy lost as Joule heat between $t = 0$ and t and compare it with the decrease in the stored energy. **d)** What is the relaxation time of this circuit and what is its quality factor? How long does it take for the amplitude of the charge to be reduced to 1% of its initial value?

Magnetic energy

P8.11 A coil has 100 turns, a section $\mathcal{S}$ = 25 cm^2 and a resistance R = 5 Ω. It is placed normal to a magnetic field B = 0.5 T and pulled out of the field in 0.2 s. Determine the average induced emf, the total induced charge, the average induced intensity, the variation of the magnetic energy and the work required to pull it.

P8.12 a) Consider the coaxial cylindrical cable of Figure 8.13. The internal cylinder of radius r_1 has a charge density $+q_l$ per unit length and the cylindrical shell of internal radius r_2 has a density $-q_l$. Applying Gauss's law to a cylindrical surface of radius r, show that the electric field between the conductors is $E = q_l/2\pi\varepsilon_o r$. Deduce that the difference of potential between them is $V = (q_l/2\pi\varepsilon_o)\ \ln(r_2/r_1)$ and that the capacitance per unit length is $C_l = 2\pi\varepsilon_o/\ln(r_2/r_1)$. Using the expression of the electric field, calculate the energy density $U_{E,v}$ and the stored energy per unit length of the cable. Using the expression $U_{E,l} = \frac{1}{2}C_l V^2$, again find the expression of C_l. **b)** To calculate the inductance, assume that a current I is carried by the internal cylinder in a direction and by the cylindrical shell in the opposite direction. Applying Ampère's law to a circle $\mathcal{C}$ of radius r (such that $r_1 < r < r_2$), show that the magnetic field between the conductors is $B = \mu_o I/2\pi r$. Show that the magnetic energy stored in the cable of length l is $U_M = (\mu_o/4\pi)I^2 l\ \ln(r_2/r_1)$. Using the fact that the magnetic energy is $\frac{1}{2}LI^2$, verify that the inductance per unit length is $L_l = (\mu_o/2\pi)\ \ln(r_2/r_1)$.

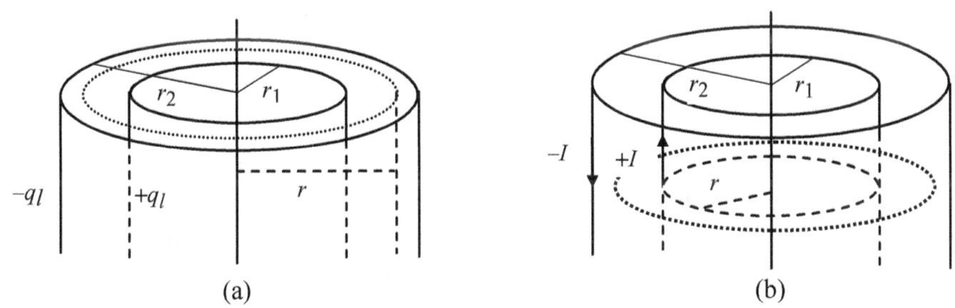

Figure 8.13. *Coaxial cable of problem 8.12*

P8.13 a) Show that the stored magnetic energy in a magnetic circuit is $U_M = \frac{1}{2}\mathcal{R}\Phi^2$, where $\mathcal{R}$ is its reluctance and Φ is the flux. **b)** Show that the inductance is $L = N^2/\mathcal{R}$ and again find the expression $U_M = \frac{1}{2}\mathcal{R}\Phi^2$ for the energy.

P8.14 A torus of permeability μ is generated by the rotation of a square of sides $2a$ about the axis Oz parallel to two sides of the square, the center of the square describing a circle of radius R. Two coils are wound around this torus: a primary of N_1 turns and a secondary of N_2 turns. **a)** Calculate the inductances L_1 and L_2 and the mutual inductance M of these coils. **b)** Assume that the secondary is connected to a resistance R_2 and that the primary of resistance R_1 is supplied by a generator of emf $\mathcal{E}_1(t)$. Write the equations of these circuits. Verify that the generator supplies a power $P = R_1I_1^2 + R_2I_2^2 + \partial_t(\frac{1}{2}L_1I_1^2 + \frac{1}{2}L_2I_2^2) + \partial_t(M\,I_1I_2)$. Interpret this expression. **c)** Assuming that the emf is $\mathcal{E}_1 = \mathcal{E}_m \cos(\omega t)$, determine the intensities I_1 and I_2.

P8.15 A coil of $N_c = 30$ turns encircles a solenoid of radius $a = 5$ cm, length $h = 50$ cm and total number of turns $N_s = 1\,000$. The coil is connected to a resistance $R = 40$ Ω. Assume that the solenoid is very long, and the self-induction of the coil is negligible. **a)** Calculate the stored energy in the solenoid if it carries a current of 10 A. How long does it take to supply this energy by a generator of 400 W? **b)** What is the induced charge in R if the intensity of the solenoid varies from $I_1 = 1$ to $I_2 = 5$ A? Does the charge depend on the radius of the coil and the required time for this variation of the intensity? Determine the direction of the induced current. **c)** Assume that the intensity in the solenoid varies according to the equation $I_s = I_m \sin(\omega t)$ where $I_m = 5$ A and $\omega = 120\pi$ rad/s. Neglecting the self-inductance of the coil and its influence on the solenoid, calculate the field B inside the solenoid. What is the induced intensity in the coil?

Magnetic forces on circuits

P8.16 a) Determine the magnetic energy of a solenoid of N turns, radius R and length h by assuming that it is very long ($R \ll h$). **b)** Two coaxial solenoids S_1 and S_2 have radii R_1 and R_2. Neglecting end effects, calculate their mutual inductance if S_2 is introduced a distance x inside S_1. Deduce their force of interaction.

P8.17 A solenoid of radius R and length $h \gg R$ is formed by using a conducting elastic wire. Calculate the magnetic force exerted by the field $\mathbf{B}$ on an element of area of the solenoid. Deduce that the solenoid is subject to a magnetic pressure $p_M = U_{M,v}$, which tends to increase its radius in agreement with the maximum flux rule. Assume that the intensity I is kept constant and R increases by δR. Show that the work of the pressure forces is equal to the variation of the magnetic energy.

Applications of the induction

P8.18 A coil of 200 turns and radius $r = 10$ cm is placed in a uniform field $B_o = 2$ T. It is connected to a circuit of resistance 50 Ω. **a)** What is the moment of the magnetic forces exerted on this coil if it carries a current I and its axis makes an angle θ with $\mathbf{B}_o$? **b)** Neglecting the self-induction, determine the moment of force that has to be exerted on the coil in order to rotate it with an angular velocity ω. What then is the induced intensity? Verify the conservation of energy. Make the same analysis, but by taking into account the self-induction.

P8.19 A metallic disk of thickness h, radius R and conductivity σ is placed in a magnetic field $B = B_m \sin(\omega t)$ parallel to its axis. Determine the induced current density at a point situated at a distance r to the axis. Calculate the average power that is dissipated as Joule heat in the disk. Numerical application: consider the case of a disk of radius 7 cm, thickness 2 mm, conductivity $10^7\ \Omega^{-1}.\text{m}^{-1}$, a magnetic field of amplitude $B_m = 0.2$ T and frequency 50 Hz.

P8.20 In a betatron, the electron moves on a circular orbit of axis Oz and radius ρ_o in a magnetic field oriented in the direction Oz. The magnitude of **B** depends on time and the distance ρ to Oz. Let $\langle B \rangle$ be the average field within the orbit and **B** the field that acts on the electron along the orbit. **a)** Show that the induced electric field is $\mathbf{E} = -\frac{1}{2}\rho\, \partial_t \langle B \rangle\, \mathbf{e}_\varphi$. Write the equation of motion and deduce that the electron moves effectively on a circular orbit of radius ρ_o if the electromagnet is conceived in such a way that $\partial_t \langle B \rangle = 2B$. The acceleration of the electron is then $\partial_t v = e\rho_o B/m$. **b)** Show that in a complete revolution, the kinetic energy varies by $\delta U_K = 2\pi e \rho_o^2 B$.

Chapter 9

Maxwell's Equations

In 1865, Maxwell unified electricity and magnetism in a single theory, called *electromagnetism*. The fields **E** and **B** cannot be considered as independent, as the variation of one in time requires the presence of the other. Thus, they constitute a single physical entity, called the *electromagnetic field.* This theory is verified by all its consequences, particularly the existence of *electromagnetic waves* that propagate in vacuum with the speed $c = 1/\sqrt{\mu_o \varepsilon_o}$ equal to the speed of light. The existence of these waves with the same properties of polarization and propagation as light waves was verified experimentally by Hertz in 1884. Electromagnetic theory also solved the old problem of the nature of light: it is an electromagnetic wave of a very short wavelength. The formulation of electromagnetism was a very important event in the history of science.

In this chapter, we write Maxwell's equations and the equations of propagation for the fields and the potentials. We discuss the questions of energy and its transfer and the radiation pressure.

9.1. Fundamental laws of electromagnetism

If electric and magnetic phenomena vary in time, all the relevant quantities (fields, free charge density q_v, conduction current density **j**, polarization **P**, magnetization **M**, etc.) may depend on time, and we expect that some of the basic equations that we have derived in previous chapters will be modified.

a) We have seen in section 3.6 that the conservation of electric charge implies that the charge flowing out of a closed surface $\mathcal{S}$ per unit time (i.e. the total intensity) is equal to the rate of decrease of the total charge Q in the enclosed volume $\mathcal{V}$ (Figure 9.1a)

$$\iint_{\mathcal{S}} d\mathcal{S}\,(\mathbf{j}.\mathbf{n}) = -\iiint_{\mathcal{V}} d\mathcal{V}\,\partial_t q_v. \qquad [9.1]$$

Using Gauss-Ostrogradsky's theorem to transform the flux of **j** into the volume integral of $\nabla.\mathbf{j}$, this equation being valid for any volume $\mathcal{V}$, we obtain the local form of the law of conservation of charge, called the *continuity equation*,

$$\nabla.\mathbf{j} + \partial_t q_v = 0. \qquad [9.2]$$

In the case of stationary phenomena, $\partial_t q_v = 0$, equation [9.2] reduces to $\nabla.\mathbf{j} = 0$ and [9.1] reduces to $\iint_{\mathcal{S}} d\mathcal{S}\,\mathbf{j}.\mathbf{n} = 0$. This implies that the intensity is the same at all points of the same branch of the circuit and Kirchhoff's nodes rule (Figure 9.1b). In the case of a surface $\mathcal{S}$, which contains the positive plate of a capacitor (Figure 9.1c), the outgoing flux $\iint_{\mathcal{S}} d\mathcal{S}\,\mathbf{j}.\mathbf{n}$ receives only a contribution $-I$ from the part of $\mathcal{S}$ that is outside the capacitor. Thus, during charging or discharging the capacitor, this flux is not equal to zero but to $-I = -\,dQ/dt$.

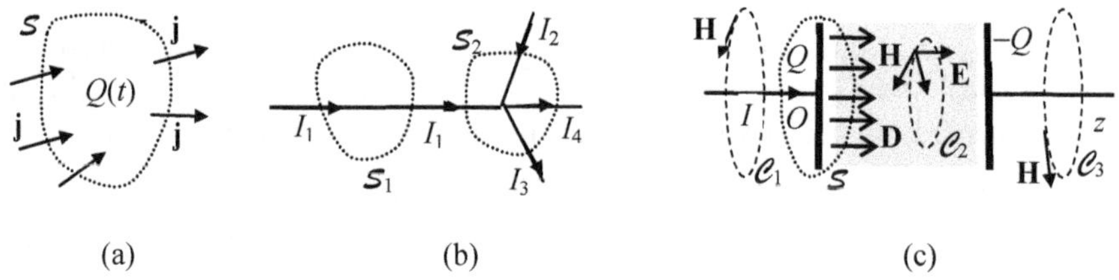

Figure 9.1. *Conservation of charge: a) in the case of time-dependent phenomena, b) in a stationary circuit, and c) displacement current and energy flux in a capacitor*

b) If the magnetic field **B** is time-dependent, there is necessarily an induced electric field **E**. The circulation of **E** over a closed path $\mathcal{C}$ is not equal to zero but to the induced e.m.f. according to the equation

$$\mathcal{E} = \int_{\mathcal{C}} d\mathbf{r}.\mathbf{E} = -\frac{d}{dt}\iint_{\mathcal{S}} d\mathcal{S}\,(\mathbf{n}.\mathbf{B}). \qquad [9.3]$$

Using Stokes's theorem to transform the circulation over $\mathcal{C}$ into the flux of $\nabla \times \mathbf{E}$ over $\mathcal{S}$, the relation being valid for any surface $\mathcal{S}$, we obtain Faraday's law in the local form

$$\nabla \times \mathbf{E} + \partial_t \mathbf{B} = 0. \qquad [9.4]$$

Thus, the electric field **E** is not conservative in the case of variable phenomena.

c) Gauss's law and the law of conservation of magnetic flux (which is the expression of the absence of magnetic charge) remain valid in the case of time-dependent phenomena. These laws may be written in the integral or local forms

$$\iint_{\mathcal{S}} d\mathcal{S}\,(\mathbf{n}.\mathbf{D}) = \iiint_{\mathcal{V}} d\mathcal{V}\, q_{\mathrm{v}} \quad \text{and} \quad \boldsymbol{\nabla}.\mathbf{D} = q_{\mathrm{v}}, \qquad [9.5]$$

$$\iint_{\mathcal{S}} d\mathcal{S}\,(\mathbf{n}.\mathbf{B}) = 0 \quad \text{and} \quad \boldsymbol{\nabla}.\mathbf{B} = 0. \qquad [9.6]$$

d) Let us consider now Ampère's law. It is obvious that equation $\boldsymbol{\nabla} \times \mathbf{H} = \mathbf{j}$ cannot hold in the case of time-dependent phenomena, as the divergence of the left-hand side is identically zero, while the divergence of the right-hand side is $-\partial_t q_{\mathrm{v}}$ according to [9.2]. Noting that $q_{\mathrm{v}} = \boldsymbol{\nabla}.\mathbf{D}$ according to Gauss's law, the continuity equation [9.2] may be written as

$$\boldsymbol{\nabla}.(\mathbf{j} + \partial_t \mathbf{D}) = 0. \qquad [9.7]$$

This equation states that the vector field $(\mathbf{j} + \partial_t\mathbf{D})$ has a conservative flux. A possible form of the modified Ampère's law is thus

$$\boldsymbol{\nabla} \times \mathbf{H} = \mathbf{j} + \partial_t \mathbf{D}. \qquad [9.8]$$

This equation is physically valid if all its consequences are verified by the experiments. Particularly, if the electric field is variable ($\partial_t\mathbf{D} \neq 0$), this equation shows that there must be necessarily a magnetic field **H**, even if $\mathbf{j} = 0$. This is the induction of a magnetic field if the electric field varies. Maxwell called $\partial_t\mathbf{D}$ the *displacement current*. To write Ampère's law in an integral form, we take the flux of both sides of [9.8] through a surface $\mathcal{S}$ bounded by a closed curve $\mathcal{C}$ and we transform the flux of $\boldsymbol{\nabla}\times\mathbf{H}$ into the circulation of **H** using Stokes's theorem. We find

$$\int_{\mathcal{C}} d\mathbf{r}.\mathbf{H} = \iint_{\mathcal{S}} dS\,\mathbf{n}.\mathbf{j} + \frac{d}{dt}\iint_{\mathcal{S}} dS\,\mathbf{n}.\mathbf{D}. \qquad [9.9]$$

It should be noted that q_{v} is the free charge density, and **j** is the free current density, which includes the conduction currents in conductors at rest, the convection currents (i.e. produced by charges that are dragged by moving bodies) and the currents that are produced by beams of charged particles. We may consider Gauss's law $\boldsymbol{\nabla}.\mathbf{D} = q_{\mathrm{v}}$ and the modified Ampère's law $\boldsymbol{\nabla} \times \mathbf{H} = \mathbf{j} + \partial_t\mathbf{D}$ as two coupled equations for **D** and **H** but, very often, they do not completely determine **D** and **H**. Similarly, equations $\boldsymbol{\nabla}.\mathbf{B} = 0$ and $\boldsymbol{\nabla}\times\mathbf{E} + \partial_t\mathbf{B} = 0$ do not completely determine **E** and **B**. The determination of all four fields necessitates the use of the relations of **D** to **E** and **B** to **H**, which involve the properties of the medium. In the general case, **E** is the sum of the field produced by the free charges, the field due to the polarization of dielectrics, and the induction field due to the variation of **B**. Similarly, **B** is the sum of the field produced by the free current density, the field produced by the magnetic bodies and the induced field due to the variation of **D**.

The electromagnetic properties of matter are determined by the polarization **P**, equivalent to a volume charge density $q'_\mathrm{v} = -\nabla.\mathbf{P}$, and the magnetization **M**, equivalent to a volume current density $\mathbf{j}'_\mathrm{v} = \nabla \times \mathbf{M}$ (besides the polarization surface charge density $q'_\mathrm{s} = \mathbf{n}.\mathbf{P}$ and the magnetization surface current density $\mathbf{j}'_\mathrm{s} = \mathbf{M} \times \mathbf{n}$). Thus, Gauss's law and Ampère's law may be written as

$$\nabla.\mathbf{E} = (q_\mathrm{v} - \nabla.\mathbf{P})/\varepsilon_\mathrm{o} \qquad \text{and} \qquad \nabla \times \mathbf{B} = \mu_\mathrm{o}(\mathbf{j} + \nabla \times \mathbf{M} + \partial_t\mathbf{D}). \qquad [9.10]$$

The displacement current density $\partial_t\mathbf{D}$ is the vector sum of a term $\varepsilon_\mathrm{o}\partial_t\mathbf{E}$, which exists in vacuum as well as in matter, and a *polarization current density* $\mathbf{j}'_\mathrm{p} = \partial_t\mathbf{P}$ due to the dielectric[1]. Thus, equations [9.10] have the same form as in vacuum, but by including the polarization charge density $q'_\mathrm{v} = -\nabla.\mathbf{P}$, the magnetization current density $\mathbf{j}'_\mathrm{M} = \nabla\times\mathbf{M}$ and the polarization current density $\mathbf{j}'_\mathrm{p} = \partial_t\mathbf{P}$.

To interpret the displacement current, let us consider, for instance, the charging of a capacitor, whose plates are disks of radius R (Figure 9.1c). The electric displacement **D** exists only between the plates, where it is uniform and given by Gauss's law $\mathbf{D} = q_\mathrm{s}\,\mathbf{e}_z$. Equation [9.7] means that the outgoing flux of the vector $\mathbf{j} + \partial_t\mathbf{D}$ from the surface $\mathcal{S}$ is equal to zero. On the part of $\mathcal{S}$ that is outside the capacitor, $\mathbf{D} = 0$, and the outgoing flux reduces to that of **j**, i.e. $-I$. On the part of $\mathcal{S}$ that is inside the capacitor, $\mathbf{j} = 0$, and the outgoing flux reduces to that of $\partial_t\mathbf{D}$, i.e. $\pi R^2\,\partial_t q_\mathrm{S}$. Thus, the conservation of the flux of $\mathbf{j} + \partial_t\mathbf{D}$ is equivalent to the evident relation $I = \pi R^2\,\partial_t q_\mathrm{S} = \partial_t Q$. The appearance of the displacement current in Ampère's law [9.8] means that it produces a magnetic field exactly like the conduction current. Indeed, applying Ampère's law [9.9] to a circular path $\mathcal{C}_2$ of radius $\rho < R$, we get $2\pi\rho H = \pi\rho^2\,\partial_t D = \pi\rho^2\,\partial_t q_\mathrm{S}$, hence $H = (\rho/2\pi R^2)\,I$. If we use Ampère's law without the term $\partial_t\mathbf{D}$, we obtain no field **H**, and this is not the case, as in the limit of a thickness of the

---

1 The polarization current density $\partial_t\mathbf{P}$ may be considered to be due to the displacement of bound charges within the atom or molecule. Indeed, **P** is the electric dipole moment of the unit volume; thus, $\mathbf{P} = \Sigma_\mathrm{k}\, q_\mathrm{k}\, \mathbf{r}_\mathrm{k}$ and $\partial_t\mathbf{P} = \Sigma_\mathrm{k}\, q_\mathrm{k}\, \mathbf{v}_\mathrm{k}$, where the sum is over the bound charges in the unit volume and this is a current density $\mathbf{j}_\mathrm{p}$. As for the term $\varepsilon_\mathrm{o}\partial_t\mathbf{E}$, Maxwell interpreted it as due to the "polarization of ether". It was widely believed at the end of the 19th Century that vacuum and matter were actually filled with a very rarefied substance called *ether* having a permittivity ε_o and permeability μ_o. It served only as a propagation medium for light and electromagnetic waves. However, no experiment has detected *ether*, and the hypothesis of its existence was completely abandoned at the beginning of the 20th Century, after Michelson's experiment and the formulation of the special theory of relativity (see Chapter 13).

capacitor equal to zero, the field must be $H = I/2\pi\rho$, identical to its expression outside the capacitor.

9.2. Maxwell's equations

A) Maxwell's equations in any medium

The electric field **E** and the magnetic induction field **B** are defined by their action on a particle of charge q and velocity **v**, called *Lorentz force:*

$$\mathbf{F} = q(\mathbf{E} + \mathbf{v} \times \mathbf{B}). \qquad [9.11]$$

In the presence of matter, we have to introduce the polarization density **P** and the magnetization density **M** or, equivalently, the electric displacement $\mathbf{D} = \varepsilon_o\mathbf{E} + \mathbf{P}$ and the magnetic field $\mathbf{H} = \mathbf{B}/\mu_o - \mathbf{M}$. The fields **E**, **D**, **B**, and **H** obey Maxwell's equations in the local form:

$$\nabla \cdot \mathbf{D} = q_v \qquad \text{(Gauss's law)} \qquad [9.12]$$

$$\nabla \times \mathbf{E} + \partial_t\mathbf{B} = 0 \qquad \text{(Faraday's law)} \qquad [9.13]$$

$$\nabla \cdot \mathbf{B} = 0 \qquad \text{(absence of magnetic charges)} \qquad [9.14]$$

$$\nabla \times \mathbf{H} = \mathbf{j} + \partial_t\mathbf{D} \qquad \text{(Ampère's law).} \qquad [9.15]$$

These are the basic equations of electromagnetic phenomena in vacuum and in matter. They are valid in insulators ($\mathbf{j} = 0$) and in conductors. In the particular case of an Ohmic conductor, $\mathbf{j} = \sigma\mathbf{E}$, where σ is the conductivity. The first three equations were established from experimental observations. The fourth equation is a generalization of Ampère's law. q_v is the charge density of free charges and **j** is the free current density (i.e. conduction, convection, and beam currents). The displacement current $\partial_t\mathbf{D}$ was introduced by Maxwell in order to preserve the equation of conservation of charge $\partial_t q_v + \nabla.\mathbf{j} = 0$. We may inverse the argument and derive this equation of conservation of charge as a consequence of equations [9.12] and [9.15].

If we know the free sources q_v and **j** and the dielectric and magnetic properties of the medium (i.e. the relations of **D** to **E** and **B** to **H**) at each point of space, Maxwell's equations have a solution but it is not unique. Indeed, we may add to it any solution ($\mathbf{E}_o$, $\mathbf{D}_o$, $\mathbf{B}_o$ and $\mathbf{H}_o$) which corresponds to $q_v = 0$ and $\mathbf{j} = 0$ and obtain another solution. We may use this property to impose initial conditions and boundary conditions at the surfaces $\mathcal{S}_i$, which limit the region $\mathcal{V}$ of the fields. If $\mathcal{V}$ is infinite but the sources q_v and **j** occupy a finite region of space, the fields (**E**, **B**, **D**, and **H**) vanish at infinity. It is always possible to find $\mathbf{E}_o$, $\mathbf{D}_o$, $\mathbf{B}_o$, and $\mathbf{H}_o$ to satisfy these initial and boundary conditions. Thus, Mawxell's equations with the initial conditions and boundary conditions always have a solution and this solution is unique. Obviously, this does not mean that it may be written in terms of simple functions. Some approximation methods or numerical methods are often necessary.

Maxwell's equations form a system of partial differential equations. They are *linear* if the dielectric and magnetic medium are linear. Thus, the solutions obey the *superposition principle:* if the sources (q_V and **j**) are multiplied by a constant k, the fields are multiplied by the same constant. Also, if a configuration of sources (q_V and **j**) produces the fields **E**, **B**, **D** and **H**, and another configuration (q'_V and **j′**) produces the fields **E′**, **B′**, **D′**, and **H′**, the superposition of sources ($q_V+q'_V$ and **j**+**j′**) produces the superposition of fields (**E** + **E′**, **B** + **B′**, **D** + **D′** and **H** + **H′**).

Sometimes, it is useful to write Maxwell's equations in an integral form instead of the local differential form. For this, we integrate both sides of [9.12] and [9.14] over a volume $\mathcal{V}$ and we use Gauss-Ostrogradsky's theorem to transform the volume integral of the divergence into the outgoing flux through the surface $\mathcal{S}$, which limits $\mathcal{V}$. As for equations [9.13] and [9.15], we calculate the flux of both sides through a surface $\mathcal{S}$ and we use Stokes's theorem to transform the flux of the curl into a circulation over the curve $\mathcal{C}$, which limits $\mathcal{S}$. Thus, we get the four equations

$$\iint_{\mathcal{S}} d\mathcal{S}\, \mathbf{n}.\mathbf{D} = Q^{(\text{in})}, \quad \text{where} \quad Q^{(\text{in})} = \iiint_{\mathcal{V}} d\mathcal{V}\, q_V \quad \text{(Gauss's law)} \quad [9.16]$$

$$\int_{\mathcal{C}} d\mathbf{r}.\mathbf{E} + \partial_t \iint_{\mathcal{S}} d\mathcal{S}\, \mathbf{n}.\mathbf{B} = 0 \quad \text{(Faraday's law)} \quad [9.17]$$

$$\iint_{\mathcal{S}} d\mathcal{S}\, \mathbf{n}.\mathbf{B} = 0 \quad \text{(absence of magnetic charges)} \quad [9.18]$$

$$\int_{\mathcal{C}} d\mathbf{r}.\mathbf{H} - \partial_t \iint_{\mathcal{S}} d\mathcal{S}\, \mathbf{n}.\mathbf{D} = I^{(\text{in})}, \quad \text{where } I^{(\text{in})} = \iint_{\mathcal{S}} d\mathcal{S}\, \mathbf{n}.\mathbf{j} \quad \text{(Ampère's law).} \quad [9.19]$$

These equations hold even if the fields have discontinuities. On the other hand, the derivatives with respect to time of the fluxes receive contributions both from the variation of the fields in time and the displacement of the surfaces. For these reasons, Maxwell's equations in the integral form have a more general validity than the differential form. In these integral forms, the fields, charges and currents are taken at the same time. For instance, if a charge enters a closed surface $\mathcal{S}$, the flux of its outgoing electric field from $\mathcal{S}$ is non-zero only when it is inside $\mathcal{S}$. However, some effects (such as propagation) depend on the local properties of the fields; they are better analyzed by using the local form of Maxwell's equations.

B) Maxwell's equations in a homogeneous, isotropic, and linear medium

If the medium is homogeneous, linear, and isotropic, **D** is proportional to **E** and **B** is proportional to **H**:

$$\mathbf{D} = \varepsilon\, \mathbf{E} \quad \text{and} \quad \mathbf{B} = \mu\, \mathbf{H}, \qquad [9.20]$$

where ε and μ are characteristic of the medium. We may then only use fields **E** and **B**, and Maxwell's equations may be written as

$$\nabla.\, \mathbf{E} = q_v/\varepsilon \qquad \text{(Gauss's law)} \qquad [9.21]$$

$$\nabla \times \mathbf{E} + \partial_t \mathbf{B} = 0 \qquad \text{(Faraday's law)} \qquad [9.22]$$

$$\nabla.\, \mathbf{B} = 0 \qquad \text{(absence of magnetic charges)} \qquad [9.23]$$

$$\nabla \times \mathbf{B} = \mu \mathbf{j} + \varepsilon\mu\, \partial_t \mathbf{E} \qquad \text{(Ampère's law).} \qquad [9.24]$$

Particularly, in vacuum ε becomes ε_o and μ becomes μ_o.

To write the integral forms of the equations in a linear but inhomogeneous medium, we must assign the corresponding constant ε and μ to each element of area $d\mathcal{S}$ or element of volume $d\mathcal{V}$; we find:

$$\iint_{\mathcal{S}} d\mathcal{S}\, \varepsilon\, \mathbf{n.E} = Q^{(in)} \qquad \text{(Gauss's law)} \qquad [9.25]$$

$$\int_{\mathcal{C}} d\mathbf{r.E} + \partial_t \iint_{\mathcal{S}} d\mathcal{S}\, \mathbf{n.B} = 0 \qquad \text{(Faraday's law)} \qquad [9.26]$$

$$\iint_{\mathcal{S}} d\mathcal{S}\, \mathbf{n.B} = 0 \qquad \text{(absence of magnetic charges)} \qquad [9.27]$$

$$\int_{\mathcal{C}} d\mathbf{r.B}/\mu = I^{(in)} + (d/dt) \iint_{\mathcal{S}} d\mathcal{S}\, \varepsilon\, \mathbf{n.D} \qquad \text{(Ampère's law).} \qquad [9.28]$$

C) Equation of propagation of the fields

Maxwell's equations form a system of coupled partial differential equations of the first order. It is possible to write uncoupled equations for each one of the fields; but they are second-order partial differential equations. Indeed, let us evaluate the curl of Faraday's equation [9.22], we find $\nabla\times(\nabla\times\mathbf{E}) + \partial_t(\nabla\times\mathbf{B}) = 0$. Using the identity $\nabla\times(\nabla\times\mathbf{E}) = \nabla(\nabla.\mathbf{E}) - \Delta\mathbf{E}$, Gauss's law [9.21] and Ampère's law [9.24], we find

$$\Delta\mathbf{E} - (1/v^2)\, \partial^2_{tt}\mathbf{E} = \mu\, \partial_t \mathbf{j} + \nabla q_v/\varepsilon, \qquad \text{where } v = 1/\sqrt{\mu\varepsilon}. \qquad [9.29]$$

Similarly, let us calculate the curl of both sides of [9.24] and use equations [9.22] and [9.23], we find the equation

$$\Delta\mathbf{B} - (1/v^2)\, \partial^2_{tt}\mathbf{B} = -\, \mu \nabla \times \mathbf{j}. \qquad [9.30]$$

The uncoupled equations, [9.29] and [9.30], for **E** and **B** determine some properties of the fields **E** and **B** if we know the charge density and the current density at each point of space. Particularly, if these densities are equal to zero everywhere, these equations become

$$\Delta\mathbf{E} - (1/v^2)\,\partial^2_{tt}\mathbf{E} = 0, \qquad \Delta\mathbf{B} - (1/v^2)\,\partial^2_{tt}\mathbf{B} = 0. \qquad [9.31]$$

These are the propagation equations (or wave equations) called *d'Alembert's equations*. The speed of propagation is

$$v = 1/\sqrt{\mu\varepsilon}. \qquad [9.32]$$

In vacuum, this speed is identical to the speed of light in vacuum

$$c = 1/\sqrt{\mu_o\varepsilon_o} = 2.997\,924\,58 \times 10^8 \text{ m/s}. \qquad [9.33]$$

9.3. Electromagnetic potentials and gauge transformation

Maxwell's equation $\boldsymbol{\nabla}.\mathbf{B} = 0$ is identically verified, if **B** is the curl of a *vector potential* **A**

$$\mathbf{B} = \boldsymbol{\nabla} \times \mathbf{A}. \qquad [9.34]$$

Substituting this expression of **B** in Faraday's equation [9.22], we get

$$\boldsymbol{\nabla} \times (\mathbf{E} + \partial_t\mathbf{A}) = 0. \qquad [9.35]$$

This equation expresses that the vector $(\mathbf{E} + \partial_t\mathbf{A})$ is conservative. Thus, we may define a *scalar potential* V such that

$$\mathbf{E} = -\,\boldsymbol{\nabla}V - \partial_t\mathbf{A}. \qquad [9.36]$$

Note that **A** and V are not uniquely determined if the fields **E** and **B** are given. Indeed, a *gauge transformation*

$$\mathbf{A} \rightarrow \mathbf{A}' = \mathbf{A} + \boldsymbol{\nabla}f, \quad V \rightarrow V' = V - \partial_t f, \qquad [9.37]$$

where f is an arbitrary scalar function of **r** and t, does not modify the fields **E** and **B**.

Substituting the expressions [9.34] and [9.36] in Maxwell's inhomogeneous equations, [9.21] and [9.24], and noting that $\boldsymbol{\nabla}\times(\nabla\times\mathbf{A}) = \boldsymbol{\nabla}(\boldsymbol{\nabla}.\mathbf{A}) - \Delta\mathbf{A}$, we find

$$\Delta V + \partial_t(\boldsymbol{\nabla}.\mathbf{A}) = -q_v/\varepsilon \quad \text{and} \quad \Delta\mathbf{A} - \varepsilon\mu\,\partial^2_{tt}\mathbf{A} - \boldsymbol{\nabla}(\boldsymbol{\nabla}.\mathbf{A} + \varepsilon\mu\,\partial_t V) = -\,\mu\mathbf{j}. \qquad [9.38]$$

These are two coupled equations for V and **A**. It is always possible to make a gauge transformation [9.37] in order to have V and **A** verify the *Lorentz condition*

$$\nabla . \mathbf{A} + \varepsilon\mu\, \partial_t V = 0. \quad [9.39]$$

Then, equations [9.38] take the uncoupled form

$$\Delta V - (1/v^2)\, \partial^2_{tt} V = -q_v/\varepsilon \qquad \text{and} \qquad \Delta \mathbf{A} - (1/v^2)\, \partial^2_{tt} \mathbf{A} = -\, \mu \mathbf{j}. \quad [9.40]$$

If the Lorentz condition [9.39] is not imposed, the propagation equations are uncoupled, hence more difficult to analyze, but the fields are evidently the same.

If the charge and current densities are known at each point of space and at any time, the propagation equations [9.40] have the particular solutions

$$V_R(\mathbf{r}, t) = (1/4\pi\varepsilon) \iiint d\mathcal{V}'\, q_v(\mathbf{r}', t - R/v)/R, \quad [9.41]$$

$$\mathbf{A}_R(\mathbf{r}, t) = (\mu/4\pi) \iiint d\mathcal{V}'\, \mathbf{j}(\mathbf{r}', t - R/v)/R, \qquad \text{where } \mathbf{R} = \mathbf{r} - \mathbf{r}', \quad [9.42]$$

These are called *retarded potentials*, because they are produced at any point **r** and time t by the sources at each space point **r′** not at the same time t but an earlier time $t - R/v$, where v is the speed of propagation. We note that there is another solution, called *advanced solution*, which is produced by the sources at the later time $t + R/v$. This solution is not acceptable as it violates the causality principle. The time delay R/v is just the time that the wave, which is emitted by the sources (q_v and **j**) at **r′**, takes to arrives at **r**. It may be shown that these potentials verify Lorentz condition [9.39]. At the limit of an infinite speed v, we find the instantaneous potentials of the *permanent regime*

$$V_P(\mathbf{r}, t) = (1/4\pi\varepsilon) \iiint d\mathcal{V}'\, q_v(\mathbf{r}', t)/R \qquad \text{and} \qquad \mathbf{A}_P(\mathbf{r}, t) = (\mu/4\pi)\iiint d\mathcal{V}'\mathbf{j}(\mathbf{r}', t)/R. \quad [9.43]$$

The expressions [9.41] and [9.42] are not the unique solutions of the equations of propagation [9.40]. We may add to them any solution of the homogeneous equations of propagation $\Delta V_o - (1/v^2)\, \partial^2_{tt} V_o = 0$ and $\Delta \mathbf{A}_o - (1/v^2)\partial^2_{tt} \mathbf{A}_o = 0$ and obtain other solutions of the equations [9.40]. If the medium is infinite and the fields must vanish at large distances, the expressions [9.41] and [9.42] are the solutions that verify this requirement. In the case of a medium bounded by surfaces $\mathcal{S}_i$, whose charges or potentials are given, the V_o and $\mathbf{A}_o$ terms must be added and they can be chosen to satisfy the boundary conditions on the surfaces $\mathcal{S}_i$.

The equations of propagation of the fields [9.29] and [9.30] may be derived from the equations of propagation of the potentials. Indeed, taking the curl of both sides of the propagation equation of **A** [9.40], we find the propagation equation of **B**

[9.30] and taking the gradient of both sides of the propagation equation of V and the time derivative of the propagation equation of **A** and adding them, we find the propagation equation of **E** [9.29].

It should be noted that the Lorentz condition does not completely fix the potentials. Indeed, it is still possible to make a gauge transformation [9.37] without affecting the Lorentz condition if the gauge function f is a solution of the equation $\Delta f - (1/v^2)\, \partial^2_{tt} f = 0$. Particularly, it is possible to choose f to impose the additional condition $\nabla.\mathbf{A} = 0$. Then, the Lorentz condition gives $\partial_t V = 0$. This special choice is called the *Coulomb gauge* (or the *transversal gauge*). Then the two equations, [9.38], may be written as

$$\Delta V = -q_v/\varepsilon \qquad \text{and} \qquad \Delta\mathbf{A} - (1/v^2)\, \partial^2_{tt}\mathbf{A} = -\,\mu\mathbf{j}. \qquad [9.44]$$

V obeys Poisson's equation and its solution is the instantaneous potential of the stationary state, but the expression of **A** should be retarded.

If we use V and **A** to analyze electromagnetic phenomena, any choice of the gauge function, f, has no effect on any physical quantities. This is referred to as *gauge invariance*. Thus, we may use any convenient gauge. This is not a drawback of the theory; on the contrary, it is of great physical importance[2].

9.4. Quasi-permanent approximation

In the general case of time-dependent q_v and **j**, the system is said to be in a *variable regime*. It obeys Maxwell's coupled equations [9.12] to [9.15]. **E** and **B** constitute the *electromagnetic field*. They are related to the potentials V and **A** by the relations $\mathbf{B} = \nabla \times \mathbf{A}$ and $\mathbf{E} = -\nabla V - \partial_t\mathbf{A}$. The potentials are given by the *retarded expressions* [9.41] and [9.42], the evaluation of which is often difficult; hence, it is necessary to use approximation methods.

a) If we completely neglect the variation of q_v in time and the motion of electric charges (thus $\mathbf{j} = 0$), the system is in the *static regime* governed by the equations of

2 Some authors consider that the potentials are simply a convenient mathematical tool to calculate the fields. Although the potentials are defined up to a gauge transformation, all the physical quantities are invariant in this transformation. It may be shown using Noether's theorem (see section 1.7) that this invariance is related to the conservation law of electric charge [9.2]. In 1959, Aharonov and Bohm showed that the vector potential has quantum mechanical observable effects. More recently, the theory of gauge fields allowed the unification of electromagnetic and weak interactions and possibly strong interactions. Thus, it seems to be the most fundamental theory of physics (the so-called *theory of everything*).

electrostatics and magnetostatics of permanent magnets, which are Maxwell's time-independent equations

$$\nabla.\mathbf{D}_S = q_v, \quad \nabla\times\mathbf{E}_S \cong 0, \quad \nabla.\mathbf{B}_S = 0 \quad \text{and} \quad \nabla\times\mathbf{H}_S \cong 0 \text{ (stationary regime).} \quad [9.45]$$

This is obviously a poor approximation with q_v, $\mathbf{E}_S$, $\mathbf{D}_S$, $\mathbf{B}_S$ and $\mathbf{H}_S$ time-independent and magnetism completely uncoupled from electrostatics.

b) A better approximation (called *permanent regime approximation*) consists of keeping the free current density $\mathbf{j}$ but neglecting Faraday's induction and the displacement current. Maxwell's equations reduce to

$$\nabla.\mathbf{D}_P = q_v, \quad \nabla\times\mathbf{E}_P \cong 0, \quad \nabla.\mathbf{B}_P = 0 \quad \text{and} \quad \nabla\times\mathbf{H}_P \cong \mathbf{j} \quad \text{(permanent regime).} \quad [9.46]$$

In the case of a linear medium, the potentials and fields may be written as

$$V_P(\mathbf{r}, t) \cong (1/4\pi\varepsilon)\iiint d\mathcal{V}'\, q_v(\mathbf{r}', t)/R \quad \text{and} \quad \mathbf{A}_P(\mathbf{r}, t) \cong (\mu/4\pi)\iiint d\mathcal{V}'\, \mathbf{j}(\mathbf{r}', t)/R, \quad [9.47]$$

$$\mathbf{E}_P(\mathbf{r}, t) \cong -\nabla V_P(\mathbf{r}, t) \cong (1/4\pi\varepsilon)\iiint d\mathcal{V}' q_v(\mathbf{r}', t)\, \mathbf{R}/R^3,$$

$$\mathbf{B}_P(\mathbf{r}, t) \cong \nabla\times\mathbf{A}_P(\mathbf{r}, t) \cong (\mu/4\pi)\iiint d\mathcal{V}\, \mathbf{j}(\mathbf{r}', t)\times\mathbf{R}/R^3, \quad [9.48]$$

where the sources are taken at the same time t as the potentials and the fields. Thus, all propagation effects are ignored. The potentials and the fields are completely uncoupled and they verify the equations

$$\Delta V_P \cong -q_v/\varepsilon, \quad \Delta\mathbf{A}_P \cong -\mu\mathbf{j}, \quad \Delta\mathbf{E}_P \cong \nabla(q_v/\varepsilon), \quad \text{and} \quad \Delta\mathbf{H}_P \cong -\nabla\times\mathbf{j}. \quad [9.49]$$

Taking the divergence of both sides of the approximate equation $\nabla\times\mathbf{H}_P \cong \mathbf{j}$, we find $\nabla.\mathbf{j} \cong 0$, which implies the continuity of the current in the branches of a circuit and Kirchhoff's rule for nodes. In this case, we also have $\nabla.\mathbf{A}_P \cong 0$.

c) The coupling of the fields $\mathbf{E}$ and $\mathbf{B}$ is due to the terms $\partial_t\mathbf{B}$ in the Maxwell-Faraday equation and the displacement current $\partial_t\mathbf{D}$ in the Maxwell-Ampère equation. If one of these terms is negligible, we expect this coupling to be weak and the solution of the problem to be easier. This is the so-called *quasi-permanent* approximation. We may consider two cases depending on whether $\partial_t\mathbf{D}$ or $\partial_t\mathbf{B}$ is small.

1) If $\partial_t\mathbf{D}$ is negligible, compared to $\mathbf{j}$, but $\partial_t\mathbf{B}$ is not small, we have the *magnetic quasi-permanent approximation* (MQ-P). Maxwell's equations reduce to

$$\nabla.\mathbf{E}_{MQ\text{-}P} = q_v/\varepsilon, \quad \nabla\times\mathbf{E}_{MQ\text{-}P} + \partial_t\mathbf{B}_{MQ\text{-}P} = 0, \quad \nabla.\mathbf{B}_{MQ\text{-}P} = 0, \quad \text{and} \ \nabla\times\mathbf{B}_{MQ\text{-}P} \cong \mu\mathbf{j}. \quad [9.50]$$

The homogeneous equations $\nabla\times\mathbf{E}_{MQ\text{-}P}+\partial_t\mathbf{B}_{MQ\text{-}P} = 0$ and $\nabla.\mathbf{B}_{MQ\text{-}P} = 0$ are verified if

$$\mathbf{E}_{MQ\text{-}P} = -\nabla V_{MQ\text{-}P} - \partial_t\mathbf{A}_{MQ\text{-}P} \quad \text{and} \quad \mathbf{B}_{MQ\text{-}P} = \nabla\times\mathbf{A}_{MQP}. \quad [9.51]$$

Making a gauge transformation [9.37], we may impose the Lorentz condition $\nabla.\mathbf{A}_{MQ\text{-}P} + \mu\varepsilon\,\partial_t V_{MQ\text{-}P} = 0$. Then, the fields and the potentials verify the equations

$$\Delta\mathbf{E}_{MQ\text{-}P} \cong \partial_t(\mu\mathbf{j}) + \nabla(q_v/\varepsilon), \qquad \Delta\mathbf{B}_{MQ\text{-}P} \cong -\nabla\times(\mu\mathbf{j}), \qquad [9.52]$$

$$\Delta V_{MQ\text{-}P} \cong -\,q_v/\varepsilon_o, \qquad \Delta\mathbf{A}_{MQ\text{-}P} \cong -\,\mu\mathbf{j}. \qquad [9.53]$$

Equations [9.53] are the same as in the permanent approximation. Their solutions have the form [9.47] with the sources taken at the same time t as the potentials

$$V_{MQ\text{-}P}(\mathbf{r}, t) \cong (1/4\pi\varepsilon)\iiint d\mathcal{V}'\, q_v(\mathbf{r}', t)/R, \quad \mathbf{A}_{MQ\text{-}P}(\mathbf{r}, t) \cong (\mu/4\pi)\iiint d\mathcal{V}'\, \mathbf{j}(\mathbf{r}', t)/R. \quad [9.54]$$

Knowing the potentials, we may evaluate the fields by using equations [9.51]:

$$\mathbf{E}_{MQ\text{-}P}(\mathbf{r}, t) = -\nabla V_{MQ\text{-}P} - \partial_t\mathbf{A}_{MQ\text{-}P} \cong (1/4\pi\varepsilon)\iiint d\mathcal{V}'\, q_V(\mathbf{r}', t)\mathbf{R}/R^3 - (\mu/4\pi)\iiint d\mathcal{V}'\partial_t\mathbf{j}(\mathbf{r}', t)/R$$

$$\mathbf{B}_{MQ\text{-}P}(\mathbf{r}, t) = \nabla\times\mathbf{A}_{MQ\text{-}P}(\mathbf{r}, t) \cong (\mu/4\pi)\iiint d\mathcal{V}'\, \mathbf{j}(\mathbf{r}', t)\times\mathbf{R}/R^3. \qquad [9.55]$$

Thus $V_{MQ\text{-}P}$, $\mathbf{A}_{MQ\text{-}P}$ and $\mathbf{B}_{MQ\text{-}P}$ are evaluated as in the permanent approximation, while $\mathbf{E}_{MQ\text{-}P}$ is given by the relation $\mathbf{E}_{MQ\text{-}P}(\mathbf{r}, t) = -\nabla V_{MQ\text{-}P}(\mathbf{r}, t) - \partial_t\mathbf{A}_{MQ\text{-}P}(\mathbf{r}, t)$.

2) In some situations, $\partial_t\mathbf{B}$ may be neglected but the displacement current $\partial_t\mathbf{D}$ cannot be neglected compared to $\mathbf{j}$ (as in the case $\mathbf{j} = 0$, for instance). This is the so-called *electrostatic quasi-permanent approximation* (EQ-P). In this case, Maxwell's equations reduce to

$$\nabla.\mathbf{E}_{EQ\text{-}P} = q_v/\varepsilon, \quad \nabla\times\mathbf{E}_{EQ\text{-}P} \cong 0, \quad \nabla.\mathbf{B}_{EQ\text{-}P} = 0, \text{ and } \nabla\times\mathbf{B}_{EQ\text{-}P} = \mu\mathbf{j} + \mu\varepsilon\partial_t\mathbf{E}_{EQ\text{-}P}. \quad [9.56]$$

Taking the divergence of both sides of the last equation and taking into account the first equation, we find the continuity equation $\nabla.\mathbf{j} + \partial_t q_v = 0$. Thus, the continuity of the intensity in the branches of a circuit and Kirchhoff's rule for nodes are not respected. On the other hand, the approximate equation $\nabla\times\mathbf{E}_{EQ\text{-}P} \cong 0$ means that $\mathbf{E}_{EA\text{-}P}$ is approximately conservative. Thus, we have the relations

$$\mathbf{E}_{EQ\text{-}P} \cong -\nabla V_{EQ\text{-}P} \qquad \text{and} \qquad \mathbf{B}_{EQ\text{-}P} = \nabla\times\mathbf{A}_{EQ\text{-}P}. \qquad [9.57]$$

Restricting the gauge transformation to time-independent gauge functions f, i.e. $V'_{EQ\text{-}P} = V_{EQ\text{-}P}$ and $\mathbf{A}'_{EQ\text{-}P} = \mathbf{A}_{EQ\text{-}P} + \nabla f(\mathbf{r})$, we may impose the condition $\nabla.\mathbf{A}_{EQ\text{-}P} = 0$. Then, the fields and the potentials verify the equations

$$\Delta V_{EQ\text{-}P} \cong -\,q_v/\varepsilon_o \quad \text{and} \quad \Delta\mathbf{A}_{EQ\text{-}P} \cong -\,\mu\mathbf{j}_T \ \text{ where } \mathbf{j}_T = \mathbf{j} + \varepsilon\,\partial_t\mathbf{E}_{EQ\text{-}P}, \quad [9.58]$$

$$\Delta\mathbf{E}_{EQ\text{-}P} \cong \nabla(q_v/\varepsilon) \quad \text{and} \quad \Delta\mathbf{B}_{EQ\text{-}P} \cong -\,\nabla\times(\mu\mathbf{j}) \cong -\,\nabla\times(\mu\mathbf{j}_T). \quad [9.59]$$

The potential $V_{\text{EQ-P}}$ and the electric field $\mathbf{E}_{\text{EQ-P}}$ verify the same equations as in the permanent approximation. Thus, they have the same expressions in terms of $q_v(\mathbf{r}, t)$:

$$V_{\text{EQ-P}}(\mathbf{r}, t) \cong (\mu/4\pi\varepsilon) \iiint d\mathcal{V}'\, q_v(\mathbf{r}', t)/R,$$
$$\mathbf{E}_{\text{EQ-P}}(\mathbf{r}, t) \cong -\nabla V_{\text{EQ-P}}(\mathbf{r}, t) \cong (1/4\pi\varepsilon) \iiint d\mathcal{V}'\, q_v(\mathbf{r}', t)\, \mathbf{R}/R^3. \qquad [9.60]$$

Knowing $\mathbf{E}_{\text{EQ-P}}$, we may evaluate $\mathbf{j}_T$, then $\mathbf{A}_{\text{EQ-P}}$ and $\mathbf{B}_{\text{EQ-P}}$ by using the same expressions as in the permanent approximation, but with the current density $\mathbf{j}_T = \mathbf{j} + \varepsilon\partial_t\mathbf{E}_{\text{EQ-P}}$:

$$\mathbf{A}_{\text{EQ-P}}(\mathbf{r}, t) \cong (\mu/4\pi) \iiint d\mathcal{V}'\, \mathbf{j}_T(\mathbf{r}', t)/R,$$
$$\mathbf{B}_{\text{EQ-P}}(\mathbf{r}, t) = \nabla \times \mathbf{A}_{\text{EQ-P}}(\mathbf{r}, t) \cong (\mu/4\pi) \iiint d\mathcal{V}'\, \mathbf{j}_T(\mathbf{r}', t)\times\mathbf{R}/R^3. \qquad [9.61]$$

To know under which conditions these approximations can be used, let us first explain what we mean by "slow variation". If a quantity G varies, we may define the *characteristic time of its variation* by $\tau \sim |G/\partial_t G|$. In the case of the sources q_v and $\mathbf{j}$, τ may be the time required to set them up or to eliminate them. It may be also their period T if they are periodic. The MQ-P approximation is valid if $|\partial_t\mathbf{D}| \ll j$. Thus, it may be used in the case of a conducting medium if the displacement current is small compared to the conduction current (see problem 9.7 and section 10.7). In this case, a time scale of the conductor is its relaxation time $\tau_c = \varepsilon/\sigma$. It is of the order of 10^{-18} s for metals. Thus, the characteristic time of variation of charges and currents must be much longer than τ_c. This is the case for electromagnetic waves of wavelength as short as X-rays. It should be noted that, to apply the MQ-P approximation, it is not necessary that $|\partial_t\mathbf{D}|$ be everywhere negligible, compared to the conduction current j. It is sufficient that its global effect be small. This is the case for electromagnetic set-ups (a solenoid for instance) where there is no accumulation of charge. The EQ-P approximation is valid if $|\partial_t\mathbf{B}| \ll |\nabla\times\mathbf{E}|$. It is not easy to know *a priori* if this condition is satisfied. If τ is the characteristic time of $\mathbf{B}$ and d is a typical distance of the system, this condition may be written as $d/\tau \ll E/B$.

Comparing the approximate quasi-permanent solutions [9.55] or [9.60] with the exact retarded solution [9.41] and [9.42], we find that the quasi-permanent approximations neglect the time delay R/v. Let d be the largest distance of the system (i.e. the distance between the positions of the sources q_v and $\mathbf{j}$ and the distance of these positions to the points, where the fields are evaluated). The variations of the sources in time are considered to be slow if their characteristic time τ_{source} is much longer than the propagation time $\tau = d/c$. This defines another scale of time, which gives a restriction on the validity of the approximations. For instance, if a circuit has a length $d = 3$ m, the propagation time is $\tau \approx 10$ ns. It is much shorter than the period of an alternating current of frequency 1 MHz ($T = 1$ μs) but certainly not 1 GHz ($T = 1$ ns).

9.5. Discontinuities on the interface of two mediums

The interface $\mathcal{S}$ of two mediums may have a charge density q_s and, if at least one of the mediums is a conductor, the interface may carry a surface current density $\mathbf{j}_s$. We assume that there is no point charge or linear current on the interface $\mathcal{S}$ and its close vicinity. Because of the discontinuity of the electric and magnetic properties of the mediums, the fields may have discontinuities; thus, their partial derivatives with respect to space coordinates are not well defined on $\mathcal{S}$. In this case, only the integral forms of Maxwell's equations may be written for the fields on the interface.

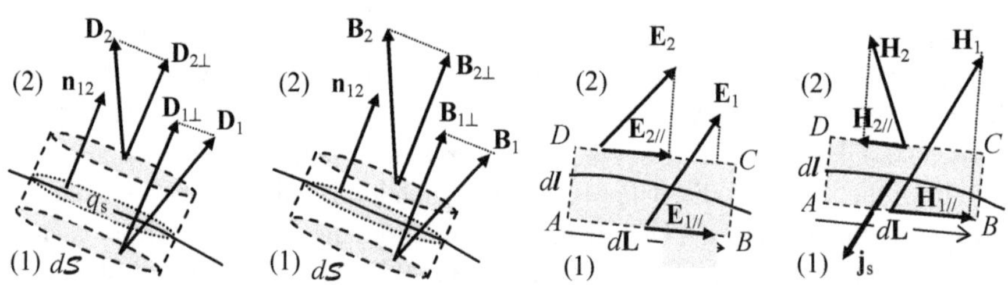

Figure 9.2. *Electromagnetic fields on the interface of two mediums*

Let us apply Gauss's law [9.25] to a thin cylinder of base $d\mathcal{S}$ parallel to the interface (Figure 9.2a). The flux of **D** outgoing from the bases situated on both sides of $\mathcal{S}$ is $(\mathbf{D}_1 - \mathbf{D}_2).\mathbf{n}_{12}\, d\mathcal{S}$, while the flux through the lateral surface tends to zero with the length of the cylinder and so is the volume charge that it contains. If q_s is the surface charge density, the internal charge is $q_s\, d\mathcal{S}$. Thus, the normal component of **D** has a discontinuity on the interface

$$\mathbf{D}_2.\mathbf{n}_{12} - \mathbf{D}_1.\mathbf{n}_{12} = q_s. \qquad [9.62]$$

The same argument may be used for equation [9.27] and the cylinder of Figure 9.2b. It shows that the normal component of **B** is continuous at the interface

$$\mathbf{B}_2.\mathbf{n}_{12} = \mathbf{B}_1.\mathbf{n}_{12}. \qquad [9.63]$$

Let us apply Faraday's induction law [9.26] to the rectangular path *ABCD* of Figure 9.2c. As the width *dl* of the rectangle tends to 0, the circulation of **E** over *BC* and *DA* tends to 0 and its circulation over *AB* and *CD* is the same as that of the tangential component of **E** in the direction of *AB*, i.e. $(\mathbf{E}_{//1} - \mathbf{E}_{//2}).d\mathbf{L}$, for any $d\mathbf{L}$. The flux of **B** through the rectangle tends to 0 with *dl*. We deduce that the tangential component of **E** is continuous on the interface

$$\mathbf{E}_{//1} = \mathbf{E}_{//2}. \quad [9.64]$$

Similarly, let us apply Ampère's law [9.28] to the rectangular path *ABCD* of Figure 9.2d. The circulation of **H** is $(H_{//1} - H_{//2})\, dL$. On the right-hand side, the flux of **D** tends to 0, while the flux of $\mathbf{j}_s$ is $j_s\, dL$ if $d\mathbf{L}$ is perpendicular to $\mathbf{j}_s$ and equal to zero if $d\mathbf{L}$ is parallel to $\mathbf{j}_s$. For any direction of $d\mathbf{L}$, it may be verified that the discontinuity of the tangential component of **H** may be written as

$$(\mathbf{H}_1 - \mathbf{H}_2) \times \mathbf{n}_{12} = \mathbf{j}_s, \quad [9.65]$$

where $\mathbf{n}_{12}$ is the unit vector normal to the interface and pointing from medium (1) toward medium (2).

As for the scalar potential *V* and the vector potential **A**, it may be shown that they are continuous unless the interface has point charges or a linear charge density and linear current densities.

9.6. Electromagnetic energy and Poynting vector

To set up an electromagnetic field, some energy must be supplied. This energy is distributed in space in electric form with a density ½(**E.D**) and in magnetic form with a density ½(**H.B**). Thus, the total electromagnetic energy density is

$$U_{\mathrm{EM,v}} = \tfrac{1}{2}\, \mathbf{E.D} + \tfrac{1}{2}\, \mathbf{H.B}. \quad [9.66]$$

If the fields are time-independent, this energy density remains constant in the course of time. The stored energy in space is initially supplied by the generators, which have emitted the fields. Once these fields are set up, the generators need to supply no more energy if the energy loss as Joule heat or others is negligible. On the contrary, if the fields depend on time, the energy density varies. The conservation of energy means that it is transmitted from one place to another. This transfer of energy is characteristic of the propagation of a wave. We define the Poynting vector as

$$\mathbf{S} = (\mathbf{E} \times \mathbf{H}). \quad [9.67]$$

To interpret this vector, using Maxwell's equations we can write

$$\boldsymbol{\nabla}.\mathbf{S} = \mathbf{H}.(\boldsymbol{\nabla}\times\mathbf{E}) - \mathbf{E}.(\boldsymbol{\nabla}\times\mathbf{H}) = -\,\mathbf{H}.\partial_t\mathbf{B} - \mathbf{E}.\,\partial_t\mathbf{D} - (\mathbf{j}.\mathbf{E}). \quad [9.68]$$

If the medium is homogeneous, linear and isotropic, we have

$$\partial_t U_{\mathrm{EM,v}} = \partial_t(\tfrac{1}{2}\varepsilon \mathbf{E}^2 + \tfrac{1}{2}\mu \mathbf{E}^2) = \varepsilon \mathbf{E}.\partial_t \mathbf{E} + \mu \mathbf{H}.\partial_t \mathbf{H} = \mathbf{E}.\partial_t \mathbf{D} + \mathbf{H}.\partial_t \mathbf{B}. \qquad [9.69]$$

Thus, we may write [9.68] in the form

$$\partial_t U_{\mathrm{EM,v}} + \mathbf{j}.\mathbf{E} = -\boldsymbol{\nabla}.\mathbf{S}. \qquad [9.70]$$

Let us integrate this equation over a volume $\mathcal{V}$ enclosed in a surface $\mathcal{S}$ and transform the volume integral of $\boldsymbol{\nabla}.\mathbf{S}$ into the flux of $\mathbf{S}$ through $\mathcal{S}$; we find the equation

$$\partial_t \iiint_{\mathcal{V}} d\mathcal{V}\, U_{\mathrm{EM,v}} + \iiint_{\mathcal{V}} d\mathcal{V}\, \mathbf{j}.\mathbf{E} = -\iint_{\mathcal{S}} d\mathcal{S}\, \mathbf{n}.\mathbf{S}. \qquad [9.71]$$

The first term is the rate of increase of the stored electromagnetic energy in the volume $\mathcal{V}$. In the second term, we have $\mathbf{j}.\mathbf{E} = q_v\mathbf{v}.\mathbf{E} = \mathbf{f}_v.\mathbf{v}$ where $\mathbf{f}_v = q_v\mathbf{E}$ is the density of electric force acting on the charges of density q_v and velocity $\mathbf{v}$. Thus, the quantity $\mathbf{j}.\mathbf{E}$ represents the work of the electric field on the charges per unit volume and per unit time. If the charges are free, they are accelerated, and this work is transformed into kinetic energy. If the medium is a conductor and the velocity of the charges remains constant, this work is dissipated as Joule heat. Thus, the second term represents the total energy supplied to the charges in the volume $\mathcal{V}$. From the principle of conservation of energy, the right-hand side must represent the energy flux inward through the surface $\mathcal{S}$. Considering an element of area $d\mathcal{S}$ normal to the unit vector $\mathbf{n}$, the flux of electromagnetic energy through $d\mathcal{S}$ in the direction of $\mathbf{n}$ in the time interval dt is

$$dU_{\mathrm{EM}} = (\mathbf{S}.\mathbf{n})\, dt\, dS. \qquad [9.72]$$

Thus, the flux of the Poynting vector $\mathbf{S}$ through a surface is the rate of flow of the radiation energy.

9.7. Electromagnetic pressure, Maxwell's tensor

Consider a volume $\mathcal{V}$ enclosed in a surface $\mathcal{S}$ and containing charges of density q_v moving with a velocity $\mathbf{v}$. The current density is $\mathbf{j} = q_v\mathbf{v}$ and the Lorentz force per unit volume of this medium is

$$\mathbf{f}_v = q_v(\mathbf{E} + \mathbf{v} \times \mathbf{B}) = q_v\mathbf{E} + \mathbf{j} \times \mathbf{B}. \qquad [9.73]$$

Using Maxwell's equations [9.12] and [9.15] to express q_v and $\mathbf{j}$ in terms of the fields, we may write

$$\mathbf{f}_v = \mathbf{E}(\boldsymbol{\nabla}.\mathbf{D}) + (\boldsymbol{\nabla} \times \mathbf{H}) \times \mathbf{B} - \dot{\mathbf{D}} \times \mathbf{B} + [(\mathbf{D}.\mathbf{B})\,\mathbf{H} + (\boldsymbol{\nabla} \times \mathbf{E}) \times \mathbf{D} + \dot{\mathbf{B}} \times \mathbf{D}],$$

where the expression in the square brackets is equal to zero, as it may be shown by using Maxwell's equations; it was added to obtain a symmetric expression. Thus, we may write in the case of a linear medium

$$\mathbf{f}_v = -\varepsilon\, \partial_t(\mathbf{E}\times\mathbf{B}) + \varepsilon[(\nabla.\mathbf{E})\mathbf{E} + (\nabla\times\mathbf{E})\times\mathbf{E}] + (1/\mu)[(\nabla.\mathbf{B})\,\mathbf{B} + (\nabla\times\mathbf{B})\times\mathbf{B}]. \quad [9.74]$$

We may write the components of $\mathbf{f}_v$ in the form

$$f_{v\,\alpha} = -\,\varepsilon\mu\, \partial_t S_\alpha + \Sigma_\beta\, \partial_\beta \tau_{\alpha\beta}, \quad [9.75]$$

where the S_α are the components of the Poynting vector and $\tau_{\alpha\beta}$ is *Maxwell's tensor*, whose components are

$$\tau_{\alpha\beta} = \varepsilon E_\alpha E_\beta + (1/\mu)\, B_\alpha B_\beta - \delta_{\alpha\beta}\, U_{EM,v}. \quad [9.76]$$

To interpret equation [9.75], we integrate it over a volume $\mathcal{V}$ enclosed in a surface $\mathcal{S}$ and convert the volume integral of the divergence of $\tau_{\alpha\beta}$ into a flux; we find

$$\iiint_{\mathcal{V}} d\mathcal{V}\, f_{v\,\alpha} + \varepsilon\mu\, (d/dt) \iiint_{\mathcal{V}} d\mathcal{V}\, S_\alpha = \iint_{\mathcal{S}} d\mathcal{S}\, \Sigma_\beta\, \tau_{\alpha\beta}\, n_\beta. \quad [9.77]$$

The first term of this equation is the total force acting on the charges in the volume $\mathcal{V}$. According to the fundamental principle of dynamics, it is equal to the rate of variation of the total momentum of matter $d\mathbf{P}_{mat}/dt$ in this volume. The second term may be interpreted as $d\mathrm{P}_{rad}/dt$ where $\mathbf{P}_{rad}$ is the momentum of the radiation contained in this volume. The radiation momentum density is

$$\mathbf{P}_{rad,v} = \varepsilon\mu\, \mathbf{S} = \mathbf{S}/v^2. \quad [9.78]$$

The right-hand side of [9.77] must be interpreted as a radiation pressure on the surface $\mathcal{S}$. Thus, Maxwell's tensor $\tau_{\alpha\beta}$ is the *electromagnetic radiation pressure tensor* similar to the mechanical pressure tensor in fluids. The pressure exerted by radiation on an element of area $d\mathcal{S}$ normal to the unit vector $\mathbf{n}$ is

$$df_{pressure,\,\alpha} = d\mathcal{S}\, \Sigma_\beta\, n_\beta\, \tau_{\alpha\beta}. \quad [9.79]$$

This allows us to write equation [9.77] in the form

$$(d/dt)\,(\mathbf{P}_{mat} + \mathbf{P}_{rad}) = \mathbf{f}_{pressure}, \quad \text{where } \mathbf{f}_{pressure,\,\alpha} = \iint_{\mathcal{S}} d\mathcal{S}\, \Sigma_\beta\, n_\beta\, \tau_{\alpha\beta}. \quad [9.80]$$

The electromagnetic field is similar to a fluid with an energy density, a momentum density (and, consequently, a density of angular momentum) and a pressure tensor, which are defined at each point in the field. When we write the laws of variation of these physical quantities, we must take into account the corresponding quantities of the field exactly as for matter.

9.8. Problems

Maxwell's equations

P9.1 a) A dielectric of polarization **P** occupies a volume $\mathcal{V}$. Admitting that the polarization charge density is $q'_{v,} = -\nabla.\mathbf{P}$, show that the total polarization charge inside a surface $\mathcal{S}$ may be written as $Q' = -\iint_{\mathcal{S}} d\mathcal{S}\, \mathbf{n}.\mathbf{P}$. If **P** is time-dependent, show that the continuity equation of the polarization charge requires the existence of a polarization current density $\mathbf{j}_p = \partial_t\mathbf{P}$. **b)** Admitting that the magnetic field of matter is the same as that of a magnetization current $\mathbf{j}_m = \nabla\times\mathbf{M}$, show that the "magnetic charge" must be equal to zero. **c)** Show that the density of total charge is $q_{v,T} = q_v - \nabla.\mathbf{P}$ and that the total current density is $\mathbf{j}_T = \mathbf{j} + \partial_t\mathbf{P} + \nabla\times\mathbf{M}$. Verify the continuity equation $\nabla.\mathbf{j}_T + \partial_t q_{v,T} = 0$ and that the Maxwell's equations may be written as

$$\nabla.\mathbf{E} = q_{v,T}/\varepsilon_o, \quad \nabla.\mathbf{B} = 0, \quad \nabla\times\mathbf{E} + \partial\mathbf{B}/\partial t = 0, \quad \nabla\times\mathbf{B} - (1/c^2)(\partial\mathbf{E}/\partial t) = \mu_o\mathbf{j}_T.$$

P9.2 a) Is it possible to have the electric field $\mathbf{E} = 10^3[(3x + y)\,\mathbf{e}_x + x\,\mathbf{e}_y + z\,\mathbf{e}_z]$ in a region that is empty of matter, electric charge, electric current, and magnetic field? If not, determine q_v, **j**, and **B**. **b)** A magnetic field $\mathbf{B} = (B_o + bt)\,\mathbf{e}_z$, where B_o and b are constants, is set up in a cylindrical region about Oz. This region is empty of charges and currents. Is it possible to determine **E**?

P9.3 Using Maxwell-Faraday's equation, show that $\nabla.\mathbf{B}$ does not depend on time everywhere. Deduce that, if the magnetic field is set up starting with a state of zero fields, the equation $\nabla.\mathbf{B} = 0$ remains valid everywhere in space. Similarly, using the Maxwell-Ampère equation, $\nabla\times\mathbf{H} - (1/c^2)\partial_t\mathbf{E} = \mu_o\mathbf{j}$, show that $\nabla.\mathbf{E} = q_v/\varepsilon_o$.

P9.4 a) A ball of radius a has a charge q uniformly distributed and independent of time. Write Maxwell's equations. Use the symmetries and determine the fields outside the ball. **b)** A cylinder of radius a carries a time-dependent current $I(t)$ uniformly distributed on the sections. Write Maxwell's equations. Discuss the directions of **E**, **B**, and **A**. Write the expressions of the retarded potentials and evaluate them in the limit of a very narrow cylinder (a rectilinear thin wire) by taking $\mathbf{j}(\mathbf{r}', t') = [I(t')/\pi a^2]\,\delta(x')\,\delta(y')\,\mathbf{e}_z$, where $\delta(x')$ and $\delta(y')$ are the Dirac delta functions. **c)** Consider the case of a constant current and the case of a constant current starting at $t = 0$.

P9.5 Consider a beam of section $\mathcal{S}$ and containing N_v particles per unit volume moving with a velocity **v**. **a)** Calculate the fields **E** and **B** outside this beam. **b)** Consider the Galilean transformation $\mathbf{r} = \mathbf{r}' + \mathbf{v}_o t$ of velocity $\mathbf{v}_o$ in the direction of the beam. Verify that Maxwell's equations are not covariant in the transformation $\mathbf{B}'(\mathbf{r}') = \mathbf{B}(\mathbf{r})$ and $\mathbf{E}'(\mathbf{r}') = \mathbf{E}(\mathbf{r}) + \mathbf{v}_o \times \mathbf{B}(\mathbf{r})$, that is required to have the invariance of the Lorentz force (see section 8.4). To simplify, consider the case $v_o = v$.

Electromagnetic potentials

P9.6 a) The potential of a stationary charge distribution of volume density $q_v(\mathbf{r})$ is

$$V(\mathbf{r}) = (1/4\pi\varepsilon_o) \iiint d\mathcal{V}' q_v(\mathbf{r}')/|\mathbf{r} - \mathbf{r}'|.$$

Admitting the relation $\Delta(1/|\mathbf{r} - \mathbf{r}'|) = -4\pi\ \delta(\mathbf{r} - \mathbf{r}')$ where $\delta(\mathbf{r} - \mathbf{r}')$ is the three-dimensional Dirac delta function (see section A.11 of the appendix A), show that $V(\mathbf{r})$ is a solution of Poisson's equation $\Delta V = -q_v/\varepsilon$. **b)** Verify that the retarded potentials [9.41] and [9.42] are solutions of the propagation equations [9.40].

Quasi-permanent approximations

P9.7 a) Using the MQ-P approximation, calculate the fields and the potentials in a solenoid carrying a variable current $I(t)$. **b)** Using the EQ-P, evaluate the fields and the potentials in a capacitor whose plates are disks of radius R, separated by a distance $d \ll R$ if the charge $q(t)$ depends on time. **c)** A metallic cylinder of conductivity σ carries a current of uniform density $\mathbf{j}(t)$. Show that an eventual charge density in this cylinder decreases according to $q_v = q_{vo}\ e^{-t/\tau_c}$ where $\tau_c = \varepsilon/\sigma$ is a characteristic time of the metal and that the current density must verify the condition that $\mathbf{j} + \tau_c\ \partial_t\mathbf{j}$ is independent of time. Write the expressions of the fields and the potentials in this case.

P9.8 Calculate the electromagnetic fields of a particle of charge q oscillating on the z axis with an angular frequency ω in the MQ-P approximation.

P9.9 An almost point-like nucleus isotropically emits electrons of velocity v_e. Assume that the nucleus starts to emit with an initial charge Q_o at $t = 0$. Let $Q(t)$ be the charge of the nucleus at time t. **a)** Determine the charge density q_v and the current density $\mathbf{j}$ everywhere and at an arbitrary time t. **b)** Determine the potentials of this charge and current distributions in the MQ-P approximation.

Electromagnetic energy and radiation pressure

P9.10 Verify that the equation of conservation of energy [9.70] is approximately verified by the MQ-P approximate solutions for the solenoid and the capacitor of problem 9.7 and exactly verified by the solution for the conducting cylinder.

P9.11 A cylindrical conductor of length L, radius a, and conductivity σ carries a current of variable density $j(t)$. Calculate the Poynting vector inside the cylinder at a distance r from the axis; what is its direction? How does the energy propagate inside the cylinder? Calculate the power that enters into this cylinder, the dissipated power as Joule heat and the stored energy. Verify the conservation of energy.

P9.12 Using Maxwell's tensor, derive the expression of the electrostatic pressure on the surface of a conductor having a constant surface charge density q_s and the expression of the magnetic pressure on the surface of a cylinder carrying a constant current density $\mathbf{j}$.

Chapter 10

Electromagnetic Waves

Maxwell's equations couple the electric field **E** and the magnetic field **B** in a single physical entity, called the *electromagnetic field.* If one of the fields varies, necessarily, the other field is induced. Similar to the relation between displacement and pressure, which are responsible for the propagation of sound waves, the coupling between **E** and **B** is responsible for the propagation of electromagnetic waves in vacuum and in matter at the speed of light. It was not possible to foresee this remarkable effect before the formulation of Maxwell's equations and electromagnetic theory. In 1884, Hertz confirmed the existence of these waves experimentally. He produced them by discharging two spheres at high potential mounted as an electric dipole. He verified that these waves propagate, interfere, diffract, and are polarized, exactly as light waves are. Today, we can produce electromagnetic waves of almost all frequencies from 10^{-2} Hz to 10^{32} Hz. They play a fundamental part in telecommunications (radio, television, radar, etc.), in medicine (X-rays, gamma rays, laser), in industry, etc.

Electromagnetic waves are emitted by variable currents in the *emitters* and they are detected by *receivers*, in which they induce currents. They are specified by the fields **E**, **D**, **B** and **H**. They may be polarized and they carry energy, momentum, and other physical quantities. Their propagation properties depend on the medium.

In this chapter, after a brief mathematical review of waves, we analyze the propagation of electromagnetic waves in vacuum, dielectrics, conductors, and plasmas, and briefly study their quantization and emission.

10.1. A short review on waves

A) One-dimensional wave equation, progressive waves

In classical physics, a wave is a mechanical or electromagnetic disturbance that propagates in a medium without transfer of matter. In modern physics, it may be any particle in motion (photons, phonons, electrons, protons, neutrinos, etc.). In general, a wave is specified by a scalar or a vector *wave function* u, which depends on space coordinates and time. Let us consider first a wave $u(x, t)$, which depends on one space dimension and time. This is the case of the displacement wave on a string for instance. In the ideal case of propagation without deformation, the disturbance at the origin $u_O = f(t)$ generates, at the point of coordinate x, the same disturbance but with a time delay x/v, that is, $u = f(t - x/v)$. The wave function u verifies the *equation of propagation*, called *d'Alembert's equation*

$$\partial^2_{xx}u(x, t) - \frac{1}{v^2}\partial^2_{tt}u(x, t) = 0. \qquad [10.1]$$

This linear, second-order partial differential equation describes the propagation of many types of wave in an ideal medium that does not dissipate energy and deform the wave. The velocity, v, which appears in this equation, is the *speed of propagation.* It is easy to verify that the d'Alembert's equation admits the solutions

$$u^{(+)} = f(t - \frac{x}{v}) \qquad \text{and} \qquad u^{(-)} = g(t + \frac{x}{v}). \qquad [10.2]$$

$u^{(+)}(x, t)$ propagates toward the positive x while $u^{(-)}(x, t)$ propagates toward the negative x with the speed v. The waves, whose physical quantities propagate from one place to another in an infinite medium, are said to be *progressive.* As the wave equation [10.1] is linear, it obeys the superposition principle. Particularly, the solutions $u^{(+)}$ and $u^{(-)}$ may be superposed to have solutions of the form

$$u(x, t) = f(t - \frac{x}{v}) + g(t + \frac{x}{v}). \qquad [10.3]$$

B) Simple harmonic progressive waves

A particular wave, which plays an important part in wave analysis, is the *simple harmonic* (or *sinusoidal*) wave, which is generated by a simple harmonic vibration of the origin $f(t) = A\cos(\omega t + \phi)$. This is obviously a mathematical concept, similar to the concept of point particle or point charge. If the wave propagates toward the positive x, for instance, the vibration at x is

$$u^{(+)}(x, t) = A\cos[\omega(t - \frac{x}{v}) + \phi]. \qquad [10.4]$$

ω is the *angular frequency* of the wave, A is its *amplitude*, and ϕ is its *phase*. It is always possible to choose ϕ $(-\pi < \phi \leq \pi)$ such that the amplitude A be positive (unless the phase varies). $u^{(+)}$ may also be written in one of the equivalent forms

$$u^{(+)}(x, t) = A \cos[\, 2\pi\tilde{\nu}(t - \frac{x}{v}) + \phi] = A \cos(\omega t - kx + \phi) = A \cos[2\pi(\frac{t}{T} - \frac{x}{\lambda}) + \phi], \quad [10.5]$$

$\tilde{\nu}$ is the *frequency*, T is the *period*, k is the *wave number*, and λ is the *wavelength*. These quantities are related by the equations

$$\tilde{\nu} = \omega/2\pi, \quad T = 1/\tilde{\nu} = 2\pi/\omega, \quad k = \omega/v, \quad \lambda = 2\pi/k = vT = v/\tilde{\nu}. \quad [10.6]$$

The expressions [10.5] show that the wave at x is the same as at the origin but with a *time delay* $\Delta t = x/v$. Considering $(\phi - \omega x/v)$ as the phase of the wave at x, we may also say that a travel of a distance x corresponds to a *phase lag*

$$\Delta\phi = \omega x/v = kx = 2\pi x/\lambda. \quad [10.7]$$

The wave is the same at points, where the phase differs by $2p\pi$ with p equal to an integer. These points are separated by a distance equal to $p\lambda$, where p is an integer. The wavelength λ is the distance that the wave travels in a period T.

It is often convenient to use the complex representation

$$\underline{u}^{(+)} = A\ e^{i(\omega t - kx + \alpha)} = \underline{A}\, e^{i(\omega t - kx)}, \qquad \underline{u}^{(-)} = B\, e^{i(\omega t + kx + \beta)} = \underline{B}\, e^{i(\omega t + kx)}, \quad [10.8]$$

where $\underline{A} = A\, e^{i\alpha}$ and $\underline{B} = B\, e^{i\beta}$ are the *complex amplitudes*. The real parts are taken at the end of the calculation if necessary.

C) Three-dimensional waves

The preceding considerations may easily be generalized to three-dimensional waves. Their d'Alembert's wave equation may be written as

$$\Delta u(x, y, z, t) - (1/v^2)\, \partial^2_{tt} u(x, y, z, t) = 0. \quad [10.9]$$

This equation has a solution of the form

$$u(\mathbf{r}, t) = f(t - \mathbf{e.r}/v) \equiv f[t - (\alpha x + \beta y + \gamma z)/v], \quad [10.10]$$

where $\mathbf{e}$ is a unit vector of components α, β and γ. The points $\mathbf{r}$ such that $(\mathbf{r.e}) = d$, where d is a constant, correspond to the same value of the wave function. These points form a plane (P) normal to $\mathbf{e}$ and situated at a distance d from the origin (Figure 10.1a). Particularly, the wave function at the points of the plane (P_o) perpendicular to $\mathbf{e}$ and containing the origin O is $u(0, t) = f(t)$. The wave at the points of (P) is thus the same as at the points of (P_o) but with a time delay d/v. The plane (P) is a *wave front* and [10.10] is a *plane wave*, which propagates in the direction **e** with a *phase velocity* v. A simple harmonic wave may be written in the real or complex forms:

$$u(\mathbf{r}, t) = A \cos[\omega(t - \mathbf{e}.\mathbf{r}/v) + \phi] = A \cos(\omega t - \mathbf{k}.\mathbf{r} + \phi), \qquad [10.11]$$

$$\underline{u}(\mathbf{r}, t) = \underline{A}\, e^{\mathrm{i}\omega(t-\mathbf{e}.\mathbf{r}/v)} = \underline{A}\, e^{\mathrm{i}(\omega t-\mathbf{k}.\mathbf{r})}. \qquad [10.12]$$

k is the *wave vector*; it points in the *direction of propagation* **e**. Its magnitude is equal to the *wave number* $k = 2\pi/\lambda$. The complex amplitude $\underline{A} = A\, e^{\mathrm{i}\phi}$ allows the combination of the real amplitude A of the wave and its phase ϕ in a single symbol. The period and the wavelength are defined by the same relations [10.6] as for the one-dimensional wave. We note, in particular, that the phase velocity is

$$v_{(\mathrm{p})} = \omega/k. \qquad [10.13]$$

The expressions [10.11] and [10.12] are solutions of the propagation equation [10.9] if the phase velocity $v_{(\mathrm{p})}$ is equal to v. This is true in the case of an infinite medium that we consider in this chapter. In the case of a bounded medium (as in the case of a waveguide), the wave must obey some conditions on the boundary surfaces and this modifies the propagation properties. Particularly, the phase velocity $v_{(\mathrm{p})}$ is no longer equal to the speed of propagation v and it may depend on the frequency of the wave. On the other hand, the properties of propagation depend on the interaction of the wave with the atoms of the medium and this interaction depends on the frequency. Thus, the phase velocity $v_{(\mathrm{p})}$ depends on the frequency, even if the medium is infinite. We say then that the medium is *dispersive* and the relation between k and ω, called *dispersion relation*

$$\omega = \omega(\mathbf{k}) \qquad [10.14]$$

is nonlinear. In spherical coordinates, ω is a function of the magnitude of **k** and the angles θ_k and ϕ_k. The phase velocity is then $v_{(\mathrm{p})}(k, \theta_k, \phi_k) = \omega(k, \theta_k, \phi_k)/k$. This relation expresses the phase velocity as a function of the wave number ($k = 2\pi/\lambda$) and the direction of propagation. If the medium is *isotropic*, the function $\omega(\mathbf{k})$ depends on the magnitude of **k** and not its direction. The phase velocity is then

$$v_{(\mathrm{p})} = \omega(k)/k. \qquad [10.15]$$

It does not depend on the direction of propagation. As $\omega = 2\pi\tilde{\nu}$ and $k = 2\pi/\lambda$, the phase velocity may be written as a function of the frequency or as a function of the wavelength (as is common in optics).

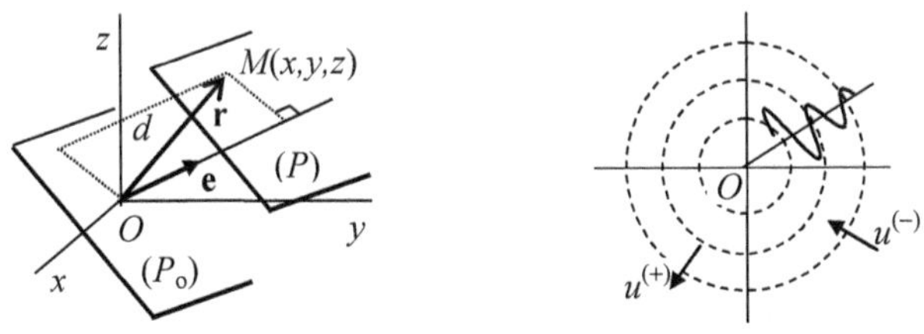

(a) (b)

Figure 10.1. *a) Three-dimensional plane wave and b) spherical wave*

D) Spherical waves

The d'Alembert's wave equation admits spherical wave solutions. These are the waves that are emitted by a point-like source or the waves that converge at a point called a *focus*. It is convenient in this case to use spherical coordinates about the source or the focus. If the wave function is isotropic, it depends only on r and t. Using the expression of the Laplacian in spherical coordinates and the fact that u does not depend on θ and φ, the equation of propagation may be written as

$$\partial^2_{rr}u + (2/r)\,\partial_r u - (1/v^2)\,\partial^2_{tt}u = 0, \qquad [10.16]$$

which admits progressive solutions of the form

$$u^{(+)} = (1/r)\,f(t - r/v), \qquad u^{(-)} = (1/r)\,g(t + r/v). \qquad [10.17]$$

$u^{(+)}$ represents a wave that is emitted from the origin (Figure 10.1b); it propagates with the speed v and it decreases like $1/r$ with the travelled distance. $u^{(-)}$ represents a wave that converges at O and increases like $1/r$ as it approaches O. We may also have a superposition of the two waves $u^{(+)}$ and $u^{(-)}$.

The study of spherical electromagnetic waves, which are transverse vector waves, began with Hansen in 1935 by analyzing antenna emission. It is quite complicated and it will not be considered further in this book.

E) Superposition of harmonic waves, Fourier analysis

As the equation of propagation is linear, it allows the superposition of solutions. The concept of simple harmonic waves is important in physics as, according to Fourier theorem, any function or wave may be written as a superposition of simple harmonic functions or waves. If a wave u, which propagates in the x direction, has a period T, it may be written as a superposition of simple harmonic waves of periods

T, $T/2$, $T/3$, ..., (i.e. angular frequencies ω, 2ω, 3ω, etc.) and wave numbers k, $2k$, $3k$,... where $\omega = 2\pi/T$ and $k = \omega/v$

$$u(x, t) = \tfrac{1}{2}\, a_o + \Sigma_{n\geq 1} A_n \cos(n\omega t - nkx + \phi_n)$$
$$\text{or} \quad \tfrac{1}{2}\, a_o + \Sigma_{n\geq 1} [a_n \cos(n\omega t - nkx) + b_n \sin(n\omega t - nkx)]. \qquad [10.18]$$

This is the *Fourier series* for the wave u. Its terms are, respectively, the *constant term, the first harmonic*, the *second harmonic*, etc. of angular frequencies 0, ω, 2ω, etc. The *Fourier coefficients* a_n are given by

$$a_n = (2/T) \int_T dt\, u(\mathbf{r}, t) \cos(n\omega t - nkx), \qquad b_n = (2/T)\int_T dt\, u(\mathbf{r}, t) \cos(n\omega t - nkx). \quad [10.19]$$

In the case of an aperiodic wave (a signal, for instance), which propagates in the x direction, the Fourier series must be replaced by a *Fourier integral*

$$u(\mathbf{r}, t) = \int_{\Delta\omega} d\omega\, [a(\omega) \cos(\omega t - kx) + b(\omega) \sin(\omega t - kx)], \qquad [10.20]$$

where $k = \omega/v$. The function $a(\omega)$ is the *cosine part* of the spectral function and $b(\omega)$ is its *sine part*. The integration is over a domain of angular frequency $\Delta\omega \equiv [\omega_1, \omega_2]$ that may be formally extended to all frequencies $[0, \infty]$ by taking $a(\omega)$ and $b(\omega)$ equal to zero outside $\Delta\omega$. We may also extend them to negative angular frequencies and use k instead of ω, then

$$u(x, t) = \int_{-\infty}^{\infty} dk\, [A(k) \cos(\omega t - kx) + B(k) \sin(\omega t - kx)$$
$$= (2\pi)^{-1/2} \int_{-\infty}^{\infty} dk\, \mathcal{U}(k)\, e^{i(\omega t - kr)}. \qquad [10.21]$$

Thus, a wave u that is aperiodic in time (i.e. a *signal*) may be considered as a superposition of simple harmonic waves, whose frequency takes continuous values in a certain band $\Delta\omega$. It may be shown that, at each point x, u has a duration in time Δt such that $\Delta\omega.\Delta t \approx 2\pi$ and, at a given time t, it takes non-negligible values in a space interval given by $\Delta x = v\, \Delta t \approx 2\pi v/\Delta\omega = 2\pi/\Delta k$. We deduce that $\Delta x.\Delta k \approx 2\pi$.

These considerations may be generalized to three-dimensional waves. They may be considered as superpositions of waves in the various directions. Thus, it is convenient to use the wave vector $\mathbf{k}$ instead of k or ω as integration variable and write for complex waves

$$\underline{u}(\mathbf{r}, t) = (2\pi)^{-3/2} \iiint d^3\mathbf{k}\, \underline{\mathcal{U}}(\mathbf{k})\, e^{i(\omega t - \mathbf{k}.\mathbf{r})}, \qquad \underline{\mathcal{U}}(\mathbf{k}) = (2\pi)^{-3/2} \iiint d^3\mathbf{r}\, \underline{u}(\mathbf{r}, t)\, e^{-i(\omega t - \mathbf{k}.\mathbf{r})}. \quad [10.22]$$

Particularly, a real wave may be written as

$$u(\mathbf{r}, t) = (2\pi)^{-3/2} \iiint d^3\mathbf{k}\; e^{-i\mathbf{k}.\mathbf{r}}\, [\underline{\mathcal{U}}(\mathbf{k})\; e^{i\omega t} + \underline{\mathcal{U}}^*(-\mathbf{k})\, e^{-i\omega t}\,]. \qquad [10.23]$$

If a signal $\underline{u}$ has an extension in space Δx, Δy, and Δz, the wave vector **k** varies in a domain Δk_x, Δk_y, and Δk_z near a certain vector $\mathbf{k}_o$, the angular frequency varies in a band $\Delta\omega$ and the signal has a time duration Δt with the *uncertainty relations*

$$\Delta x.\Delta k_x \approx 2\pi, \quad \Delta y.\Delta k_y \approx 2\pi, \quad \Delta z.\Delta k_z \approx 2\pi, \quad \text{and} \quad \Delta\omega.\Delta t \approx 2\pi. \quad [10.24]$$

F) Dispersion

If the medium of propagation is dispersive or if the wave is guided, each spectral component of angular frequency ω propagates with the corresponding phase velocity. Thus, the signal is deformed as it propagates (it spreads in space and time; see problem 10.2).

Equations [10.24] show that an exactly monochromatic wave (i.e. $\Delta\omega = 0$) cannot be emitted during a finite interval of time Δt and occupy a finite region of space. As the energy density is proportional to $|u|^2$ the emission of such a monochromatic wave requires an infinite amount of energy; thus, it is impossible. A wave of finite duration, and which occupies a finite region of space, is a *wave packet* of band width $\Delta\omega$. In the case of light emitted by an atom undergoing a transition from an excited state to the ground state, the excited state is characterized by a mean lifetime τ. Thus, the emitted light has a minimum band width $\Delta\omega \approx 2\pi/\tau$ and it is widened by the Doppler effect due to the thermal agitation of atoms and the fact that light is emitted by a multitude of atoms that emit spontaneously in an uncoordinated way. The duration of a wave packet $\Delta t = 2\pi/\Delta\omega$ is called *coherence time* and the space extension of the wave packet $\Delta x = v\,\Delta t$ is called *coherence length.* In the case of laser light, the atoms are stimulated by the wave itself; thus, they emit in a coordinated way and the wave is quite coherent. The analysis of the superposition of waves is important, especially in studying interference and diffraction.

Some mediums, such as crystals, are anisotropic. Then, the phase velocity (i.e. the velocity of the wave front of a simple harmonic wave) depends on the frequency and the direction of propagation. The dispersion relation of these mediums may be written as $\omega = \omega(\mathbf{k})$ and the phase velocity is $v_{(p)}(\mathbf{k}) = \omega(\mathbf{k})/k$ in the direction of **k**. The center of a wave packet moves in this medium with a velocity, called *group velocity* $\mathbf{v}_{(g)}$. In the case of a one-dimensional wave, it may be shown that the group velocity is given by

$$v_{(g)} = d\omega/dk. \quad [10.25]$$

In the case of a three-dimensional wave, writing the wave function as a Fourier integral [10.22], $u(\mathbf{r}, t)$ has significant values if the phase $\varphi = \omega t - \mathbf{k.r}$ varies very little near $\mathbf{k}_o$. Otherwise, the real part and the imaginary part of the exponential oscillate rapidly between -1 and $+1$ and the integral is negligible. Thus, at a given time t, the wave is concentrated near a point $\mathbf{r}$ which verifies the condition $\partial\varphi/\partial k_x \approx 0$, $\partial\varphi/\partial k_y \approx 0$ and $\partial\varphi/\partial k_z \approx 0$, i.e. $x \approx (\partial\omega/\partial k_x)t$, $y \approx (\partial\omega/\partial k_y)t$ and $z \approx (\partial\omega/\partial k_z)t$. This shows that the center of the packet moves with a velocity equal to the *group velocity*

$$\mathbf{v}_{(g)} = \nabla_k\omega \equiv (d\omega/dk_x)\,\mathbf{e}_x + (d\omega/dk_y)\,\mathbf{e}_y + (d\omega/dk_z)\,\mathbf{e}_z. \quad [10.26]$$

If the propagation medium is infinite and non-dispersive, the speed of propagation, the phase velocity and the group velocity are equal ($v_{(p)} = v_{(g)} = v$). On the contrary, if the medium is bounded or dispersive, the three velocities may not be equal. The group velocity is effectively the velocity of signals and of all the physical quantities attached to the signal, such as energy, momentum, etc.

To transmit information (sound, image, etc.), a simple harmonic wave is not very useful; it must be modulated by varying its amplitude, frequency, or phase according to the information to be transmitted. The emitted wave is then a superposition of simple harmonic waves in a certain band. Each of these spectral components propagates with its proper velocity. This causes a deformation of the signal, while propagating with the group velocity.

G) Standing waves

Let us consider the one-dimensional equation of propagation [10.1]. Using the *method of separation of variables*, we look to solutions of the form

$$u(x, t) = f(x)\,g(t). \quad [10.27]$$

Substituting this expression in equation [10.1] and dividing by $f(x)\,g(t)$, we obtain

$$(v^2/f)\,\partial^2_{xx}f = (1/g)\,\partial^2_{tt}g. \quad [10.28]$$

The left-hand side is a function of x and the right-hand side is a function of t. The equation may be identically verified (i.e. for any x and t) only if both sides are equal to a constant C, hence

$$\partial^2_{tt}g - Cg = 0, \qquad \partial^2_{xx}f - Cf/v^2 = 0. \quad [10.29]$$

The mathematical form of the solution depends on the sign of the constant C.

– If C is negative, we set $C = -\omega^2$ and $k = \omega/v$. The solutions of [10.29] may be written as $g = A\cos(\omega t + \alpha)$ and $f = B\cos(kx + \beta)$, hence

$$u = A\cos(\omega t + \alpha)\cos(kx + \beta). \quad [10.30]$$

This is a *standing wave*, which may exist notably in a bounded medium. The integration constants A, α and β, as well as the frequency ω, are determined by the vibration of the source and the boundary conditions of the medium of propagation. Then, the wave number k is determined by the dispersion relation of the medium.

– If C is positive, we set $C = \delta^2 v^2$. The solution of equations [10.29] may be written as $g = A\,e^{\delta vt} + B\,e^{-\delta vt}$ and $f = P\,e^{\delta x} + Q\,e^{-\delta x}$, hence

$$u = (P\,e^{\delta x} + Q\,e^{-\delta x})(A\,e^{\delta vt} + B\,e^{-\delta vt}). \qquad [10.31]$$

This is an *exponential wave*. The constants A, B, P and Q, as well as δ, are determined by the source and the boundary conditions of the medium. The exponential waves cannot be stationary and will not be considered in this book.

10.2. Electromagnetic waves in infinite vacuum and dielectrics

A) Equation of propagation of the fields and the potentials

We consider the electromagnetic fields in infinite vacuum or in an infinite linear, isotropic and homogeneous dielectric of permittivity ε and permeability μ. In the absence of charge and currents, Maxwell's equations are

$$\nabla.\mathbf{E} = 0, \qquad \nabla \times \mathbf{E} + \partial_t \mathbf{B} = 0, \qquad [10.32]$$

$$\nabla.\mathbf{B} = 0, \qquad \nabla \times \mathbf{B} - \varepsilon\mu\, \partial_t \mathbf{E} = 0. \qquad [10.33]$$

We have shown in section 9.2c the equations of propagation of the fields

$$\Delta\mathbf{E} - \mu\varepsilon\, \partial^2_{tt}\mathbf{E} = 0, \qquad \Delta\mathbf{B} - \mu\varepsilon\, \partial^2_{tt}\mathbf{B} = 0. \qquad [10.34]$$

The speed of propagation is $v = 1/\sqrt{\mu\varepsilon}$ in the medium and $c = 1/\sqrt{\mu_o\varepsilon_o}$ in vacuum.

It is evident that equations [10.34] do not contain all the information of the four Maxwell's equations. Indeed, equations [10.34] express no relation between **E** and **B**. Thus, to [10.34] we must add one of the Maxwell-Faraday or Maxwell-Ampère equations and one of the equations $\nabla.\mathbf{E} = 0$ and $\nabla.\mathbf{B} = 0$. On the other hand, the wave equations [10.34] are partial differential equations with an infinite number of solutions. To determine the solution corresponding to a given physical situation, we need the initial conditions and eventually the boundary conditions on the surfaces of the medium.

It is possible to specify the wave by the potentials V and **A** (see section 9.3) such that

$$\mathbf{E} = -\nabla V - \partial_t \mathbf{A}, \qquad \mathbf{B} = \nabla \times \mathbf{A}, \qquad \text{and } \nabla.\mathbf{A} + \varepsilon\mu\, \partial_t V = 0. \qquad [10.35]$$

The potentials obey d'Alembert's equation of propagation similar to [10.34]

$$\Delta V - \mu\varepsilon\, \partial^2_{tt} V = 0, \qquad \Delta \mathbf{A} - \mu\varepsilon\, \partial^2_{tt} \mathbf{A} = 0. \qquad [10.36]$$

B) Simple harmonic plane waves in dielectrics

Consider a simple harmonic plane wave propagating in the direction of $\mathbf{e} = \mathbf{k}/k$ (Figure 10.2a), and which is specified by the complex electric field

$$\underline{\mathbf{E}} = \underline{\mathbf{E}}_m\, e^{i(\omega t - \mathbf{k}.\mathbf{r} + \phi)}. \qquad [10.37]$$

This expression is a solution of the propagation equation $\Delta \mathbf{E} - \mu\varepsilon\, \partial^2_{tt}\mathbf{E} = 0$ if ω and k are related by the dispersion relation

$$\omega = vk. \qquad [10.38]$$

Thus, the field [10.37] propagates with the phase velocity $v_{(p)} = \omega/k = v = 1/\sqrt{\mu\varepsilon}$. Particularly, the phase velocity is c in vacuum. We define the *index of refraction* of the medium by

$$n = c/v = \sqrt{\mu_r \varepsilon_r}, \qquad [10.39]$$

where ε_r and μ_r are the *relative electric permittivity* of the medium and its *relative magnetic permeability*, respectively. Often, the medium is not magnetic, then $\mu_r = 1$. Particularly, the vacuum is non-dispersive for electromagnetic waves. Its index is $n = 1$ by definition. As for matter, the constants μ and ε depend on the interaction of matter with the wave (which depends on ω in general); thus, matter is always dispersive.

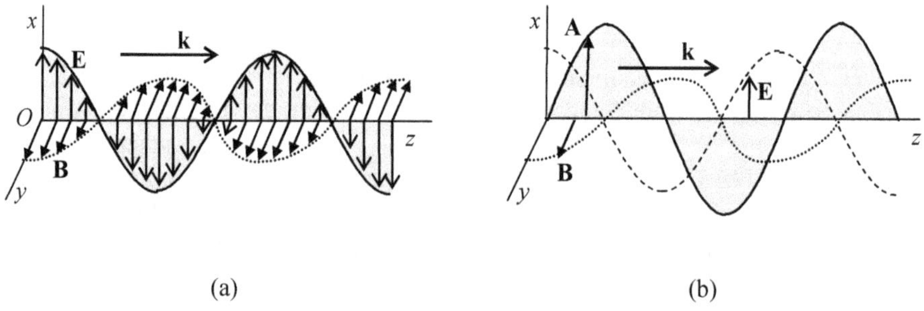

Figure 10.2. *a) Plane electromagnetic wave that is polarized linearly in the direction Ox, propagating in the direction Oz if it is represented by the fields* **E** *and* **B**. *b) The same wave represented by the vector potential* **A** *in the Coulomb gauge*

We note that differentiating the plane wave [10.37] with respect to time is equivalent to multiplying it by $i\omega$, and acting with the operator $\boldsymbol{\nabla}$ is equivalent to multiplying it by $-i\mathbf{k}$. Thus, substituting the expression [10.37] in equations [10.32], we get

$$\mathbf{k}.\mathbf{E}_{\mathrm{m}} = 0, \qquad \text{and} \qquad \underline{\mathbf{B}} = (1/\omega)\,(\mathbf{k} \times \underline{\mathbf{E}}_{\mathrm{m}})\; e^{i(\omega t - \mathbf{k}.\mathbf{r} + \phi)}. \tag{10.40}$$

The first equation expresses the orthogonality of $\mathbf{E}$ to $\mathbf{k}$ and the second equation determines $\mathbf{B}$ in terms of $\mathbf{E}$. As for the equations [10.33], they may be written as

$$\mathbf{k}.\mathbf{B}_{\mathrm{m}} = 0, \qquad \text{and} \qquad \mathbf{k} \times \mathbf{B}_{\mathrm{m}} + \varepsilon\mu\,\omega\,\mathbf{E}_{\mathbf{m}} = 0. \tag{10.41}$$

They are verified by the expression of $\mathbf{B}$ [10.40] if we take into account the dispersion relation [10.38]. The equation $\mathbf{k}.\mathbf{B}_{\mathrm{m}} = 0$ means that $\mathbf{B}$ is orthogonal to $\mathbf{k}$.

We may also consider the real part of the fields

$$\begin{aligned} &\mathbf{E} = \mathbf{E}_{\mathrm{m}} \cos(\omega t - \mathbf{k}.\mathbf{r} + \phi), && \mathbf{B} = \mathbf{B}_{\mathrm{m}} \cos(\omega t - \mathbf{k}.\mathbf{r} + \phi), \\ &\mathbf{B}_{\mathrm{m}} = (1/v)\,(\mathbf{e} \times \mathbf{E}_{\mathrm{m}}), && \mathbf{E}_{\mathrm{m}} = -v\,(\mathbf{e} \times \mathbf{B}_{\mathrm{m}}). \end{aligned} \tag{10.42}$$

Some properties of the fields of a plane electromagnetic wave in dielectrics may be deduced from equations [10.40] and [10.41] or equations [10.42]:

– The fields $\mathbf{E}$ and $\mathbf{B}$ are orthogonal to each other and both are orthogonal to the direction of propagation $\mathbf{e} = \mathbf{k}/k$. We say that the electromagnetic wave is *transverse*. The trihedron of vectors $\mathbf{e}$, $\mathbf{E}$, and $\mathbf{B}$ is right-handed.

– The fields $\mathbf{E}$ and $\mathbf{B}$ are in phase.

– The amplitudes of the fields $\mathbf{E}$ and $\mathbf{B}$ are related by the equation

$$E_{\mathrm{m}} = v\,B_{\mathrm{m}}. \tag{10.43}$$

We note that these properties hold for waves that propagate in an isotropic, linear, homogeneous, and infinite dielectric. In the case of a guided wave, the phase velocity depends on the geometry of the waveguide and the frequency of the wave, even if the medium is not dispersive. Then, the phase velocity $v_{(\mathrm{p})} = \omega/k$ and the group velocity $v_{(\mathrm{g})} = \partial\omega/\partial k$ are both different from $v = 1/\sqrt{\mu\varepsilon}$. On the other hand, the fields are not always transverse.

If we use the potentials to specify the plane electromagnetic wave, they may be written in the complex forms

$$\mathbf{A} = i\underline{\mathbf{A}}_{\mathrm{m}}\; e^{i(\omega t - \mathbf{k}.\mathbf{r} + \phi)}, \qquad \text{and} \qquad V = i\underline{V}_{\mathrm{m}}\; e^{i(\omega t - \mathbf{k}.\mathbf{r} + \phi)}, \tag{10.44}$$

where $i\underline{\mathbf{A}}_m$ and $i\underline{V}_m$ are the complex amplitudes and the factor i is explicitly written in order to make the amplitudes of **E** and **B** real. Thus, we find

$$\nabla V = -i\mathbf{k}V, \quad \nabla.\mathbf{A} = -i\mathbf{k}.\mathbf{A}, \quad \nabla\times\mathbf{A} = -i\mathbf{k}\times\mathbf{A}, \quad \Delta V = -k^2 V, \quad \text{and } \Delta\mathbf{A} = -k^2\mathbf{A}.$$

If we impose the Lorentz condition $\nabla.\mathbf{A} + \mu\varepsilon\,\partial_t V = 0$, we must have the relation

$$\underline{V}_m = (\mathbf{k}.\underline{\mathbf{A}}_m)/\varepsilon\mu\omega = v\,(\mathbf{e}.\underline{\mathbf{A}}_m). \qquad [10.45]$$

Then, the fields are given by

$$\mathbf{B} = \nabla\times\mathbf{A} = (\mathbf{k}\times\underline{\mathbf{A}}_m)\,e^{i(\omega t-\mathbf{k}.\mathbf{r}+\phi)} \equiv \underline{\mathbf{B}}_m\,e^{i(\omega t-\mathbf{k}.\mathbf{r}+\phi)},$$

$$\mathbf{E} = -\nabla V - \partial_t\mathbf{A} = \omega\,[\underline{\mathbf{A}}_m - \mathbf{e}\,(\mathbf{e}.\underline{\mathbf{A}}_m)]\,e^{i(\omega t-\mathbf{k}.\mathbf{r}+\phi)} \equiv \underline{\mathbf{E}}_m\,e^{i(\omega t-\mathbf{k}.\mathbf{r}+\phi)}. \qquad [10.46]$$

We note that the Lorentz condition does not completely fix the choice of the potentials. Indeed, if we make a new gauge transformation

$$\mathbf{A}\to\mathbf{A}' = \mathbf{A} + \nabla f \quad \text{and} \quad V\to V' = V - \partial_t f, \qquad [10.47]$$

The potentials V' and $\mathbf{A}'$ still verify the Lorentz condition if the gauge function f is a solution of the equation

$$\Delta f - \mu\varepsilon\,\partial^2_{tt} f = 0. \qquad [10.48]$$

Particularly, taking $f = -i\underline{f}_m\,e^{i(\omega t-\mathbf{k}.\mathbf{r}+\phi)}$ with $\omega = kv$, the new potentials V' and $\mathbf{A}'$ are also plane waves with the same angular frequency ω and the same wave vector **k** as V and **A**, but their amplitudes are

$$\underline{\mathbf{A}}'_m = \underline{\mathbf{A}}_m - \mathbf{k}\,\underline{f}_m, \qquad \underline{V}'_m = \underline{V}_m - \omega\underline{f}_m. \qquad [10.49]$$

Thus we subtract from the amplitude $\underline{\mathbf{A}}_m$ a longitudinal vector $\mathbf{k}\,\underline{f}_m$ (i.e. parallel to the direction of propagation **e**) and we subtract $\omega\underline{f}_m$ from the amplitude $\underline{V}_m$. This does not modify the expressions [10.46] of the fields. Particularly, if we take $\underline{f}_m = \underline{V}_m/\omega$, we find $\underline{V}'_m = 0$ and $\underline{\mathbf{A}}'_m = \underline{\mathbf{A}}_m - \mathbf{e}(\mathbf{e}.\mathbf{A}_m)$, i.e. $V' = 0$ and $\underline{\mathbf{A}}'$ transverse. As we have seen, this particular choice of the Lorentz gauge is called *Coulomb's gauge.* This allows the plane wave to be represented by the vector potential **A** (Figure 10.2b):

$$V = 0, \quad \underline{\mathbf{A}} = i\underline{\mathbf{A}}'_m\,e^{i(\omega t-\mathbf{k}.\mathbf{r}+\phi)} \quad \text{with } \mathbf{k}.\underline{\mathbf{A}}'_m = 0,$$

$$\underline{\mathbf{E}} = -\partial_t\underline{\mathbf{A}} = \omega\underline{\mathbf{A}}'_m\,e^{i(\omega t-\mathbf{k}.\mathbf{r}+\phi)} \quad \text{and } \underline{\mathbf{B}} = \nabla\times\underline{\mathbf{A}} = \mathbf{k}\times\underline{\mathbf{A}}'\,e^{i(\omega t-\mathbf{k}.\mathbf{r}+\phi)}. \qquad [10.50]$$

We may also use the real potentials and fields. Taking $\underline{\mathbf{A}}_m$ real, we find

$$V = 0, \qquad \mathbf{A} = -\mathbf{A}_m \sin(\omega t - \mathbf{k.r} + \phi), \qquad \text{with } \mathbf{k}.\underline{\mathbf{A}}_m = 0,$$
$$\mathbf{E} = \omega \mathbf{A}_m \cos(\omega t - \mathbf{k.r} + \phi), \quad \text{and} \quad \mathbf{B} = \mathbf{k} \times \mathbf{A}_m \cos(\omega t - \mathbf{k.r} + \phi). \quad [10.51]$$

10.3. Polarization of electromagnetic waves

Most of the physical and chemical actions of electromagnetic waves are exerted by the electric field. Indeed, the magnitude of **B** is of the order of E/c where c is the speed of light in vacuum and the speed of the charged particles in the medium is often much less than c. Thus, the electric force $q\mathbf{E}$ is much more important than the magnetic force $q\mathbf{v}_q \times \mathbf{B}$. Furthermore, only the electric force exerts work and contributes to the energy transfer. For these reasons, the electromagnetic wave is often specified by the field **E**. Then, the Maxwell-Faraday equation determines $\partial_t\mathbf{B}$ and the integration with respect to time gives **B**. However, in the case of a nonlinear, anisotropic or bounded medium, the analysis is much more complicated.

In the case of an infinite, homogeneous, linear and isotropic medium, **E** and **B** are perpendicular to the direction of propagation **e**. The direction of **E** specifies the *polarization* of the wave. Figure 10.2a represents a plane electromagnetic wave, which propagates in the direction *Oz* and such that **E** is everywhere parallel to *Ox*. The field **B** is then oriented in the direction *Oy*. We say that the wave is *polarized in the plane Oxz* (or *linearly polarized in the direction Ox*). If the wave is specified by the vector potential **A**, the wave is polarized in the plane of **k** and **A**, while **B** is perpendicular to this plane. If we use Coulomb's gauge, **A** points in the same direction as **E** and the polarization may be also specified by the direction of **A**.

Let us consider a simple harmonic wave of angular frequency ω propagating in the direction $\mathbf{e} = \mathbf{k}/k$. As the field **E** is transverse, it lies in the plane perpendicular to **e**. We choose, as basis vectors in this plane, two orthogonal unit vectors $\mathbf{e}_1$ and $\mathbf{e}_2$ such that the trihedron $(\mathbf{e}_1, \mathbf{e}_2, \mathbf{e})$ is right-handed. The superposition of two waves $\mathbf{E}_1$ and $\mathbf{E}_2$ of angular frequency ω, polarized linearly in the directions $\mathbf{e}_1$ and $\mathbf{e}_2$ respectively and such that $\mathbf{E}_2$ has a phase lead[1] ϕ over $\mathbf{E}_1$, is

$$\mathbf{E} = \mathbf{E}_1 + \mathbf{E}_2 = A_1 \cos(\omega t - \mathbf{k.r})\, \mathbf{e}_1 + A_2 \cos(\omega t - \mathbf{k.r} + \phi)\, \mathbf{e}_2. \qquad [10.52]$$

The polarization of the resultant wave **E** depends on the amplitudes A_1 and A_2 and ϕ.

1 The amplitudes A_1 and A_2 are positive. The waves may have the phases ϕ_1 and ϕ_2, then $\phi = \phi_2 - \phi_1$. Taking $\phi_1 = 0$ is equivalent to changing the origin of time.

a) If the waves $\mathbf{E}_1$ and $\mathbf{E}_2$ are in phase ($\phi = 2n\pi$ where n is an integer), the resultant wave may be written as

$$\mathbf{E}' = E_m \cos(\omega t - \mathbf{k.r})\, \mathbf{e}_p, \quad \text{where } E_m = \sqrt{A_1^2 + A_2^2} \text{ and } \mathbf{e}_p = (A_1\mathbf{e}_1 + A_2\mathbf{e}_2)/E_m \quad [10.53]$$

$\mathbf{E}'$ has an amplitude E_m and it is polarized linearly in the direction $\mathbf{e}'_p$ of the first diagonal of the rectangle of sides $2A_1$ and $2A_2$ in the directions $\mathbf{e}_1$ and $\mathbf{e}_2$, respectively, thus making an angle $\theta = \text{Arctan}(A_2/A_1)$ with $\mathbf{e}_1$ (Figure 10.3).

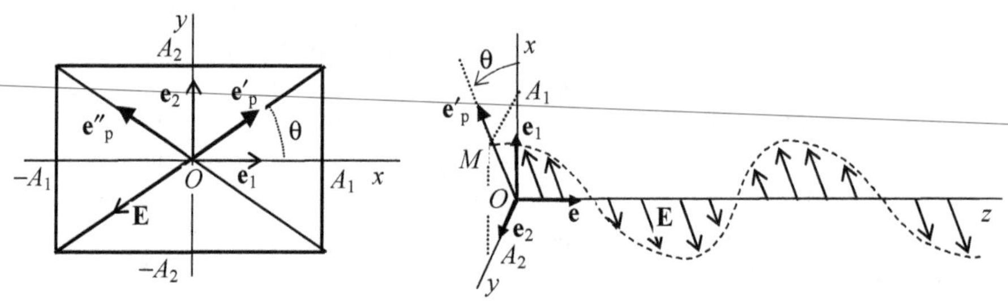

Figure 10.3. *a) The superposition of two waves, which are in phase and polarized linearly in the directions* $\mathbf{e}_1$ *and* $\mathbf{e}_2$*, is a linearly polarized wave in the direction* $\mathbf{e}'_p$*, b) the field* $\mathbf{E}$ *in the three-dimensional space. If the waves are in phase opposition, the resultant wave is linearly polarized in the direction of the second diagonal* $\mathbf{e}''_p$ *of the rectangle*

b) If the waves $\mathbf{E}_1$ and $\mathbf{E}_2$ are in phase opposition ($\phi = \pm\pi + 2n\pi$), the resultant wave may be written as

$$\mathbf{E}'' = E_m \cos(\omega t - \mathbf{k.r})\, \mathbf{e}''_p, \quad \text{where } E_m = \sqrt{A_1^2 + A_2^2} \text{ and } \mathbf{e}''_p = (A_1\mathbf{e}_1 - A_2\mathbf{e}_2)/E_m . \quad [10.54]$$

$\mathbf{E}''$ has an amplitude E_m and it is polarized linearly in the direction $\mathbf{e}''_p$ of the second diagonal of the rectangle of sides $2A_1$ and $2A_2$ in the directions $\mathbf{e}_1$ and $\mathbf{e}_2$, respectively, thus making an angle $\theta = -\text{Arctan}(A_2/A_1)$ with $\mathbf{e}_1$.

c) If the waves $\mathbf{E}_1$ and $\mathbf{E}_2$ have the same amplitude ($A_1 = A_2 \equiv E_o$) and a phase shift $\phi = -\pi/2 + 2n\pi$, the resultant wave is

$$\mathbf{E}^{(-)} = E_o\,[\cos(\omega t - \mathbf{k.r} + \alpha)\, \mathbf{e}_1 + \sin(\omega t - \mathbf{k.r} + \alpha)\, \mathbf{e}_2] \equiv E_o\, \mathbf{e}^{(-)} . \quad [10.55]$$

This is a vector of magnitude E_o and making an angle $\theta = (\omega t - \mathbf{k.r} + \alpha)$ with $\mathbf{e}_1$ (taken in the direction $\mathbf{e}_x$ in Figure 10.4a). An observer who receives this wave, sees the field $\mathbf{E}^{(-)}$ pointing in the direction of $\mathbf{e}^{(-)}$. At a given point, $\mathbf{e}^{(-)}$ rotates clockwise with the angular velocity ω about $\mathbf{e}$ taken in the direction Oz. At the various points of Oz, the tip of $\mathbf{E}^{(-)}$ moves on a helix about Oz rolled in the direction of the fingers of the left-hand if the thumb points in the direction of propagation (Figure 10.4b). We say that the wave is *left-handed circularly polarized*[2].

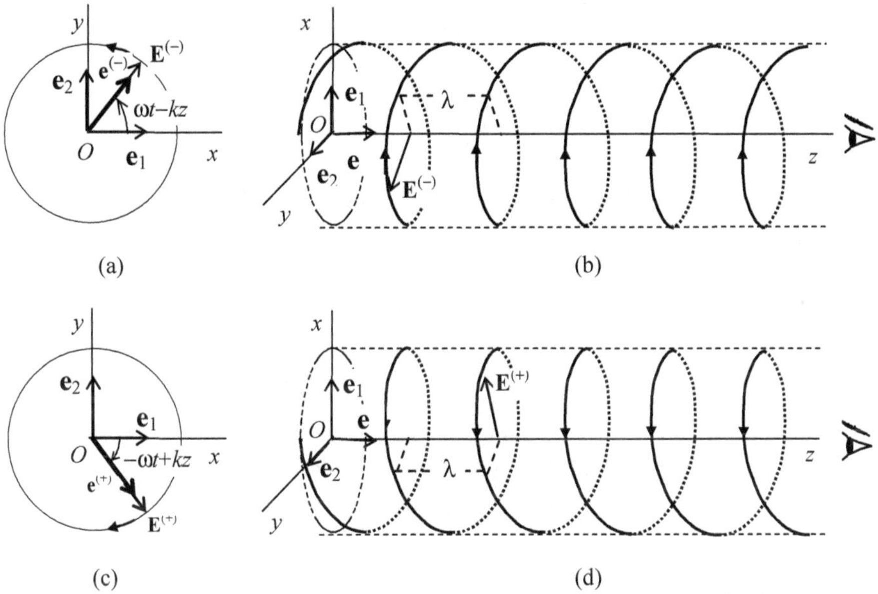

Figure 10.4. *Left-handed circularly polarized wave* $\mathbf{E}^{(-)}$*: a) if the wave propagates in the direction Oz, the observer sees the field at a given space point z move clockwise on a circle of radius* E_o *parallel to the* ($\mathbf{e}_1$, $\mathbf{e}_2$) *plane, and b) at a given time, the tip of* $\mathbf{E}^{(-)}$ *at the various points z are located on a helix of radius* E_o *in the direction of the left hand fingers if the thumb points in the direction of propagation. c) and d) correspond to a right-handed circularly polarized wave* $\mathbf{E}^{(+)}$*: the field* $\mathbf{E}^{(+)}$ *rotates in the opposite direction to that of* $\mathbf{E}^{(-)}$

d) If the waves $\mathbf{E}_1$ and $\mathbf{E}_2$ have the same amplitude ($A_1 = A_2 \equiv E_o$) and a phase shift $\phi = \pi/2 + 2n\pi$, the resultant wave may be written as

$$\mathbf{E}^{(+)} = E_o\,[\cos(\omega t - \mathbf{k.r} + \alpha)\,\mathbf{e}_1 - \sin(\omega t - \mathbf{k.r} + \alpha)\,\mathbf{e}_2] \equiv E_o\,\mathbf{e}^{(+)}. \qquad [10.56]$$

This is a vector of magnitude E_o and making with $\mathbf{e}_1$ an angle $\theta = (-\omega t + \mathbf{k.r} - \alpha)$ (Figure 10.4c). The field $\mathbf{E}^{(+)}$ points in the direction of $\mathbf{e}^{(+)}$ and rotates in the opposite direction to $\mathbf{E}^{(-)}$. At a given space point z, it rotates anticlockwise with the angular

2 In some domains of physics, reversed conventions are used for right-handed or left-handed polarization.

velocity ω and, at the various points z, its tip is located on a helix rolled like the fingers of the right-hand (Figure 10.4d). $\mathbf{E}^{(+)}$ is a *right-handed circularly polarized wave.*

We may also use the complex representation and introduce a complex basis $\underline{\mathbf{e}}_{(\pm)} = (\mathbf{e}_1 \pm \mathrm{i}\mathbf{e}_2)/\sqrt{2}$ and conversely, $\mathbf{e}_1 = (\underline{\mathbf{e}}_{(+)} + \underline{\mathbf{e}}_{(-)})/\sqrt{2}$ and $\mathbf{e}_2 = (\underline{\mathbf{e}}_{(+)} - \underline{\mathbf{e}}_{(-)})/\sqrt{2}\,\mathrm{i}$. Setting $\underline{E}_{\mathrm{o}} = E_{\mathrm{o}}\, e^{\mathrm{i}\alpha}$, the right-handed and left-handed circularly polarized waves may be written as

$$\underline{\mathbf{E}}^{(\pm)} = \underline{E}_{\mathrm{o}}\; e^{\mathrm{i}(\omega t-\mathbf{k.r})}\,\mathbf{e}_1 + \underline{E}_{\mathrm{o}}\; e^{\mathrm{i}(\omega t-\mathbf{k.r}\pm\pi/2)}\,\mathbf{e}_2 = \underline{E}_{\mathrm{o}}\; e^{\mathrm{i}(\omega t-\mathbf{k.r})}\,(\mathbf{e}_1 \pm \mathrm{i}\mathbf{e}_2) \equiv \underline{E}^{(\pm)}\underline{\mathbf{e}}_{(\pm)}. \qquad [10.57]$$

A wave that is linearly polarized in the direction $\mathbf{e}_{\mathrm{p}} = \cos\theta\,\mathbf{e}_1 + \sin\theta\,\mathbf{e}_2$ may be written as

$$\underline{\mathbf{E}} = \underline{E}_{\mathrm{o}}\; e^{\mathrm{i}(\omega t-\mathbf{k.r})}\,\mathbf{e}_{\mathrm{p}} = \underline{E}_{\mathrm{o}}\; e^{\mathrm{i}(\omega t-\mathbf{k.r})}\,(\cos\theta\,\mathbf{e}_1 + \sin\theta\,\mathbf{e}_2)$$

$$= (\underline{E}_{\mathrm{o}}/\sqrt{2}\,)[\,e^{\mathrm{i}(\omega t-\mathbf{k.r}-\theta)}\,\underline{\mathbf{e}}_{(+)} + e^{\mathrm{i}(\omega t-\mathbf{k.r}+\theta)}\,\underline{\mathbf{e}}_{(-)}\,] \equiv \underline{E}^{(+)}\,\underline{\mathbf{e}}_{(+)} + \underline{E}^{(-)}\,\underline{\mathbf{e}}_{(-)}\,. \qquad [10.58]$$

Thus, any linearly polarized wave in the direction making an angle θ with $\mathbf{e}_1$ may be considered as a superposition of a right-handed circularly polarized wave and a left-handed circularly polarized wave with opposite phases $\pm\theta$.

e) In a more general case, if the phase shift of the waves $\mathbf{E}_1$ and $\mathbf{E}_2$ is ϕ and they have any amplitude, the resultant wave may be written as

$$\mathbf{E} = A_1\cos(\omega t - \mathbf{k.r})\,\mathbf{e}_1 + A_2\cos(\omega t - \mathbf{k.r} + \phi)\,\mathbf{e}_2. \qquad [10.59]$$

At a given space point $\mathbf{r}$, the tip of $\mathbf{E}$ moves on an ellipse that is inscribed in a rectangle of sides $2A_1$ and $2A_2$. The elliptical motion is anticlockwise if $-\pi < \phi < 0$ (the wave is then *left-handed elliptically polarized*) or clockwise if $0 < \phi < \pi$ (the wave is then *right-handed elliptically polarized*).

f) A wave may be *unpolarized* if ϕ changes arbitrarily (then, the direction of $\mathbf{E}$ varies arbitrarily in time) or if it is a superposition of non-coherent waves having various polarizations. This is the case of light waves that are emitted by thermal sources (incandescence lamps, flames, etc.). The wave may be *partially polarized* if it is the superposition of a polarized wave and an unpolarized wave. On the contrary, lasers emit coherent and polarized light. A *polarizer* is a device that produces polarized waves from an unpolarized wave. This is the case of a Polaroid sheet, which transmits light that is polarized in a specific direction Ox and absorbs light that is polarized in the perpendicular direction Oy. If a wave of amplitude E_{o} and polarized linearly in a direction making an angle θ with Ox is incident on the

Polaroid, the amplitude of the transmitted wave is $E_o \cos\theta$. There also exist polarizers that produce circularly polarized waves.

10.4. Energy and intensity of plane electromagnetic waves

Consider a progressive electromagnetic wave that propagates in the direction $\mathbf{e} = \mathbf{k}/k$, whose fields are

$$\mathbf{E} = \mathbf{E}_m \cos(\omega t - \mathbf{k.r}), \qquad \mathbf{B} = \mathbf{B}_m \cos(\omega t - \mathbf{k.r}), \tag{10.60}$$

where $E_m = vB_m = B_m/\sqrt{\mu\varepsilon}$. We define the *impedance* Z of the medium per unit area and the impedance of vacuum Z_o per unit area as

$$Z = E/H = \mu E/B = \mu v = \sqrt{\mu/\varepsilon} \quad \text{and} \quad Z_o = \mu_o c = 376.73\ \Omega/\text{m}^2. \tag{10.61}$$

At each point of space, the densities of electric and magnetic energies are

$$U_{E,v} = \tfrac{1}{2}\varepsilon\mathbf{E}^2 = \tfrac{1}{2}\varepsilon E_m^2 \cos^2(\omega t - \mathbf{k.r}), \quad U_{M,v} = \mathbf{B}^2/2\mu = (B_m^2/2\mu)\cos^2(\omega t - \mathbf{k.r}). \tag{10.62}$$

Thus, the two densities of energy are equal and we may write the total density of electromagnetic energy

$$U_{EM,v} = U_{E,v} + U_{M,v} = \varepsilon E_m^2 \cos^2(\omega t - \mathbf{k.r}). \tag{10.63}$$

The average value of the energy density taken in time over a period $T = 2\pi/\omega$ or taken in space over a wavelength is

$$< U_{EM,v} > = \tfrac{1}{2}\varepsilon E_m^2. \tag{10.64}$$

The Poynting vector of this wave is

$$\mathbf{S} = (\mathbf{E} \times \mathbf{B})/\mu = (E_m^2/Z)\cos^2(\omega t - \mathbf{k.r})\,\mathbf{e}. \tag{10.65}$$

It points in the direction of propagation $\mathbf{e}$. Thus, the energy is localized in the region of the fields and it propagates from one place to another in the direction of propagation, starting from the emitter. To sustain this permanent flux of radiation energy, the emitter must continuously supply energy.

The energy that is received during an interval of time Δt by an element of area $\delta\mathcal{S}$ perpendicular to the Poynting vector $\mathbf{S}$ (Figure 10.5a) is $(\mathbf{S.n})\,\delta\mathcal{S}\,\Delta t = S\,\delta\mathcal{S}\,\Delta t$. This is the energy contained in a cylinder of base $\delta\mathcal{S}$ and length $v\,\Delta t$ parallel to the direction of propagation $\mathbf{e}$. Its volume is $v\ \Delta t\ \delta\mathcal{S}$ and the energy that it contains is $U_{\mathrm{EM,v}}\, v\ \Delta t\ \delta\mathcal{S}$. Comparing the two expressions, we get

$$S(\mathbf{r},t) = v\ U_{\mathrm{EM,v}}(\mathbf{r},t), \quad \text{i.e. } \mathbf{S}(\mathbf{r},t) = vU_{\mathrm{EM,v}}(\mathbf{r},t)\ \mathbf{e} \quad \text{(progressive waves).} \qquad [10.66]$$

This relation holds only in the case of progressive waves of any profile. It is verified by the expressions [10.63] and [10.65] for simple harmonic waves.

The power, which is received by $d\mathcal{S}$, is the flux of the Poynting vector over $d\mathcal{S}$

$$dP = \mathbf{n}.\mathbf{S}\ d\mathcal{S}. \qquad [10.67]$$

The average power that is received by the unit area placed perpendicularly to the direction of propagation is the *intensity* of the wave

$$\mathcal{I} = \langle dP/d\mathcal{S} \rangle = \langle S \rangle = E_{\mathrm{m}}^2/2\mu v = E_{\mathrm{m}}^2/2Z. \qquad [10.68]$$

If we use the potentials to specify the electromagnetic wave, the average energy density and the intensity of the wave may be written as

$$\langle U_{\mathrm{EM,v}} \rangle = \tfrac{1}{2}\varepsilon\omega^2\,[A_{\mathrm{m}}^2 - (\mathbf{e}.\mathbf{A}_{\mathrm{m}})^2] \quad \text{and} \quad \mathcal{I} = \tfrac{1}{2}\,v\varepsilon\omega^2[A_{\mathrm{m}}^2 - (\mathbf{e}.\mathbf{A}_{\mathrm{m}})^2]. \qquad [10.69]$$

Particularly, if we use Lorentz's gauge, we find

$$\langle U_{\mathrm{EM,v}} \rangle = \tfrac{1}{2}\varepsilon\omega^2 A_{\mathrm{m}}^2 \quad \text{and} \quad \mathcal{I} = \tfrac{1}{2}\,v\varepsilon\omega^2 A_{\mathrm{m}}^2. \qquad [10.70]$$

If we use the complex representation, we cannot calculate the energy density and the Poynting vector by using directly the complex fields. The superposition principle, which justifies the use of the complex fields, does not hold if the quantities are nonlinear. Thus, the real part of the fields must be taken before calculating the energy density and the Poynting vector. If the fields are $\mathbf{E} = \underline{\mathbf{E}}_{\mathrm{m}}\, e^{\mathrm{i}(\omega t - \mathbf{k}.\mathbf{r})}$ and $\mathbf{B} = \underline{\mathbf{B}}_{\mathrm{m}}\, e^{\mathrm{i}(\omega t - \mathbf{k}.\mathbf{r})}$, it is possible to write the average values directly

$$\langle U_{\mathrm{E,v}} \rangle = (\varepsilon/4)(\mathbf{E}.\mathbf{E}^*), \quad \langle U_{\mathrm{M,v}} \rangle = (1/4\mu)(\mathbf{B}.\mathbf{B}^*), \quad \mathcal{I} = \langle S \rangle = (1/2Z)(\mathbf{E}.\mathbf{E}^*). \qquad [10.71]$$

As the relation of the energy and the Poynting vector to the fields is quadratic, if the wave $\mathbf{E}$ is the superposition of two waves $\mathbf{E}_1$ and $\mathbf{E}_2$, the energy density and the Poynting vector of $\mathbf{E}$ are not always the sum of those of $\mathbf{E}_1$ and $\mathbf{E}_2$. For instance, if $\mathbf{E}_1$ and $\mathbf{E}_2$ propagate in the same direction $\mathbf{e}$, their superposition $\mathbf{E} = \mathbf{E}_1 + \mathbf{E}_2$ propagates in the same direction. Thus, we have $\mathbf{B} = \mathbf{e} \times \mathbf{E}/v$ and $\mathbf{B}^2 = \mathbf{E}^2/v^2$ and, consequently, $\mathbf{E}^2 = \mathbf{E}_1^2 + \mathbf{E}_2^2 + 2\mathbf{E}_1.\mathbf{E}_2$. The total energy density is $U_{\mathrm{EM,v}} = \varepsilon\mathbf{E}^2 = U_{\mathrm{EM,v1}} + U_{\mathrm{EM,v2}} + 2\varepsilon\mathbf{E}_1.\mathbf{E}_2$. Only if $\mathbf{E}_1.\mathbf{E}_2 = 0$, we find $U_{\mathrm{EM,v}} = U_{\mathrm{EM,v1}} + U_{\mathrm{EM,v2}}$, $\mathbf{S} = \mathbf{S}_1 + \mathbf{S}_2$ and $\mathcal{I} = \mathcal{I}_1 + \mathcal{I}_2$. This is the case if $\mathbf{E}_1$ and $\mathbf{E}_2$ are polarized linearly in orthogonal directions. If the waves are unpolarized (the angle of $\mathbf{E}_1$ and $\mathbf{E}_2$ varies rapidly or at random) or if they have different frequencies, the relations hold only in average values. We define the *degree of polarization* of the wave in the direction $\mathbf{e}_1$ as the ratio

$$\mathcal{P}_1 = \frac{\mathcal{I}_1 - \mathcal{I}_2}{\mathcal{I}_1 + \mathcal{I}_2}. \qquad [10.72]$$

If the wave is totally polarized in the direction $\mathbf{e}_1$ (then, $E_2 = 0$ and $\mathcal{I}_2 = 0$), we find $\mathcal{P}_1 = 1$. On the contrary, if the wave is totally polarized in the direction $\mathbf{e}_2$ perpendicular to $\mathbf{e}_1$ (then, $E_1 = 0$ and $\mathcal{I}_1 = 0$), we find $\mathcal{P}_1 = -1$. If the wave is polarized circularly, $\underline{E}_1$ and $\underline{E}_2$ have the same amplitude but a phase shift $\pm\,\pi/2$, then $\mathcal{P}_1 = 0$. If the wave is unpolarized, it is a superposition of waves $\underline{E}_1$ and $\underline{E}_2$ of randomly varying amplitudes or phase shift ϕ or a superposition of right-handed and left-handed circularly polarized waves with a randomly varying phase shift ϕ, we find $\mathcal{P}_1 = 0$.

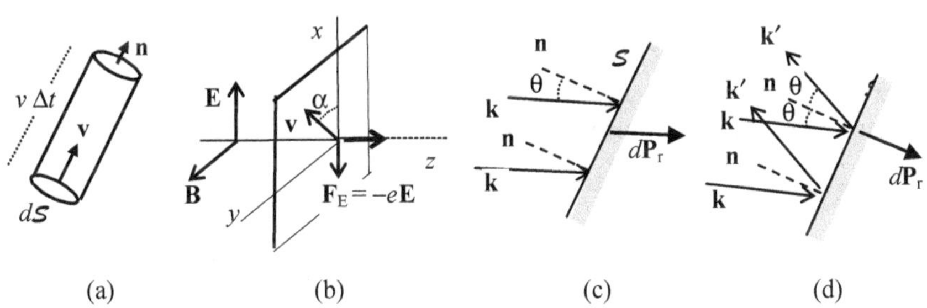

Figure 10.5. *a) Interpretation of the relation* $\mathbf{S} = vU_{EMv}\,\mathbf{e}$. *Radiation pressure of a wave that is b) incident normally on a totally absorbing plate, c) incident obliquely on a totally absorbing body, and d) incident obliquely on a totally reflecting body*

10.5. Momentum and angular momentum densities, radiation pressure

Besides energy and momentum, a body may have *orbital angular momentum* $\mathbf{L} = \mathbf{r}\times\mathbf{P}$ and an *intrinsic angular momentum* or *spin* $\mathbf{s}$. The *total angular momentum* $\mathbf{J} = \Sigma_i(\mathbf{L}_i + \mathbf{s}_i)$ of all the bodies of an isolated system is conserved. A continuous medium (a fluid, for instance) has a *density of momentum* $\mathbf{P}_v$, a *density of orbital angular momentum* $\mathbf{L}_v = \mathbf{r} \times \mathbf{P}_v$ and a *density of intrinsic angular momentum* $\mathbf{s}_v$. We expect the electromagnetic field to have similar densities. The surface of a body that intercepts the electromagnetic wave, receives a certain amount of energy, momentum, and angular momentum. The received momentum is equivalent to a *radiation pressure* on the body and the received angular momentum may set the body in rotation.

Let us consider an electromagnetic wave that propagates in the direction Oz and which is linearly polarized in the direction Ox. We assume that this wave is incident normally upon a metallic plate, which lies in the Oxy plane and absorbs all the radiation (Figure 10.5b). The fields of the wave act on the conduction electrons with a Lorentz force $\mathbf{F} = -e(\mathbf{E} + \mathbf{v}_e \times \mathbf{B})$ and set them in motion. **E** points in the direction Ox and **B** in the direction Oy. If the plate is thin, the velocity $\mathbf{v}_e$ cannot have a significant component in the direction Oz normal to the plate. In the interval of time dt, an electron receives an energy

$$dW = \mathbf{F}.\mathbf{v}_e\, dt = -\, e(\mathbf{E}.\mathbf{v}_e)\, dt = -\, e\, v_e E \cos\alpha\, dt, \qquad [10.73]$$

where α is the angle of $\mathbf{v}_e$ with Ox. The electron receives also a momentum $d\mathbf{P} = \mathbf{F}\, dt = -\, e\, dt\, (\mathbf{E} + \mathbf{v}_e \times \mathbf{B})$. The first term, proportional to **E**, is a sinusoidal function of time; its average value over a period is equal to zero. Thus the effective received momentum is

$$d\mathbf{P}_z = -\, e\, (\mathbf{v}_e \times \mathbf{B})\, dt = -\, e\, v_e B \cos\alpha\, dt\, \mathbf{e}_z. \qquad [10.74]$$

The expressions [10.73] and [10.74] show that $dW/dP = E/B = v$. Thus, the electromagnetic wave carries momentum oriented in the direction of propagation. Its energy density $U_{EM,v}$ and momentum density $P_{EM,v}$ are related by the equation

$$\mathbf{P}_{EM,v} = (U_{EM,v}/v)\, \mathbf{e} = \mathbf{S}/v^2. \qquad [10.75]$$

This is the same relation, [9.78], that we have deduced from Maxwell's tensor.

If the wave is incident on a body in a direction making an angle θ with the normal to the body, an element of area $\mathcal{S}$ of its surface receives during dt the energy $dU_{EM} = U_{EM,v}\, \mathcal{S} v \cos\theta\, dt$ and the momentum $dP_{EM} = P_{EM,v}\, \mathcal{S}\, v \cos\theta\, dt$ contained in the cylinder of base area $\mathcal{S}$ and length $v\, dt$ in the direction of propagation. If the body is totally absorbing (Figure 10.5c), this transfer of momentum is equivalent to a force per unit area $F_{EM,s} = vP_{EM,v} \cos\theta$ pointing in the direction of propagation. Particularly, if the wave is incident normally ($\theta = 0$), the absorbing surface is under a *radiation pressure*

$$p_r = vP_{EM,v} = U_{EM,v} \qquad \text{(totally absorbing surface).} \qquad [10.76]$$

If the surface is totally reflecting (Figure 10.5d), there is no transfer of energy to the body but it receives momentum $2\mathcal{S}v\, P_{EM,v} \cos\theta$ per unit time, which is normal to its surface. This transfer of momentum is equivalent to a radiation pressure

$$p_r = 2vP_{EM,v} \cos\theta = 2U_{EM,v} \cos\theta \qquad \text{(totally reflecting surface).} \qquad [10.77]$$

Although the radiation pressure is small, it may have significant effects if it acts on small particles (such as electrons or dust) or if it acts for a long interval of time. It may become very important in the case of intense sources such as lasers. In astrophysics, mediums at very high temperature (as the core of stars) may produce radiation whose pressure may counterbalance gravity pressure. It may even exceed it, provoking the explosion of a star.

An electromagnetic wave also carries angular momentum. Let us consider again a wave that is incident on a metallic plate (Figure 10.5b). As **E** and **B** are orthogonal to Oz, the Lorentz force $-e(\mathbf{E} + \mathbf{v} \times \mathbf{B})$ acting on an electron has a component $-e\mathbf{E}$ in the plane of the plate. If the wave is polarized linearly in a direction $\mathbf{e}_P$, the electron moves only in this direction and its mean angular momentum remains equal to zero. Thus, a linearly polarized electromagnetic wave carries no angular momentum. A circularly polarized wave may be written as

$$\mathbf{E}^{(\pm)} = E_o\,[\cos(\omega t - kz)\,\mathbf{e}_x + \cos(\omega t - kz \pm \pi/2)\,\mathbf{e}_y],$$

$$\mathbf{B}^{(\pm)} = -\,(E_o/v)\,[\cos(\omega t - kz \pm \pi/2)\,\mathbf{e}_x - \cos(\omega t - kz)\,\mathbf{e}_y]. \qquad [10.78]$$

Neglecting the magnetic force, the electron is subject to an electric force

$$\mathbf{F}^{(\pm)}_{//} = -e\mathbf{E}^{(\pm)} = -eE_o\,[\cos(\omega t - kz)\,\mathbf{e}_x + \cos(\omega t - kz \pm \pi/2)\,\mathbf{e}_y] \qquad [10.79]$$

lying in the plane Oxy of the plate. The equation of motion of the electron is $m\,\ddot{\mathbf{r}}^{(\pm)}_{//} = \mathbf{F}^{(\pm)}_{//}$ and we find by integration

$$\mathbf{r}^{(\pm)}_{//} = (eE_o/m\omega^2)\,[\cos(\omega t - kz)\,\mathbf{e}_x + \cos(\omega t - kz \pm \pi/2)\,\mathbf{e}_y]. \qquad [10.80]$$

The electron moves on a circle of radius $eE/m\omega^2$ in the direction of rotation of $\mathbf{E}^{(\pm)}$. Neglecting the variation of the electron spin, the received angular momentum is

$$\mathbf{J}^{(\pm)} = m\mathbf{r}^{(\pm)} \times \dot{\mathbf{r}}^{(\pm)} = \mp\,(e^2E_o^2/m\omega^3)\,\mathbf{e}_z = \mp\,(ve^2/m\omega^3)\,\mathbf{E}^{(\pm)} \times \mathbf{B}^{(\pm)}. \qquad [10.81]$$

As the received energy by the electron is $U_{EM} = e^2E_o^2/m\omega^2$, we deduce that a progressive circularly polarized wave carries angular momentum of density $\mathbf{s}_{EM,v}^{(\pm)} = \mp\,U_{EM,v}^{(\pm)}/\omega$. In the general case of a progressive wave, a standing wave or a superposition of both types of waves, we write

$$\mathbf{s}_{EM,v}^{(\pm)} = \mp\,\mathbf{S}/v\omega. \qquad [10.82]$$

In the case of a purely progressive wave, the Poynting vector is $\mathbf{S} = vU_{EM,v}\,\mathbf{e}$. The momentum of the wave being directed toward the electron, the orbital angular momentum carried by the wave is equal to zero and $\mathbf{s}_{EM,v}^{(\pm)}$ is effectively the *density of intrinsic angular momentum* (or *spin*) of the wave.

In the case of a left-handed circularly polarized wave $\mathbf{E}^{(-)}$, $\mathbf{s}_v^{(-)}$ points in the direction of propagation (positive helicity), while in the case of the right-handed circularly polarized wave $\mathbf{E}^{(+)}$, $\mathbf{s}_v^{(+)}$ points in the opposite direction of the direction of propagation (negative helicity).

Generally, an electromagnetic wave has *orbital angular momentum*, which adds to its intrinsic angular momentum; this increases the effect of angular momentum transfer. The total angular momentum density may be written as

$$\mathbf{J}_{EM,v}^{(\pm)} = \mathbf{L}_{EM,v} + \mathbf{s}_{EM,v}^{(\pm)}, \quad \text{where } \mathbf{L}_{EM,v} = \mathbf{r} \times \mathbf{P}_{EM,v} = \mu\varepsilon\, \mathbf{r} \times \mathbf{S}. \qquad [10.83]$$

If a body receives an electromagnetic wave, the transfer of angular momentum may set the body in rotation. This effect was verified experimentally by Berth in 1936 by observing the rotation of a plate of quartz if it intercepts a circularly polarized wave. Actually, the effect may be easily observed by using a laser beam.

10.6. A simple model of dispersion

The speed of propagation $v = 1/\sqrt{\mu\varepsilon}$ is independent of the signal profile if the magnetic permeability μ and the electric permittivity ε are constant characteristics of the medium. The reality is more complex: if an electromagnetic wave is incident on a dielectric, a fraction of it penetrates as a *primary wave*. It acts on the charged particles in the medium. These particles (especially the electrons) emit *secondary* electromagnetic waves. The superposition of the primary wave and secondary waves is the transmitted wave, while the reflected wave is a superposition of secondary waves emitted back, toward the incidence medium. We first note that on the microscopic scale, the electrons, atoms, and molecules are in permanent thermal agitation, the fields undergo large fluctuations and they become infinite at the positions of the particles. These are the *microscopic fields*. On the contrary, the Maxwell's equations are written in terms of the *macroscopic* fields, charge densities, and current densities that are experimentally observable. These are mean values taken over time intervals and elements of volume and they are regular functions of the position and time.

If the medium is anisotropic, the directions of **E** and **P** (or **D**) are different in general (see section 4.14). We assume in this chapter that the medium is isotropic. The speed of the charged particles in the dielectric is extremely small. Thus, the force exerted by the magnetic field $q\,\mathbf{v} \times \mathbf{B}$ is completely negligible. By symmetry, the electrons move in the direction of the electric field. Thus, the polarization **P** of the dielectric is also in this direction. On the other hand, if there is no hysteresis effect in the medium, **P** vanishes if the electric field is removed. Writing $P = f(E)$ and assuming that the field **E** is not very strong, we may always write $f(E)$ as a power series in E. If only the first order term, $P = \varepsilon_o \chi_e E$, is important for the considered phenomenon, we say that the medium is *linear*. In this chapter, we do not consider the variation of the electric susceptibility χ_e with temperature.

The properties of propagation of electromagnetic waves in a dielectric depend on the interaction of the wave with the particles of the dielectric. This interaction can be correctly formulated only in the framework of quantum mechanics. In this section, we adopt a simple classical model, which considers the medium as containing classical oscillators that may be excited by the electromagnetic wave. These oscillators are electrons, bound to their equilibrium positions by elastic forces $-m\omega_j^2 \mathbf{u}_j$, where m is the electron mass, $\mathbf{u}_j$ is its displacement from equilibrium and ω_j is its proper angular frequency. The index (j) labels the different possible bindings of the electrons to the atoms. We also assume that the electrons are subject to dissipative forces of the form $-2m\beta_j \dot{\mathbf{u}}_j$, where the β_j are constant. An electromagnetic wave propagating in the medium acts on the electron with a force $-e\mathbf{E}_l$, where $\mathbf{E}_l$ is the *local electric field*, which is the resultant of the primary field **E** and the field of the charges of the medium other than the considered electron. If ε is the permittivity of the dielectric, the density of polarization is

$$\mathbf{P} = (\varepsilon - \varepsilon_o)\,\mathbf{E} = \Sigma_j\,(-eN_j)\,\mathbf{u}_j, \qquad [10.84]$$

where N_j is the number of electrons of the type (j) per unit volume. We have seen in section 4.12 that the local field is

$$\mathbf{E}_l = \mathbf{E} + \mathbf{P}/3\varepsilon_o \equiv \gamma\mathbf{E}, \qquad \text{where } \gamma = 2/3 + \varepsilon/3\varepsilon_o. \qquad [10.85]$$

Thus, the equation of motion of the electron of the type (j) may be written as

$$\ddot{\mathbf{u}}_j + 2\beta_j \dot{\mathbf{u}}_j + \omega_j^2 \mathbf{u}_j = -(e\gamma/m)\,\mathbf{E}. \qquad [10.86]$$

If the medium is linear and isotropic and the primary wave in the medium has an angular frequency ω, the electrons undergo sustained oscillations with the same frequency. By symmetry, the polarization **P**, the secondary electric field (that is emitted by the excited atoms and molecules) and, consequently, the total field **E** all have the same angular frequency ω as the primary field, and they are polarized in the same direction. Thus, we write the solution in the dielectric in the form

$$\mathbf{E} = \mathbf{E}_o\, e^{i(\omega t - \mathbf{k}.\mathbf{r})}, \qquad \mathbf{u}_j = \mathbf{u}_{jo}\, e^{i(\omega t - \mathbf{k}.\mathbf{r})}. \qquad [10.87]$$

Substituting these expressions in equations [10.84] and [10.86], we find the relations

$$\frac{3(\varepsilon-\varepsilon_o)}{2\varepsilon_o+\varepsilon}=\frac{e^2}{m\varepsilon_o}\Sigma_j\frac{N_j}{\omega_j^2-\omega^2+2i\beta_j\omega}, \quad [10.88]$$

$$\mathbf{u}_{jo}=-\frac{\gamma e}{m}\frac{1}{\omega_j^2-\omega^2+2i\beta_j\omega}\mathbf{E}_o. \quad [10.89]$$

Then equation [10.84] gives the amplitude of the polarization, which results from these displacements

$$\mathbf{P}=\Sigma_j(-eN_j)\,\mathbf{u}_{j,o}\equiv\varepsilon_o\chi_e\mathbf{E}_o \quad \text{where } \chi_e=\frac{\gamma e^2}{m\varepsilon_o}\Sigma_j\frac{N_j}{\omega_j^2-\omega^2+2i\beta_j\omega}. \quad [10.90]$$

$\chi_e=\varepsilon/\varepsilon_o-1$ is the susceptibility of the dielectric. Thus, the index of refraction is

$$n=c/v=\sqrt{\varepsilon\mu/\varepsilon_o\mu_o}=\sqrt{(1+\chi_e)\mu/\mu_o}. \quad [10.91]$$

In particular, if the medium is non-magnetic ($\mu\approx\mu_o$ and $\varepsilon=\varepsilon_o n^2$), equation [10.88] becomes Clausius-Mossoti's equation (called also Lorentz-Lorenz formula)

$$\frac{n^2-1}{n^2+2}\frac{1}{N}=\frac{e^2}{3m\varepsilon_o}\Sigma_j\frac{f_j}{\omega_j^2-\omega^2+2i\beta_j\omega}, \quad [10.92]$$

where N is the number of bound electrons per unit volume and f_j is the fraction of these electrons, which are of the type (j). The right-hand side does not depend on the temperature, if the molecules are non-polar. If the quantity [10.88] is small, compared to 1, we find

$$n^2=\frac{\varepsilon}{\varepsilon_o}=1+\frac{e^2}{m\varepsilon_o}\Sigma_j\frac{N_j}{\omega_j^2-\omega^2+2i\beta_j\omega}. \quad [10.93]$$

This expression shows that n is complex, of the form

$$n=n_{(r)}-in_{(i)}. \quad [10.94]$$

The expressions of the real part and the imaginary part of n are

$$n_{(r)}=1+\frac{e^2}{2m\varepsilon_o}\Sigma_j\frac{N_j(\omega_j^2-\omega^2)}{(\omega_j^2-\omega^2)^2+4\beta_j^2\omega^2}, \quad n_{(i)}=\frac{e^2}{m\varepsilon_o}\Sigma_j\frac{N_j\beta_j\omega}{(\omega_j^2-\omega^2)^2+4\beta_j^2\omega^2}. \quad [10.95]$$

The imaginary part of the index corresponds to an absorption of the wave and, consequently, to its attenuation. Indeed, the phase of the wave is

$$\omega(t - \frac{\mathbf{e.r}}{v}) = \omega[t - \frac{\mathbf{e.r}}{c(n_r - in_i)}] = \omega[t - \frac{n_r\mathbf{e.r}}{c(n_r^2 + n_i^2)}] - i\frac{\omega n_i\mathbf{e.r}}{c(n_r^2 + n_i^2)}.$$

The wave propagates with a phase velocity $v_{(p)}$ and a free path l, given by

$$v_{(p)} = c[n_{(r)} + \frac{n_{(i)}^2}{n_{(r)}}] \quad \text{and} \quad l = \frac{1}{\delta} = \frac{c}{\omega}[\frac{n_{(r)}^2}{n_{(i)}} + n_{(i)}]. \qquad [10.96]$$

We note that the imaginary part of the index vanishes and there is no attenuation of the wave if there are no dissipative forces ($\beta_j = 0$).

Figure 10.6 illustrates typical variations of $n_{(r)}$ and $n_{(i)}$ versus the frequency of the wave. The curve representing $n_{(i)}$ is a succession of resonance curves at the characteristic frequencies ω_j of the medium. For a wave frequency close to one of these frequencies, the wave is very damped.

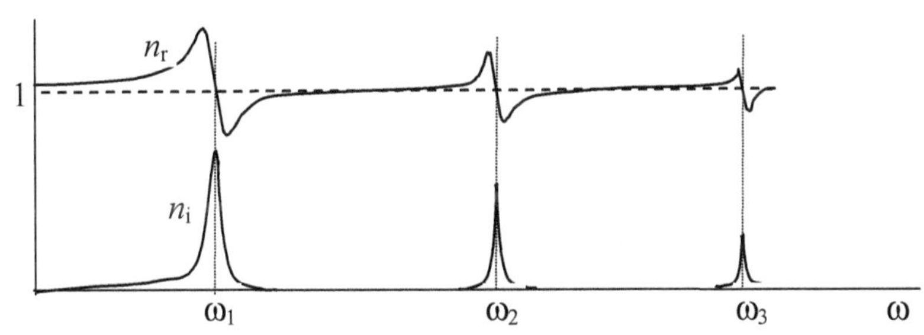

Figure 10.6. *Typical variations of the real part of the index of refraction and its imaginary part*

In quantum theory, the energy of the photon $h\tilde{\nu}_j$ is equal to the difference of energy levels $E_j - E_o$, where E_j is the energy of an excited state of the atom and E_o is that of the ground state. The real part $n_{(r)}$ is an increasing function of ω (*normal dispersion*) except in a small frequency band near ω_j in which the dispersion is *abnormal*. $n_{(r)}$ has a maximum and a minimum below and above each one of the ω_j. At high frequencies ($\omega >> \omega_j$), $n_{(r)}$ approaches 1 and $n_{(i)}$ approaches zero. In this limit, the medium behaves as the vacuum (since the atoms cannot respond to an excitation of very high frequency). This behavior of the index is generally valid in quantum theory with other interpretations of the constants β and N_j.

10.7. Electromagnetic waves in conductors

In a conductor at electrostatic equilibrium, the charge density q_v, the electric field **E**, and the current density **j** are all equal to zero. The conductor is then equipotential. In the case of a quasi-permanent regime with a certain difference of potential maintained between the ends of the conductor, any eventual charge density vanishes rapidly, but **E** and **j** do not necessarily vanish. The magnetic field is related to the current by Ampère's law. Thus, it may not vanish inside the conductor. In the following, we consider only Ohmic conductors carrying an alternating current or exposed to a simple harmonic electromagnetic wave. We show that the fields and the current density decrease exponentially with depth, the higher the frequency the lesser the penetration depth. This effect is due to Eddy currents and the dissipation of energy as Joule heat.

A) Equations of propagation and plane wave solutions

In an Ohmic conductor, the current density is related to the electric field by Ohm's law $\mathbf{j} = \sigma\mathbf{E}$. Thus, Maxwell's equations may be written as

$$\nabla.\mathbf{E} = q_v/\varepsilon \qquad [10.97]$$

$$\nabla \times \mathbf{E} + \dot{\mathbf{B}} = 0 \qquad [10.98]$$

$$\nabla.\mathbf{B} = 0 \qquad [10.99]$$

$$\nabla \times \mathbf{B} = \mu\sigma\mathbf{E} + \mu\varepsilon\,\dot{\mathbf{E}}. \qquad [10.100]$$

We know that, at electrostatic equilibrium, the charge density q_v is equal to zero because of the repulsion of like charges. This property remains often valid in the variable regimes. Indeed, by taking the divergence of both sides of [10.100] and using [10.97], we find the equation of evolution of q_v

$$\partial_t q_v + q_v/\tau_c = 0, \qquad \text{where } \tau_c = \varepsilon/\sigma. \qquad [10.101]$$

The solution of this equation is $q_v = q_{vo}\, e^{-t/\tau_c}$. Thus, an eventual charge density, equal to q_{vo} at $t = 0$, decreases exponentially in time. In the case of a typical metallic conductor, such as copper, whose conductivity is $\sigma = 5.98 \times 10^7\ \Omega^{-1}.\text{m}^{-1}$, the *characteristic time* (or *relaxation time*) is $\tau_c = 1.5 \times 10^{-19}$ s and, in the case of a semiconductor, τ_c may be of the order of 10^{-11} s. The charge density decreases to $6.7 \times 10^{-3}\ q_{vo}$ after a time equal to $5\tau_c$ and to $0.45 \times 10^{-6}\, q_{vo}$ after $10\tau_c$. Thus, the charge density q_v is practically equal to zero in the case of low-frequency phenomena compared to $1/\tau_c$ (quasi-permanents approximation).

Setting $q_v = 0$ in Maxwell's equations [10.97] to [10.100] and making the same analysis as in section 9.2C, we find the equations of propagation of the fields in conductors

$$\Delta\mathbf{E} - \varepsilon\mu\partial^2_{tt}\mathbf{E} - \mu\sigma\,\partial_t\mathbf{E} = 0, \qquad \Delta\mathbf{B} - \varepsilon\mu\partial^2_{tt}\mathbf{B} - \mu\sigma\,\partial_t\mathbf{B} = 0. \qquad [10.102]$$

In the limit of a non-conducting medium ($\sigma = 0$), we find d'Alembert's equation of propagation.

In the following, we introduce the characteristic angular frequency of the conducting medium $\omega_c = 1/\tau_c = \sigma/\varepsilon$. The equation of $\mathbf{E}$ has a plane wave solution $\underline{\mathbf{E}} = \mathbf{E}_m e^{i(\omega t - p\mathbf{e}.\mathbf{r}+\phi)}$ if p verifies the equation $v^2p^2 = \omega^2 - i\omega\omega_c$, where $v = 1/\sqrt{\mu\varepsilon}$. The root p of this equation is complex of the form $p = \pm(k - i\eta)$. Taking the sign (+), the wave may be written as

$$\underline{\mathbf{E}} = \mathbf{E}_m\, e^{-\eta\mathbf{e}.\mathbf{r}}\, e^{i(\omega t - k\mathbf{e}.\mathbf{r}+\phi)}$$

$$k = \frac{\omega}{v\sqrt{2}}\{\sqrt{1+Q^2} + 1\}^{1/2},\ \eta = \frac{\omega}{v\sqrt{2}}\{\sqrt{1+Q^2} - 1\}^{1/2}, \quad \text{where } Q = \omega_c/\omega. \quad [10.103]$$

The Maxwell-Gauss equation [10.97] (with $q_v = 0$) is verified if $\mathbf{E}.\mathbf{e} = 0$, which means the transversality of $\underline{\mathbf{E}}$. Then, the Maxwell-Faraday equation [10.98] gives

$$\begin{aligned}\underline{\mathbf{B}} &= (k/\omega - i\eta/\omega)\,\mathbf{e}\times\mathbf{E}_m e^{-\eta\mathbf{e}.\mathbf{r}}\, e^{i(\omega t - k\mathbf{e}.\mathbf{r}+\phi)}\\ &= (1/v)\,(1+Q^2)^{1/4}\,\mathbf{e}\times\mathbf{E}_m\, e^{-\eta\mathbf{e}.\mathbf{r}}\, e^{i(\omega t - k\mathbf{e}.\mathbf{r}+\phi-\alpha)}.\end{aligned} \qquad [10.104]$$

The other two Maxwell equations [10.99] and [10.100] are verified. Thus, the fields $\mathbf{E}$ and $\mathbf{B}$ in the conductor are orthogonal to each other and to the direction of propagation $\mathbf{e}$ and the trihedron of the vectors $\mathbf{e}$, $\mathbf{E}$ and $\mathbf{B}$ is right-handed. The real amplitudes of $\mathbf{B}$ and $\mathbf{E}$ are related by the equation

$$\mathbf{B}_m = (1/v)(1+Q^2)^{1/4}(\mathbf{e}\times\mathbf{E}_m). \qquad [10.105]$$

The phase lag of the magnetic field over the electric field is

$$\alpha = \mathrm{Arctan}(\eta/k) = \mathrm{Arctan}[\sqrt{1+1/Q^2} + 1/Q] \qquad (0 < \alpha < \pi/2). \qquad [10.106]$$

To interpret k and η, we note that the real fields may be written as

$$\begin{aligned}\mathbf{E} &= \mathbf{E}_m e^{-\eta\mathbf{e}.\mathbf{r}}\cos(\omega t - k\mathbf{e}.\mathbf{r} + \phi),\\ \mathbf{B} &= (1/v)(1+Q^2)^{1/4}\,\mathbf{e}\times\mathbf{E}_m\, e^{-\eta\mathbf{e}.\mathbf{r}}\cos(\omega t - k\mathbf{e}.\mathbf{r} + \phi - \alpha).\end{aligned} \qquad [10.107]$$

The cosine functions express that the wave is simple harmonic of angular frequency ω propagating in the direction **e** with a wave number k. Thus, its phase velocity is $v_p = \omega/k$. Because of the factor $e^{-\eta \mathbf{e.r}}$, the amplitudes of **E** and **B** decrease exponentially while propagating. Note that choosing the root $p' = -(k - i\eta)$ is equivalent to change the direction of propagation **e** to **–e**, we find a function $\cos(\omega t + k\mathbf{e.r})$, which expresses that the wave propagates in the direction **–e** with the wave number k and a amplitude $\mathbf{E}_m e^{-\eta(-\mathbf{e.r})}$, which decreases exponentially in the direction of propagation. If a wave of angular frequency ω is incident normally on the face *Oxy* of an infinite conductor toward the positive z (Figure 10.7a), a part of the wave penetrates in the conductor and continue to propagate toward the positive z ($\mathbf{e} = \mathbf{e}_z$). If it does not meet an obstacle, on which it may be reflected, no wave propagates in the conductor toward the negative z. While propagating, the wave is attenuated with an *attenuation coefficient* η. Its amplitude decreases like $e^{-z/\delta}$ where $\delta = 1/\eta$ is called *skin depth* or *penetration depth.* After each travel of a distance δ, the amplitude is divided by e. If the wave is incident on the first face of a thin conducting plate (Figure 10.7b), the wave may reach the second face, be reflected on it and propagate toward the negative z with an amplitude, which varies like $e^{z/\delta}$. A part of this wave crosses the first face back to the incidence medium.

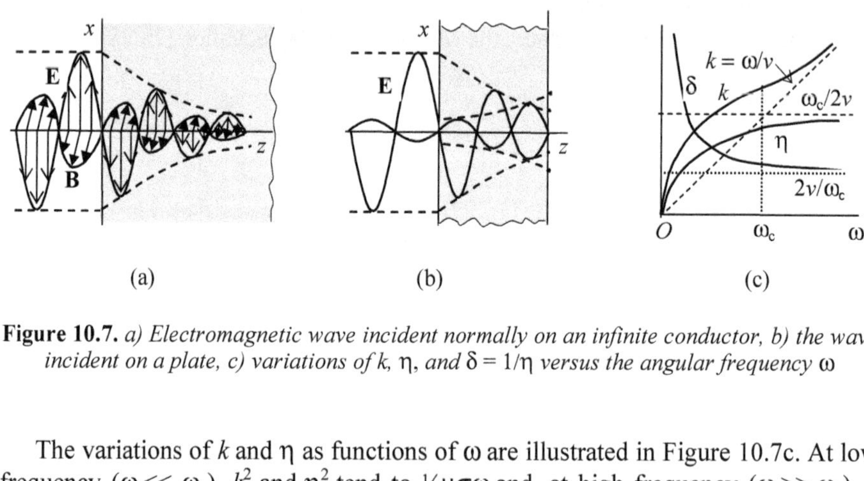

Figure 10.7. *a) Electromagnetic wave incident normally on an infinite conductor, b) the wave incident on a plate, c) variations of k, η, and δ = 1/η versus the angular frequency ω*

The variations of k and η as functions of ω are illustrated in Figure 10.7c. At low frequency ($\omega << \omega_c$), k^2 and η^2 tend to $\frac{1}{2}\mu\sigma\omega$ and, at high frequency ($\omega >> \omega_c$), k approaches asymptotically ω/v and η approaches $\omega_c/2v$. For $\omega = \omega_c$, the expressions [10.103] give $k \approx 1.1\ \omega_c/v$ and $\eta \approx 0.45\ \omega_c/v$. For instance, in the case of copper ($\sigma = 5.98 \times 10^7\ \Omega^{-1}.\text{m}^{-1}$), the characteristic angular frequency is $\omega_c \approx \sigma/\varepsilon_0 = 6.77 \times 10^{18}$ rad.s^{-1}. Using equations [10.103], we find $\eta = 1.54 \times 10^4\ \text{m}^{-1}$ and $\delta = 65\ \mu\text{m}$ for 1 MHz radiowaves, $\eta = 3.76 \times 10^8\ \text{m}^{-1}$ and $\delta = 2.7$ nm for $\lambda = 500$ nm visible light and $\eta = 1.19 \times 10^9\ \text{m}^{-1}$ and $\delta = 0.84$ nm for $\lambda = 50$ nm ultraviolet.

B) Skin effect and magnetic shielding

In a cylindrical conductor, for instance, a time-independent current I is usually uniformly distributed on the sections with a uniform density $j = I/\pi R^2$. The magnetic field is then $B^{(\text{in})} = (\mu_o I/2\pi)(r/R^2)$ inside the cylinder and $B^{(\text{ex})} = \mu_o I/2\pi r$ outside it (Figure 10.8a and b). If this conductor is used to carry an alternating current or if it receives an electromagnetic wave, an oscillating field is set up inside it. Then, a field and a current are induced and, according to Lenz law, they oppose the variations, which produce them. This induction is so important at high frequency that it cancels the current and the fields at sufficient depth in the conductor (Figure 10.8c) and produces a reflected wave on the conductor. Thus, the current and the fields are restricted to a narrow layer near the surface of the conductor. The consequences of this *skin effect* are important for the analysis and conception of high-frequency circuits, transmission lines, and antennas. It must be taken into account even in the distribution of electric energy at 50 Hz.

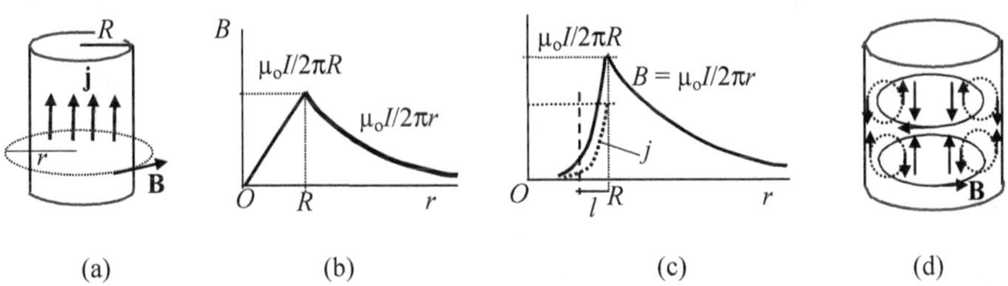

Figure 10.8. *Skin effect in a cylinder: a) the field* **B** *in the case of a constant direct current and b) variation of B versus r in this case, c) the same variation for an alternating current, and d) concentration of j and* **B** *near the surface because of Eddy currents*

a) Quasi-permanent regimes

The term $-\varepsilon\mu\partial^2_{tt}\mathbf{E}$ in the propagation equation of **E** is due to the displacement current $\varepsilon\,\partial_t\mathbf{E}$ in the Maxwell-Ampère equation $\nabla\times\mathbf{B} = \mu(\mathbf{j} + \varepsilon\,\partial_t\mathbf{E})$, while the term $-\mu\sigma\,\partial_t\mathbf{E}$ is due to the conduction current $\mathbf{j} = \sigma\mathbf{E}$. In the case of a sinusoidal field $\mathbf{E} = \mathbf{E}_m e^{i\omega t}$, the ratio of the amplitudes of these currents is $\varepsilon\,\partial_t E/j = \omega\varepsilon/\sigma = \omega/\omega_c \equiv 1/Q$. Thus, the displacement current is negligible, compared to the conduction current if $Q = \omega_c/\omega >> 1$, this is the case also in the magnetic quasi-permanent approximation. Then, Maxwell's equations may be written as

$$\nabla.\mathbf{E} \cong 0, \qquad \nabla\times\mathbf{E} + \partial_t\mathbf{B} = 0, \qquad \nabla.\mathbf{B} = 0, \qquad \nabla\times\mathbf{B} \cong \mu\mathbf{j}. \qquad [10.108]$$

From the last equation, we deduce that $\nabla.\mathbf{j} \cong 0$ and, replacing **j** by $\sigma\mathbf{E}$, we deduce the so-called *diffusion equations*

$$\Delta\mathbf{E} - \mu\sigma\,\partial_t\mathbf{E} \cong 0 \quad \text{and} \qquad \Delta\mathbf{B} - \mu\sigma\,\partial_t\mathbf{B} \cong 0. \qquad [10.109]$$

Consider, for instance, a plane wave of angular frequency ω, polarized in the direction Ox and propagating in the direction Oz. In complex notation, it may be written as $\mathbf{E} = f(z)\, e^{i\omega t}\, \mathbf{e}_x$. Substituting this expression in equations [10.108], we find that $\mathbf{B} = (i/\omega)\partial_z f\, e^{i\omega t}\, \mathbf{e}_y$ and f must be a solution of the equation $\partial^2_{zz} f \cong i\mu\sigma\omega f$, hence $f = E_m\, e^{-ipz}$ provided that $p^2 + i\mu\sigma\omega = 0$, thus $p = (1-i)\eta$ where $\eta = \sqrt{\mu\sigma\omega/2}$. Thus, the fields may be written as

$$\mathbf{E} = E_m\, e^{-\eta z}\, e^{i(\omega t - \eta z)}\, \mathbf{e}_x \quad \text{and} \quad \mathbf{B} = \sqrt{\mu\sigma/\omega}\, E_m\, e^{-\eta z}\, e^{i(\omega t - \eta z - \pi/4)}\, \mathbf{e}_y. \qquad [10.110]$$

These are the limits of [10.103] and [10.104] for $\omega \ll \omega_c$, i.e. $Q \gg 1$. The characteristic angular frequency ω_c of good conductors being very high (of the order of 10^{18}), the skin depth in this case may be written as

$$\delta = 1/\eta = \sqrt{2/\mu\sigma\omega}. \qquad [10.111]$$

The expressions [10.110] represents a simple harmonic wave of wave vector $k = \eta$ but its amplitude decreases exponentially with an attenuation coefficient η. As $\mathbf{j} = \sigma\mathbf{E}$, the current density decreases with the penetrated distance according to the same exponential law. This decrease of the current density may be explained by the apparition of Eddy currents (Figure 10.8d). These currents reduce the current density $\mathbf{j}$ that produces them in the depth of the conductor and reinforces $\mathbf{j}$ near the surface.

b) Effect on the resistance of conductors

One of the important consequences of the skin effect is to increase the resistance of the conductor, as it reduces the section area of the conductor that carries the current. The exact analysis of this effect is complicated but, if it is assumed that the current has only a depth δ near the surface, the effective section in the case of a cylindrical conductor is $2\pi r\delta$ and the resistance of the conductor is $R \approx h/2\pi\sigma r\delta = (h/2\pi r)\sqrt{\mu\omega/2\sigma}$. Thus, the higher the frequency, the smaller the depth and the higher the resistance. For instance, the resistance of a copper wire of length 1 m and section 0.1 mm^2 is 0.17 Ω for constant direct current, 0.23 Ω at 1 MHz and 0.72 Ω at 10 MHz. On the other hand, the skin effect depends on the magnetic permeability of the conductor. Thus, iron is not useful as conductor to carry currents of frequencies higher than 1 kHz as its resistance becomes very high.

The skin effect may be provoked also by nearby conductors carrying currents, in the case of a cable formed by several conductors, for instance. This proximity skin effect must be added to the own skin effect of each conductor. It may be reduced by keeping the conductors apart. However, in doing so, the inductance is increased and this is not desirable. To reduce the resistance, it is not useful to increase the diameter of conductors to values much higher than δ because of the skin effect. Tube-like conductors have the same resistance as full conductors. One may also use several thin, parallel, isolated conductors with a convenient geometry to reduce proximity effects. As silver has the highest conductivity of all metals, it has the largest skin depth. Thus, it helps to plate a good conductor such as copper with a thin layer of silver. The current circulates over all the section at low frequency and mostly in the silver layer at high frequency.

c) Good conductors and superconductors

The equations of the quasi-permanent regime [10.108] are valid if the conduction current density σE is higher than the displacement current $\varepsilon\omega E$, thus $Q = \sigma/\varepsilon\omega$ is much larger than 1 (say, $Q > 100$). The medium is then considered as a *good conductor* for waves (i.e. signals) of frequency ω. So, the ratio Q may be considered as a *quality factor* for conduction. For instance, copper has a characteristic frequency $\omega_c = 6.8 \times 10^{18}$ rad.s^{-1}. Thus, it is considered as a good conductor if $\omega < \omega_c/100 \approx 10^{16}$ Hz, i.e. up to X-rays.

In the case of a good conductor, the wave number is $k = \eta = (1/v)\sqrt{\omega\omega_c/2}$. The phase velocity is $v_{(p)} = \omega/k = v\sqrt{2/Q}$ while the group velocity is $v_{(g)} = \partial\omega/\partial k = 2v\sqrt{2/Q}$. In the limit of a superconductor ($\sigma = \infty$), the skin depth is equal to zero. Thus, there are no fields and current in the superconductor and there is no energy loss. Hence, the use of superconductors to construct powerful electromagnets (used, for instance, in the powerful particle accelerators). Superconductivity also allows the conception of extremely precise measurement instruments.

d) Magnetic shielding

If a conductor is exposed to a time-independent magnetic field ($\omega = 0$), the attenuation coefficient η is equal to zero and the field may penetrate without attenuation in the entire conductor and in the cavities that it may contain. If the field depends on time, it may always be considered as a superposition of simple harmonic waves of various angular frequencies ω (Fourier theorem). These waves are attenuated inside the conductor and a large part is reflected, especially the high-frequency spectral components. If a magnetic disturbance is incident on a metallic plate of thickness d, it may partially cross the plate if d is comparable to the skin depth δ. Conversely, if d exceeds about 5δ, the disturbance is almost totally reflected. The points P, which are behind the plate, are electromagnetically protected. According to equation [10.111], this protection is improved if the product $\sigma\mu\omega$ is large.

C) Energy dissipation in conductors

The attenuation of electromagnetic waves in conductors is due to the dissipation of energy as Joule heat. Let us consider a wave whose real fields are [10.107]. Contrary to the case of a non-conducting medium, we find that the density of electric energy $U_{E,v} = \frac{1}{2}\varepsilon \mathbf{E}^2$ and the density of magnetic energy $U_{M,v} = B^2/2\mu$ are not equal. The ratio of their average values over a period of time is

$$< U_{M,v} > / < U_{E,v} > = \sqrt{1+Q^2}\ . \qquad [10.112]$$

It tends to 1 if $Q << 1$ (i.e. $\omega >> \omega_c$). In the case of a good conductor ($Q >> 1$), we find $U_{M,v}/U_{E,v} \approx Q >> 1$. The energy is then essentially magnetic. The total electromagnetic energy density $U_{EM,v} = \frac{1}{2}\varepsilon E^2 + B^2/2\mu$ may be written as

$$U_{EM,v} = \tfrac{1}{2}\varepsilon E_m^2\, e^{-2\eta \mathbf{e.r}}\, \{\cos^2(\omega t - k\mathbf{e.r} + \phi)$$
$$+ (v/\omega)^2\, [k\cos(\omega t - k\,\mathbf{e.r} + \phi) + \eta \sin(\omega t - k\,\mathbf{e.r} + \phi)]^2\}. \qquad [10.113]$$

The Poynting vector and the intensity of the wave may be written as

$$\mathbf{S} = \mathbf{e}\,(E_m^2/\mu v)(1 + Q^2)^{1/4}\, e^{-2\eta \mathbf{e.r}}\, \cos(\omega t - k\,\mathbf{e.r} + \phi)\cos(\omega t - k\,\mathbf{e.r} + \phi - \alpha)$$

$$\mathcal{I} = \langle S \rangle = (k/2\mu\omega)\, E_m^2\, e^{-2\eta \mathbf{e.r}} \equiv \mathcal{I}_o\, e^{-2\eta \mathbf{e.r}}\,, \qquad [10.114]$$

where $\mathcal{I}_o$ is the intensity at the entry ($\mathbf{r} = 0$). The intensity $\mathcal{I}$ decreases exponentially with distance travelled. In copper, for instance, after a distance $x = 0.1$ mm, the intensity is divided by a factor $e^{2x/\delta} \cong 22$ in the case of a 1 MHz radio wave, and it becomes practically zero in the cases of visible light and ultraviolet.

10.8. Electromagnetic waves in plasmas

A plasma is a partially or totally ionized gas, formed from a mixture of electrons and heavy positive ions. Although it is rare on Earth, plasmas constitute a very large part of matter in the Universe. Ionization may be provoked at high temperatures (several thousands of degrees) by the energetic collisions of the molecules. Partial ionization may be also provoked at low temperatures if the gas is bombarded by particles. This occurs effectively in the ionosphere (i.e. the upper atmosphere at an altitude varying between 200 and 400 km). In these layers, low ionization of the order of 1% is produced by the absorption of ultraviolet radiation from the Sun and the collisions of cosmic rays. At a higher altitude, the ionization is lower because air is more rarefied and, at lower altitude, the ionization is less probable because the number of ionizing particles is fewer.

We assume that the plasma is at low pressure, in order to neglect all collisions between ions and molecules and we neglect the thermal agitation. This allows us to consider only motion due to electromagnetic waves.

A) Equation of propagation in a plasma

At equilibrium, the positive and negative charges counterbalance at each point. Thus, there are no global charge density, current density, and field. Under the influence of an electromagnetic disturbance, the very heavy positive ions move very little but the electrons move and produce a change in the charge density, a current density, and, consequently, electric and magnetic fields. These fields tend to bring the electrons back to their equilibrium positions. However, because of the electric acceleration, they come back with some kinetic energy. Thus, they continue their motion in the opposite direction and so on. Then, we have a *plasma oscillation* similar to the oscillation of a gas producing a sound wave.

Consider a plasma containing at equilibrium N_v electrons per unit volume and an electromagnetic wave that is polarized linearly in the direction Ox and incident normally from the side $z < 0$ on the face Oxy of the plasma (Figure 10.9a). During motion, the electrons emit secondary waves, which superpose backward to form a reflected wave and they superpose with the incident wave in the plasma to form a transmitted wave. The very heavy positive ions do not move and do not contribute to the emission of secondary waves. Let $\mathbf{u}(\mathbf{r}, t)$ be the average displacement of electrons at the point $\mathbf{r}$ and at time t, measured from their equilibrium positions. An element of volume $\delta\mathcal{V} = \delta x\, \delta y\, \delta z$ at equilibrium becomes $\delta\mathcal{V}' \cong (1 + \boldsymbol{\nabla}.\mathbf{u})\, \delta\mathcal{V}$ after the displacement $\mathbf{u}$. The number of electrons in this volume $N_v\, \delta\mathcal{V}$ being unchanged, the density of electrons becomes $N'_v = N_v\, \delta\mathcal{V}/\delta\mathcal{V}' \cong N_v(1 - \boldsymbol{\nabla}.\mathbf{u})$. The density of positive ions remains almost unchanged and equal to N_v. Thus, the total density of charge becomes $q_v = N_v e(\boldsymbol{\nabla}.\mathbf{u})$ and the current density (due to the displacement of electrons of charge $-e$ and velocity $\partial_t\mathbf{u}$) is $\mathbf{j} = -N_v e\, \partial_t\mathbf{u}$. Thus, Maxwell's equations in the plasma take the form

$$\boldsymbol{\nabla}.\mathbf{E} = (N_v e/\varepsilon)\,(\boldsymbol{\nabla}.\mathbf{u}), \qquad [10.115]$$

$$\boldsymbol{\nabla} \times \mathbf{E} + \partial_t\mathbf{B} = 0, \qquad [10.116]$$

$$\boldsymbol{\nabla}.\mathbf{B} = 0, \qquad [10.117]$$

$$\boldsymbol{\nabla} \times \mathbf{B} - \varepsilon\mu\, \partial_t\mathbf{E} = -\mu N_v e\, \partial_t\mathbf{u}. \qquad [10.118]$$

To these equations, we must add the equation of motion of the electrons

$$m\,\partial^2_{tt}\mathbf{u} = -\,e\mathbf{E}. \qquad [10.119]$$

By writing this equation, we neglect the magnetic force compared to the electric force. Indeed, if v is the speed of propagation, we have $E \approx vB \approx cB$, thus $f_M/f_E = |\mathbf{v} \times \mathbf{B}|/E \approx (B/E)|\partial_t\mathbf{u}| = |\partial_t\mathbf{u}|/c$ and the electrons velocity $\partial_t\mathbf{u}$ is much smaller than c (non-relativistic plasma).

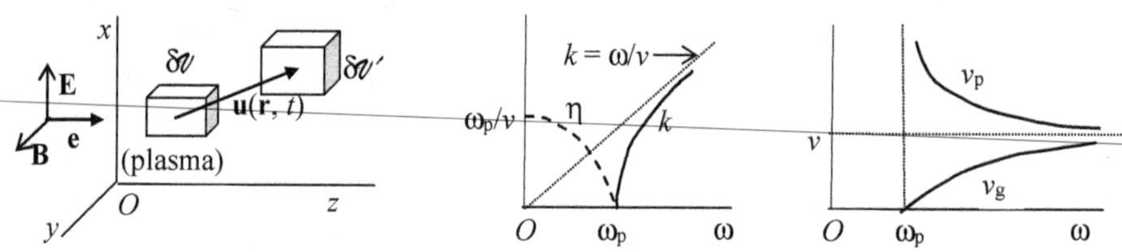

Figure 10.9. *a) Displacement in a plasma, b) attenuation coefficient* η *(if* $\omega < \omega_p$*) and wave number* k *(if* $\omega > \omega_p$*), c) phase velocity and group velocity if* $\omega > \omega_p$

B) Electromagnetic plane waves in a plasma

In the case of a plane wave of angular frequency ω and wave vector $\mathbf{k} = k\mathbf{e}$, the fields and the displacement $\mathbf{u}$ may be represented by the complex expressions

$$\mathbf{E} = \underline{\mathbf{E}}_m\, e^{i(\omega t - k\mathbf{e}.\mathbf{r})}, \qquad \mathbf{B} = \underline{\mathbf{B}}_m\, e^{i(\omega t - k\mathbf{e}.\mathbf{r})}, \qquad \mathbf{u} = \underline{\mathbf{u}}_m \exp e^{i(\omega t - k\mathbf{e}.\mathbf{r})}. \qquad [10.120]$$

These expressions verify Maxwell's equations and equation [10.119] if

$$\mathbf{e}.\underline{\mathbf{E}}_m = 0, \quad \mathbf{e}.\underline{\mathbf{B}}_m = 0, \quad \underline{\mathbf{B}}_m = (k/\omega)\,(\mathbf{e} \times \underline{\mathbf{E}}_m), \quad \underline{\mathbf{u}}_m = (e/m\omega^2)\,\underline{\mathbf{E}}_m \qquad [10.121]$$

provided that ω and k be related by the dispersion relation

$$\omega^2 = \omega_p^{\,2} + v^2k^2, \qquad \text{where } \omega_p^{\,2} = N_v e^2/m\varepsilon \text{ and } v = 1/\sqrt{\mu\varepsilon}. \qquad [10.122]$$

v is the speed of propagation in the plasma (different from the phase velocity of the wave $v_p = \omega/k$) and ω_p is the *plasma angular frequency*. Thus, the wave in the plasma is transverse ($\mathbf{E}_m$ and $\mathbf{B}_m$ orthogonal to the direction $\mathbf{e}$ of propagation) and they are orthogonal to each other. The electrons oscillate in the direction of the electric field (which is perpendicular to the direction of propagation) and this oscillation is in phase with $\mathbf{E}$.

It is also possible to write an equation of propagation for the fields in the plasma. Indeed, differentiating equation [10.118] with respect to time and using the equation of motion [10.119], we find

$$\nabla \times \partial_t \mathbf{B} - \mu\varepsilon\, \partial^2_{tt}\mathbf{E} = \mu\varepsilon\, \omega_p\, \mathbf{E}. \qquad [10.123]$$

Using equation [10.116], we get the equation

$$v^2\, \Delta\mathbf{E} - \partial^2_{tt}\mathbf{E} - \omega_p^{\,2}\mathbf{E} - v^2\, \nabla(\nabla.\mathbf{E}) = 0. \qquad [10.124]$$

If we take the divergence of this equation, we get $[\partial^2_{tt} + \omega_p^{\,2}](\nabla.\mathbf{E}) = 0$. If $\mathbf{E}$ is simple harmonic of angular frequency ω different from ω_p, we must have necessarily $\nabla.\mathbf{E} = 0$ and the equation of propagation may be simplified to

$$v^2\, \Delta\mathbf{E} - \partial^2_{tt}\mathbf{E} - \omega_p^{\,2}\mathbf{E} = 0. \qquad [10.125]$$

The magnetic field **B** and the displacement **u** obey similar equations. These equations of propagation are called *Klein-Gordon equations*. In the limit of a non-ionized medium ($N_v = 0$ and $\omega_p = 0$), we find d'Alembert's equation of propagation.

The dispersion relation [10.122] allows k to be determined as a function of ω. As this relation is nonlinear, the properties of propagation in the plasma depend heavily on the frequency. For this reason, we must distinguish two cases, depending on whether the angular frequency ω is higher or lower than ω_p.

1) If ω is higher than the plasma angular frequency ω_p, the dispersion relation [10.122] gives a real value of k. This means that the plasma is dispersive and the wave propagates without attenuation with a wave number k, a phase velocity $v_{(p)}$ and a group velocity $v_{(g)}$ given by

$$k = (\omega/v)\,\gamma, \quad v_{(p)} = \omega/k = v/\gamma, \quad \text{and} \quad v_{(g)} = v\gamma \text{ where } \gamma \equiv [1 - \omega_p^{\,2}/\omega^2]^{1/2} < 1. \qquad [10.126]$$

The variations of k and $v_{(p)}$ as functions of the frequency are illustrated in Figure 10.9. We note that $v_{(p)}$ is always higher than the speed of propagation in the non-ionized gas $v = (\varepsilon\mu)^{-1/2}$. It tends asymptotically to infinity if ω tends to ω_p and it tends to v at very high frequency. The ratio of the amplitudes of the fields in the plasma is

$$B_m/E_m = k/\omega = \gamma/v. \qquad [10.127]$$

To evaluate the total energy density, we must include the density of the electrons kinetic energy to the energy density of the fields. Its average value over a period is

$$< U_v > = < \tfrac{1}{2}\varepsilon E^2 > + < \tfrac{1}{2}B^2/\mu > + < \tfrac{1}{2}m(\partial_t\mathbf{u})^2 > = \tfrac{1}{2}\varepsilon E_m^{\,2}. \qquad [10.128]$$

The intensity of the electromagnetic wave is

$$\mathcal{I} = < S > = (1/2\mu) <|\mathbf{E}\times\mathbf{B}^*|> = (k/2\mu v)\, E_m^{\ 2} = (1/2\mu v)\, \gamma E_m^{\ 2} = \gamma v <U_v>. \quad [10.129]$$

This result means that the energy propagates in the plasma with the group velocity $v_{(g)} = v\gamma < v$. The variation of $v_{(g)}$ versus ω is illustrated in Figure 10.9c. It increases from 0 at $\omega = \omega_p$ to v at very high frequency. The fact that $v_{(g)}$ is less than the speed of propagation in the non-ionized gas v, which is itself less than the speed of light in vacuum, agrees with the special theory of relativity, which requires that the velocity of any particle or signal be less than the speed of light in vacuum. Note that the phase velocity is not associated with a displacement of physical quantities; thus, it has not to be necessarily less than the speed of light. At the frequency ω_p, the group velocity vanishes; this means that there is no transfer of energy or other physical quantities. If $\omega > \omega_p$, the energy propagates without any loss as Joule heat or other. This is due to our starting assumption that the plasma is at very low pressure and at a very low temperature, which allows us to neglect collisions and, consequently, the resistivity.

2) If ω is less than the plasma frequency ω_p, the dispersion relation [10.122] gives an imaginary value for k. Writing $k = -i\eta$, we find

$$\eta = (1/v)\sqrt{\omega_p^{\ 2} - \omega^2}\,. \quad [10.130]$$

In this case, the fields and the displacement may be written as

$$\mathbf{E} = \underline{\mathbf{E}}_m\, e^{-\eta\mathbf{e}.\mathbf{r}}\, e^{i\omega t}, \quad \mathbf{B} = \underline{\mathbf{B}}_m\, e^{-\eta\mathbf{e}.\mathbf{r}}\, e^{i\omega t}, \quad \mathbf{u} = \underline{\mathbf{u}}_m\, e^{-\eta\mathbf{e}.\mathbf{r}}\, e^{i\omega t}. \quad [10.131]$$

This is an attenuated wave with an attenuation coefficient η, whose variation versus ω is illustrated in Figure 10.9b. It decreases from $\omega_p/v = (\mu n e^2/m)^{½}$ for $\omega = 0$ to 0 for $\omega = \omega_p$. If we choose the root $k = +i\eta$ instead of $-i\eta$, the solution would be exponentially increasing with the travelled distance; this is physically impossible in an infinite plasma.

Substituting the expressions [10.131] in Maxwell's equations [10.115] to [10.118] and in [10.119], we find that they are verified if

$$\mathbf{e}.\underline{\mathbf{E}}_m = 0, \quad \mathbf{e}.\underline{\mathbf{B}}_m = 0, \quad \underline{\mathbf{B}}_m = -i(\eta/\omega)(\mathbf{e}\times\underline{\mathbf{E}}_m), \quad \underline{\mathbf{u}}_m = (e/m\omega^2)\,\underline{\mathbf{E}}_m. \quad [10.132]$$

$\mathbf{E}_m$ and $\mathbf{B}_m$ are orthogonal to each other and to $\mathbf{e}$, and $\mathbf{B}$ has a phase lag of $\pi/2$ over $\mathbf{E}$.

Taking the real parts of the fields [10.131] and evaluating the mean values over a period for the electric, magnetic and kinetic energy densities, we find the total energy density

$$< U_v> = <\tfrac{1}{2}\varepsilon E^2> + <\tfrac{1}{2}B^2/\mu> + <\tfrac{1}{2}m(\partial_t \mathbf{u})^2> = \tfrac{1}{2}\varepsilon\,(\omega_p^2/\omega^2)E_m^2\, e^{-2\boldsymbol{\delta}\mathbf{.r}}\,. \quad [10.133]$$

The intensity of the wave is

$$\mathcal{I} = < S > = 0. \quad [10.134]$$

In this case, there is no transfer of energy or any other physical quantity. Thus, we cannot consider that there is a wave in the plasma. If an electromagnetic wave of frequency less than ω_p is incident on the plasma, it penetrates the plasma to a depth of the order of $\delta = 1/\eta$. Initially there is some small transfer of energy to set up the energy density [10.133] in this layer. However, after this transient regime, no more energy is transferred to the plasma. The wave is then totally reflected on the surface of the plasma exactly as on a perfect mirror (see section 11.5).

In the ionosphere, N_v is of the order of 10^{12} to 10^{13} free electrons per m^3 depending on the hour of the day, seasons, and solar activity. The corresponding plasma frequency is about $\tilde{\nu}_p$ = 10 to 30 MHz (which corresponds to a wavelength of 10 m to 30 m). If $\tilde{\nu} < \tilde{\nu}_p$ (thus, $\lambda > \lambda_p$), the incident wave on the ionosphere is totally reflected. On the contrary, if $\tilde{\nu} > \tilde{\nu}_p$ (i.e. $\lambda < \lambda_p$), the wave propagates in the ionosphere without attenuation. For instance, if $\tilde{\nu}_p$ = 20 MHz ($\omega_p = 1.3 \times 10^8$rad.s^{-1}), a radio wave of wavelength 1 m ($\tilde{\nu} = 3 \times 10^8$ Hz) propagates with a phase velocity $v_{(p)} = 3.20 \times 10^8$ m/s and a group velocity $v_{(g)} = 2.72 \times 10^8$ m/s while a radio wave of 200 m ($\tilde{\nu} = 1.5 \times 10^6$ Hz) has an attenuation constant $\eta = 0.42$ m^{-1} and a penetration depth $\delta = 1/\eta = 2.3$ m. Thus, it is totally reflected. AM radio waves of wavelength longer than 30 m allow the connection of points on the Earth' surface that are not in line of sight. On the contrary, FM and TV emissions, which use wavelengths shorter than 30 m, propagate without attenuation in the ionosphere.

The displacement of electrons in a plasma produces a current of density

$$\mathbf{j} = -Ne\,\partial_t\mathbf{u} = -\mathrm{i}\,(Ne^2/m\omega)\,\underline{\mathbf{E}}_m\, e^{\mathrm{i}(\omega t - \mathbf{k}\mathbf{e.r})} = -\mathrm{i}\varepsilon(\omega_p^2/\omega)\,\underline{\mathbf{E}}. \quad [10.135]$$

This relation is similar to Ohm's law but with an imaginary conductivity

$$\sigma_p = -\mathrm{i}\varepsilon\omega_p^2/\omega. \quad [10.136]$$

A 1 m^3 cube of plasma has an impedance

$$Z = 1/\sigma_p = \mathrm{i}\omega/\varepsilon\omega_p^2. \quad [10.137]$$

This expression is similar to the impedance of a self-inductance ($Z = i\omega L$). Thus, the equivalent inductance of the plasma per unit volume is

$$L_p = 1/\varepsilon\omega_p^2 = m/Ne^2 = 3.55 \times 10^7/N \text{ H/m}^3. \quad [10.138]$$

The plasma behaves as an inductance that dissipates no energy. If we do not neglect the collisions of electrons with themselves and with the positive ions at rest, a certain amount of energy is dissipated and this corresponds to a supplementary resistance of the plasma.

10.9. Quantization of electromagnetic waves

According to classical concepts, a wave is extended in space and time with a certain continuous distribution of physical quantities, such as energy and momentum and a continuous flux of these quantities. However, some effects of emission and absorption of electromagnetic waves, such as the *photoelectric effect*, discovered by Hertz in 1887 and interpreted by Einstein in 1905, can be understood only if the wave is really constituted of "packets" of energy; we say that the wave is *quantized*. The quantum of radiation, called *photon*, of frequency $\tilde{\nu}$ has an energy

$$E_\gamma = h\tilde{\nu}, \quad \text{where } h = 6.626\,176 \times 10^{-34} \text{ J.s} = 4.135\,669 \times 10^{-15} \text{ eV.s.} \quad [10.139]$$

h is *Planck's constant*. The effects of radiation are fundamentally due to the interaction of a single photon with matter. Thus, they depend on its energy, i.e. its frequency $\tilde{\nu}$ or wavelength $\lambda = c/\tilde{\nu}$. A shorter wavelength corresponds to a more energetic photon and, consequently, a more important effect of the radiation.

If N_v is the number of photons per unit volume of the wave, its energy density is $U_{EM,v} = N_v h\tilde{\nu}$. According to the relation [10.75], the wave also has a momentum density $P_{EM,v} = U_{EM,v}/c = N_v h\tilde{\nu}/c$. Thus, the momentum of each photon is

$$p_\gamma = h/\lambda, \quad [10.140]$$

where λ is the wavelength. De Broglie generalized this relation to all particles.

According to the special theory of relativity, the energy and the momentum of a particle of mass m are related by the equation $E = \sqrt{p^2c^2 + m^2c^4}$. Thus, the relation $E_\gamma = cp_\gamma$ implies that the photon has no mass. The expressions [10.139] and [10.140] for the energy and momentum of the photon are confirmed by all experiments, notably the *Compton effect* (1923).

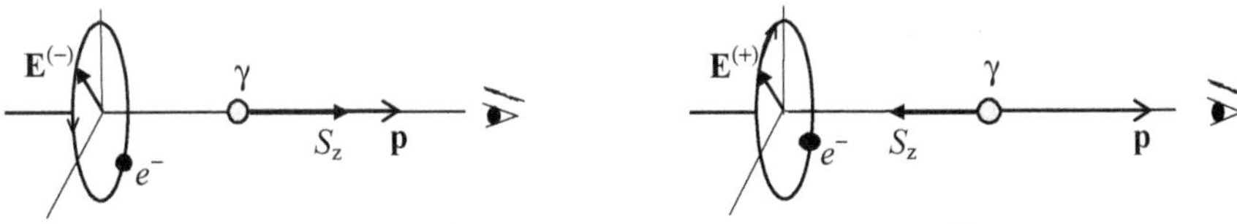

Figure 10.10. *a) A left-handed circularly polarized wave is constituted of photons of helicity* +1 *(forward spin), b) a right-handed circularly polarized wave is constituted of photons of helicity* −1 *(backward spin)*

On the other hand, the electromagnetic wave has a density of intrinsic angular momentum $\mathbf{s}_{EM,v}{}^{(\pm)}$. As it contains N_v photons per unit volume, each photon has an intrinsic angular momentum (or spin)

$$s^{(\pm)} = \mathbf{s}_{EM,v}{}^{(\pm)}/N_v = \mp \hbar\, \mathbf{e}_z , \qquad \text{where } \hbar = h/2\pi. \qquad [10.141]$$

Thus, the spin of the photon is quantized: it may be either $+\hbar$ if the photon is polarized forward (positive helicity, Figure 10.10a) or $s = -\hbar$ if the photon is polarized backward (negative helicity, Figure 10.10b). A linearly polarized wave may be considered as a superposition of two right-handed and left-handed circularly polarized waves having the same amplitude. Thus, it is constituted of an equal number of photons that are polarized forward and backward. The intrinsic angular momentum of this wave is thus equal to zero. This is also the case of an unpolarized wave.

10.10. Electromagnetic spectrum

An electromagnetic wave carries energy; thus, it cannot be emitted by charges at rest or in uniform motion: only accelerated charges (or variable currents) may emit them. If the current is a simple harmonic function of time, the emitted wave is simple harmonic of equal frequency. The emitter of a wavelength λ is always a system whose dimensions are comparable to λ. For instance, waves of wavelengths roughly more than 1 mm are emitted by macroscopic systems (electronic or electric).

While propagating, an electromagnetic wave is subject to the various effects due to the propagation medium and eventual obstacles. The most important effects are the decrease of intensity with the travelled distance like $1/r^2$ (in the case of a spherical wave), reflection on surfaces of large dimensions compared to the wavelength, diffraction by apertures and obstacles of dimensions comparable to λ, absorption by the medium, etc.

The most important applications of electromagnetic waves are in the domain of telecommunications. Information (sound, image, etc.) may be transmitted by *modulated waves*, that is, of amplitude, frequency or phase that vary according to the information. Modulation is realized by using an *input transducer*, which transforms the information into an electric signal. Microphones and photoelectric cells are examples of input transducers. In the receiver, an *output transducer* transforms the electrical signal of the modulated wave into a non-electrical signal. A loudspeaker and a liquid crystal screen are examples of output transducers. The extracted signal reproduces the original one with some deformation depending on the quality of the system.

The following is a classification of electromagnetic waves according to their frequency $\tilde{\nu}$ (or their wavelength $\lambda = c/\tilde{\nu}$) and some of their principal uses.

1) Waves of *industrial frequencies* (30 Hz $< \tilde{\nu} <$ 3 kHz and 10^5 m $< \lambda < 10^7$ m): these are emitted by alternating current generators and electrical setups. They are used in traditional telephony and for the transport of electrical energy.

2) *Radio waves* (or Hertzian waves): these include:

– very low-frequency (*VLF*) waves (3 kHz $< \tilde{\nu} <$ 30 kHz and 10^4 m $< \lambda < 10^5$ m);

– low-frequency (*LF*) waves (30 kHz $< \tilde{\nu} <$ 300 kHz and 10^3 m $< \lambda < 10^4$ m);

– medium-frequency (*MF*) waves (0.3 MHz $< \tilde{\nu} <$ 3 MHz and 10^2 m $< \lambda < 10^3$ m);

– high-frequency waves (*HF*) (3 MHz $< \tilde{\nu} <$ 30 MHz and 10 m $< \lambda < 10^2$ m);

– very high-frequency waves (*VHF*) (30 MHz $< \tilde{\nu} <$ 0.3 GHz and 1 m $< \lambda <$ 10 m).

Hertzian waves are emitted by macroscopic antennas. They are particularly used with amplitude modulation in telephony and AM radio emissions with a band width of the order of 10 kHz (which is the frequency band of the usual audible sounds). Frequency modulation is used for FM radio emissions and for television emissions.

3) *Microwaves* or ultra-high frequency (*UHF*) waves (0.3 GHz $< \tilde{\nu} <$ 300 GHz and 1 mm $< \lambda <$ 1 m) are emitted by some atoms and molecules (vibration spectrum and rotation spectrum). The ground state of the cesium atom consists of two very close energy levels separated by 4.14×10^{-15} eV. In the transition from the higher to the lower level, the atom emits a precise frequency of $9.192\,631\,77 \times 10^9$ Hz, which is the basis of the atomic clocks. Polar water molecules absorb microwaves easily; hence their use in microwave ovens (λ = 12.2 cm) to cook food that contains a large amount of water. They are also used by diathermy machines (to warm muscles and

joints in order to relieve soreness), and telecommunications with aircraft and satellites, radio-astronomy, and *radar* systems.

4) *Infrared* radiations ($3 \times 10^{11} < \tilde{\nu} < 3 \times 10^{14}$ Hz and 1 μm < λ < 1 mm) are emitted and absorbed by molecules. Most hot materials emit heat in the form of infrared radiations. They have many applications in medicine (diagnostics, thermotherapy, and thermography), industry, and infrared teledetection.

5) *Light* ($4 \times 10^{14} < \tilde{\nu} < 7.5 \times 10^{14}$ Hz and 400 nm < λ < 750 nm) is a major agent for the transport of solar energy to Earth, photosynthesis, and life itself. Its color is associated with its frequency, and it is emitted by atoms and molecules.

6) *Ultraviolet* radiations ($3 \times 10^{15} < \tilde{\nu} < 3 \times 10^{16}$ Hz and 10 nm < λ < 100 nm) are emitted by atoms. They are used to study atoms and efficiently initiate some photochemical reactions (such as the combination of chloride and hydrogen, the breaking of the carbon-carbon bond, etc.). They depolymerize nucleic acids, destroy proteins and inhibit the body's immune system, they have adverse effects on the skin and may cause skin cancer. They constitute a part of solar radiation and most of them are absorbed by the atmosphere, particularly the ozone (O_3) layer.

7) *X-rays* (20 pm < λ < 10 nm and $3 \times 10^{16} < \tilde{\nu} < 15 \times 10^{20}$ Hz) are emitted by atoms and charged particles (especially electrons) as they collide with other atoms. They are used in the study of crystalline structures and in medicine (radioscopy and radiotherapy). *Gamma rays* (λ < 20 pm and $\tilde{\nu} > 15 \times 10^{20}$) are emitted by atomic nuclei and by decelerating charged particles. They are used in scientific research and in medicine (gammascopy and gammatherapy).

10.11. Emission of electromagnetic radiations

One of the basic ideas of quantum theory is that atoms and molecules can only be in discrete states of well-defined energies. Normally, they are in the lowest energy level, called the *ground state*, and they emit no radiation. If they are in an excited energy state E_j, they undergo a transition to a lower energy level E_i by emitting a photon of energy

$$E_j - E_i = h\tilde{\nu}_{ji}, \qquad [10.142]$$

where $\tilde{\nu}_{ji}$ is the frequency of the emitted radiation. For instance, the ground state energy of the hydrogen atom is $E_1 = -13.6$ eV and that of the first excited level is $E_2 = -3.40$ eV. Thus, the wavelength of the emitted radiation in the transition $2 \to 1$ is $\lambda_{21} = hc/(E_2 - E_1) = 0.122$ μm. The emission spectrum of atoms and molecules is discrete and it depends only on their energy levels. It is almost independent of the physical or chemical conditions. This is true if the atoms or the molecules are well separated (in the gaseous state, for instance). In the case of dense mediums (liquids or solids), the emission is rather collective and the spectrum is almost continuous.

In reality, the waves, which are emitted in atomic transitions, are not monochromatic. The reason is that the exact transition time cannot be predicted. Quantum theory asserts only that, if the atom is excited at $t = 0$, the probability of staying in the excited state decreases exponentially according to $e^{-t/\tau}$, where τ is the *mean life* of the excited state. τ may be considered to be the duration of the emission. It is usually of the order of 10^{-9} to 10^{-8} s. According to Fourier theory (section 10.1E), this wave packet may be considered as a superposition of monochromatic waves of frequencies $\tilde{\nu}$ and wave numbers $k = \omega/v$. A wave packet, which propagates in the direction of Oz, may be written as

$$\underline{E}(z, t) = \int_0^\infty d\tilde{\nu}\; \underline{\mathcal{E}}(\tilde{\nu})\; e^{i(\omega t - kz)} = (1/2\pi) \int_0^\infty d\omega\; \underline{\mathcal{E}}(\omega)\; e^{i(\omega t - kz)}. \qquad [10.143]$$

In the case of a monochromatic wave $a_0\, e^{i(\omega_0 t - k_0 z)}$, the spectral amplitude may be written as $\underline{\mathcal{E}}(\omega) = 2\pi a_0\, \delta(\omega - \omega_0) = a_0\, \delta(\tilde{\nu} - \tilde{\nu}_0)$ and the spectral intensity is $\mathcal{I}(\omega) = \mathcal{I}_0\, \delta(\omega - \omega_0)$, where δ is the Dirac delta-function (see section A.11 of the appendix A). A three-dimensional wave may be written as

$$\underline{\mathbf{E}}(\mathbf{r}, t) = (2\pi)^{-3/2} \iiint d^3\mathbf{k}\; \hat{\mathbf{E}}(\mathbf{k})\; e^{i(\omega t - \mathbf{k}.\mathbf{r})}, \qquad \text{where } \omega = ck. \qquad [10.144]$$

We may easily show the Parseval relation

$$\iiint d^3\mathbf{r}\, |\underline{\mathbf{E}}(\mathbf{r}, t)|^2 = \iiint d^3\mathbf{k}\, |\hat{\mathbf{E}}(\mathbf{k})|^2, \qquad [10.145]$$

which expresses the intensity of the packet as the integral of its spectral energy.

The spectral function $\hat{\mathbf{E}}(\mathbf{k})$ is important in a band of widths Δk_x, Δk_y, and Δk_z related to the extensions Δx, Δy, and Δz of the wave packet in the direction of Ox, Oy, and Oz by the uncertainty relations [10.24]. The band width $\Delta\omega$ is related to the duration Δt of the packet by the uncertainty relation $\Delta\omega.\Delta t \approx 2\pi$. If we identify Δt with τ, we find a band width $\Delta\tilde{\nu}_0 \approx 1/\tau \approx 10^8$ to 10^9 Hz. This *quantum width* is very small, compared to the frequency of the radiation ($\approx 10^{15}$ Hz for visible light). In this case, the radiation is said to be *quasi-monochromatic*. In fact, $\Delta\tilde{\nu}$ is wider than $\Delta\tilde{\nu}_0$ because of the thermal agitation of atoms (producing collisions and a widening of the band by the Doppler effect).

A light wave that is emitted by a macroscopic body consists of many wave packets, which are emitted by the atoms of the body. There are two types of sources:

– *Incoherent sources*, in which the atoms emit independently, thus the emitted packets have no relation of polarization, amplitude, frequency, and phase. In this case, the total wave is $\mathbf{E} = \Sigma_j\, \underline{\mathbf{E}}_j(\mathbf{r}, t)$, where $\underline{\mathbf{E}}_j$ are packets of the form of [10.144] whose random phases ϕ_j are related to the time of emission by the atoms. Thus, the total wave $\mathbf{E}(\mathbf{r}, t)$ is neither polarized, nor coherent (Figure 10.11a). The intensity of the total wave is the sum of intensities that are emitted by the individual atoms $\mathcal{I}(\mathbf{r}, t) = \Sigma_j\, \mathcal{I}_j(\mathbf{r}, t)$. This is the case of traditional (non-laser) sources, such as thermal sources (flames and incandescent lamp) and electric discharge sources (electric arcs, neon discharge tubes, and spectral lamps).

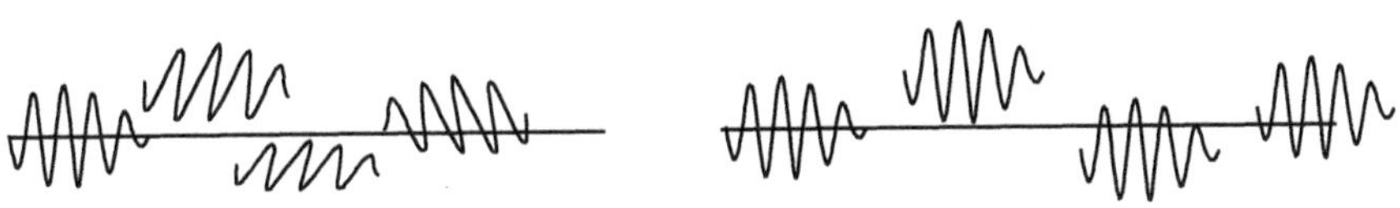

(a) (b)

Figure 10.11. *a) Non-coherent wave and b) coherent wave*

– *Coherent sources*, such as lasers, in which the wave packets $\mathbf{E}_j(\mathbf{r}, t)$ that are emitted by the atoms have well-defined polarization and phase relations. The atoms in these sources cannot be considered as completely independent, because the wave that is emitted by one atom acts on the others. The total emitted wave $\underline{\mathbf{E}} = \Sigma_j\, \underline{\mathbf{E}}_j(\mathbf{r}, t)$ is polarized and coherent (Figure 10.11b). Its intensity is given by

$$\mathcal{I}(\mathbf{r}, t) = (1/2\mu v)\, |\underline{\mathbf{E}}(\mathbf{r}, t)|^2 = (1/2\mu v)\, |\Sigma_j\, \underline{\mathbf{E}}_j(\mathbf{r}, t)|^2. \qquad [10.146]$$

It may be much larger (or smaller) than the sum of the intensities $\Sigma_j\, \mathcal{I}_j$ of the individual atoms.

10.12. Spontaneous and stimulated emissions

The emission of radiation by an atom as it undergoes the transition from an energy level E_j to a lower level E_i may be spontaneous, if it occurs without the influence of an incident wave. This is the type of emission by the common natural or artificial sources. Atoms emit independently according to a probability law without any amplitude, polarization and phase correlation. The emitted radiation is *incoherent*. Conversely, if an atom is in an energy level E_i and it receives a photon of energy $E_j - E_i = h\tilde{\nu}_{ji}$, it may undergo an excitation to a higher level E_j followed by the transition back to the level E_i with the emission of a photon of frequency $\tilde{\nu}_{ji}$ but generally without any relation to the initial photon. Thus, a good radiation emitter is also a good absorber of this radiation.

In 1917, Einstein discovered that if an atom is in an excited level $E_j > E_i$ and it receives a photon of energy $E_j - E_i$, instead of being excited to a higher level E_j, there is a large probability that it undergoes a transition to the lower level E_i with the emission of a photon that is identical to the incident photon. This *stimulated emission* amplifies the incident wave. It is the basic principle of *lasers*. The emitted photon may excite another atom and so on, producing an avalanche of emissions. Noting that, if N identical waves $\mathbf{E}_j$ of the same intensity $\mathcal{I}_1$ superpose, the resulting intensity is, according to [10.146], $\mathcal{I}(\mathbf{r}, t) = (1/2\mu v)\,|N\underline{E}_1(\mathbf{r}, t)|^2 = N^2\,\mathcal{I}_1$; thus, it may be very large. This large energy is evidently supplied by the external system, which excites the atoms coherently. The word "laser" is an acronym for "*light amplification by stimulated emission of radiation*". The emitted wave is extremely phase-coherent and directional with a very narrow spectral band.

In conventional light emitters, the stimulated emission is not significant because of the extremely small probability that an atom is in an excited state. The situation is different in the case of *population inversion*, i.e. many atoms in an excited state caused by a process of *optical pumping*. For this, the excited state must have a sufficiently long lifetime τ (of the order of 10^{-3} s or more); it is then said to be a *metastable state*. To reinforce the process of light amplification, the lasing medium is placed between two plane or spherical mirrors (Figure 10.12a). The wave propagates back and forth with a period of $2en/c$, where n is the index of refraction of the medium and e is the distance between the mirrors. The system constitutes an *optical cavity* with standing waves of frequency $\tilde{\nu}_p = pc/2ne$ similar to the waves on a stretched string. Here p is an integer, which labels the mode of the standing wave. The waves, which are emitted in the oblique directions with respect to the mirrors axis or whose frequencies are different from $\tilde{\nu}_p$, are taken out of the beam by reflection on the mirrors or are rapidly attenuated. Only the waves propagating perpendicularly to the mirrors remain in the beam after many reflections. The beam intensity increases by a resonance phenomenon. Two parallel plates placed at the ends of the cavity receive the wave at the Brewster incidence and reflect only the wave that is polarized perpendicularly to the incidence plane (see the section 11.2d). One of the mirrors (the *output coupler*) is partially transparent; it transmits a laser beam that is highly directional (if the cavity is long enough), polarized, and quasi-monochromatic. The high energy of the beam is provided by the optical pumping system, which maintains the population inversion.

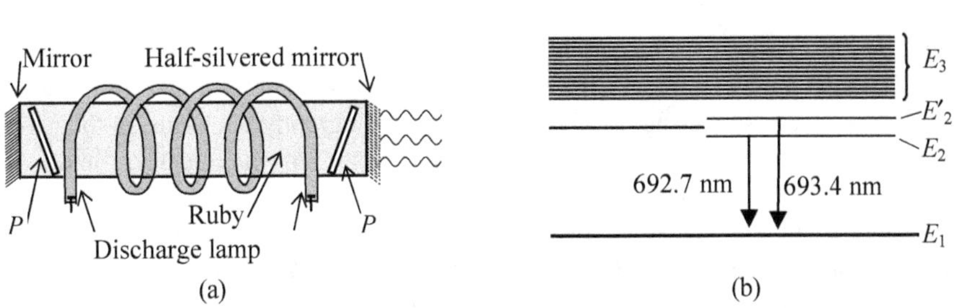

Figure 10.12. *a) Schematic representation of a ruby laser. The Brewster polarizers P polarize the beam, and b) the energy levels*

A lasing medium with only two levels E_i and E_j is not possible. A typical three level laser (and the first laser to be realized) is the *ruby laser*. This is a crystal of aluminum oxide (Al_2O_3) where some aluminum atoms are replaced with chromium atoms in a ratio of $1/10^4$ (Figure 10.12a). The energy levels are illustrated in Figure 10.12b. The intermediary level is in fact two very close levels E_2 and E'_2. The levels E_3 form a wide band covering the entire visible spectrum. The atoms of the lasing medium absorb any visible light and get excited to one level of the continuous band. Then they undergo a rapid transition to one of the levels E_2 and E'_2, which have a relatively long lifetime ($\approx$ 2 ms); this favors the inversion. Afterwards, they undergo transitions to the ground state E_1 with the emission of 692.7 and 693.4 nm radiations, respectively. The difference between the absorbed energy in the excitation and the emitted radiation energy is dissipated as heat in the medium and must be evacuated.

The output of a laser is a very narrow and intense beam. For instance, a beam of power 10 mW on a section of 1 mm^2 corresponds to an intensity of 10^4 $W.m^{-2}$, compared to about 150 $W.m^{-2}$ for sunlight. Being very coherent, this beam can travel a very long distance without significantly spreading or it can be focused to a very tiny spot of very high irradiance. The laser may be operating continuously or in a *pulsed mode*, emitting flashes of very short duration (of the order of 1 μs). In the so-called Q-switched laser, the emission is held-up while the population inversion is allowed to build up to the maximum level; then, a rapid lasing is allowed, producing a pulse of very short duration (of the order of a nanosecond) but with very high power (up to 10^9 W).

Actually, we have many types of lasers (*gas lasers, solid-state lasers, semiconductor lasers, optical fiber lasers*, etc.), for virtually all electromagnetic radiation: visible light, infrared laser, ultraviolet laser, X-ray laser, and so on. Similar devices operating at microwave and radio frequencies are called *masers* rather than lasers. The power of lasers varies between several milliwatts for the common types to considerable values for pulsed lasers.

Due to the almost monochromatism of their radiation, coherence, polarization and high intensity, laser beams have many applications in research. They include spectroscopy, laser scattering and interferometry, holography, etc. Due to their high intensity, they may be used to study the interaction of the electric field with nonlinear materials, produce very hot plasma, realize nuclear fusion, etc. Laser beams enable high-precision measurements of small angles, small lengths (for machine tools, for instance), and large distances (for instance, the distance from the Earth to the Moon), and to achieve precise alignment for mechanical construction, etc. In industry, well-focused laser beams are used for cutting, welding, marking parts, micro-piercing of metals and very hard materials, realization of semi-conductors, non-contact measurements, etc. Military applications for lasers include guiding munitions, missile defense, an alternative to radar, etc. In medicine, they are used in bloodless surgery, laser healing, kidney stone treatment, eye treatment, etc. Their daily use includes compact disc players and engravers, laser printers, barcode scanners in supermarkets. In telecommunications, we are in the first stages of a new era of optical communications combining lasers and fiber optics with an incredible increase in data-handling capacity for television transmission, phone conversations, etc.

10.13. Problems

Propagation of waves

P10.1 Consider the one-dimensional wave equation $\partial^2_{tt}u - v^2\partial^2_{xx}u = 0$. Introduce the variables $\xi = t - x/v$ and $\eta = t + x/v$. **a)** Show that the wave equation becomes $\partial^2 u/\partial\xi\partial\eta = 0$. **b)** Deduce that the solution of this equation may be written in the form $u = f(\xi) + g(\eta)$ where f and g are two arbitrary functions.

P10.2 **a)** Two waves of the same amplitude and the same linear polarization in the direction Ox propagate in the direction Oz in a dispersive medium. Let ω_1 and ω_2 be their frequencies and k_1 and k_2 their wave numbers. Determine the points where the amplitude of their superposition is maximum at a given time t. Show that, over the course of time, these points move with a velocity $v_{max} = (\omega_1-\omega_2)/(k_1-k_2)$ that is close to the group velocity $v_{(g)} = d\omega/dk$ if ω_2 is close to ω_1. **b)** Consider now the general case of a signal of space extension Δz. By Fourier theorem, this signal may be considered as a superposition of waves of angular frequencies ω and wave numbers k, of the form $u(z, t) = \int_{\Delta\omega} d\omega\, f(\omega)\, e^{i(\omega t - kz)}$, where k is related to ω by the dispersion relation of the medium. The integral is non-negligible at z and t if $(\omega t - kz)$ remains almost constant as ω varies in the frequency band $\Delta\omega \equiv [\omega_1, \omega_2]$. Deduce that the points, where $u(z, t)$ is large, are such that $t - (dk/d\omega)z = 0$. Deduce that, over the course of time, these points move with the group velocity $v_{(g)} = d\omega/dk$. To study the effect of dispersion on the propagation of wave packets, assume that the amplitude

$A(k)$ is real and Gaussian of the form $A(k) = A_{\text{m}} e^{-(k-k_o)^2/4\kappa^2}$, where κ = constant. $A(k)$ takes the value A_{m}/e for $k = k_o \pm 2\kappa$. Thus, $\Delta k = 4\kappa$ may be considered as the band width of k. Show that we may write the first two terms of the Taylor series $\omega = \omega_o + v_{(g)}(k-k_o) + \frac{1}{2}\beta(k-k_o)^2 + \ldots$ Using the integral $\int_{-\infty}^{+\infty} du\, e^{au-bu^2} = \sqrt{\pi/b}\, e^{a^2/4b}$ and neglecting β, show that $u(0,t) = 2\sqrt{\pi}\,\kappa A_{\text{m}} e^{-(\kappa v_g t)^2}\, e^{i\omega_o t}$. Deduce that the time width is $\Delta t = 2/\kappa v_{(g)}$ such that $\Delta t.\Delta\omega = 8$. Show that $u(x,0) = 2\sqrt{\pi}\,\kappa\, A\, e^{-\kappa^2 x^2}\, e^{-ik_o x}$, thus $\Delta x = 2/\kappa$ and $\Delta x.\Delta k = 8$. Keeping the term β in the expansion of ω, evaluate $u(x, t)$ and verify that its space extension is $\Delta x = (2/\kappa)\sqrt{1+4\beta^2\kappa^4 t^2}$; thus, it widens as it propagates.

P10.3 **a)** Knowing the expression of the phase velocity as a function of the wavelength λ, show that the group velocity may be written in the form $v_{(g)} = v_{(p)} - \lambda(dv_{(p)}/d\lambda)$. **b)** In optics, the index of refraction n is usually expressed as a function of the wavelength. Show that the group velocity may be written in the form $v_{(g)} = c/n + (c\lambda/n^2)(dn/d\lambda)$. **c)** Knowing the expression of the phase velocity as a function of the frequency, show that the group velocity is given by the equation $1/v_{(g)} = 1/v_{(p)} - (\tilde{\nu}/v_{(p)}^2)(dv_{(p)}/d\tilde{\nu})$. **d)** The refraction index of glass depends on the light wavelength in vacuum, according to Cauchy empirical formula $n = A + B/\lambda^2$. Calculate the phase velocity and the group velocity in a glass, whose parameters are $A = 1.584$ and $B = 1.270 \times 10^4$ nm^2 for the wavelength $\lambda = 600$ nm (in vacuum).

P10.4 Assume that the transverse waves propagate on a string with a speed v. Look to waves of the form $u(x, t) = f(x) \cos(\omega t)$. Show that $f(x)$ obeys a differential equation. Write its general solution. If the string is fixed at its ends, the wave must verify the boundary conditions $u(0, t) = u(L, t) = 0$. Show that the solution depends on an integer n. What are the possible values of the frequency and the wavelength?

Electromagnetic waves in dielectrics, polarization

P10.5 Determine the polarization of the following waves:

a) $\mathbf{E} = A\cos(\omega t - kz)\,\mathbf{e}_x + B\cos(\omega t - kz)\,\mathbf{e}_y$,
b) $\mathbf{E} = A\cos(\omega t - kz)\,\mathbf{e}_x + A\sin(\omega t - kz)\,\mathbf{e}_y$,
c) $\mathbf{E} = A\cos(\omega t - kx)\,\mathbf{e}_y + A\sin(-\omega t + kx)\,\mathbf{e}_z$,
d) $\mathbf{B} = A\cos(\omega t - kz + \pi/3)\,\mathbf{e}_x + A\sin(\omega t - kz)\,\mathbf{e}_y$,
e) $\mathbf{B} = -A\cos(\omega t - ky)\,\mathbf{e}_x + A\sin(\omega t - ky)\,\mathbf{e}_z$,
f) $\mathbf{B} = A\cos(\omega t + kx)\,\mathbf{e}_z - A\sin(\omega t + kx)\,\mathbf{e}_y$.

P10.6 Write down the expression for a plane wave propagating in a direction **e**, which lies in the plane Oxz making an angle θ with Oz if: **a)** it is polarized in the plane Oxz,

b) polarized perpendicularly to this plane, **c)** polarized right-circularly, and **d)** polarized left-circularly.

P10.7 Consider the superposition of two waves, which propagate in the direction Oz and which are polarized linearly in the directions $\mathbf{e}_x$ and $\mathbf{e}_y$. The components of $\mathbf{E}$ in the Oxy plane are $x = E_x \cos(\omega t - kz)$ and $y = E_y \cos(\omega t - kz + \phi)$, where ϕ is the waves phase shift. **a)** By considering the derivative $\partial_t y$ at $t = 0$, show that the tip of $\mathbf{E}$ moves on an ellipse clockwise if $0 < \phi < \pi$ and anticlockwise if $-\pi < \phi < 0$. **b)** Let Ox' and Oy' be the axes obtained from Ox and Oy by rotation through an angle α. Write $\mathbf{E}$ as a superposition of two waves polarized in the directions Ox' and Oy'. **c)** Let E'_x and E'_y be the amplitude of these waves and ϕ' be their phase shift. Show that $\frac{1}{2}E_x E_y \sin\phi = \frac{1}{2} E'_x E'_y \sin\phi'$. Interpret this relation. **d)** What should α be in order to have the ellipse axes in the directions of Ox' and Oy'?

P10.8 Consider the superposition of two waves $\underline{\mathbf{E}}_j = E_j \mathbf{e}_j$, which propagate in the same direction $\mathbf{e}$ and which are polarized linearly in the directions $\mathbf{e}_j$, where $j = 1.2$. They are not necessarily monochromatic. We define the 2×2 tensor, whose elements are $I_{ij} = \langle \underline{E}_i \underline{E}_j^* \rangle$. Here $\langle f \rangle$ stands for the time average of f. **a)** We define the *Stokes parameters*: $s_o = I_{11} + I_{22}$, $s_1 = 2\,\mathcal{Re}\, I_{21}$, $s_2 = 2\,\mathcal{Im}\, I_{21}$ and $s_3 = I_{11} - I_{22}$. Interpret s_o. Write the tensor I_{ij} as a 2×2 matrix $[I]$ in terms of the parameters s_i. What is its determinant and what is its trace? We define the *polarization* of the wave as $P = \mathbf{s}^2/s_o$, where $\mathbf{s}^2 = s_1^2 + s_2^2 + s_3^2$. Write $[I]$ in a new basis $\mathbf{e}'_1$ and $\mathbf{e}'_2$ obtained from $\mathbf{e}_1$ and $\mathbf{e}_2$ through a rotation of 45° and in the basis formed by the complex vectors $\mathbf{e}_{(\pm)} = (\mathbf{e}_1 \pm i\mathbf{e}_2)/\sqrt{2}$. Verify that the trace of $[I]$ and its determinant do not depend on the used basis. **b)** What are the values of the parameters s_i if the wave is completely unpolarized? **c)** Verify that $s_o^2 = \mathbf{s}^2$, thus $P = 1$ if the wave is monochromatic and polarized (elliptically in the general case).

P10.9 In an isotropic but nonlinear crystal, the electric displacement $\mathbf{D}$ is in the same direction as $\mathbf{E}$ but with a magnitude $D = \varepsilon E + \gamma E^2$. A plane harmonic wave of angular frequency ω is incident normally on this crystal. Assume that the wave is polarized in the direction Ox and it propagates in the direction Oz. **a)** Show that the symmetries imply that the wave in the crystal is polarized in the direction Ox and it propagates in the direction Oz. **b)** Nothing requires that the wave in the crystal be harmonic with the same frequency ω, but it must be repeated at intervals of time equal to the period of the excitation wave. Thus, harmonic waves of frequencies 2ω, 3ω, etc., may be generated in the crystal. Write the solution in the form:

$$\mathbf{E} = \Sigma_p E_p \cos(p\omega t - k_p z + \phi_p)\, \mathbf{e}_x.$$

Write the expression of **D**. Show that Faraday's induction law gives:

$$\mathbf{B} = \Sigma_p\,(k_p/p\omega)\,E_p \cos(p\omega t - k_p z + \phi_p)\,\mathbf{e}_y.$$

Show that Maxwell's equations $\nabla.\mathbf{B} = 0$ and $\nabla.\mathbf{D} = 0$ are verified. **c)** Impose Ampère's equation and deduce that we must have k_p and ϕ_p proportional to p, of the forms $k_p = pk$ and $\phi_p = p\phi$. Keeping only two terms in the power series, show that $E_1 = E_2$ and that the phase velocity depends on the field amplitude according to the relation $1/v_{(p)}^2 = \mu\varepsilon + \frac{1}{2}\mu\gamma E_1$.

Energy and momentum of plane electromagnetic waves

P10.10 a) Two waves propagate in the direction Oz. In which case, the energy density and the intensity of their superposition are the sum of the corresponding quantities for the waves? **b)** Consider a wave $\underline{\mathbf{E}} = \underline{E}\, e^{i(\omega t - kz)}\,\mathbf{e}_x$. Calculate its field $\underline{\mathbf{B}}$. Verify that the intensity of this wave may be written as $|\underline{\mathbf{E}} \times \underline{\mathbf{B}}^*|/2\mu$. **c)** Natural unpolarized light may be considered as the superposition of two linearly polarized waves in the directions Ox and Oy, respectively, with a phase shift between them varying randomly. What is the transmitted intensity if this light in incident on a plate that absorbs completely the y component?

P10.11 A wave of angular frequency ω propagates in a non-magnetic medium of index n. Its direction of propagation **e** lies in the Oyz plane and makes an angle θ with Oz. The wave is polarized in the Oyz plane. **a)** Write the expressions of its fields **E** and **B**. **b)** Calculate its energy density, its Poynting vector and its intensity. What is the average power received by the unit area of the Oxy plane?

P10.12 A spacecraft of mass 10^3 kg moves in free space propelled by reaction to a beam of light of 10^6 W that it emits in the opposite direction to its motion. Neglecting the gravitational force, determine the exerted force on this spacecraft and its velocity after 24 hours, starting from rest.

P10.13 A He-Ne laser beam has a wavelength $\lambda = 632.8$ nm, a section $\mathcal{S} = 0.10$ cm^2 and a power $P = 1$ W. Calculate the amplitude of the corresponding electric field and that of the magnetic field. What should the power of an incandescence lamp of efficiency 10% be to produce the same light intensity at 1 m from the lamp? What should be the radius of a particle of mass density 10^3 kg.m^{-3} in order for it to be suspended by this upward laser beam? Assume that the particle absorbs light totally.

P10.14 A light beam of intensity $\mathcal{I}$ falls on a surface $\mathcal{S}$ with an angle of incidence θ. **a)** Assuming that all light is diffused isotropically back to the incidence medium, calculate the rate of momentum transfer to the surface. **b)** Assume now that the surface absorbs a fraction α of this energy and diffuse the remaining isotropically.

Determine the rate of momentum transfer to the surface and the radiation pressure. **c)** Assuming that the surface is a perfect mirror, calculate the force and the exerted pressure on this mirror if $\mathcal{I}$ = 20 W/cm^2, θ = 30° and $\mathcal{S}$ = 100 cm^2.

P10.15 The *solar constant* is the flux of the solar electromagnetic energy that is incident normally on the unit area of the surface of the Earth. It is about 1340 W/m^2. **a)** Calculate the amplitude of the corresponding electric field. **b)** The average distance of the Earth to the Sun is 1.49×10^8 km. Calculate the total power that the Sun emits as radiation. Using the relativity relation $\Delta U_o = \Delta m\, c^2$, calculate the mass that the Sun loses per second because of this emission. **c)** One project to convert solar energy into electric energy. What should be the surface of the solar panel in order to produce 1 kW, assuming 30% efficiency? **d)** Compare the force exerted by the radiation on a particle of radius r_o with the gravitational attraction of the Sun, knowing that the mass of the Sun is about 2×10^{30} kg and the gravitational constant is $G = 6.67 \times 10^{-11}$ N.m^2/kg^2. Assume that the particle density is m_V = 10^3 kg/m^3 and that it absorbs all the radiation. Show that, if r_o is less than a certain value, the radiation pressure may be larger than the gravitational attraction. The particle is then repulsed by the Sun instead of being attracted. This may explain the comet's tail in the opposite direction of the Sun.

P10.16 The emitter of a radar station has a power P and it consists of a small electric dipole located at the focus of a parabolic antenna whose aperture radius is R. As we shall see in section 11.8, the emitted wave of wavelength λ is diffracted in a cone of half-angle θ, such that $\sin\theta = 0.6\ \lambda/R$. **a)** Assuming that the wave is emitted isotropically in this cone, calculate the intensity of the wave $\mathcal{I}(r)$ at a distance r from the antenna. **b)** The wave encounters an obstacle of area $\mathcal{S}$ at a distance D. The reflected wave is assumed to be isotropic in all directions on one side of this surface. It is intercepted by the same parabolic antenna. Calculate the intercepted power P_i.

Momentum and angular momentum densities, radiation pressure

P10.17 An electromagnetic wave propagates in the direction Oz and is polarized linearly in the direction Ox. It is intercepted by a plane plate of area $\mathcal{S}$ initially lying in the Oxy plane and then rotated through an angle θ about Oy. Calculate the Maxwell's tensor. Deduce the force exerted by the radiation on this surface $\mathcal{S}$, assuming that it absorb totally the radiation.

P10.18 a) Using the model of section 10.6 with a single binding force $-m\omega_o^2 u$ for electrons and neglecting the friction force, show that the dispersion relation in this medium is $c^2k^2 = \omega^2 + \omega_p^2\omega^2/(\omega_o^2 - \omega^2)$ with $\omega_p^2 = Ne^2/\varepsilon_o m$. Calculate the group velocity. **b)** Plot the index of refraction versus the angular frequency, ω. How can we interpret the negative values of n^2 and the values of the index that are positive but less than 1? According to special relativity, the speed of particles, of energy, or any physical quantity is always less than the speed of light in vacuum c. Does an index of refraction less than 1 contradict special relativity? **c)** Plot the group velocity versus the angular frequency of the wave. Show that, if $|n|^2$ increases as a function of ω (normal dispersion), $v_{(g)}$ is less than c, in agreement with special relativity. But, if $|n|^2$ decreases as a function of ω (abnormal dispersion), $v_{(g)}^2$ is negative. In this case $v_{(g)}$ cannot be interpreted as the speed of energy or signals.

Electromagnetic waves in conductors

P10.19 Analyze the equation of propagation of **E** in a conductor by using the method of separation of variables. Deduce that $\mathbf{E} = \mathbf{F}(\mathbf{r})\, e^{-t/\tau_c}$, where $\mathbf{F}(\mathbf{r})$ verifies the equation $\Delta\mathbf{F}(\mathbf{r}) = \kappa\, \mathbf{F}(\mathbf{r})$ with κ contant. Using the continuity equation of the charge, show that $\tau_c = \varepsilon/\sigma$ and $\kappa = 0$. Deduce that an electric field, which may exist at $t = 0$, decreases exponentially in time with the same relaxation time τ_c as the charge density.

P10.20 Show that, if we use a gauge such that $\boldsymbol{\nabla}.\mathbf{A} + \mu\varepsilon\, \partial_t V + \mu_o\sigma V = 0$, the equations of propagation of V and **A** may be written in an Ohmic conductor in the uncoupled forms

$$\Delta V - \mu\varepsilon\, \partial_{tt} V - \mu_o\sigma\, \partial_t V = -q_v/\varepsilon, \qquad \Delta\mathbf{A} - \mu\varepsilon\, \partial_{tt}\mathbf{A} - \mu_o\sigma\partial_t\mathbf{A} = 0.$$

P10.21 Let us assume that the electric field is $\mathbf{E} = \mathbf{F}(\mathbf{r})\, e^{i\omega t}$ in a conductor, whose conductivity is σ. We assume that $\mu = \mu_o$ and $\varepsilon = \varepsilon_o$. **a)** Using Maxwell's equations, determine the field **B**, the current density **j** and the charge density q_v. **b)** Assuming that $q_v = 0$, verify that $\Delta\mathbf{j} + \omega\mu_o(\omega\varepsilon_o - i\sigma)\, \mathbf{j} = 0$. Compare the displacement current with the conduction current.

P10.22 A wave is polarized in the direction Ox and it propagates in the direction Oz in an infinite conductor. Write the expression of the Poynting vector. Calculate the energy loss between the planes z and $z + dz$ and verify that it is equal to the energy dissipated as Joule heat.

P10.23 a) Show that the Maxwell's equations in a linear, homogeneous and isotropic conductor imply that $\partial_t U_{EM,v} + \mathbf{j}.\mathbf{E} = -\, \boldsymbol{\nabla}.\mathbf{S}$. Integrate this equation over a volume $\mathcal{V}$ bounded by a surface $\mathcal{S}$ and transform the volume integral of $\boldsymbol{\nabla}.\mathbf{S}$ into the flux of **S** over $\mathcal{S}$. Interpret the two terms of the left-hand side. Deduce that the vector **S** is the *density of the energy flux*. **b)** Determine the Poynting vector **S** and analyze the energy flow inside a conducting cylinder carrying a current of uniform density **j**.

P10.24 A simple harmonic wave of angular frequency ω propagates in the direction Oz and is polarized in the direction Ox in a conductor of conductivity σ. **a)** Write the expressions of the fields **E** and **B**. **b)** In the usual metallic conductors, the charge density is negligible. At which frequencies is this condition verified? What then are the expressions of the fields and the current density?

P10.25 Let us consider a good conductor and a wave that propagates in the direction Oz and suppose that the charge density is negligible. **a)** Show that **E** is transverse. Let us assume that $\underline{\mathbf{E}} = \underline{E}_{\mathrm{m}} e^{\mathrm{i}(\omega t - pz)} \mathbf{e}_{\mathrm{x}}$. Show that $\mathbf{B} = (p/\omega)\underline{E}_{\mathrm{m}} e^{\mathrm{i}(\omega t - pz)} \mathbf{e}_{\mathrm{y}}$. Show that the Maxwell's equations are verified if $p = k - \mathrm{i}\eta$ where k and η are given by the expressions [10.103]. **b)** Calculate the equivalent surface current density and the fields just outside the surface. **c)** Write the expressions of the real fields.

Electromagnetic waves in plasmas

P10.26 If we neglect the friction force exerted on the conduction electrons in a metal and, consequently, the dissipation of energy as Joule heat, the metal may be treated as a plasma. Assume that the number of conduction electrons is 1 electron per atom in silver. **a)** Calculate the number of conduction electrons per unit volume and the cut-off frequency $\tilde{\nu}_{\mathrm{p}}$. Can visible light propagate in solid silver? **b)** What is the attenuation coefficient in silver for light of wavelength $\lambda = 580$ nm in vacuum? Verify that visible light penetrates silver only a fraction of a micron. **c)** What should the thickness of silver on a glass plate be in order to have a mirror that reflects half the light intensity at 99%? At what wavelength does silver become transparent?

P10.27 An ionized gas fills a parallel plate capacitor of thickness d and area $\mathcal{S}$. The total current is the sum of the displacement current and the conduction current due to the motion of electrons. Assuming that the electric field in the plasma is $\mathbf{E} = \underline{\mathbf{E}}_{\mathrm{m}} e^{\mathrm{i}(\omega t - kz)}$, show that the total current density is $\mathbf{j} = \mathrm{i}(\varepsilon_0\omega - N_{\mathrm{v}}e^2/m\omega)\mathbf{E}$, where m is the mass of the electron, N_{v} is the number of electrons per unit volume and $-e$ is the charge of the electron. Deduce that this set-up is equivalent to a capacitor of capacitance $C = \varepsilon_0\mathcal{S}/d$ and a solenoid of inductance $L = md/\mathcal{S}N_{\mathrm{v}}e^2$ connected in parallel.

P10.28 In section 3.7, we have interpreted Ohm's law in the case of a stationary current by assuming that each conduction electron is subject to a friction force $\mathbf{f} = -b\mathbf{v}$. This force is due to the collision of the electron with the other electrons and with the positive ions. **a)** Let us assume first that the electric field **E** in the conductor is time-independent. Show that the electrons are accelerated by the field and that their velocity tends to a limit $\mathbf{v} = -e\mathbf{E}/m$. Deduce that the electric current is given by Ohm's law $\mathbf{j} = \sigma\mathbf{E}$ where $\sigma = N_{\mathrm{v}}e^2/b$ and N_{v} is the number of electrons per unit volume. **b)** If the field **E** is sinusoidal with angular frequency ω, we expect that the result of question (a) holds at low frequency. Neglecting the magnetic force, write the equation of motion of the electron and its solution. Deduce that the conductor obeys Ohm's law with a complex conductivity $\underline{\sigma} = N_v e^2/(b + im\omega)$. **c)** Calculate the dissipated power per unit volume in this conductor. **d)** A medium may be considered as "metallic" if $b >> m\omega$ and as a "plasma" if $b << m\omega$. At what frequencies may copper be considered a plasma? **e)** In a dense plasma, the electrons collide frequently with other electrons and the positive ions. Thus, they are subject to a friction force $\mathbf{f} = -b\mathbf{v}$ exactly as an Ohmic conductor. Write Maxwell's equations in this plasma. Deduce the relations

$$\mathbf{e}.\underline{\mathbf{E}}_m = 0, \quad \mathbf{e}.\underline{\mathbf{B}}_m = 0, \quad \underline{\mathbf{B}}_m = (k/\omega)\,(\mathbf{e} \times \underline{\mathbf{E}}_m), \quad \text{and} \quad \underline{\mathbf{u}}_m = e\underline{\mathbf{E}}_m/(m\omega^2 - ib\omega).$$

What is the dispersion relation of this medium?

Quantization of electromagnetic waves

P10.29 A wave is specified by the electric field $\mathbf{E} = \mathbf{E}_m \cos(\omega t - kz)$ with the amplitude $\mathbf{E}_m$ pointing in the direction Ox. **a)** Calculate the corresponding energy density and Poynting vector. **b)** Determine the energy of the photons of this wave, their momentum, and their average number per unit volume. These photons are intercepted by a totally absorbing surface $\mathcal{S}$, which lies in the Oxy plane. Determine the average number of photons, which are absorbed per unit area of this plate and per unit time, the momentum that it receives and the force exerted by the radiation on it. **c)** Assume that the plate is metallic with the conduction electrons subject to a friction force $-b\mathbf{v}$. Show that, at low frequency ($\omega << b/m_e$), they tend to a terminal velocity $\mathbf{v} = -e\mathbf{E}/b$. Show that the electrons are subject to a magnetic force $F_M = (e^2/2b)\,EB$ pointing in the direction Oz. **d)** This force may be assimilated to a radiation pressure. Show that it is equivalent to a rate of transfer of momentum $\Delta p = (e^2/b)\,EB$ per electron and per unit time. Show that the work exerted by the electric field on the electron in motion is $\Delta W = (e^2/b)\,E^2$. Verify that, in the case of an electromagnetic wave in vacuum, $\Delta W = c\,\Delta p$. This is the relationship between the energy density and the momentum density of the wave or between the energy and the momentum of the photons that are associated with the wave and absorbed by the metallic plate.

P10.30 The ground state energy of the hydrogen atom is $E_1 = -13.6057$ eV and its first excited level is at $E_2 = -3.4014$ eV. What is the wavelength of the emitted radiation in the transition from E_2 to E_1? What should the kinetic energy of an electron, which collides with the atom, be in order to excite it from the ground state to the level E_2? What should this kinetic energy be in order to ionize the atom from the ground state?

Chapter 11

Reflection, Interference, Diffraction and Diffusion

In this chapter we study the laws of reflection and transmission of electromagnetic waves at the interface of two mediums (one of them may be the vacuum). The wave may be totally or partially reflected, a part being transmitted across the interface. The amplitudes of the reflected wave and the transmitted wave, as well as their possible phase shifts, are determined by the *boundary conditions* at the interface. First, we formulate the laws determining the direction of propagation of the reflected wave and the transmitted wave in the case of two dispersive mediums. These laws hold for any type of wave. Then, we study the reflection and refraction of electromagnetic waves on the interface of two dielectrics, of a dielectric with a conductor and of a dielectric with a plasma. In the second part of this chapter, we study the interference of two and several waves, the diffraction, and the diffusion of waves.

11.1. General laws of reflection and refraction

The laws of reflection and refraction of light, which were initially established experimentally, were interpreted by Huygens for any type of wave. To simplify, we consider a scalar wave $\underline{u}$, which propagates in a medium (1). If it meets the interface of this medium with another medium (2), generally there is a reflected wave $\underline{u}'$ back toward the medium (1) and a transmitted wave $\underline{u}''$ in the medium (2) (Figure 11.1). Huygens' principle assumes that each point P of a wavefront Σ_0 at time t_0 behaves like a point source, emitting a secondary wavelet S_P. The envelope Σ of these wavelets is the wavefront later. It is not necessary for the sources to be material ones; the principle holds even in the case of the propagation of light in vacuum. From this general principle, it is possible to deduce the following laws of reflection and refraction, which are verified experimentally:

– The direction of propagation of the incident wave, that of the reflected wave, and that of the refracted wave lie in the same plane containing the normal to the interface $\mathcal{S}$ at each point of incidence.

– The angle of reflection θ' (between the direction of propagation of the reflected wave and the normal Oz' to $\mathcal{S}$) is equal to the angle of incidence θ (between the direction of propagation of the incident wave and the normal Oz to $\mathcal{S}$). The angle of refraction θ'' (between the direction of propagation of the refracted wave and the normal Oz to $\mathcal{S}$) is related to θ by Snell's law

$$n_1 \sin\theta = n_2 \sin\theta''. \qquad [11.1]$$

$n_1 = c/v_1$ and $n_2 = c/v_2$ are the *indices of refraction* of mediums (1) and (2), where c is the speed of propagation in a medium of reference (the vacuum in the case of electromagnetic waves).

Equation [11.1] determines the angle of refraction θ'', if $(n_1/n_2)\sin\theta < 1$. This condition can be always satisfied if $n_1 < n_2$. In the case $n_1 > n_2$, we must have $\sin\theta < n_1/n_2$. Thus, θ must be less than a *critical angle* (or *limiting angle*) i_L, given by

$$\sin i_L = n_2/n_1. \qquad [11.2]$$

At the angle of incidence equal to i_L the angle of refraction is $\theta'' = 90°$. If the angle of incidence θ is larger than i_L, the wave undergoes *total reflection*.

The wave theory enables us to establish the laws of reflection and refraction using the *boundary conditions* (or *continuity equations*) at the interface $\mathcal{S}$. We have only to assume that these conditions are expressed as *linear relations* between the incident wave $\underline{u}$, the reflected wave $\underline{u}'$, and the transmitted wave $\underline{u}''$, and their partial derivatives with respect to time or space coordinates. In the case of simple harmonic waves $u = u_m\, e^{i(\omega t - \mathbf{k.r})}$, the time derivative is simply $i\omega u$ and the derivative with respect to x, for instance, is $-ik_x u$. Thus, the boundary conditions are linear, of the general form $\underline{au} + \underline{bu'} + \underline{cu''} = 0$, that is,

$$\underline{a}\,\underline{u}_m\, e^{i(\omega t - \mathbf{k.r})}\, e^{i(\omega t - \mathbf{k.r})} + \underline{b}\,\underline{u}'_m\, e^{i(\omega' t - \mathbf{k}'.\mathbf{r})} + \underline{c}\,\underline{u}''_m\, e^{i(\omega'' t - \mathbf{k}''.\mathbf{r})} = 0. \qquad [11.3]$$

A relation of this form can be verified at any t and at any point $\mathbf{r}$ of the interface $\mathcal{S}$ only if the angular frequency and the scalar product $(\mathbf{k.r})$ on $\mathcal{S}$ are the same for the three waves, u, u' and u''. We deduce that the angular frequency undergoes no change in the reflection and transmission ($\omega = \omega' = \omega''$) and the tangential component of **k**, **k**′, and **k**" (i.e. their components which are parallel to $\mathcal{S}$) are the same. The normal component of the wave vectors are obtained using the dispersion relation $\omega = \omega(\mathbf{k})$ for each medium.

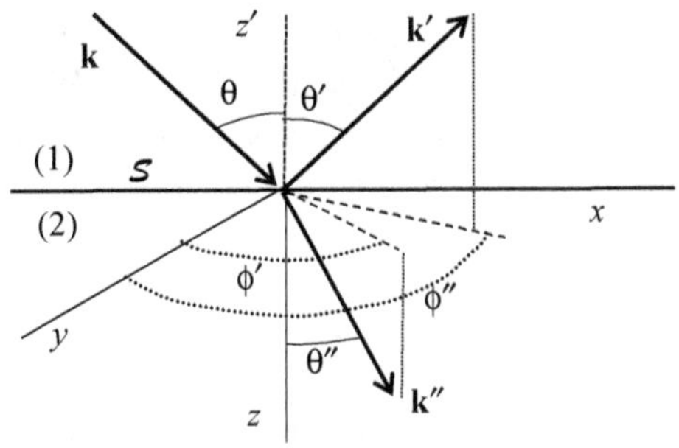

Figure 11.1. *Laws of reflection and refraction*

We choose the axes of coordinates in such a way that *Oxy* is tangent to the interface $\mathcal{S}$ with *Oz* oriented from medium (1) toward medium (2) and the x axis in the plane of incidence (formed by **k** and the normal axis *Oz*) (Figure 11.1). Due to the symmetry with respect to the plane of incidence *Oxz*, the vectors **k**′ and **k**" are in this plane. This same result may also be obtained by analysis. Indeed, let θ be the angle of **k** with *Oz*, θ′ the angle of **k**′ with *Oz*′, and θ" the angle of **k**" with *Oz* (so they lie between 0 and 90°). Let us assume that the azimuthal angles of the reflection plane and the refraction plane are ϕ′ and ϕ", respectively. Thus, we have

$$k'(x \sin\theta' \cos\phi' + y \sin\theta' \sin\phi') = k''(x \sin\theta'' \cos\phi'' + y \sin\theta'' \sin\phi'') = k\, x \sin\theta.$$

These relations are satisfied for all values of the coordinates x and y if

$$k' \sin\theta' \cos\phi' = k'' \sin\theta'' \cos\phi'' = k \sin\theta, \qquad k' \sin\theta' \sin\phi' = k'' \sin\theta'' \sin\phi'' = 0.$$

As $k = k' = \omega/v_1$ and $k'' = \omega/v_2$, we obtain the relations

$$\phi' = \phi'' = 0, \qquad \theta = \theta', \qquad (1/v_1)\sin\theta = (1/v_2)\sin\theta''. \qquad [11.4]$$

The equations $\phi' = 0$ and $\phi'' = 0$ are the expressions that the vectors **k**, **k**′, and **k**" lie in the plane of incidence *Oxz*. The equality $\theta = \theta'$ is the law of reflection and the last relationship [11.4] is the expression of Snell's law [11.1]. Consequently, the wave vectors may be written as:

$$\mathbf{k} = k(\sin\theta\, \mathbf{e}_x + \cos\theta\, \mathbf{e}_z), \quad \mathbf{k}' = k'(\sin\theta'\, \mathbf{e}_x - \cos\theta'\, \mathbf{e}_z),$$
$$\mathbf{k}'' = k''(\sin\theta''\, \mathbf{e}_x + \cos\theta''\, \mathbf{e}_z). \qquad [11.5]$$

Taking into account the equality of the phases ($\omega t - \mathbf{k.r}$) on $\mathcal{S}$, the boundary conditions [11.3] reduce to linear relationships between the complex amplitudes $\underline{a}\, \underline{u}_m + \underline{b}\, \underline{u}'_m + \underline{c}\, \underline{u}''_m = 0$. We must have two relations of this type to determine the complex amplitudes $\underline{u}'_m$ of the reflected wave and $\underline{u}''_m$ of the refracted wave in terms of the amplitude $\underline{u}_m$ of the incident wave. This is equivalent to determining the amplitudes and the phase shifts of $\underline{u}'_m$ and $\underline{u}''_m$. The continuity relationships being linear, we find that $\underline{u}'_m$ and $\underline{u}''_m$ are proportional to $\underline{u}_m$. We define the *reflection coefficient* $\mathcal{R}$ and the *transmission coefficient* $\mathcal{T}$ as the ratios of the complex amplitudes of the reflected wave and the transmitted wave, respectively, to that of the incident wave

$$\mathcal{R} = \underline{u}'_m/\underline{u}_m, \qquad \mathcal{T} = \underline{u}''_m/\underline{u}_m. \qquad [11.6]$$

The law of *conservation of energy* requires that the normal components of the vectors energy flux density (or Poynting vector) **S** verify the continuity condition

$$S_z + S'_z = S''_z \qquad \text{(at } z = 0\text{)}. \qquad [11.7]$$

The tangential components of **S**, **S′**, and **S''** play no part in the transfer of energy across the interface, as they correspond to a propagation of energy in a direction that is parallel to the interface in each medium. **S** being quadratic in **u**, equation [11.7] can be verified at each point of the interface $\mathcal{S}$ only if the phase ($\omega t - \mathbf{k.r}$) is the same for the three waves at the interface $\mathcal{S}$. Thus, we must have $\omega = \omega' = \omega''$ and $\mathbf{k.r} = \mathbf{k'.r} = \mathbf{k''.r}$ for $z = 0$. We conclude that the laws of reflection and refraction are closely related to the principle of conservation of energy.

The concepts of simple harmonic wave and plane wave are useful mathematical models. A real wave is always a superposition of waves in a certain band of frequency $\Delta\omega$ and a certain band of wave vector $\Delta\mathbf{k}$. The reflected wave has the same band of frequency and a band of wave vector $\Delta\mathbf{k}'$ (such that $\Delta\mathbf{k}'_{//} = \Delta\mathbf{k}_{//}$ and $\Delta k'_z = -\Delta k_z$), while the refracted wave has the same band of frequency $\Delta\omega$ and a band of wave vector $\Delta\mathbf{k}''$ (such that $\Delta\mathbf{k}''_{//} = \Delta\mathbf{k}_{//}$ and $|\Delta k'_z| \neq |\Delta k_z|$).

11.2. Reflection and refraction on the interface of two dielectrics

Consider an electromagnetic plane wave that is incident from a dielectric (1) at an angle θ on the face $\mathcal{S}$ of a dielectric (2). We assume that the dielectrics are linear and isotropic; thus, the electromagnetic field is specified by two fields **E** and **B** (since $\mathbf{D} = \varepsilon\mathbf{E}$ and $\mathbf{H} = \mathbf{B}/\mu$). The primary electromagnetic fields act on the electric charges; they oscillate and emit secondary waves of the same frequency. The superposition of the secondary waves constitutes the reflected wave backward toward medium (1) and their superposition with the primary wave constitutes the refracted wave forward in medium (2). The amplitudes of the reflected wave and the refracted wave depend on the interaction of the electromagnetic field with matter, which is related to the electric susceptibility and the magnetic susceptibility of the medium. Globally, this process produces the boundary conditions on the interface $\mathcal{S}$. As the interface of the dielectrics carries no charge and current densities, these conditions may be written as (see section 9.5)

$$\mathbf{E}_{//1} = \mathbf{E}_{//2}, \qquad \mathbf{B}_{//1}/\mu_1 = \mathbf{B}_{//2}/\mu_2, \qquad \varepsilon_1 E_{1\perp} = \varepsilon_2 E_{2\perp}, \qquad B_{1\perp} = B_{2\perp}. \qquad [11.8]$$

An electromagnetic wave has two independent states of linear polarization determined by the direction of the electric field **E**. Any wave is a superposition of these two states. Thus, it is sufficient to study the reflection and refraction of plane waves in these two states of polarization. As the mediums are assumed to be isotropic, if we choose the first direction of polarization in the plane of incidence *Oxz* and the second perpendicular to this plane, the incident wave is symmetric with respect to the plane of incidence (for the direction of propagation and the direction of polarization). The reflected wave and the transmitted wave must have the same symmetry. Thus their wave vectors are in the plane of incidence (and given by [11.5]) and they must have the same linear polarization as the incident wave. We write the fields in the form $\mathbf{E} = \underline{\mathbf{E}}_m e^{i(\omega t - \mathbf{k.r})}$, $\mathbf{B} = \underline{\mathbf{B}}_m e^{i(\omega t - \mathbf{k.r})}$, etc., and, for simplicity, we omit the exponentials, which have the same value on the interface.

A) Case of linear polarization in the plane of incidence

In this case, the fields **E**, **E′**, and **E″** of the incident, reflected, and refracted waves are in the plane of incidence *Oxz* (Figure 11.2a), while **B**, **B′**, and **B″** are parallel to *Oy*. Thus, it is convenient to determine the waves by the magnetic fields. The electric fields are then $\mathbf{E} = \mathbf{B}\times\mathbf{k}/\omega\mu_1\varepsilon_1$, $\mathbf{E}' = \mathbf{B}'\times\mathbf{k}'/\omega\mu_1\varepsilon_1$, and $\mathbf{E}'' = \mathbf{B}''\times\mathbf{k}''/\omega\mu_2\varepsilon_2$. The complex amplitudes of the fields may be written as

$$\begin{aligned}
&\underline{\mathbf{B}}_m = \underline{B}_m\,\mathbf{e}_y, && \underline{\mathbf{E}}_m = v_1\underline{B}_m(\cos\theta\,\mathbf{e}_x - \sin\theta\,\mathbf{e}_z),\\
&\underline{\mathbf{B}}'_m = -\underline{B}'_m\,\mathbf{e}_y, && \underline{\mathbf{E}}'_m = v_1\underline{B}'_m(\cos\theta'\,\mathbf{e}_x + \sin\theta'\,\mathbf{e}_z),\\
&\underline{\mathbf{B}}''_m = \underline{B}''_m\,\mathbf{e}_y, && \underline{\mathbf{E}}''_m = v_2\underline{B}''_m(\cos\theta''\,\mathbf{e}_x - \sin\theta''\,\mathbf{e}_z),
\end{aligned} \qquad [11.9]$$

where a (–) sign was introduced into the expression of $\mathbf{B}'_m$ in order for $\mathbf{E}'_m$ and $\mathbf{E}_m$ to have the same sign at the limit $\theta = 0$. At the interface $\mathcal{S}$ ($z = 0$), the boundary conditions [11.8] may be written as

$$\varepsilon_1 v_1(\underline{B}_m \sin\theta - \underline{B}'_m \sin\theta') = \varepsilon_2 v_2\,\underline{B}''_m \sin\theta'',$$
$$(\underline{B}_m - \underline{B}'_m)/\mu_1 = \underline{B}''_m/\mu_2, \qquad v_1(\underline{B}_m\cos\theta + \underline{B}'_m\cos\theta') = v_2\,\underline{B}''_m\cos\theta''.$$

These equations determine the amplitudes B'_m and B''_m of the reflected wave and the transmitted wave and, consequently, the electric fields as functions of B_m. We obtain the reflection and transmission coefficients

$$\mathcal{R}_{//} = \frac{\underline{E}'_m}{\underline{E}_m} = \frac{\underline{B}'_m}{\underline{B}_m} = \frac{\mu_2 \sin(2\theta'') - \mu_1 \sin(2\theta)}{\mu_1 \sin(2\theta) + \mu_2 \sin(2\theta'')} = \frac{Z_2 \cos\theta'' - Z_1 \cos\theta}{Z_1 \cos\theta + Z_2 \cos\theta''},$$

$$\mathcal{T}_{//} = \frac{\underline{E}''_m}{\underline{E}_m} = \frac{v_2}{v_1}\frac{\underline{B}''_m}{\underline{B}_m} = \frac{4\mu_2 \sin\theta'' \cos\theta}{\mu_1 \sin(2\theta) + \mu_2 \sin(2\theta'')} = \frac{2Z_2 \cos\theta}{Z_1 \cos\theta + Z_2 \cos\theta''}, \qquad [11.10]$$

where we have used Snell's law and the impedances of the mediums $Z_i = \sqrt{\mu_i/\varepsilon_i}$. We find that $\underline{B}_m$, $\underline{B}'_m$, and $\underline{B}''_m$ are in phase or in opposite phase; thus, we may take them to be real. If the mediums are non-magnetic ($\mu_i = \mu_o$, $\varepsilon_i = \varepsilon_o n_i^2$ and $Z_i = Z_o/n_i$), we get *Fresnel's formulas for p-waves*

$$\mathcal{R}_{//} = \frac{n_1 \cos\theta'' - n_2 \cos\theta}{n_1 \cos\theta'' + n_2 \cos\theta} = \frac{\tan(\theta'' - \theta)}{\tan(\theta + \theta'')}, \qquad \mathcal{T}_{//} = \frac{2\cos\theta \sin\theta''}{\sin(\theta + \theta'') \cos(\theta - \theta'')}. \qquad [11.11]$$

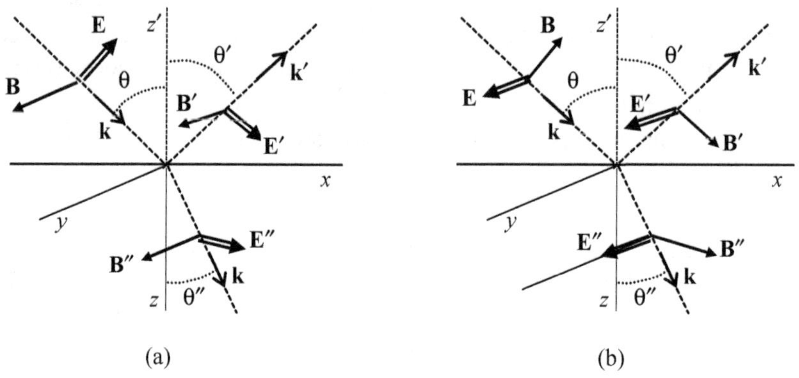

Figure 11.2. *Reflection and refraction of an electromagnetic wave that is linearly polarized: a) in the plane of incidence, and b) perpendicularly to the plane of incidence*

B) Case of linear polarization perpendicular to the plane of incidence

In this case, **E** is parallel to Oy and $\mathbf{B} = (\mathbf{k} \times \mathbf{E})/vk$ is in the plane of incidence (Figure 11.2b). Thus, it is convenient to determine the waves directly by the electric fields and write the amplitudes

$$\begin{aligned} &\underline{\mathbf{E}}_m = \underline{E}_m \mathbf{e}_y, && \underline{\mathbf{B}}_m = (\underline{E}_m/v_1)(-\cos\theta\, \mathbf{e}_x + \sin\theta\, \mathbf{e}_z), \\ &\underline{\mathbf{E}}'_m = \underline{E}'_m \mathbf{e}_y, && \underline{\mathbf{B}}'_m = (\underline{E}'_m/v_1)(\cos\theta'\, \mathbf{e}_x + \sin\theta'\, \mathbf{e}_z), \\ &\underline{\mathbf{E}}''_m = \underline{E}''_m \mathbf{e}_y, && \underline{\mathbf{B}}''_m = (\underline{B}''_m/v_1)(-\cos\theta''\, \mathbf{e}_x + \sin\theta''\, \mathbf{e}_z). \end{aligned} \qquad [11.12]$$

The boundary conditions of [11.8] give

$$E_{\mathrm{m}} + E'_{\mathrm{m}} = E'', \qquad (n_1/\mu_1)\,(E_{\mathrm{m}} \cos\theta - E'_{\mathrm{m}} \cos\theta') = (n_2/\mu_2)\, E''_{\mathrm{m}} \cos\theta. \quad [11.13]$$

These equations determine the reflection and transmission coefficients

$$\mathcal{R}_{\perp} = \frac{\underline{E}'_{\mathrm{m}}}{\underline{E}_{\mathrm{m}}} = \frac{\underline{B}'_{\mathrm{m}}}{\underline{B}_{\mathrm{m}}} = \frac{\mu_2 \,\mathrm{tg}\,\theta'' - \mu_1 \,\mathrm{tg}\,\theta}{\mu_2 \,\mathrm{tg}\,\theta'' + \mu_1 \,\mathrm{tg}\,\theta} = \frac{Z_2 \cos\theta - Z_1 \cos\theta''}{Z_2 \cos\theta + Z_1 \cos\theta''},$$

$$\mathcal{T}_{\perp} = \frac{\underline{E}''_{\mathrm{m}}}{\underline{E}_{\mathrm{m}}} = \frac{v_2}{v_1}\frac{\underline{B}''_{\mathrm{m}}}{\underline{B}_{\mathrm{m}}} = \frac{2\mu_2 \tan\theta''}{\mu_2 \tan\theta'' + \mu_1 \tan\theta} = \frac{2 Z_2 \cos\theta}{Z_2 \cos\theta + Z_1 \cos\theta''}. \quad [11.14]$$

Particularly, if the mediums are non-magnetic, we get *Fresnel's formulas for s-waves*

$$\mathcal{R}_{\perp} = \frac{n_1 \cos\theta - n_2 \cos\theta''}{n_1 \cos\theta + n_2 \cos\theta''} = \frac{\sin(\theta'' - \theta)}{\sin(\theta'' + \theta)}, \quad \mathcal{T}_{\perp} = \frac{2 n_1 \cos\theta}{n_1 \cos\theta + n_2 \cos\theta''} = \frac{2\cos\theta \sin\theta''}{\sin(\theta'' + \theta)}. \quad [11.15]$$

Note that the reflection and transmission coefficients in the case of incidence from medium (1) on medium (2) and from medium (2) on medium (1), for any state of polarization, verify the relations

$$\mathcal{R}_{12} = -\mathcal{R}_{21} \qquad \text{and} \qquad \mathcal{T}_{12}\,\mathcal{T}_{21} = 1 - \mathcal{R}_{12}\,\mathcal{R}_{21}. \quad [11.16]$$

If the angle of incidence is small ($\theta \approx \theta' \approx \theta'' \approx 0$), $\mathcal{R}_{/\!/}$ and $\mathcal{R}_{\perp}$ approach the same limit $\mathcal{R}_{\mathrm{o}}$, while $\mathcal{T}_{/\!/}$ and $\mathcal{T}_{\perp}$ approach the same limit $\mathcal{T}_{\mathrm{o}}$ given by

$$\mathcal{R}_{\mathrm{o}} \to \frac{\theta'' - \theta}{\theta'' + \theta} = \frac{n_1 - n_2}{n_1 + n_2} \qquad \text{and} \qquad \mathcal{T}_{\mathrm{o}} \to \frac{2\theta''}{\theta'' + \theta} = \frac{2 n_1}{n_1 + n_2}. \quad [11.17]$$

The variations of the reflection and transmission coefficients as functions of the angle of incidence θ are illustrated in Figure 11.3. $\mathcal{R}_{\perp}$ decreases from $\mathcal{R}_{\mathrm{o}}$ to -1 and $\mathcal{R}_{/\!/}$ increases from $\mathcal{R}_{\mathrm{o}}$ to $+1$. If $n_1 < n_2$, $\mathcal{T}_{\perp}$ and $\mathcal{T}_{/\!/}$ are positive and decrease from $\mathcal{T}_{\mathrm{o}}$ for $\theta = 0$ to 0 for $\theta = \pi/2$. If $n_1 > n_2$, we have similar variations, but θ cannot be larger than the critical angle i_{L} given by [11.2]. $\mathcal{R}_{/\!/}$ vanishes for an angle of incidence $\theta = \theta_{\mathrm{B}}$ (called the Brewster angle) such that $\theta + \theta_{\mathrm{B}} = \pi/2$. For $\theta < \theta_{\mathrm{B}}$ and for any state of polarization, the ratio $E'_{\mathrm{m}}/E_{\mathrm{m}}$ is negative if $n_1 < n_2$ and positive if $n_1 > n_2$. The reflection on a more refringent medium ($n_2 > n_1$) occurs with a change of sign (i.e. a phase shift of π). The reflection on a less refringent medium ($n_2 < n_1$) occurs without a change of sign (i.e. without phase shift). As the transmission coefficient is always positive, the transmitted wave has no phase shift.

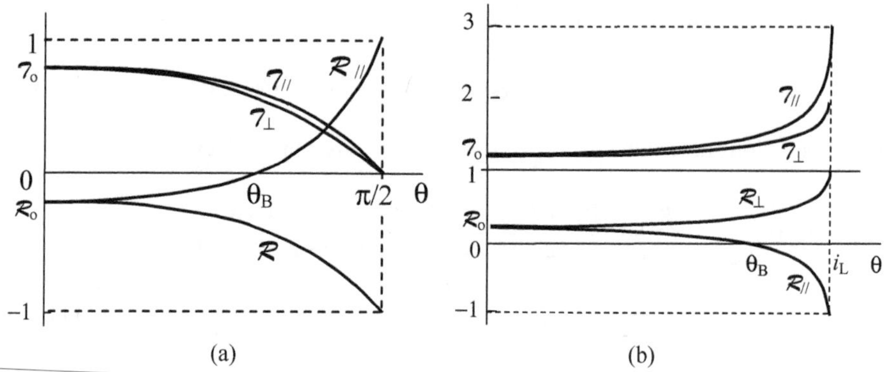

Figure 11.3. *Variations of $\mathcal{R}_\perp$, $\mathcal{R}_{//}$, $\mathcal{T}_\perp$ and $\mathcal{T}_{//}$ versus θ: a) if $n_1 < n_2$ (the graph corresponds to $n_1 = 1$ and $n_2 = 1.5$), and b) if $n_2 < n_1$ (the graph corresponds to $n_1 = 1.5$ and $n_2 = 1$)*

C) Conservation of energy

The Poynting vector of an electromagnetic wave is

$$\mathbf{S} = \frac{1}{\mu}\mathbf{E} \times \mathbf{B} = \frac{1}{\mu v} E^2 \mathbf{e} = \frac{1}{Z} E^2 \mathbf{e}. \qquad [11.18]$$

The conservation of energy on the interface $\mathcal{S}$ of two mediums requires that the power that any element of area of the interface $d\mathcal{S}$ receives from medium (1) is equal to the transmitted power toward medium (2). If the axis Oz is normal to $\mathcal{S}$ and oriented from medium (1) toward medium (2), this condition can be expressed by the equation $S_z + S'_z = S''_z$, that is, in the case of the interface of two dielectrics

$$E^2 \frac{\cos\theta}{Z_1^2} - E'^2 \frac{\cos\theta}{Z_1^2} = E''^2 \frac{\cos\theta''}{Z_2^2}. \qquad [11.19]$$

Using the reflection and transmission coefficients, this relation may be written as

$$(1 - \mathcal{R}^2) \frac{\cos\theta}{Z_1} = \mathcal{T}^2 \frac{\cos\theta''}{Z_2}. \qquad [11.20]$$

Any wave may be considered as the superposition of two waves, which are polarized in the plane of incidence and perpendicularly to this plane, respectively. The Poynting vector is the sum of the corresponding Poynting vectors (see section 10.4). The equation of conservation of energy [11.20] holds for each of them, as may easily be verified using the expressions [11.10] and [11.14]. Thus, [11.20] holds for any state of polarization and for non-polarized waves.

We define the *energy reflection factor* as the ratio of the reflected power to the incident power and the *energy transmission factor* as the ratio of the transmitted power to the incident power per unit area of the interface $\mathcal{S}$

$$f_R \equiv \frac{S'_z}{S_z} = \frac{E'^2}{E^2} = \mathcal{R}^2, \qquad f_T \equiv \frac{S''_z}{S_z} = \frac{E''^2}{E^2}\frac{Z_1}{Z_2}\frac{\cos\theta''}{\cos\theta} = \mathcal{T}^2\frac{Z_1}{Z_2}\frac{\cos\theta''}{\cos\theta}. \qquad [11.21]$$

The conservation of energy on the interface $\mathcal{S}$ requires that

$$f_R + f_T = 1. \qquad [11.22]$$

The variations of f_R and f_T as functions of the angle of incidence are illustrated in Figure 11.4a. f_R and f_T vary very little for small angles of incidence and, in the case of non-magnetic mediums, their values are given by

$$f_{R\,o} = \left(\frac{n_1 - n_2}{n_1 + n_2}\right)^2, \qquad f_{T\,o} = \frac{4n_1 n_2}{(n_1 + n_2)^2}. \qquad [11.23]$$

f_{Ro} is usually small. For instance, it is 4% at the air-glass interface and 2% at the air-water interface.

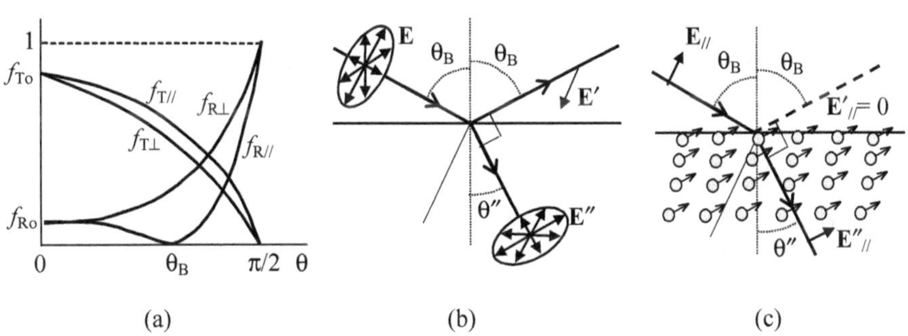

Figure 11.4. *a) Variations of the energy reflection factors $f_{R\perp}$ and $f_{R//}$ and the transmission factors $f_{T\perp}$ and $f_{T//}$ versus θ, b) Brewster polarization, and c) its interpretation*

D) Brewster's law

For any values of the indices n_1 and n_2, the coefficient $\mathcal{R}_{//}$ is equal to zero if $\tan(\theta + \theta'') = \infty$, that is, $\theta + \theta'' = \pi/2$. Thus, the intensity of the reflected wave vanishes if it is polarized in the plane of incidence. Using Snell's law, we deduce that the Brewster angle verifies the relationship

$$\tan\theta_B = n_2/n_1. \qquad [11.24]$$

This result may be used to obtain a linearly polarized wave from waves of any polarization or from non-polarized waves. Writing such waves as a superposition of a wave polarized in the plane of incidence and a wave polarized perpendicularly to this plane, at Brewster incidence only the component that is polarized perpendicularly to the plane of incidence produces a reflected wave with the same polarization (Figure 11.4b). This phenomenon was first observed by Malus in 1807 and was analyzed by Brewster in 1815. It may be explained by the polarization of the molecules of the second medium in the direction of **E** and the vanishing of the dipole radiation in the direction of the dipole. As the reflected wave is the superposition of these dipole radiations, it vanishes in this direction (Figure 11.4c).

In the case of a wave that is incident from the air side on the air-glass interface ($n_1 = 1$ and $n_2 = 1.5$), we find $2n_1/(n_1 + n_2) = 0.8$, $(n_1 - n_2)/(n_1 + n_2) = -0.2$ and $\theta_B = 56.31°$. If the wave is incident from the glass side, we find $2n_1/(n_1+n_2) = 1.2$ and $(n_1 - n_2)/(n_1 + n_2) = 0.2$ and $\theta_B = 18.43°$.

11.3. Total reflection

To analyze what happens if the angle of incidence exceeds the critical angle i_L, we consider again the boundary conditions at the interface of the mediums, which determine the laws of reflection and refraction. For instance, we consider the case of an electromagnetic wave that is polarized in the plane of incidence (Figure 11.2a). Let us assume that the incident wave has the form

$$\underline{\mathbf{B}} = \underline{B}_m e^{i(\omega t - \mathbf{k.r})} \mathbf{e}_y, \quad \text{and} \quad \underline{\mathbf{E}} = v_1 \underline{B}_m (\cos\theta\, \mathbf{e}_x - \sin\theta\, \mathbf{e}_z)\, e^{i(\omega t - \mathbf{k.r})}, \quad [11.25]$$

where we have used Maxwell's equation $\boldsymbol{\nabla} \times \mathbf{B} = \mu_o \varepsilon_o\, \partial_t \mathbf{E}$ to relate **E** to **B**. Like the incident wave, the reflected wave in the medium (1) is progressive of the form

$$\underline{\mathbf{B}}' = -\underline{B}'_m\, e^{i(\omega t - \mathbf{k}'.\mathbf{r})} \mathbf{e}_y, \qquad \underline{\mathbf{E}}' = v_1 \underline{B}'_m (\cos\theta'\, \mathbf{e}_x + \sin\theta'\, \mathbf{e}_z)\, e^{i(\omega t - \mathbf{k}'.\mathbf{r})}. \quad [11.26]$$

Using the symmetries or a proof similar to that of section 11.1, we may show that **k** and **k′** lie in the plane of incidence Oxz and they are of the form [11.5] with a magnitude ω/v_1. In the second medium, we may always write the fields in the form

$$\underline{\mathbf{B}}'' = \underline{B}''_m\, e^{i(\omega t - px - qy - sz)} \mathbf{e}_y, \qquad \underline{\mathbf{E}}'' = (v_2^2/\omega)\, \underline{B}''_m (s\, \mathbf{e}_x - p\mathbf{e}_z)\, e^{i(\omega t - px - qy - sz)}. \quad [11.27]$$

Writing the continuity conditions of the normal components of **B** and $\mathbf{D} = \varepsilon\mathbf{E}$ and the tangential components of **E** and $\mathbf{H} = \mathbf{B}/\mu$ at the interface, the equality of the phases for $z = 0$ gives the relations

$$\theta' = \theta, \qquad k' = k, \qquad q = 0, \qquad p = k \sin\theta = (\omega/v_1)\sin\theta. \quad [11.28]$$

On the other hand, the fields [11.27] must verify the equations of propagation in the second medium

$$\Delta\mathbf{E} - (1/v_2^2)\,\partial^2_{tt}\mathbf{E} = 0 \quad \text{and} \quad \Delta\mathbf{B} - (1/v_2^2)\,\partial^2_{tt}\mathbf{B} = 0. \qquad [11.29]$$

Thus, we must have

$$p^2 + q^2 + s^2 = \omega^2/v_2^2. \qquad [11.30]$$

Using equations [11.28], we find

$$s^2 = \frac{\omega^2}{v_2^2}\,[1 - \frac{v_2^2}{v_1^2}\sin^2\theta\,] = \frac{\omega^2}{v_2^2}\,[1 - \frac{\sin^2\theta}{\sin^2 i_L}\,]. \qquad [11.31]$$

As long as $\theta < i_L$, we have $\sin\theta < \sin i_L$ and s is real. In this case, the three parameters (p, q, s) are the components of a wave vector in the second medium and the wave is progressive. On the other hand, if $\theta > i_L$, s is imaginary, in the form

$$s = \pm\, i\eta \qquad \text{with} \qquad \eta = (2\pi/\lambda_2)\,\sqrt{\sin^2\theta/\sin^2 i_L - 1}\,, \qquad [11.32]$$

where $\lambda_2 \equiv 2\pi v_2/\omega$ is the wavelength in the second medium. In the general case, the wave in the second medium is a superposition of a solution with $s = +i\eta$ and a solution with $s = -i\eta$, in the form

$$\underline{\mathbf{B}}'' = [\underline{B}''_m\, e^{-\eta z} + \underline{C}''_m\, e^{\eta z}\,]\; e^{i(\omega t - px)}\,\mathbf{e}_y\,,$$
$$\mathbf{E}'' = (v_2^2/\omega)\,[B''_m\, e^{-\eta z}\,(i\eta\mathbf{e}_x + p\,\mathbf{e}_z) + C''_m\, e^{\eta z}\,(-i\eta\mathbf{e}_x + p\,\mathbf{e}_z)]\; e^{i(\omega t - px)}. \qquad [11.33]$$

The oscillatory factor of the fields is $e^{i(\omega t - px)}$. This indicates that the wave propagates in the direction Ox parallel to the interface with a wave vector p (Figure 11.5a). Its phase velocity is

$$v_p = \omega/p = v_1/\sin\theta. \qquad [11.34]$$

The amplitude of this wave depends on the distance z to the interface: it is the sum of a term $\underline{B}''_m\, e^{-\eta z}$, which decreases exponentially with z and a term $\underline{C}''_m\, e^{\eta z}$, which increases exponentially with z. If the second medium is finite in the direction Oz, the two terms are possible (and even necessary in order to respect the boundary conditions on the other face of the second medium). If the second medium is infinite in the direction Oz, the second term is not physically possible. Thus, the wave decreases exponentially according to the expressions

$$\mathbf{B}'' = B''_m e^{-\eta z} e^{i(\omega t - px)} \mathbf{e}_y, \qquad \mathbf{E}'' = (v_2^2/\omega) B''_m e^{-\eta z} (s\,\mathbf{e}_x - p\,\mathbf{e}_z) e^{i(\omega t - px)}. \quad [11.35]$$

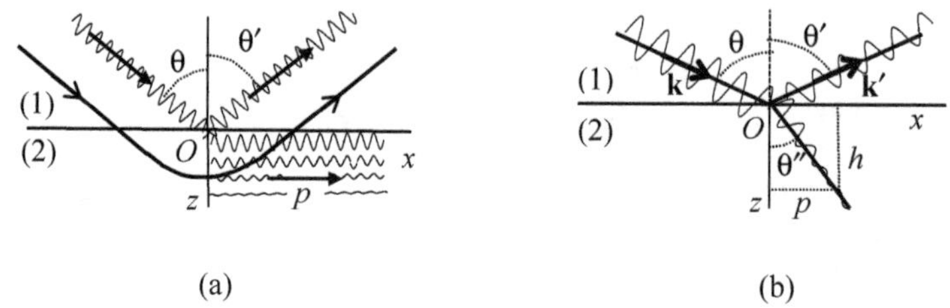

Figure 11.5. *Reflection on a medium in which the wave is attenuated: a) total internal reflection, b) reflection on a resistive or a reactive medium*

The attenuation coefficient η of the wave in the direction of Oz is proportional to the inverse of the wavelength λ_2 in the second medium. For instance, in the case of total reflection on the interface water-air, $\delta = 3.6\,\lambda_2^{-1}$ for $\theta = 60°$. The wave only penetrates a distance of the order of λ_2 in the second medium.

In the case of an infinite second medium, the boundary conditions at the interface may be written as

$$B_m/\mu_1 - B'_m/\mu_1 = B''_m/\mu_2, \qquad (\underline{B}_m + \underline{B}'_m)\, v_1 \cos\theta = -i\underline{B}''_m \delta/\mu_2 \varepsilon_2 \omega. \quad [11.36]$$

These equations determine the amplitudes of the reflected and the transmitted waves

$$\frac{\underline{B}'_m}{\underline{B}_m} = -\frac{\mu_1 v_1 \varepsilon_2 \omega \cos\theta + i\eta}{\mu_1 v_1 \varepsilon_2 \omega \cos\theta - i\eta}, \qquad \frac{\underline{B}''_m}{\underline{B}_m} = \frac{2\mu_2 v_1 \varepsilon_2 \omega \cos\theta}{\mu_1 v_1 \varepsilon_2 \omega \cos\theta - i\eta}. \quad [11.37]$$

Setting $\phi = \arctan(\eta/\mu_1 v_1 \varepsilon_2 \cos\theta)$, we may write

$$\underline{B}'_m = -\underline{B}_m\, e^{2i\phi} \quad \text{and} \quad B''_m = 2\underline{B}_m (\mu_2/\mu_1) \cos\phi\, e^{i\phi}. \quad [11.38]$$

The physical fields are the real parts of the expressions [11.25], [11.26] and [11.27]; that is, taking $\underline{B}_m$ as real

$$\mathbf{B} = B_m \cos(\omega t - \mathbf{k.r})\,\mathbf{e}_y, \qquad \mathbf{E} = v_1 B_m (\cos\theta\,\mathbf{e}_x - \sin\theta\,\mathbf{e}_z)\cos(\omega t - \mathbf{k.r}),$$

$$\mathbf{B}' = -B_m \cos(\omega t - \mathbf{k}'.\mathbf{r} + 2\phi)\,\mathbf{e}_y, \quad \mathbf{E}' = v_1 B_m (\cos\theta\,\mathbf{e}_x + \sin\theta\,\mathbf{e}_z)\cos(\omega t - \mathbf{k}'.\mathbf{r} + 2\phi),$$

$$\mathbf{B}'' = 2(\mu_2/\mu_1)\, B_m e^{-\eta z} \cos\phi \cos(\omega t - px + \phi)\,\mathbf{e}_y,$$

$$\mathbf{E}'' = 2(\mu_2 v_2^2/\mu_1 \omega)\, B_m e^{-\eta z} \cos\phi\, [\delta \sin(\omega t - px + \phi)\,\mathbf{e}_x - p\cos(\omega t - px + \phi)\mathbf{e}_z]. [11.39]$$

We find that the reflected **E** wave has the same amplitude as the incident wave but with a phase shift 2ϕ. The average values of the Poynting vectors over a period are

$$<\mathbf{S}> = \frac{v_1}{2\mu_1} B_m^2 (\sin\theta\, \mathbf{e}_x + \cos\theta\, \mathbf{e}_z), \quad <\mathbf{S}'> = \frac{v_1}{2\mu_1} B_m^2 (\sin\theta\, \mathbf{e}_x - \cos\theta\, \mathbf{e}_z),$$

$$<\mathbf{S}''> = 2\frac{v_2^2}{v_1}\frac{\mu_2}{\mu_1^2} B_m^2\, e^{-2\eta z} \cos^2\phi \sin\theta\, \mathbf{e}_x. \qquad [11.40]$$

In the second medium, we have a propagation of energy in the direction Ox parallel to the surface but only at a depth not exceeding a few wavelengths (*evanescent wave*). The energy propagates as if it slightly penetrates the second medium where it propagates near the surface and returns to the first medium without any loss.

11.4. Reflection on a conductor

Consider a wave that is incident from a dielectric medium (1) on a conducting medium (2) (Figure 11.5b). The wave equations in these mediums are respectively

$$\Delta\mathbf{E}_1 - (1/v_1^2)\, \partial^2_{tt}\mathbf{E}_1 = 0 \quad \text{and} \quad \Delta\mathbf{E}_2 - (1/v_2^2)\partial^2_{tt}\mathbf{E}_2 - \mu_2\sigma\, \partial_t\mathbf{E}_2 = 0. \quad [11.41]$$

As we have done in the preceding section, we write the fields of the incident wave and those of the reflected wave as progressive waves

$$\mathbf{E} = \underline{\mathbf{E}}_m\, e^{i(\omega t - \mathbf{k}.\mathbf{r})}, \quad \mathbf{B} = \underline{\mathbf{B}}_m\, e^{i(\omega t - \mathbf{k}.\mathbf{r})}, \quad \mathbf{E}' = \underline{\mathbf{E}}'_m\, e^{i(\omega t - \mathbf{k}'.\mathbf{r})}, \quad \text{and } \mathbf{B}' = \underline{\mathbf{B}}'_m\, e^{i(\omega t - \mathbf{k}'.\mathbf{r})}.$$

We write the transmitted fields in the general form $\mathbf{E}'' = \underline{\mathbf{E}}''_m\, e^{i(\omega t - px - qy - sz)}$ and $\mathbf{B}'' = \underline{\mathbf{B}}''_m\, e^{i(\omega t - px - qy - sz)}$. The continuity conditions at the interface $z = 0$ give the relations [11.28]. On the other hand, $\mathbf{E}''$ and $\mathbf{B}''$ verify the wave equations in the conductor if

$$p^2 + q^2 + s^2 = (\omega/v_2)^2\,(1 - iQ), \qquad [11.42]$$

where we introduced the quality factor $Q = \omega_c/\omega$, and $\omega_c = \sigma/\varepsilon_2$ is the characteristic angular frequency of the conductor (see section 10.7); hence,

$$s^2 = \frac{\omega^2}{v_2^2}(\chi - iQ) \qquad \text{with} \quad \chi = 1 - \frac{v_2^2}{v_1^2}\sin^2\theta. \qquad [11.43]$$

This equation shows that s is always complex of the form

$$\underline{s} = h - i\eta, \qquad [11.44]$$

$$h = \frac{\omega}{v_2\sqrt{2}}\,[\sqrt{\chi^2 + Q^2} + \chi]^{½} \quad \text{and } \eta = \frac{\omega}{v_2\sqrt{2}}\,[\sqrt{\chi^2 + Q^2} - \chi]^{½}. \qquad [11.45]$$

The wave in the conductor has the form $\underline{\mathbf{E}}'' = \underline{E}''_{\mathrm{m}}\, e^{-\eta z}\, e^{i(\omega t - px - qy - sz)}\, \mathbf{e}_y$. It propagates with an attenuation coefficient η in the direction Oz and a wave vector $p\,\mathbf{e}_x + h\,\mathbf{e}_z$, which makes with Oz an angle θ'' given by

$$\tan\theta'' = p/h. \qquad [11.46]$$

In the case of an electromagnetic wave that is polarized perpendicularly to the plane of incidence, for instance, the fields have the forms

$$\underline{\mathbf{E}} = \underline{E}_{\mathrm{m}}\,\mathbf{e}_y\, e^{i(\omega t - \mathbf{k}.\mathbf{r})}, \qquad \underline{\mathbf{B}} = (\underline{E}_{\mathrm{m}}/v_1)(-\cos\theta\,\mathbf{e}_x + \sin\theta\,\mathbf{e}_z)\, e^{i(\omega t - \mathbf{k}.\mathbf{r})},$$
$$\underline{\mathbf{E}}' = \underline{E}'_{\mathrm{m}}\,\mathbf{e}_y\, e^{i(\omega t - \mathbf{k}'.\mathbf{r})}, \qquad \underline{\mathbf{B}}' = (\underline{E}'_{\mathrm{m}}/v_1)(\cos\theta'\,\mathbf{e}_x + \sin\theta'\,\mathbf{e}_z)\, e^{i(\omega t - \mathbf{k}'.\mathbf{r})},$$
$$\underline{\mathbf{E}}'' = \underline{E}''_{\mathrm{m}}\mathbf{e}_y\, e^{-\eta t} e^{i(\omega t - px - qy - sz)}, \qquad \underline{\mathbf{B}}'' = (\underline{E}''_{\mathrm{m}}/\omega)(-\underline{s}\mathbf{e}_x + p\mathbf{e}_z)\, e^{-\eta t} e^{i(\omega t - px - qy - sz)}. \qquad [11.47]$$

Assuming that the surface of the conductor carries no surface charge density and no surface current density (although this is not always justified), the continuity conditions of the normal components of **B** and ε**E** and the tangential components of **B**/μ and **E** may be written as

$$\underline{E}_{\mathrm{m}} + \underline{E}'_{\mathrm{m}} = \underline{E}''_{\mathrm{m}}, \qquad (\underline{E}_{\mathrm{m}} - \underline{E}'_{\mathrm{m}})\cos\theta = \frac{\mu_1 v_1}{\mu_2\omega}\,\underline{s}\,\underline{E}''. \qquad [11.48]$$

We deduce from these equations the reflection and transmission coefficients

$$\frac{\underline{E}'_m}{\underline{E}_m} = \frac{\mu_2\omega\cos\theta - \mu_1 v_1(h - i\eta)}{\mu_2\omega\cos\theta + \mu_1 v_1(h - i\eta)}, \qquad \frac{\underline{E}''_m}{\underline{E}_m} = \frac{2\mu_2\omega\cos\theta}{\mu_2\omega\cos\theta + \mu_1 v_1(h - i\eta)}. \qquad [11.49]$$

Thus, the reflected wave has a phase lead $\phi_+ + \phi_-$, while the transmitted wave gets a phase lead ϕ_+. The phases $\phi_\pm$ are given by the expressions

$$\tan\phi_\pm = \frac{\eta\mu_1 v_1}{\mu_2\omega\cos\theta \pm \mu_1 v_1 h} \qquad (0 < \phi_\pm < \pi). \qquad [11.50]$$

In the case of typical metals, σ is of the order of $5 \times 10^7\ \Omega^{-1}\ \mathrm{m}^{-1}$; thus, the characteristic angular frequency ω_c is of the order of 10^{18} rad/s. If the frequency is not very high, only the conduction electrons contribute to the secondary waves and the term ω_c/ω is much larger than 1; hence, $h \approx \eta \approx (\omega/v_2)\sqrt{Q/2}$. The term $\mu_1 v_1(h - i\eta)$ has a much larger magnitude than $\mu_2\,\omega\cos\theta$ and we may write

$$\mathcal{R}_\perp = \frac{E'_m}{E_m} \approx -1 + \gamma, \qquad \mathcal{T}_\perp = \frac{E''_m}{E_m} \approx \gamma, \qquad \text{where } \gamma = 2\frac{\mu_2 v_2}{\mu_1 v_1 \sqrt{Q}}\, e^{i\pi/4} \cos\theta. \quad [11.51]$$

The wave is almost totally reflected with a phase shift close to π and there is a small energy transfer across the surface of the conductor. The transferred energy is totally dissipated as Joule heat.

At very high frequency (ultraviolet and X-rays), the positive ions in the metal contribute with the conduction electrons to the emission of secondary waves. This increases the dissipated energy in the metal and reduces the reflection coefficient to values as low as 0.2 in the case of the reflection of light on silver. In this case, a thin metallic film transmits a certain fraction of the wave.

The mean values of the Poynting vectors of the incident wave, the reflected wave, and the transmitted wave, evaluated over a period, are respectively

$$\langle \mathbf{S} \rangle = \mathcal{R}e\, \mathbf{E} \times \mathcal{R}e\, \mathbf{B}/\mu = (1/2\mu_1 v_1)\, |\underline{E}_m|^2 (\cos\theta\, \mathbf{e}_z + \sin\theta\, \mathbf{e}_x),$$

$$\langle \mathbf{S}' \rangle = \mathcal{R}e\, \mathbf{E}' \times \mathcal{R}e\, \mathbf{B}'/\mu = (1/2\mu_1 v_1)\, |\mathcal{R}|^2 |\underline{E}_m|^2 (-\cos\theta\, \mathbf{e}_z + \sin\theta\, \mathbf{e}_x),$$

$$\langle \mathbf{S}'' \rangle = \mathcal{R}e\, \mathbf{E}'' \times \mathcal{R}e\, \mathbf{B}''/\mu = (1/2\mu_2 \omega)\, e^{-2\eta z}\, |\underline{E}''_m|^2 (\mathcal{R}e\, \underline{s}\, \mathbf{e}_z + p\, \mathbf{e}_x)]. \quad [11.52]$$

The transferred intensity to the conductor is

$$\mathcal{I}'' = \langle S_z \rangle - \langle S'_z \rangle = \frac{\cos\theta}{2\mu_1 v_1}(1 - |\mathcal{R}|^2)\, |\underline{E}_m|^2 = \frac{2\mu_2 h\omega \cos^2\theta}{\left|\mu_2\omega\cos\theta + \mu_1 v_1 (h - i\eta)\right|^2}\, |\underline{E}_m|^2. \quad [11.53]$$

In the case of a good conductor and a relatively low frequency ($\omega << \omega_c$), we find $\eta \approx h \approx \sqrt{\mu_2 \sigma\omega/2} = (\omega/v_2)\sqrt{Q/2} >> \mu_2\, \omega \cos\theta$, thus

$$\mathcal{I}'' = \varepsilon_1 v_2 \frac{\mu_2}{\mu_1}\sqrt{\frac{2}{Q}}\, |\underline{E}_m|^2 \cos^2\theta = 2\frac{v_2\mu_2}{v_1\mu_1}\sqrt{\frac{2}{Q}}\, \mathcal{I} \cos^2\theta. \quad [11.54]$$

On the other hand, in the case of a relatively high frequency ($\omega >> \omega_c$) and $\sin^2\theta << v_1^2/v_1^2$, we find $\omega_c^2/\omega^2 << \chi^2$, $\delta \approx \omega_c/2v_2\sqrt{\chi}$ and $h \approx \omega\sqrt{\chi}/v_2$, thus

$$\mathcal{I}'' = 2\frac{\sqrt{\chi}}{\mu_2 v_2}\, \underline{E}_m|^2 = 4\frac{\mu_1 v_1 \sqrt{\chi}}{\mu_2 v_2}\, \mathcal{I}. \quad [11.55]$$

In the case of a superconductor, $\omega_c \to \infty$, $Q \to \infty$ and $\delta \to \infty$. Thus, the wave cannot penetrate in the conductor. The reflection coefficient is then $E'_m/E_m = -1$. The reflected electric field has the same amplitude as the incident electric field but they are in phase opposition.

11.5. Reflection on a plasma

Let us consider an electromagnetic wave incident from a dielectric medium (1) on a plasma (medium 2), such as the ionosphere. The wave equations in these mediums are, respectively,

$$\Delta \mathbf{E}_1 - (1/v_1^2)\,\partial^2_{tt}\mathbf{E}_1 = 0 \qquad \text{and} \qquad \Delta \mathbf{E}_2 - (1/v_2^2)\partial^2_{tt}\mathbf{E}_2 - (\omega_p^2/v_2^2)\mathbf{E}_2 = 0. \quad [11.56]$$

We assume, for instance, that the wave is polarized in the plane of incidence. The incident fields and the reflected fields have the forms [11.25] and [11.26], respectively, while the fields transmitted to the plasma may be written in the forms $\mathbf{E}'' = \underline{\mathbf{E}}''_m e^{i(\omega t - px - qy - sz)}$ and $\mathbf{B}'' = \underline{\mathbf{B}}''_m e^{i(\omega t - px - qy - sz)}$. The equality of the phase on the interface $z = 0$ gives the equations [11.28]. On the other hand, $\mathbf{E}''$ and $\mathbf{B}''$ verify the wave equations in the plasma if

$$p^2 + q^2 + s^2 = (\omega^2 - \omega_p^2)/v_2^2, \quad \text{thus } s = (\omega/v_2)\sqrt{1 - \omega_p^2/\omega^2 - (v_2^2/v_1^2)\sin^2\theta}\,. \quad [11.57]$$

If the quantity inside the square root in [11.57] is negative, s is imaginary:

$$s = \pm\, i\eta \qquad \text{with} \quad \eta = \frac{2\pi}{\lambda_2}\sqrt{\left(\frac{v_2}{v_1}\right)^2 \sin^2\theta + \left(\frac{\omega_p}{\omega}\right)^2 - 1}\,. \qquad [11.58]$$

This is effectively the case if $\omega < \omega_p$ for any angle θ or if $\omega > \omega_p$ and θ larger than a critical angle θ_L given by

$$\sin\theta_L = (v_1/v_2)\sqrt{1 - \omega_p^2/\omega^2}\,. \qquad [11.59]$$

This case is similar to that of an incident wave on a dielectric at an angle of incidence larger than the critical angle i_L. If the plasma is infinite in the normal direction Oz, the energy is totally reflected (with an amplitude equal to that of the incident wave and a phase lead equal to 2ϕ where $\phi = \arctan(\eta/\mu_1 v_1 \varepsilon_2 \cos\theta)$, and η is given by [11.58]. The wave penetrates in the plasma to a depth of the order of λ_2. In the limit $\omega_p = 0$, we find obviously the results of the reflection and refraction on a dielectric.

If the quantity inside the square root of [11.57] is positive, s is real. In this case, the wave in the plasma is progressive with a wave vector $\mathbf{k}'' = p\mathbf{e}_x + s\mathbf{e}_z$. This is effectively the case if $\omega > \omega_p$ and θ is less than the critical angle θ_L. In this case, there is a transfer of energy to the plasma as a progressive wave. This wave forms with the normal an angle θ'' such that

$$\sin\theta'' = p/\sqrt{p^2 + s^2} = (v_2/v_1)\sin\theta \,/\sqrt{1 - \omega_p^2/\omega^2} = \sin\theta / \sin\theta_L. \qquad [11.60]$$

The reflection and transmission coefficients are determined from the boundary conditions at the interface. To write them, we use the expressions [11.25] and [11.26] in the incidence medium and [11.27] in the plasma.

11.6. Interference of two electromagnetic waves

If two waves superpose in a region of space, the energy, momentum, and other physical quantities, which are received at a given point, are not necessarily the sums of the corresponding quantities for each wave if it is intercepted separately. We say that the waves *interfere*.

Consider two electromagnetic waves, specified by their electric fields, with equal frequency and, to simplify, equal amplitudes and linearly polarized in the same direction. Then, we may treat the fields as two scalar quantities

$$E_1 = a\cos[\omega(t - r_1/v) - \phi_1], \qquad E_2 = a\cos[\omega(t - r_2/v) - \phi_2]. \qquad [11.61]$$

r_1 and r_2 are the distances of the observation point M to the sources and v is the speed of propagation. ϕ_1 and ϕ_2 are the phases of the sources chosen in such a way that the amplitudes are positive. We assume that the sources are at a sufficiently long distance from the region of observation, so that the amplitudes of the waves vary little in this region. The resultant wave at M is

$$E(M) = E_1 + E_2 = 2a\cos(\tfrac{1}{2}\Delta\phi)\cos(\omega t - \phi_1 - \tfrac{1}{2}\Delta\phi), \qquad [11.62]$$

where $\Delta\phi$ is the phase lag of the wave (2) over the wave (1) at M. It is given by

$$\Delta\phi = \Delta_o\phi + \Delta_m\phi \quad \text{with} \quad \Delta_o\phi \equiv \phi_2 - \phi_1 \quad \text{and} \quad \Delta_m\phi \equiv (\omega/v)\,\Delta r = 2\pi\,(\Delta r/\lambda). \qquad [11.63]$$

$\Delta_o\phi$ is the initial phase lag (at $t = 0$ and $r_1 = r_2 = 0$) and $\Delta_s\phi$ is the phase lag due to the *path difference* $\Delta r \equiv r_2 - r_1$ at M. The total wave at M is a sinusoidal function of time, whose angular frequency is ω and amplitude is

$$A(M) = 2a\cos(\tfrac{1}{2}\Delta\phi). \qquad [11.64]$$

If $\Delta\phi$ is an integer multiple of 2π, the amplitude at M is maximum in absolute value and equal to $2a$; the waves at M are *in phase* and the interference is *constructive*. On the contrary, if $\Delta\phi$ is a half-integer multiple of 2π, the amplitude $A(M)$ vanishes; the waves at M are *in opposite phase* and the interference is *destructive*:

$$\Delta\phi = 2p\pi, \qquad p = \text{integer} \qquad \text{(maximums)}, \qquad [11.65]$$

$$\Delta\phi = 2p\pi, \qquad p = \text{half-integer} \qquad \text{(minimums)}. \qquad [11.66]$$

Thus, the state of interference at M is determined by the phase shift between the waves at M; We may define the *order of interference* p by

$$p \equiv \frac{\Delta\phi}{2\pi} = \frac{\Delta r}{\lambda} + \frac{\Delta_0\phi}{2\pi}. \qquad [11.67]$$

The interference is constructive if the order p is an integer number (positive, negative, or zero) and destructive if the order p is a half-integer number. The locus of the maxima of given order p is an *interference fringe.* In particular, the order 0 corresponds to the waves reaching M with a phase shift equal to 0. Each increase of the path difference by $\Delta r = \lambda$ increases the order of interference by 1.

It is often convenient to use the complex representation $\underline{E} = a e^{i(\omega t - \omega r/v - \phi)}$ of the wave, then $E = \mathcal{Re}\, \underline{E}$. The intensity $\mathcal{I}$ of the wave [11.62] may be written as

$$\mathcal{I} = \underline{E}\,\underline{E}^*/2\mu v = (2a^2/\mu v)\cos^2(\tfrac{1}{2}\Delta\phi). \qquad [11.68]$$

As a function of $\Delta\phi$, $\mathcal{I}(M)$ oscillates between minimums equal to zero and maximums equal to $2a^2/\mu v$ (Figure 11.6a). As the average value of $\cos^2(\tfrac{1}{2}\Delta\phi)$ is ½, the average value of $\mathcal{I}(M)$ over space is $\langle\mathcal{I}(M)\rangle = a^2/\mu v = \mathcal{I}_1 + \mathcal{I}_2$. Thus, the interference produces a redistribution of the intensity, the average intensity being the sum of the intensities emitted by the sources.

In general, the amplitudes of the waves may be different and they may depend on the observation point M. Using the complex representation, the waves at M are

$$\underline{E}_1 = a_1(M)\, e^{i(\omega t - \phi_1 - \omega r_1/v)} \equiv \underline{a}_1(M)\, e^{i\omega t}, \qquad \underline{E}_2 = a_2(M)\, e^{i(\omega t - \phi_2 - \omega r_2/v)} \equiv \underline{a}_2(M)\, e^{i\omega t}.$$

The resultant wave may be written as $\underline{E}(M) = E_1 + E_2 = [\underline{a}_1(M) + \underline{a}_2(M)]\, e^{i\omega t}$. Using the phase shift [11.63], the intensity at M can then be written as

$$\begin{aligned}\mathcal{I}(M) &= (1/2\mu v)\,|\underline{a}_1(M) + \underline{a}_2(M)|^2 = (1/2\mu v)[a_1(M)^2 + a_2(M)^2 + 2a_1(M)\, a_2(M)\cos(\Delta\phi)]\\ &= \mathcal{I}_1(M) + \mathcal{I}_2(M) + 2\sqrt{\mathcal{I}_1(M)\mathcal{I}_2(M)}\,\cos(\Delta\phi). \end{aligned} \qquad [11.69]$$

$\mathcal{I}_1(M)$ and $\mathcal{I}_2(M)$ are the intensities of the waves if they are observed separately. The third term is the *interference term.* It may be positive or negative. Thus, the resultant intensity varies between a minimum $\mathcal{I}_{min}$ and a maximum $\mathcal{I}_{max}$ given by

$$\mathcal{I}_{min} = [\sqrt{\mathcal{I}_1(M)} - \sqrt{\mathcal{I}_2(M)}\,]^2 \quad \text{and} \quad \mathcal{I}_{max} = [\sqrt{\mathcal{I}_1(M)} + \sqrt{\mathcal{I}_2(M)}\,]^2. \qquad [11.70]$$

Figure 11.6b is the *phasor diagram* for the interference of the waves [11.61]. The waves are represented by the projection over Ox of two vectors making with Ox the angles $(\omega t - \omega r_1/v - \phi_1)$ and $(\omega t - \omega r_2/v - \phi_2)$. The parallelogram rotates about O at the angular velocity ω without deformation. If the waves have different amplitudes but the same frequency and a fixed phase shift, the resultant amplitude varies between a minimum $|a_1 - a_2|$ if the order $p_M = \Delta\phi/2\pi$ is equal to a half-integer (then the vectors point in opposite directions), and a maximum $(a_1 + a_2)$ if the order p_M is equal to an integer (then, the vectors point in the same direction). We define the *contrast* or *visibility factor* by

$$\mathcal{C} = \frac{\mathcal{I}_{max} - \mathcal{I}_{min}}{\mathcal{I}_{max} + \mathcal{I}_{min}} = \frac{2\sqrt{\mathcal{I}_1(M)\mathcal{I}_2(M)}}{\mathcal{I}_1(M) + \mathcal{I}_2(M)}. \qquad [11.71]$$

$\mathcal{C}$ always lies between 0 and 1. The maximum value $\mathcal{C} = 1$ corresponds to $\mathcal{I}_{min} = 0$, that is, the waves having equal amplitudes. The minimum value $\mathcal{C} = 0$ corresponds to $\mathcal{I}_{max} = \mathcal{I}_{min}$, that is, the total absence of interference fringes near the point M.

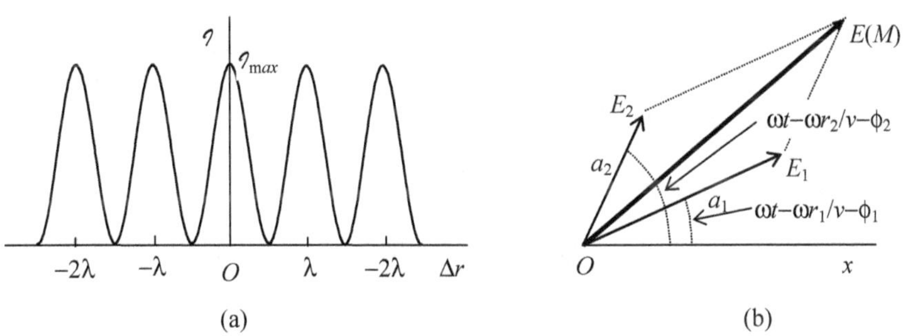

Figure 11.6. *a) The intensity versus Δr, and b) phasor diagram for interference*

The most useful interference phenomena are those of light waves. The historic experiment by Young confirmed the wave nature of light and even measured its wavelength (see problem 11.11). Another interference effect occurs between the reflected waves at the faces of thin films (see problem 11.12).

11.7. Superposition of several waves, conditions for observable interference

Consider several waves $E_j = a_j \cos(\omega_j t - \phi_j)$ that superpose at points M. Their phases ϕ_j may depend on M. We assume that the waves are polarized in the same direction, so they can be treated as scalar waves. The resultant wave is then their algebraic sum $E = \Sigma_i \, a_j \cos(\omega_j t - \phi_j)$. We deduce that

$$\begin{aligned} E^2 &= \Sigma_{k,j} \; a_k \, a_j \cos(\omega_j t - \phi_j) \cos(\omega_k t - \phi_k) \\ &= \Sigma_j \, a_j^2 \cos^2(\omega_j t - \phi_j) + \tfrac{1}{2} \Sigma_{k \neq j} \, a_k a_j \cos[(\omega_k + \omega_j)t - \phi_k - \phi_j] \\ &\quad + \tfrac{1}{2} \Sigma_{k \neq j} \, a_k a_j \cos[(\omega_k - \omega_j)t - \phi_k + \phi_j], \end{aligned}$$

where we separated the terms $k = j$ from the terms $k \neq j$. To evaluate the resultant intensity, we take the average over the time of observation. The first term gives the sum of the intensities of the waves taken separately $\mathcal{I}_j = a_j^2 / 2\mu v$. The second term has an average equal to zero. Thus, we obtain

$$\mathcal{I} = \Sigma_j \, \mathcal{I}_j + \Sigma_{k \neq j} \sqrt{\mathcal{I}_k \mathcal{I}_j} < \cos[(\omega_k - \omega_j)t + (\phi_j - \phi_k)] >, \qquad [11.72]$$

where $< f(t) >$ designates the average of $f(t)$ over the time of observation. The second term is the interference term. As an application of [11.72] we consider the following cases:

a) The waves have different angular frequencies: in this case, the time-average of the interference term is zero and the total intensity is equal to the sum of the intensities of the individual waves

$$\mathcal{I} = \Sigma_j \, \mathcal{I}_j \qquad \text{(waves of different frequencies).} \qquad [11.73]$$

b) The waves have equal frequencies but different phases: in this case, we obtain a generalization of equation [11.69]

$$\mathcal{I} = \Sigma_j \mathcal{I}_j + \Sigma_{k \neq j} \sqrt{\mathcal{I}_k \mathcal{I}_j} \cos(\phi_k - \phi_j) \qquad \text{(waves of equal frequencies);} \qquad [11.74]$$

c) The number of waves N is large with different phases, or N is small but the phases ϕ_j change at random during the observation time (*non-coherent waves*). In this case, the average of the function $\cos[(\omega_k - \omega_j)t + (\phi_j - \phi_k)]$ for all the waves is equal to zero and the total intensity is equal to the sum of the intensities $\mathcal{I}_i$

$$\mathcal{I} = \Sigma_j \, \mathcal{I}_j \qquad \text{(large number of waves or non-coherent waves).} \qquad [11.75]$$

Only in case (b) does the intensity $\mathcal{I}$ depend on the observation point via the phases ϕ_k. In the other cases, the intensity is uniform. Thus, the interference is observable under the following conditions:

– the waves must be synchronous (i.e. they have the same frequency);

– the waves must have constant phases (i.e. independent of time), we say that they are *coherent temporally*;

– the waves must have the same polarization, otherwise the minimums are not zero and the contrast is reduced;

– in the case of light, the *spatial coherence* requires that the sources are not very large in order for its different points to be approximately coherent.

The only way to have coherent light waves from traditional sources is to use an *aperture splitting* setup or an *amplitude splitting* setup. In setups of the first type (such as Young's double slit experiment), two or several secondary waves originate from different parts of the same primary wave front. In setups of the second type, the secondary waves originate from the same part of the primary wave front (the half-silvered mirror is an example of such setups). In all cases, we have twin wave packets originating from the same atom, which have a short duration τ_c, called the *coherence time*.

11.8. Huygens-Fresnel's principle and diffraction by an aperture

Diffraction refers to the bending of waves near obstacles, i.e. their deviation from rectilinear propagation in homogeneous mediums, contrarily to the displacement of free particles. The mathematical problem consists of determining the solution of the wave equation, which verifies some given boundary conditions according to the nature of the obstacle. This is in fact a considerably difficult problem; hence, approximation methods must be used.

Huygens' principle qualitatively explains the diffraction by assuming that the points of the aperture behave like sources, emitting spherical wavelets whose envelope at a later time is the wave front of the diffracted wave (Figure 11.7a). However, this simple geometrical formulation does not allow the determination of the angular distribution of the diffracted intensity and, in particular, does not explain the absence of a diffracted wave in the backward direction, i.e. the opposite direction to the incident wave. In order to quantitatively analyze diffraction, Fresnel assumed that the diffracted wave results from the interference of the spherical wavelets $d\underline{u}_d$ emitted by the elements of area $d\mathcal{S}$ of the aperture. According to the *Fresnel-Huygens principle*, these wavelets may be written as

$$d\underline{E}_d(M) = d\mathcal{S}\ \mathcal{T}(P)\eta_P(\theta', \theta)\underline{E}(P)\frac{1}{PM}\, e^{-ik\,PM}, \quad \text{where } k = 2\pi/\lambda. \qquad [11.76]$$

In this expression, we have omitted the factor $e^{i\omega t}$ that is common to all the wavelets. $\underline{E}(P)$ is the incident wave at P and $(1/PM)\,e^{-ik\,PM}$ characterizes the spherical wavelet that is emitted by the element of area $d\mathcal{S}$. The function $\mathcal{T}(P)$ is the coefficient of transmission equal to 1 if the aperture is completely transparent at P and equal to 0 if it is completely opaque. θ and θ' are, respectively, the angles that the direction of the diffracted wave and the incident wave make with the normal to the aperture oriented in the direction of propagation. Intuitively, Fresnel chose the *inclination factor* $\eta_P(\theta', \theta)$ equal to $(1/2\lambda)(\cos\theta + \cos\theta')$ to favor the direction of geometrical optics. The factor $1/2\lambda$ is necessary in order for η_P to have the right dimensions of the inverse of a length, which can only be λ. Later on, Kirchhoff explained the origin of this factor but with a phase of $\pi/2$ (see problem 11.16); thus, it may be written as

$$\eta_P(\theta', \theta) = \frac{i}{2\lambda}(\cos\theta + \cos\theta'). \qquad [11.77]$$

The diffracted wave is obtained by integrating [11.76] over the aperture

$$\underline{E}_d(M) = \iint_{\mathcal{S}} d\mathcal{S}\, \eta_P(\theta', \theta)\, \mathcal{T}(P)\, \underline{E}(P). \qquad [11.78]$$

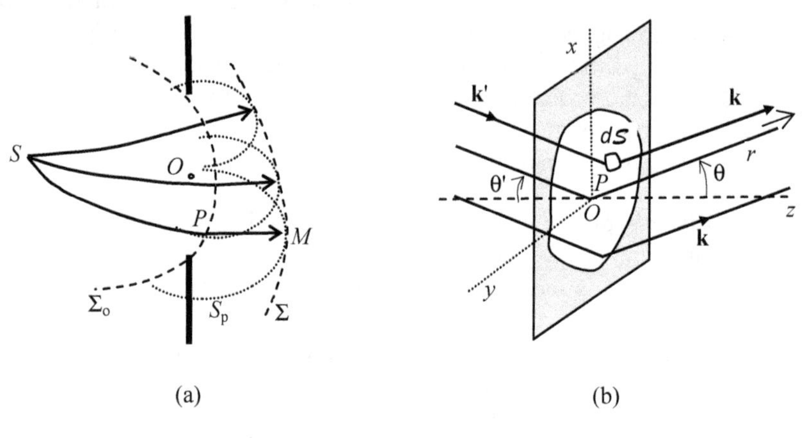

Figure 11.7. *a) Huygens' principle, and b) Fraunhofer diffraction by an aperture*

In general, the mathematical analysis of diffraction using [11.78] is very complicated. Indeed, it contains four scales of length: the wavelength λ, the dimensions of the aperture d, and the distances $r \equiv OM$ and $r' \equiv OS$ of the center of the aperture to the observation point and to the source. It is simplified if the source S and the point of observation are far from the aperture and the angles θ and θ' are small. This is the so-called *Fraunhofer diffraction*. In this case, the incident wave and the total diffracted wave are plane waves (Figure 11.7b). In the expression [11.78], the distance PM and the inclination factor η_P are approximately independent of P and the incident wave is a plane wave $\underline{u}(P) = A\, e^{-i\mathbf{k}'.\overrightarrow{OP}}$, where $\mathbf{k}'$ is the incident wave vector. Setting $\overrightarrow{PM} = \mathbf{r} - \overrightarrow{OP}$ (where $\mathbf{r} \equiv \overrightarrow{OM}$ and $OP << r$) and $\mathbf{k}$ for the wave vector in the direction of $\overrightarrow{PM}$ (thus, $k\,PM = \mathbf{k}.\overrightarrow{PM}$), we may write

$$\frac{1}{PM}\, e^{-ik\,PM} \approx \frac{1}{PM}\, e^{-i\mathbf{k}.\overrightarrow{PM}} \approx \frac{1}{r}\, e^{-i\mathbf{k}.\mathbf{r}}\, e^{i\mathbf{k}.\overrightarrow{OP}} .$$

Similarly, if the source S is at large distance r', it may be assimilated to a point source emitting a spherical wave $\underline{E}(P)$. Setting $\mathbf{r}' \equiv \overrightarrow{OS}$, we may write

$$\underline{E}(P) = \frac{1}{SP}\, e^{-ik'SP} \approx \frac{1}{SP}\, e^{-i\mathbf{k}.\overrightarrow{SP}} \approx \frac{1}{r'}\, e^{i\mathbf{k}'.\mathbf{r}'}\, e^{-i\mathbf{k}'.\overrightarrow{OP}} .$$

Thus, the incident wave and diffracted wave are approximately plane waves. Their wave vectors $\mathbf{k}'$ and $\mathbf{k}$ have equal magnitudes $k = k' = 2\pi/\lambda$. On the other hand, if the aperture is plane, the inclination factor $\eta_P(\theta', \theta)$ does not depend on P and, for small angles θ' and θ, $\eta_P(\theta', \theta)$ may be considered as a constant. If x and y are the coordinates of the point P of the aperture, we may write the total diffracted wave as

$$\underline{E}_d(\theta) = \chi I_{\mathcal{S}}(\mathbf{K}), \quad \text{where } I_{\mathcal{S}}(\mathbf{K}) = \iint_{\mathcal{S}} d\mathcal{S}\; \mathcal{T}(x,y)\, e^{i(xK_x + yK_y)} \text{ and } \mathbf{K} = \mathbf{k} - \mathbf{k}'. \quad [11.79]$$

In this expression, χ is a constant, which is independent of the observation angle θ if it is small. The total diffracted wave is obtained by integration over the aperture. Using an appropriate transmission factor $\mathcal{T}(P)$, it is possible to extend the integration to the entire aperture screen. This allows us to consider the *aperture function* $I_{\mathcal{S}}(\mathbf{K})$ as the Fourier transform of the transmittance. In the following, we apply [11.79] to the case of a narrow slit, a rectangular aperture, and a circular aperture.

a) Consider the case of a narrow slit of width d in the direction Ox with length L in the direction Oy (Figure 11.8a). If L is large and the wave vector $\mathbf{k}'$ of the incident wave is parallel to the plane Oxz, the system has translational symmetry in the direction Oy. Thus, we consider the observation points M in the $O'Xz$ plane in the direction making an angle θ with Oz (then, $k_x = k \sin\theta$ and $k_y = 0$). If the incident wave is normal to the aperture, $k'_x = k'_y = 0$, we have $K_x = k \sin\theta$ and $K_y = 0$ and the expression [11.79] may be written as

$$\underline{E}_d(M) = \chi \int_{-d/2}^{d/2} dx \int_{-L/2}^{L/2} dy\; e^{ixK_x} = \chi L d\, \frac{\sin(\Phi/2)}{\Phi/2} \quad \text{with } \Phi = k_x d = 2\pi \frac{d}{\lambda} \sin\theta. \quad [11.80]$$

As $d \sin\theta$ is the path difference between the wavelets that are emitted from the extreme points of the slit in the width direction Ox, Φ is the phase shift between these wavelets. The diffracted intensity may be written as

$$\mathcal{I}(\theta) = \mathcal{I}_o\, \mathcal{F}_d(\Phi), \qquad \text{where} \quad \mathcal{F}_d(\Phi) = \left[\frac{\sin(\Phi/2)}{\Phi/2}\right]^2. \qquad [11.81]$$

The diffracted intensity $\mathcal{I}(\theta)$ is illustrated in Figure 11.8b. For $\Phi = 0$, $\mathcal{F}_d(\Phi) = 1$ and the intensity has a sharp principal maximum equal to $\mathcal{I}_o$. The intensity $\mathcal{I}(\theta)$ vanishes for $\Phi = 2p\pi$, i.e.

$$\sin\theta = p\frac{\lambda}{d}, \qquad p = \pm 1, \pm 2, \pm 3 \ldots \text{ (zeros of intensity)}. \qquad [11.82]$$

Approximately, halfway between two zeros of $\mathcal{I}(\theta)$, i.e. for $\sin\theta \approx (p + \frac{1}{2})(\lambda/d)$, the intensity has a secondary maximum equal to $\mathcal{I}_o^2/[(p + \frac{1}{2})\pi]^2$, which decreases if p increases. Most of the diffracted intensity is concentrated in the principal maximum and, in the case of light, precise photometric measurements using photomultipliers confirm this distribution of the intensity. We may define the half-width as the distance between the center of the principal maximum ($\Phi = 0$) and the first zero of $\mathcal{I}$ ($\Phi = 2\pi$, i.e. $\sin\theta = \lambda/d$). Note that the width of the principal maximum is twice the width of the secondary maximums.

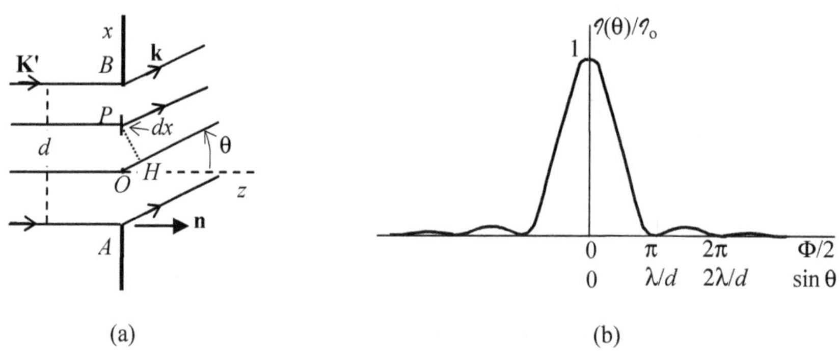

Figure 11.8. *a) Diffraction by a slit AB of width d in the case of a normal incident wave, and b) variation of the intensity as a function of* $\Phi/2$ *or* $\sin\theta$

If the angle of incidence on the aperture is θ' (algebraic and measured from the normal **n** to the aperture in the direction of the incident wave), the previous expressions remain valid but with a phase shift $\Phi = 2\pi\,(d/\lambda)(\sin\theta - \sin\theta')$. Then, the direction of the principal maximum is $\theta = \theta'$.

The first zero of the intensity corresponds to $\Phi = 2\pi$, i.e. the wavelets emitted by the end points A and B in phase. To interpret this result, imagine that the slit is divided into strips of infinitesimal width in the direction AB. If the direction θ_1 is such that the phase shift between the extreme wavelets emitted by A and B is 2π, the phase shift between the wavelet emitted by A and that which is emitted by the middle O of the slit is π. Their interference is destructive and will be so for all the elements of the upper half of the slit and the corresponding elements of the lower half. The diffracted intensity is thus equal to zero in the direction θ_1.

This argument may be repeated for the other zeros of the intensity: it is sufficient that the slit be formed by an even number of zones, such that the wavelets emitted by the points of a zone and the corresponding points of the next zone are in the opposite phase in the direction θ_p. Conversely, if the direction θ is such that the slit is formed by an odd number of zones, the wavelets emitted by the unpaired last zone do not interfere destructively with any other zone, and we have a secondary maximum. The emitted intensity is that of this unpaired zone and it decreases as the number of zones increases because the area of the unpaired zone decreases.

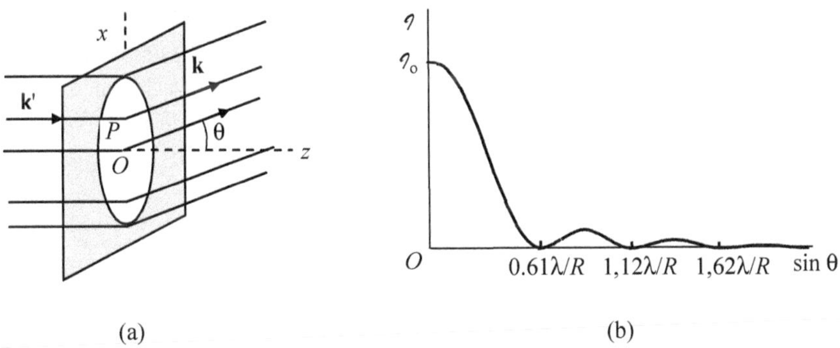

Figure 11.9. *a) Diffraction by a circular aperture, and b) variation of the intensity as a function of* sin θ, *where* θ *is the angle of observation*

b) Consider the case of a circular aperture of radius R illuminated by a plane wave incident normally (Figure 11.9a). This setup has a rotational symmetry around the axis of the aperture Oz; thus, the intensity of the diffracted wave has the same symmetry. We divide the aperture into small elements of area and we use the polar coordinates (ρ, ϕ) for a running point P of the aperture. In this case, the diffracted wave [11.79] may be written as

$$\underline{E}_d(M) = \chi \int_0^R d\rho\, \rho \int_0^{2\pi} d\phi\; e^{iK_x \rho \cos\phi}, \qquad \text{where } K_x = (2\pi/\lambda)\sin\theta. \qquad [11.83]$$

The integral over ϕ may be expressed in terms of the Bessel function J_o (see section 5.7)

$$\int_0^{2\pi} d\phi\; e^{is\cos\phi} = 2\pi\, J_o(s), \qquad \text{where } s = k_x R = 2\pi(R/\lambda)\sin\theta. \qquad [11.84]$$

As we have $(d/ds)[sJ_1(s)] = s\,J_o(s)$, where J_1 is Bessel function of order 1, we get

$$E_d(M) = 2\pi\chi \int_0^R d\rho\, \rho\; J_o(k_x\rho) = 2\pi\chi \frac{1}{k_x{}^2} \int_0^{s_o} ds\; s\; J_o(s) = 2\pi\chi R^2 \frac{1}{s} J_1(s). \quad [11.85]$$

The intensity in the direction θ may be written as

$$\mathcal{I}(M) = \mathcal{I}_o \left[\frac{2}{s} J_1(s)\right]^2. \qquad [11.86]$$

The function $J_1(s)$ oscillates similarly to a sine function but its amplitude decreases and its zeros s_{1p} are not equally spaced. The ratio $J_1(s)/s$ has a principal maximum equal to ½ for $s = 0$. Figure 11.9b illustrates the variation of $\mathcal{I}$ as a function of $\sin\theta$. It has a sharp maximum equal to some value $\mathcal{I}_o$ for $\sin\theta = 0$, vanishes for a series of values $\sin\theta_{min} = (s_{1p}/2\pi)(\lambda/R)$ and it has a secondary maximum at $\sin\theta_{max}$ between each two consecutive zeros:

$$\sin\theta_{max} = p_{max}\frac{\lambda}{R}, \qquad \sin\theta_{min} = p_{min}\frac{\lambda}{R}. \qquad [11.87]$$

The first values of p_{max}, p_{min}, and the relative intensities of the first maximums, are

$p_{min} =$	0.61,	1.12,	1.62,	2.12,	etc.
$p_{max} =$	0,	0.82,	1.32,	1.84,	etc.
$I_{max} =$	1,	0.0175,	0.0042,	0.0016,	etc. [11.88]

If the wave is intercepted on a screen normal to the axis, the diffraction pattern is a luminous disk, called the *Airy disk*, of radius given by the relation $\sin\theta = 0.61\,\lambda/R$ and followed by circular rings of decreasing intensity.

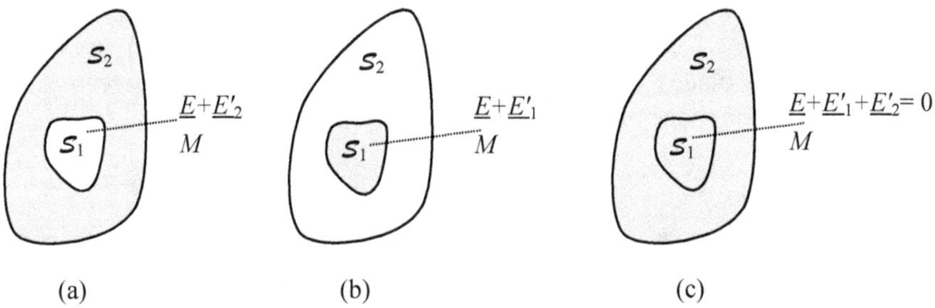

Figure 11.10. *Babinet's theorem: a) an aperture in a screen and (b) its complementary screen*

11.9. Diffraction by an obstacle, Babinet's theorem

The diffracted wave may be considered as resulting from the superposition of the primary wave $\underline{E}$ and the secondary wavelets emitted by the atoms of the obstacle. Consider a wave $\underline{E}$ incident on an opaque screen with an aperture $\mathcal{S}_1$ (Figure 11.10a). Let $\mathcal{S}_2$ be the opaque part of this screen. The diffracted wave received at a point M is $\underline{E}_{(a)} = \underline{E} + \underline{E}'_2$, where $\underline{E}'_2$ is the resultant of the waves that are emitted by the atoms of $\mathcal{S}_2$ if they are excited by $\underline{E}$. Figure 11.10b illustrates the *complementary screen* of Figure 11.10a (i.e. the opaque parts of one are the transparent parts of the other). The wave at the same point M is $\underline{E}_{(b)} = \underline{E} + \underline{E}'_1$, where $\underline{E}'_1$ is the wave emitted by the atoms of $\mathcal{S}_1$. If the two parts $\mathcal{S}_1$ and $\mathcal{S}_2$ were opaque (Figure 11.10c), the wave at M would be $\underline{E}_{(c)} = \underline{E} + \underline{E}'_1 + \underline{E}'_2$ and this should be equal to zero, thus $\underline{E}'_1 + \underline{E}'_2 = -\underline{E}$. By writing these expressions, we assume that the waves emitted by the atoms are the same in Figures 11.10a, b and c; this means that an atom is excited only by the primary wave (and not by the waves emitted by the other atoms). Comparing these relationships, we find that

$$\underline{E}_{(a)} + \underline{E}_{(b)} = \underline{E}, \quad \text{hence } <|\underline{E}_{(a)}|^2> = <|\underline{E}_{(b)}|^2> + <|\underline{E}|^2> - 2\,\mathcal{R}e <\underline{E}^*\underline{E}_{(b)}> . \quad [11.89]$$

From this result, we draw the following conclusions:

– If $\underline{E}_{(a)} = 0$, then $\underline{E}_{(b)} = \underline{E}$. This means that the dark points in the case (a) receive the total wave in the case (b) (that is, the wave received if there is no screen).

– If $\underline{E} = 0$, then $\underline{E}_{(a)} = -\underline{E}_{(b)}$. This means that at the points M, that receive no wave if there is no screen, the diffracted waves by the complementary screens are opposite; thus, they receive the same diffracted intensity. For instance, if a wave converges ideally at a point A of an observation screen P, no wave is received at the other points of P. If this wave is diffracted by an aperture (a) or by its complementary (b) placed before P, the intensity observed on this screen is the same, whatever is the shape of the aperture, except at the point A.

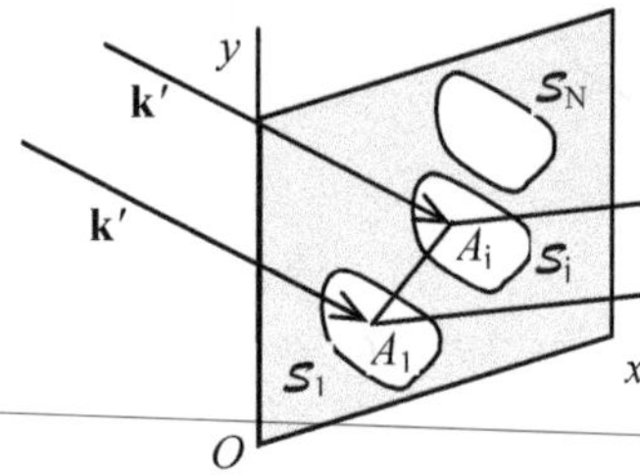

Figure 11.11. *Diffraction by randomly distributed apertures*

11.10. Diffraction by several randomly distributed identical apertures

Suppose that a wave of wave vector $\mathbf{k}'$ is incident on a large number N of identical apertures. If they are randomly distributed in an opaque screen without any overlap, they are obtained from one of them (aperture 1, for instance) by translations $\overrightarrow{A_1 A'_j} = \xi_j \, \mathbf{e}_x + \eta_j \, \mathbf{e}_y$ (Figure 11.11). The diffracted wave at a point M situated at large distance in the direction of the wave vector $\mathbf{k}$ is the superposition of the waves that are diffracted by all the apertures, thus

$$\underline{E}_d(M) = \chi \int_{S_1} dx_1 dy_1 \; e^{i(x_1 K_x + y_1 K_y)} + \chi \, \Sigma_{j \neq 1} \int_{S_j} dx_j \; dy_j \; e^{i(x_j K_x + y_j K_y)}, \qquad [11.90]$$

where we have set $\mathbf{K} = \mathbf{k} - \mathbf{k}'$. The first term is the wave $\underline{E}_{d1}(M)$ diffracted by the aperture (1). If we make a change of integration variables $x_j = x_1 + \xi_j$ and $y_j = y_1 + \eta_j$ in the integrals over S_j, we find that they are equal to the first integral multiplied by a phase factor $e^{i\phi_j}$, where $\phi_j = \xi_j K_x + \eta_j K_y$. Thus, the total diffracted wave is

$$\underline{E}_d(M) = \underline{E}_{d1}(M) \, [1 + \Sigma_{1<j \leq N} \, e^{i\phi_j}]. \qquad [11.91]$$

As the number of apertures is large and they are randomly distributed, the phases ϕ_j take all values between 0 and 2π; thus, $\Sigma_{j \neq 1} \, e^{i\phi_j} = 0$ and the total diffracted wave is the same as the diffracted wave by a single aperture. However, this argument fails if

$\mathbf{k} = \mathbf{k}'$, that is, in the direction of the incident wave, as all the phase shifts ϕ_j are then equal to 0, and we find

$$\underline{E}_d = N\,\underline{E}_{d1}. \qquad [11.92]$$

Thus, the diffraction pattern produced by a large number N of randomly distributed identical apertures is the same as that of a single aperture except in the direction of the principal maximum whose intensity is N^2 times the intensity diffracted by a single aperture. The same result holds if the wave is diffracted by randomly distributed obstacles (dust or powder, for instance).

11.11 Diffraction grating

A *diffraction grating* is usually a series of N parallel and identical slits with a spacing d between adjacent slits. More generally, it is an optical structure that is repeated N times in the x direction with a period d of the order of the wavelength λ (Figure 11.12a). They produce, by reflection or transmission, N coherent waves of equal amplitude a with the same phase-shift between consecutive waves. Their diffraction properties are essentially determined by the spacing d and the number N of slits. It is possible to conceive diffraction gratings for light waves, infrared and ultraviolet waves, microwaves, acoustic waves, etc. The characteristic effects of diffraction gratings are easily observed on a compact disk, a fish scale, a bird's fine feather, etc.

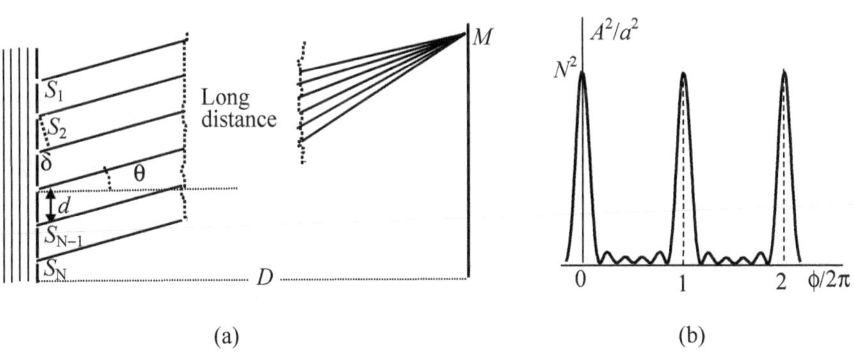

Figure 11.12. *a) Interference of several waves, and b) variation of A^2/a^2 versus $\phi/2\pi$*

The resultant wave at an observation point M is the superposition of N waves emitted or diffracted by the slits. If the distance D from the diffraction grating to the observation screen is large, compared to Nd, the rays S_iM are nearly parallel. The setup is called *Fraunhofer diffraction* (or *diffraction at infinity*). If the observation point M is in the direction θ (with the normal Oz to the diffraction grating), the path difference between two consecutive waves is $\delta = d \sin\theta$ and their phase shift at M is $\phi = 2\pi(d/\lambda)\sin\theta$. Using the complex representation for the N waves, the resultant wave is

$$\underline{E} = a\,e^{i(\omega t-\mathbf{k.r})} + a\,e^{i(\omega t-\mathbf{k.r}-\phi)} + \ldots + a\,e^{i[\omega t-\mathbf{k.r}-(N-1)\phi]}$$
$$= a\,e^{i(\omega t-\mathbf{k.r})}\,[1 + e^{-i\phi} + e^{-2i\phi} + \ldots + e^{-i(N-1)\phi}].$$

The expression in brackets is a geometrical progression; it may be written as

$$[\,] = a\,e^{i(\omega t-\mathbf{k.r})}\,\frac{1-e^{-iN\phi}}{1-e^{-i\phi}} = A\,e^{i[\omega t-\mathbf{k.r}-(N-1)\phi/2]}.$$

Thus, the real part of $\underline{E}$ is

$$\mathcal{Re}\,\underline{E} = A(\phi)\cos[\omega t - \mathbf{k.r} - \tfrac{1}{2}(N-1)\phi] \quad \text{where } A(\phi) = a\,\frac{\sin(N\phi/2)}{\sin(\phi/2)}. \qquad [11.93]$$

and the resultant intensity is

$$\mathcal{I} = \mathcal{I}_o\,\mathcal{F}_{dg}(\phi), \quad \text{where } \mathcal{I}_o = \alpha N^2 a^2 \text{ and } \mathcal{F}_{dg}(\phi) \equiv \frac{\sin^2(N\phi/2)}{N^2\sin^2(\phi/2)}. \qquad [11.94]$$

$\mathcal{F}_{dg}(\phi)$ is the *diffraction grating function.* The amplitude $A(\phi)$ has principal maximums equal to Na and $\mathcal{F}_{dg}(\phi)$ has principal maximums equal to 1 for $\phi = 2p\pi$. Between two consecutive principal maximums, $A(\phi)$ and $\mathcal{F}_{dg}(\phi)$ have $(N-1)$ zeros (corresponding to directions such that $\sin(N\phi/2) = 0$, other than those of the principal maximums) and $(N-2)$ secondary maximums of relatively small amplitudes. The principal maximums correspond to all the slit waves in phase. The zeros correspond to phase shifts, such that the phasor representation is a closed regular polygon formed one or several times. The larger N is, the sharper the principal maximums are, and the more numerous and smaller the secondary maximums are. This makes diffraction grating very useful in spectroscopy. Figure 11.12b illustrates the function $\mathcal{F}_{dg}$ as a function of the phase shift $\phi/2\pi$.

If the incident wave makes an angle θ' with the normal $\mathbf{n}$ to the diffraction grating (θ' is algebraic measured from $\mathbf{n}$ pointing in the direction of the incident wave), the previous expressions remain valid but with a phase shift $\phi = 2\pi\,(d/\lambda)(\sin\theta - \sin\theta')$. Then, the direction of the principal maximum of order $p = 0$ is $\theta_o = \theta'$.

Instead of having the superposition of waves that are diffracted by N slits, we may have the superposition of waves received by N receptors. For instance, a radio telescope is a set of highly adjustable antennas (often parabolic), which may be directed toward the sky to receive radio waves emitted by celestial bodies and to measure their wavelength (going from 1 mm to more than 1 km) and their intensity. The parabolic antenna must have a large aperture area to detect as much energy as possible and reduce the diffraction by the aperture of each antenna (see section 11.8b). The largest interferometric radio telescope in operation is the Allen Telescope Array in California. It is composed of 42 antennas of diameter 6.1 m and distributed over an area of diameter 1 km. It was conceived to detect waves of frequencies from 0.5 to 11 GHz (i.e. a wavelength going from 2.7 to 60 cm).

One of the important uses of a diffraction grating is as a dispersive element for spectroscopic instruments, enabling very precise measurement of wavelengths. The direction θ_p of the order p is such that $\phi_p = 2\pi(d/\lambda)\sin\theta_p = 2p\pi$. Thus, a measurement of θ_p allows us to determine the wavelength $\lambda = (d/p)\sin\theta_p$. The first zeros of intensity correspond to $\phi_p \pm \delta\phi = 2p\pi \pm 2\pi/N$. Thus, the half-width of the principal maximum of order p is $\delta\theta_p = \lambda/Nd\cos\theta_p$. If the incident wave is a superposition of two wavelengths λ and $\lambda' = \lambda + \Delta\lambda$, the principal maximum of order 0 is in the same direction $\theta = 0$ for both waves; while the other principal maximums are in slightly different directions such that $\lambda = (d/p)\sin\theta_p$. By differentiating this relationship we obtain the angular separation $\Delta\theta_p = p\,\Delta\lambda/d\cos\theta_p$ for their principal maximums of order p. According to the *Rayleigh criterion*, the two radiations can be distinguished by observing the order p if their principal maximums are separated by an angle $\Delta\theta_p$ larger than the half-width of these principal maximums (Figure 11.13). Thus, we must have $\Delta\theta_p > \delta\theta$, that is, $\Delta\lambda > \lambda/pN$. This is the smallest wavelength difference that may be measured with this instrument. The *resolving power* in the pth-order is

$$\mathcal{R}_p \equiv \lambda/\Delta\lambda = pN. \qquad [11.95]$$

The higher the order and the larger the number of lines N, the higher the resolving power.

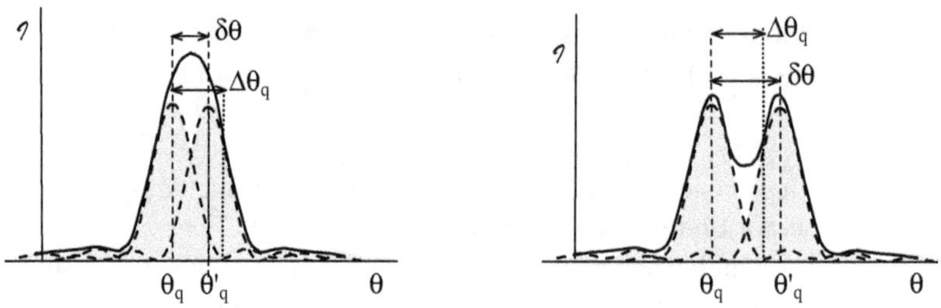

Figure 11.13. *Rayleigh criterion and resolving power*

11.12. X-ray diffraction

A diffraction grating may be two-dimensional or three-dimensional if the diffracting centers are distributed periodically over a plane or in a volume with spacing of the order of λ in all directions. X-rays are electromagnetic waves of very short wavelength, of the order of 0.1 nm (i.e. the inter-atomic distances in solids). It is impossible to have a one-dimensional diffraction grating with such small spacing. In 1912, Max von Laue observed the diffraction of X-rays by a crystal. He proposed that a crystal is a regular array of atoms, which may act as a three-dimensional diffraction grating for X-rays.

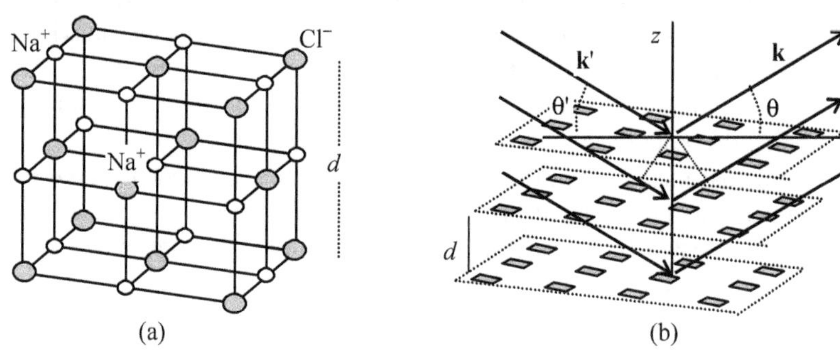

Figure 11.14. *a) Unit cell of NaCl, b) the atomic planes reflect the wave like a mirror. The direction of the principal maximums verify Bragg condition*

The fundamental structure of a crystal is the *unit cell*, which is an arrangement of a certain number of atoms (or ions) in a characteristic geometrical configuration. A macroscopic crystal is a periodical three-dimensional juxtaposition of unit cells in all directions. For instance, the unit cell of sodium chloride is face-centered cubic of sides $d = 0.562737$ nm. It is formed by eight juxtaposed cubes with alternating Cl^- and Na^+ ions occupying their summits (Figure 11.14a). Note that all the ions of the unit cell, except the central Na^+, are shared by two, four or eight neighboring cells. Thus, on average, each unit cell is formed by only four Cl^- ions and four Na^+ ions. Each of these cells is a diffracting center for X-rays. The crystal is then equivalent to equally spaced diffraction centers in any direction (Figure 11.14b). The total diffracted wave results from the interference of these partial waves. The direction of the principal maximums of interference is determined by the geometry of the configuration of the unit cells, while the intensity of the diffracted wave in a given direction is determined by the distribution of the electric charge within the unit cell. Thus, by observing the direction of the diffraction maximums, we obtain information on the geometry of the diffraction centers, and by observing the intensity distribution, we obtain information on the distribution of charge within the unit cell.

Analyzing the diffraction by a linear chain of unit cells, then by all the linear chains of the atomic plane *Oxy* (or any atomic plane parallel to *Oxy*), it may be shown that the intensity has a principal maximum in the direction of the reflected wave ($\theta' = -\theta$) and in the direction of the incident wave ($\theta' = \theta$). Here θ and θ' are the angles of the incident wave vector **k** and the diffracted wave vector **k**' with the atomic plane (Figure 11.14b). If d is the spacing between the atomic planes, the path difference of the waves diffracted by two consecutive planes is $2d \sin\theta$. Thus, the total diffracted wave has a principal maximum if this path difference is an integer multiple of λ; hence, Bragg's law

$$2d \sin\theta_p = p\,\lambda, \qquad \text{where } p \text{ is an integer.} \qquad [11.96]$$

This condition must be verified by the angle θ_p of **k** with the atomic plane (which is equal to the angle of **k**' with this plane) in order to have a principal maximum. In the other directions, there is some phase shift between the waves diffracted by the planes. As the number of these planes is very large, the reflected intensity in these directions is negligible. We conclude that, if a wave (X-rays, electronic wave, etc.) is incident on the crystal in a direction making an angle θ_p with the atomic plane, there is a diffracted wave only in the directions verifying Bragg's law [11.96]. For instance, in the case of X-rays of wavelength $\lambda = 0.200$ nm incident on a NaCl crystal, the Bragg condition gives $\sin\theta = p\lambda/2d = 0.178\,p$, thus, $\theta = 0 - 10.2° - 20.8°$, etc.

The diffraction of X-rays of known wavelength enables the study of crystalline structures. If the spacing of a particular crystal (NaCl or $CaCO_3$, for instance) is determined by a method other than X-rays, this crystal may then be used as a spectrometer for X-rays. Observation of the diffraction of a beam of X-rays by this crystal allows us to determine the wavelength of the X-rays. Then, the diffraction of this beam of known λ by other crystals can be used to study their structures.

The angular distribution of the diffracted wave depends on the structure of the unit cell. Let x_j, y_j and z_j (j = 1, 2, …, n) be the coordinates of the n atoms of the cell. If the crystal is formed by N unit cells, we may consider it as the superposition of n simple crystals C_j, whose N atoms of the type j form parallel planes of spacing d. If the incident wave has an amplitude E_o and a wave vector **k**', the emitted wave by an atom (j) with a wave vector **k** is of the form $E_o A_j e^{i\phi_j}$. As ϕ_j is the same for all the N atoms of type j and given by $\phi_j = K_x x_j + K_y y_j + K_z z_j$ with $\mathbf{K} = \mathbf{k} - \mathbf{k}'$, the total wave that is emitted by these atoms is $\underline{E}_j = N E_o A_j e^{i\phi_j}$. The total wave that is diffracted by all the crystal is the superposition of the waves $\underline{E}_j$ diffracted by the simple crystals C_j, that is,

$$\underline{E} = \Sigma_{1\le j\le n} \underline{E}_j = N E_o \underline{F}(\mathbf{K}) \qquad \text{with} \quad \underline{F}(\mathbf{K}) = \Sigma_{1\le j\le n} A_j e^{i\phi_j}, \qquad [11.97]$$

where the summation runs over the atoms of a single unit cell. The complex function $\underline{F}(\mathbf{K})$ is called *structure factor of the crystal*. It is a characteristic of its unit cell (type of atoms, their number, and their geometrical configuration), not on the crystal, as a whole. A measurement of the intensity that is diffracted in the various directions allows the determination of $|\underline{F}(\mathbf{K})|$ and provides information on the cell structure.

11.13. Diffusion of waves*

If a light beam propagates in a medium, waves of small amplitude may be emitted in directions, and sometimes with frequencies, different from those of the primary wave. This effect, called *diffusion*, is due to the scattering of the wave by the particles of the medium designated as the *scatterers*. It may be also interpreted as due to the collision of individual photons with the scatterers and the subsequent emission of photons in different directions and eventually different energies. The study of the diffusion by atoms and molecules must inevitably use quantum mechanics and this is beyond the scope of this book. The laws of classical physics are sufficient to study the diffusion by macroscopic scatterers (large molecules, dust, smoke, fog, density fluctuations, roughness of surfaces, etc.). The difference between the diffusion and diffraction patterns is due to the randomly distributed scatterers; thus the absence of any coherent phase relation between the scattered waves. Several effects of propagation are consequences of diffusion.

In classical theory, the electric field of the wave acts on the charged particles of the scatterers and the energy of the wave may be absorbed and subsequently emitted in all directions according to the laws of electromagnetism. In quantum theory, a photon collides with a target particle and it may be absorbed or scattered according to the laws of quantum mechanics. The duration of the collision is very short and the wave or photon are scattered in various directions with a certain law of probability within certain limits imposed by the laws of conservation of energy and momentum.

If the medium is dense, the scatterers are not completely independent, and the analysis becomes more complicated. In some cases, the direction of the resultant diffused wave is more restricted and it may be even in a single direction, because of the constructive interference in this direction and the destructive interference in the other directions. This is the case of the reflection and refraction on polished surfaces.

Diffusion is not an exclusive property of electromagnetic waves: all waves and particles (electrons, neutrons, alpha particles, etc.) undergo collisions with other particles. A collision is said to be *elastic* if there is little exchange of energy. Then, the incident and scattered photons have equal energy and, consequently, equal frequency and wavelength. The wave is simply deviated with a certain angular distribution, which may depend on energy. In principle, the complete knowledge of the properties of the medium enable the determination of the laws of diffusion (this is the *direct problem*). Conversely, the analysis of the diffusion in a medium allows us to study some properties of the scatterers (this is the *inverse problem*). The analysis of diffusion is a very important means to study these interactions, and it has many applications in physics, biology, physics of the atmosphere, etc.

A) Resonance scattering

An atom (or a molecule) is usually in its ground state of energy E_1. If an electromagnetic wave is incident on the medium, it acts on the electrons and excites the atom to a state of energy E_i. According to quantum theory, this interaction is a process of absorption and subsequent emission of a single photon of energy $E_\gamma = h\tilde{\nu}$. If E_γ is close to $E_i - E_1$, that is, $\tilde{\nu} \cong \tilde{\nu}_{i,1} = (E_i - E_1)/h$, there will be a resonance effect. The atom then has a high probability of absorbing the photon, being excited to the state of energy E_i, and subsequently, returning to the ground state by emitting a photon of the same frequency but in a direction often different from that of the incident photon. The emitted photon may again be scattered by other atoms and finally emerge out of the medium in an arbitrary direction or be absorbed as heat. This scattering is said to be *resonant.* A beam having a frequency $\tilde{\nu}_{i,1}$ is almost totally diffused in all directions, giving to the body its characteristic color if it is observed by diffusion or reflection. The medium is almost opaque to this beam and the body appears black if it is observed by transmission. If a beam of continuous spectrum (white light, for instance) traverses this medium, the frequencies $\tilde{\nu}_{i,1}$ will be absent from the emerging beam, producing dark lines in the beam spectrum (this is the so-called *absorption spectrum*). A red filter, for instance, is transparent to the part of the spectrum close to red and it absorbs the part that is close to the blue.

On the contrary, if E_γ is not close to the excitation energy $E_i - E_1$ of one of the energy levels, the wave propagates through the medium almost without change of intensity and frequency; the medium is transparent to this frequency. We say in this case that the diffusion is *non-resonant*. If E_γ is lower than all the excitation energies $E_i - E_1$ of the atom, the photon cannot be absorbed by the atom. It may only make the electrons oscillate slightly and emit a secondary wave with a very small intensity. On the contrary, if E_γ is higher than the excitation $E_i - E_1$, the photon may be absorbed and the exceeding energy transformed into kinetic energy and ultimately to heat. The excited atom subsequently emits a photon of frequency $\tilde{\nu}_{i,1}$ in any direction.

If a body has no resonance frequency $\nu_{i,1}$ in the spectrum, it is transparent to all frequencies and it produces no diffusion; thus, it is completely invisible. However, if it is pulverized into small pieces, these parts diffuse all frequencies and it appears white if it is illuminated with daylight.

The optical properties of a medium depend strongly on the degree of order of the scatterers and on the mean distance between them, compared to the wavelength. If the scatterers are distributed at random, there will be no phase shift relation between the secondary waves that they emit; the wave is then diffused in all directions. On the contrary, if the scatterers are exactly periodic (as in the case of a diffraction grating or a crystal) there will be no diffused wave but diffraction in specific directions. On the other hand, if the spacing between the scatterers is of the order of λ, the emitted wave emitted by one of them acts on the others; then, the scatterers cannot be considered as independent. For this reason, a strong correlation exists between the waves that are emitted by the molecules of a liquid or a solid that contains no impurities. This makes a plane wave propagate in these mediums in a single direction according to the laws of reflection and refraction. This direction is that of the constructive interference of the secondary emitted waves by these molecules. On the contrary, if the scatterers are distributed at random with spacing much larger than λ, we usually find the other extreme case of completely independent and incoherent sources. Then, the interference is completely negligible and the resultant intensity is the sum of the intensities that are emitted by the sources. This is the case of the molecules of a very rarefied gas.

Assume that a primary wave is polarized in the direction Ox and propagates in the direction Oz in a medium. Under the influence of this wave, the scatterers emit secondary waves. The motion of electrons is proportional to the electric field and, by symmetry, it is in the direction Ox of $\mathbf{E}$. Then, the scatterers are equivalent to electric dipoles that are polarized in the direction of $\mathbf{E}$ and oscillate with the same frequency. According to the laws of electromagnetism, the secondary wavelets, that they emit, are polarized also in the direction Ox and they propagate essentially in the normal direction to this axis.

B) Propagation in dense mediums

A medium is considered as *dense*, if the spacing between scatterers is of the order of the wavelength λ; thus, much shorter than the coherence length of the wave. This is the case of liquids, solids, and a gas at high pressure. The wavelets emitted by neighboring scatterers interfere with the primary wave to produce the total wave, which propagates in the medium in a well-defined direction with a speed that is characteristic of the medium. To understand this, we first recall that the interference of a large number of waves whose phase randomly takes all values is completely destructive (see section 11.10). Consider a cylindrical beam of light in the direction Ox (Figure 11.15a). The scatterers of a plane wave front $\mathcal{S}_1$ are excited simultaneously. Thus, they emit wavelets in phase. However, the scatterers being very numerous and distributed randomly in the plane $\mathcal{S}_1$, emit wavelets that arrive to an observation point M_1 outside the cylindrical beam with random phases. Their interference is destructive at this point (Figure 11.15b). Conversely, at points situated inside the beam, such as M_2 following $\mathcal{S}_1$ or M_3 preceding $\mathcal{S}_1$, many of these wavelets arrive almost in phase; thus, their interference is constructive, at least partially.

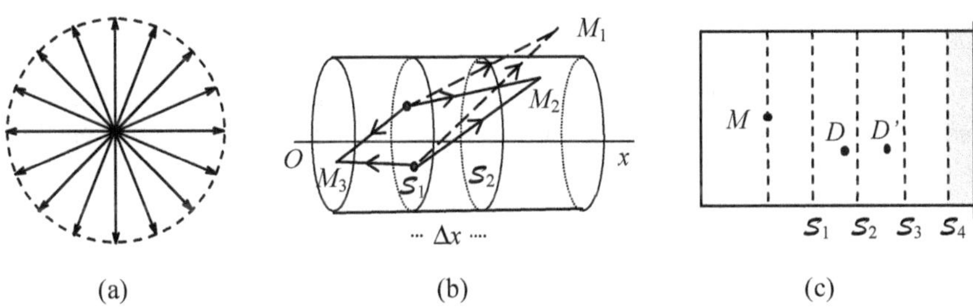

Figure 11.15. *Diffusion of a wave: a) the superposition of randomly phased waves is completely destructive, b) propagation of a wave as interference of the diffused waves and the primary wave, and c) reflection as interference of diffused wavelets*

To understand why this interference produces a wave only in the forward direction, we have to consider the waves emitted by different planes $\mathcal{S}_i$. If the wave propagates from $\mathcal{S}_1$ toward $\mathcal{S}_2$ situated at a distance Δx, the scatterers of $\mathcal{S}_2$ start to emit with a time delay $\Delta x/v$ over those of $\mathcal{S}_1$. However, to reach M_2, the wave emitted by $\mathcal{S}_2$ must travel a distance Δx less than the wave emitted by $\mathcal{S}_1$. Thus, the two waves arrive in phase and they interfere constructively in the forward direction. On the other hand, if we consider a point M_3 that precedes $\mathcal{S}_1$, the wavelets diffused by the scatterers of $\mathcal{S}_1$ interfere constructively at this point, but the wavelets diffused by $\mathcal{S}_2$ must travel a distance Δx more than those of $\mathcal{S}_1$. They interfere at M_3 with a time delay $2\Delta x/v$ (i.e. a phase shift $4\pi\ \Delta x/\lambda$). Thus, the waves diffused by the various sections have random phase shifts in the backward direction if the medium is infinite; their interference is destructive. This explains why the wave propagates only in the forward direction if the medium is infinite toward the positive x.

In the case of a dense medium bounded by a plane surface $\mathcal{S}$ (Figure 11.15c), let us assume that the region situated between M and $\mathcal{S}$ is divided into slices of thickness $\lambda/4$. The waves diffused by two consecutive slices interfere destructively at M, as the waves diffused backward by the corresponding scatterers D and D' are out of phase. Thus, the reflected wave at M results only from the interference of the wavelets that are diffused by the scatterers of the unpaired slice (or part of a slice) that precedes $\mathcal{S}$. The interference of these wavelets can be constructive only if the surface $\mathcal{S}$ is polished, that is, if its irregularities do not exceed a small fraction of the wavelength. Otherwise, the wave is diffused backward in all directions.

We note that the propagation of electromagnetic waves in vacuum is a very particular phenomenon with no diffusion at all. In modern physics, the vacuum is not completely devoid of electromagnetic properties, such as *vacuum polarization, vacuum fluctuations*, etc.

C) Diffusion by rarefied mediums: Rayleigh diffusion

A medium is said to be rarefied if the average spacing of the scatterers is much larger than the wavelength λ and they are non-coherent. This is the case of a gas at a very low pressure (as in the upper atmosphere) or rare impurities in a dense and homogeneous medium (such as smog, dust, density and temperature fluctuations of the atmosphere, etc.). In this case, the interference of the scattered wavelets is negligible and we have to add their intensities instead of their amplitudes.

In vacuum, a beam of light progresses forward and it may be observed only in the direction of the beam, as there is no lateral diffusion. This is what is observed outside the Earth's atmosphere: the Sun is visible only if it is looked at directly and the sky is black in the other directions. On the other hand, if we look to the sky from the ground, a pure atmosphere is almost completely transparent, as nitrogen and oxygen molecules have no resonances in the visible spectrum. Thus, light propagates in a pure atmosphere with very little absorption (provoked mostly by the ozone and vapor molecules). However, in the rarefied upper atmosphere, light undergoes some non-resonant diffusion by the nitrogen and oxygen molecules. Each molecule behaves as a small electric dipole, absorbing light energy and subsequently emitting mostly perpendicularly to the dipole. Being at relatively large distances apart, these molecules behave as independent sources of light. The rotation and vibration frequency spectrums of these molecules are lower than visible light while their electronic spectrums are higher. Thus, blue light is more diffused than red light and this is the reason for the blue sky, if we do not look in the direction of the Sun.

On the other hand, because of this more significant diffusion of blue, if light from the Sun crosses a larger thickness of the atmosphere (as at sunrise and at sundown), it tends to be red.

If the medium is homogeneous, the interference is practically destructive in all non-forward directions. The lateral diffusion becomes more significant if the medium is less homogeneous, containing some impurities or fluctuations. For instance, if drops of milk are poured in a large amount of water in a glass container illuminated with white light, the diffused light is bluish while the transmitted light is reddish. Amorphous and transparent solids, such as glass and plastics, diffuse some light, in contrast to crystals (because of the perfect periodic structure of crystals). Non-ordered and well-spaced irregularities are very good scatterers, as their waves are less coherent. Thus, they do not interfere destructively in the lateral directions.

The non-resonant diffusion depends on the size of the scatterers compared to the wavelength. If the dimensions of the scatterers are much smaller than λ, the intensity of the diffused wave depends on the frequency like $\tilde{\nu}^4$. This is the so-called *Rayleigh scattering* (see section 11.14). This is the case of O_2 and N_2 molecules in the atmosphere. They have a diameter of about 0.2 nm, thus much smaller than the wavelength of visible light. The $\tilde{\nu}^4$ dependence explains why violet and blue are much more diffused laterally than red.

If the size of the scatterers is comparable to λ or larger, the diffused intensity depends on the frequency like $\tilde{\nu}^p$ with $p < 4$ and it depends on the size and the shape of the scatterers. This is the case of a cloudy sky; the droplets of water have a much larger size than λ. All wavelengths are diffused with almost the same intensity and the cloud appears white or grayish. This is the so-called *Mie diffusion.* Both Rayleigh and Mie scatterings are elastic, that is, without change of frequency. If the size of the scatterers exceeds about 10 λ, we may use the laws of geometrical optics. Then, the deviation of light depends on the wavelength because of dispersion (as in the case of rainbow). This effect is not usually considered as diffusion.

11.14. Cross-section*

The first study of diffusion or scattering was attributed to Rayleigh (1871) who analyzed the scattering of sound by a spherical body and the scattering of light by a gas (the blue sky). The scattering of an electron beam started with Faxen and Holtsmark in 1927. Actually, the scattering of beams of particles, X-rays and γ-rays by atomic nuclei is a very important means to study the structure of matter and the interaction of particles. In optics, the scattering of light by molecules, atoms, and ions allows the study of the optical properties of matter (opacity and transparency, color, index of refraction, etc.).

Let us assume that a wave $u_{\mathrm{i}} = A\, e^{\mathrm{i}(\omega t - kz)}$, comes from far away in the direction Oz to meet a small-sized target (Figure 11.16). In optics, u may represent the electric field. The total wave is then

$$u = A\, e^{\mathrm{i}\omega t}\, [\, e^{-\mathrm{i}kz} + f(\theta)\frac{1}{r}\, e^{-\mathrm{i}kr}\,], \qquad [11.98]$$

where the second term represents the scattered spherical wave. The function $f(\theta)$ is the *scattering amplitude* in the direction θ; it depends on the interaction of the wave with the target. If the target and the wave are symmetric about Oz, f does not depend on the angle φ about this axis.

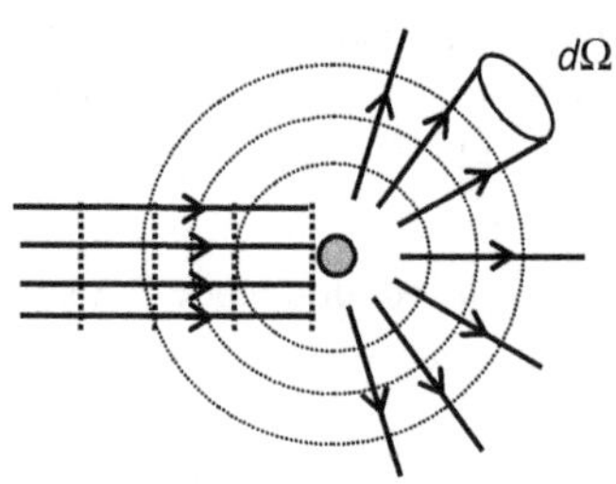

Figure 11.16. *Scattering experiment*

In a scattering experiment, a beam of particles is incident on a target and one measures the number of scattered particles by using appropriate detectors placed around the target (Figure 11.16). Let us assume that the incident beam contains N_{i} particles per unit volume moving with velocity v_{i}. The *flux of particles* is the number of particles crossing, per unit time, the unit area placed perpendicularly to the beam, i.e. $F_{\mathrm{i}} = N_{\mathrm{i}} v_{\mathrm{i}}$. If the target opposes to the beam a transverse area σ, each particle that passes through this area is absorbed or scattered in a direction other than that of the beam. Thus, the total number of particles extracted from the beam per unit time is the number of particles intercepted by the area σ, that is, $N_{\mathrm{s}} = F_{\mathrm{i}}\sigma$. This concept may be generalized to any target. The number of particles that are scattered per unit time depends on the interaction of the incident particles with the target. The *cross-section* is defined by

$$\sigma = N_{\mathrm{s}}/F_{\mathrm{i}}. \qquad [11.99]$$

In fact, the concept of cross section is not simple. If the particles of the beam do not interact with the target, there will be no scattering and the cross section vanishes, exactly as if there is no target. On the other hand, the interaction may depend on the energy of the particles; thus, the cross section depends on the energy and the target seems to be "smaller" or "larger" if the energy varies.

It is also possible to measure the number dN_s of scattered particles in a solid angle $d\Omega$ around the target, by setting a counter in this direction; $d\Omega$ is then the solid angle subtended by the entry of the counter as seen from the target. The *differential cross section* is defined by

$$\sigma(\Omega)\, d\Omega = dN_s/F. \qquad [11.100]$$

The cross section σ is obviously the integral of $\sigma(\Omega)$ over all directions. In modern physics, the wave-particle duality allows the interpretation of $|u|^2$, in the case of a beam, as the number of particles per unit volume. Then, it may be shown that

$$\sigma(\Omega) = |f(\Omega)|^2. \qquad [11.101]$$

Let us consider, for instance, an incident wave on a medium containing electrons that are bound to the atoms with a force $-m\omega_o^2 x$. Using the equation [10.89], we find that the electrons oscillate with an amplitude

$$A = -\frac{e}{m}\frac{1}{\omega_o^2 - \omega^2}E_o. \qquad [11.102]$$

The average power of the radiation that is emitted by this electron is (see section 15.8)

$$<P> = \frac{\mu_o e^2}{12\pi c}A^2\omega^4 = \frac{\mu_o e^4}{12\pi c m^2}\frac{\omega^4}{(\omega_o^2 - \omega^2)^2}E_o^2 = \frac{\mu_o^2 e^4}{6\pi m^2}\frac{\omega^4}{(\omega_o^2 - \omega^2)^2}\mathcal{I}, \qquad [11.103]$$

where we have used the intensity of the incident wave, $\mathcal{I} = E_o^2/2\mu_o c$, which corresponds to a flux of photons $F_i = \mathcal{I}/h\nu$. The power $<P>$ corresponds to $N_s = <P>/h\tilde{\nu}$ scattered photons. Thus, the cross section is

$$\sigma = N_s/F_i = <P>/\mathcal{I} = \frac{\mu_o^2 e^4}{6\pi m^2}\frac{\omega^4}{(\omega_o^2 - \omega^2)^2} = \frac{8\pi r_o^2}{3}\frac{\omega^4}{(\omega_o^2 - \omega^2)^2}. \qquad [11.104]$$

The quantity $r_o = e^2/4\pi\varepsilon_o mc^2 = 2.81794 \times 10^{-15}$ m is the so-called *classical radius of the electron*. In particular, at high frequency ($\omega >> \omega_o$), we find *Thomson cross section*

$$\sigma_{Th} = (8/3)\,\pi r_o^2. \qquad [11.105]$$

In the case of air molecules, the characteristic angular frequency ω_o is much higher than the angular frequencies of visible light; thus, in the first approximation,

$$\sigma = (8/3)\,\pi r_o^2\,(\omega/\omega_o)^4. \qquad [11.106]$$

σ varies like the fourth power of the frequency (see Rayleigh scattering).

The cross section of the scatterers is also related to the absorption coefficient of the wave in the medium. First, let us note that the intensity $\mathcal{I}$ of the wave is related to the flux F (defined as the number of particles that are received by the unit area and per unit time) by the relationship $\mathcal{I} = FE$, where E is the energy of each particle. Assume that the medium contains N_d scatterers per unit volume with cross section σ. If a beam of section $\mathcal{S}$ and flux F travels in this medium a distance dx, it meets $N_d\mathcal{S}\,dx$ scatterers. Each scatterer extracts σF particles per unit time from the beam. Thus, the beam loses $\sigma F N_d \mathcal{S}\,dx$ particles per unit time and its flux varies by

$$dF = -\,F\sigma N_d\,dx\,. \qquad [11.107]$$

We deduce the variation law of the flux or the intensity

$$F(x) = F_o\,e^{-\mu x}\,, \quad \text{i.e.} \quad \mathcal{I}(x) = \mathcal{I}_o\,e^{-\mu x} \quad \text{with} \;\; \mu = \sigma N_d\,. \qquad [11.108]$$

Like σ, the *absorption coefficient* μ depends on the energy of the particles (or the frequency of the wave). On the other hand, μ is proportional to the number of scatterers per unit volume. Thus, the attenuation in a gas increases if it is compressed. Finally, if the medium is formed by several types (i) of scatterers, its absorption coefficient μ is the sum of the μ_i:

$$\mu = \Sigma_i\,\sigma_i\,N_{di} = \Sigma_i\,\mu_i. \qquad [11.109]$$

11.15. Problems

Reflection and refraction on the interface of two dielectrics

P11.1 A non-polarized electromagnetic wave is incident at an angle θ on the interface of two dielectrics of indices n_1 and n_2. **a)** Calculate the energy reflection and transmission factors. Verify the conservation of energy. **b)** Assuming that $n_1 = 1$, $n_2 = 1.5$ and $\theta = 45°$, calculate the reflection and transmission coefficients and the reflection and transmission factors. Determine the degree of polarization in the plane of incidence for the reflected and the transmitted waves.

P11.2 An electromagnetic plane wave is polarized in a direction that makes an angle α with the plane of incidence. **a)** Show that the reflected and transmitted electric fields form with the plane of incidence, respectively, the angles α' and α'' given by the relations $\tan \alpha' = (\mathcal{R}_\perp / \mathcal{R}_{//}) \tan \alpha$ and $\tan \alpha'' = (\mathcal{T}_\perp / \mathcal{T}_{//}) \tan \alpha$, where $\mathcal{R}_\perp$ and $\mathcal{R}_{//}$ are the reflection coefficients, while $\mathcal{T}_\perp$ and $\mathcal{T}_{//}$ are the transmission coefficients for parallel polarization and perpendicular polarization to the plane of incidence. **b)** Calculate the coefficients $\mathcal{R}_\perp$, $\mathcal{T}_\perp$, $\mathcal{R}_{//}$ and $\mathcal{T}_{//}$ for light that is incident on water ($n = 1.33$) at 60°. Calculate α' and α'' if $\alpha = 45°$.

P11.3 An electromagnetic wave is incident at an angle θ on the interface of two dielectrics of indices n_1 and n_2. The Fresnel formulas show that the amplitude of the transmitted electric field may be larger than that of the incident field. **a)** Verify that the incident power is always equal to the sum of the reflected and transmitted powers. **b)** A light wave is incident normally on a plate of index 1.5. Neglecting multiple reflections, calculate the global energy transmission factor. What should this factor be for a set-up of four parallel but separated plates?

P11.4 A wave is incident on a glass plate of thickness L and index n at an angle of incidence θ. It is polarized perpendicularly to the plane of incidence. Calculate the global transmission factor f_T. Note that, in the plate, there are a transmitted wave across the first face and a reflected wave on the second face. What is the numerical value of f_T if $L = 3$ mm, $n = 1.5$ for light of wavelength $\lambda = 0.6\ \mu$ incident normally?

P11.5 Any wave that is incident on the interface of two dielectrics may be written in the form $\mathbf{E} = \underline{E}_{//}\,\mathbf{e}_{//} + \underline{E}_\perp \mathbf{e}_\perp$, where $\mathbf{e}_{//}$ and $\mathbf{e}_\perp$ are unit vectors that are perpendicular to the direction of propagation. $\mathbf{e}_{//}$ is in the plane of incidence and $\mathbf{e}_\perp$ is perpendicular to this plane. **a)** Write in this form the expression of a plane wave that is incident at an angle θ and polarized linearly, polarized circularly, and non-polarized. **b)** Consider the case of a non-polarized incident wave. Calculate the energy reflection and transmission factors. It will be convenient to set $n = n_2/n_1$. Determine the degree of polarization perpendicularly to the incidence plane for the reflected wave and for the transmitted wave. What are their values in the case of the incidence at the Brewster angle and at $\theta = 30°$ on the surface of water?

P11.6 a) An electromagnetic wave that is polarized perpendicularly to the plane of incidence falls on a glass plate of index 1.5. Show that the intensity of the reflected wave varies from 4% to 100% as θ varies from 0 to 90°. Calculate the reflected intensity at $\theta = 45°$. **b)** Show that, in the case of a polarized wave in the plane of incidence, the reflected intensity decreases from 4% at $\theta = 0$ to zero at Brewster's incidence, then it increases to 100% at $\theta = 90°$. What is the reflected intensity at $\theta = 45°$? **c)** A wave is incident on the glass-air interface from the glass side. Analyze the reflected intensity if the wave is polarized in the plane of incidence and if it is polarized perpendicularly to this plane. **d)** A loss of 4% is sometimes intolerable if light crosses the plate many times. Show that, at the Brewster incidence, after a large number of crossings, the light component that is polarized perpendicularly to the plane of incidence is eliminated and only the component that is polarized in this plane remains with an intensity nearly equal to the incident intensity.

Total reflection

P11.7 a) What is the critical angle on the glass-air interface if the glass index is 1.5? A wave is incident at 50° on the glass-air interface from the glass side. Calculate the attenuation coefficient in air for light of wavelength 0.6 μm in vacuum. How deeply does it penetrate the air? **b)** Consider a thin layer of air between the planes $z = 0$ and $z = L$ separating two glass plates, where L is of the order of the wavelength of light. A wave propagates in the first plate and falls on the surface $z = 0$ at an angle of incidence $\theta > i_L$. Assume that the wave is polarized in the plane of incidence. Write the boundary conditions on the faces of the air layer. Show that the wave may cross this layer, produce a wave in the second plate and propagate parallel to the direction of propagation in the first plate. Calculate the amplitude of the reflected wave and that of the transmitted wave for $L = 1$ μm. Note that, in the air layer, the wave has the form $f\,e^{-\delta z} + g\,e^{\delta z}$.

Reflection on conductors and plasmas

P11.8 A wave is incident normally on the interface of a dielectric and a good conductor. **a)** Show that, at low frequency ($\omega << \sigma/2\varepsilon_2$), the reflection and the transmission coefficients are approximately $\mathcal{R} \approx -1 + 4i\kappa$ and $\mathcal{T} \approx 4i\kappa$, where $\kappa = (\mu_2 v_2/\mu_1 v_1)(\varepsilon_2\omega/\sigma)$. Deduce that the wave is almost totally reflected with a phase lead ϕ given by $\tan\phi = -4\kappa$ with $\pi/2 < \phi < \pi$). **b)** Consider the case of a good non-magnetic conductor, such as silver (of conductivity $\sigma = 6.29 \times 10^7\ \Omega.\mathrm{m}^{-1}$). Using the exact expression of $\mathcal{R}$ and assuming that $v_2 = v_1 = c$, determine the frequency of the electromagnetic waves that are reflected at more than 95% in intensity. **c)** Estimate the magnitude and the phase of the reflection coefficient of silver for visible light of wavelength $\lambda = 500$ nm, ultraviolet of $\lambda = 100$ nm and X-rays of $\lambda = 0.1$ nm. **d)** What should the thickness of a film of silver be in order to transmit less than 10 % of the light intensity that crosses the entry face? Assume that $\lambda = 500$ nm.

P11.9 An electromagnetic wave is incident from a dielectric (medium 1) normally on the surface of a good conductor (medium 2). **a)** Show that Maxwell's equations in the conductor are the same as those in a dielectric with a complex dielectric constant $\underline{\varepsilon} = \varepsilon_2 - i\sigma/\omega$. Write the solution representing a plane wave, and deduce the usual properties of waves in conductors. **b)** Is it possible to define a complex index using $\underline{\varepsilon}$, and write Snell's law to determine the direction of propagation in the conductor? **c)** Consider a good conductor that is non-magnetic (like silver with $\sigma = 6.29 \times 10^7\ \Omega.\mathrm{m}^{-1}$). Assume that $v_2 = v_1 = c$ and use the exact expression of $\mathcal{R}$. For which frequencies is the electromagnetic wave reflected with more than 95% in intensity? Estimate the modulus and the phase of the reflection coefficient on silver in the case of visible light of wavelength 0.5 μm.

P11.10 Show that an electromagnetic wave of frequency $\tilde{\nu}$ is totally reflected on the ionosphere if its angle of incidence θ is such that $\cos\theta < \tilde{\nu}_p/\tilde{\nu}$, where $\tilde{\nu}_p$ is the plasma frequency of the ionosphere. This means that, for an angle of incidence θ, the effective cut-off frequency is $\tilde{\nu}_p/\cos\theta$.

Interference of two electromagnetic waves

P11.11 In Young's historic experiment, light was incident normally on two parallel slits separated by a distance d and the interference of the diffracted waves was observed on a screen situated at a large distance D from the slits (Figure 11.17). **a)** Show that the phase shift between the waves at a point M of the screen is $\phi \cong 2\pi(\delta/\lambda)\sin\theta \cong 2\pi\,(d/\lambda)(x/D)$, where x is the distance of M to the axis Oz and θ is the angle of OM with Oz. Deduce that the bright fringes correspond to $x = p\,Dd/\lambda$. **b)** Taking $d = 0.150 \pm 0.002$ mm and $D = 1.000 \pm 0.003$ m and using quasi-monochromatic light, the fourth bright fringe is found at a distance $x_4 = 12.7 \pm 0.1$ mm from the central fringe at O'. What is the value of the wavelength? Estimate the precision of this measurement. Using sunlight, a bright white fringe is found at O'. A pinhole is made in the screen at a distance of 1 cm from O'. Show that the light spectrum passing through this hole has dark lines (channeled spectrum). Determine the corresponding wavelengths in the visible spectrum 390 nm $< \lambda <$ 760 nm. **c)** A light source is at equal distances from the slits. It emits two coherent waves of equal amplitudes and close wavelengths λ and $\lambda + \delta\lambda$. Determine the intensity at a point M. Calculate the contrast $\mathcal{C}$ at M and study its variation as a function of x. Is it possible to use these results to determine λ and $\delta\lambda$?

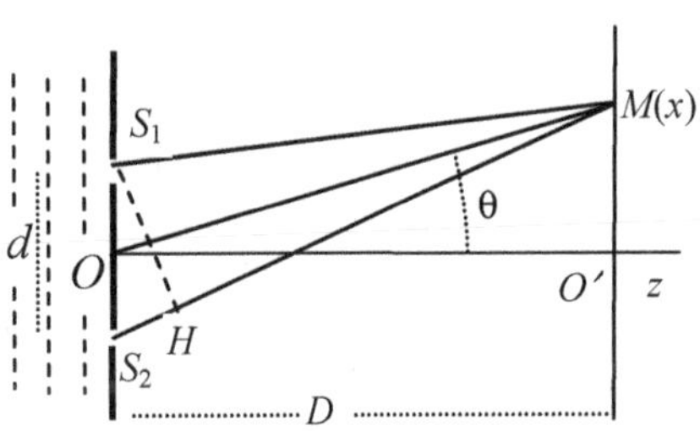

Figure 11.17. *Young's experiment*

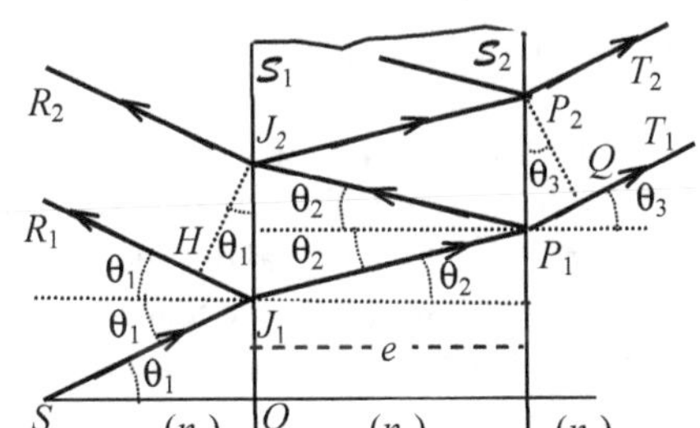

Figure 11.18. *Interference on a thin film*

P11.12 The interference of light on a thin film of thickness e produces the intense colors of a soap bubble, for instance. It occurs between the reflected waves R_1 and R_2 (or the transmitted waves T_1 and T_2) on both faces of the film (Figure 11.18).

a) Show that the phase shift between R_1 and R_2 is $\phi_R = \phi + \hat{\phi}$ and between T_1 and T_2 is $\phi_T = \phi + \hat{\phi} + \pi$, where $\phi = 4\pi(e/\lambda_2)\cos\theta_2$ and $\hat{\phi} = 0$ if n_2 lies between n_1 and n_3 and $\hat{\phi} = \pm\pi$ otherwise. Deduce that, in the case of normal incidence, the reflected intensity is maximal if $e = \frac{1}{4}\lambda_2(1 + 2q)$ where q is an integer. **b)** Consider a wedge-shaped air film situated between two plates, that make a small angle α and illuminated normally. Show that the bright fringes are parallel to the wedge at positions $x_q = (1 + 2q)(\lambda_2/4\alpha)$. **c)** A light wave is incident normally on the air film situated between the plane surface of a glass plate and the spherical surface of a planar convex lens whose radius of curvature is R. Show that the bright fringes are circular of radii $r_p \approx \sqrt{(p+1/2)\lambda R}$ called Newton's rings. Describe what one observes if this setup is illuminated with sunlight.

Multi-slit interference

P11.13 Consider a set-up similar to that of Young's experiment but with six parallel slits emitting waves of amplitude a in phase. **a)** Draw a phasor diagram for the resultant wave in the direction θ. Verify that it has principal maximums $A_{max} = 6a$ in the directions θ such that $\sin\theta = q\lambda/d$ and that two principal maximums are separated by four secondary maximums and five minimums. Verify that the first minimum is in the direction given by $\sin\theta_{min} = \lambda/6d$. **b)** A radar station uses an emitter that consists of six rectilinear parallel wires separated by a distance $d = 20$ cm between consecutive wires. The station uses a wavelength of 10 cm. What are the directions of the principal maximums of order 0 and 1 and what is their angular width? Instead of rotating the array, it is possible to use a *phase command*, which consists of producing a phase shift $\phi_i(t)$ between adjacent antennas. What then is the direction of the principal maximum of order 0? How should we choose $\phi_i(t)$ in order to have sweeping at an angular speed ω?

P11.14 a) A light wave of wavelength $\lambda = 0.6$ μm, is incident normally on a slit of width 0.2 mm. The diffraction pattern is observed on a screen that coincides with the focal plane of a lens of focal distance $f = 100$ cm. Determine the distances of the first minimum and the first secondary maximum from the center of the diffraction pattern? **b)** Describe the diffraction pattern if the incident light is formed by two wavelengths $\lambda_1 = 0.6$ μm and $\lambda_2 = 0.5$ μm. **c)** The slit is illuminated with an ideally monochromatic light of wavelength $\lambda = 0.6$ μm, but at an angle $\theta' = 30°$ with the normal to the slit. Show that the light intensity in the direction θ is $\mathcal{I} = \mathcal{I}_0\mathcal{F}_d(\Phi)$ with $\Phi = 2\pi(d/\lambda)(\sin\theta - \sin\theta')$ and $\mathcal{F}_d(\Phi) = \sin^2(\Phi/2)/(\Phi/2)^2$. Deduce that the principal maximum is in the direction of geometrical optics. Determine the positions of the first minimums on both sides of the principal maximum.

P11.15 The *transmittance* of an aperture is the ratio of the transmitted amplitude to the incident amplitude. Describe the Fraunhofer diffraction pattern produced by a slit of width d and transmittance depending on the distance x to the longitudinal axis of the slit according to the expression $\mathcal{T}(x) = \cos^2(\pi x/d)$ for $|x| < d/2$ and $\mathcal{T}(x) = 0$ for $|x| > d/2$.

P11.16 To simplify, we consider the propagation of a scalar wave $\underline{E} = f(\mathbf{r})\,e^{i\omega t}$. **a)** Show that $f(\mathbf{r})$ obeys Helmholtz equation $\Delta f + k^2 f = 0$, where $\mathbf{k}$ is the wave vector of magnitude $k = \omega/v$ and v is the speed of propagation in the medium. **b)** Consider the vector field $(\Psi_1\nabla\Psi_2 - \Psi_2\,\nabla\Psi_1)$. Applying Gauss-Ostrogradsky's theorem to a volume $\mathcal{V}$ bounded by a surface $\mathcal{S}$, show *Green's identity*

$$\iiint_{\mathcal{V}} d\mathcal{V}\,'\,[\Psi_1(\mathbf{r}')\Delta'\Psi_2(\mathbf{r}') - \Psi_2(\mathbf{r}')\Delta'\Psi_1(\mathbf{r}')]$$

$$= \iint_{\mathcal{S}} d\mathcal{S}(\mathbf{r}')\;\mathbf{n}(\mathbf{r}').[\Psi_1(\mathbf{r}')\nabla'\Psi_2(\mathbf{r}') - \Psi_2(\mathbf{r}')\;\nabla'\Psi_1(\mathbf{r}')], \qquad \text{[P11.1]}$$

where Δ' and ∇' are the Laplacian and the vector differential operators with respect to the coordinates of $\mathbf{r}'$ and $\mathbf{n}'$ is the normal unit vector pointing outward from $\mathcal{S}$ at the point $\mathbf{r}'$. **c)** We would like to write a representation of the solution $f(\mathbf{r})$ of the Helmholtz equation $\Delta f + k^2 f = 0$ at a point M in terms of the boundary conditions on a given surface $\mathcal{S}_1$ surrounding M (Figure 11.19). We take $\Psi_1(\mathbf{r}') = f(\mathbf{r}')$ and $\Psi_2(\mathbf{r}) = (1/R)\,e^{-ikr}$, where $\mathbf{R} = \overrightarrow{M'M}$. $\Psi_2(\mathbf{r})$ may be interpreted as a spherical wave that is emitted by a point-source at M and evaluated at M' on the surface $\mathcal{S}_1$. Let $\mathcal{S}_2$ be a sphere of center M and radius ρ. Apply Green's identity [P11.1] to the volume $\mathcal{V}$ bounded by the surfaces $\mathcal{S}_1$ and $\mathcal{S}_2$. Verify first that $\nabla'\Psi_2 = (\mathbf{R}/R^3)(1 + ikR)\,e^{-ikr}$ and that Ψ_2 is a solution of Helmholtz's equation at any point $\mathbf{r}'$ except the points where $R = 0$, therefore Ψ_2 is singular. No point M' of $\mathcal{V}$ corresponds to $R = 0$. Deduce that the left-hand side of equation [P11.1] is equal to 0. Evaluating the integral over $\mathcal{S}_2$, show that

$$\iint_{\mathcal{S}_2} d\mathcal{S}'\;\mathbf{n}'.[f(\mathbf{r}')\nabla'\Psi_2(\mathbf{r}') - \Psi_2(\mathbf{r}')\;\nabla' f(\mathbf{r}')] = e^{-ik\rho}\iint_{\mathcal{S}_2} d\Omega'\; f(1+ik\rho) + \boldsymbol{\rho}.\nabla f],$$

where $\boldsymbol{\rho} = \overrightarrow{MM'_2}$. In the limit $\rho \to 0$, the right-hand side tends to $4\pi f(M)$. Deduce Kirchhoff's representation

$$f(M) = \frac{1}{4\pi}\iint_{\mathcal{S}_1} d\mathcal{S}'\,\frac{e^{-ikR}}{R}\,\{\mathbf{n}'.\nabla' f(\mathbf{r}') - f(\mathbf{r}')\frac{\mathbf{R}.\mathbf{n}'}{R^2}(1+ik)\}$$

$$= \frac{1}{4\pi}\iint_{\mathcal{S}_1} d\mathcal{S}'\,\frac{e^{-ikR}}{R}\,\{\frac{\partial f}{\partial x_n} - f(\mathbf{r}')\frac{\cos\theta'}{R}(1+ikR)\}, \qquad \text{[P11.2]}$$

where θ is the angle that $\mathbf{n}'$ forms with $\mathbf{R}$ and x_n is the normal coordinate to $\mathcal{S}_1$. This relation allows f to be determined at each point M, if we know f and its normal derivative $\partial f/\partial x_n$ on a closed surface $\mathcal{S}_1$. Note that, if M is outside $\mathcal{S}_1$, we consider the volume $\mathcal{V}$, which is outside $\mathcal{S}_1$ and $\mathcal{S}_2$ and bounded by a closed surface $\mathcal{S}_3$, which we may take at infinity. If the function f decreases rapidly to 0 at large distance, the integral over $\mathcal{S}_3$ tends to 0. Thus, the relation [P11.2] holds in all cases with $\mathcal{S}_1$ being any closed surface. **d)** In the case of a point-source S, the wave on $\mathcal{S}_1$ is of the form $f(\mathbf{r}') = (A/r')e^{-ikr'}$, where $r' = SM'$. Show that the relationship [P11.2] may be written as

$$f(M) = \frac{i}{\lambda} \iint_{\mathcal{S}_1} d\mathcal{S}' \, \frac{e^{-ikR}}{R} \, \frac{e^{-ikr'}}{r'} \, \frac{\cos\theta + \cos\theta'}{2}. \quad \text{[P11.3]}$$

We find the Fresnel inclination factor ½(cos θ + cos θ') with the right factor of proportionality $1/\lambda$ but with an additional phase shift $\pi/2$.

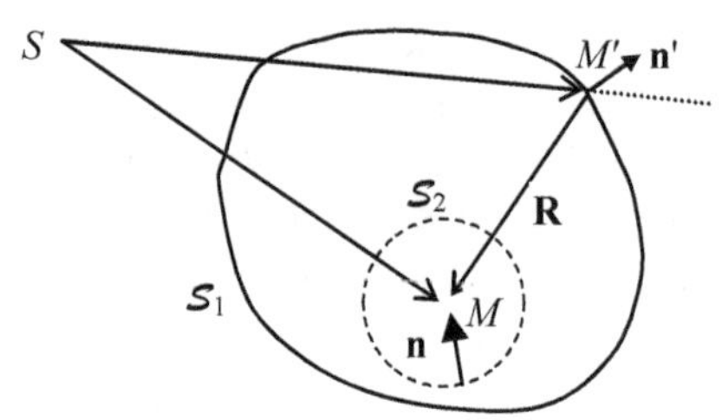

Figure 11.19. *Problem 11.16*

Diffraction by randomly distributed identical apertures or obstacles

P11.17 Identical opaque disks of radius R are randomly distributed on a plate of glass and illuminated normally. The diffraction pattern is observed in the focal plane of a converging lens parallel to the plate. **a)** Show that, at any point except the lens focal point, the diffraction pattern is the same as that of a single disk of radius R. **b)** A powder, assimilated to small disks, is spread randomly on a plate of glass. Using light of wavelength 600 nm and a lens of focal distance 2 m, the radius of the first dark ring is 4.90 cm. What is the radius of the powder grains? **c)** The Sun and the Moon appear to be surrounded by a halo when the atmosphere is slightly cloudy or dusty. Explain why. What can you conclude if the angular radius of this halo is 3° and light has an average wavelength of 0.6 μm?

Diffraction grating

P11.18 In the case of radio waves, it is easy to have sources with spacing of the order of the wavelength. A rectilinear antenna, carrying an oscillating current, emits an electromagnetic wave especially in radial directions of its median plane. Consider two parallel antennas with a spacing d (Figure 11.20). Let $E_1 = E_m \sin(\omega t)$ and $E_2 = E_m \sin(\omega t + \phi)$ be the waves that they emit. **a)** Write the expressions of the waves at a long distance r from the center O and in the direction θ in the median plane of the antennas. What is the resultant wave? Determine the directions corresponding to the maximum intensity of the wave and to the minimum intensity. **b)** Assuming that $d = \lambda/2$ and $\phi = 0$, determine the intensity for $\theta = 0$, 30°, 60° and 90°. What can you deduce concerning the direction of the emitted wave? **c)** Now assume that $d = \lambda/2$ and $\phi = 30°$. What is the new direction of the emitted wave? **d)** Is it possible to narrow the direction of emission by increasing the number of antennas?

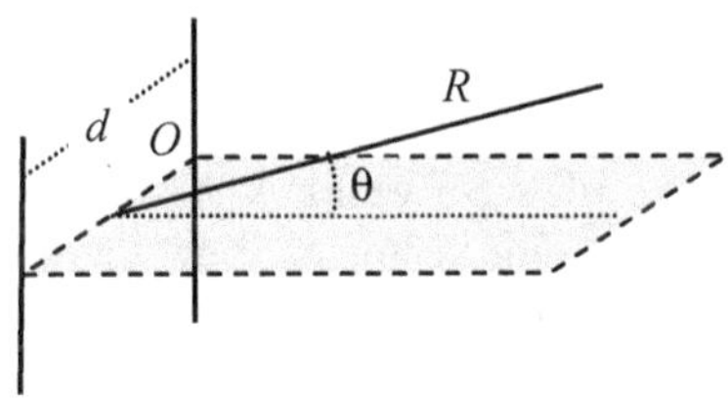

Figure 11.20. *Problem 11.18*

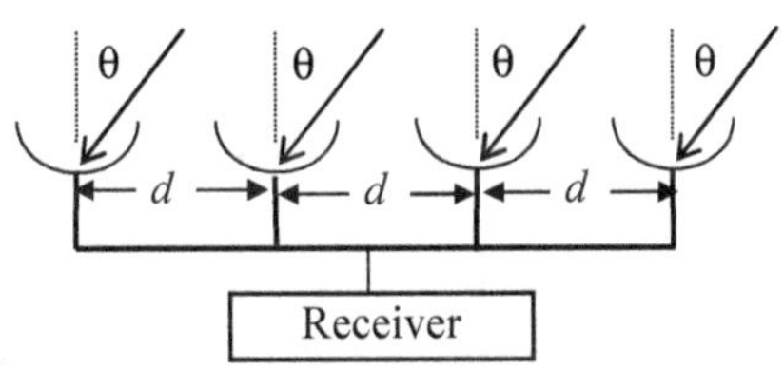

Figure 11.21. *Problem 11.19*

P11.19 Radio-interferometers are used in radioastronomy. They consist of two or more antennas at a distance d apart (Figure 11.21). The radio signals are transmitted by cables to a global receiver, where they interfere. **a)** Let us consider first the case of two antennas. Show that the angles of incidence that correspond to maximums of intensity are given by $\sin\theta = p\lambda/d$. Plot the interference intensity versus θ. The Green Bank (West Virginia) interferometer has two antennas separated by a distance d, which may be as long as 2,700 m and it uses a wavelength $\lambda = 11$ cm. What is the angular separation of two maximums of interference? **b)** A radio-interferometer in Australia is formed by 32 antennas aligned and 7 m apart. What is the angular width of the central maximum and the angular separation of two consecutive principal maximums if $\lambda = 21$ cm?

P11.20 a) Analyze the Fraunhofer diffraction by three identical slits of width d at a distance a apart, if they are illuminated normally. **b)** The middle slit is covered with a plate of thickness e, which produces a phase shift ϕ in the transmitted wave without modifying its intensity. Analyze the new distribution of intensity on a screen placed at large distance and parallel to the slits. Consider successively the case of a quarter-wave plate producing a phase shift $\phi = \pi/2$, a half-wave plate producing a phase shift $\phi = \pi$, and a plate producing a phase shift $\phi = \pi/2 + \varepsilon$ ($\varepsilon << \pi/2$).

P11.21 a) In order to determine the spacing of a diffraction grating, we observe the diffraction by transmission for a beam of wavelength 589 nm. One of the principal maximums is in the direction of the incident beam and the third is in the direction of 45.0°. What is the spacing of the diffraction grating? **b)** This diffraction grating is illuminated with light of unknown wavelength. We find that the third-order maximum is in the direction of 35.5°. What is the value of λ? **c)** Hydrogen and deuterium have two lines that differ by 0.18 nm close to $\lambda = 656.3$ nm. What should be the minimum number of diffraction grating lines in order to separate these two lines? **d)** This diffraction grating is illuminated normally with sunlight. The transmitted light is received on a screen located at 60 cm from the diffraction grating. In this screen, we make a slit between the distances 10 and 11 cm from the central maximum. What is the band of the wavelength passing by this slit?

X-rays diffraction

P11.22 a) Consider the diffraction of a plane wave of wave vector **k'** by a linear chain of unit cells of spacing d in the direction of the unit vector **e**. Show that the interference is constructive in the directions of wave vector $\mathbf{k}_q$ such that $\mathbf{k}_q.\mathbf{e} - \mathbf{k}'.\mathbf{e} = 2\pi q/d$ where q is an integer. **b)** Consider now the waves that are diffracted by the parallel chains of an atomic plane. Take the origin O at the position of one of the cells and two axes (oblique in general) of unit vectors $\mathbf{e}_1$ and $\mathbf{e}_2$ in the directions of two chains. Any cell of this plane occupies the position $\mathbf{r}_{m,n} = md_1\mathbf{e}_1 + nd_2\,\mathbf{e}_2$, where m and n are integers and d_1 and d_2 are the spacing in the directions $\mathbf{e}_1$ and $\mathbf{e}_2$ respectively. Show that the phase shift of the diffused wave by this cell in a direction of wave vector $\mathbf{k} = k_1\,\mathbf{e}_1 + k_2\,\mathbf{e}_2 + k_z\,\mathbf{e}_z$ is $\phi_{m,n} = m\,d_1\,\mathbf{e}_1.(\mathbf{k} - \mathbf{k}') + n\,d_2\,\mathbf{e}_2.(\mathbf{k} - \mathbf{k}')$. We have a principal maximum if $\phi_{m,n}$ is an integer multiple of 2π for any m and n. This is possible if $d_1\,\mathbf{e}_1.(\mathbf{k} - \mathbf{k}') = 2q_1\pi$ and $d_2\,\mathbf{e}_2.(\mathbf{k} - \mathbf{k}') = 2q_2\pi$. As the directions of the chains $\mathbf{e}_1$ and $\mathbf{e}_2$ may be chosen in many ways in the atomic plane, show that this implies that $\mathbf{k}'_{//} = \mathbf{k}_{//}$, thus $k_\perp = k'_\perp$ (transmitted wave in the direction of incidence) or $k_\perp = -k'_\perp$ (reflected wave on the atomic plane). **c)** Finally, consider the waves that are diffracted by the various atomic parallel planes. Show that they are in phase if Bragg's law is verified. **d)** X-rays, of wavelength $\lambda = 0.200$ nm, are incident on a NaCl crystal. Determine the angles θ formed by the incident and reflected rays with the atomic planes, which are at a distance $d = 0.5627$ nm apart.

Cross section

P11.23 Consider a beam of N_i photons per unit volume, incident in the direction Oz on a sphere of radius R (Figure 11.22). **a)** What is the photons' flux? Assuming that the photons are reflected on the surface of the sphere according to the usual laws of reflection, determine the angle of deviation θ for the photons that are reflected at the point *I*, situated at a distance *b* from *Oz*. **b)** A counter, which is placed at large distance, counts the scattered photons in the solid angle $d\Omega$. What is the count per second? Deduce the differential cross section $\sigma(\Omega)$. Calculate the total cross section σ and verify that it is πR^2. **c)** Calculate the absorption coefficient of light in water if it contains 100 particles per cm^3 and the diameter of these particles is 10 μm.

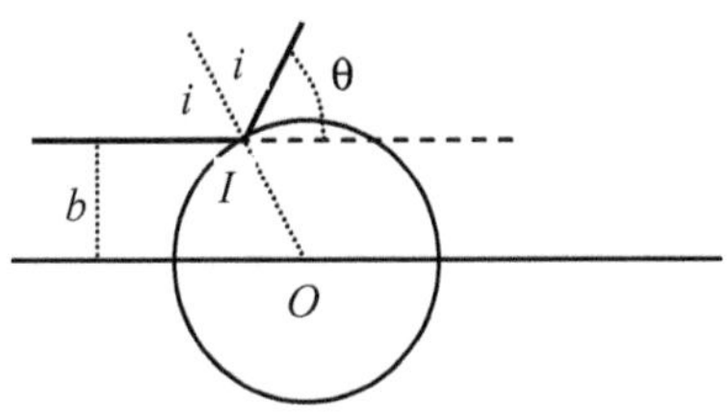

Figure 11.22. *Problem 11.23*

Chapter 12

Guided Waves

If a wave propagates in an infinite and non-dispersive medium, its phase velocity $v_{(p)}$ and its group velocity $v_{(g)}$ are equal to the speed of propagation v, which appears in the wave equation for any frequency of the wave. All physical quantities associated with the wave are transferred with the group velocity. A wave is *guided* if it is canalized between surfaces, which limit the propagation medium in one transverse direction or in both of them. Guided waves propagate in specific *modes*. Each mode is characterized by a cut-off frequency, a phase velocity and a group velocity, which depend on the frequency of the wave and on the geometry of the waveguide. The propagation properties in the infinite medium are recovered if the transverse dimensions of the waveguide are much larger than the wavelength. We may analyze their propagation by studying the successive reflections on the guide walls. However, a more practical and general method consists of directly finding the solutions of the wave equation that satisfy the boundary conditions.

If the medium is bounded in the direction of propagation, it can support only *standing* (or *stationary*) *waves* in *normal modes* of discrete frequencies (called *normal frequencies*). The modes are determined from the wave equation and the boundary conditions of the medium. In each mode, the propagation medium is a juxtaposition of wave zones with points called *antinodes*, where the amplitude of the wave is large, and points called *nodes*, where the amplitude is equal to zero. The physical quantities oscillate at each space point with no transfer from one zone to the other. More generally, guided or standing waves may be superpositions of modes.

In this chapter, we study the propagation of the potential and the current along an electric line. Then we consider the propagation of electromagnetic waves, guided by two conductors or within a hollow conductor. We evoke some applications of waveguides, namely in telecommunications, allowing the transmission of energy and information over long distances.

12.1. Transmission lines

Let us consider a transmission line constituted by two long and parallel conductors, carrying the electric current there and back. Eventually one of the conductors may be the ground. The current produces a magnetic field; thus, the line has a certain *inductance* L_l per unit length. The conductors also constitute a long capacitor with a certain capacitance C_l per unit length. In the following, we neglect the resistance of the conductors and an eventual current leak between them (*ideal line*).

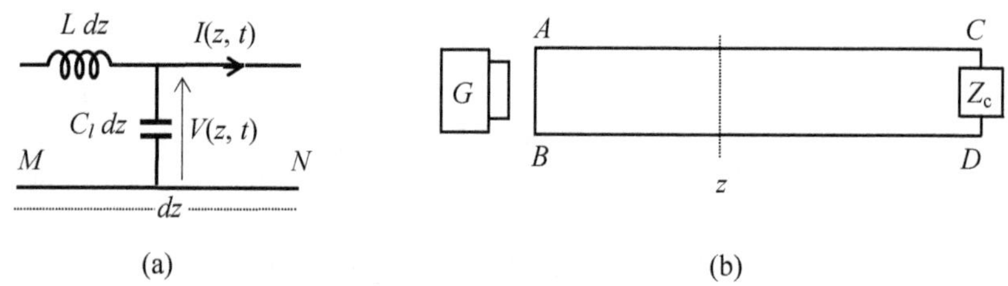

Figure 12.1. *a) Element of length dz of a transmission line, and b) a transmission line supplied by a generator G and connected to an impedance Z_C*

An element of the line MN of length dz consists of a *self-inductance* $L_l\,dz$ in the direction of the line and a transverse capacitor of capacitance $C_l\,dz$ (Figure 12.1a). The potential difference V between the two conductors and the current intensity I are functions of z and t. The potential drop between M and N is then $-\,dV = \partial_t I\, L_l\, dz$. The charge of the equivalent capacitor of length dz is $VC_l\,dz$ and the current intensity in the transverse branch of the capacitor is $\partial_t V\, C_l\, dz$. The decrease of the intensity between M and N is thus $-\,dI = \partial_t V\, C_l\, dz$. Dividing by dz, we get the equations

$$L_l\,\partial_t I + \partial_z V = 0, \qquad C_l\,\partial_t V + \partial_z I = 0. \qquad [12.1]$$

Differentiating one of these equations with respect to t and the other with respect to z and making linear combinations, we find the equations of propagation

$$\partial^2_{tt} I - v^2\partial^2_{zz} I = 0, \qquad \partial^2_{tt} V - v^2\partial^2_{zz} V = 0, \qquad \text{where } v = 1/\sqrt{C_l L_l}\,. \qquad [12.2]$$

In the case of a line constituted by two long, plane and parallel plates of width D, we find $C_l = \varepsilon D/d$ and $L_l = \mu d/D$, where d is the distance between the plates. In the case of a coaxial cable constituted by a cylindrical wire of radius r_1 surrounded by a cylindrical shell of internal radius r_2, we find $C_l = 4\pi\varepsilon/\ln(r_2/r_1)$ and $L_l = (\mu/4\pi)\ln(r_2/r_1)$ (see problem 8.12). In these cases, the speed of propagation is $v = 1/\sqrt{C_l L_l} = 1/\sqrt{\mu\varepsilon}$. This is also the speed of light in the medium that separates the conductors. This result holds for any geometry of the line and, in the limit, of non-guided waves. This result means that the potential and the current are produced by the electromagnetic wave that propagates in the medium that separates the conductors. The phase velocity $v_{(p)} = \omega/k = 1/\sqrt{\mu\varepsilon}$ depends only on the nature of this medium and it is independent of the frequency. The ideal line is thus *non-dispersive*; it transmits the electric signals without deformation or attenuation.

If a generator G of electromotive force (emf) $\mathcal{E} = \mathcal{E}_m \cos(\omega t)$ is connected at the entry of an infinite line toward the positive z, the d'Alembert's equation for the potential $\partial^2_{tt}V - v^2\partial^2_{zz}V = 0$ has solutions $V(z, t) = A\cos(\omega' t - k'z + \phi')$ if $\omega'/k' = v$. Imposing the condition $V(0, t) = \mathcal{E}_m \cos(\omega t)$ at the entry, we find that $A = \mathcal{E}_m$ and $\phi' = 0$, $\omega' = \omega$ and, consequently, $k' = \omega/v$. Substituting V in the first equation [12.1], we find $L_l\,\partial_t I = -k\mathcal{E}_m \sin(\omega t - kz)$ and, consequently, $I = (k\mathcal{E}_m/\omega L_l)\cos(\omega t - kz)$. As $k/\omega = 1/v = \sqrt{C_l L_l}$, we deduce that

$$V(z, t) = \mathcal{E}_m \cos(\omega t - kz), \quad I(z, t) = (\mathcal{E}_m/Z)\cos(\omega t - kz), \tag{12.3}$$

where $Z = \sqrt{L_l/C_l}$ is the *impedance* of the line, that is, the ratio (V_m/I_m) of the amplitude of the voltage to that of the intensity at any point of the line.

The electric energy and the magnetic energy are distributed in the dielectric that separates the conductors. Their densities per unit length of the line are

$$U_{E\,l} = U_{M\,l} = \tfrac{1}{2} C_l\, \mathcal{E}_m^2 \cos^2(\omega t - kz). \tag{12.4}$$

This energy is not stationary; it propagates along the line. Indeed, the power that flows through the section at z may be written as

$$P(z, t) = V(z, t)\, I(z, t) = (\mathcal{E}_m^2/Z)\cos^2(\omega t - kz). \tag{12.5}$$

In particular, the power at the entry is the power that is supplied by the generator

$$P_{(g)} = V(0, t)\, I(0,t) = \mathcal{E}(t)\, I(0,t). \tag{12.6}$$

The average power, taken over a period, is independent of z and t

$$\langle P \rangle = \tfrac{1}{2}\sqrt{C_l/L_l}\; \mathcal{E}_m^2 = \tfrac{1}{2}\mathcal{E}_m^2/Z. \tag{12.7}$$

In the case of a line constituted by two long, plane and parallel plates of width D and separated by a distance d, we find $Z = \sqrt{\mu/\varepsilon}\,(d/D)$. Particularly, if $D = d$, we get the *impedance of the medium* $Z = \sqrt{\mu/\varepsilon}$. Its value in the case of vacuum is $Z_o = \sqrt{\mu_o/\varepsilon_o} = 377\ \Omega$.

Consider an ideal line of length d, capacitance C_l and inductance L_l per unit length (Figure 12.1b). Assume that it is short-circuited at the entry AB and connected at its terminal CD to a circuit of impedance Z_c. If it is excited at the entry by a generator of angular frequency ω, the wave on the line is, in general, a superposition of a wave of the form [12.3] propagating toward the positive z and a wave that propagates in the opposite direction. To manage the phase shifts, it is convenient to use the complex representation and write

$$\underline{V}(z,t) = \underline{V}_1\, e^{i(\omega t - kz)} + \underline{V}_2\, e^{i(\omega t + kz)}, \quad \underline{I}(z,t) = (1/Z)[\,\underline{V}_1\, e^{i(\omega t - kz)} - \underline{V}_2\, e^{i(\omega t + kz)}\,]. \quad [12.8]$$

The line being short-circuited at AB, we must have $V(0, t) = 0$ and its end being connected to an impedance $\underline{Z}_c$, we must have $V(d, t) = \underline{Z}_c\ I(d, t)$. These two conditions give two equations, which allow $\underline{V}_1$ and $\underline{V}_2$ to be determined:

$$\underline{V}_1 + \underline{V}_2 = 0, \qquad [\underline{V}_1\, e^{-ikd} + \underline{V}_2\, e^{ikd}\,] = (\underline{Z}_c/Z)\,[\underline{V}_1\, e^{-ikd} - \underline{V}_2\, e^{ikd}\,]. \quad [12.9]$$

In the particular case $\underline{Z}_c = Z$, these equations are verified if $V_2 = 0$, that is, no reflected wave; we say that the impedance is *matched* (or *adapted*). The line of length d is thus equivalent to an infinite line toward the positive z.

In the general case $\underline{Z}_c \neq Z$, equations [12.9] have a non-trivial solution (i.e. non-zero solution) only if

$$e^{-ikd} - e^{ikd} = (\underline{Z}_c/Z)[\ e^{-ikd} + e^{ikd}\,], \qquad \text{hence } \tan(2\pi d/\lambda) = (\underline{Z}_c/Z). \quad [12.10]$$

a) If the line is short-circuited at its end ($\underline{Z}_c = 0$), the possible wavelengths are $\lambda_n = 2d/n$, the corresponding angular frequency is $\omega_n = \pi n v/d$ and the wave is

$$I = \mathcal{R}e(\underline{V}_1/Z)\ e^{i\omega_n t}\ [\ e^{-ik_n z} + e^{ik_n z}\,] = (A/Z)\cos(\omega_n t + \phi + \pi/2)\cos(n\pi z/d),$$
$$V = \mathcal{R}e\, \underline{V}_1\ e^{i\omega_n t}\ [\ e^{-ik_n z} - e^{ik_n z}\,] = A\cos(\omega_n t + \phi)\sin(n\pi z/d). \quad [12.11]$$

Thus, the line can support waves only according to the discrete *modes* (n) of frequencies $\tilde{\nu}_n = nv/2d$. The entry and the end of the line are *nodes* for the voltage, i.e. $V(0, t) = V(d, t) = 0$, and *antinodes* for the intensity, i.e. I_m has the maximum value A/Z. In the mode (n), the length of the line is equal to n times the half-wavelength ($d = \frac{1}{2} n\lambda_n$).

b) If the line is open at its end ($Z_c = \infty$), then $\tan(2\pi d/\lambda) = \infty$, i.e. $d = ½\lambda(n+½)$. The mode (n) has the frequency $\tilde{\nu}_{n+1/2} = (n+½)v/2d$. The corresponding wave is

$$I = \mathcal{R}e\ (\underline{V}_1/Z)\ e^{i\omega_n t}\ [\ e^{-ik_n z} + e^{ik_n z}\ ] = (A/Z)\ \cos(\omega_n t + \phi + \pi/2)\ \cos[\pi(n + ½)z/d],$$

$$V = \mathcal{R}e\ \underline{V}_1\ e^{i\omega_n t}\ [\ e^{-ik_n z} - e^{ik_n z}\ ] = A\cos(\omega_n t + \phi)\sin[\pi(n + ½)\ z/d]. \qquad [12.12]$$

The entry of the line is a node of V and an antinode of I, while its end is an antinode of V and a node of I. The length of the line is $d = (n + ½)(\lambda/2)$.

The existence of these standing waves was verified by Lecher in 1890 by exciting at high frequency the entry of a line open at its end. A measurement of V between the conductors at the various points z confirmed the existence of nodes and antinodes with the open end being an antinode of V. The distance between consecutive nodes is $\lambda/2$. Knowing the frequency, Lecher verified that the speed of propagation of the electromagnetic wave on the line is equal to the speed of light.

If the two points C and D are disconnected and the conductors AC and BD aligned, we get a *dipole-antenna* of length $2d = \lambda(n + ½)$, whose extremities are nodes of I and antinodes of V. The shortest antenna of this type has a length $2d = ½\lambda$ (*half-wave antenna,* Figure 12.2a). If the points C and D are grounded, we find a dipole-antenna of length $2d = n\lambda$, whose extremities are nodes of V and antinodes of I. The shortest antenna of this type has a length ($2d = \lambda$, Figure 12.2b).

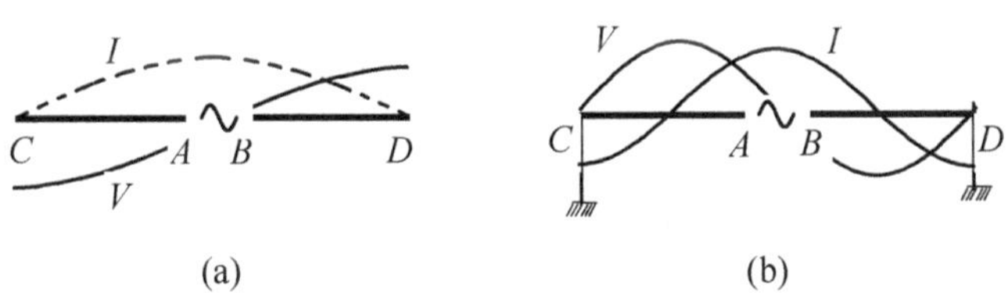

Figure 12.2. *Standing electromagnetic waves: a) a dipole-antenna whose ends are free (nodes of I), and b) dipole-antenna whose ends are grounded (nodes of V)*

A standing electromagnetic wave on a line may be considered as an interference of a progressive wave and the corresponding reflected wave on both ends of the line. In general, a wave is stationary if it is confined in a finite region of space. If there is no energy dissipation, once the wave is established, it stays indefinitely and the generator needs to supply no more energy. In the case of a partial reflection on the boundary, a part of the wave may escape and the wave dies out. To maintain it, the generator must continuously supply energy. In this case, the wave on the line is a superposition of a standing wave and a progressive wave.

The modes are determined by the boundary conditions. If the electromagnetic wave meets the polished surface of a conductor, a part of the wave is reflected and the other is dissipated within the conductor as Joule heat. In the case of an ideal superconductor, there is no field inside it. The continuity of the tangential component of **E** and the normal component of **B** imply that the sum of the incident wave E_i and the reflected wave E_j vanishes on the surface of the conductor ($E_i + E_r = 0$). Thus, we have total reflection with a phase shift of π for E.

12.2. Guided waves

We have seen in section 10.7 that the attenuation coefficient of an electromagnetic wave in a good conductor is large at high frequency; this means that the wave cannot propagate in conductors but it can in dielectrics (or vacuum). An ordinary electric circuit behaves as an antenna, emitting a large part of its energy as radiation. The amplitude of the emitted wave decreases with the travelled distance as, besides absorption, the wave spreads over a larger and larger wave front. To reduce this energy loss and the decrease in intensity, the electromagnetic wave must be guided by a metallic structure. The waveguide may be a hollow metallic pipe that is empty or filled with a dielectric (Figure 12.3a) or a two-conductor transmission line (coaxial cable, two-wire, or microstrip transmission lines) to conduct the current in both directions. Figure 12.3b shows the lines of **E** and **B** in the case of a transmission line constituted by two parallel wires usually used for the transport of electric energy and in traditional telephony. The two wires are isolated and set very close to each other in order to reduce the magnetic field and the inductance of the circuit and to reduce the interference of signals that are transmitted by nearby pairs of lines in cables containing several pairs. Figure 12.3c shows a coaxial cable constituted by a cylindrical conductor of radius R_1, surrounded by an insulator and a cylindrical conducting shell of internal radius R_2. In this case, the fields **E** and **B** are perpendicular to the direction of propagation parallel to the axis Oz of the line. We say that the wave is *transverse electromagnetic* (*TEM*).

The propagation in the coaxial cable is easier to analyze. Using cylindrical coordinates, the symmetries require that the fields be of the form $\mathbf{E} = F(\rho)\, e^{i(\omega t - kz)}\, \mathbf{e}_\rho$ and $\mathbf{B} = G(\rho)\, e^{i(\omega t - kz)}\, \mathbf{e}_\varphi$. The Maxwell-Gauss equation $\boldsymbol{\nabla}.\mathbf{E} = 0$ in the dielectric implies that $F = (E_o/\rho)$ and the Maxwell-Faraday equation implies that $G = F/v$. The real fields may be written as

$$\mathbf{E} = (E_m/\rho)\cos(\omega t - kz)\,\mathbf{e}_\rho, \qquad \mathbf{B} = (E_m/v\rho)\cos(\omega t - kz)\,\mathbf{e}_\varphi. \qquad [12.13]$$

The Poynting vector is $\mathbf{S} = \mathbf{E} \times \mathbf{B}/\mu = (E_m^2/v\rho^2)\cos^2(\omega t - kz)\,\mathbf{e}_z$ and the transmitted power is the flux of **S** over the section of the insulator, that is,

$$P = \int_{R_1}^{R_2} d\rho\,\rho \int_0^{2\pi} d\varphi\,(E_m^2/v\rho^2)\cos^2(\omega t - kz) = (2\pi E_m^2/v)\cos^2(\omega t - kz)\ln(R_2/R_1). \quad [12.14]$$

Its average value is $\langle P \rangle = (\pi E_m^2/v)\ln(R_2/R_1)$. In reality, there is always a certain energy loss in the dielectric and in the conductors.

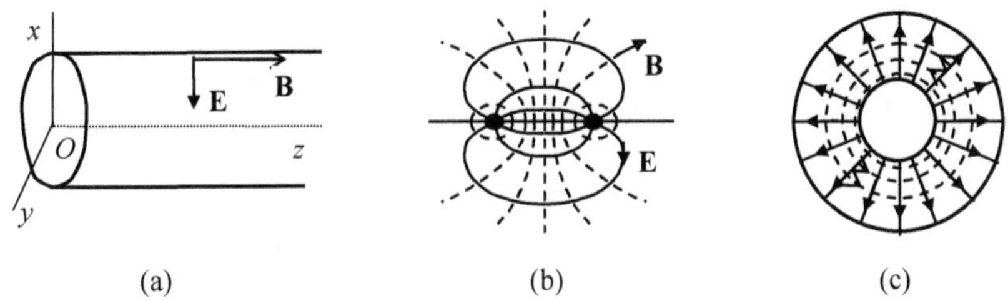

Figure 12.3. *a) Electromagnetic waveguide in the form of a hollow conductor, b) waveguide formed by two parallel conducting wires, and c) coaxial cable. The lines of* **E** *are drawn as solid lines and those of* **B** *are drawn as dashed lines*

In any waveguide, the wave must obey some boundary conditions at the surface of the conductors. In general, these conditions are satisfied if the wave propagates in specific modes, which depend on the geometry of the waveguide. As we have seen in section 9.5, the tangential component of **E** and the normal component of **B** vanish on the surface of an ideal conductor, while the normal component of **E** and the tangential component of **B** may not vanish, as this surface may carry charge and current densities. In the case of a good conductor, these properties are approximately valid. The energy loss as Joule heat is then negligible. In the following, we assume that the conductors are ideal and that the waveguide has translational symmetry in the direction Oz (Figure 12.3). At each point of the dielectric, the fields verify Maxwell's equations with no charge and current densities:

$$\nabla.\mathbf{E} = 0, \quad [12.15]$$

$$\nabla.\mathbf{B} = 0, \quad [12.16]$$

$$\nabla \times \mathbf{B} = (1/v^2)\,\partial_t \mathbf{E}, \quad [12.17]$$

$$\nabla \times \mathbf{E} = -\,\partial_t \mathbf{B}, \quad [12.18]$$

where $v = 1/\sqrt{\mu\varepsilon}$. We deduce the equations of propagation

$$v^2\Delta\mathbf{E} - \partial^2_{tt}\mathbf{E} = 0, \qquad v^2\Delta\mathbf{B} - \partial^2_{tt}\mathbf{B} = 0. \quad [12.19]$$

On the other hand, on the surface of the conductors, the fields verify the conditions:

$$\mathbf{B}_{\perp} = 0 \qquad \text{and} \qquad \mathbf{E}_{//} = 0, \tag{12.20}$$

where $\mathbf{B}_{\perp}$ is the component of **B**, which is normal to the surface of the conductor, and $\mathbf{E}_{//}$ is the component of **E**, which is parallel to this surface. The conditions [12.20] together with Maxwell's equations impose that the wave can propagate in the guide only in certain modes with a characteristic lower cut-off frequency. In each mode, the fields **E** and **B** have specific orientations with respect to the walls and they propagate with a specific phase velocity and group velocity, which depend on the frequency and the geometry of the guide. In general, the conditions of [12.20] are compatible with Maxwell's equations for three types of waves:

a) *Transverse magnetic* (*TM*) waves such that

$$B_z = 0 \ \text{(everywhere)} \quad \text{and} \quad \mathbf{E}_{//} = 0 \ \text{(on the conductors)}. \tag{12.21}$$

b) *Transverse electric* (*TE*) waves such that

$$E_z = 0 \ \text{(everywhere)} \quad \text{and} \quad \mathbf{B}_{\perp} = 0 \ \text{(on the conductors)}. \tag{12.22}$$

c) *Transverse electromagnetic* (*TEM*) waves such that

$$E_z = 0 \ \text{(everywhere)} \quad \text{and} \quad B_z = 0 \ \text{(everywhere)}. \tag{12.23}$$

The *TEM* waves propagate only in free space or along transmission lines formed by two conductors, such as two parallel wires (Figure 12.3b) or a coaxial cable (Figure 12.3c). The fields **E** and **B** lie in the normal sections and are perpendicular to each other. Near the conductor, **E** is perpendicular to the conductor and **B** is tangential.

In the case of a wave of angular frequency ω, which propagates in the direction of the axis *Oz* of a waveguide, the fields are of the form

$$\underline{\mathbf{E}} = \hat{\underline{\mathbf{E}}}(x,y)\ e^{i(\omega t - kz)}, \qquad \underline{\mathbf{B}} = \hat{\underline{\mathbf{B}}}(x,y)\ e^{i(\omega t - kz)}, \tag{12.24}$$

where $\hat{\underline{\mathbf{E}}}(x,y)$ and $\hat{\underline{\mathbf{B}}}(x,y)$ are two vector functions. Substituting [12.24] to $\underline{\mathbf{E}}$ and $\underline{\mathbf{B}}$ in the wave equations [12.19], we find that $\hat{\underline{\mathbf{E}}}$ and $\hat{\underline{\mathbf{B}}}$ verify Helmholtz equations

$$\partial^2_{xx}\hat{\underline{\mathbf{E}}} + \partial^2_{xx}\hat{\underline{\mathbf{E}}} + (\omega^2/v^2 - k^2)\hat{\underline{\mathbf{E}}} = 0, \qquad \partial^2_{xx}\hat{\underline{\mathbf{B}}} + \partial^2_{xx}\hat{\underline{\mathbf{B}}} + (\omega^2/v^2 - k^2)\hat{\underline{\mathbf{B}}} = 0. \tag{12.25}$$

We note that the functions $\hat{\underline{\mathbf{E}}}(x,y)$ and $\hat{\underline{\mathbf{B}}}(x,y)$ are not independent, because $\underline{\mathbf{E}}$ and $\underline{\mathbf{B}}$ are related by Maxwell's equations [12.17] and [12.18].

12.3. Waveguides formed by two plane and parallel plates

This is the simplest waveguide to analyze mathematically. We chose the axes of coordinates so that one of the plates lies in the plane *Oyz* and the other in the plane $x = a$ (Figure 12.4). The fields do not depend on *y* because of the translational symmetry in this direction. Thus, the fields may be written as

$$\underline{\mathbf{E}} = \hat{\underline{\mathbf{E}}}(x)\ e^{\mathrm{i}(\omega t-kz)} \quad \text{and} \quad \underline{\mathbf{B}} = \hat{\underline{\mathbf{B}}}(x)\ e^{\mathrm{i}(\omega t-kz)}, \qquad [12.26]$$

where $\hat{\underline{\mathbf{E}}}(x)$ and $\hat{\underline{\mathbf{B}}}(x)$ are two vector functions of *x* only. In this case, the Helmholtz equations of [12.25] become the simple differential equations

$$\partial^2{}_{xx}\hat{\underline{\mathbf{E}}} + q^2\hat{\underline{\mathbf{E}}} = 0 \quad \text{and} \quad \partial^2{}_{xx}\hat{\underline{\mathbf{B}}} + q^2\hat{\underline{\mathbf{B}}} = 0, \quad \text{where } q^2 = \omega^2/v^2 - k^2. \quad [12.27]$$

The equation for $\hat{\underline{\mathbf{B}}}(x)$ has the general solution

$$\hat{\underline{\mathbf{B}}}(x) = (B_1\,\mathbf{e}_x + B_2\,\mathbf{e}_y + B_3\,\mathbf{e}_z)\ e^{\mathrm{i}qx} + (B'_1\,\mathbf{e}_x + B'_2\,\mathbf{e}_y + B'_3\,\mathbf{e}_z)\ e^{-\mathrm{i}qx}. \qquad [12.28]$$

The corresponding magnetic field is

$$\underline{\mathbf{B}} = [(B_1\,\mathbf{e}_x + B_2\,\mathbf{e}_y + B_3\,\mathbf{e}_z)\ e^{\mathrm{i}qx} + (B'_1\,\mathbf{e}_x + B'_2\,\mathbf{e}_y + B'_3\,\mathbf{e}_z)\ e^{-\mathrm{i}qx}]\ e^{\mathrm{i}(\omega t-kz)}. \quad [12.29]$$

Maxwell's equation $\boldsymbol{\nabla}.\underline{\mathbf{B}} = 0$ is verified if

$$qB_1 = kB_3, \qquad qB'_1 = -kB'_3. \qquad [12.30]$$

The electric field is related to the magnetic field by the third Maxwell-Ampère equation $\boldsymbol{\nabla}\times\underline{\mathbf{B}} = \mathrm{i}\omega\underline{\mathbf{E}}/v^2$, which gives in this case

$$\underline{\mathbf{E}} = (v^2/\omega)\{[kB_2\mathbf{e}_x - (kB_1 + qB_3)\mathbf{e}_y + qB_2\mathbf{e}_z]\ e^{\mathrm{i}qx} + [kB'_2\mathbf{e}_x - (kB'_1 - qB'_3)\mathbf{e}_y - qB'_2\,\mathbf{e}_z]\ e^{-\mathrm{i}qx}\}\ e^{\mathrm{i}(\omega t-kz)}. \qquad [12.31]$$

The other Maxwell's equations $\boldsymbol{\nabla}.\underline{\mathbf{E}} = 0$ and $\boldsymbol{\nabla}\times\underline{\mathbf{E}} = -\dot{\underline{\mathbf{B}}}$ are identically verified. Using [12.30] and $q^2 + k^2 = \omega^2/v^2$, we may write

$$\underline{\mathbf{B}} = \{[B_1\mathbf{e}_x + B_2\mathbf{e}_y + (q/k)B_1\mathbf{e}_z]\ e^{\mathrm{i}qx} + [B'_1\mathbf{e}_x + B'_2\mathbf{e}_y - (q/k)\,B'_1\,\mathbf{e}_z]\ e^{-\mathrm{i}qx}\}\ e^{\mathrm{i}(\omega t-kz)},$$

$$\underline{\mathbf{E}} = (v^2/\omega)\{[kB_2\,\mathbf{e}_x - (\omega^2/kv^2)\,B_1\,\mathbf{e}_y + qB_2\,\mathbf{e}_z]\,e^{iqx} + [kB'_2\,\mathbf{e}_x - (\omega^2/kv^2)\,B'_1\,\mathbf{e}_y - qB'_2\mathbf{e}_z]\,e^{-iqx}\}\;e^{i(\omega t-kz)}. \quad [12.32]$$

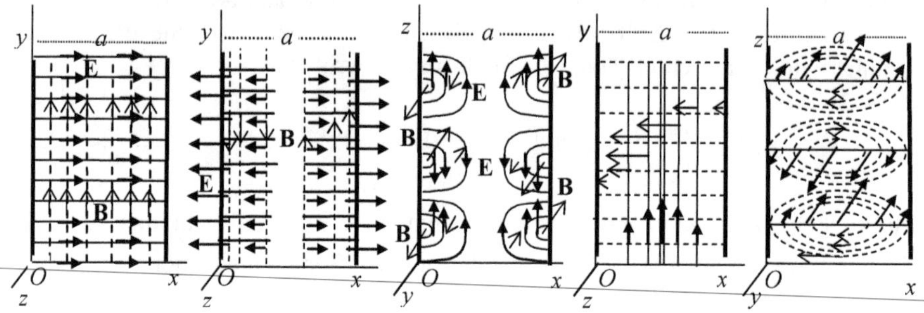

(a) (b) (c) (d) (e)

Figure 12.4. *Fields* **E** *and* **B** *in a waveguide formed by two plane parallel plates: a) fields of a TEM wave, b) fields of a TM wave* ($m = 1$) *in a transverse section, and c) in a longitudinal section, d) fields of a TE wave* ($m = 1$) *in a transverse section, and e) in a longitudinal section. The lines of* **E** *are solid lines and those of* **B** *are dashed lines*

The boundary conditions [12.20] are verified on the plate $x = 0$ if $B'_1 = -B_1$ and $B'_2 = B_2$. Thus, we may write the fields in the form

$$\underline{\mathbf{B}} = 2[iB_1\sin(qx)\,\mathbf{e}_x + B_2\cos(qx)\,\mathbf{e}_y + (q/k)B_1\cos(qx)\,\mathbf{e}_z]\;e^{i(\omega t-kz)},$$
$$\underline{\mathbf{E}} = 2(v^2/\omega)[kB_2\cos(qx)\,\mathbf{e}_x - i(\omega^2/kv^2)B_1\sin(qx)\,\mathbf{e}_y + iqB_2\sin(qx)\,\mathbf{e}_z]\;e^{i(\omega t-kz)} \quad [12.33]$$

The boundary conditions [12.20] on the plate $x = a$ are verified if

$$B_1 \sin qa = 0 \qquad \text{and} \qquad qB_2 \sin qa = 0. \quad [12.34]$$

These equations may be verified in the following three cases:

i) $q = 0$, then $k = \omega/v$ and the fields have the form

$$\underline{\mathbf{E}} = vB_o\,\mathbf{e}_x\;e^{i(\omega t-kz)} \qquad \text{and} \qquad \underline{\mathbf{B}} = B_o\,\mathbf{e}_y\;e^{i(\omega t-kz)}, \quad [12.35]$$

where we have redefined the amplitude $\underline{B}_o = 2\underline{B}_2$. This is a *TEM* wave with k related to ω by the dispersion relation

$$\omega = vk. \quad [12.36]$$

The phase velocity and the group velocity are

$$v_{(p)} = \omega/k = v \qquad \text{and} \qquad v_{(g)} = \partial\omega/\partial k = v. \tag{12.37}$$

Thus, this waveguide transmits *TEM* waves at any frequency with a phase velocity and a group velocity equal to the speed of propagation v, exactly as if the medium were infinite. Figure 12.4a shows the lines of the fields **E** and **B** for this *TEM* wave.

ii) $\sin(qa) = 0$ (with $q \neq 0$) and $\underline{B}_1 = 0$. Then q may have one of the values

$$q_m = m\pi/a, \qquad \text{where } m = 1, 2, 3, \ldots \tag{12.38}$$

The integer number m specifies the mode. The corresponding value of k is given by the dispersion relation

$$k^2 = \omega^2/v^2 - q_m^{\ 2}. \tag{12.39}$$

We may write also

$$k = k_m = \gamma_m\omega/v, \quad \text{where } \gamma_m = \sqrt{1 - \omega_m^{\ 2}/\omega^2} \text{ and } \omega_m = m\pi v/a. \tag{12.40}$$

If $\omega > \omega_m$, k_m is real; the wave is progressive and the guide is *dispersive*. On the other hand, if $\omega < \omega_m$, k_m is imaginary ($k = -i\eta_m$), the guide is *reactive* and the wave is attenuated with an attenuation coefficient in the direction Oz

$$\eta_m = (1/v)\sqrt{\omega_m^{\ 2} - \omega^2}\,. \tag{12.41}$$

Thus, ω_m is a lower *cut-off frequency* for the mode m. The fields in this mode are

$$\underline{\mathbf{E}} = v\underline{B}_o[\gamma_m \cos(m\pi x/a)\,\mathbf{e}_x + i(\omega_m/\omega)\sin(m\pi x/a)\,\mathbf{e}_z]\; e^{i(\omega t - kz)},$$

$$\underline{\mathbf{B}} = \underline{B}_o \cos(m\pi x/a)\; e^{i(\omega t - kz)}\,\mathbf{e}_y, \tag{12.42}$$

where we have redefined the amplitude $\underline{B}_o = 2\underline{B}_2$. The corresponding real fields are

$$\mathbf{E} = vB_o[\gamma_m \cos(m\pi x/a)\cos(\omega t - kz + \phi)\,\mathbf{e}_x - (\omega_m/\omega)\sin(m\pi x/a)\sin(\omega t - kz + \phi)\mathbf{e}_z]$$

$$\mathbf{B} = B_o \cos(m\pi x/a)\cos(\omega t - kz + \phi)\,\mathbf{e}_y. \tag{12.43}$$

As $B_z = 0$ everywhere, this is a *TM* wave. Figures 12.4b and 12.4c illustrate the fields of a *TM* wave ($m = 1$) between the plates.

iii) $\sin(qa) = 0$ (with $q \neq 0$) and $B_2 = 0$. q may have one of the values of [12.38] with the same dispersion relation as [12.39] and a cut-off frequency $\omega_m = m\pi v/a$. The fields in the mode m are

$$\underline{\mathbf{E}} = B_o (v/\gamma_m) \sin(m\pi\, x/a)\; e^{i(\omega t - kz)}\; \mathbf{e}_y.$$

$$\underline{\mathbf{B}} = B_o[-\sin(m\pi\, x/a)\, \mathbf{e}_x + i(\omega_m/\omega\gamma_m) \cos(m\pi\, x/a)\, \mathbf{e}_z]\; e^{i(\omega t - kz)}, \qquad [12.44]$$

or the real fields

$$\mathbf{E} = B_o(v/\gamma_m) \sin(m\pi\, x/a) \cos(\omega t - kz + \phi)\, \mathbf{e}_y,$$
$$\mathbf{B} = B_o[-\sin(m\pi x/a) \cos(\omega t - kz + \phi)\, \mathbf{e}_x - (\omega_m/\omega\gamma_m) \cos(m\pi\, x/a) \sin(\omega t - kz + \phi)\mathbf{e}_z]. \qquad [12.45]$$

As $E_z = 0$ everywhere, this is a *TE* wave. Figures 12.4d and 12.4e illustrate the fields **E** and **B** of a *TE* (m = 1) wave.

Contrary to *TEM* waves, *TM* and *TE* waves can only propagate in the guide in a mode m if its angular frequency ω is higher than the cut-off angular frequency of this mode $\omega_m = m\pi v/a$. Using the dispersion relation [12.39], we obtain the phase velocity and the group velocity of the mode m

$$v_{(p)\,m} = \omega/k = v/\gamma_m > v \qquad \text{and} \qquad v_{(g)\,m} = \partial\omega/\partial k = v\gamma_m < v. \qquad [12.46]$$

We always have $v_{(p)m} > v$, $v_{(g)m} < v$ and $v_{(p)m}v_{(g)m} = v^2$.

12.4. Guided electromagnetic waves in a hollow conductor

At hyper-frequencies ($\tilde{\nu}$ of the order of the GHz), two-conductor waveguides are not practical. A single conductor waveguide (i.e. a hollow conductor) may be used. This type of waveguide cannot support *TEM* waves at any frequency. It may support a *TE* wave or a *TM* wave in a given mode if the frequency of the wave is higher than the cut-off frequency of the mode, which is determined by the geometry of the waveguide. The *TE* or *TM* mode with the lowest cut-off frequency is called the *dominant mode*. Thus, the frequency of the dominant mode is the minimum frequency of a wave that can propagate in the waveguide without attenuation. In this section, we study the propagation in a rectangular waveguide (Figure 12.5a) and we provide some results for circular waveguides (Figure 12.5b).

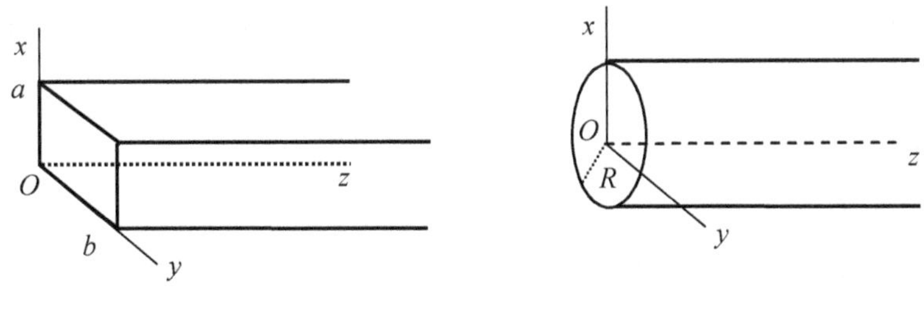

Figure 12.5. *a) Rectangular waveguide, and b) circular waveguide*

The simplest waveguide to analyze has a rectangular cross-section with sides *a* and *b* (Figure 12.5a). Such a guide is used for the transmission of linearly polarized waves. If we write the fields in the form [12.24], each component of the vector functions $\underline{\hat{\mathbf{E}}}(x,y)$ and $\underline{\hat{\mathbf{B}}}(x,y)$ obeys Helmholtz equation [12.25]. Let us write, for instance

$$\underline{\hat{E}}_x(x,y) = X(x)\,Y(y). \qquad [12.47]$$

Substituting this expression into Helmholtz equation and dividing by XY, we get

$$(1/X)\partial^2_{xx}X + (1/Y)\partial^2_{yy}Y + \omega^2/v^2 - k^2 = 0. \qquad [12.48]$$

The first term being a function of x and the second a function of y, the equation is identically verified only if each term is constant, that is,

$$\partial^2_{xx}X = -p^2X, \quad \partial^2_{xx}Y = -q^2Y \qquad \text{with} \quad p^2 + q^2 + k^2 = \omega^2/v^2, \qquad [12.49]$$

where we chose the (–) sign in order not to have exponential solutions, which cannot respect the boundary conditions. Thus, we have

$$X = A\sin(px + \phi), \qquad Y = B\sin(qy + \psi) \qquad [12.50]$$

and similarly for the other components $\underline{\hat{E}}_y(x,y)$ and $\underline{\hat{E}}_z(x,y)$. The condition $\boldsymbol{\nabla}.\mathbf{E} = 0$ may be verified only if the components of $\mathbf{E}$ are simple harmonic functions of x and y with the same parameters p and q. Thus, omitting the global factor $e^{i(\omega t - kz)}$, we may write

$$\underline{\hat{E}}_x(x,y) = A_1\sin(px + \phi_1)\sin(qy + \psi_1),\ \underline{\hat{E}}_y(x,y) = A_2\sin(px + \phi_2)\sin(qy + \psi_2),$$
$$\underline{\hat{E}}_z(x,y) = A_3\sin(px + \phi_3)\sin(qy + \psi_3), \qquad [12.51]$$

where p and the ϕ_i are not all equal to zero, this is also the case for q and the ψ_i (because this is equivalent to a field equal to zero). We may take the phases lying between 0 and π (π excluded) and the amplitudes to be positive, negative or zero. The boundary condition $\mathbf{E}_{//} = 0$ on the surfaces $x = 0$ and $y = 0$ gives the conditions

$$A_1 \sin \psi_1 = A_2 \sin \phi_2 = A_3 \sin \phi_3 = A_3 \sin \psi_3 = 0. \quad [12.52]$$

Redefining the amplitudes, we may write the electric field in the form

$$E_x = A_1 \sin(px + \phi_1) \sin(qy), \qquad E_y = A_2 \sin(px) \sin(qy + \psi_2),$$

$$E_z = A_3 \sin(px) \sin(qy). \quad [12.53]$$

The condition $\mathbf{E}_{//} = 0$ on the surfaces $x = a$ and $y = b$ is verified if

$$A_1 \sin qb = A_2 \sin pa = A_3 \sin pa = A_3 \sin qb = 0. \quad [12.54]$$

The magnetic field is given by the equation $\boldsymbol{\nabla} \times \mathbf{E} = -\partial_t \mathbf{B}$, that is,

$$\underline{B}_x = (1/\omega)\sin(px)\, [\mathrm{i}qA_3 \cos(qy) - kA_2 \sin(qy + \psi_2)],$$

$$\underline{B}_y = (1/\omega) \sin(qy)\, [\, kA_1 \sin(px + \phi_1) - \mathrm{i}pA_3 \cos(px)],$$

$$\underline{B}_z = (1/\omega)\, [\, \mathrm{i}pA_2 \cos(px) \sin(qy + \psi_2) - \mathrm{i}qA_1 \sin(px + \phi_1) \cos(qy)]. \quad [12.55]$$

The condition $\mathbf{B}_\perp = 0$ on the surfaces $x = 0$, $x = a$, $y = 0$ and $y = b$ is verified if equations [12.52] and [12.54] are verified. Maxwell's equation $\boldsymbol{\nabla}.\mathbf{E} = 0$ is verified if

$$pA_1 \cos \phi_1 = qA_2 \cos \psi_2 = pA_1 \sin \phi_1 + qA_2 \sin \psi_2 + \mathrm{i}kA_3 = 0, \quad [12.56]$$

while the other Maxwell equations are verified. All conditions [12.52], [12.54], and [12.56] can only be verified for the following types of waves:

a) *TE* waves:

$$\hat{\underline{\mathbf{B}}} = (A/\omega)\, [pk \sin(px) \cos(qy)\mathbf{e}_x + kq \sin(qy) \cos(px)\mathbf{e}_y - \mathrm{i}(p^2+q^2) \cos(px) \cos(qy)\mathbf{e}_z],$$
$$\hat{\underline{\mathbf{E}}} = A\, [q \cos(px) \sin(qy)\, \mathbf{e}_x - p \sin(px) \cos(qy)\, \mathbf{e}_y]; \quad [12.57]$$

b) *TM* waves:

$$\hat{\underline{\mathbf{E}}} = (Cv^2/\omega)[pk \cos(px) \sin(qy)\mathbf{e}_x + qk \sin(px) \cos(qy)\mathbf{e}_y + \mathrm{i}(p^2+q^2) \sin(px) \sin(qy)\, \mathbf{e}_z],$$
$$\hat{\underline{\mathbf{B}}} = C[- q \sin(px) \cos(qy)\, \mathbf{e}_x + p \sin(qy) \cos(px)\, \mathbf{e}_y]. \quad [12.58]$$

In these expressions, p and q only take the values

$$p = m\pi/a, \qquad q = n\pi/b, \qquad \text{where } m \text{ and } n = 0, 1, 2, \ldots \quad [12.59]$$

The integers m and n specifying the mode cannot both be equal to zero. Then, the last relation [12.49] gives

$$k = k_{m,n} = \gamma_{m,n}\,\omega/v, \qquad \gamma_{m,n} = \sqrt{1 - \omega_{m,n}{}^2/\omega^2}\,, \tag{12.60}$$

where $\omega_{m,n}$ is the angular cut-off frequency given by

$$\omega_{m,n} = \pi v\sqrt{m^2/a^2 + n^2/b^2}\,. \tag{12.61}$$

If the angular frequency ω of the wave is higher than $\omega_{m,n}$, $k_{m,n}$ is real; the wave is then progressive without attenuation with the dispersion relation [12.60]. The phase velocity and the group velocity are

$$v_{(p)\,m,n} = \omega/k = v/\gamma_{m,n} > v \qquad \text{and} \qquad v_{(g)\,m,n} = \partial\omega/\partial k = \gamma_{m,n}v < v. \tag{12.62}$$

Conversely, if ω is lower than $\omega_{m,n}$, $k_{m,n}$ is imaginary of the form $-i\eta_{m,n}$; the wave is then attenuated in the Oz direction with an attenuation coefficient

$$\eta_{m,n} = (1/v)\sqrt{\omega_{m,n}{}^2 - \omega^2}\,. \tag{12.63}$$

Figure 12.6a illustrates the dispersion relation for the mode (m, n): ω increases from the cut-off angular frequency $\omega_{m,n}$ and asymptotically approaches the straight-line $\omega = vk$, which is the dispersion relation in an infinite medium. Figure 12.6b illustrates the attenuation coefficient $\eta_{m,n}$ and the wave number $k_{m,n}$ versus ω: $\eta_{m,n}$ and $k_{m,n}$ are equal to zero at the cut-off frequency $\omega_{m,n}$. Figure 12.6c illustrates the phase velocity and the group velocity versus ω: we always have $v_{(p)m,n} > v$, $v_{(g)m,n} < v$ and $v_{(p)m,n}\,v_{(g)m,n} = v^2$. At high frequency, $v_{(p)m,n}$ and $v_{(g)m,n}$ asymptotically approach the speed of propagation v in the infinite medium.

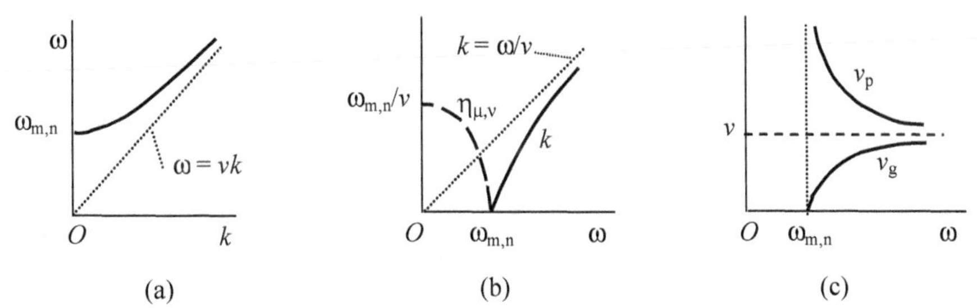

Figure 12.6. *a) Dispersion relation of a waveguide, b) attenuation coefficient and wave number versus ω, and c) phase velocity and group velocity versus ω for a mode (m, n)*

For instance, if $a = 3$ cm and $b = 7$ cm, the cut-off frequency of the modes are $\tilde{\nu}_{m,n} = 15\,(m^2/9 + n^2/49)^{½}$ GHz. The first two frequencies are $\tilde{\nu}_{0,1} = 2.14$ GHz and $\tilde{\nu}_{0,2} = 4.29$ GHz. To reduce the deformation of signals, the waveguide must be used only in the first mode, using only frequencies $\tilde{\nu}$ lying between 2.14 and 4.29 GHz. A wave of 3 GHz has a factor $\gamma_{0,1} = 0.7$; thus, it propagates with phase velocity $v_{(p)\,0,1} = 4.29 \times 10^8$ m/s and a group velocity $v_{(g)\,0,1} = 2.1 \times 10^8$ m/s. A wave of 1.5 GHz entering the guide according to the first mode has an attenuation coefficient $\eta = 32\ \text{m}^{-1}$. Thus, this wave can penetrate only few centimeters.

In the case of a circular waveguide of radius R, the fields can be expressed in terms of Bessel functions; the cut-off frequencies are not the same for the TE modes and the TM modes; they are given by

$$\omega_{m,j}(TM) = x_{m,j}\, v/R \qquad \text{and} \qquad \omega_{m,j}\,(TE) = x'_{m,j}\, v/R, \qquad [12.64]$$

where $x_{m,\,j}$ is the jth zero of the Bessel function $J_m(x)$. The first zeros are

$$\begin{array}{llll} \text{for } m = 0, & x_{0,1} = 2.405, & x_{0,2} = 5.520\,, & x_{0,3} = 8.654.... \\ \text{for } m = 1, & x_{1,1} = 3.832, & x_{1,2} = 7.076, & x_{1,3} = 10.173... \\ \text{for } m = 2, & x_{2,1} = 5.136, & x_{2,2} = 8.417, & x_{2,3} = 12.620... \end{array} \qquad [12.65]$$

$x'_{m,j}$ is the jth zero of thc function $J'_m(x) \equiv dJ_m/dx$. The first zeros are

$$\begin{array}{llll} \text{for } m = 0, & x'_{0,1} = 3.832, & x'_{0,2} = 7.016, & x'_{0,3} = 10.174... \\ \text{for } m = 1, & x'_{1,1} = 1.841, & x'_{1,2} = 5.331, & x'_{1,3} = 8.536... \\ \text{for } m = 2, & x'_{2,1} = 3.054, & x'_{2,2} = 6.706, & x'_{2,3} = 9.970... \end{array} \qquad [12.66]$$

The lowest cut-off frequency is that of the TE mode $m = 1$ and $j = 1$, i.e. $\omega_{1,1}(TE) = x'_{1,1}\, v/R = 1.84\, v/R$.

12.5. Energy propagation in waveguides

Let us calculate the energy density in a rectangular waveguide in the case of a transverse electric wave in the mode (m,n) [12.57], for instance. Taking the real part of the fields, we find the density of electric and magnetic energy per unit volume

$$\begin{aligned} U_{E,v} &= ½\varepsilon \mathbf{E}^2 = ½\varepsilon A^2[q^2 \cos^2(px)\sin^2(qy) + p^2\sin^2(px)\cos^2(qy)]\cos^2(\omega t - kz), \\ U_{M,v} &= \mathbf{B}^2/2\mu = ½(A^2/\mu\omega^2)\{[p^2k^2\sin^2(px)\cos^2(qy) + k^2q^2\cos^2(px)\sin^2(qy)]\cos^2(\omega t - kz) \\ &\quad + (p^2 + q^2)^2\cos^2(px)\cos^2(qy)\sin^2(\omega t - kz)\}. \end{aligned} \qquad [12.67]$$

Contrary to the case of progressive waves in infinite space, these energy densities are not equal. Taking the average values over a period and integrating over the transverse dimensions x from 0 to a and y from 0 to b and using the integrals

$$\int_0^a dx \ \sin^2(px) = (1/2p)\,[ap - \sin(ap)] = a/2 \quad (\text{si } p = m\pi/a). \qquad [12.68]$$

We find that the stored energy between the sections z and $z + dz$ is

$$dU_E = dU_M = (\varepsilon/16)\, abA^2(q^2 + p^2)\, dz = (\varepsilon/16v^2)abA^2\, \omega_{m,n}^2\, dz, \qquad [12.69]$$

where we have used the relation $p^2 + q^2 + k^2 = \omega^2/v^2$. Thus, we find that the two time-averaged densities of energy per unit length are equal. The total electromagnetic energy density per unit length is

$$U_{EM,l} = (\varepsilon/8v^2)\, abA^2\, \omega_{m,n}^2. \qquad [12.70]$$

The Poynting vector $\mathbf{S} = \mathbf{E} \times \mathbf{B}/\mu$ of the wave may be written as

$$\mathbf{S} = (A^2/4\mu\omega)\{-(p^2+q^2)[p \sin(2px)\cos^2(qy)\mathbf{e}_x + q\, \cos^2(px)\sin(2qy)\ \mathbf{e}_y]\sin(2\omega t - 2kz) + 4k\,[q^2\cos^2(px)\sin^2(qy) + p^2\sin^2(px)\cos^2(qy)]\cos^2(\omega t - kz)\,\mathbf{e}_z\}. \qquad [12.71]$$

Evaluating its average over a period, the transverse components of $<\mathbf{S}>$ are equal to zero; thus, the energy propagates effectively only in the direction of the waveguide

$$<\mathbf{S}> = (A^2k/2\mu\omega)\,[q^2\cos^2(px)\sin^2(qy) + p^2\sin^2(px)\cos^2(qy)]\,\mathbf{e}_z. \qquad [12.72]$$

Evaluating the flux of $<\mathbf{S}>$ over a section, we get the transferred average power

$$<P> = (A^2k/8\mu\omega)\, ab\,(q^2 + p^2) = (\varepsilon k/8\omega)\, ab\, A^2\, \omega_{m,n}^2 = v_{(g)\,m,n} U_{EM,l}. \qquad [12.73]$$

This expression shows that the electromagnetic energy propagates in the waveguide with the group velocity $v_{(g)\,m,n}$.

The preceding analysis assumes that the waveguide walls are superconductors (i.e. having infinite conductivity σ). In which case, there is no electric field in the superconducting wall and the continuity of the component of **E** parallel to the wall requires that it is equal to zero near the wall. Thus, the Poynting vector **S** has no component perpendicular to the wall. In other words, there is no energy flow through the wall. In the case of usual conductors, σ is finite and **E** varies with the traveled distance in the conductor according to the relations [10.110], that is, by taking the real part, $E = E_m\, e^{-\eta z}\cos(\omega t - \eta z)$ where $\eta \approx \sqrt{\mu\sigma\omega/2}$. The time-averaged power dissipated as Joule heat in the unit volume of the conductor is

$$< P_J > = < \sigma E^2 > = \tfrac{1}{2}\, \sigma\, E_m^2\, e^{-2\eta z}. \qquad [12.74]$$

Thus, the time-averaged power that is dissipated per unit area of the walls is

$$P_s = \int_0^\infty dz\ \tfrac{1}{2}\sigma\, E_m^2\, e^{-2\eta z} = (\eta/2\mu\omega)\, E_m^2. \qquad [12.75]$$

In general, the power loss in the waveguide depends on the wall material, its geometry, the mode of propagation, and the dielectric that fills the guide. While propagating from z to $z + dz$, power loss dP is proportional to dz and P. Thus, we write $dP = -\,\zeta\, P(z)\, dz$. By integration, we find that the transferred power varies with the traveled distance according to the law

$$P(z) = P_o\, e^{-\zeta z}, \qquad [12.76]$$

where P_o is the power at the entry.

12.6. Cavities

An electromagnetic cavity is a finite volume bounded by a conductor in all directions. If there is no energy loss, a wave remains indefinitely confined in the cavity as a standing wave in *modes* having well-defined discrete frequencies. The modes are determined by Maxwell's equations and the boundary conditions on the cavity walls. The analysis is complicated in the case of arbitrary shape. Let us consider a cavity obtained by closing a waveguide at both ends. In order to respect the boundary conditions on the lateral surfaces, the waves must propagate according to the modes of the waveguide, that is, with a wavelength of the guided wave $\lambda' = v_{(p)}/\tilde{\nu} = 2\pi/k$ where $v_{(p)}$ is the phase velocity of the mode. The boundary conditions on the end surfaces, are respected if the waveguide has a length L equal to an integer multiple of $\lambda'/2$; thus, we must have $L = p\lambda'/2 = p\pi/k$. In the case of a rectangular cavity, using the relations [12.60] and [12.61], this condition may be written as $L\gamma_{m,n}\, \omega/v = p\pi$. The discrete angular frequencies that verify this condition are

$$\omega_{m,n,p} = \pi v \sqrt{m^2/a^2 + n^2/b^2 + p^2/L^2}\ . \qquad [12.77]$$

Thus, a wave can be stationary in a cavity if its frequency is equal to one of the *normal frequencies* of the cavity, which depend on its geometrical form and dimensions. The cavity may be excited by sending a wave through a small hole in the walls. Resonance occurs if the frequency of excitation is close to one of the normal frequencies of the cavity.

In reality, there is always some energy loss (absorption in the dielectric, Joule heat in the conductors, etc.). This gives each mode a certain characteristic relaxation time τ and this corresponds to a *resonance width* $\Delta\omega \approx 2\pi/\tau$. If the excitation frequency falls within the bandwidth of a normal frequency, the normal mode is excited. In principle, all the normal modes may be excited. However, the excitation of a mode of high frequency $\tilde{\nu}_n$ is less likely because it requires higher energy. Figure 12.7 illustrates the variation of the absorbed power as a function of the excitation frequency. It is formed by a series of resonance peaks with a certain width. At low excitation frequencies, the resonances are well separated. At high frequencies, the resonances are closer and their width becomes comparable to their frequency spacing. The response of the system to the excitation becomes a slowly varying function of the frequency.

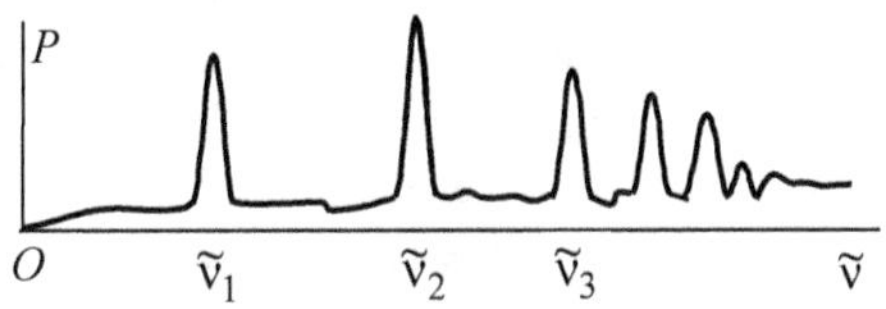

Figure 12.7. *Absorbed power*

A quantum system, such as atoms or molecules, has an infinite number of stationary modes, exactly like a cavity. If it is exposed to an electromagnetic field, it may absorb energy. The variation of the absorbed power versus the wave frequency is similar to that of Figure 12.7 with sharp peaks at the normal frequencies $\tilde{\nu}_n$. If, for instance, a gas is exposed to a beam of light of bandwidth $\Delta\tilde{\nu}$, the frequencies of this beam, which are equal to the normal frequencies of the gas molecules, are absorbed. Analyzing the spectrum of the emerging beam by using a spectrometer, the absorbed frequencies appear as dark lines (absorption spectrum).

12.7. Applications of waveguides

An electromagnetic wave of any frequency $\tilde{\nu}$ may propagate in the *TEM* mode only in waveguides formed by two conductors. It may propagate in a *TE* or *TM* mode if $\tilde{\nu}$ is higher than the cut-off frequency $\tilde{\nu}_i$ of this mode. Thus, a wave with a frequency lower than the frequency $\tilde{\nu}_1$ of the dominant mode cannot propagate in any *TE* or *TM* mode. If $\tilde{\nu}_1 < \tilde{\nu} < \tilde{\nu}_2$, the wave may propagate only in the dominant mode and, if $\tilde{\nu}_2 < \tilde{\nu} < \tilde{\nu}_3$, it may propagate in the dominant mode and the second mode and so on. As the velocity depends on the frequency and the mode, dispersion occurs. If a signal is a superposition of several frequencies, each spectral component propagates with its proper speed and the signal is inevitably deformed, even if a single mode is used for the transmission. The deformation is more important if more than one mode is used. Thus, to transmit a signal in a waveguide constituted by a hollow conductor with minimum deformation, only the dominant mode must be used. This is achieved by using frequencies that lie between the cut-off frequency of the dominant mode and that of the second mode. Other frequency bands may be used with appropriate filters to enable a single mode.

As the cut-off frequencies depend on the geometry of the waveguide, the transverse dimensions must be chosen so that the wave propagates in the dominant mode. They are comparable to the wavelength. For instance, the mode m of a waveguide formed by two plane and parallel plates has a cut-off frequency $\tilde{\nu}_m = mv/2a$. A wave of frequency $\tilde{\nu}$ may only propagate in the dominant mode if $v/2a < \tilde{\nu} < v/a$, i.e. $a < \lambda < 2a$, where $\lambda = v/\tilde{\nu}$ is the wavelength in the infinite medium. Similarly, to transmit light in an optical fiber, its radius must be of the order of the micrometer. As the transverse dimensions of waveguides are of the order of λ, an important part of the wave is diffracted at its extremity and another part is reflected back. If the waveguide ends in an antenna, the transmitted wave may be directed and the reflection is reduced at its end.

One of the most important uses of waveguides is to transmit signals and information. In order to only allow the dominant mode, the fields at the entry of the guide must be established in this mode configuration by using appropriate mode filters. We may transmit waves of frequency between several dozen kilohertz and a few hundred megahertz by coaxial cables formed by a wire surrounded by an insulating medium and a braided copper sheathing. This also protects the signal from the interference of other unwanted signals. In such a cable, the *TEM* wave of any frequency can propagate besides *TM* and *TE* waves if their frequency is higher that the cut-off frequency of the modes. Coaxial cables are used, for instance, to connect the parabolic antenna to a television set and the amplifier of a radar system to its antenna. In telephony, a bundle of some 20 cables of this type allows more than 1.3×10^5 simultaneous telephone calls to be transmitted. At hyper-frequencies (several gigahertz), we may use the *TE* or *TM* modes if the *TEM* can be eliminated. This can be achieved with a rectangular or cylindrical single-conductor waveguide. For instance, a circular waveguide of 5 cm has a useful bandwidth of several gigahertz near 40 GHz. This enables the transmission of approximately 5×10^4 telephone calls or 50 television channels.

A transparent fiber is an optical waveguide. This is a very thin cylindrical fiber made of a transparent material with a high refraction index. It is surrounded by a layer with a lower index, which is specially chosen to increase the reflection coefficient for the light used and to protect the fiber. For short distance transmissions, we may use plastic fibers (polystyrene, for instance), which are economical, light, and flexible, but have high attenuation. We also use fibers fabricated from a core of pure silica surrounded by a silicone layer. They are stiffer and more costly than plastic fibers, but they have lower attenuation. We may also use fluoride glass, which is more difficult to manufacture but has extremely low attenuation.

A large number of fibers may be assembled to form an optical fiber cable, which may be bent, with a radius as small as 1 cm, as light propagates separately in each fiber. This type of cable may be used to transmit images, thus to observe normally inaccessible places, such as internal organs of the body (endoscopy). It may be used to transmit an intense laser beam to be used in medical surgery and in industry. Because of the high frequency of light waves, fibers and lasers may be associated for the transmission of an enormous quantity of information as optical signals for distances of up to thousands of kilometers.

12.8. Problems

Transmission lines

P12.1 A line has a capacitance C_l and an inductance L_l per unit length. **a)** Verify that a wave that propagates toward the positive z, may be written as $V = V_m e^{i(\omega t - kz)}$ and $I = (V_m/Z)\, e^{i(\omega t - kz)}$, while a wave that propagates toward the negative z, may be written as $V = V_m e^{i(\omega t + kz)}$ and $I = -(V_m/Z)\, e^{i(\omega t + kz)}$. **b)** If a wave propagates toward a junction point $z = D$, where Z has a discontinuity, show that a part of the wave is reflected and another part is transmitted. Knowing the incident wave V, write the expressions of the reflected wave V' and the transmitted wave V'', as well as the expressions of the intensities I' and I''. **c)** If the line ends at the point $z = D$, show that it is equivalent to assuming that it is joined to another line of infinite impedance Z_2. Then, there is a total reflection with a change of sign for I and without change of sign for V. **d)** If the line is short-circuited ($Z_2 = 0$) at $z = D$, show that the wave is reflected totally with a change of sign for V and without change of sign for I. **e)** Is it possible to connect the end to an impedance $\underline{Z}_c$ in order to have no reflected wave?

P12.2 An air-filled coaxial cable of length L is formed by a cylindrical conductor of radius 1 mm surrounded by a cylindrical shell of internal radius 5 mm. Assume that the resistance of the conductors is negligible. A potential wave of amplitude V_m and frequency 10 MHz is sent along this cable. **a)** Calculate the impedance at the entry. **b)** Calculate the reflection coefficient, if the cable is connected to a circuit of impedance $\underline{Z}_c$. **c)** Calculate the current intensity and the average power that is supplied to this circuit and verify the conservation of energy. **d)** What should the value of $\underline{Z}_c$ be in order for the wave to be totally reflected? What should the value of $\underline{Z}_c$ be in order not to have a reflected wave? What should the value of $\underline{Z}_c$ be in order to have the maximum supplied average power?

P12.3 A real line of inductance L_l and capacitance C_l per unit length has a small longitudinal resistance r_l per unit length (for both conductors) and a small leak conductance G_l per unit length (i.e. reciprocal of a large resistance). Let $V(z, t)$ and $I(z, t)$ be the voltage and the intensity at point z and time t. **a)** Show that V and I obey the equations $C_l\,\partial_t V + G_l\,V + \partial_z I = 0$ and $L_l\,\partial_t I + r_l\,I + \partial_z V = 0$. Deduce the equation of propagation $\partial^2_{tt}V - v^2\partial^2_{zz}V + 2\beta\partial_t V + \gamma V = 0$ and the same for I, where we have set $v = (C_l\,L_l)^{-1/2}$, $\beta = \frac{1}{2}(G_l/C_l + r_l/L_l)$ and $\gamma = r_l G_l/C_l L_l$. **b)** A generator of electromotive force $\underline{\mathcal{E}} = \mathcal{E}_m\,e^{i\omega t}$ is connected to the entry. Show that the voltage and the intensity may be written as $\underline{V} = \mathcal{E}_m\,e^{-\eta x}\,e^{i(\omega t - kz)}$ and $\underline{I} = \underline{I}_m\,e^{-\eta x}\,e^{i(\omega t - kz)}$ where k and η are given by the equations $k^2 = \frac{1}{2}(A + B)$ and $\eta^2 = \frac{1}{2}(B - A)$ with $A = (\omega^2/v^2 - r_l G_l)$ and $B = [A^2 + \omega^2(G_l L_l + r_l C_l)^2]^{1/2}$. **c)** Show that $\underline{I}_m = \mathcal{E}_m/\underline{Z}$ where $Z = (\omega L_l - ir_l)/(k - i\eta)$ is the impedance of the line. Is it possible to choose the line parameters in order that there is no dispersion? Does the attenuation then depend on the frequency?

Electromagnetic standing waves

P12.4 The entry of an electric line is short-circuited and excited at high frequency while its end ($z = D$) is connected to an impedance $\underline{Z}_c$. **a)** Write the general solutions for $\underline{V}$ and $\underline{I}$, impose the boundary conditions and show that the modes may be written as $V = A\,\cos(\omega t + \phi)\,\cos(\omega z/v)$ and $I = (A/Z_l)\,\sin(\omega t + \phi)\,\sin(\omega z/v)$, where the normal frequencies are given by the condition $\tan(2\pi D/\lambda) = i(\underline{Z}_c/Z_l)$. **b)** Determine the normal modes if the line is short-circuited and if it is open-ended.

P12.5 A linearly polarized wave of angular frequency ω is incident normally on the face Oxy of an ideal conductor. **a)** Write the expression of the total field in front of the conductor and show that the wave is stationary. Determine the nodal planes of **E**. **b)** Calculate the electromagnetic energy density and the Poynting vector of this wave. **c)** Calculate the surface charge density and the surface current density on the conductor surface. Deduce the radiation pressure on this surface. **d)** Using the quantization of radiation, what is the number of photons that are intercepted by this surface? What are the energy and the momentum that are transferred to the conductor per unit time and per unit area? Deduce the radiation pressure. **e)** Assume now that the incident wave is right-handed circularly polarized. Write the expression of the reflected wave and that of the total wave. Determine its Poynting vector and radiation pressure. **f)** Assume that a linearly polarized wave is confined between two metallic surfaces at $z = 0$ and $z = d$. What are the frequencies that correspond to standing waves? Write the expressions of **E** and **B** for these modes.

P12.6 Assume that an electric field $\mathbf{E} = A\,\cos(kz)\,\cos(\omega t)\,\mathbf{e}_x$ is established in a region of a dielectric. **a)** Calculate the magnetic field. **b)** Calculate the electric energy density, the magnetic energy density, and the Poynting vector. **c)** Describe the fields **E** and **B** as functions of z in a wave zone situated between two nodal planes of **E** at the instants $t = 0$, $T/8$, $T/4$, $3T/8$ and $T/2$. Analyze the distribution of the electric energy, magnetic energy, and total energy, the magnitude and the direction of the Poynting vector at these instants of time. **d)** Show that the total energy that is stored in a half-zone (of length $\lambda/4$ between a nodal plane and an antinodal plane) remains constant in the course of time. Thus, there is no energy transfer from one half-zone to the adjacent half-zone. Calculate the total electromagnetic energy that is stored in a half-zone.

P12.7 A circularly polarized standing wave may be considered as the superposition of two linearly polarized waves that are in quadrature

$$\mathbf{E}^{\pm} = E_o[\cos(\omega t)\mathbf{e}_x + \cos(\omega t \pm \pi/2)\,\mathbf{e}_y]\,\sin(kz).$$

a) Show that, at points whose z coordinates differ by $\lambda/2$, the tips of **E** move on equal circles. The nodes of **E** correspond to a radius equal to 0 and the antinodes to a radius equal to E_o. What is the direction of rotation? **b)** Show that, if a circularly polarized progressive wave is reflected totally on a fixed obstacle, the reflected wave is polarized circularly in the opposite direction to that of the incident wave. Show that the wave, which results from the superposition of the incident wave and the reflected wave, is a circularly polarized standing wave. **c)** Show that the reversal of the direction of circular polarization is a consequence of conservation of angular momentum.

Guided waves

P12.8 An electromagnetic wave is incident at an angle θ on an ideal metallic plate M_1 lying in the Oyz plane (Figure 12.8). Assume that the wave is polarized in the direction Oy. **a)** Write the expressions of the incident wave **E**, the reflected wave **E′**, and those of the corresponding magnetic fields. **b)** Write the expression of the resultant field **E**. Show that the total wave propagates in the direction Oz with a phase velocity $v_{(p)} = \omega/k \sin\theta$ and that it has nodes and antinodes. **c)** Calculate the resulting energy density, Poynting vector, and intensity. **d)** The wave is reflected also on a metallic plate M_2 parallel to M_1 at a distance d. Write the continuity conditions on the plates. Deduce the cut-off frequency of the modes. **e)** Determine the charge densities and the densities of the currents that are induced on the plates. What is the radiation pressure on them?

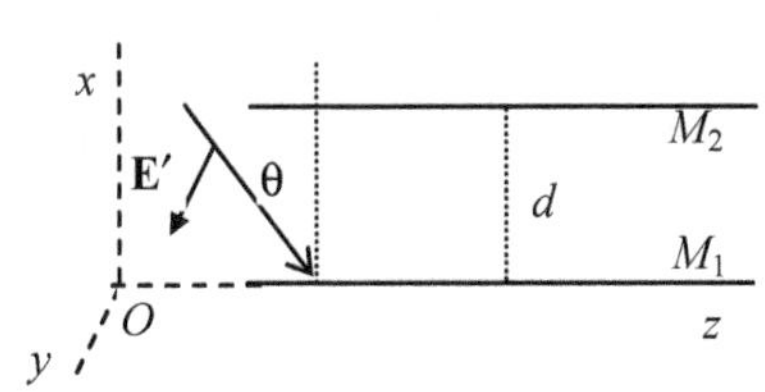

Figure 12.8. *Problem 12.8*

P12.9 A waveguide is formed by two plane and parallel plates separated by a distance a. Verify that a TE wave may be written as

$$\mathbf{E} = -\mathrm{i}vB_o\,\mathbf{e}_y\,[\,e^{\mathrm{i}(\omega t+px-kz)} - e^{\mathrm{i}(\omega t-px-kz)}\,,$$

$$\mathbf{B} = \mathrm{i}B_o\{[\mathbf{e}_x + (\omega_m/\gamma_m\omega)\,\mathbf{e}_z]\,e^{\mathrm{i}(\omega t+px-kz)} + [-\mathbf{e}_x + (\omega_m/\gamma_m\omega)\,\mathbf{e}_z]\,e^{\mathrm{i}(\omega t-px-kz)}\,\}.$$

Deduce that this wave may be considered as the superposition of an oblique wave and the corresponding reflected wave on the plate (zigzag wave).

Hollow waveguides and cavities

P12.10 An electromagnetic waveguide has a rectangular section of sides $a = 2$ cm and $b = 1$ cm. **a)** What is the lowest cut-off frequency for TE waves? Write down the expressions of the electric and magnetic fields in this mode. Determine the frequency of the first five modes. **b)** Calculate the phase velocity and the group velocity for a wave with a frequency of 10 GHz that propagates in the dominant mode.

P12.11 What is the cut-off frequency of the dominant mode in a waveguide whose section is a square of side a? Consider a fiber of this type with index 1.5. If light has a frequency that is less than the cut-off frequency, it cannot propagate. What should the minimum value a_m of the side a be to allow the propagation of light, whose wavelength is $\lambda = 500$ nm, in vacuum? Determine the phase velocity and the group velocity for the dominant mode if $a = 2a_m$.

P12.12 a) What are the three first frequencies of a cubic cavity of side a? What is their degeneracy? **b)** Determine the cut-off frequencies of the first three modes of a cylindrical waveguide of radius $R = 10$ cm. What are the lowest frequencies of a cylindrical cavity of radius $R = 10$ cm and length 10 cm?

Applications of waveguides

P12.13 Suppose that a cylindrical hollow conductor of internal radius 1.5 cm is used as a waveguide to transmit electromagnetic waves. Determine the cut-off frequency of the first three modes. To reduce the deformation of signals, only the frequencies that lie between the cut-off frequency of the dominant mode and that of the second mode must be used. What are the limits of this frequency band? Estimate the number of simultaneous telephone calls that may be transmitted by this waveguide. Estimate the number of TV channels that may be transmitted by this waveguide.

Chapter 13

Special Relativity and Electrodynamics

Until the end of the 19th Century, classical mechanics was confirmed by all experiments and nobody dared to think that this might not be the case in electromagnetism. However, several experiments have shown some contradictions between classical mechanics and electromagnetic phenomena, especially the propagation of light. In fact, as we shall see in this chapter, Maxwell's equations, which are the basic laws of electromagnetism, are not in accordance with the Galilean invariance, which is one of the basic principles of classical mechanics. Several attempts have been made, without success, to modify Maxwell's equations in order to make them agree with classical mechanics. Lorentz adopted the opposite strategy and proposed to modify classical mechanics by replacing the Galilean transformation by the now-called *Lorentz transformation.* In 1905, Einstein analyzed the basic concepts of space and time, and formulated the *special theory of relativity*. The Lorentz transformation resulted straightforwardly from this analysis. Up to now, all the consequences of this theory have been verified experimentally.

The *special theory of relativity* and the *general theory of relativity*, both formulated by Einstein, are new perceptions of physics and the Universe with very important consequences. Special relativity is used to study high-velocity (thus high-energy) phenomena. All fundamental physical theories must be formulated in accordance with relativity in order to be covariant (that is, independent of the observation frame). In this chapter we introduce the basic ideas of this theory and analyze some of its consequences in mechanics and in electromagnetism.

13.1. Galilean relativity in mechanics

One of the basic principles of physics is the *principle of relativity*, according to which the *laws of physics can be formulated in a way that is independent of the frame of reference*. Without this principle, physics would not be a universal science. This principle, first discussed by Galileo at the beginning of the 17th Century, is valid in mechanics for a class of frames, called *inertial frames*. They are in relative uniform motion, one with respect to the other. They are assumed to exist but nobody can say whether a frame (such as the frame of the Earth) is inertial or not! One of these frames is fixed with respect to the stars assumed to be "at rest". If we prefer to use a non-inertial frame (i.e. an accelerated frame), so-called *inertial forces* must be introduced in order to maintain the validity of the laws of mechanics.

Some physical quantities (such as mass, charge, etc.) have the same values in all inertial frames; they are said to be *invariant*. Other physical quantities, A, B, etc., may depend on the frame of reference. The *theory of relativity* requires that any physical law may be written as a mathematical relationship that holds in any inertial frame: $f(A, B, ...) = g(C, D, ...)$ in S and $f'(A', B', ...) = g'(C', D', ...)$ in S'. We say that the law is *covariant*.

Obviously, covariance does not require that $A' = A$, $B' = B$, etc., or $f' = f$ (i.e. invariance). If the frames of reference are fixed with respect to the same material support, they can be obtained from one another by a time-independent translation of the origin or rotation of the axes of coordinates. The validity of the physical laws in these frames requires that the physical quantities A, B, etc., as well as the functions $f(A, B...)$ and $g(C, D, ...)$ have well-defined transformation rules in translations and rotations. They may be scalars, *vectors*, or *tensors* of various ranks. The covariance of the law $f = g$ requires that f and g have the same nature (both are scalars, vectors, etc.).

Things are not so evident if we consider physical quantities measured in different inertial frames S and S' (that is, defined with respect to different material supports in relative uniform motion). For this reason, the transformation of coordinates and time plays an important part in the theory.

Let us assume that S' is moving with a constant velocity $\mathbf{v}_0$ with respect to S. If necessary, by making appropriate translation and rotation of the axes of coordinates in one frame or the other and an appropriate time shift, we may always assume that the axes of coordinates coincide at $t = 0$ and that the velocity $\mathbf{v}_0$ is in the direction Oz. In classical mechanics, the coordinates of the same point in space, measured in two frames of reference, are related by the *Galilean transformation*

$$x = x', \qquad y = y', \qquad z = z' + v_0 t, \qquad \text{i.e.} \qquad \mathbf{r} = \mathbf{r}' + \mathbf{v}_0 t. \qquad [13.1]$$

Although, the origin of time and that of the positions are arbitrary in the inertial frames, the time interval Δt between two events is the same in all frames. This is also true for distances $\Delta r = |\mathbf{r}_1 - \mathbf{r}_2|$, angles, volumes, etc. Thus, in classical mechanics, these quantities are *invariant* (or *absolute*). Other quantities, such as the velocity, the energy, etc., are *relative*. Differentiating the relations [13.1] with respect to time (assumed to be the same in both frames), we obtain the *Galilean transformation of the velocity* of a body $\mathbf{v}$ in S to $\mathbf{v}'$ in S'

$$\mathbf{v} = \mathbf{v}' + \mathbf{v}_o. \qquad [13.2]$$

Differentiating once more, we find that the acceleration is the same in both frames. The fundamental law $\mathbf{f} = m\mathbf{a}$ holds in all inertial frames if the force $\mathbf{f}$ is independent of the frame (this is just the definition of inertial frames).

If we use spherical coordinates around the origin, we specify the velocity by its magnitude v, its angle θ with Oz and its azimuthal angle φ around Oz. The vector relation [13.2] is equivalent to the relations

$$v = \sqrt{v'^2 + v_o^2 + 2\,v'v_o \cos\theta'}\,, \qquad [13.3]$$

$$\tan\theta = \sin\theta'/[\cos\theta' + v_o/v'], \qquad \varphi = \varphi'. \qquad [13.4]$$

We note that the equality of the azimuthal angles φ and φ' is expected from symmetry: if $\mathbf{v}_o$ is in the direction of Oz, we have a rotational symmetry about Oz; the plane P containing $\mathbf{v}'$ and Oz is a plane of reflection symmetry; thus, $\mathbf{v}$ is in the same plane.

As any mechanical quantity may be expressed in terms of positions and velocities, we deduce its transformation law from equations [13.1] and [13.2]. For instance, the momentum and kinetic energy of a particle transform according to the relationships

$$\mathbf{p} = m\mathbf{v} = m(\mathbf{v}' + \mathbf{v}_o) = \mathbf{p}' + m\mathbf{v}_o,$$
$$U_K = \tfrac{1}{2}m\mathbf{v}^2 = \tfrac{1}{2}m(\mathbf{v}+\mathbf{v}_o)^2 = \tfrac{1}{2}m\mathbf{v}'^2 + \tfrac{1}{2}m\mathbf{v}_o^2 + m(\mathbf{v}'.\mathbf{v}_o) = U'_K + \tfrac{1}{2}m\mathbf{v}_o^2 + m(\mathbf{v}'.\mathbf{v}_o).$$

13.2. Galilean relativity and wave theory*

A simple harmonic plane wave is characterized by its *angular frequency* ω, its phase velocity v_p, its *direction of propagation*, and its *amplitude* (related to the wave intensity, i.e. the energy that it carries). Note that, generally, v_p is different from the speed of propagation v, which appears in the wave equation (that we may write only in the rest frame of the medium). Taking Oz and $O'z'$ in the direction of the velocity $\mathbf{v}_o$ (Figure 13.1a), the wave function may be written in S and S'

$$u(\mathbf{r},t) = u_m \cos[\omega(t - \mathbf{e}.\mathbf{r}/v_p)] \equiv u_m \cos(\omega t - \mathbf{k}.\mathbf{r}), \quad [13.5]$$
$$u'(\mathbf{r}',t') = u'_m \cos[\omega'(t' - \mathbf{e}'.\mathbf{r}'/v'_p)] \equiv u'_m \cos(\omega' t' - \mathbf{k}'.\mathbf{r}'). \quad [13.6]$$

$\mathbf{e} = \mathbf{k}/k$ and $\mathbf{e}' = \mathbf{k}'/k'$ are the unit vectors in the direction of propagation. Figure 13.1b illustrates the directions of propagation in S and S'.

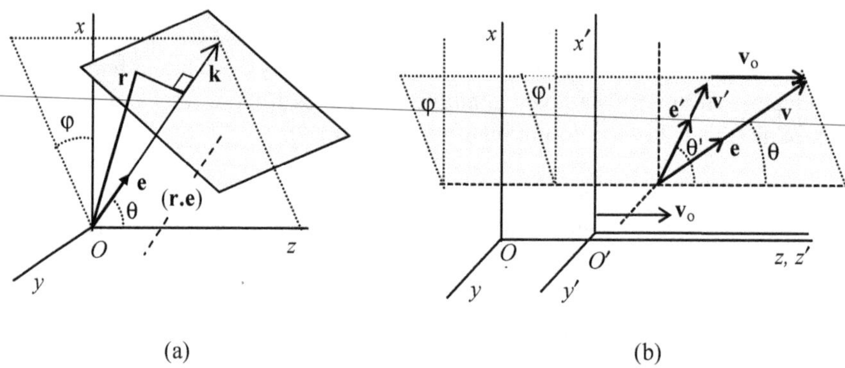

Figure 13.1. *Transformation of a wave: a) the wave in the frame S, and b) directions of propagation in S and S'*

The laws of transformation of the wave must verify the condition that the phase at each point M of space and at any time is the same in both frames of reference. The reason is that, if the wave reaches a maximum or a minimum S, it must be so in any other frame S'. Thus, we have

$$\omega(t - \mathbf{e}.\mathbf{r}/v_p) = \omega'(t' - \mathbf{e}'.\mathbf{r}'/v'_p) \quad [13.7]$$

at any time ($t = t'$) and at the same point in space of position $\mathbf{r}$ and $\mathbf{r}'$ in S and S'. The direction of propagation $\mathbf{e}$ is determined by its angle θ with $\mathbf{v}_o$ and its azimuthal angle φ about $\mathbf{v}_o$ taken in the direction Oz (Figure 13.1b). Expressing x, y and z in terms of x', y' and z' by using the Galilean transformation [13.1], the relationship [13.7] may be written as

$$\omega'\{t - [x' \sin\theta' \cos\varphi' + y' \sin\theta' \cos\varphi' + z' \cos\theta']/v'_p\}$$
$$= \omega\{t - [x' \sin\theta \cos\varphi + y' \sin\theta \cos\varphi + (z' + v_o t) \cos\theta]/v_p\}. \quad [13.8]$$

This relation is identically verified, that is, for any x', y', z' and t, if the coefficients of these quantities on both sides are equal; thus, we find the equations

$$(\omega'/v'_p)\sin\theta'\cos\varphi' = (\omega/v_p)\sin\theta\cos\varphi, \qquad (\omega'/v'_p)\cos\theta' = (\omega/v_p)\cos\theta,$$
$$\omega' = \omega\,[1-(v_o/v_p)\cos\theta]\,.$$

These equations imply that

$$\theta' = \theta, \qquad \varphi' = \varphi, \tag{13.9}$$
$$v'_p = v_p - v_o\cos\theta, \tag{13.10}$$
$$\omega' = \omega[1-(v_o/v_p)\cos\theta]. \tag{13.11}$$

Obviously, the inverse transformations are obtained by changing v_o into $-v_o$. These are the Galilean transformations for the characteristics of a wave. The relations [13.10] and [13.11] imply that the wavelength $\lambda = 2\pi v_p/\omega$ is the same in both frames. This result agrees with the invariance of distances in classical physics and the interpretation of the wavelength as the distance separating two crests and measured in interference experiments (see Chapter 11).

The equations [13.9] state that the direction of propagation of a wave is the same in all reference frames. The aberration of light, discovered by Bradley in 1727, confirms the transformation property for the direction of the velocity of light in the corpuscular model [13.4], but not for the direction of propagation of a wave [13.9]. This aberration is due to the motion of the Earth on its orbit, which gives a deviation $\alpha \cong v_o/c \cong 10^{-4}$ rad for the starlight (see problem 13.1).

Equation [13.11] is the transformation law of frequency. It agrees with the *Doppler effect* discovered in 1842 for sound waves and later generalized by Fizeau to light waves. To analyze this effect, we designate the proper frames of the medium of propagation, the observer, and the source by S_M, S_O and S_S, respectively. The velocities of the source and of the observer with respect to the medium are designated $\mathbf{v}_S$ and $\mathbf{v}_O$. The angles of these velocities with the direction of propagation $\mathbf{e}$ of the wave (Figure 13.2a) are designated by θ_O and θ_S. The physical quantities are measured in S_O, the phase velocity $v_{p,M}$ in S_M is known, and in S_S the proper angular frequency ω_S of the source is known. Using equation [13.11] to transform from S_M to S_O and then to transform from S_M to S_S, we obtain

$$\omega_O = \omega_S\,\frac{1-(v_O/v_{p,M})\cos\theta_O}{1-(v_S/v_{p,M})\cos\theta_S} = \omega_S\,\frac{1-(\mathbf{v}_O.\mathbf{e})/v_{p,M}}{1-(\mathbf{v}_S.\mathbf{e})/v_{p,M}}\,. \tag{13.12}$$

If the observer is at rest in the medium ($S_O \equiv S_M$ and $\mathbf{v}_O = 0$) and the source is moving with a velocity $\mathbf{v}_S$, we obtain $\omega_O = \omega_S/[1-\mathbf{v}_S.\mathbf{e}/v_{p(M)}]$ and if the source is at rest in the medium ($S_M = S_S$ and $\mathbf{v}_S = 0$) and the observer is moving with a velocity $\mathbf{v}_O$, we obtain $\omega_O = \omega_S[1-\mathbf{v}_S.\mathbf{e}/v_{p(M)}]$. If the observer is moving away from the source θ_O is acute, therefore $\omega_O < \omega_S$, and if the source moves toward the source θ_S

is obtuse, therefore $\omega_O > \omega_S$. We note that the Doppler shift does not have the same expression if the source is moving toward the observer, or the observer is moving with the same relative velocity toward the source. The wave also undergoes a Doppler-Fizeau effect if it is reflected on a moving body (Figure 13.2b); the effect is used in this case to measure the velocity of the moving obstacle (see problem 13.2b).

Sometimes, in the case of electromagnetic waves and especially light waves, the Doppler effect is expressed in terms of the "the wavelength in vacuum" $\lambda_o = 2\pi c/\omega$ instead of the frequency. In non-relativistic theory, λ_o is shifted while the wavelength in the medium $\lambda = 2\pi v/\omega$ is not.

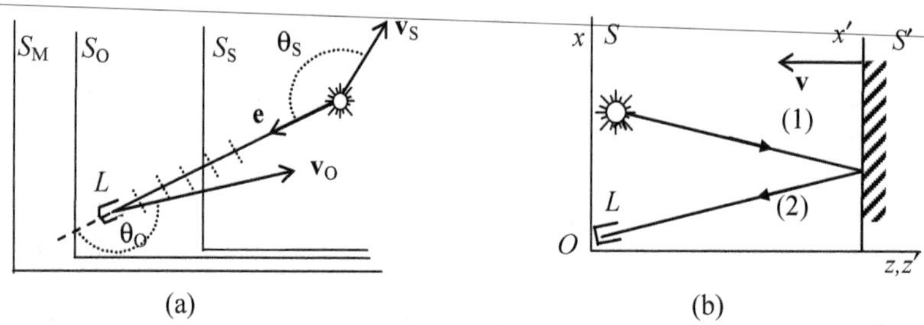

Figure 13.2. *The Doppler-Fizeau effect: a) in the case of an observer and a source in motion with respect to the medium, and b) in the case of the reflection of a wave on a moving mirror*

The radiation spectrum emitted by an atom is characteristic of that atom, independently of the physical or chemical conditions. This allows the determination of the elements that constitute the stars (mostly hydrogen, helium, and other light elements in small proportions). The measurement of the Doppler shift allows the determination of the velocity of celestial bodies, for instance, the rotation velocity of the Sun, and the approach velocity of celestial bodies. The experiment shows a red shift of light emitted by all galaxies. In 1929, Edwin Hubble proposed that this red shift is a Doppler effect, which shows that the galaxies move away. In other words, the Universe is expanding from the initial *big bang* explosion at the formation of the Universe. The velocity v of a moving away galaxy is proportional to its distance D, according to the law $v = HD$, where *H is the Hubble constant.* Its reciprocal $T = 1/H$ is considered in certain cosmological models as the age of the Universe (about 16×10^9 years).

Equation [13.10] is the law of transformation of the phase velocity (identical to the group velocity in the case of propagation of electromagnetic waves in vacuum). It is not the same as the transformation law of the velocity of a particle [13.3], unless the motion of the particle and the propagation of the wave are in the direction of the velocity of transformation $\mathbf{v}_o$ (then, $\theta = \theta' = 0$ and $v = v' + v_o$) or in the opposite direction (then, $\theta = \theta' = \pi$ and $v = v' - v_o$).

To derive the law of transformation of the group velocity, we note that the invariance of the wavelength ($\lambda = \lambda'$) implies the invariance of the wave number ($k = k'$). As the direction of $\mathbf{k}$ is the same in both frames of reference (see [13.9]), we deduce that the components of $\mathbf{k}$ are invariant

$$k_x = k'_x, \qquad k_y = k'_y, \qquad k_z = k'_z. \qquad [13.13]$$

This result may also be obtained by writing $u_m \cos(\omega t - \mathbf{k.r}) \equiv u'_m \cos(\omega' t' - \mathbf{k}'.\mathbf{r}')$. On the other hand, the phase velocity is $v_p = \omega/k$. Thus, we may write the law of transformation of the angular frequency as

$$\omega' = \omega - kv_o \cos\theta = \omega - k_z v_o. \qquad [13.14]$$

This relation shows that, in a frame S' moving with respect to the medium (frame S), the propagation of the wave is not isotropic and there is necessarily a dispersion effect.

Differentiating both sides of [13.14] with respect to k'_x, for instance, and noting that ω' is related to k'_x, k'_y, and k'_z by the dispersion relation in S' and ω is related to k_x k_y and k_z by the dispersion relation in S, we find the components of the group velocity

$$v'_{gx} = \partial\omega'/\partial k'_x = \partial\omega/\partial k_x = v_{gx}, \qquad v'_{gy} = \partial\omega'/\partial k'_y = \partial\omega/\partial k_y = v_{gy},$$
$$v'_{gz} = \partial\omega'/\partial k'_z = \partial\omega/\partial k_z - v_o = v_{gz} - v_o. \qquad [13.15]$$

Thus, the group velocity transforms exactly like the velocity of a particle [13.2]. Assimilating the direction of light with that of the group velocity, we may explain the aberration of starlight in the wave theory, exactly as in the corpuscular theory.

If a particle is moving in a medium, its velocity $\mathbf{v}$ does not depend on the properties of the medium if they do not interact. The physical quantities, associated with this motion (energy, momentum, etc.) are transferred with the same velocity as the particle. On the other hand, the phase velocity of a wave does not depend on the source but essentially on the frequency and the medium (its nature, physical conditions, its dimensions if it is bounded, etc.). The physical quantities associated with the wave are transferred with the group velocity (not the phase velocity). This is evident if we consider the case of a wave packet, which moves with the group velocity and obviously contains all the physical quantities.

13.3. The 19th Century experiments on the velocity of light

Before the formulation of electromagnetism by Maxwell, light was widely considered as a mechanical wave, similar to elastic waves and sound. However, contrarily to mechanical waves, light propagates in vacuum. The physicists of that period assumed that a very rarefied medium, called ether, fills the vacuum and the transparent bodies and serves to transmit light. Thus, Earth sails in a "sea of ether" assumed to be at rest and to fill all the space. The phase velocity of light (c in vacuum and c/n in transparent mediums) is the phase velocity in the ether.

If the observer is moving with respect to the ether with a velocity v_o, the phase velocity of light must be modified according to the Galilean law of transformation of the velocity [13.2] in the corpuscular theory of light and [13.10] in the wave theory. In practice, v_o is much less than v and this modification of light speed is too small to be observed. The highest speed v_o for macroscopic bodies at disposal on Earth is the velocity of Earth itself on its orbit, which is about 30 km/s. This produces a relative modification of only 10^{-4} in the velocity of light (and only 10^{-8} in experiments measuring back-and-forth time of propagation). Only very high precision experiments (notably using interference) can detect it. Several experiments have tried to detect this modification (and indirectly detect the ether). We discuss in the following only the historic Michelson-Morley experiment in 1881, which has been repeated several times. It has shown without any doubt that the speed of light in vacuum does not depend on the motion of the observer.

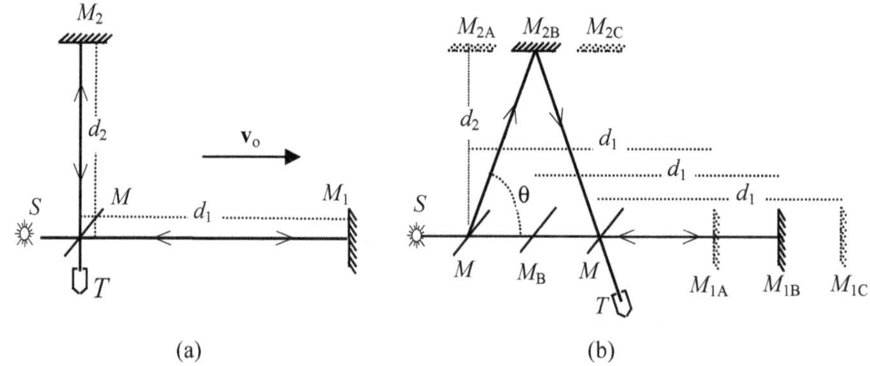

Figure 13.3. *Michelson-Morley experiment: a) the light beams in the frame of the Earth, and b) the beams in the frame of the hypothetical ether*

Michelson's interferometer is illustrated in Figure 13.3. A monochromatic light beam of wavelength λ is split into two beams by a half-silvered mirror M. They propagate on two perpendicular axes from M to the mirrors M_1 and M_2 and back to

M. Their interference pattern is observed through a telescope *T*. The interferometer is set on a horizontal turntable. Figure 13.3a illustrates the light beams as observed in the Earth reference frame and Figure 13.3b illustrates them in the frame of the ether. As time is assumed absolute in classical physics, the time of travel of these beams may be calculated in either frame, by using the appropriate speed. The speed of light is c in all directions in the ether frame and $(\mathbf{c} - \mathbf{v}_o)$ in the Earth frame. In the following, we set $\mathbf{v}_o = c\boldsymbol{\beta}$ for the velocity of the frame S' (of the Earth) with respect to the frame S (of the ether).

First, the arm MM_1 was oriented in the direction of the Earth velocity $\mathbf{v}_o$. The travel times along MM_1 and M_1M are easily calculated in the frame of the Earth (Figure 13.3a). The speeds of light in the directions MM_1 and M_1M are $c - v_o$ and $c + v_o$, respectively. The travel time for MM_1M is $t_1 = d_1/(c-v_o) + d_1/(c+v_o) = 2d_1\gamma^2/c$, where $\gamma = 1/\sqrt{1-\beta^2}$. The travel times for MM_2 and M_2M are calculated easily in the frame of the ether (Figure 13.3b). The path MM_2 makes with $\mathbf{v}_o$ an angle θ such that $\cos\theta = v_o/c$. The length of the path MM_2M is $2d_2/\sin\theta = 2d_2\gamma$. As the speed on this path is c, the travel time on this path is $t_2 = 2d_2\gamma/c$. The interference pattern is determined by the difference $\delta t = t_1 - t_2 = 2d_1\gamma^2/c - 2d_2\gamma/c$. A rotation of the turntable through 90° brings the arm MM_2 to the direction of $\mathbf{v}_o$ of the Earth's velocity and the arm MM_1 to the perpendicular direction. The difference in time becomes $\delta t' = t'_1 - t'_2 = 2d_1\gamma/c - 2d_2\gamma^2/c$. If i is the interference fringe width, the rotation produces a displacement of the fringes by a distance

$$x = (i/\lambda)\,(c\,\delta t - c\,\delta t') = 2(i/\lambda)(d_1 + d_2)\,(\gamma^2 - \gamma) \approx (i/\lambda)\,(d_1 + d_2)\beta^2\,. \qquad [13.16]$$

Taking $d_1 + d_2 = 11$ m and $\lambda = 300$ nm, this analysis expects a displacement of $0.37i$. The experiment has shown no displacement, although the set-up was able to detect a displacement as small as 0.01 i. This negative result has shown that the Galilean transformation of velocity is not valid in optics, neither in the corpuscular theory of light, nor in wave theory. It has shown also that it is impossible to detect experimentally the ether. Moreover, this hypothetical medium should have some contradictory properties.

13.4. Special theory of relativity

The discrepancy of the results of the 19th Century experiments with the law of transformation of velocities has a deep theoretical origin. Indeed, light waves are electromagnetic. The fundamental laws of electromagnetism are Maxwell's equations, which are non-covariant in the Galilean transformation. In particular, these equations predict a velocity of electromagnetic waves in vacuum equal to c

independently of the inertial frame of reference. Comparing with sound, for instance, it is possible to formulate a theory for the speed of sound in the rest frame of the medium. In the case of electromagnetic waves in vacuum, it does not make sense to speak of the "rest frame of vacuum". All attempts to modify Maxwell's equations to make them covariant in the Galilean transformation have failed. In 1892, Lorentz proposed to modify the Galilean transformation to make Maxwell's equations covariant. He obtained the famous *Lorentz transformation*, according to which, both space coordinates and the time of events depend on the inertial frame of reference. An important consequence of this transformation is the universality (or invariance) of the speed of light in vacuum. This explains the negative result of Michelson-Morley experiment.

In 1905, Einstein followed another and more radical approach. Analyzing the basic physical concepts of space and time, starting with the simple case of the simultaneity of events, he concluded that time cannot be absolute, but relative. To measure time, synchronized clocks are assumed to be distributed everywhere in the Universe and the synchronization may be carried out at different places only by the exchange of light signals, which can propagate even in vacuum. This definition of time led him to the formulation of the *special theory of relativity*, based on two principles:

– *the principle of relativity*: any physical law can be written in a covariant mathematical form (i.e. valid in all inertial frames);

– *the principle of universality of the speed of light in vacuum*: in vacuum, light propagates isotropically with the same speed c in all inertial frames of reference.

The first principle generalizes the principle of relativity to all physics (not just mechanics). According to our discussion of section 13.1, the physical quantities must have well-defined transformation properties from one inertial frame to another and similarly for the equations, which formulate the physical law. In Einstein's formulation, the universality of the speed of light in vacuum must be considered as a logically required "principle" that allows time to be defined. Actually, there are several direct experimental verifications of this universality.

Using these two principles, Einstein derived Lorentz transformation. In the case of an inertial frame S and S' such that their axes are parallel and they coincide at $t = t' = 0$ and S' is moving with a velocity $v_0 \equiv \beta c$ in the direction Oz parallel to Oz, this transformation is

$$x = x', \qquad y = y', \qquad z = \gamma(z' + v_0 t'), \qquad t = \gamma(t' + v_0 z'/c^2)\,, \qquad [13.17]$$

where $\gamma = 1/\sqrt{1-\beta^2}$. If the transformation velocity v_o is small compared to c, neglecting terms of the order of $1/c^2$, the Lorentz transformation [13.17] reduces to the Galilean transformation. The inverse transformation of [13.17] is obtained by changing v_o into $-v_o$:

$$x' = x, \qquad y' = y, \qquad z' = \gamma(z - v_o t), \qquad t' = \gamma(t - v_o z/c^2). \qquad [13.18]$$

If the velocity of S' with respect to S is in an arbitrary direction of unit vector $\mathbf{e}_o$, we write $\mathbf{v}_o = v_o\mathbf{e}_o$. It may be shown that the Lorentz transformation may be written as

$$\mathbf{r} = \mathbf{r}' + (\gamma - 1)(\mathbf{r}'.\mathbf{e}_o)\mathbf{e}_o + \gamma\mathbf{v}_o t', \qquad t = \gamma(t' + \mathbf{v}_o.\mathbf{r}'/c^2). \qquad [13.19]$$

The Lorentz transformation treats space and time as a single entity called *space-time*. A point of space-time represents an *event* specified by its four coordinates x, y, z, and ct. Two events (1) and (2) are separated by an *interval*

$$\Delta S^2 = (x_1 - x_2)^2 + (y_1 - y_2)^2 + (z_1 - z_2)^2 - c^2(t_1 - t_2)^2 \equiv \Delta\mathbf{r}^2 - c^2\,\Delta t^2. \qquad [13.20]$$

An event may be the position of a particle at a given time. Note that the interval ΔS^2 may be positive, negative, or equal to zero.

The Lorentz transformation is linear; thus, it holds for the difference of the coordinates of two events Δx, Δy, Δz, and Δt. It implies that the period of time Δt and the distance Δr separating two events are relative quantities but the interval ΔS^2 between them is invariant (that is, the same in all inertial frames) contrarily to Galilean physics, where both Δt and Δr are invariant. In particular, if the events represent the space coordinates and the timing of a photon, which moves at the velocity c, we must have $\Delta r = c\Delta t$ and the interval $\Delta S^2 = \Delta r^2 - c^2\Delta t^2$ is equal to zero in all frames.

As a direct consequence of Lorentz transformation, we mention the *time dilation*. If two events occur at the same space point in S' ($\Delta\mathbf{r}' = 0$) and they are separated by a time $\Delta t'$, the transformation [13.17] gives $\Delta x = 0$, $\Delta y = 0$, $\Delta z = \gamma v_o \Delta t'$ and

$$\Delta t = \gamma\,\Delta t' > \Delta t'. \qquad [13.21]$$

The time dilation holds for all phenomena, in particular the readings of clocks and the lifetime of particles (and even the life of human beings). The time dilation is verified experimentally to very high precision even in our daily telephone call and use of the global positioning system (GPS) via satellites.

Another consequence of the Lorentz transformation is the contraction of length in the direction of motion. Consider a rod, that is moving with a velocity v_o in the direction of its length taken along Oz. The length of the rod in the frame S of the observer is the distance Δz separating two events, which occur simultaneously at its ends ($\Delta t = 0$). The Lorentz transformation [13.18] gives $\Delta x' = \Delta x = 0$, $\Delta y' = \Delta y = 0$ and $\Delta z' = \gamma(\Delta z - c\beta\Delta t) = \gamma\Delta z$ for the difference of coordinates of these events in the proper frame S' of the rod (i.e. the frame in which the rod is at rest). The length of the rod in S is related to its proper length by the relationship

$$\Delta z = \Delta z'/\gamma < \Delta z'. \qquad [13.22]$$

Thus, it is contracted in the ratio $1/\gamma$. A similar analysis shows that, if the length of the rod is perpendicular to its velocity, it is not contracted. In general, an extended object is contracted in the direction of its motion and not in the transverse directions. Thus, its volume is contracted in the same ratio $1/\gamma$, its angles and its shape are consequently relative, its mass density is increased, etc. The length contraction explains the negative result of Michelson-Morley experiment.

As our concepts (such as the time duration and distances) and, consequently, the principles of the Galilean and Newtonian mechanics are based on our daily observation of objects whose velocity is much lower than c, the relativistic effects seem to be unusual or paradoxical. Actually, it is firmly established that any physical theory concerning phenomena that involve velocities comparable to c (in other words, high-energy effects) must be formulated in accordance with the special theory of relativity.

13.5. Four-dimensional formalism

In a change of the frame of reference, the transformation of the space-time coordinates of an event is similar to the transformation of the ordinary three-dimensional space coordinates. Thus, it is convenient to treat space-time as a four-dimensional vector space. An event is then represented by a *four-vector* that we write as a bold-faced, overlined symbol

$$\overline{\mathbf{x}} = (x, y, z, ct) \equiv (x_1, x_2, x_3, x_4). \qquad [13.23]$$

Sometimes, we write explicitly the coordinates x_μ with Greek indices (μ, α, etc.) taking the values 1, 2, 3 and 4. Ordinary three-dimensional vectors are represented by non-overlined, bold-faced symbols ($\mathbf{x}$, $\mathbf{A}$, etc.) of components x_i, A_i, etc., with Latin indices taking the values 1, 2 and 3.

A Lorentz transformation is represented by a 4×4 matrix $[\overline{\mathbf{L}}]$ of elements $L_{\mu\nu}$, such that the transformation of the four-coordinates x_μ may be written in the four-vector notation

$$\overline{\mathbf{x}} = \overline{\mathbf{L}}\,\overline{\mathbf{x}'}, \quad \text{i.e.} \quad x_\mu = \Sigma_\alpha L_{\mu\alpha} x'_\alpha \quad \text{with} \quad [\overline{\mathbf{L}}] = \begin{bmatrix} 1 & 0 & 0 & 0 \\ 0 & 1 & 0 & 0 \\ 0 & 0 & \gamma & \gamma\beta \\ 0 & 0 & \gamma\beta & \gamma \end{bmatrix}. \qquad [13.24]$$

In the equation $x_\mu = \Sigma_\alpha L_{\mu\alpha} x'_\alpha$, the repeated summation index α is a dummy (or a contraction) index; it may be renamed ν, β, etc. The μ index is fixed for the equation; it may be renamed only in both sides of the equation.

Four quantities A_1, A_2, A_3, A_4 are the components of a four-vector $\overline{\mathbf{A}}$, if they transform like the four coordinates x_μ in a Lorentz transformation. In the case of a transformation of velocity $v_o = c\beta$ in the direction Oz, for instance, we must have

$$A'_1 = A_1, \qquad A'_2 = A_2, \qquad A'_3 = \gamma(A_3 - \beta A_4), \qquad A'_4 = \gamma(\beta A_3 - A_4). \quad [13.25]$$

We define the *norm* of a four-vector $\overline{\mathbf{A}}$ and the *four-scalar product* of two four-vectors $\overline{\mathbf{A}}$ and $\overline{\mathbf{B}}$ as

$$(\overline{\mathbf{A}}\,.\,\overline{\mathbf{A}}) \equiv A_\mu A_\mu = \mathbf{A}^2 - (A_4)^2, \qquad (\overline{\mathbf{A}}\,.\,\overline{\mathbf{B}}) \equiv A_\mu B_\mu = (\mathbf{A.B}) - A_4 B_4. \quad [13.26]$$

It is easy to verify that they are invariant in a Lorentz transformation. We say that they are *four-scalars*. For instance, the norm of the four-vector Δx_μ separating two events is the interval of the events and it is invariant. Contrarily to the three-dimensional space, the norm of a four-vector may be positive, negative, or equal to zero.

A *four-scalar field* is represented by a single function $f(\mathbf{r}, t)$ that we write as $f(x_\mu)$ or $f(\overline{\mathbf{x}})$. It is invariant in a Lorentz transformation, that is,

$$f'(x'_\mu) = f(x_\mu) \qquad \text{with} \qquad x_\mu = L_{\mu\nu} x'_\nu. \qquad [13.27]$$

Similarly, a *four-vector field* is represented by four components $F_\mu(\mathbf{r}, t) \equiv F_\mu(\overline{\mathbf{x}})$ that transform exactly like the x_μ, that is,

$$F_\mu(\overline{\mathbf{x}}) = L_{\mu\nu}\, F'_\nu(\overline{\mathbf{x}'}) \qquad \text{with} \quad x_\mu = L_{\mu\nu} x'_\nu. \qquad [13.28]$$

Note that the components $F_\mu(\bar{\mathbf{x}})$ are linear combinations of the components $F'_\nu(\bar{\mathbf{x}}')$ where $\bar{\mathbf{x}}$ and $\bar{\mathbf{x}}'$ are related by the same Lorentz transformation as F_μ. We may define also *four-tensor fields* of various ranks $T_{\mu\nu\alpha\ldots}$. It is possible to add or compare tensors of the same rank. The tensorial rank of a tensor or a tensorial expression is the number on non-dummy indices: a four-scalar has no non-dummy index (rank 0), a four-vector has a single non-dummy index (rank 1), etc.

The derivatives with respect to the space-time coordinates x_μ

$$\partial_\mu \equiv \left(\frac{\partial}{\partial x_1}, \frac{\partial}{\partial x_2}, \frac{\partial}{\partial x_3}, \frac{\partial}{\partial x_4}\right) \equiv \left[\nabla, \frac{1}{c}\frac{\partial}{\partial t}\right]. \qquad [13.29]$$

are the components of a four-vector operator. Acting on a four-scalar field $f(\bar{\mathbf{x}})$, it gives a four-vector field $\partial_\mu f$, called the *four-gradient* of *f*. Acting on the components of a four-vector field $F_\mu(\bar{\mathbf{x}})$ with contraction of μ, we get a four-scalar $\partial_\mu F_\mu(\bar{\mathbf{x}})$, called *four-divergence*. A repeated action of this operator (with contraction on μ) defines the *d'Alembertian*:

$$\Box \equiv \partial_\mu\partial_\mu \equiv \nabla^2 - \frac{1}{c^2}\frac{\partial^2}{\partial t^2}. \qquad [13.30]$$

This is a four-scalar operator. Acting on a four-scalar field $f(\bar{\mathbf{x}})$, it gives another four-scalar field $\partial_\mu\partial_\mu f(\bar{\mathbf{x}})$ and acting on a four-vector field, it gives another four-vector field $\partial_\mu\partial_\mu F_\nu(\bar{\mathbf{x}})$.

Consider the integrals of four-scalar, four-vector, or four-tensor fields over a four-volume of space-time $d^4x \equiv dx\, dy\, dz\, dt$, of the form

$$A = \iiiint d^4x\ f(\bar{\mathbf{x}}), \qquad A_\nu = \iiiint d^4x\ F_\nu(\bar{\mathbf{x}}), \quad \text{etc.} \qquad [13.31]$$

They are a four-scalar A, a four-vector A_μ, etc. We note that the Jacobian of the Lorentz transformation is equal to 1, hence

$$d^4x = \frac{D(x,y,z,t)}{D(x',y',z',t')}\ d^4x' = d^4x'. \qquad [13.32]$$

A physical theory is said to be *relativistic* if it is formulated in accordance with the principles of the special theory of relativity. For this, it is necessary that all its physical quantities have well-defined transformation properties under Lorentz transformations; these quantities must be four-scalars, four-vectors or four-tensors. In relativistic quantum field theory, we may also have *spinors*. If the equations of the theory are expressed in terms of such quantities, they are *covariant* in Lorentz transformations. We say that the theory is *invariant* in Lorentz transformations.

13.6. Elements of relativistic mechanics

From the Lorentz transformation [13.17], we deduce the law of transformation of the velocity of a particle defined by its components dx/dt, dy/dt, and dz/dt in a frame S and dx'/dt', dy'/dt', and dz'/dt' in a frame S'

$$v_x = \frac{dx}{dt} = \frac{v'_x}{\gamma(1+\beta v'_z/c)}, \qquad v_y = \frac{dy}{dt} = \frac{v'_y}{\gamma(1+\beta v'_z/c)}, \qquad v_z = \frac{dz}{dt} = \frac{v'_z + c\beta}{1+\beta v'_z/c}. \quad [13.33]$$

Particularly, if $\mathbf{v}'$ is parallel to $\mathbf{v}_o$, we find that $\mathbf{v}$ is parallel to $\mathbf{v}_o$ and the law of transformation of velocity (also called *law of addition of velocities*) reduces to

$$v = \frac{v' + v_o}{1 + \beta v'/c}. \quad [13.34]$$

For small velocities compared to c, equations [13.33] reduce to the Galilean equation of addition of velocities [13.2] if we neglect terms of the second order in $1/c$. However, the relativistic law gives completely different results than the Galilean law, if the velocities are comparable to c. In particular, in the limit $v' \rightarrow c$ or $v_o \rightarrow c$, we find $v \rightarrow c$. On the other hand, Lorentz transformation becomes meaningless if $v_o > c$ as γ becomes imaginary. Thus, the speed of light is the upper limit of the velocity of particles. This is also the upper limit for the speed of transfer of any physical quantity (energy, momentum, etc.), information, or interaction.

The law of transformation [13.33], being different from that of x, y, and z, the components of the velocity are not the first three components of a four-vector. This is due to the fact that dt in the definition of the velocity $v_i = dx_i/dt$ is not a scalar quantity. We may define a *four-vector velocity* V_μ, if we replace dt by an infinitesimal quantity $d\tau$, which is a four-scalar having the dimension of time. Recalling that the norm $ds^2 = dx_\mu\, dx_\mu$, is a four-scalar, we may calculate it in any inertial frame. In the proper frame (S') of the particle (in which $d\mathbf{r}' = 0$) and the frame S of the observer (in which $d\mathbf{r} = \mathbf{v}\, dt$, where $\mathbf{v}$ is the velocity of the particle), we may write

$$(dS^2)_{S'} = -c^2\, dt'^2, \qquad (dS^2)_S = d\mathbf{r}^2 - c^2 dt^2 = (\mathbf{v}^2 - c^2)dt^2. \quad [13.35]$$

The time t' is the *proper time* of the particle, usually called τ. Comparing the two expressions of dS^2, we deduce that

$$d\tau = \sqrt{1 - v^2/c^2}\; dt. \quad [13.36]$$

As $dS^2 = -c^2\, d\tau^2$ is a four-scalar, we deduce that the infinitesimal proper time $d\tau$ is a four-scalar. Thus, we define the four-vector velocity as

$$V_\mu = \frac{dx_\mu}{d\tau}. \qquad [13.37]$$

Writing explicitly the space and time components of V_μ, we find

$$V_i = \frac{dx_i}{dt\sqrt{1-v^2/c^2}} = \frac{v_i}{\sqrt{1-v^2/c^2}}, \qquad V_4 = \frac{dx_4}{dt\sqrt{1-v^2/c^2}} = \frac{c}{\sqrt{1-v^2/c^2}}. \qquad [13.38]$$

The norm of this four-vector may be calculated in any frame of reference. Particularly, in the proper frame of the particle, we find $V'_i = 0$ and $V'_4 = c$, hence

$$V_\mu\, V_\mu = \mathbf{V}'^2 - (V'_4)^2 = \mathbf{V}^2 - (V_4)^2 = -\,c^2. \qquad [13.39]$$

Admitting that the mass m is the same in all inertial frames (i.e. m is a four-scalar), the four quantities

$$P_\mu = mV_\mu \qquad [13.40]$$

are the components of a four-vector, whose space and time components are

$$P_i = \frac{mv_i}{\sqrt{1-v^2/c^2}}, \qquad P_4 = \frac{mc}{\sqrt{1-v^2/c^2}}. \qquad [13.41]$$

To interpret these quantities, let us write them as power series of (v/c), we find

$$P_i = mv_i + \tfrac{1}{2}(v/c)^2\, mv_i + \ldots, \qquad cP_4 = mc^2 + \tfrac{1}{2}\, mv^2 + \ldots \qquad [13.42]$$

We recognize in the expression of P_i, the components of the classical momentum $p_i = mv_i$. This allows P_i to be interpreted as the three components of the *relativistic momentum* of the particle. In the series of cP_4, we recognize the kinetic energy $\tfrac{1}{2}mv^2$. Thus, we interpret cP_4 as the *relativistic energy* of the particle and we write

$$W = cP_4 = \frac{mc^2}{\sqrt{1-v^2/c^2}}. \qquad [13.43]$$

In particular, if the particle is at rest

$$W_o = mc^2. \qquad [13.44]$$

Contrarily to classical physics, a particle at rest has energy. In fact this is an enormous energy; the rest energy of 1 g of matter is 9×10^{13} J! The *relativistic kinetic energy* U_K is defined as the difference between the energy W of the body in motion and its energy at rest:

$$U_K = \frac{mc^2}{\sqrt{1-v^2/c^2}} - mc^2. \qquad [13.45]$$

All experiments in nuclear physics and particle physics show that the mass may be transformed into energy according to [13.44] and that the relativistic momentum and energy of isolated systems as defined by [13.41] and [13.43] are conserved.

The velocity of a particle may be expressed in terms of $\mathbf{P}$ and W by the relation

$$\mathbf{v} = c^2\mathbf{P}/W. \qquad [13.46]$$

Substituting this expression of $\mathbf{v}$ in the expression of W, we find the relation between the energy and the momentum of a particle of mass m

$$W = \sqrt{m^2c^4 + c^2\mathbf{P}^2} \quad \text{or} \quad W^2 = m^2c^4 + c^2\mathbf{P}^2. \qquad [13.47]$$

We may introduce the concept of force as the rate of variation of the relativistic momentum in time

$$\mathbf{f} = d\mathbf{P}/dt. \qquad [13.48]$$

Using the expression [13.47], we deduce that $W\,dW = c^2\,\mathbf{P}.d\mathbf{P}$, hence a variation of the relativistic energy (or the kinetic energy),

$$dW = dU_K = c^2\,\mathbf{P}.\mathrm{d}\mathbf{P}/W = d\mathbf{P}.\mathbf{v} = dt\,\mathbf{f}.\mathbf{v} = \mathbf{f}.d\mathbf{r}, \qquad [13.49]$$

where we have used [13.46] and [13.48]. Thus, the variation of W (or U_K in the case of constant masses) is equal to the work of the force $\mathbf{f}$.

Using the law of transformation of $\mathbf{P}$, we deduce the law of transformation of $\mathbf{f}$

$$\begin{aligned}\mathbf{f} = d\mathbf{P}/dt &= [d\mathbf{P}'/dt' + (\gamma-1)(\mathbf{e}_o.d\mathbf{P}'/dt')\,\mathbf{e}_o + \gamma(\mathbf{v}_o/c^2)\,dW'/dt']\,/\,(dt/dt') \\ &= [\,\mathbf{f}' + (\gamma-1)(\mathbf{e}_o.\mathbf{f}')\,\mathbf{e}_o + \gamma\mathbf{v}_o(\mathbf{f}'.\mathbf{v}')/c^2]/\gamma[1 + (\mathbf{v}_o.\mathbf{v}')/c^2]\end{aligned} \qquad [13.50]$$

since $d\mathbf{P}'/dt' = \mathbf{f}'$, $dW'/dt' = \mathbf{f}'.\mathbf{v}'$ and $dt = \gamma[dt' + (\mathbf{v}_o.d\mathbf{r}')/c^2] = \gamma\,dt'\,[1 + (\mathbf{v}_o.\mathbf{v}')/c^2]$.

The three components of the force, defined by [13.48], are not the first three components of a four-vector. For this reason, $\mathbf{f}$ is called *Newton's force*. It should not be confused with the "classical Newton's force" $\mathbf{f}_c = d\mathbf{p}/dt$, where $\mathbf{p} = m\mathbf{v}$. Using the proper time τ of the particle, we may define a *four-vector force* $\overline{\mathbf{F}}$ of components

$$F_\mu = dP_\mu/d\tau \,. \tag{13.51}$$

Explicitly, the components of this four-vector are

$$\mathbf{F} = \frac{d\mathbf{P}}{d\tau} = \frac{d\mathbf{P}/dt}{\sqrt{1-v^2/c^2}} = \frac{\mathbf{f}}{\sqrt{1-v^2/c^2}}, \quad F_4 = \frac{dP_4}{d\tau} = \frac{dW/dt}{c\sqrt{1-v^2/c^2}} = \frac{\mathbf{f.v}}{c\sqrt{1-v^2/c^2}}. \tag{13.52}$$

Now, let us consider the case of a massless particle. Apparently, the expressions [13.41] seem to indicate that $\mathbf{P} = 0$ and $W = 0$, which mean that the particle carries no momentum and no energy, contrarily to experiment (in the case of photons, for instance). In fact, $\mathbf{P}$ and W may not be equal to zero if the velocity of the particle is equal to c. Then the denominator of $\mathbf{P}$ and W is equal to zero and these expressions are meaningless. Thus a massless particle must be always moving with the speed of light in vacuum c. Note also that, if v exceeds c, $\mathbf{P}$ and W become imaginary; Thus, the velocity of a particle cannot exceed c. On the other hand, setting $m = 0$ in [13.47], we find the relation

$$W = cP \qquad (\text{if } m = 0). \tag{13.53}$$

In the case of photons, this relation may be obtained from the energy and momentum densities for electromagnetic waves (see section 10.5). In quantum theory, a wave of frequency $\tilde{\nu}$ is associated with photons of energy $W = h\tilde{\nu}$, where $h = 6.626\ 075 \times 10^{-34}$ J.s is Planck's constant. According to equation [13.53], the momentum of the photons is $p = W/c = h\tilde{\nu}/c = h/\lambda$. This is de Broglie's relation.

13.7. Special relativity and wave theory*

Using spherical coordinates, the velocity of a particle is specified by its magnitude v, its angle θ with Oz (i.e. with $\mathbf{v}_o$) and its azimuthal angle φ about Oz. The Lorentz transformation for these quantities is

$$\varphi = \varphi', \quad \tan\theta = \frac{v'\sin\theta'}{\gamma(v'\sin\theta' + c\beta)}, \quad v = \frac{\sqrt{v'^2 + v_o^2 + 2v_o v'\cos\theta' - (v'\beta)^2\sin^2\theta'}}{1+\beta(v'/c)\sin\theta'}. \tag{13.54}$$

In the case of a wave, the invariance of the phase may be written as

$$\omega(t - \mathbf{e.r}/v_p) = \omega'(t' - \mathbf{e}'.\mathbf{r}'/v'_p). \tag{13.55}$$

Writing the scalar products **e.r** and **e′.r′** explicitly, and expressing x, y, z and t in terms of x', y', z', and t' by using Lorentz transformation [13.17], we find

$$\omega'\{t' - (1/v'_p)[x' \sin\theta' \cos\varphi' + y' \sin\theta' \cos\varphi' + z' \cos\theta']\}$$
$$= \omega\{\gamma(t' + \beta z'/c) - (1/v_p)[x' \sin\theta \cos\varphi + y' \sin\theta \cos\varphi + \gamma(z' + c\beta t') \cos\theta]\}. \quad [13.56]$$

The coefficients of x', y', z' and t' must be equal on both sides, hence the relations

$$\varphi' = \varphi, \qquad \tan\theta' = \frac{\sin\theta}{\gamma(\cos\theta - \beta v_p/c)}, \quad [13.57]$$

$$v'_p = \frac{v_p - v_o \cos\theta}{\sqrt{(\beta v_p/c - \cos\theta)^2 + (1-\beta^2)\sin^2\theta}}, \quad [13.58]$$

$$\omega' = \omega\gamma\,[\,1 - (c\beta/v_p)\cos\theta\,]. \quad [13.59]$$

Contrarily to classical wave theory, the relations [13.57] imply a modification of the direction of propagation. In the case of starlight aberration, if $\theta = 90°$, we get

$$\tan\theta' = -\,1/\gamma\beta, \quad [13.60]$$

which corresponds to an inclination $\alpha = \theta' - 90°$ given by

$$\tan\alpha = -\,\gamma\beta. \quad [13.61]$$

This result may be also obtained by using the relativistic law of transformation of the direction of the velocity of a particle [13.54]. Thus, the equivalence between the wave theory and the corpuscular theory of light is recovered for the aberration phenomenon.

Equation [13.59] is the expression of the *relativistic Doppler effect*; we may write it as

$$\omega' = \omega\gamma\,[1 - \mathbf{e}.\mathbf{v}_o/v_p\,]. \quad [13.62]$$

Particularly, in the case of propagation of light in vacuum ($v_p = c$), this relation shows that the Doppler effect depends only on the velocity $\mathbf{v}_o$ of S' with respect to S. Thus, the effect depends only on the relative velocity of the source with respect to the observer, not on the velocity of each one, contrarily to the classical expression [13.12]. Indeed, the concept of the ether being abandoned, the velocity of the source or the observer with respect to the medium of propagation is meaningless in this case. An experiment by Ives and Stilwell in 1938 has shown that the Doppler shift of light emitted by atoms in motion is in perfect agreement with the relativistic expression [13.62] and not with the classical expression [13.12].

We note that the Doppler effect multiplies the angular frequency of a simple harmonic wave by a factor $\rho_D = \gamma\,[1 - (\mathbf{e}.\mathbf{v}_o)/v_p]$, thus the period by $1/\rho_D$. If the wave has a period T without being necessarily simple harmonic, Fourier theory allows it to be considered as a superposition of simple harmonic waves of angular frequencies $n\omega$. The Doppler effect makes these angular frequencies $n\omega\rho_D$. If the medium is non-dispersive, ρ_D does not depend on the frequency, and the profile of the wave is not affected by the Doppler effect, only its period becomes T/ρ_D. In the case of a signal, its central frequency and the band width of its spectrum are multiplied by ρ_D. Thus, its time duration is divided by ρ_D and so is the time interval separating two signals. The Doppler effect is thus related to the transformation of time in special relativity. Contrarily to the classical theory, if the relative motion is perpendicular to the direction of propagation ($\mathbf{v}_o.\mathbf{e} = 0$), the relativistic theory predicts a *transverse Doppler effect* $\omega' = \omega\gamma$, which is identical to time dilation.

The law of transformation of the angular frequency [13.59] and the law of transformation of the phase velocity [13.58] imply a transformation of the wavelength according to the relation

$$\lambda' = \frac{v'_p}{\nu'} = \frac{\lambda}{\gamma\sqrt{(\beta v_p/c - \cos\theta)^2 + (1-\beta^2)\sin^2\theta}}. \qquad [13.63]$$

In particular, in the case of a light wave in vacuum, $v_p = c$, this equation reduces to

$$\lambda' = \frac{\lambda}{\gamma(1 - \beta\cos\theta)}. \qquad [13.64]$$

This result may be obtained easily from the law of transformation of the angular frequency by using the relation $\omega = 2\pi c/\lambda$ and $\omega' = 2\pi c/\lambda'$. Contrarily to the Galilean transformation, the wavelength is not the same in all inertial frames. In special relativity, time duration and distances depend on the frame.

The law of transformation of the phase velocity [13.58] is not similar to the law of transformation of the velocity of a particle [13.33]. Thus, contrarily to the velocity of a particle, nothing requires that the phase velocity be less than the velocity of light in vacuum. This is not surprising, since the phase velocity does not correspond to the propagation of physical quantities, such as energy and momentum. Physical quantities are transferred at the group velocity, which must always be less than the velocity of light in vacuum, in order to respect the causality principle.

To derive the law of transformation of the group velocity, we write the invariance of the phase as

$$\omega t - k_x x - k_y y - k_z z = \omega' t' - k'_x x' - k'_x y' - k'_x z' \,. \qquad [13.65]$$

Using the transformation [13.17] to express x, y, z, and t in terms of x', y', z', and t' and identifying the coefficients, we get the law of transformation of the wave vector **k** and the angular frequency ω:

$$k'_x = k_x, \qquad k'_y = k_y, \qquad k'_z = \gamma\,(k_z - \beta\omega/c), \qquad \omega' = \gamma\,(\omega - c\beta k_z)\,. \qquad [13.66]$$

These relations express that k_x, k_y, k_z, and ω/c are the components of a four-vector, just to make $\mathbf{k.r} - \omega t$ a scalar product of two four-vectors, thus invariant in Lorentz transformation. Writing the inverse of the last equation [13.66], i.e. $\omega = \gamma(\omega' + c\beta k'_z)$ and using the dispersion relation to express ω as a function of **k** and ω' as a function of $\mathbf{k}'$, the relations [13.66] may be written as

$$k'_x = k_x, \quad k'_y = k_y, \quad k'_z = \gamma\,[k_z - \beta\omega(\mathbf{k})/c], \quad \omega = \gamma[\omega'(\mathbf{k}') + c\beta k'_z]\,. \qquad [13.67]$$

Differentiating ω with respect to k_z, for instance, we find

$$\frac{\partial\omega}{\partial k_z} = \gamma[\frac{\partial\omega'}{\partial k'_x}\frac{\partial k'_x}{\partial k_z} + \frac{\partial\omega'}{\partial k'_y}\frac{\partial k'_y}{\partial k_z} + \frac{\partial\omega'}{\partial k'_z}\frac{\partial k'_z}{\partial k_z} + v_o\frac{\partial k'_z}{\partial k_z}] = \gamma[v'_{gz}\frac{\partial k'_z}{\partial k_z} + v_o\frac{\partial k'_z}{\partial k_z}].$$

Thus we can write

$$v_{gz} = \gamma^2(v'_{gz} + v_o)[1 - \beta v_{gz}/c], \qquad \text{i.e. } v_{gz}(1-\beta^2) = (v'_{gz} + v_o)[1 - \beta v_{gz}/c]\,. \qquad [13.68]$$

We deduce that

$$v_{gz} = \frac{v'_{gz} + v_o}{(1 + \beta v'_{gz}/c)}\,. \qquad [13.69]$$

We find a law of transformation similar to that of the third component of the velocity of a particle (see equation [13.33]). Similarly, we may derive the transformation laws for v_{gx} and v_{gy}. Thus, the group velocity of a wave transforms exactly like the velocity of a particle.

Considering a wave packet, the energy and the other physical quantities of the wave are obviously localized in the same region as the wave packet. Thus, they are transferred with the same group velocity $\mathbf{v}_g$ as the wave packet. Similarly, if a modulated wave is used to transmit information, the velocity of this transmission is the group velocity. As the group velocity transforms like the velocity of a particle, the group velocity cannot exceed the velocity of light in vacuum.

13.8. Elements of relativistic electrodynamics

We have seen that the instantaneous interaction-at-a-distance disagrees with the principle of causality. Thus, all interactions must be local, that is, mediated by fields propagating with finite velocities. On the other hand, for velocities (of particles and waves) comparable to the velocity of light, relativistic theories must be used. Electromagnetism is a typical theory, which is formulated in terms of fields and it analyzes phenomena with velocities comparable (or equal) to the velocity of light. Thus, it is not surprising that special relativity was created from the contradictions of electromagnetism with the principles of the Galilean-Newtonian mechanics. For a theory to be relativistic, it is necessary that its laws be formulated in terms of quantities and functions that are four-scalars, four-vectors or four-tensors. In fact, electromagnetism was a relativistic theory before the formulation of special relativity, although the laws of electromagnetism are not usually written in a manifestly covariant form. In this section we introduce some elements of the relativistic formulation of electromagnetism, leaving many details to specialized texts. The association of relativistic electromagnetism with quantum theory produces quantum electrodynamics, which agrees perfectly with experiment.

A) Transformation laws of charge and current densities

Exactly like mass, electric charge is a characteristic quantity of particles. Experiments show that it is independent of their motion or the motion of the observer; thus, it is independent of the inertial frame. The charge contained in an element of volume $d^3\mathbf{r} = dx\ dy\ dz$ near a point $\mathbf{r}$ and time t is $dq = q_V(\mathbf{r}, t)\ dx\ dy\ dz$. This charge being invariant in a Lorentz transformation, comparing it with the invariant $d^4x \equiv dt\ dx\ dy\ dz$ (see [13.32]), we deduce that the charge density q_V transforms like dt, i.e. the fourth component of a four-vector.

The current density $\mathbf{j}(\mathbf{r}, t)$ is related to the charge that is intercepted by an element of area $d\mathcal{S}$ normal to the unit vector $\mathbf{n}$ by the relation $dq = (\mathbf{n}.\mathbf{j})\ d\mathcal{S}\ dt$. Consider, for instance, an element of area $d\mathcal{S} = dx\ dy$ parallel to the (x, y) plane. The intercepted charge during dt is $dq = (\mathbf{e}_z.\mathbf{j})\ d\mathcal{S}\ dt = j_3\ dx\ dy\ dt$ and it is four-invariant. Comparing it with the invariant d^4x, we deduce that j_3 is the third component of a four-vector. Similarly, j_1 and j_2 are, respectively, the first and the second components of the same four-vector. Thus, the four quantities j_1, j_2, j_3 and cq_V are the four components of the *four-vector current density* J_μ

$$J_k \equiv j_k, \qquad J_4 \equiv cq_v \qquad (k = 1, 2, 3)\,. \qquad [13.70]$$

This implies that the four-vector ($\mathbf{j}$, cq_v) transforms exactly like ($\mathbf{r}$, ct). Using the transformation [13.19], the law of transformation of $\mathbf{j}$ and q_v may be written as

$$\mathbf{J} = \mathbf{J}' + (\gamma-1)(\mathbf{J}'.\mathbf{e}_o)\,\mathbf{e}_o + \gamma\, q'_v\,\mathbf{v}_o, \qquad q_v = \gamma\,[q'_v + (\mathbf{J}'.\mathbf{v}_o)/c^2]\,. \qquad [13.71]$$

The inverse transformation is obtained by changing $\mathbf{v}_o$ into $-\mathbf{v}_o$:

$$\mathbf{J}' = \mathbf{J} + (\gamma - 1)(\mathbf{J}.\mathbf{e}_o)\,\mathbf{e}_o - \gamma q_v \mathbf{v}_o, \qquad q'_v = \gamma\,[q_v - (\mathbf{J}.\mathbf{v}_o)/c^2]\,. \qquad [13.72]$$

Particularly, if the charges are at rest in the frame S', the current density $\mathbf{J}'$ is equal to zero. We find in the frame S of the observer

$$\mathbf{J} = \gamma q'_v \mathbf{v}_o = q_v \mathbf{v}_o\,. \qquad q_v = \gamma q'_v. \qquad [13.73]$$

Thus, in the frame of the observer S, we have both charge and current densities. This is the case of a rigid body carrying fixed charges (with respect to the body) and moving with a velocity $\mathbf{v}_o$ with respect to the observer. S' is then the proper frame of the body. This is also the case of a beam of charged particles of velocity $\mathbf{v}_o$ (S' is then the proper frame of the particles). Note that the relation $q_v = \gamma q'_v$ can be understood because of the contraction of volumes in the ratio $1/\gamma$, thus an increase of the charge density in the ratio γ. Also the relation $\mathbf{J} = q_v \mathbf{v}_o$ is expected because charge density q_v in S is moving with the velocity $\mathbf{v}_o$. This current is not due to the displacement of charges with respect to the body (i.e. conduction current) but to the displacement of the body itself, whether it is a dielectric or a conductor (i.e. convection current). In general, if there is a conduction current density $\mathbf{j}'$ as well as a charge density q'_v in S', using [13.71], we may write the current density is S as

$$\mathbf{j} = [\,\mathbf{j}' + (\gamma^{-1} - 1)(\mathbf{j}'.\mathbf{e}_o)\,\mathbf{e}_o] + q_v \mathbf{v}_o \equiv \mathbf{j}_c + q_v\,\mathbf{v}_o\,. \qquad [13.74]$$

This is the sum of a *conduction current density* $\mathbf{j}_c$ and a convection *current density* $q_v \mathbf{v}_o$. Both currents have some common physical properties (production of a magnetic field for instance), but they may have some different properties (for instance, the production of Joule heat by the conduction current but not by the convection current).

Using the four-vector current density, the local equation of conservation of electric charge $\partial_t q_v + \boldsymbol{\nabla}.\mathbf{J} = 0$ may be written as $\partial_t q_v + \Sigma_k\, \partial_k J_k = 0$, i.e.

$$\partial_\mu J_\mu = 0\,. \qquad [13.75]$$

Note that $\partial_\mu J_\mu$ is the contracted product of the four-vector operator ∂_μ and the four-vector J_μ. Thus, it is invariant in Lorentz transformations; if it vanishes in one inertial frame, it must vanish in any other inertial frame.

B) Transformation law of fields and potentials

Assume that we have an electric field $\mathbf{E}(\mathbf{r}, t)$ and a magnetic field $\mathbf{B}(\mathbf{r}, t)$ in an inertial frame S. Let S' be another inertial frame moving with a velocity $\mathbf{v}_o$ with respect to S. To write the expressions of the fields $\mathbf{E}'(\mathbf{r}', t')$ and $\mathbf{B}'(\mathbf{r}', t')$ in S'; we start with the definition of the fields in relation to the relativistic Newton force $\mathbf{f} = d\mathbf{P}/dt$ that the fields exert on a charge q:

$$\mathbf{f} = q[\mathbf{E} + (\mathbf{v} \times \mathbf{B})], \qquad \mathbf{f}' = q[\mathbf{E}' + (\mathbf{v}' \times \mathbf{B}')], \qquad [13.76]$$

where we have used the fact that the charge q is the same in both frames. Using the laws of transformation of velocity and force, we may express $\mathbf{f}$ and $\mathbf{v}$ in the first equation [13.76] in terms of $\mathbf{f}'$ and $\mathbf{v}'$, then substitute the expression $q[\mathbf{E}' + (\mathbf{v}' \times \mathbf{B}')]$ for $\mathbf{f}'$. We get an equation that must be identically verified for any velocity $\mathbf{v}'$. We deduce the law of transformation of the fields:

$$\mathbf{E} = \gamma\mathbf{E}' + (1-\gamma)(\mathbf{E}'.\mathbf{e}_o)\,\mathbf{e}_o - c\gamma(\boldsymbol{\beta}\times\mathbf{B}'), \quad \mathbf{B} = \gamma\mathbf{B}' + (1-\gamma)(\mathbf{B}'.\mathbf{e}_o)\,\mathbf{e}_o + (\gamma/c)(\boldsymbol{\beta}\times\mathbf{E}') \quad [13.77]$$

The inverse transformation is obtained by changing $\mathbf{v}_o$ into $-\mathbf{v}_o$:

$$\mathbf{E}' = \gamma\mathbf{E} + (1-\gamma)(\mathbf{E}.\mathbf{e}_o)\mathbf{e}_o + c\gamma(\boldsymbol{\beta}\times\mathbf{B}), \quad \mathbf{B}' = \gamma\mathbf{B} + (1-\gamma)(\mathbf{B}.\mathbf{e}_o)\mathbf{e}_o - (\gamma/c)(\boldsymbol{\beta}\times\mathbf{E}). \quad [13.78]$$

Decomposing the fields $\mathbf{E}$ and $\mathbf{B}$ into longitudinal components $\mathbf{E}_{//} = (\mathbf{E}.\mathbf{e}_o)\,\mathbf{e}_o$ and $\mathbf{B}_{//} = (\mathbf{B}.\mathbf{e}_o)\,\mathbf{e}_o$ (parallel to $\mathbf{v}_o$) and transverse components $\mathbf{E}_\perp = \mathbf{E} - (\mathbf{E}.\mathbf{e}_o)\,\mathbf{e}_o$ and $\mathbf{B}_\perp = \mathbf{B} - (\mathbf{B}.\mathbf{e}_o)\,\mathbf{e}_o$ (orthogonal to $\mathbf{v}_o$), their laws of transformation are

$$\begin{aligned} &\mathbf{E}_{//} = \mathbf{E}'_{//}, \qquad && \mathbf{E}_\perp = \gamma[\mathbf{E}'_\perp - (\mathbf{v}_o\times\mathbf{B}')], \\ &\mathbf{B}_{//} = \mathbf{B}'_{//}, && \mathbf{B}_\perp = \gamma[\mathbf{B}'_\perp + (\mathbf{v}_o\times\mathbf{E}')/c^2]. \end{aligned} \qquad [13.79]$$

Thus, the longitudinal components of the fields are not modified in the transformation. In particular, if the velocity of transformation $\mathbf{v}_o$ is in the direction of Oz, we find

$$\begin{aligned} &E_1 = \gamma[E'_1 + c\beta B'_2], \qquad && E_2 = \gamma[E'_2 - c\beta B'_1], \qquad && E_3 = E'_3, \\ &B_1 = \gamma[B'_1 - (\beta/c)E'_2], && B_2 = \gamma[B'_2 + (\beta/c)E'_1], && B_3 = B'_3. \end{aligned} \qquad [13.80]$$

These transformation laws show that the fields **E** and **B** cannot be considered as independent fields. For instance, if the field in S is purely electric ($\mathbf{B} = 0$), we find both electric and magnetic fields in S' given by

$$\mathbf{E}' = \gamma\, \mathbf{E} + (1-\gamma)\,(\mathbf{E}.\mathbf{e}_o)\mathbf{e}_o, \qquad \mathbf{B}' = -\,(\beta\gamma/c)\,(\mathbf{e}_o \times \mathbf{E})\,. \qquad [13.81]$$

Similarly, if the field in S is purely magnetic ($\mathbf{E} = 0$), we have both electric and magnetic fields in S' given by

$$\mathbf{E}' = c\gamma\beta\,(\mathbf{e}_o \times \mathbf{B}), \qquad \mathbf{B}' = \gamma\mathbf{B} + (1-\gamma)(\mathbf{B}.\mathbf{e}_o)\,\mathbf{e}_o\,. \qquad [13.82]$$

We verify easily that **E**' and **B**' in [13.81] and [13.82] are orthogonal.

In the non-relativistic limit ($v_o << c$, i.e. $\beta << 1$), we have $\gamma \approx 1 + ½\beta^2$. The law of transformation becomes (to the first order in β):

$$\mathbf{E} \approx \mathbf{E}' - \mathbf{v}_o \times \mathbf{B}', \qquad \mathbf{B} \approx \mathbf{B}' + (\mathbf{v}_o \times \mathbf{E}')/c^2\,. \qquad [13.83]$$

The equations of propagation of the scalar potential V and the vector potential **A** may be written in the Lorentz gauge as

$$\Box V = -\frac{q_v}{\varepsilon_0}, \qquad \Box\mathbf{A} = -\mu_o\mathbf{j} \qquad \text{with} \quad \Box \equiv \partial_\mu\partial_\mu \equiv \nabla^2 - \frac{1}{c^2}\frac{\partial^2}{\partial t^2}\,. \qquad [13.84]$$

Using the relation $\varepsilon_o\mu_o = 1/c^2$ and the four-vector current density $J_v = (\mathbf{j}, cq_v)$, these equations of propagation may be written in a single equation

$$\Box A_v = -\mu_o J_v\,, \qquad [13.85]$$

where we have grouped the three components of **A** and V/c to form a single entity A_v. As with any physical law, the relation [13.85] must be valid in any inertial frame. The d'Alembertian being a four-scalar operator and J_v a four-vector, we deduce that

$$A_\mu = (\mathbf{A}, V/c) \qquad [13.86]$$

is a four-vector called *four-vector potential.* We may then write its law of transformation:

$$\mathbf{A} = \mathbf{A}' + (\gamma-1)(\mathbf{A}'.\mathbf{e}_o)\,\mathbf{e}_o + \gamma V'\boldsymbol{\beta}/c\,, \qquad V = \gamma[V' + c(\mathbf{A}'.\boldsymbol{\beta})]\,. \qquad [13.87]$$

In particular, in the case of a transformation of velocity $\mathbf{v}_o = c\boldsymbol{\beta}$ in the direction Oz, this law of transformation can be explicitly written as

$$A_1 = A'_1, \quad A_2 = A'_2, \quad A_3 = \gamma[A'_3 + V'\beta/c] \quad \text{and} \quad V = \gamma[V' + c\beta A'_3]\,. \qquad [13.88]$$

If the field in S' is purely electric (as in the case of the field of a charge at rest), we may take $\mathbf{A}' = 0$ and find in S

$$\mathbf{A} = \gamma V'\boldsymbol{\beta}/c, \qquad V = \gamma V'. \qquad [13.89]$$

Thus, we find in S both an electric and a magnetic field. The electromagnetic induction is closely related to the transformation law of the fields (see section 8.4).

It may be shown that Maxwell's equations are covariant in Lorentz transformations (problem 13.23). This shows the covariance of electromagnetism.

13.9. Problems

Galilean relativity and waves

P13.1 Let S be the reference frame of the fixed stars (and the Sun) and S' be the Earth's frame moving with a speed v_o = 30 km/s on the ecliptic. For simplicity, assume that the starlight falls normally to the ecliptic and the velocity $\mathbf{v}_o$ is in the direction $O'z'$. In this problem, we use Galilean transformation. **a)** Determine the velocity and the direction of the starlight as seen by the observer on Earth. Verify that the apparent direction of the star makes an angle $\alpha \approx 10^{-4}$ with the normal to the ecliptic, the star describing an ellipse in 1 year. **b)** Considering the wave fronts in both the stars and the Earth's frame, show that the deviation of light cannot be explained by using the phase velocity. Show that it may be explained by using the group velocity.

P13.2 a) Choose the right frames of reference and show the Doppler-Fizeau formula $\omega_O = \omega_S/[1 - (v_S/v_{p(M)}) \cos \theta_O]$ in the case of a moving source and an observer at rest and $\omega_O = \omega_S\,[1 - (v_O/v_{p(M)}) \cos \theta_O]$ in the case of a moving observer and a source at rest. **b)** Show that, if a wave of frequency ω_1 is emitted by an observer and reflected on a moving mirror toward the observer, the reflected wave has a frequency $\omega_2 = \omega_1(1+ v/v_p)/(1- v/v_p)$.

Special relativity

P13.3 An event (1) occurs at the origin at time $t_1 = 0$ and an event (2) occurs at a time t_2 at the point $(0, 0, z_2)$. What are the positions and times of these events as seen by an observer travelling at a speed v_o in the direction Oz? Can he see event (2) precede event (1)? What is the condition on z_2 and t_2 for this does not happen? Interpret this result in terms of causality.

P13.4 To derive Lorentz transformation, we consider an event of space-time coordinates $(x_1, x_2, x_3, x_4 = ct)$ in a frame S and (x'_1, x'_2, x'_3, x'_4) in a frame S'.

According to the principle of inertia, a free particle has a uniform motion in all frames. This is possible if the transformation is linear, of the form

$$x^1 = a^1{}_1 x'^1 + a^1{}_2 x'^2 + a^1{}_3 x'^3 + a^1{}_4 x'^4, \quad x^2 = a^2{}_1 x'^1 + a^2{}_2 x'^2 + a^2{}_3 x'^3 + a^2{}_4 x'^4,$$
$$x^3 = a^3{}_1 x'^1 + a^3{}_2 x'^2 + a^3{}_3 x'^3 + a^3{}_4 x'^4, \quad x^4 = a^4{}_1 x'^1 + a^4{}_2 x'^2 + a^4{}_3 x'^3 + a^4{}_4 x'^4.$$

a) Show that the reflection symmetry with respect to the planes $O'y'z'$ and $Ox'z'$ requires that $a_{12} = a_{13} = a_{14} = a_{21} = a_{23} = a_{24} = 0$. Show that the 90° rotational symmetry about Oz implies that $a_{11} = a_{22}$ and $a_{31} = a_{32} = a_{41} = a_{42} = 0$. **b)** Writing that O' has the coordinates $x'_1 = x'_2 = x'_3 = 0$ in S', $x_1 = x_2 = 0$ and $x_3 = v_o t$ in S, show that $a_{34} = \beta\, a_{44}$, where $\beta = v_o/c$. Thus, the transformation is of the form

$$x_1 = a_{11}\, x'_1, \quad x_2 = a_{11}\, x'_2, \quad x_3 = a_{33}\, x'_3 + \beta\, a_{44}\, x'_4, \quad x_4 = a_{43}\, x'_3 + a_{44}\, x'_4.$$

c) Consider two events ($\mathbf{r}$, t) and ($\mathbf{r} + d\mathbf{r}$, $t + dt$), which correspond to two positions of a photon and the interval $dS^2 = d\mathbf{r}^2 - c^2\, dt^2$. The invariance of the velocity of light requires that $dS'^2 = 0$ if $dS^2 = 0$. Write $dS^2 = f(x', y', z', t', v_o)\, dS'^2$. It may be shown that the homogeneity of space and time and other requirements imply that $f = 1$. Show that the invariance of the interval ($dS^2 = dS'^2$) implies that $a_{11}{}^2 = 1$, $a_{33}{}^2 - a_{43}{}^2 = 1$, $a_{44}{}^2 = \gamma^2$ and $\beta a_{33} = a_{43}$, where $\gamma = 1/(1-\beta^2)^{1/2}$. **d)** As the transformation reduces to the identity at the limit $\beta \to 0$, deduce that

$$x_1 = x'_1, \qquad x_2 = x'_2, \qquad x_3 = \gamma(x'_3 + \beta x'_4), \qquad x_4 = \gamma(x'_3 + \beta x'_4)\,.$$

P13.5 a) In order to write the Lorentz transformation in the case of an arbitrary velocity $\mathbf{v}_o \equiv c\boldsymbol{\beta}$ in the direction of the unit vector $\mathbf{e}_o$, we decompose the position vector $\mathbf{r}'$ into a longitudinal component $(\mathbf{r}'.\mathbf{e}_o)\mathbf{e}_o$ (parallel to $\mathbf{v}_o$) and a transverse component $\mathbf{r}' - (\mathbf{r}'.\mathbf{e}_o)\mathbf{e}_o$ (normal to $\mathbf{v}_o$). Noting that the transverse component is not modified in the transformation, while the longitudinal component transforms as z in [13.17], show that $\mathbf{r} = \mathbf{r}' + (\gamma - 1)(\mathbf{r}'.\mathbf{e}_o)\mathbf{e}_o + \gamma x'_4 \boldsymbol{\beta}$ and $x_4 = \gamma[x'_4 + (\boldsymbol{\beta}.\mathbf{r})]$. **b)** Derive the transformation laws of the velocity and the force.

$$\mathbf{v} = [\mathbf{v}' + (\gamma - 1)(\mathbf{v}'.\mathbf{e}_o)\mathbf{e}_o + \gamma c\boldsymbol{\beta}]/\gamma(1 + \boldsymbol{\beta}.\mathbf{v}'/c)\,,$$
$$\mathbf{f} = [\,\mathbf{f}' + (\gamma - 1)(\mathbf{f}'.\mathbf{e}_o)\mathbf{e}_o + \gamma(\boldsymbol{\beta}/c)(\mathbf{f}'.\mathbf{v}')]/\gamma[1 + (\boldsymbol{\beta}/c)\mathbf{v}']\,.$$

Four-dimensional formalism

P13.6 Verify that the norm of a four-vector $\overline{\mathbf{A}}$ and the four-scalar product of two four-vectors $\overline{\mathbf{A}}$ and $\overline{\mathbf{B}}$ defined by $(\overline{\mathbf{A}}\,.\,\overline{\mathbf{A}}) \equiv \mathbf{A}^2 - (A_4)^2$ and $(\overline{\mathbf{A}}\,.\,\overline{\mathbf{B}}) \equiv (\mathbf{A}.\mathbf{B}) - A_4 B_4$ are invariant in a Lorentz transformation in the direction Oz. This may also be shown for a Lorentz transformation in an arbitrary direction. In fact, the general

Lorentz transformation (including rotations) is defined by the condition of the invariance of the scalar product of four-vectors.

P13.7 Using Lorentz transformation in the direction *Oz*, show that the d'Alembertian is a four-scalar operator, that is, it gives a four-scalar if it acts on a four-scalar $f(x, y, z, t)$. It may be shown that this property holds for Lorentz transformations in any direction. Deduce that, if a wave propagates with a speed *c* isotropically in a medium at rest in a frame *S'*, it will be so in any frame *S*. Does this result hold for waves that propagate with a speed less than *c*?

Elements of relativistic mechanics

P13.8 A galaxy subtends an angle of 20' as seen from Earth. It rotates about itself with a period of 8×10^4 years. Requiring that the velocity of the stars at its periphery does not exceed *c*, what is the upper limit on its distance to Earth?

P.13.9 a) Two electrons move toward each other with a speed $v = c/2$. Knowing that $m_e = 0.51$ MeV/c^2 or 9.11×10^{-31} kg, calculate their momentum, their energy and their kinetic energy and compare with the classical values. **b)** What is the relative velocity (i.e. the velocity of one of them in the rest frame of the other)?

P13.10 Compare the expressions of the invariant $(\overline{\mathbf{P}}.\overline{\mathbf{P}})$ in the proper frame of a particle and another frame, derive the relation of *W* to **P**.

P13.11 The mass of the electron is $m = 9.11 \times 10^{-31}$ kg. **a)** What is its rest energy in joules and in MeV? **b)** Assume that it is accelerated by a voltage of 3×10^5 V. Calculate its velocity and its momentum. What do you find if you use non-relativistic mechanics? **c)** What is the produced energy in the pair (e^+e^-) annihilation if the initial particles are at rest and if they move toward each other with the velocities of 10^8 m/s?

Special relativity and waves

P13.12 In 1851, Fizeau tried to measure an eventual modification of light speed in moving water. A half-silvered mirror *M* splits a beam of light into two parts, which propagate in opposite directions in water that flows in a glass tube. Their interference is observed through a telescope *T* (Figure 13.4). The speed of light with respect to the hypothetical ether of water is c/n where *n* is the index of water. If water is moving with a speed v_o, assume that its ether is dragged with a speed αv_o where $0 < \alpha < 1$. Let *d* be the total distance travelled in water. **a)** Show that the difference in the travel time of these beams is $t_2(v_o) - t_1(v_o) \approx 2\alpha n^2(dv_o/c^2)$. Thus, by observing the interference figure before and after setting water in motion, Fizeau's was able to determine the drag coefficient α. Using a particular mechanical model of the ether, Fresnel has found $\alpha = 1 - 1/n^2$ in agreement with Fizeau result. **b)** Show that this "partial drag" can be obtained by using the relativistic transformation of the velocity both in the particle and the wave model of light.

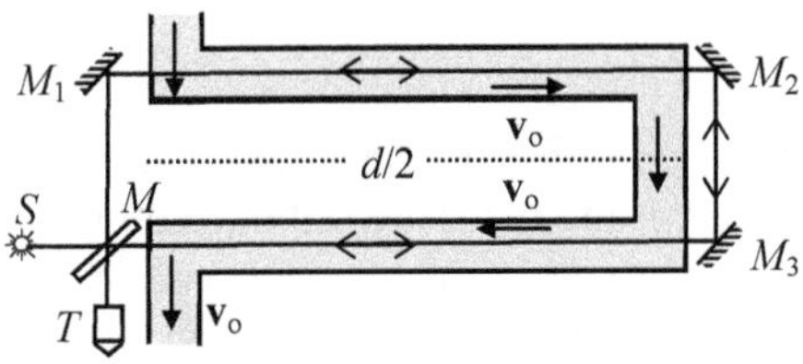

Figure 13.4. *Fizeau experiment, P13.12*

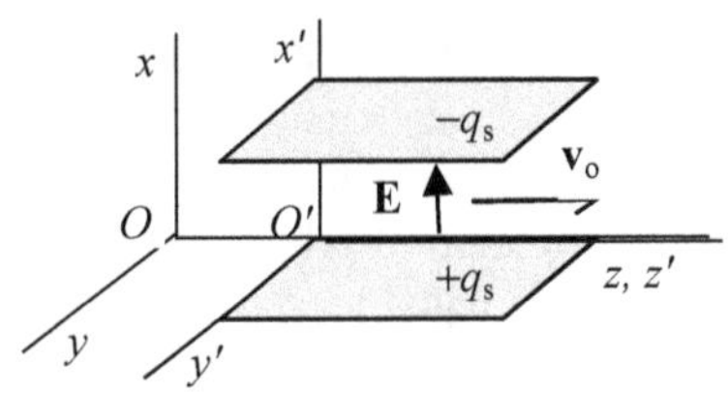

Figure 13.5. *Problem 13.15*

P13.13 A source of light moves toward the observer and emits red light of wavelength $\lambda_o = 700$ nm. What should be the source velocity v_o in order that this light appears blue of wavelength $\lambda'_o = 400$ nm?

P13.14 A quasar is a very far away celestial body. It has the size of a star but it radiates as much energy as a thousand galaxies. The emitted radiation by the farthest known quasar has a wavelength 4.4 times its value on Earth. What is the velocity of this quasar? Hubble law states that the velocity of this body is proportional to its distance D according to the relation $v = HD$, where $H \approx 20$ km.s^{-1} /10^6 light-year is *Hubble constant*. What is the distance D to this quasar?

Elements of relativistic electromagnetism

P13.15 The plane parallel plates of a capacitor have charge densities $\pm q'_s$ in their proper frame S' (Figure 13.5). The capacitor moves with a velocity v_o with respect to the observer (whose proper frame is S). Assume that v_o is in the direction Oz and that the plates are parallel to Oyz. Calculate the fields **E** and **B** in S' and deduce the fields in S. Calculate the fields directly in S.

P13.16 In a frame S, an electron moves with a velocity **v** in the direction Ox in a field **E** parallel to the plane Oyz and making an angle α with Oy. An observer is moving with a velocity $\mathbf{v}_o$ in the direction Oz. **a)** Calculate the velocity of the electron in the frame S' of the observer. Using the force exerted on the electron in S and the transformation law of the force, calculate the force exerted on the electron in S'. **b)** Calculate the electric and magnetic fields in S' and verify the expression of the force.

P13.17 A particle of charge q moves with respect to the observer with a velocity $\mathbf{v}_o$ in the direction Oz. In the proper frame S' of the particle, its field is purely electrostatic. Using the transformation law of the electromagnetic field, determine the fields in the frame S of the observer. Show that these fields verify the equations $\nabla\times\mathbf{E} + \partial_t\mathbf{B} = 0$ and $\nabla.\mathbf{B} = 0$.

P13.18 **a)** Show that the quantities $(\mathbf{E}.\mathbf{B})$ and $(E^2 - c^2B^2)$ are invariant in Lorentz transformation of arbitrary velocity. Deduce that, if the fields are orthogonal or one of them is equal to zero in a frame S, the fields are necessarily orthogonal in any other frame S'. **b)** Given the fields $\mathbf{E}$ and $\mathbf{B}$ in a frame S, is it possible to find a frame S' such that $\mathbf{B}' = 0$ or a frame S'' such that $\mathbf{E}'' = 0$?

P13.19 **a)** Two particles of charges q_1 and q_2 are moving with the same velocity $\mathbf{v}_o = c\boldsymbol{\beta}$ and they are separated by a distance $\mathbf{r}$. Show that the force exerted by q_1 on q_2 is $\mathbf{F}_{1\to2} = (q_1q_2\,K_o/r^3)\,\eta[\mathbf{r}/\gamma^2 + \boldsymbol{\beta}\,\beta r\cos\theta]$, where θ is the angle that $\mathbf{r}$ forms with $\mathbf{v}_o$ and $\eta = (1-\beta^2)(1-\beta^2\sin^2\theta)^{-3/2}$. Verify that this force is not in the direction of the line joining the two particles but $\mathbf{F}_{1\to2} = -\,\mathbf{F}_{2\to1}$. Compare the electric force with the magnetic force depending on the values of $\mathbf{v}_o$ and θ. **b)** Assume now that the particles are situated at points $\mathbf{r}_1$ and $\mathbf{r}_2$ and they are moving with different velocities $\mathbf{v}_1$ and $\mathbf{v}_2$. Calculate the forces exerted by the one on the other. Is the principle of action and reaction verified? **c)** A very long linear conductor carries a current I. What is the force that it exerts on a charge q situated at a distance r from this conductor?

P13.20 Consider a very thin cylindrical beam of particles having charge q and velocity $\mathbf{v}_o$ in the frame S of the observer. Let q'_L be the charge per unit length in the frame S' in which the charges are at rest. Calculate the fields and the potentials in S' and deduce the fields and the potentials in S. Verify Ampère's law.

P13.21 Using the transformation laws [13.87] for the potentials $\mathbf{A}$ and V and the expressions of the fields in terms of the potentials, derive the laws of transformation of the fields.

P13.22 Assume that, in a frame S', the electromagnetic field is static and that S' is moving with respect to S with a velocity $\mathbf{v}_o$ in an arbitrary direction. Verify that Maxwell's equations are valid in S.

P13.23 Show the covariance of Maxwell's equations in a Lorentz transformation of velocity $\mathbf{v}_o$ in an arbitrary direction.

Chapter 14

Motion of Charged Particles in an Electromagnetic Field

The motion of charged particles in an electromagnetic field is of great practical importance. It is used in observation instruments (oscilloscopes, electron microscopes etc.), accelerators, mass spectroscopy, the investigation of nuclear and particle reactions, etc. It is also important in some other fields of physics: plasma physics, astrophysics, cosmic ray physics, electronics, etc. In this chapter we analyze the simplest problems of motion in uniform electric and magnetic fields both in Newtonian and relativistic mechanics. We also consider some simple applications.

14.1. Motion of a charged particle in an electric field

In this section we consider the motion of a charged particle in a uniform electric field $\mathbf{E}$. The relevant effect is its acceleration and deviation, which is widely used in observation and measurement instruments (oscilloscopes, television sets, etc.).

A) Non-relativistic analysis

An electric field $\mathbf{E}$ exerts a force $\mathbf{F} = q\mathbf{E}$ on a particle of mass m and charge q. Thus, its equation of motion is

$$\mathbf{a} = \mathbf{F}/m = (q/m)\,\mathbf{E}. \qquad [14.1]$$

If the field $\mathbf{E}$ is uniform, the acceleration $\mathbf{a}$ is constant. The equation of motion [14.1] may be integrated to give the velocity and the position

$$\mathbf{v} = (q/m)\mathbf{E}\, t + \mathbf{v}_o, \qquad \mathbf{r} = (q/2m)\mathbf{E}\, t^2 + \mathbf{v}_o t + \mathbf{r}_o, \qquad [14.2]$$

where $\mathbf{v}_o$ and $\mathbf{r}_o$ are the initial velocity and the initial position, respectively.

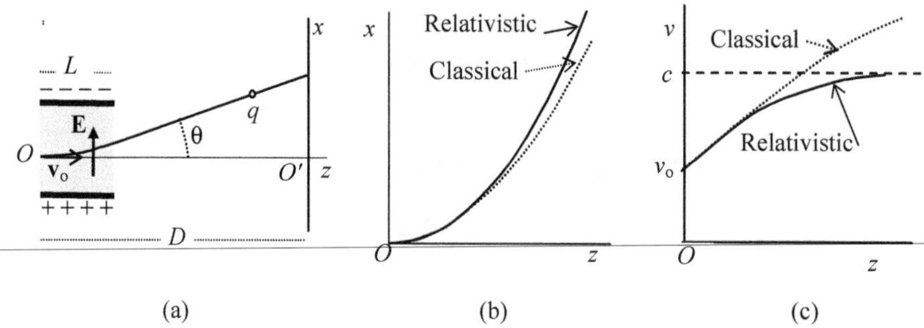

Figure 14.1. *a) Trajectory of a charged particle in a uniform field* **E**, *b) deviation of the particle versus z, and c) variation of the velocity versus z*

Assume that the particle enters at the origin O with a velocity $\mathbf{v}_o$ pointing in the direction Oz in a uniform field $\mathbf{E} = E\mathbf{e}_x$, for instance between the plates of a parallel plate capacitor (Figure 14.1a). Using [14.2] and, taking into account the initial conditions, the components of the velocity of the particle and its coordinates may be written as

$$v_x = (q/m)Et, \qquad v_y = 0, \qquad v_z = v_o, \qquad [14.3]$$

$$x = (q/2m)Et^2, \qquad y = 0, \qquad z = v_o t. \qquad [14.4]$$

The particle remains in the plane Oxz. Its motion is uniform in the direction Oz and accelerated in the direction Ox of the field. The equation of the trajectory is obtained by eliminating t between x and z. We find a parabola whose apex is at O

$$x = (qE/2mv_o^2)\, z^2. \qquad [14.5]$$

If the length of the capacitor in the direction Oz is L, the particle leaves the field at time $t = L/v_o$ with a velocity having the components $v_x = qEL/mv_o$ and $v_z = v_o$. Outside the field, the particle is free; its velocity remains constant. Its path is rectilinear and it makes with the initial direction Oz an angle θ such that

$$\tan\theta = (v_x/v_z) = qEL/mv_o^2 = qVL/mdv_o^2, \qquad [14.6]$$

where $V = Ed$ is the voltage of the capacitor and d is the plates spacing. The impact point on a screen perpendicular to Oz and situated at a distance D is displaced by

$$x = D\tan\theta = k\, V, \qquad \text{where } k = qLD/mdv_o^2. \qquad [14.7]$$

Thus, this displacement is proportional to the voltage V of the capacitor. For instance, if the electron is first accelerated by a potential V_o = 10 kV, its initial kinetic energy is $U_K = eV_o = 1.602\times10^{-15}$ J and its initial velocity is $v_o = (2U_K/m)^{1/2}$ = 0.593×10^8 m/s (thus, almost non-relativistic). If the capacitor has a length L = 1 cm and a thickness d = 2 mm, a voltage V = 100 V produces a displacement x = 5 mm on a screen situated at a distance D = 20 cm from the capacitor.

The variation of the kinetic energy in the interval of time dt in the field is

$$dU_K(t) = m\,\mathbf{v}.d\mathbf{v} = m\mathbf{v}.\mathbf{a}\,dt = q\mathbf{E}.\mathbf{v}\,dt = -\,q\,\boldsymbol{\nabla}V.d\mathbf{r} = -q\,dV = -\,dU_E. \qquad [14.8]$$

It is the opposite of the variation of the potential energy $U_E = qV$ of the charge in the field. This is valid also for the variation in a finite interval of time independently of the path

$$\Delta U_K = -\Delta U_E. \qquad [14.9]$$

This equation is the basis of electrostatic accelerators: charged particles are produced almost at rest by a source at a point O. A potential V_o is applied between O and another point P. The particles move from O to P and gain a kinetic energy $|qV_o|$. Their velocity at P is then $v = \sqrt{2|qV_o|/m}$.

B) Relativistic analysis

In the non-relativistic treatment, the velocity of a charged particle in an electric field increases indefinitely if the particle travels a large distance or if the field is very strong. However, if the particle velocity becomes comparable to the speed of light in vacuum c, the relativistic mechanics must be used and the particle speed will never reach c. The relativistic corrections are approximately of the order of v^2/c^2 at low velocities. For instance, if the velocity of the particle is 0.1 c, the relativistic corrections are roughly 1%. Another way to see whether the non-relativistic expressions are valid consists of comparing the kinetic energy of the particle to its rest energy. For instance, the electron rest energy is mc^2 = 511 keV. Thus, if an electron is accelerated by a potential of 20 kV, its kinetic energy becomes 20 keV and $U_K/mc^2 \approx 4$ %. The same correction is reached for protons (which are 1836 times heavier than electrons) if they are accelerated by a potential of $V \approx 40$ MV.

Consider the problem of a particle fired with a velocity $\mathbf{v}_o = v_o\,\mathbf{e}_z$ in a field $\mathbf{E} = E\,\mathbf{e}_x$. The relativistic equation of motion $d\mathbf{P}/dt = \mathbf{f} = q\mathbf{E}$ gives the equations

$$dP_x/dt = qE, \qquad dP_y/dt = 0, \qquad dP_z/dt = 0. \qquad [14.10]$$

The initial momentum is $P_o = mv_o/(1 - v_o^2/c^2)^{½}$ and the initial relativistic energy is $W_o = mc^2/(1 - v_o^2/c^2)^{½}$. Equations [14.10] may be solved with these initial conditions to give

$$P_x = qEt, \qquad P_y = 0, \qquad P_z = P_o. \qquad [14.11]$$

The energy of the particle is then $W = \sqrt{c^2P^2 + m^2c^4}$, that is,

$$W(t) = \sqrt{W_o^2 + q^2c^2E^2t^2}\ , \quad \text{where} \quad W_o = \sqrt{c^2P_o^2 + m^2c^4}\ . \qquad [14.12]$$

Knowing the momentum and the energy, we may determine the velocity by using the equation $\mathbf{v} = c^2\mathbf{P}/W$, hence

$$v_x = dx/dt = c^2qEt/W(t), \quad v_y = dy/dt = 0, \quad v_z = dz/dt = c^2P_o/W(t). \qquad [14.13]$$

The coordinates of the particle are obtained by integrating these equations with respect to time. Taking into account the initial position at O, we find

$$x = [W(t) - W_o]/qE, \quad y = 0, \quad z = (cP_o/qE)\sinh^{-1}(cqEt/W_o^2). \qquad [14.14]$$

The equation of the trajectory in the Oxz plane is

$$z = (W_o/qE)\,[\cosh^{-1}(qEx/cP_o) - 1]. \qquad [14.15]$$

The speed of the particle is

$$v = c^2\sqrt{P_o^2 + q^2E^2t^2}\ /W(t) = c\,[\,1 - m^2c^4/(qEz + W_o)^2]^{½}\ . \qquad [14.16]$$

The variation of the particle energy between $t = 0$ and t is given by

$$W - W_o = qEz = -q(V - V_o), \qquad [14.17]$$

where $(V - V_o)$ is the variation of the electric potential.

The non-relativistic limits of the preceding expressions correspond to small velocities compared to c. For this reason, it is necessary that the initial velocity be small and that the field E be sufficiently weak. In this limit, we may write $P \cong mv$ and $W \cong mc^2 + ½mv^2$. Expanding the cosh function [14.15] as a power series and keeping the second order, we find

$$z \approx (mc^2/qE)\,[1 + (qEx/cmv_o)^2 - 1] = (qE/2mv_o^2)\,x^2. \qquad [14.18]$$

Thus, we find again the classical parabola. The relativistic trajectory and the classical trajectory are illustrated in Figure 14.1b. The speed [14.16], has the non-relativistic limit $v = (v_o^2 + 2qEz/m)^{1/2}$. The relativistic treatment shows that the speed of the particle never exceeds the speed of light in vacuum c, while it increases indefinitely in the non-relativistic treatment (see Figure 14.1c).

C) Applications

Many types of equipment use the deflection of charged particles in an electric field **E**. In an ink jet printer, for instance, fine, charged droplets of ink are fired and pass between the parallel plates of a capacitor under the control potential of the signal. The resulting electric field directs the beam of ink as the paper moves in front of it.

The deflection of charged particles in an electric field **E** is used in cathode-ray tubes, which were the basic elements in oscilloscopes and television sets before the use of liquid crystal screens. In these vacuum tubes, a beam of electrons is emitted from the cathode and accelerated toward the anode. They are deflected by vertical and horizontal electric fields between the plates of capacitors (Figure 14.2) and they strike a fluorescent screen producing a bright spot. The spot is deviated vertically and horizontally proportionally to the applied voltages on the capacitors. An oscilloscope allows voltages to be compared and the shape of a voltage signal $V(t)$ to be observed. The instrument may be used to observe any physical quantity that may be transformed into an electric signal (sound waves, variation of temperature, heartbeat, etc.). In television sets, the deviation of the spot is controlled by the modulated electromagnetic wave, which is picked up and amplified.

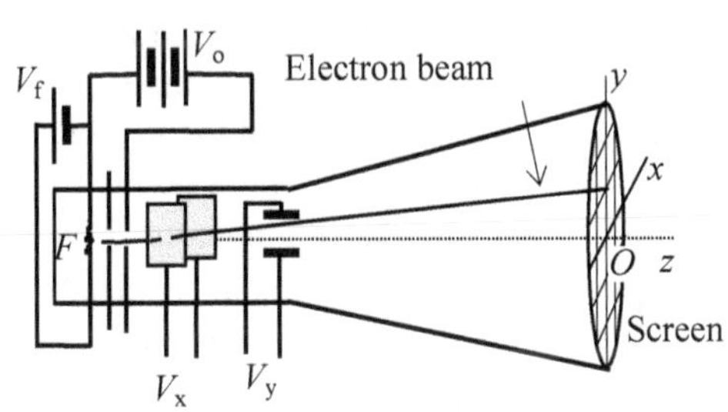

Figure 14.2. *Cathode-ray tube*

14.2. Bohr model for the hydrogen atom*

By 1905, the quantization of radiation was established. A light wave of frequency $\tilde{\nu}$ is formed by photons of energy $E_\gamma = h\tilde{\nu}$, where h is Planck's constant.

One of the long-standing problems was the interpretation of the hydrogen spectrum: the wavelengths of the emitted radiations by the hydrogen atom are given by the empirical formula

$$1/\lambda_{n,p} = R_H (1/p^2 - 1/n^2), \qquad \text{where } R_H = 1.097 \times 10^7 \text{ m}^{-1}. \qquad [14.19]$$

R_H is the *Rydberg constant* for hydrogen, while n and p are positive integers.

The analysis of section 2.9 shows that, in classical mechanics, a system of charged particles can be in a stable equilibrium configuration only if the positive charge and negative charge are distributed with vanishing total charge density. This led Thomson to propose a model, according to which the protons are distributed in the whole volume of the atom of radius 10^{-10} m, and in this way, neutralize the electrons. However, Rutherford's experiment showed that the protons are concentrated in a small nucleus of radius ~10^{-15} m. If the electrons are at rest, they cannot be stable, as nothing prevents them from being attracted and neutralized by the nucleus. In 1913, Bohr tried to interpret the empirical formula [14.19]. By analogy to the solar system, he proposed a model of the atom, according to which the electrons move on orbits around the nucleus at distances of the order of 10^{-10} m. Thus, the orbital motion of the electrons prevents them from being captured by the nucleus.

Consider a system formed by a nucleus of charge Ze and an electron of charge $-e$. The nucleus being much heavier than the electron, its displacement is negligible. The electron is subject to Coulomb's force $F = -K_o Ze^2/r^2$, which corresponds to a potential energy $U_E = -K_o Ze^2/r$. The study of this motion shows that the electron may follow elliptical, circular, parabolic, or hyperbolic orbits. The two last types correspond to collisions, while the closed orbits of the first two types correspond to a bound electron in the atom. Let us consider the simplest case of a circular orbit of radius r. The conservation of energy requires that the velocity be constant and given by the radial equation of motion $F = -K_o Ze^2/r^2 = -m_e v^2/r$. This gives the electron velocity $v = (K_o Ze^2/m_e r^2)^{1/2}$ and angular momentum $L = m_e rv = (K_o Ze^2 m_e r)^{1/2}$. According to electromagnetic theory, an accelerated electron radiates energy. To have the atom in a stationary state, Bohr supposed (without apparent raison) that it radiates no energy and its angular momentum L is an integer multiple of $\hbar = h/2\pi$ (called *reduced planck's constant*)

$$L = n\hbar\,, \qquad \text{where } n = 0, 1, 2, \ldots \qquad [14.20]$$

He deduced that the radius of the orbit that corresponds to a stationary state takes only the discrete values

$$r_n = n^2 (\varepsilon_o h^2/\pi Z m_e e^2). \qquad [14.21]$$

The total energy of this state is conserved and given by

$$E_n = \tfrac{1}{2}\, m_e v^2 + U_E = -\frac{1}{2}\frac{Ze^2}{4\pi\varepsilon_o r_n} = -\frac{RhcZ^2}{n^2}, \qquad [14.22]$$

where the constant R is given by

$$R = \frac{me^4}{8c\varepsilon_o{}^2 h^3} = 1.097\ 373\ 20 \times 10^7\ \mathrm{m}^{-1}, \qquad [14.23]$$

which is very close to Rydberg's constant R_H [14.19] and in perfect agreement with the experimental value $1.097\ 373\ 76 \pm 1.2) \times 10^7\ \mathrm{m}^{-1}$.

The state of lowest energy, called the *ground state*, corresponds to $n = 1$. In the case of the hydrogen atom, the corresponding energy E_1 and orbit radius (called *Bohr radius*) are

$$E_1 = -hcR = -2.18 \times 10^{-18}\ \mathrm{J} = -13.61\ \mathrm{eV}, \qquad [14.24]$$

$$r_1 = \varepsilon_o h^2 / \pi m_e e^2 = 0.529\ 175 \times 10^{-10}\ \mathrm{m}. \qquad [14.25]$$

The states $n = 2, 3...$ are *excited states.* The higher the *principal quantum number n*, the longer the orbit radius r_n (proportional to n^2) and the higher the energy level (proportional to $-1/n^2$). Figure 14.3a illustrates the Bohr orbits for hydrogen and Figure 14.3b illustrates the energy levels. The orbit $n = \infty$ has an energy $E_\infty = 0$ and a radius $r_\infty = \infty$, it corresponds to an electron separated from the nucleus. The energy required to separate the electron from the state n is $E_\infty - E_n = -E_n$. In particular, the energy necessary to separate the electron from the ground state is the *ionization energy* of the atom $-E_1 = 13.61\ \mathrm{eV} = 2.1806 \times 10^{-18}\ \mathrm{J}$.

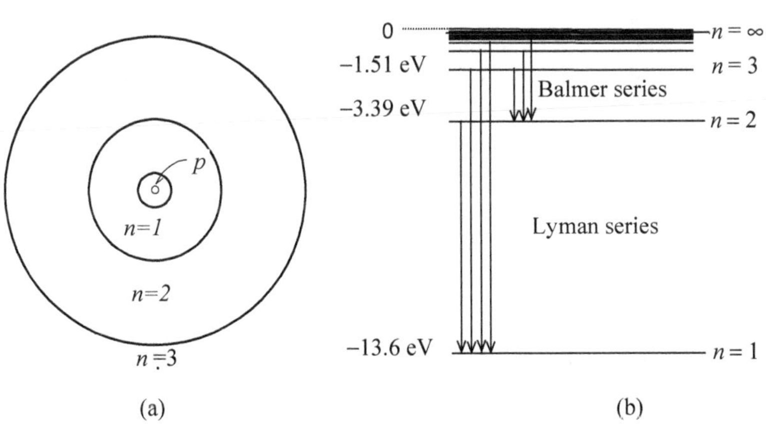

Figure 14.3. *Bohr model for the hydrogen atom: a) the orbits, and b) the energy levels*

To interpret the hydrogen spectrum, Bohr assumed that the atom is usually in the ground state E_1. If it is excited to a state of energy E_n, it may undergo a transition to a lower energy state E_p with the emission of a photon of energy

$$h\tilde{\nu}_{n,p} = E_n - E_p = hcRZ^2 (1/p^2 - 1/n^2). \quad [14.26]$$

The corresponding wavelength is given by the Rydberg relation

$$1/\lambda_{n,p} = \tilde{\nu}_{n,p}/c = Z^2 R_H (1/p^2 - 1/n^2). \quad [14.27]$$

Conversely, if the atom is in a state E_p, it may undergo a transition to a higher energy level E_n if it absorbs a photon of energy $h\tilde{\nu}_{n,p} = E_n - E_p$. The frequencies of the radiations that an atom may emit are thus the same as those of the radiations that it may absorb. According to Bohr's model, in a given stationary state, the atom obeys the laws of classical mechanics but it violates the laws of electromagnetism as the accelerated electron emits no radiation. The transitions from one state to another are jumps in a very short time interval, during which the laws of motion are not known, thus classical mechanics is probably violated.

Bohr's model is considered as *semi-classical* as the motion of the electron is in accordance with the laws of classical mechanics but subject to the quantization rule [14.20]. This model is not satisfactory because it does not justify the origin of this quantization rule and the absence of radiation in the stationary states. The properties of the atom can be understood only in the framework of quantum mechanics, which abandons the concept of orbit in favor of a distribution of probability and the duality of the wave-particle. The electrons form an electronic cloud around the nucleus, even in the hydrogen atom, which has a single electron. The atom may be only in quantized states completely determined by quantum theory. It undergoes transitions from a state to another not by a continuous variation of the physical quantities but a quantum jump with the emission of a photon. The ground state is the stable configuration because there is no lower energy state. In quantum mechanics, the concepts of position and motion do not have the same significance as in classical physics. Particularly, they cannot be determined simultaneously with precision.

14.3. Rutherford's scattering *

While studying the scattering of alpha particles by a thin gold leaf, Rutherford and his collaborators, Marsden and Geiger, observed that a good many particles were scattered backward. Rutherford interpreted this unexpected result by proposing that the positive charges in the gold atom are concentrated in a very small nucleus of radius $R_n \approx 7 \times 10^{-15}$ m instead of being spread in all the atom of radius $R_a \approx 10^{-10}$ m as in Thomson's model. Indeed, we saw in section 2.7 that the maximum of the electric field of a charged ball is at its surface. Thus, it is $E_{Th} = K_oZe/R_a^2 = 1.14 \times 10^{13}$ V/m in Thomson's model and $E_{Ru} = K_oZe/R_n^2 = 2.32 \times 10^{21}$ V/m in Rutherford's model. The alpha particles are scattered in Rutherford's model like small bullets fired on a very hard metallic ball instead of a large cotton ball.

The alpha particle of charge $Z'e$ (where $Z' = 2$) moves on a hyperbolic trajectory, whose focus is at the position of the nucleus of charge Ze. It approaches from far away with a velocity v_o initially on a straight line at a normal distance b from the nucleus (Figure 14.4). b is the *impact parameter* and it is related to the angular momentum L by the relation $b = L/mv_o$. After the collision, it moves away on a straight line, which makes an angle θ with its initial direction. The conservation of energy and angular momentum implies that it is scattered with the same velocity v_o and the same impact parameter b.

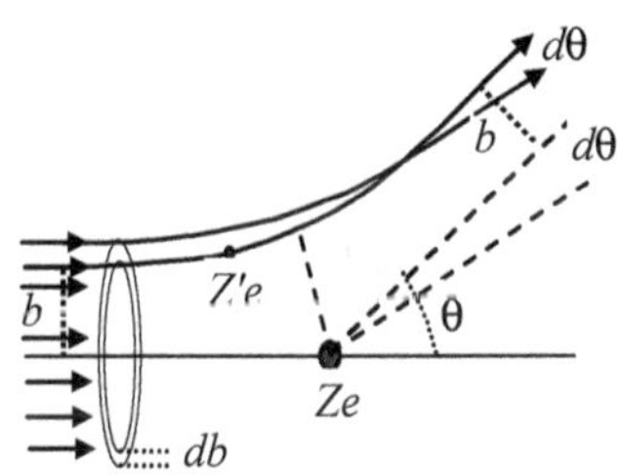

Figure 14.4. *Rutherford's scattering*

The atom being globally neutral, the electrons constitute a screen, which reduces the electric field of the nucleus. The alpha particle is strongly repulsed only if it penetrates deeply in the atom. It is then subject to the repulsion of the almost point-like nucleus. It may be shown that its deviation θ is related to b by the relation $\tan \theta/2 = K_oZZ'e^2/2Eb$ (see problem 14.10).

14.4. Motion of a charged particle in a magnetic field

A) Non-relativistic analysis

Let us assume that a particle of mass m and charge q is fired from the origin O with a velocity $\mathbf{u}$ in a uniform magnetic field $\mathbf{B}$ pointing in the direction Oz (Figures 14.5a and 14.5b). The equation of motion of this particle is

$$\dot{\mathbf{v}} = (q/m)\,(\mathbf{v} \times \mathbf{B}). \quad [14.28]$$

Projecting this equation on the axes, we find

$$\dot{v}_x = (qB/m)\, v_y, \qquad \dot{v}_y = -(qB/m)\, v_x, \qquad \dot{v}_z = 0. \qquad [14.29]$$

Let u_y and u_z be the components of **u** in the directions Oy and Oz. The equation $\dot{v}_z = 0$ implies that v_z be constant, thus equal to its initial value

$$v_z = u_z. \qquad [14.30]$$

The first two equations [14.29] are coupled. To uncouple them, we differentiate the second with respect to time and use the first equation, we find

$$\ddot{v}_y + \omega_c^2\, v_y = 0 \quad \text{where} \quad \omega_c = |q|B/m. \qquad [14.31]$$

The general solution of this equation is

$$v_y = A\cos(\omega_c t + \alpha). \qquad [14.32]$$

Using the second equation [14.29], we find

$$v_x = \pm A\sin(\omega_c t + \alpha), \qquad [14.33]$$

where ± is the sign of the charge. Imposing now the initial conditions $v_x = 0$, $v_y = u_y$, we find $\alpha = 0$ and $A = u_y$. Thus, the components of the velocity of the particle are

$$v_x = \pm u_y \sin(\omega_c t), \qquad v_y = u_y \cos(\omega_c t), \qquad v_z = u_z. \qquad [14.34]$$

Integrating once more with respect to time and taking into account the initial conditions $x = y = z = 0$, we find the coordinates

$$x = \pm R\,[1 - \cos(\omega_c t)], \qquad y = R\sin(\omega_c t), \qquad z = u_z t, \qquad [14.35]$$

where we have set

$$R = u_y/\omega_c = mu_y/|q|B. \qquad [14.36]$$

a) If the charge q is positive, the coordinates are given by

$$x = R\,[1 - \cos(\omega_c t)], \qquad y = R\sin(\omega_c t), \qquad z = u_z t. \qquad [14.37]$$

The trajectory is a helix of radius R around the axis of equation $x = R$ and $y = 0$ parallel to **B**. The particle has about this axis an angular velocity ω_c, called *cyclotron* *frequency* (Figure 14.5a). The motion about this axis is left-handed and the velocity in the direction of the field is constant (equal to the component of the initial velocity in the direction of Oz). The pitch of the helix, i.e. the difference of the z coordinates of two positions that are separated by a period $T = 2\pi/\omega_c$, is

$$\Delta z = u_z\,(2\pi/\omega_c) = 2\pi m u_z/|q|B. \qquad [14.38]$$

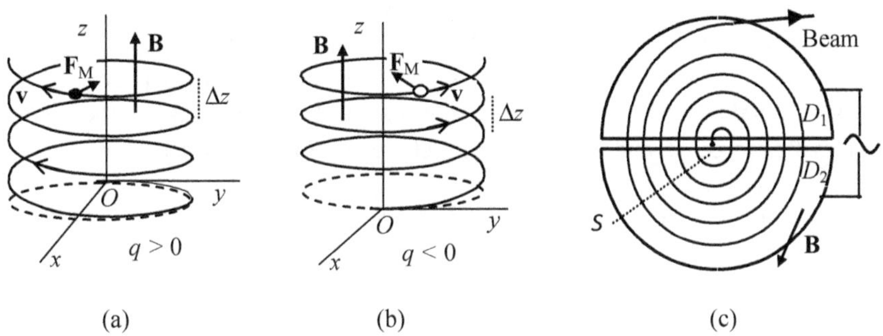

Figure 14.5. *a) Motion of a positively charged particle in a field* **B**, *b) motion of a negatively charged particle in a field* **B**, *and c) schematic representation of the cyclotron*

b) If the charge q is negative, the coordinates are given by

$$x = R\,[\cos(\omega_c t) - 1], \qquad y = R\sin(\omega_c t), \qquad z = u_z t. \qquad [14.39]$$

The trajectory is a helix with the same characteristics, except that the axis is at $x = -R$ and $y = 0$ and the motion about this axis is right-handed (Figure 14.5b).

In particular, if the particle is fired with an initial velocity perpendicular to the magnetic field ($u_z = 0$), the motion is circular of radius R and angular velocity ω_c in a plane perpendicular to the field.

In the case of the motion in a constant magnetic field, we have two important conserved quantities:

– The kinetic energy of the particle is constant since the work of the magnetic force on the particle $q\,\mathbf{v} \times \mathbf{B}$ is zero (as it is orthogonal to $\mathbf{v}$). Consequently, the speed of the particle remains constant. In fact both the longitudinal component $v_{//}$ (parallel to **B**) and the transverse component $v_\perp$ remain constant. On the other hand, the cyclotron angular frequency ω_c does not depend on the particle speed, while the radius of the helix is proportional to the transverse component of the velocity.

– The longitudinal component of the angular momentum is conserved, i.e.

$$L_z = -(q/|q|)\,mR^2\omega_c = -\,qR^2\omega_c, \qquad [14.40]$$

where the (–) sign must be taken since, for instance, negative charges have a right-handed motion. Consequently, the orbital magnetic moment is conserved

$$\mathcal{M} = qL_z/2m = -\tfrac{1}{2}\,qR^2\omega_c, \qquad |\mathcal{M}| = \tfrac{1}{2}\,mv_\perp^{\,2}/B. \qquad [14.41]$$

The motion in a time-dependent or a non-uniform field is much more complicated. If the field is uniform but slowly varying in time, in the sense that during the cyclotron period $T_c = 2\pi/\omega_c$, the field varies very little, $T_c\,|\partial_t\mathbf{B}| << |\mathbf{B}|$, the helicoidal motion varies slowly. The variation of **B** induces a field **E** according to Faraday's equation $\nabla\times\mathbf{E} = -\partial_t\mathbf{B}$. This induced electric field acts on the charged particle with a force $q\mathbf{E}$ and, during a complete revolution, it produces a work

$$\Delta W = \oint_{\mathcal{C}} d\mathbf{r}.\mathbf{F} = q\oint_{\mathcal{C}} d\mathbf{r}.\mathbf{E} = q\iint_{\mathcal{S}} d\mathcal{S}\;\mathbf{n}.\nabla\times\mathbf{E} = -q\iint_{\mathcal{S}} d\mathcal{S}\;\mathbf{n}.\partial_t\mathbf{B} = -q\,\partial_t\Phi = -q\,\pi R^2\partial_t\mathbf{B},$$

where $\mathcal{S}$ is a surface bounded by $\mathcal{C}$, and Φ is the magnetic flux through $\mathcal{S}$. This work produces a variation of the particule kinetic energy at a rate

$$dU_K/dt \cong \Delta W/T_c = -\tfrac{1}{2}\,q\omega_c R^2\partial_t B \cong \mathcal{M}\,\partial_t B. \qquad [14.42]$$

This variation of the particle energy by varying the magnetic field is used in the betatron (see section 8.9E). If B increases, as the particle progresses in the direction of **B**, $\omega_c = |q|B/m$ increases and $R = (2\mathcal{M}/|q|\,\omega_c)^{1/2}$ decreases.

Let us now consider the motion of a particle in a slowly varying **B** field in space. We assume that the field is symmetric about Oz and B_z is independent of ρ (Figure 14.6). Using cylindrical coordinates, the equation $\nabla.\mathbf{B} = \rho^{-1}\partial_\rho(\rho B_\rho) + \partial_z B_z = 0$ implies that $B_\rho = -\tfrac{1}{2}(\partial_z B_z)\rho$. Assuming that B_z varies slowly, such that $R\,|\partial_z B_z| << |B_z|$, we deduce that $|B_\rho|/|B_z| << \tfrac{1}{2}\rho/R$. The small component B_ρ acts on the charge with a force parallel to Oz producing a drift in this direction according to the equation of motion

$$m\,(\partial_t v_z) = -qv_\varphi B_\rho = \tfrac{1}{2}qv_\varphi R(\partial_z B_z) = (q/|q|)\,\mathcal{M}\,(\partial_z B_z). \qquad [14.43]$$

As the kinetic energy $U_K = \tfrac{1}{2}m(R^2\omega_c^{\,2} + v_z^{\,2}) = -(m/q)\mathcal{M}\omega_c + \tfrac{1}{2}\,mv_z^{\,2}$ is conserved and $\mathcal{M}$ is conserved, we deduce by differentiation that

$$m\,(\partial_t\omega_c) = v_z\,|q|(\partial_z B_z). \qquad [14.44]$$

If, for instance, the particle moves toward a stronger field, the right-hand side is positive; thus, ω_c increases. As $\mathcal{M} = -\tfrac{1}{2}qR^2\omega_c$ is constant, R decreases and the particle slows down. If the field is strong enough, the particle may return back as if it is reflected from a *magnetic mirror*. It may also be confined between two regions where the field **B** is very strong, forming so-called *magnetic bottles*. This effect is used to confine very hot plasmas, which cannot be confined in an ordinary container, in order to study thermonuclear fusion.

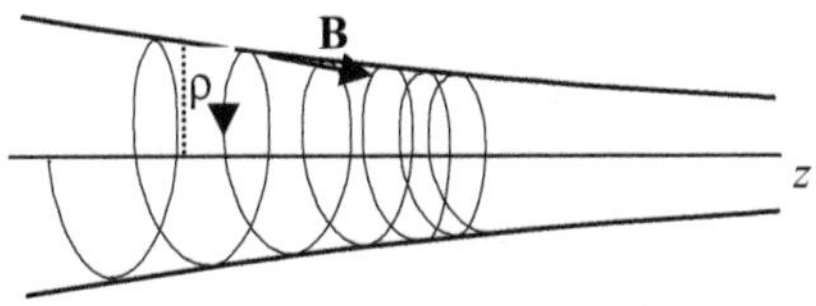

Figure 14.6. *Motion of a charged particle in a slowly varying magnetic field*

B) Relativistic analysis

If the particle has high velocity, its momentum and energy vary according to

$$d\mathbf{P}/dt = \mathbf{f} = q\,\mathbf{v} \times \mathbf{B}, \qquad dW/dt = \mathbf{f}.\mathbf{v} = q\,(\mathbf{v} \times \mathbf{B}).\mathbf{v} = 0. \qquad [14.45]$$

The second equation shows that the energy of the particle remains constant (equal to its initial energy W_o). Using the relation $\mathbf{v} = c^2\mathbf{P}/W$, the first equation becomes

$$d\mathbf{v}/dt = (qc^2/W_o)\,(\mathbf{v} \times \mathbf{B}). \qquad [14.46]$$

The only difference with the non-relativistic equation [14.28] is the replacement of $1/m$ by the constant factor c^2/W_o. We find a helicoidal motion with a cyclotron angular frequency and radius

$$\omega_c = |q|c^2B/W_o, \qquad R = u_y/\omega_c = P_{oy}/|q|B, \qquad [14.47]$$

where $P_{oy} = W_o v_{oy}/c^2$ is the transverse component of the initial relativistic momentum (in the direction Oy). The non-relativistic limit is obtained by taking the limits $P_o \to mv_o$ and $W_o \to mc^2$, thus neglecting the kinetic energy compared to the rest energy.

C) Applications

1) The *cyclotron* is a particle accelerator using the combined action of a constant magnetic field and an alternating electric field. The first cyclotron was constructed by Lawrence in 1932. It is essentially formed by two hollow, copper, D-shaped, half-cylinders (also know as "*dees*"), D_1 and D_2, separated by a gap and immersed in a magnetic field **B** parallel to the axes of the dees (Figure 14.5c). An alternating electric field is set up in the gap by applying an alternating potential between the dees. The particles of positive charge q, for instance, are produced with a small

velocity by a source S near the center. They are injected in D_1 perpendicularly to **B**. Each particle follows a half-circle path inside D_1 with the cyclotron angular frequency ω_c. As it reaches the gap, the alternating potential V is at its maximum V_m. The particle receives an energy qV_m and enters in D_2 where it follows a half-circle of larger radius but with the same frequency. By reaching the gap again, the alternating potential becomes $-V_m$ if it has a frequency exactly equal to the cyclotron frequency. The particle gains again an energy qV_m and enters D_1 and so on.

As $P = |q|BR$, to accelerate the particle to high energy (thus, high momentum P) the field B and R must be as large as possible. Practically, there is a limit as it is very difficult to have a strong uniform magnetic field over a large area. On the other hand, at high energy, relativistic relations must be used. As the cyclotron frequency decreases with increasing energy, the frequency of the alternating potential must be decreased accordingly. Such accelerators are called *synchro-cyclotrons*. It is also possible to vary B and the frequency in order to have an orbit of given radius, practically in a circular ring, whose radius may be very large (hundreds of meters); such accelerators are called *synchrotrons*. To give an idea of the order of magnitude in the case of a cyclotron of radius $R = 0.5$ m and a field $B = 2$ T, it is possible to accelerate protons to about 50 MeV, deuterons to 25 MeV, and alpha particles to 12 MeV.

2) Frequently, the charged particles (electrons, protons, α particles, ions, etc.) are emitted by sources with various velocities. A *velocity selector* allows a focalized beam of mono-energetic particles to be selected (Figure 14.7a). The initial beam is sent through two holes O_1 and O_2. Between them, an electric field **E** and a magnetic field **B** are set up in perpendicular directions by a capacitor and an electromagnet. The particles pass through both holes if they are not deviated by the fields. As in the case of Thomson's experiment, the condition for this is $qE = qvB$. This determines the velocity of the particles $v = E/B$.

3) A *mass spectrometer* uses the motion of charged particles in a magnetic field **B** to measure its mass or to separate particles of different masses m_i. It is used, for instance, to measure the mass of isotopes and separate them. The atoms are ionized in a source and accelerated by a potential V (Figure 14.7b). They form a beam of ions of velocity $v_i = (2qV/m_i)^{1/2}$. They normally enter a uniform field **B** and travel in half-circles, whose radii depend on the masses of the particles according to

$$R_i = m_i v_i/|q|B = \sqrt{2m_i V/qB^2} \,. \qquad [14.48]$$

After traveling in half-circles, they are intercepted by a plate at M_i at a distance $OM_i = D_i = 2R_i$. Thus the mass of the atoms are given by the relation $m_i = |q|B^2D_i^2/8V$. In the case of a mixture of isotopes, the M_i form mass spectral lines.

Thus, the mass spectrometer allows the separation of isotopes that cannot be separated by chemical methods; however, this method is very slow and costly.

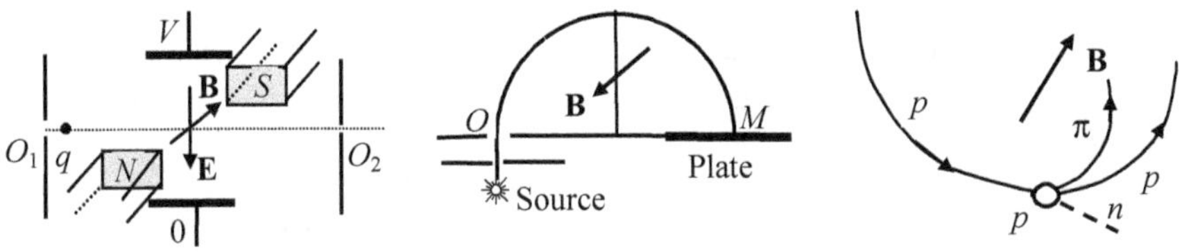

Figure 14.7. *Applications of the motion of a charged particle in a field* **B***: a) velocity selector, b) mass spectrometer, and c) visualization of the reaction* $p + p \to p\, \pi^+ n$

4) A charged particle, which moves in a liquid (often hydrogen or propane) maintained under pressure at a temperature slightly higher than the boiling point, produces bubbles around the ions along its trajectory. The *bubble chamber* uses this effect to detect charged particles and to visualize their circular trajectories in a strong field **B**. For instance, the production of a π^+ meson in the proton-proton collision $p + p \to p + n + \pi^+$ may be observed and analyzed (Figure 14.7c). The incident proton arrives along a circular path and collides with a proton of the hydrogen liquid initially at rest (thus, it has no trajectory). The photograph shows two circular trajectories that are curved in the same direction as the incident proton; thus, they are positively charged particles. A measurement of the radii of these trajectories allows their momentums to be determined. The conservation of momentum implies the existence of another particle, the neutron, which is invisible because it is neutral. The experiments undertaken using bubble chambers played an important part in the development of particle physics before the discovery of more efficient methods. Instead of bubbles in a liquid, the *Charpak chamber* (1968) is formed by a grid of high-voltage parallel wires. If a particle passes near one of them, it provokes a detectable signal. If the chamber is immersed in a strong magnetic field, the curved trajectory may be reconstituted with the help of computers.

14.5. Motion in crossed electric and magnetic fields

Let us assume that, in a frame S, a charged particle is subject to the fields $\mathbf{E} = E\,\mathbf{e}_x$ and $\mathbf{B} = B\,\mathbf{e}_y$, where E and B are positive. To simplify, we assume that the motion is non-relativistic and that the particle is fired with an initial velocity $\mathbf{v}_o$ in the plane Oxz. The equation of motion $m\dot{\mathbf{v}} = q(\mathbf{E} + \mathbf{v} \times \mathbf{B})$ is equivalent to the equations

$$m\dot{v}_x = qE - qBv_z, \qquad m\dot{v}_y = 0, \qquad m\dot{v}_z = qBv_x. \qquad [14.49]$$

As the initial velocity lies in the plane Oxz, the motion remains in this plane. It is convenient to use complex notation and set $\underline{u} = x + \mathrm{i}z$. This complex variable verifies the equation $m\ddot{\underline{u}} = qE + \mathrm{i}qB\dot{\underline{u}}$, whose solution is $\underline{u} = \mathrm{i}(E/B)t + \underline{b} + \underline{a}\exp(\mathrm{i}\eta\omega_c t)$, where $\underline{a}$ and $\underline{b}$ are arbitrary complex constants, $\omega_c = |q|B/m$ is the cyclotron angular frequency and $\eta = \pm 1$ according to the sign of the charge. Setting $\underline{a} = A\exp(\mathrm{i}\phi)$ and $\underline{b} = b_x + \mathrm{i}b_z$ we find

$$x = A\cos(\eta\omega_c t + \phi) + b_x \quad \text{and} \quad z = (E/B)t + A\sin(\eta\omega_c t + \phi) + b_z.$$

Setting $v_{oz} - E/B = v'_z$ and taking into account the initial conditions $x = z = 0$ and $\mathbf{v}_o = v_{ox}\mathbf{e}_x + v_{oz}\mathbf{e}_z$, we find

$$\omega_c x = -\eta v'_z[1 - \cos(\omega_c t)] + v_{o,x}\sin(\omega_c t),$$
$$\omega_c z = \omega_c(E/B)t + v'_z\sin(\omega_c t) + \eta v_{o,x}[1 - \cos(\omega_c t)]. \qquad [14.50]$$

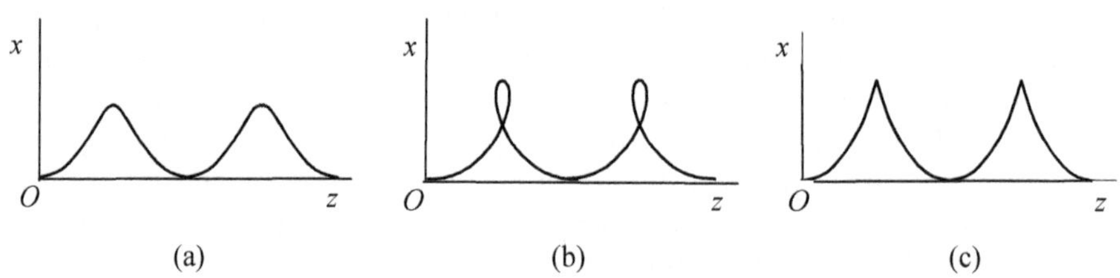

Figure 14.8. *Trajectory in the plane Oxz, respectively, in the cases: a) $v'_z < E/B$, b) $v'_z > E/B$, and c) $v'_z = E/B$*

Particularly, if the initial velocity points in the direction Oz, we find

$$\omega_c x = -\eta v'_z[1 - \cos(\omega_c t)] \quad \text{and} \quad \omega_c z = \omega_c(E/B)t + v'_z\sin(\omega_c t). \qquad [14.51]$$

Differenciating these equations with respect to time, we obtain the velocity of the particle $\mathbf{v} = -\eta v'_z\sin(\omega_c t)\,\mathbf{e}_x + [E/B + v'_z\cos(\omega_c t)]\mathbf{e}_z$. The average velocity over a period is the *drift velocity* $\langle\mathbf{v}\rangle = (E/B)\,\mathbf{e}_z$. The motion of the charged particle is the superposition of a helicoidal motion of radius $R = |v'_z| = |v_{oz} - E/B|$ and angular frequency $\omega_c = |q|B/m$ and a translational motion in the direction Oz with a velocity equal to the *drift velocity* E/B. If $|v'_z| < E/B$ (that is, $0 < v_{oz} < 2E/B$), the sign of the velocity in the direction $\mathbf{e}_z$ does not change and the trajectory is illustrated in Figures 14.8a. If $|v'_z| > E/B$ (that is, $v_{oz} < 0$ or $v_{oz} > 2E/B$) the sign of the velocity in the direction $\mathbf{e}_z$ changes at certain times and we find the trajectory of Figure 14.8b. In the particular case $v'_z = -E/B$, we find the trajectory of Figure 14.8c.

14.6. Magnetic moment in a magnetic field

An electron, in orbit around a nucleus at O, has an orbital magnetic moment $\mathcal{M}_o = -e\boldsymbol{l}/2m$, where m is the electron mass and $\boldsymbol{l}$ is its orbital angular momentum with respect to O. If a magnetic field $\mathbf{B}$ acts on the atom, the orbit changes continuously in time. This change is conveniently analyzed as a variation of the angular momentum $\boldsymbol{l}$ or that of the magnetic moment $\mathcal{M}_o$. The moment of force acting on the atom may be written as $\boldsymbol{\Gamma} = \mathcal{M}_o \times \mathbf{B}$ and this produces a variation of the angular momentum given by the equation

$$\frac{d\boldsymbol{l}}{dt} = \boldsymbol{\Gamma} = -\frac{e}{2m}\boldsymbol{l} \times \mathbf{B}. \qquad [14.52]$$

Taking the axis Oz in the direction of $\mathbf{B}$, we may write

$$dl_x/dt = -\omega_L l_y, \qquad dl_y/dt = \omega_L l_x, \qquad dl_z/dt = 0, \qquad [14.53]$$

where we have set $\omega_L = eB/2m$. Combining the equations [14.53], we may show that l^2 remains constant in the course of the motion. The three equations [14.53] have the solution

$$l_x = l_o \cos(\omega_L t + \alpha), \qquad l_y = l_o \sin(\omega_L t + \alpha), \qquad l_z = \text{Cte}. \qquad [14.54]$$

This shows that the vector $\boldsymbol{l}$ undergoes a precession about $\mathbf{B}$ with an angular frequency ω_L called *Larmor precession.* Thus, the orbit of the electron undergoes the same precession about the magnetic field without modification of shape (Figure 14.9a). The moment of force $\boldsymbol{\Gamma} = \mathcal{M}_o \times \mathbf{B}$ is equivalent to a magnetic interaction energy

$$U_{MO} = -\mathcal{M}_o.\mathbf{B} = \frac{e}{2m}(\boldsymbol{l}.\mathbf{B}). \qquad [14.55]$$

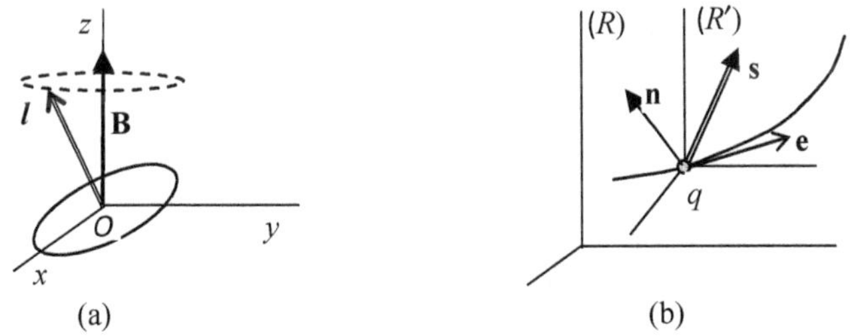

Figure 14.9. *a) Larmor precession, and b) spin precession in a magnetic field*

The electron also has an intrinsic angular momentum or *spin* **s** associated with a magnetic moment $\boldsymbol{\mathcal{M}}_s = -(ge/2m)\mathbf{s}$, where g is the *gyromagnetic ratio* of the electron (very close to 2). A magnetic field $\mathbf{B}'$ (measured in the rest frame S' of the electron) acts on this magnetic moment with a moment of force $\mathbf{\Gamma}_s = \boldsymbol{\mathcal{M}}_s \times \mathbf{B}'$. Thus, the variation of the spin is given by

$$\frac{d\mathbf{s}}{dt} = \mathbf{\Gamma}_s = -\frac{ge}{2m}\,\mathbf{s} \times \mathbf{B}'. \quad [14.56]$$

The corresponding interaction energy is

$$U_s = -\boldsymbol{\mathcal{M}}_s.\mathbf{B}' = \frac{ge}{2m}(\mathbf{s}.\mathbf{B}'). \quad [14.57]$$

Let us consider an electron, which moves with a velocity **v** in the fields **E** and **B** measured in the frame of reference S of the observer (Figure 14.9b). The magnetic field in the proper frame of the electron is

$$\mathbf{B}' = \mathbf{B} - (\mathbf{v} \times \mathbf{E})/c^2 + \mathcal{O}(\beta^2). \quad [14.58]$$

The interaction energy of this spin with the magnetic field is

$$U_{Ms} = \frac{ge}{2m}(\mathbf{s}.\mathbf{B}) - \frac{ge}{2mc^2}\,\mathbf{s}.(\mathbf{v} \times \mathbf{E}). \quad [14.59]$$

The first term is responsible for the *Zeeman effect*, that is, the splitting of the spectral lines if a magnetic field acts of the atom.

Assume that the electron is moving in the central electrostatic potential $V(r)$ of a nucleus. Its interaction energy is $U(r) = -eV(r)$ and the electric field acting on the electron is

$$\mathbf{E} = -\nabla V(r) = \frac{1}{e}\frac{dU}{dr}\frac{\mathbf{r}}{r}. \quad [14.60]$$

In the proper frame of the electron, we have both electric and magnetic fields. The induced magnetic field acts on the electron spin. The energy of this interaction is

$$U'_s = -\frac{ge}{2mc^2}\,\mathbf{s}.(\mathbf{v} \times \mathbf{E}) = \frac{g}{2m^2c^2}\frac{1}{r}\frac{dU}{dr}(\mathbf{s}.\boldsymbol{l}), \quad [14.61]$$

where we have used the relation $\boldsymbol{l} = m\,\mathbf{r} \times \mathbf{v}$. This interaction, which is proportional to $(\mathbf{s}.\boldsymbol{l})$, is called *spin-orbit* coupling. It contributes to the fine structure of atoms. However, the corresponding experimental values are only half of [14.61]. This discrepancy is due to the fact that the proper frame of the electron is not an inertial frame. Because of the instantaneous velocity **v** and acceleration $\mathbf{a} = d\mathbf{v}/dt$ of the electron, its proper frame has not only a translational motion of velocity **v** but also the so-called *Thomas precession* of angular frequency $\mathbf{\Omega}_\mathrm{T} = (\mathbf{a} \times \mathbf{v})/2c^2$. It may be shown that, in this non-inertial frame, equation [14.56] must be replaced by

$$\frac{d\mathbf{s}}{dt} + (\Omega_\mathrm{T} \times \mathbf{s}) = \mathbf{\Gamma}_\mathrm{s} = -\frac{ge}{2m}\mathbf{s} \times \mathbf{B}'. \qquad [14.62]$$

This is equivalent to replace the magnetic field $\mathbf{B}'$ by $\mathbf{B}' + m\,\mathbf{\Omega}_\mathrm{T}/e$ in the spin equation of motion and to an additional interaction energy

$$U''_\mathrm{s} = \frac{ge}{4mc^2}\mathbf{s}.(\mathbf{a} \times \mathbf{v}). \qquad [14.63]$$

The acceleration of the electron being $\mathbf{a} = \mathbf{F}/m = -(e/m)\mathbf{E} = -(1/mr)(dU/dr)\mathbf{r}$, equation [14.63] becomes $U''_\mathrm{s} = -(g/4m^2c^2r)(dU/dr)(\mathbf{s}.\boldsymbol{l})$, which is exactly half of the expression [14.61] with opposite sign. Thus, the spin-orbit coupling corrected by Thomas precession may be written as

$$U_\mathrm{s} = \frac{g}{4m^2c^2}\frac{1}{r}\frac{dU}{dr}(\mathbf{s}.\boldsymbol{l}). \qquad [14.64]$$

14.7. Problems

Motion of a charged particle in an electric field

P14.1 a) Write the equation of motion for an electron in a uniform electric field E pointing in the direction Ox and determine its motion if it is fired from the origin with a velocity v_o in the direction Ox. **b)** Assume that $E = 10^5$ V/m and $v_o = 0$. What is the acceleration of the electron and the distance that it travels to attain the speed $c/10$, where c is the speed of light? **c)** Assume that $v_o = 10^6$ m/s and $E = 10^5$ V/m, both of them pointing in the positive x direction. What is the distance that the electron travels before it changes direction and the time it takes to reach this position?

P14.2 An electron is fired with a kinetic energy of 10^4 eV toward a sphere of radius 1 cm and charge $Q = -5$ μC. From what distance must it be fired in order to hit the sphere?

P14.3 In 1913 Millikan's experiment established the quantization of electric charge and determined the elementary charge with reasonable precision. The set-up is illustrated in Figure 14.10. A uniform electric field **E** is set up between the plane parallel plates of a capacitor. Fine droplets of oil from a spray fall on this field through a hole in the upper plate. A droplet may acquire a charge q by friction in the spray or by collision with the air molecules. In the absence of electric field, the droplet is subject to its weight $m\mathbf{g}$, the Archimedes' buoyancy in air $-m'\mathbf{g}$ (negligible in this experiment), and the viscosity force given by Stokes's law $\mathbf{f}_v = -6\pi\eta r\mathbf{v}$, where η is the viscosity of air, r is the radius of the droplet and $\mathbf{v}$ is its velocity. Let μ be the mass density of the oil and μ' that of air. Write the equation of motion of the droplet if the plates have a difference of potential V_0. Show that the velocity may be written as $v = A\,e^{-t/\tau} + (mgd + qV_0)/6\pi\eta rd$, where $\tau = m/6\pi\eta r$. What is the limit velocity? It was found that, in the absence of electric field, the droplets fall a distance of 1 mm in 27.4 s and that they stay in equilibrium in a field $E = 8.50 \times 10^3$ V/m. How many excess electrons does it contain knowing that $\eta = 1.8 \times 10^{-5}$ N.s/m, $\mu = 950$ kg/m^3 and $\mu' = 1.29$ kg/m^3?

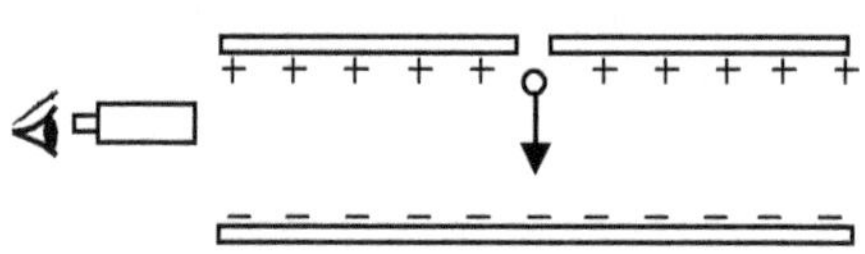

Figure 14.10. *Millikan experiment, problem 14.3*

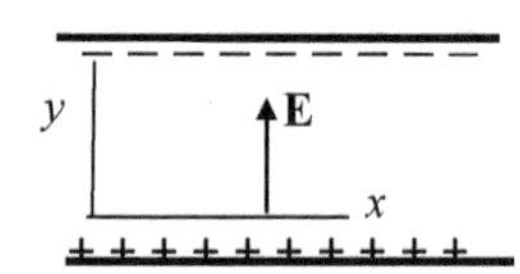

Figure 14.11. *Problem 14.4*

P14.4 An electron is fired with a kinetic energy of 3×10^{-16} J in a uniform electric field $E = 2 \times 10^4$ V/m set up between the plates of a parallel plate capacitor. The initial velocity of the electron is in the direction Ox parallel to the plates (Figure 14.11). **a)** Calculate the displacement of the electron in the normal direction Oy as it leaves the capacitor, whose length in the direction Ox is of 1 cm. Determine its velocity, direction, and energy upon leaving. **b)** The electron is intercepted on a screen situated at 20 cm from the capacitor. What is the displacement of the impact point on this screen due to the electric field?

P14.5 An electrostatic precipitator is formed by a rod of radius 1 cm surrounded by a metallic cylindrical shell of radius 50 cm with a difference of potential of 60 kV. Calculate the electric field near the cylinder. Determine the force that acts on a singly ionized dust particle of mass 0.1 μg and its acceleration if it is near the rod and if it is near the cylindrical shell.

P14.6 In a cathode-ray tube, the electrons are emitted with negligible velocity by a heated filament and attracted by an anode at a potential of 20 kV higher than the cathode. **a)** What is the velocity of the electrons as they arrive to the anode? **b)** As this velocity is not small compared to c, the relativistic expressions of the kinetic energy $U_K = mc^2(1 - v^2/c^2)^{-1/2} - mc^2$ must be used. What is the exact value of the velocity?

Bohr model for the hydrogen atom

P14.7 In an ionized atom, the point-like nucleus has a charge Ze and the Z' electrons are distributed uniformly in a sphere of radius R around the nucleus. Calculate the force acting on an electron situated at a distance r from the nucleus ($r < R$). Calculate the electrostatic interaction energy of this system and discuss its stability.

P14.8 a) According to Thomson's model, protons and electrons are distributed uniformly in all the volume of the atom of radius R. What is the force acting on an electron as a function of r in the case of a neutral atom? Determine the motion of this electron and its frequency if $R = 10^{-10}$ m. **b)** Using Bohr's model for the hydrogen atom, calculate the frequency of the electron on its ground state orbit of radius 0.5×10^{-10} m. Compare the results of both models with visible light frequencies.

Rutherford's scattering

P14.9 An alpha particle of mass 6.6×10^{-27} kg and energy 5 MeV is incident on a gold nucleus ($Z = 79$). What is the velocity of this particle? What is its shortest distance of approach to the nucleus? Determine the electric field and potential at this distance. What is the maximum acceleration of this particle?

P14.10 The trajectory of a particle of mass m subject to a central force K/r^2 exerted by a second particle of large mass M may be written in polar coordinates in the form $1/r = C\,[1 + \eta \cos(\phi - \phi_o)]$ where $C = -mK/L^2$, $\eta = \sqrt{1 + 2L^2E/mK^2}$ and ϕ_o is the polar angle of the major axis. E is the energy and L is the angular momentum, which are conserved quantities. If E is positive (as in the case of a collision), the eccentricity η is larger than 1 and the trajectory is a hyperbola. In the case of Rutherford's scattering (Figure 14.4), a particle of charge $Z'e$ is fired from large distance along an axis situated at a distance b (called *impact parameter*) from a nucleus of charge Ze. At large distance after the collision, the particle moves away along a straight line that makes an angle θ with the initial direction. **a)** Express the constants L, η and C in terms of E and b. By considering the asymptotic directions (as $r \to \infty$), show that $\tan \theta/2 = (\varepsilon^2 - 1)^{-1/2} = ZZ'e^2/8\pi\varepsilon_o Eb$. **b)** Consider a beam of intensity I (that is, the number of particles that are incident per unit time on the unit area placed normal to the beam). Show that the number of incident particles per unit time with an impact parameter lying between h and $b + db$ is $dN = 2\pi Ib\, db$. Show that these particles are scattered with an angle lying between θ and $\theta + d\theta$, where $d\theta = -(db/b) \sin \theta$, that is, in a solid angle $d\Omega = 2\pi \sin\theta\, d\theta$. The differential cross-section $\sigma(\Omega)$ is defined by the relation $\sigma(\Omega)\, d\Omega = dN/I$. Deduce Rutherford's formula

$$\sigma(\Omega) = (ZZ'e^2/16\pi\varepsilon_0 E)^2\, [1/\sin^4(\theta/2)].$$

Motion of a charged particle in a magnetic field

P14.11 In the cathode-ray tube of a television set, the electron beam is accelerated by a potential $V = 50$ kV. **a)** Calculate the velocity of the electrons. **b)** Assume that the Earth magnetic field has a vertical component $B = 4 \times 10^{-5}$ T and that the beam travels from west to east. In which direction is the beam deviated by this magnetic field and what is this deviation if the beam travels 25 cm in the tube?

P14.12 a) A proton moves on a circle of radius 0.8 m in a field $B = 2$ T. Calculate its velocity, energy, and period. What is the voltage that may accelerate the proton to this velocity? **b)** What is the energy of an electron in order to have the same period as the proton in this field?

P14.13 Assume that, within the atom, an electron moves on a circle of radius r. A magnetic field **B** is applied perpendicularly to this orbit. Does the frequency of this motion increase or decrease? Verify that the variation of the frequency is $\delta\tilde{\nu} = \pm eB/4\pi m$. This variation of the frequency is responsible for the Zeeman effect in this classical model.

P14.14 A parallel plate capacitor of thickness d is under a voltage V. If an energetic photon is incident on the negative plate, an electron may be extracted and attracted by the positive plate. A magnetic field **B** is applied perpendicularly to **E**. Neglecting the initial velocity of the extracted electron, analyze its motion and show that it may reach the positive plate only if $B > [2mV/ed^2]^{1/2}$.

P14.15 An electron of energy 50 keV enters a magnetic field $B = 2$T with its velocity making 60° with **B**. Analyze its motion.

P14.16 In a cyclotron, the magnetic field is $B = 0.5$ T and the dees have a radius of 0.5 m. What is the frequency of the accelerating potential for protons? What is the energy of the protons as they leave at the periphery of a dee? How many turns will the protons have travelled in order to be accelerated to this energy if the amplitude of the accelerating potential is 20 kV? Is it possible to use this cyclotron to accelerate electrons?

P14.17 An electron of energy 15 keV enters an electric field $E = 200$ V/cm horizontally pointing downward. How should a magnetic field **B** be oriented in order that the electron suffers no deviation? What should the orbit of the electron be if **E** is removed?

P14.18 In a mass spectrometer, carbon ions are accelerated by a potential of 10 kV. They enter a magnetic field of 0.3 T. What is the mass of these ions if they move on half-circles of radius R = 16.62 cm? In fact, the ionized gas is a mixture of ^{12}C and ^{13}C of masses 12.00000 u and 13.00336 u where u = 1.66055×10^{-27} kg is the unit of atomic mass. What is the spacing between the mass spectral lines of these ions?

Motion in crossed electric and magnetic fields

P14.19 We consider the motion of a charged particle in a field **E** pointing in the direction Ox and a field **B** pointing in the direction Oy in a frame S. In order to study this motion, it is convenient to make a Lorentz transformation to a frame S' where there is only one field. **a)** If $E < cB$, show that it is possible to find a frame S' whose velocity with respect to S is $\mathbf{v}_o = c^2(\mathbf{E}\times\mathbf{B})/E^2$ and such that

$$\mathbf{E}' = 0 \quad \text{and} \quad \mathbf{B}' = \gamma\,[\mathbf{B} - (\mathbf{v}_o \times \mathbf{E})/c^2\,] = [1 - E^2/B^2c^2]^{1/2}\mathbf{e}_y.$$

Study the motion in S' and deduce the motion in S. **b)** If $E > cB$, show that it is possible to find a frame S' whose velocity with respect to S is $\mathbf{v}_o = (\mathbf{E}\times\mathbf{B})/B^2$ and such that

$$\mathbf{B}' = 0 \quad \text{and} \quad \mathbf{E}' = \gamma\,[\mathbf{E} - (\mathbf{v}_o \times \mathbf{B})] = [1 - E^2/B^2c^2]^{1/2}\,\mathbf{e}_x.$$

Chapter 15

Emission of Radiation

The purpose of this last chapter is to study the emission of waves by time-dependent sources: moving charges and simple harmonic currents in antennas. As in the case of a sustained oscillator, the source is taken into account by a term $f(\mathbf{r}, t)$ on the right-hand side of the equation of propagation. In mechanics, knowledge of the forces and the equations of motion is not sufficient to determine the motion; the *initial conditions* are needed. In wave theory, knowledge of the equation of propagation and the sources $f(\mathbf{r}, t)$ at each point $\mathbf{r}$ and at any time t is not sufficient to determine the wave. We need the initial conditions, i.e. the values of u and its time derivative at the initial time $t = 0$ and at each point in space. If the medium is bounded, we need also the *boundary conditions*. In this chapter, we assume that the medium is infinite, linear, and isotropic of electric susceptibility ε and magnetic permeability μ, and that the source is restricted to a small region, so that the solution and its gradient vanish rapidly at large distances.

15.1. Retarded potentials and fields

We have seen that the fundamental laws of electromagnetic phenomena are Maxwell's equations [9.12] to [9.15]. We have also seen in section 9.3 that it is often practical to use the *vector potential* $\mathbf{A}$ and the *scalar potential* V such that

$$\mathbf{B} = \nabla \times \mathbf{A} \qquad \text{and} \qquad \mathbf{E} = -\nabla V - \partial_t \mathbf{A}. \qquad [15.1]$$

The homogeneous Maxwell equations $\nabla.\mathbf{B} = 0$ and $\nabla\times\mathbf{E} + \partial_t\mathbf{B} = 0$ are then identically verified. If we use potentials that verify Lorentz condition

$$\nabla.\mathbf{A} + \mu\varepsilon\, \partial_t V = 0, \qquad [15.2]$$

the other two Maxwell equations, $\nabla.\mathbf{E} = q_v/\varepsilon$ and $\nabla \times \mathbf{B} = \mu\mathbf{j} + \varepsilon\mu\, \partial_t\mathbf{E}$, are verified if V and $\mathbf{A}$ are solutions to the equations of propagation

$$\Delta V - \mu\varepsilon\, \partial^2_{tt} V = -\, q_v/\varepsilon \qquad \text{and} \qquad \Delta \mathbf{A} - \mu\varepsilon\, \partial^2_{tt}\mathbf{A} = -\mu\mathbf{j}, \qquad [15.3]$$

where $v = 1/\sqrt{\mu\varepsilon}$ is the *speed of propagation*. These equations have particular solutions called *retarded potentials*

$$V_{\text{ret}}(\mathbf{r}, t) = (1/4\pi\varepsilon) \iiint d\mathcal{V}'\, q_v(\mathbf{r}', t - R/v)/R \quad \text{where} \quad \mathbf{R} = \mathbf{r} - \mathbf{r}', \qquad [15.4]$$

$$\mathbf{A}_{\text{ret}}(\mathbf{r}, t) = (\mu/4\pi) \iiint d\mathcal{V}'\, \mathbf{j}(\mathbf{r}', t - R/v)/R. \qquad [15.5]$$

The charge density $q_v(\mathbf{r}, t)$ appears to be the source of V and the current density $\mathbf{j}(\mathbf{r}, t)$ the source of $\mathbf{A}$. In fact, as V and $\mathbf{A}$ are coupled by the Lorentz condition, each one of these densities may be a source for both V and $\mathbf{A}$.

Knowing the retarded potentials, the relations [15.1] determine the retarded fields. We may write them in the form (which is not very useful)

$$\mathbf{E}_{\text{ret}}(\mathbf{r}, t) = -(1/4\pi\varepsilon) \iiint d\mathcal{V}'\, (1/R)\, [\mu\varepsilon\, \partial_{t'}\mathbf{j}(\mathbf{r}', t) + \nabla' q_v(\mathbf{r}', t')]_{t' = t - R/v}, \qquad [15.6]$$

$$\mathbf{B}_{\text{ret}}(\mathbf{r}, t) = (\mu/4\pi) \iiint d\mathcal{V}'\, (1/R)\, [\nabla' \times \mathbf{j}(\mathbf{r}', t')]_{t' = t - R/v}. \qquad [15.7]$$

To the solutions $V_{\text{ret}}(\mathbf{r}, t)$ and $\mathbf{A}_{\text{ret}}(\mathbf{r}, t)$ we may add any solutions $V_o(\mathbf{r}, t)$ and $\mathbf{A}_o(\mathbf{r}, t)$ of the homogeneous equations

$$\Delta V_o - \mu\varepsilon\, \partial^2_{tt} V_o = 0 \qquad \text{and} \qquad \Delta \mathbf{A}_o - \mu\varepsilon\partial^2_{tt}\mathbf{A}_o = 0 \qquad [15.8]$$

and obtain another solution. Accordingly, we add the terms $\mathbf{E}_o = -\nabla V_o - \partial_t \mathbf{A}_o$ and $\mathbf{B}_o = \nabla \times \mathbf{A}_o$ to the retarded fields [15.6] and [15.7]. It is always possible to choose V_o and $\mathbf{A}_o$ in order that the solutions $V = V_{\text{ret}} + V_o$ and $\mathbf{A} = \mathbf{A}_{\text{ret}} + \mathbf{A}_o$ satisfy the initial and the boundary conditions. The retarded potentials [15.4] and [15.5] and the corresponding fields correspond to an infinite medium with $V_{\text{ret}} \to 0$ and $\mathbf{A}_{\text{ret}} \to 0$ as $r \to \infty$ if q_v and $\mathbf{j}$ occupy a finite region of space.

In the particular case of time-independent charge density and current density, we find the time-independent solutions

$$V_P(\mathbf{r}) = (1/4\pi\varepsilon) \iiint d\mathcal{V}'\, q_v(\mathbf{r}')/R, \qquad \mathbf{A}_P(\mathbf{r}) = (\mu/4\pi)\iiint d\mathcal{V}'\, \mathbf{j}(\mathbf{r}')/R, \qquad [15.9]$$

$$\mathbf{E}_P(\mathbf{r}) = (1/4\pi\varepsilon) \iiint d\mathcal{V}'\, q_v(\mathbf{r}')\, \mathbf{R}/R^3, \quad \mathbf{B}_P(\mathbf{r}) = (\mu/4\pi)\iiint d\mathcal{V}'\, \mathbf{R} \times \mathbf{j}(\mathbf{r}')/R^3. \quad [15.10]$$

In the following, we consider only the particular retarded solutions. To simplify, we omit the indices (ret) and we designate the solutions simply by V, $\mathbf{A}$, $\mathbf{E}$, and $\mathbf{B}$.

15.2. Dipole radiation

Consider a charge and current distribution in a small region $\mathcal{V}$ that constitutes the emitter. The retarded potentials [15.4] and [15.5] at a point **r** may be written as

$$V(\mathbf{r}, t) = \frac{1}{4\pi\varepsilon} \iiint_{\mathcal{V}} d\mathcal{V}' \frac{1}{R} q_v(\mathbf{r}', t'), \qquad \mathbf{A}(\mathbf{r}, t) = \frac{\mu}{4\pi} \iiint_{\mathcal{V}} d\mathcal{V}' \frac{1}{R} \mathbf{j}(\mathbf{r}', t'), \quad [15.11]$$

where we have set $\mathbf{R} = \mathbf{r} - \mathbf{r}'$ and $t' = t - R/v$. Here $\mathbf{r}'$ is a running point in the volume $\mathcal{V}$ of the source. In general, these expressions are very difficult to use, as the charge and current densities must be taken at an earlier time t' with a time delay R/v that depends on $\mathbf{r}'$. Thus, some approximations are necessary. If we are interested only in the potentials at large distance from the volume $\mathcal{V}$, we may write up to the first order in $\mathbf{r}'/r$

$$R = |\mathbf{r} - \mathbf{r}'| \cong r[1 - \frac{\mathbf{r}.\mathbf{r}'}{r^2}], \qquad \frac{1}{R} = \frac{1}{|\mathbf{r} - \mathbf{r}'|} \cong \frac{1}{r} + \frac{\mathbf{e}_r.\mathbf{r}'}{r^2}, \qquad \text{where } \mathbf{e}_r = \frac{\mathbf{r}}{r}.$$

Within the same approximation, the time of the densities at $\mathbf{r}'$ may be written as

$$t' \cong t_o + (\mathbf{r}.\mathbf{r}')/vr, \qquad \text{where } t_o \equiv t - r/v. \qquad [15.12]$$

Writing all quantities as power series in $\mathbf{r}'/r$ and keeping only the first order, we find

$$q_v(\mathbf{r}', t') \cong q_v(\mathbf{r}', t_o) + \partial_{t_o} q_v(r'.t_o)\ (\mathbf{e}_r.\mathbf{r}')/v,$$

$$\mathbf{j}(\mathbf{r}', t') \cong \mathbf{j}(\mathbf{r}', t_o) + \partial_{t_o} \mathbf{j}(r'.t_o)\,(\mathbf{e}_r.\mathbf{r}')/v. \qquad [15.13]$$

Thus, the potentials may be written as

$$V(\mathbf{r}, t) \cong V_1(\mathbf{r}, t) + V_2(\mathbf{r}, t), \qquad \mathbf{A}(\mathbf{r}, t) \cong \mathbf{A}_1(\mathbf{r}, t) + \mathbf{A}_2(\mathbf{r}, t), \qquad [15.14]$$

$$V_1(\mathbf{r}, t) \cong \frac{1}{4\pi\varepsilon} \frac{q}{r} \Big|_{\text{ret}}, \qquad V_2(\mathbf{r}, t) \cong \frac{1}{4\pi\varepsilon} \frac{1}{r^2} [(\mathbf{e}_r.\mathbf{p}) + \frac{r}{v} (\mathbf{e}_r.\, \dot{\mathbf{p}})] \Big|_{\text{ret}}, \qquad [15.15]$$

$$\mathbf{A}_1(\mathbf{r}, t) \cong \frac{\mu}{4\pi} \frac{\dot{\mathbf{p}}}{r} \Big|_{\text{ret}}, \qquad \mathbf{A}_2(\mathbf{r}, t) \cong \frac{\mu}{4\pi} \frac{1}{r^3} [(\mathcal{M} \times \mathbf{r}) + \frac{r}{v} (\dot{\mathcal{M}} \times \mathbf{r})]_{\text{ret}}, \quad [15.16]$$

where $f|_{\text{ret}}$ means that f must be evaluated at time $t_o = t - r/v$, that is,

$$q|_{\text{ret}} = \iiint_{\mathcal{V}} d\mathcal{V}' q_v(\mathbf{r}', t_o),$$

$$\mathbf{p}|_{\text{ret}} = \iiint_{\mathcal{V}} d\mathcal{V}'\, \mathbf{r}' q_v(\mathbf{r}', t_o), \qquad \mathcal{M}|_{\text{ret}} = \tfrac{1}{2} \iiint_{\mathcal{V}} d\mathcal{V}'\, \mathbf{r}' \times \mathbf{j}_v(\mathbf{r}', t_o). \qquad [15.17]$$

V_1 involves the total charge q, V_2 involves the *electric dipole moment* $\mathbf{p}$, and its time-derivative $\dot{\mathbf{p}}$, $\mathbf{A}_1$ involves $\dot{\mathbf{p}}$, and $\mathbf{A}_2$ involves the magnetic dipole moment $\mathcal{M}$ and its time-derivative $\dot{\mathcal{M}}$.

The total charge q of atoms, molecules, and macroscopic systems is often equal to zero. The approximation $R \cong r(1 - \mathbf{r}.\mathbf{r}'/r^2)$, which consists of keeping only the first order r'/r, is equivalent to neglecting the contributions of electric and magnetic multipoles that are higher than the electric dipole moment $\mathbf{p}$ and the magnetic dipole moment $\mathcal{M}$. For this reason, it is called *dipole approximation.* If the sources are simple harmonic functions of angular frequency ω, all the quantities depend on time by the intermediary of a factor $e^{i\omega t}$. In this case, we may replace the differentiation with respect to time by the factor $i\omega$. Designating by $k = \omega/v$ the wave number, we find:

$$V(\mathbf{r}, t) \cong (1/4\pi\varepsilon r^2)\{(1 + ikr)(\mathbf{e}_r.\mathbf{p})\}|_{ret},$$

$$\mathbf{A}(\mathbf{r}, t) \cong (\mu/4\pi r^2)\ \{ivkr\mathbf{p} + (1 + ikr)(\mathcal{M} \times \mathbf{e}_r)\}|_{ret}, \qquad [15.18]$$

$$\mathbf{E}(\mathbf{r}, t) \cong (1/4\pi\varepsilon r^3)\{\ (3 + 3ikr - k^2r^2)(\mathbf{e}_r.\mathbf{p})\ \mathbf{e}_r - (1 + ikr - k^2r^2)\ \mathbf{p} + i(kr/v)\ (1 + ikr)(\mathbf{e}_r \times \mathcal{M})\ \}|_{ret}, \qquad [15.19]$$

$$\mathbf{B}(\mathbf{r}, t) \cong (\mu/4\pi r^3)\{\ ivkr(1 + ikr)(\mathbf{p} \times \mathbf{e}_r) + (3 + 3ikr - k^2r^2)(\mathbf{e}_r.\mathcal{M})\mathbf{e}_r - (1 + ikr - k^2r^2)\ \mathcal{M}\ \}|_{ret}. \qquad [15.20]$$

Note that the potentials V and $\mathbf{A}$ are related by Lorentz condition [15.2], which may be written as $\nabla.\mathbf{A} + ikV/v = 0$. Thus, in this case, V and the fields $\mathbf{E}$ and $\mathbf{B}$ are known if we know the vector potential $\mathbf{A}$:

$$V = i(v/k)\ (\nabla.\mathbf{A}), \qquad \mathbf{E} = -i(v/k)\ [k^2\,\mathbf{A} + \nabla(\nabla.\mathbf{A})], \qquad \mathbf{B} = \nabla \times \mathbf{A}. \qquad [15.21]$$

15.3. Electric dipole radiation

Consider an electric dipole $\mathbf{p}(t)$ situated at the origin and oriented in the direction Oz and assume that it is a simple harmonic function of time

$$\mathbf{p}(t) = p_m\ e^{i\omega t}\ \mathbf{e}_z. \qquad [15.22]$$

This dipole produces a time-dependent electric field that induces a magnetic field. Thus, it emits an electromagnetic wave of angular frequency ω. Such a dipole may be modeled as two balls located at A and B at a distance d apart with charges

$\pm q_m e^{i\omega t}$ such that $p_m = q_m d$ (Figure 15.1). These charges are supplied by an oscillating current

$$I(t) = \dot{q} = i\omega q_m e^{i\omega t} = I_m e^{i\omega t} \qquad \text{with } p_m = -i(d/\omega)I_m. \qquad [15.23]$$

We assume in this section that the distance d is small enough, so that the electric current may be considered uniform on the segment AB. To get an idea of the order of magnitude, we note that an electric current propagates in an electric circuit with the speed of the electromagnetic wave in the surrounding dielectric or vacuum. The wave takes a time d/c to travel the distance d. The intensity I may be considered as uniform over $AB = d$, if the time d/c is much shorter than the period $2\pi/\omega$ of oscillation of the dipole. This is equivalent to the condition

$$d << \lambda, \qquad [15.24]$$

where $\lambda = v/\tilde{v} = 2\pi v/\omega$ is the wavelength. The case where d is of the order of λ will be considered in section 15.5 in studying emission by antennas.

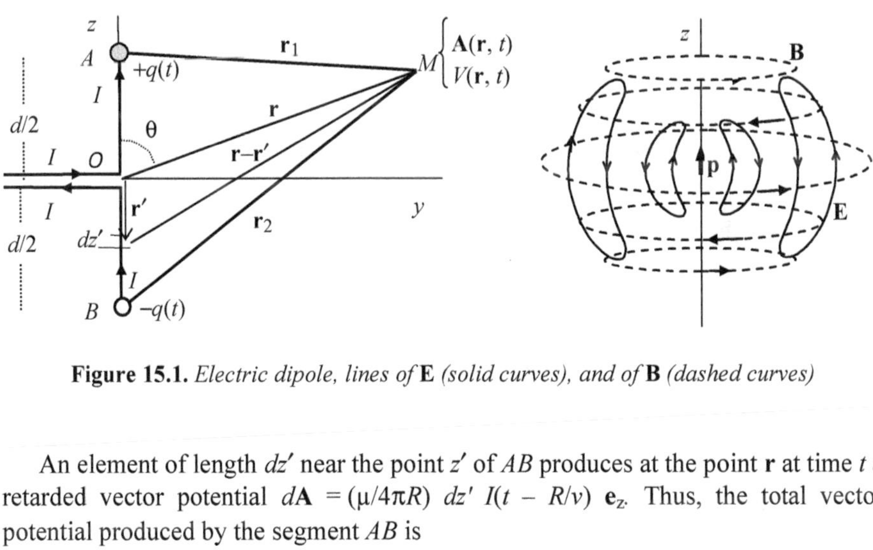

Figure 15.1. *Electric dipole, lines of* **E** *(solid curves), and of* **B** *(dashed curves)*

An element of length dz' near the point z' of AB produces at the point $\mathbf{r}$ at time t a retarded vector potential $d\mathbf{A} = (\mu/4\pi R)\, dz'\, I(t - R/v)\, \mathbf{e}_z$. Thus, the total vector potential produced by the segment AB is

$$\mathbf{A}(\mathbf{r}, t) = \frac{\mu}{4\pi} \mathbf{e}_z \int_{-d/2}^{d/2} \frac{dz'}{|\mathbf{r}-\mathbf{r}'|} I(t - |\mathbf{r}-\mathbf{r}'|/v). \qquad [15.25]$$

Using the expression [15.23] for I, we may write in the limit of large distance compared to d ($r >> d$) and large wavelength compared to d ($kd << 1$)

$$|\mathbf{r}-\mathbf{r}'| \cong r - z'\cos\theta, \quad e^{i\omega(t-|\mathbf{r}-\mathbf{r}'|/v)} \cong e^{i\omega(t-r/v)}\, e^{ikz'\cos\theta} \cong (1 + ikz'\cos\theta)\, e^{i\omega(t-r/v)}. [15.26]$$

Thus, to the first order in d

$$\mathbf{A}(\mathbf{r},t)=\frac{\mu}{4\pi}\frac{ip_{\mathrm{m}}\omega}{r}e^{\mathrm{i}(\omega t-kr)}\mathbf{e}_z\equiv\frac{\mu}{4\pi}\frac{\mathrm{i}\omega}{r}\mathbf{p}\big|_{\mathrm{ret}},\qquad[15.27]$$

where $\mathbf{p}\big|_{\mathrm{ret}}$ means that $\mathbf{p}$ must be evaluated at time $t'=t-r/v$.

We note that $\mathbf{A}$ propagates with a phase velocity equal to v. The equations [15.18], [15.19], and [15.20] give the scalar potential and the fields

$$V=\frac{1}{4\pi\varepsilon}\frac{1}{r^2}(1+\mathrm{i}kr)(\mathbf{p}.\mathbf{e}_r)_{\mathrm{ret}},\qquad[15.28]$$

$$\mathbf{E}=\frac{1}{4\pi\varepsilon}\frac{1}{r^3}[(3+3\mathrm{i}kr-k^2r^2)(\mathbf{p}.\mathbf{e}_r)\mathbf{e}_r-(1+\mathrm{i}kr-k^2r^2)\mathbf{p}]_{\mathrm{ret}},$$

$$\mathbf{B}=\frac{\mu}{4\pi}\frac{1}{r^3}\mathrm{i}\omega(1+\mathrm{i}kr)(\mathbf{r}\times\mathbf{p})_{\mathrm{ret}}.\qquad[15.29]$$

Explicitly, using spherical coordinates about Oz, we find

$$\mathbf{A}(\mathbf{r},t)=\frac{\mu}{4\pi}\frac{i\omega p}{r}(\cos\theta\,\mathbf{e}_r-\sin\theta\,\mathbf{e}_\theta)\big|_{\mathrm{ret}},\quad V=\frac{1}{4\pi\varepsilon}\frac{p}{r^2}(1+ikr)\cos\theta\big|_{\mathrm{ret}},\qquad[15.30]$$

$$\mathbf{E}=\frac{1}{4\pi\varepsilon}\frac{p}{r^3}[(2+2\mathrm{i}kr)\cos\theta\,\mathbf{e}_r+(1+\mathrm{i}kr-k^2r^2)\sin\theta\,\mathbf{e}_\theta)]_{\mathrm{ret}},$$

$$\mathbf{B}=\frac{\mu}{4\pi}\frac{i\omega p}{r^2}(1+ikr)\sin\theta\,\mathbf{e}_\varphi\big|_{\mathrm{ret}}.\qquad[15.31]$$

The lines of the field $\mathbf{E}$ and $\mathbf{B}$ are illustrated in Figure 15.1.

If the wavelength is very long ($\lambda >> r >> d$), the terms $1/r^3$ in the expression of $\mathbf{E}$ and $1/r^2$ in the expression of $\mathbf{B}$ are dominant. So, the potentials and the fields may be written as

$$\mathbf{A}(\mathbf{r},t)=\frac{\mu}{4\pi}\frac{i\omega p}{r}(\cos\theta\,\mathbf{e}_r-\sin\theta\,\mathbf{e}_\theta)\big|_{\mathrm{ret}},\qquad V(\mathbf{r},t)=\frac{1}{4\pi\varepsilon}\frac{p}{r^2}\cos\theta\big|_{\mathrm{ret}},\qquad[15.32]$$

$$\mathbf{E}(\mathbf{r},t)=\frac{1}{4\pi\varepsilon}\frac{p}{r^3}[(2\cos\theta\,\mathbf{e}_r+\sin\theta\,\mathbf{e}_\theta]_{\mathrm{ret}},\quad \mathbf{B}(\mathbf{r},t)=\frac{\mu}{4\pi}\frac{i\omega p}{r^2}\sin\theta\,\mathbf{e}_\varphi\big|_{\mathrm{ret}}.\qquad[15.33]$$

But, at large distances compared to λ ($r >> \lambda >> d$), i.e. in the so-called *wave zone*, the terms $1/r$ in the potentials and the fields are dominant and we find

$$\mathbf{A}(\mathbf{r},t)=\frac{\mu}{4\pi}\frac{i\omega p}{r}(\cos\theta\,\mathbf{e}_r-\sin\theta\,\mathbf{e}_\theta)\big|_{\mathrm{ret}}=\frac{\mu}{4\pi}\frac{d}{r}(\cos\theta\,\mathbf{e}_r-\sin\theta\,\mathbf{e}_\theta)\,I(t-\frac{r}{v})\big|_{\mathrm{ret}},\qquad[15.34]$$

$$V(\mathbf{r}, t) = \frac{1}{4\pi\varepsilon} \frac{d}{vr} I(t - \frac{r}{v}) \cos\theta |_{\text{ret}}, \qquad [15.35]$$

$$\mathbf{E} = \frac{1}{4\pi\varepsilon} \frac{ikd}{vr} I(t - \frac{r}{v}) \sin\theta\, \mathbf{e}_\theta)|_{\text{ret}}, \quad \mathbf{B} = \frac{\mu}{4\pi} \frac{ikd}{r} I(t - \frac{r}{v}) \sin\theta\, \mathbf{e}_\varphi|_{\text{ret}}. \qquad [15.36]$$

In this zone, the wave propagates in the radial direction $\mathbf{e}_r$ with the phase velocity v, but it does not have spherical symmetry. The fields **E** and **B** are orthogonal to the direction of propagation, the trihedron ($\mathbf{E}$, $\mathbf{B}$, $\mathbf{e}_r$) is right-handed, and the ratio E/B is equal v. These are the usual properties of electromagnetic plane waves.

The energy density and the Poynting vector associated with this wave and averaged over a time equal to a period of the wave are

$$< U_{(\text{DE})\,\text{v}} > = \frac{\mu}{32\pi^2 v^2} \frac{1}{r^2} p_{\text{m}}^2\, \omega^4 \sin^2\theta, \qquad [15.37]$$

$$< \mathbf{S}_{(\text{DE})} > = \frac{1}{2\mu} (\mathbf{E}\times\mathbf{B}^*) = \frac{\mu}{32\pi^2 v} \frac{1}{r^2} p_{\text{m}}^2 \omega^4 \sin^2\theta\, \mathbf{e}_r = < U_{(\text{DE})\,\text{v}} > v\, \mathbf{e}_r. \qquad [15.38]$$

This means that the energy propagates in the radial direction with the velocity v. We note that the intensity of the electric dipole radiation $\mathcal{I}_{(\text{DE})} = < S_{(\text{DE})} >$ decreases like $1/r^2$, this can be explained by the distribution of the energy over a sphere of radius r.

The power of the electric dipole radiation, emitted in the solid angle $d\Omega$ in the direction of the angles θ and φ, is

$$dP_{(\text{DE})} = \mathcal{I}_{(\text{DE})} r^2 d\Omega = \frac{\mu}{32\pi^2 v} p_{\text{m}}^2 \omega^4 \sin^2\theta\, d\Omega = \frac{v}{32\pi^2 \varepsilon} p_{\text{m}}^2 k^4 \sin^2\theta\, d\Omega. \qquad [15.39]$$

The radiation is not isotropic. The intensity of radiation vanishes in the direction of the dipole ($\theta = 0$ or $\theta = \pi$), since the fields **E** and **B** vanish in this direction. The averaged total emitted power is the flux of <**S**> over a sphere of radius r:

$$\begin{aligned} P_{(\text{DE})} &= \iint d\boldsymbol{s} < \mathbf{S}_{(\text{DE})} > \mathbf{e}_r = \iint_{\text{sphere}} \mathcal{I}_{(\text{DE})} r^2 \sin\theta\, d\theta\, d\varphi \\ &= \frac{\mu}{32\pi^2 v} p_{\text{m}}^2 \omega^4 \int_0^\pi d\theta \sin^3\theta \int_0^{2\pi} d\varphi = \frac{\mu}{12\pi v} p_{\text{m}}^2 \omega^4. \end{aligned} \qquad [15.40]$$

It may be expressed in terms of the effective intensity $I_{\text{eff}} = I_{\text{m}}/\sqrt{2} = \omega p_{\text{m}}/d\sqrt{2}$ as

$$P_{(\text{DE})} = \frac{\mu v}{6\pi} (kd)^2 I_{\text{eff}}^2 = \frac{2}{3} \pi\mu v \left(\frac{d}{\lambda}\right)^2 I_{\text{eff}}^2. \qquad [15.41]$$

Particularly, for waves in vacuum we find $P_{(DE)} \cong 80\pi^2 (d/\lambda)^2 I_{eff}^2$. This power is obviously supplied by the generator of the electric current. The electric dipole consumes as much energy as if it has a *radiation resistance*

$$R_{(DE)} = (2/3)\,\pi\mu v\,(d/\lambda)^2 = 80\pi^2 (d/\lambda)^2 \qquad \text{(in ohms)}. \qquad [15.42]$$

For a given intensity I_{eff}, the total emitted power varies as the frequency squared, at least for large wavelengths such that $kd = 2\pi d/\lambda << 1$.

15.4. Magnetic dipole radiation

The magnetic moment of a loop of area $\mathcal{S}$ carrying a current $I(t)$ is $\mathcal{M}(t) = \mathcal{S}I(t)$. If the loop lies in the plane Oxy, $\mathcal{M}$ points in the direction Oz (Figure 15.2a). We assume that the loop is circular with a small radius, compared to the wavelength λ. The current is then approximately the same at all the points of the loop. If the current is simple harmonic of angular frequency ω, the magnetic moment is a simple harmonic function with the same angular frequency:

$$I = I_m\, e^{i\omega t}, \qquad \mathcal{M} = \mathcal{M}_m\, e^{i\omega t}\, \mathbf{e}_z, \quad \text{where} \quad \mathcal{M}_m = \mathcal{S} I_m. \qquad [15.43]$$

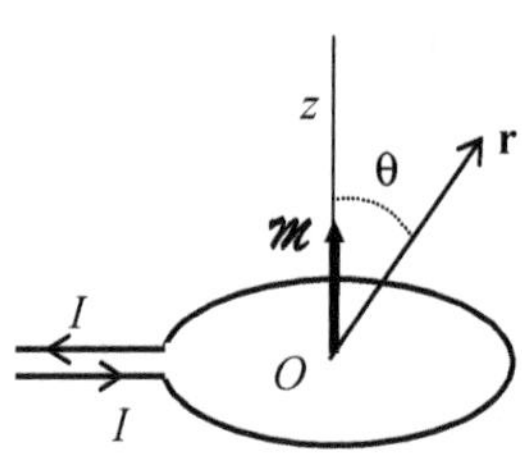

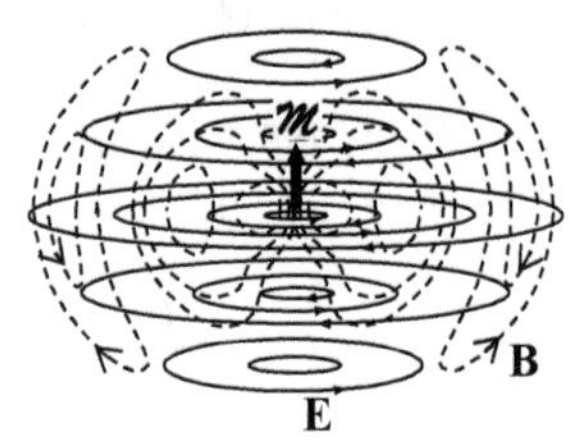

Figure 15.2. *Magnetic dipole moment and its lines of field*

The retarded vector potential, produced by this circuit at a point **r** and a time t, is

$$\mathbf{A}(\mathbf{r}, t) = \frac{\mu}{4\pi} \int_C R\, d\varphi' \frac{1}{|\mathbf{r} - \mathbf{r}'|} I(t - |\mathbf{r} - \mathbf{r}'|/v)\, \mathbf{e}'_\varphi, \qquad [15.44]$$

where we have considered an element of the circuit of length $dr' = R\, d\varphi'$. As $R << r$, we may write, to the first order in R/r:

$$|\mathbf{r} - \mathbf{r}'| \cong r\,[1 - (R/r) \sin\theta \cos(\varphi - \varphi')]. \qquad [15.45]$$

Using the expression [15.43] of $I(t)$ and assuming that $kR \ll 1$ (that is, $\lambda \gg R$), we may write, to the first order in kR and R/r:

$$\mathbf{A}(\mathbf{r}, t) \cong \frac{\mu}{4\pi}\frac{R}{r} I(t-\frac{r}{v}) \int_0^{2\pi} d\varphi' \,[1 + (\frac{R}{r} + ikR) \sin\theta \cos(\varphi-\varphi')][-\sin\varphi' \mathbf{e}_x + \cos\varphi' \mathbf{e}_y]$$

$$= \frac{\mu}{4\pi}\frac{R^2}{r^2} I(t-\frac{r}{v})(1+ikr) \sin\theta[-\sin\varphi\, \mathbf{e}_x + \cos\varphi\, \mathbf{e}_y] = \frac{\mu}{4\pi r^3}(1+ikr)(\mathcal{M}\times\mathbf{r})\big|_{\text{ret}}. \quad [15.46]$$

As there is no accumulation of electric charge ($q_v = 0$), the retarded scalar potential $V(\mathbf{r}, t)$ is equal to zero at all points of space. On the other hand, using [15.46], we may verify easily that $\nabla.\mathbf{A} = 0$. The Lorentz condition [15.2] implies that $\partial_t V = 0$. As V is necessarily of the form $V_m\, e^{i\omega t}$, we deduce that $V_m = 0$ and, consequently, that $V = 0$. Knowing V and $\mathbf{A}$, the fields may be written as

$$\mathbf{E} = -\nabla V - \partial_t \mathbf{A} = \frac{\mu}{4\pi}\frac{\omega}{r^3}(-i + kr)(\mathcal{M}\times\mathbf{r})\big|_{\text{ret}},$$

$$\mathbf{B} = \nabla\times\mathbf{A} = \frac{\mu}{4\pi}\frac{1}{r^3}[\frac{\mathbf{r}}{r^2}(\mathbf{r}.\mathcal{M})(3 + 3ikr - k^2r^{2]} - \mathcal{M}(1 + ikr - k^2r^{2]}]_{\text{ret}}. \quad [15.47]$$

Explicitly, using spherical coordinates, we find

$$\mathbf{E} = \frac{\mu}{4\pi}\frac{vk}{r^2}(-i + kr)\,\mathcal{M}(t - \frac{r}{v}) \sin\theta\, \mathbf{e}_\varphi,$$

$$\mathbf{B} = \frac{\mu}{4\pi}\frac{1}{r^3}\mathcal{M}(t - \frac{r}{v})[\,(2 + 2ikr)\cos\theta\, \mathbf{e}_r + (1 + ikr - k^2r^2 \sin\theta\, \mathbf{e}_\theta]. \quad [15.48]$$

Particularly, we find at large distances

$$\mathbf{E} = \frac{\mu}{4\pi}\frac{vk^2}{r}\mathcal{M}(t - \frac{r}{v}) \sin\theta\, \mathbf{e}_\varphi, \qquad \mathbf{B} = -\frac{\mu}{4\pi}\frac{k^2}{r}\mathcal{M}(t-\frac{r}{v}) \sin\theta\, \mathbf{e}_\theta. \quad [15.49]$$

E and **B** are orthogonal to the direction of propagation $\mathbf{e}_r$, the trihedron ($\mathbf{E}$, $\mathbf{B}$, $\mathbf{e}_r$) is right-handed and the ratio E/B is equal to v. These are the usual properties of electromagnetic plane waves. Both fields decrease like $1/r$ at large distances.

The energy density and the Poynting vector associated with this wave and averaged over a time equal to a period of the wave are

$$< U_{(DM)\,v} > = \frac{\mu}{32\pi^2}\frac{k^4}{r^2}\mathcal{M}_m^{\;2}\sin^2\theta, \quad [15.50]$$

$$< \mathbf{S}_{(DM)} > = \frac{1}{2\mu}(\mathbf{E}\times\mathbf{B}^*) = \frac{\mu}{32\pi^2}\,\mathcal{M}_m^{\;2}\sin^2\theta\, \mathbf{e}_r = <U_{(DM)\,v}>\, v\, \mathbf{e}_r. \quad [15.51]$$

The energy propagates in the radial direction with the velocity v and the intensity decreases like $1/r^2$. The power that is emitted in a solid angle $d\Omega$ is

$$dP_{(DM)} = \mathcal{I}_{(DM)}\, r^2\, d\Omega = \frac{\mu}{32\pi^2}\, v k^4 \mathcal{M}_m^{\,2} \sin^2\theta\, d\Omega\,. \qquad [15.52]$$

The radiation is not isotropic. The intensity of radiation vanishes in the direction of $\mathcal{M}$ ($\theta = 0$ or $\theta = \pi$), since the fields **E** and **B** vanish in this direction. The averaged total emitted power intercepted by a sphere of radius r is

$$P_{(DM)} = \iint_{\mathcal{S}} \langle \mathbf{S}_{(DM)} \rangle\, \mathbf{e}_r\, d\mathcal{S} = \iint_{\mathcal{S}} \mathcal{I}_{(DM)}\, r^2 \sin\theta\, d\theta\, d\varphi$$

$$= \frac{\mu}{16\pi}\, v k^4 \mathcal{M}_m^{\,2} \iint_{\mathcal{S}} \sin^3\theta\, d\theta = \frac{\mu}{12\pi}\, v k^4 \mathcal{M}_m^{\,2}. \qquad [15.53]$$

Expressing this power in terms of the effective intensity $I_m/\sqrt{2}$ in the loop, we find

$$P_{(DM)} = \frac{1}{6}\, \pi v \mu (kR)^4\, I_{eff}^{\,2}. \qquad [15.54]$$

Thus, the magnetic dipole consumes energy as if it has a *radiation resistance*

$$R_{(DM)} = (1/6)\pi v \mu\, (kR)^4 = (8/3)\pi^5 v \mu\, (R/\lambda)^4, \qquad [15.55]$$

that is, $R_{(DM)} = 3.074 \times 10^5\, (R/\lambda)^4$ (in ohms). Comparing the power that is emitted by an electric dipole [15.40] and by a magnetic dipole [15.54] having the same current intensity, we find

$$\frac{P_{(DE)}}{P_{(DM)}} = \left(\frac{d}{\pi k R^2}\right)^2 = \frac{1}{4\pi^4}\left(\frac{d}{R}\right)^2 \left(\frac{\lambda}{R}\right)^2. \qquad [15.56]$$

As λ is usually much larger than R and the length of the dipole d for an antenna may be taken much larger than the radius of the loop, the emission of an electric dipole is often much more efficient than that of a magnetic dipole.

15.5. Antennas

We consider an antenna constituted by two rectilinear conductors OA and OB of length $d/2$ each and supplied at O by a sinusoidal current of angular frequency ω (Figure 15.1a). We suppose that d is comparable to the wavelength λ. The electric current is established in the conductors under the influence of the electromagnetic wave produced by the current itself. The wave propagates in the medium surrounding the antenna with the speed of light in this medium and the electric current is established in the conductor with this speed. The current is reflected at the end points A and B. It is a sinusoidal function of time of angular frequency ω and a function of z ($-d/2 < z < d/2$) along the antenna with a wave number $k = \omega/v$. Assuming that $I(t, z)$ is symmetric in z, it may be written as

$$I(t, z) = \underline{I}_{1\,\mathrm{m}}\, e^{\mathrm{i}(\omega t - k|z'|)} + \underline{I}_{2\,\mathrm{m}}\, e^{\mathrm{i}(\omega t + k|z'|)} . \qquad [15.57]$$

The intensity being equal to zero at the ends A and B ($z = \pm\, d/2$) at any time, we must have $\underline{I}_{2\,\mathrm{m}} = -\,\underline{I}_{1\,\mathrm{m}}\, e^{-ikd}$. Redefining the amplitude, we may write:

$$I(t, z) = \underline{I}_{\mathrm{m}}\, e^{\mathrm{i}\omega t} \sin(k|z'| - \tfrac{1}{2}kd) . \qquad [15.58]$$

Each element dz' of the antenna emits like a small electric dipole of length dz' carrying a current of amplitude $\underline{I}_{\mathrm{m}} \sin(k|z'| - \tfrac{1}{2}kd)$. According to equation [15.23], it is equivalent to an electric dipole $d\mathbf{p}_{\mathrm{m}} = (1/\mathrm{i}\omega)\, \underline{I}_{\mathrm{m}} \sin(k|z'| - \tfrac{1}{2}kd)\, dz'\, \mathbf{e}_z$. Using the expression [15.34], the vector potential of this element at the point M of position $\mathbf{r}$ is

$$d\mathbf{A}(\mathbf{r}, t) = \frac{\mu}{4\pi} \frac{I_{\mathrm{m}}}{R}\, e^{\mathrm{i}(\omega t - kR)} \sin(\tfrac{1}{2}\, kd - k|z'|)\, dz'\, \mathbf{e}_z, \qquad [15.59]$$

where $R \approx r[1 - (z'/r) \cos\theta]$ is the distance from the element dz' to M and r is the distance from the center O of the antenna to M. The total potential at M is obtained by integration over z'. The dominant term at large distance r is

$$\begin{aligned} \mathbf{A}(\mathbf{r}, t) &= \frac{\mu}{2\pi} \frac{I_{\mathrm{m}}}{r}\, e^{\mathrm{i}(\omega t - kr)}\, \mathbf{e}_z \int_{-d/2}^{d/2} dz'\; e^{ikz'\cos\theta} \sin(kd/2 - k|z'|) \\ &= \frac{\mu}{2\pi} \frac{I_{\mathrm{m}}}{kr} \frac{1}{\sin^2\theta} [\cos(\tfrac{1}{2}\, kd \cos\theta) - \cos(\tfrac{1}{2}kd)]\, e^{\mathrm{i}(\omega t - kr)}\, \mathbf{e}_z . \end{aligned} \qquad [15.60]$$

The corresponding scalar potential and fields at large distances are

$$\begin{aligned} V &= \frac{\mu}{2\pi} \frac{vI_{\mathrm{m}}}{kr} \frac{\cos\theta}{\sin^2\theta} [\cos(\tfrac{1}{2}kd \cos\theta) - \cos(\tfrac{1}{2}kd)]\, e^{\mathrm{i}(\omega t - kr)} , \\ \mathbf{E} &= \mathrm{i}\frac{\mu}{2\pi} \frac{vI_{\mathrm{m}}}{r} \frac{1}{\sin\theta}\, e^{\mathrm{i}(\omega t - kR)} [\cos(\tfrac{1}{2}kd \cos\theta) - \cos(\tfrac{1}{2}kd)]\, \mathbf{e}_\theta, \\ \mathbf{B} &= \mathrm{i}\frac{\mu}{2\pi} \frac{I_{\mathrm{m}}}{r} \frac{1}{\sin\theta}\, e^{\mathrm{i}(\omega t - kR)} [\cos(\tfrac{1}{2}kd \cos\theta) - \cos(\tfrac{1}{2}kd)]\, \mathbf{e}_\phi . \end{aligned} \qquad [15.61]$$

The fields decrease like $1/r$. They are orthogonal to the direction of propagation, the trihedron ($\mathbf{E}$, $\mathbf{B}$, $\mathbf{e}_r$) is right-handed, and the ratio E/B is equal to v. On the other hand, the field $\mathbf{E}$ is in the plane containing the antenna and the radial direction from O to the observation point M; thus the wave is polarized in the azimuthal plane. The fields vanish if OM is in the direction of the antenna. Taking the real part of the fields, we may calculate the Poynting vector:

$$\mathbf{S} = \frac{\mathbf{E}\times\mathbf{B}}{\mu} = \frac{\mu}{4\pi^2}\,\frac{vI_{\mathrm{m}}^{\;2}}{r^2}\,\frac{1}{\sin^2\theta}\sin^2(\omega t - kr)\,[\cos(\tfrac{1}{2}kd\cos\theta) - \cos(\tfrac{1}{2}kd)]^2\,\mathbf{e}_{\mathrm{r}}. \quad [15.62]$$

Thus, the intensity of the wave in the direction θ is

$$\mathcal{I} = \langle S\rangle = \frac{\mu}{8\pi^2}\,\frac{vI_{\mathrm{m}}^{\;2}}{r^2}\,\frac{1}{\sin^2\theta}[\cos(\tfrac{1}{2}kd\cos\theta) - \cos(\tfrac{1}{2}kd)]^2. \quad [15.63]$$

The radiation power that is emitted in the solid angle $d\Omega$ is

$$dP_{(\mathrm{ray})} = \langle S\rangle r^2\,d\Omega = \frac{\mu}{8\pi^2}\,vI_{\mathrm{m}}^{\;2}\,\frac{1}{\sin^2\theta}\,[\cos(\tfrac{1}{2}kd\cos\theta) - \cos(\tfrac{1}{2}kd)]^2\,d\Omega. \quad [15.64]$$

The most efficient antenna is such that $kd = n\pi$, i.e. $d = n\lambda/2$. The shortest antenna verifying this condition has a length $d = \lambda/2$ (half-wave antenna). The corresponding angular distribution of the emitted power is

$$\frac{dP_{(\mathrm{ray})}}{d\Omega}\Big|_{d=\lambda/2} = \frac{\mu}{8\pi^2}\,vI_{\mathrm{m}}^{\;2}\,\frac{1}{\sin^2\theta}\,\cos^2(\tfrac{1}{2}\pi\cos\theta). \quad [15.65]$$

In the case of a full-wave antenna ($d = \lambda$, i.e. $kd = 2\pi$), we find:

$$\frac{dP_{(\mathrm{ray})}}{d\Omega}\Big|_{d=\lambda} = \frac{\mu}{8\pi^2}\,vI_{\mathrm{m}}^{\;2}\,\frac{4}{\sin^2\theta}\cos^4(\tfrac{1}{2}\pi\cos\theta). \quad [15.66]$$

The angular distribution of the radiation in the case of a half-wave antenna is similar to that of an electric dipole but that of a full wave antenna is more concentrated in the normal direction to the antenna ($\theta = \pi/2$).

The time-averaged power intercepted by the sphere of radius r is

$$P_{(\mathrm{ray})} = \iint_S dP_{(\mathrm{ray})} = \frac{\mu}{4\pi}\,cI_{\mathrm{m}}^{\;2}\int d\theta\,\frac{1}{\sin\theta}\,[\cos(\tfrac{1}{2}kd) - \cos(\tfrac{1}{2}kd\cos\theta)]^2. \quad [15.67]$$

Making the successive change of variables $u = \cos\theta$, $1\pm u = x$, then $kdx/2 = \xi$, we may write:

$$P_{(\mathrm{ray})} = \frac{\mu}{4\pi}\,cI_{\mathrm{m}}^{\;2}\int_0^{kd} d\xi\,\frac{1}{\xi}\,[(1-\cos\xi)\cos(\tfrac{1}{2}kd) - \sin(\tfrac{1}{2}kd)\sin\xi]. \quad [15.68]$$

These integrals may be expressed in terms of the so-called *cosine integral* and *sine integral*:

$$\mathrm{Ci}(x) \equiv -\int_x^{\infty} d\xi\,(1/\xi)\cos\xi, \qquad \mathrm{Si}(x) \equiv \int_0^x d\xi\,(1/\xi)\sin\xi,$$
$$\int_0^x d\xi\,(1/\xi)(1-\cos\xi) = \gamma + \ln(x) - \mathrm{Ci}(x), \tag{15.69}$$

where $\gamma = 0.57721$ is Euler's constant. We find

$$P_{(\mathrm{ray})} = \frac{\mu}{4\pi}\, cI_{\mathrm{m}}^{2}\{\cos(\tfrac{1}{2}kd)\,[\gamma + \ln(kd) - \mathrm{Ci}(kd)] - \sin(\tfrac{1}{2}kd)\,\mathrm{Si}(kd)\}, \tag{15.70}$$

$$P_{(\mathrm{ray})} = 1.21\,\frac{\mu}{4\pi}\, cI_{\mathrm{m}}^{2} \qquad \text{(half-wave antenna)},$$
$$P_{(\mathrm{ray})} = 3.35\,\frac{\mu}{4\pi}\, cI_{\mathrm{m}}^{2} \qquad \text{(full-wave antenna)}. \tag{15.71}$$

Thus, the full-wave antenna emits almost three times more energy than a half-wave antenna with the same current.

15.6. Potentials and fields of a charged particle*

Let us consider a particle of charge q, position $\mathbf{r}_{\mathrm{q}}(t')$ and velocity $\dot{\mathbf{r}}_{\mathrm{q}}(t')$. This particle is equivalent to a charge density and a current density, at each point $\mathbf{r}'$, proportional to three-dimensional Dirac functions around the point $\mathbf{r}_{\mathrm{q}}(t')$:

$$q_{\mathrm{v}}(\mathbf{r}', t') = q\,\delta^3[\mathbf{r}' - \mathbf{r}_{\mathrm{q}}(t')], \qquad \mathbf{j}(\mathbf{r}', t') = q\,\dot{\mathbf{r}}_{\mathrm{q}}(t')\,\delta^3[\mathbf{r}' - \mathbf{r}_{\mathrm{q}}(t')]. \tag{15.72}$$

Substituting these expressions to q_{v} and $\mathbf{j}$ in equations [15.4] and [15.5], we find the scalar potential, for instance

$$V(\mathbf{r}, t) = \frac{q}{4\pi\varepsilon}\iiint \frac{d^3r'}{|\mathbf{r}-\mathbf{r}'|}\int dt'\,\delta(t' - t + \frac{|\mathbf{r}-\mathbf{r}'|}{v})\,\delta^3[\mathbf{r}' - \mathbf{r}_{\mathrm{q}}(t')]$$
$$= \frac{q}{4\pi\varepsilon}\int dt'\,\frac{1}{R(t')}\,\delta[t' - t + \frac{R(t')}{v}], \tag{15.73}$$

where we have integrated over $\mathbf{r}'$ by using the Dirac function $\delta^3[\mathbf{r}' - \mathbf{r}_{\mathrm{q}}(t')]$ and set

$$\mathbf{R}(t') = \mathbf{r} - \mathbf{r}_{\mathrm{q}}(t'), \qquad \text{thus} \quad R(t')^2 \equiv \mathbf{r}^2 + \mathbf{r}_{\mathrm{q}}(t')^2 - 2\,\mathbf{r}.\mathbf{r}_{\mathrm{q}}(t'). \tag{15.74}$$

Using the property of Dirac delta function $\delta[F(t')] = \Sigma_j\, \delta(t'- t_i)/|F'(t_j)|$, where the sum is over the roots t_j of $F(t') = 0$ (see the section A.11 of the Appendix A) and assuming that the argument of $\delta[t'- t + R(t')/v]$ has a single root, we may write

$$V(\mathbf{r}, t) = \frac{q}{4\pi\varepsilon}\, \frac{1}{g(t')R(t')}\Big|_{\text{ret}}, \qquad [15.75]$$

where

$$g(t') \equiv 1 - \mathbf{e}_q(t').\boldsymbol{\beta}_q(t'), \qquad \mathbf{e}_q(t') = \mathbf{R}(t')/R(t') \text{ and } \boldsymbol{\beta}_q(t') = \dot{\mathbf{r}}_q(t')/v. \qquad [15.76]$$

$\mathbf{e}_q(t')$ is the unit vector in the direction that joins the position $\mathbf{r}_q(t')$ of the charge at time t' to the point $\mathbf{r}$, where the potential is calculated (see Figure 15.3a). $\boldsymbol{\beta}_q(t')$ is the vector velocity of the particle, measured in units of the speed of propagation v of the wave (i.e. $\boldsymbol{\beta}_q = \mathbf{v}_q/v$). The symbol (ret) means that the preceding expression is to be calculated at a time t', which is a root of the equation

$$t' - t + R(t')/v = 0, \qquad \text{i.e. } \mathbf{r}^2 + \mathbf{r}_q(t')^2 - 2\, \mathbf{r}.\mathbf{r}_q(t') = v^2(t' - t)^2. \qquad [15.77]$$

Similarly, the vector potential may be written as:

$$\mathbf{A}(\mathbf{r}, t) = \frac{\mu q}{4\pi}\, \frac{\dot{\mathbf{r}}_q(t')}{g(t')R(t')}\Big|_{\text{ret}}. \qquad [15.78]$$

The potentials [15.75] and [15.78] are the *Lienard-Wieckert retarded potentials*. These expressions are valid for any motion of the charge. They show that we must take into account the time delay due to the propagation if the charge q or its position varies. We note that equation [15.77], which determines t', may be very complicated. It may have no roots; then, the potentials vanish at the considered position and time. If it has several roots t_i, the potentials are the superpositions of the potentials produced by the particle at the corresponding positions and arriving at M at the same time t.

To understand the physical meaning of the retarded potentials and fields, let us consider the following simple example: assume that a charge q is initially at rest at the origin O and it is suddenly accelerated at $t = 0$ to have a velocity v_q. At $t < 0$, the field configuration is that of a particle at rest, i.e. $\mathbf{B} = 0$ and $\mathbf{E} = q\mathbf{r}/4\pi\varepsilon r^3$ (spherically symmetric). After $t = 0$, the field is deformed. It is evident that a test charge placed close to O will feel this deformation before a test charge placed far away. A test charge located at a distance R from O feels this deformation only after the time $t = R/v$. At this time, the source charge q is no longer at O but a point O' such that $OO' = v_q t$. At a given time t, the sphere of radius $R = ct$ divides the space into two regions: outside the sphere, where there is only the spherical symmetric electric field $\mathbf{E} = q\mathbf{r}/4\pi\varepsilon r^3$ of the charge q at rest and inside the sphere, where we have both electric and magnetic fields with the electric field having no spherical symmetry.

If the charge velocity is small, compared to the propagation speed of the wave ($\beta_q << 1$), the correction factor $g(t')$ is almost equal to 1. Thus, we find

$$V(\mathbf{r}, t) = \frac{q}{4\pi\varepsilon}\frac{1}{R(t')}\Big|_{\text{ret}}, \qquad \mathbf{A}(\mathbf{r}, t) = \frac{\mu q}{4\pi}\frac{\dot{\mathbf{r}}_q(t')}{R(t')}\Big|_{\text{ret}}. \qquad [15.79]$$

If, in addition, the distance R from the position of the charge q to the point $\mathbf{r}$, where the potentials are evaluated, is always small in the sense that the delay R/v is much smaller than the characteristic time of the charge motion (such as the period of its motion if it is periodic), the propagation effects are small and we find the usual expressions of the permanent regime:

$$V(\mathbf{r}, t) = \frac{q}{4\pi\varepsilon}\frac{1}{R(t)}, \quad \mathbf{A}(\mathbf{r}, t) = \frac{\mu q}{4\pi}\frac{\dot{\mathbf{r}}_q(t)}{R(t)}, \quad \text{with } R(t) = |\mathbf{r} - \mathbf{r}_q(t)|. \quad [15.80]$$

For instance, if an antenna of length 1 m emits a wave of 10 MHz, the effects of the time delay are negligible if $R << cT = 30$ m.

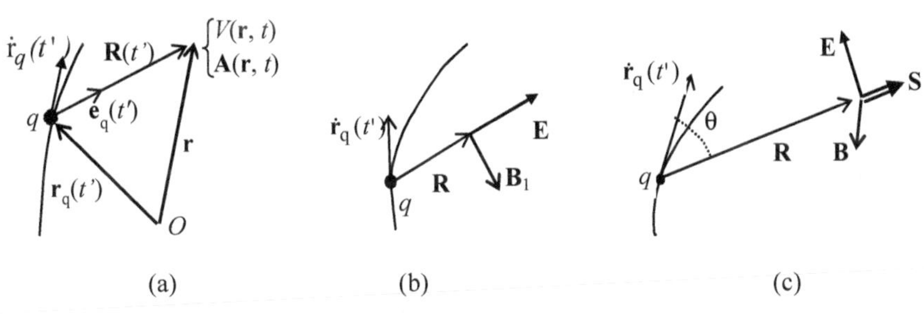

Figure 15.3. *a) Potentials of a charge q, b) field of a point charge at small distance, and c) fields at large distance (wave zone)*

Knowing the retarded potentials, we may calculate the retarded fields, by using the relation [15.1]. Using the expression [15.74] of $R(t')^2$, we find

$$\partial R/\partial x_\alpha = R_\alpha/R = e_{q,\alpha}, \quad \partial R/\partial t' = -v\,\mathbf{e}_q(t').\boldsymbol{\beta}_q(t') \quad \text{and } \partial t/\partial t' = g(t').$$

Thus, the retarded electric field may be written as

$$\mathbf{E} = -\nabla V - \partial_t \mathbf{A} \equiv \mathbf{E}_1 + \mathbf{E}_2, \qquad \mathbf{B} = \nabla \times \mathbf{A} \equiv \mathbf{B}_1 + \mathbf{B}_2. \qquad [15.81]$$

The fields $\mathbf{E}_1$ and $\mathbf{E}_2$ are given by:

$$\mathbf{E}_1(\mathbf{r}, t) = \frac{q}{4\pi\varepsilon} \frac{1}{\gamma^2 g^3 R^2} (\mathbf{e}_q - \boldsymbol{\beta}_q)\,|_{ret},$$

$$\mathbf{E}_2(\mathbf{r}, t) = \frac{q}{4\pi\varepsilon} \frac{1}{v g^3 R} \{\mathbf{e}_q \times [(\mathbf{e}_q - \boldsymbol{\beta}_q) \times \boldsymbol{\alpha}_q]\}\,|_{ret}, \qquad [15.82]$$

where we have set $\gamma = (1-\beta_q^2)^{-1/2}$ and $\boldsymbol{\alpha}_q(t) = d\boldsymbol{\beta}_q/dt = \ddot{\mathbf{r}}_q/v$, $\ddot{\mathbf{r}}_q(t)$ being the acceleration of the charged particle. The magnetic fields $\mathbf{B}_1$ and $\mathbf{B}_2$ are given by

$$\mathbf{B}_1(\mathbf{r}, t) = \frac{1}{v}(\mathbf{e}_q \times \mathbf{E}_1) = \frac{\mu q}{4\pi} \frac{1}{\gamma^2 g^3 R^3} (\dot{\mathbf{r}}_q \times \mathbf{R})_{ret},$$

$$\mathbf{B}_2(\mathbf{r}, t) = \frac{1}{v}(\mathbf{e}_q \times \mathbf{E}_2) = \frac{\mu q}{4\pi} \frac{1}{g^3 R} \mathbf{e}_q \times \{\mathbf{e}_q \times [(\mathbf{e}_q - \boldsymbol{\beta}_q) \times \boldsymbol{\alpha}_q]\}_{ret}. \qquad [15.83]$$

We note that the fields $\mathbf{E}_1$ and $\mathbf{B}_1$ decrease as functions of the distance like $1/R^2$, while $\mathbf{E}_2$ and $\mathbf{B}_2$ decrease like $1/R$. At short distance from the charge (Figure 15.3b), the fields $\mathbf{E}_1$ and $\mathbf{B}_1$ dominate, with $\mathbf{E}_1$ almost longitudinal (in the direction of $\mathbf{e}_q$) and $\mathbf{B}_1$ transverse. On the contrary, at large distances, i.e. the wave zone, (Figure 15.3c), the fields $\mathbf{E}_2$ and $\mathbf{B}_2$ dominate. They are transverse (perpendicular to $\mathbf{e}_q$) and they vanish if the particle is not accelerated at instant t'. If the velocity of the particle is small and the fields are evaluated at short distance, the time delay R/v is negligible and we find the permanent regime expressions

$$\mathbf{E}(\mathbf{r}, t) = \frac{q}{4\pi\varepsilon} \frac{\mathbf{R}(t)}{R(t)^3}, \qquad \mathbf{B}(\mathbf{r}, t) = \frac{\mu q}{4\pi} \frac{\dot{\mathbf{r}}_q(t) \times \mathbf{R}(t)}{R(t)^3}. \qquad [15.84]$$

15.7. Case of a charged particle with constant velocity *

Consider a particle of charge q and constant velocity $\mathbf{v}_q = v\boldsymbol{\beta}_q$. We may assume, without loss of generality, that the particle is at the origin at time $t' = 0$ and that $\mathbf{v}_q$ is in the direction Oz (Figure 15.4a). Thus, we have $\mathbf{r}_q(t') = v\boldsymbol{\beta}_q t'$. Equation [15.77] is quadratic in t'; its retarded root is

$$t' = t - (\gamma^2/v)\,[\,\mathbf{R}_q.\boldsymbol{\beta}_q + D(t)\,], \qquad \text{where } \gamma = 1/\sqrt{1-\beta_q^2}, \qquad [15.85]$$

where $\mathbf{R}_q(t) = \mathbf{r} - v\boldsymbol{\beta}_q t$ is the vector that joins the position $v\boldsymbol{\beta}_q t$ of the particle at time t (at which we evaluate the fields) to the point $\mathbf{r}$ (where we calculate the field) and

$$D(t) = \sqrt{\mathbf{R}_q(t)^2(1-\beta_q^2) + [\mathbf{R}_q(t).\boldsymbol{\beta}_q]^2} = \sqrt{(\mathbf{r}^2 - v^2t^2)(1-\beta_q^2) + (\mathbf{r}.\boldsymbol{\beta}_q - vt)^2}\ . \quad [15.86]$$

Thus, the retarded potentials may be written as

$$V(\mathbf{r}, t) = \frac{q}{4\pi\varepsilon}\frac{1}{D(t)} \quad \text{and} \quad \mathbf{A}(\mathbf{r}, t) = \frac{\mu q}{4\pi}\frac{\dot{\mathbf{r}}_q}{D(t)} \quad [15.87]$$

and the fields may be written as

$$\mathbf{E} = \frac{q}{4\pi\varepsilon}\frac{\mathbf{R}_q}{\gamma^2 D^3}, \qquad \mathbf{B} = \frac{\mu q}{4\pi}\frac{(\dot{\mathbf{r}}_q \times \mathbf{R}_q)}{\gamma^2 D^3}. \quad [15.88]$$

We note that these fields are in fact $\mathbf{E}_1$ and $\mathbf{B}_1$, which decrease like $1/R_q^2$ since the acceleration of the charged particle is equal to zero.

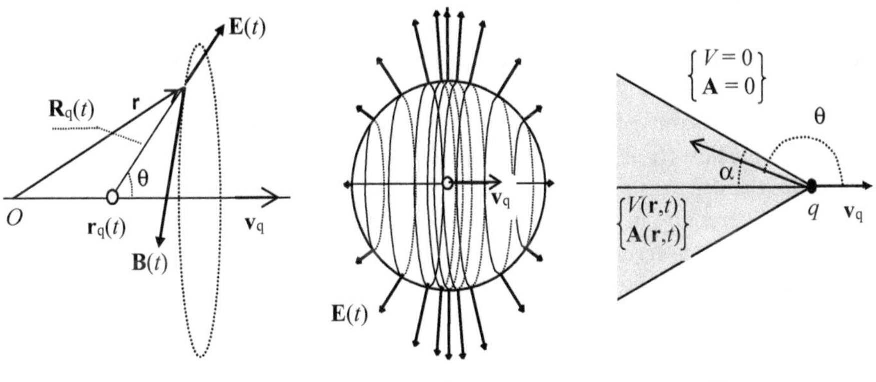

Figure 15.4. *a) Fields of a charged particle in uniform motion, b) the field lines of* **E** *are radial and those of* **B** *are circular about the trajectory. The fields are the most intense in the plane containing the particle and normal to the trajectory* ($\theta = \pi/2$)*, c) Cherenkov effect*

Designating by $\theta(t)$ the angle that $\mathbf{R}_q(t)$ form with $\boldsymbol{\beta}$, we may write

$$V(\mathbf{r}, t) = \frac{q}{4\pi\varepsilon}\frac{\eta}{R_q(t)}, \qquad \mathbf{A}(\mathbf{r}, t) = \frac{\mu q}{4\pi}\frac{\eta\dot{\mathbf{r}}_q(t)}{R_q(t)}, \qquad \text{where } \eta = (1-\beta_q^2\sin^2\theta)^{-1/2}, \quad [15.89]$$

$$\mathbf{E} = \frac{q}{4\pi\varepsilon}\frac{\eta'\mathbf{R}_q}{R_q^3}, \qquad \mathbf{B} = \frac{\mu q}{4\pi}\frac{\eta'\dot{\mathbf{r}}_q \times \mathbf{R}_q}{R_q^3}, \qquad \text{where } \eta' = (1-\beta_q^2)(1-\beta_q^2\sin^2\theta)^{-3/2}. \quad [15.90]$$

The potentials and the fields depend on time through $\mathbf{R}_q(t)$ and $\theta(t)$. V and $\mathbf{E}$ are not isotropic as in the case of a charge at rest but more concentrated in the directions normal to the velocity (θ close to $\pi/2$), and this concentration becomes more and more accentuated as the velocity of the charged particle increases. They have a rotational symmetry about the rectilinear trajectory of the particle and reflection symmetry with respect to the normal plane containing the particle (Figure 15.4b).

According to the special theory of relativity, the velocity of a particle v_q is always less than c, hence $\beta_q < 1$. If the charged particle is moving in vacuum ($v = c$), the expressions of the potentials and the fields make sense for any angle θ. However, if the particle is moving in matter, where $v < c$, its velocity may exceed v, the potentials and the fields become infinite in the directions $\theta = \alpha$ and $\pi - \alpha$ such that

$$|\sin \alpha| = 1/\beta_q = v/v_q \quad (\alpha < \pi/2). \qquad [15.91]$$

If the angle θ lies between α and $\pi - \alpha$, $(1 - \beta_q^2 \sin^2\theta)^{1/2}$ is imaginary and the expressions become meaningless. Thus, the wave is concentrated in a cone whose vertex is at the position of the particle and situated behind the particle with a half-angle α (Figure 15.4c). The cone situated in front of the particle, such that $\theta < \alpha$, is forbidden by causality. This effect is the electromagnetic equivalent of the shock wave in acoustics (α is then called Mach angle). It was initially predicted by Sommerfeld and it was observed in water by the French doctor Mallet in 1926 and studied by Cherenkov in 1934. The Cherenkov emission has a continuous spectrum and the wave is strongly polarized. It is produced by high-energy electrons. The emitted radiation may be detected by photomultipliers, and it may be used to study nuclear reactions and reactions of particles.

The potentials and the fields of a charged particle may be evaluated using the laws of transformation of the potentials and the fields from the proper frame of the particle S' to the observer's frame S. In S', we have $\mathbf{A}' = 0$, $V'(\mathbf{r}', t') = q/4\pi\varepsilon_0 R'$, $\mathbf{B}' = 0$ and $\mathbf{E}'(\mathbf{r}', t') = q\mathbf{R}'/4\pi\varepsilon_0 R'^3$. Using the Lorentz transformation for the coordinates, the potentials and the fields, we obtain the same results [15.87] and [15.88] (see problem 15.12).

15.8. Radiated energy by a moving charge

Consider the electromagnetic fields [15.82] and [15.83] of a moving charged particle. Assuming that the particle is non-relativistic ($\beta \ll 1$) and keeping only the first order in β, we may replace the factors g and $(1 - \beta^2)$ by 1 and write

$$\mathbf{E}_1 = \frac{q}{4\pi\varepsilon}\frac{\mathbf{R}}{R^3}\Big|_{\text{ret}}, \quad \mathbf{B}_1 = \frac{\mu q}{4\pi}\frac{1}{R^3}(\dot{\mathbf{r}}_q \times \mathbf{R})\Big|_{\text{ret}},$$

$$\mathbf{E}_2 = \frac{q}{4\pi\varepsilon}\frac{1}{vR^3}[\mathbf{R}\times(\mathbf{R}\times\boldsymbol{\alpha})]\Big|_{\text{ret}} = -\frac{q}{4\pi\varepsilon}\frac{1}{vR}\boldsymbol{\alpha}_\perp\Big|_{\text{ret}},$$

$$\mathbf{B}_2 = \frac{\mu q}{4\pi}\frac{1}{R^4}\{\mathbf{R}\times[\mathbf{R}\times(\mathbf{R}\times\boldsymbol{\alpha})]\}\Big|_{\text{ret}} = -\frac{\mu q}{4\pi}\frac{1}{R^2}(\mathbf{R}\times\boldsymbol{\alpha}_\perp)\Big|_{\text{ret}}, \qquad [15.92]$$

where $\boldsymbol{\alpha}_{q\perp}$ is the component of $\boldsymbol{\alpha}_q = \ddot{\mathbf{r}}_q/v$ that is normal to $\mathbf{R}$. The Poynting vector of this wave may be written as

$$\mathbf{S} = \frac{q^2}{16\pi^2\varepsilon}\Big[\frac{v}{R^6}\mathbf{R}\times(\boldsymbol{\beta}\times\mathbf{R}) - \frac{1}{R^5}\mathbf{R}\times(\mathbf{R}\times\boldsymbol{\alpha}_{q\perp})$$

$$-\frac{1}{R^4}\boldsymbol{\alpha}_{q\perp}\times(\boldsymbol{\beta}\times\mathbf{R}) + \frac{1}{vR^3}\boldsymbol{\alpha}_{q\perp}\times(\mathbf{R}\times\boldsymbol{\alpha}_{q\perp})\Big]\Big|_{\text{ret}}. \qquad [15.93]$$

As we have seen in section 15.6, the fields $\mathbf{E}_2$ and $\mathbf{B}_2$ are dominant in the wave zone. $\mathbf{E}_2$ is orthogonal to $\mathbf{R}(t')$ and $\mathbf{B}_2$ is orthogonal to both $\mathbf{R}(t')$ and $\mathbf{E}_2$ (Figure 15.3c). The Poynting vector in this region is

$$\mathbf{S} \cong \frac{1}{\mu}\mathbf{E}_2\times\mathbf{B}_2 = \frac{1}{\mu v}(E_2)^2\mathbf{e}_q = \frac{q^2}{16\pi^2\varepsilon v R^2}\alpha_{q\perp}{}^2\mathbf{e}_q = \frac{\mu q^2}{16\pi^2 v R^2}\ddot{\mathbf{r}}_{q\perp}^2\,\mathbf{e}_q\Big|_{\text{ret}}. \qquad [15.94]$$

Thus, the energy propagates in the radial direction $\mathbf{e}_q = \mathbf{R}_q/R_q$. The intensity of the wave being proportional to $\ddot{\mathbf{r}}_{q\perp}^2$, there is no emitted energy in the direction of acceleration. The total radiated power is the flux of $\mathbf{S}$ through a large surface $\mathcal{S}$ surrounding the charge:

$$P_{(\text{ray})} = \iint_{\mathcal{S}} d\mathcal{S}\,\mathbf{n}.\mathbf{S} = \int_0^{2\pi} d\varphi \int_0^{\pi} d\theta\; r^2 \sin\theta\, S.$$

Taking for $\mathcal{S}$ a sphere of radius R with a polar axis in the direction of $\boldsymbol{\alpha}$, we find $\alpha_\perp = \alpha\sin\theta$ and the radiated power is given by *Larmor formula*

$$P_{(\text{ray})} = \frac{\mu q^2}{6\pi v}|\ddot{\mathbf{r}}|^2. \qquad [15.95]$$

As an application, consider an electron bound to the atom by a force $-m\omega^2 x$ where x is the displacement from the equilibrium position. For simplicity, we assume that the electron oscillates on the axis $x'Ox$ with an amplitude A. Its position is $x = A\cos(\omega t)$ and its acceleration is $\ddot{x} = -A\omega^2\cos(\omega t)$. As it is emitting in vacuum, $v = c$ and the time-averaged radiated power is

$$<P_{(\text{ray})}> = \frac{1}{T}\int_0^T dt\; P = \frac{\mu_0 e^2}{12\pi c} A^2\omega^4. \qquad [15.96]$$

As the energy of this oscillator is $U = ½m\omega^2A^2$, it decreases according to the equation

$$\frac{dU}{dt} = -<P_{(\text{ray})}> = -\frac{\mu_0 e^2}{6\pi cm}\,\omega^2 U, \qquad [15.97]$$

whose solution is

$$U = U_0\, e^{-2t/\tau} \qquad \text{with } \tau = \frac{12\pi cm}{\mu_0 e^2\omega^2}. \qquad [15.98]$$

τ is the *relaxation time* in this classical model of emission. In the case of light emission ($\omega \approx 3 \times 10^{15}$ rad/s), we find $\tau \approx 10^{-8}$ s. This corresponds to a band width given by the uncertainty relation

$$\Delta\omega \approx \frac{1}{\tau} = \frac{\mu_0 e^2}{12\pi cm}\,\omega^2 \approx 10^8 \text{ Hz}. \qquad [15.99]$$

15.9. Problems

Electric dipole radiation

P15.1 Consider an electric dipole $\mathbf{p}(t)$ variable but at rest. It is modeled by two charges $-q(t)$ and $+q(t)$ at the points A and B of axis Oz at a distance d apart. Using the expressions [15.4] and [15.5], show directly the expression [15.15] for the retarded scalar potential.

P15.2 Using the definitions [15.17] and the expressions of the charge density of a point-like charge and the current density of a thin current (in terms of Dirac delta functions), evaluate the electric dipole moment of two point charges q and $-q$ at a distance d apart and the magnetic moment of a plane loop made of a thin wire.

P15.3 In this problem we review the emission of radiation by an oscillating electric dipole. We consider two small balls situated at A and B of coordinates $d/2$ and $-d/2$ on the axis Oz and connected by a wire of negligible resistance (Figure 15.1a). Let $+Q(t)$ and $-Q(t)$ be the charges of these balls at the time t. The current intensity is then $I = \partial_t Q$ fed at the origin O. **a)** Show that, at a point **r** situated at a large distance ($r >> d$), the retarded vector potential may be written as

$$\mathbf{A}(\mathbf{r}, t) = \frac{\mu}{4\pi} \int_{-d/2}^{d/2} dz' \frac{1}{|\mathbf{r} - \mathbf{r}'|} I(z', t - |\mathbf{r} - \mathbf{r}'|/v)\, \mathbf{e}_z .$$

Show that $|\mathbf{r} - \mathbf{r}'| = r - z' \cos\theta$. Assume that I is a sinusoidal function of angular frequency ω, such that $d << \lambda$, where λ is the wavelength of the radiation. Show that $\mathbf{A}(\mathbf{r}, t) = (\mu/4\pi)(d/r)\, I(t - r/v)\, \mathbf{e}_z$. **b)** Using the Lorentz condition, show that the scalar potential may be written as

$$V(\mathbf{r}, t) = \frac{1}{4\pi\varepsilon} \frac{zd}{r^3} [Q(t - \frac{r}{v}) + \frac{r}{v} I(t - \frac{r}{v})].$$

c) Consider the case $Q = -Q_m \cos(\omega t)$. Calculate the fields **E** and **B** and show that the time-averaged radiated power is $<P> = \mu d^2 \omega^2 I_m^2/12\pi\, v$. Deduce that the dipole consumes the same energy as a resistance $R = (2\pi/3)\sqrt{\mu/\varepsilon}\,(d/\lambda)^2$. Calculate R if the dipole emits in vacuum.

Magnetic dipole radiation

P15.4 A small electric circuit of area S lies in the plane Oxy around O. It carries a current $I(t)$. **a)** Determine the vector potential $\mathbf{A}(\mathbf{r}, t)$ and the scalar potential $V(\mathbf{r}, t)$ that it produces at large distance. **b)** Deduce the expressions of the fields $\mathbf{E}(\mathbf{r}, t)$ and $\mathbf{B}(\mathbf{r}, t)$. **c)** Write the expressions of **A**, **E** and **B**, if I is a sinusoidal function of angular frequency ω. **d)** Calculate the radiated power in the solid angle $d\Omega$ and the total radiated power.

P15.5 a) What is the length of a half-wave antenna emitting an *FM* wave of frequency $\tilde{\nu}$ = 100 MHz? What is the emitted power if this antenna is fed with an effective current of 10 A? **b)** Assume that a circular loop of the same length is fed with the same current. What is the emitted power?

Potentials and fields of a point charge

P15.6 Work out the details of the calculation leading from the expression [15.73] to the expression [15.75] for the retarded potential of a point-like particle of charge q.

P15.7 a) The Fourier transform of $1/r$ is $F(\mathbf{k}) = (2\pi)^{-3/2} \iiint d^3\mathbf{r}\, (1/r) e^{-i\mathbf{k}.\mathbf{r}}$. The Fourier theory allows to write the inverse relation $1/r = (2\pi)^{-3/2} \iiint d^3\mathbf{k}\, F(\mathbf{k}) e^{i\mathbf{k}.\mathbf{r}}$. Show that $F(\mathbf{k}) = 2(2\pi)^{-1/2} k^2$. Deduce that $\Delta(1/r) = -4\pi\, \delta(\mathbf{r})$ where $\delta(\mathbf{r})$ is the three-dimensional Dirac function (see section A.11 of appendix A). Generalize this relation in the form

$\Delta(1/|\mathbf{r}-\mathbf{r}'|) = -4\pi\,\delta(\mathbf{r}-\mathbf{r}')$. **b)** Deduce that the retarded potentials [15.4] and [15.5] verify the equations of propagation [15.3].

P15.8 An element $d\mathbf{L}$ of a conductor points in the direction Oz and its center is at the origin O. Assume that the conductor contains negative conduction charges of density $-q_l$ per unit length, moving with a velocity $-\mathbf{v}$ in the direction Oz and producing a current I. Using spherical coordinates, show that the fields $\mathbf{E}$ and $\mathbf{B}$ produced by these charges at a point $\mathbf{r}$ are

$$\mathbf{E}^- = -(q_l\, dL/4\pi\varepsilon_0)[(1-\beta^2)/(1-\beta^2\sin^2\theta)^{3/2}]\,(\mathbf{e}_r/r^2),$$
$$\mathbf{B}^- = (\mu_0 q_l\, dL\, v\sin\theta/4\pi)[(1-\beta^2)/(1-\beta^2\sin^2\theta)^{3/2}]\,(\mathbf{e}_\varphi/r^2)\,.$$

The positive charges in the conductor are at rest with a density $+q_l$ per unit length. They produce the fields $\mathbf{E}^+ = (q_l/4\pi\varepsilon_0)(\mathbf{e}_r/r^2)$ and $\mathbf{B}^+ = 0$. Deduce the expressions of the total electric and magnetic fields produced by the conductor. Show that the global field $\mathbf{E}$ is not exactly equal to zero but it points in a certain direction for some values of θ and in the opposite direction for other values of θ and its average value is equal to zero. What are the approximate expressions of $\mathbf{E}$ for $\beta \ll 1$? Verify that we find Biot-Savart law in this limit.

P15.9 Using the retarded potentials of a charged particle in uniform motion, derive the expressions of the fields.

P15.10 A charge q oscillates on the z axis with an angular frequency ω and a amplitude z_m. **a)** Calculate its electric dipole moment and show that the average power of the electric dipole radiation that it emits in the solid angle $d\Omega$ around the angles θ and φ is

$$dP_{(DE)} = \mathcal{I}_{(DE)}\, r^2\, d\Omega = \frac{\mu}{32\pi^2 v}\, p_m^{\,2}\omega^4 \sin^2\theta\, d\Omega = \frac{v}{32\pi^2\varepsilon}\, p_m^{\,2}\, k^4 \sin^2\theta\, d\Omega..$$

Calculate the total emitted power. Draw the radiation diagram, that is, a plot of $\mathcal{I}(\theta)$ in polar coordinates. What is the fraction of the power that is emitted in the directions making an angle less than 45° with the Oxy plane?

P15.11 Using the expressions [15.87] of the potentials of a point-like charge q moving with constant velocity, derive the expressions [15.88] of the fields. For this, first derive the relations $\boldsymbol{\nabla} D^{-1} = -D^{-3}[\mathbf{R}_q(1-\beta_q^{\,2})+(\mathbf{R}_q.\boldsymbol{\beta}_q)\boldsymbol{\beta}_q]$ and $\partial_t D^{-1} = D^{-3}(\mathbf{R}_q.\boldsymbol{\beta}_q)$. Show that V and $\mathbf{A}$ verify the Lorentz condition $\boldsymbol{\nabla}.\mathbf{A} + \varepsilon_0\mu_0\,\partial_t V = 0$. Calculate the flux of $\mathbf{B}$ through a sphere surrounding this charge, thus verifying Gauss's law. Show directly that $\mathbf{B}$ verifies the equation div $\mathbf{B} = 0$.

P15.12 A particle of charge q is moving with a velocity $\mathbf{v}_q = c\boldsymbol{\beta}_q$. Let S' be the frame of the particle and S that of the observer. Assume that the particle is at the origin of both frames. **a)** Write the Lorentz transformation ($\mathbf{r}$, t) to ($\mathbf{r}'$, t') for the space-time coordinates, for the potentials and for the fields. Set $\mathbf{R}_q = \mathbf{r} - \mathbf{v}_q t$ for the relative distance of $\mathbf{r}$ from the position of the charge at time t. Show that

$$\mathbf{r}' = \mathbf{R}_q + (\gamma-1)(\mathbf{R}_q.\mathbf{e}_q)\mathbf{e}_q \text{ and } r' = \gamma D(t),$$

where $\gamma_q = (1 - \beta_q^2)^{\frac{1}{2}}$, $\mathbf{e}_q = \mathbf{v}_q/v_q$ and $D(t) = [R_q^2(1 - \beta_q^2) + (\mathbf{R}_q.\boldsymbol{\beta}_q)^2]^{\frac{1}{2}}$. **b)** Write the expressions of the potentials and the fields in the frame S'. **c)** Using the transformation of the potentials and the fields, show that

$$V(\mathbf{r}, t) = K_o\gamma q/r' = K_o q/D(t) \qquad \text{and} \qquad \mathbf{A}(\mathbf{r}, t) = K_o\gamma_q q\mathbf{v}_q/c^2 r' = \mu_o q/4\pi D(t'),$$
$$\mathbf{E}(\mathbf{r}, t) = (K_o q/r'^3)[\gamma_q\mathbf{r}' + (1 - \gamma_q)(\mathbf{r}'.\mathbf{e}_q)\,\mathbf{e}_q] = K_o q\mathbf{R}_q/\gamma_q^2 D^3,$$
$$\mathbf{B}(\mathbf{r}, t) = (K_o q/c^2 r'^3)\,\gamma_q(\mathbf{v}_q \times \mathbf{r}') = (\mu_o q/4\pi\gamma_q^2 D^3)\,(\mathbf{v}_q \times \mathbf{R}_q).$$

Derive the expressions [15.89] and [15.90]. **d)** Show that in the non-relativistic limit, we obtain the expressions $\mathbf{E}(\mathbf{r}, t) = K_o q\mathbf{R}_q/R_q^3$ and $\mathbf{B}(\mathbf{r}, t) = (\mu_o q/4\pi R_q^3)(\mathbf{v}_q \times \mathbf{R}_q)$.

Radiated energy by a point charge

P15.13 a) Calculate the Poynting vector of the radiation emitted by a charged particle. **b)** Calculate its limit at large distances. **c)** Deduce the expression [15.94].

P15.14 a) Work out the derivation of the expression of the emitted power by an accelerated charge q. Assume that the charge is moving along the axis Oz with a velocity $\dot{\mathbf{r}}_q = v\boldsymbol{\beta}(t)$. **b)** Show that, because of this radiation, the particle is subject to a braking force and that its velocity decreases according to the law $\dot{r} = \dot{r}_o\, e^{-t/\tau}$, where $\tau = \mu q^2/6\pi m v$. Calculate τ in the case of the electron. **c)** In a linear accelerator, a proton is uniformly accelerated from rest to a final kinetic energy U_K over a distance L. Calculate the energy that the proton radiates in this process if $L = 10$ m and $U_K = 10$ MeV, $m_p = 1.67 \times 10^{-27}$ kg and $e = 1.60 \times 10^{-19}$ C.

P15.15 *Green function method.* Consider the emission equation

$$\Delta u - (1/v^2)\partial^2_{tt}u = -f(\mathbf{r}, t),$$

where u stands for the potential and the source f is, in general, non-sinusoidal. We may always write a Fourier representation of u and f,

$$u(\mathbf{r}, t) = \int d\omega\; e^{i\omega t}\,\mathcal{U}(\mathbf{r},\omega) \qquad \text{and} \qquad f(\mathbf{r}, t) = \int d\omega\; e^{i\omega t}\,\mathcal{F}(\mathbf{r},\omega).$$

Show that $\mathcal{U}$ verifies Helmholtz equation $\Delta\mathcal{U} + k^2\mathcal{U} = -\mathcal{F}$, where $k = \omega/v$. **b)** We define the *Green function* as the solution of the equation $\Delta\mathcal{G}(\mathbf{r}) + k^2\mathcal{G}(\mathbf{r}) = -\delta(\mathbf{r})$, where δ is the three-dimensional Dirac function. Verify that $\mathcal{G}(\mathbf{r}) = e^{-ikr}/4\pi r$. Deduce that

$$\mathcal{U}(\mathbf{r},\omega) = \int d^3\mathbf{r}'\, \mathcal{G}(\mathbf{r}-\mathbf{r}')\, \mathcal{F}(\mathbf{r}',\omega) \quad \text{and} \quad u(\mathbf{r},t) = (1/4\pi)\int d^3\mathbf{r}'\, f(\mathbf{r}', t - |\mathbf{r}-\mathbf{r}'|/v)/|\mathbf{r}-\mathbf{r}'|.$$

Answers to Some Problems

Chapter 1

P1.2 a) $\mathbf{e}'_1 = \cos\varphi\, \mathbf{e}_1 + \sin\varphi\, \mathbf{e}_2$, $\mathbf{e}'_2 = -\sin\varphi\, \mathbf{e}_1 + \cos\varphi\, \mathbf{e}_2$, $\mathbf{e}'_3 = \mathbf{e}_3$
$A'_1 = A_1 \cos\varphi + A_2 \sin\varphi$, $A'_2 = -A_1 \sin\varphi + A_2 \cos\varphi$, $A'_3 = A_3$

$$[R_z(\varphi)] = \begin{bmatrix} \cos\varphi & \sin\varphi & 0 \\ -\sin\varphi & \cos\varphi & 0 \\ 0 & 0 & 1 \end{bmatrix}, \quad R_z(\varphi)^{-1} = \widetilde{R}_z(\varphi) = [R_z(-\varphi)] = \begin{bmatrix} \cos\varphi & -\sin\varphi & 0 \\ \sin\varphi & \cos\varphi & 0 \\ 0 & 0 & 1 \end{bmatrix}.$$

b) $\mathbf{B} = (\mu_o I/2\pi r'^2)(-y'\,\mathbf{e}'_x + x'\,\mathbf{e}'_y)$ is a vector equal to $(\mu_o I/2\pi r'^2)(\mathbf{e}'_z \times \mathbf{r}')$, while $\mathbf{B}' = (\mu_o I/2\pi r'^2)\,[(x' \sin 2\varphi + y' \cos 2\varphi)\,\mathbf{e}'_1 + (x' \cos 2\varphi - y' \sin 2\varphi)\,\mathbf{e}'_2]$ is not a vector.

P1.4 $\iint_{\mathcal{S}} d\mathcal{S}\, \mathbf{n}.\mathbf{E}(\mathbf{r}) = 4\pi R^2 f(R)$, $\nabla.\mathbf{E} = 2f/r + \partial_r f$ and $\iiint_{\mathcal{V}} d\mathcal{V}\, \nabla.\mathbf{E} = 4\pi R^2 f(R)$.

P1.5 Let $x'_\alpha = \Sigma_\beta R_{\beta\alpha} x_\beta$. We find $\partial_\alpha = \Sigma_\beta\, \partial'_\beta.(\partial x'_\beta/\partial x_\alpha) = \Sigma_\beta\, \partial'_\beta R_{\alpha\beta} = \Sigma_\beta R_{\alpha\beta}\, \partial'_\beta$.

P1.6 b) If $d\mathbf{r}$ is on V = Constant, $dV = \Sigma_\alpha\, \partial_\alpha V\, dx_\alpha = \nabla V.d\mathbf{r} = 0$, hence ∇V is normal to $d\mathbf{r}$. **c)** If only x varies, as $r^2 = x^2 + y^2 + z^2$, we get $2r\, dr = 2x\, dx$, hence $\partial_\alpha r = x_\alpha/r$ and $\partial_\alpha f(r) = (df/dr)(\partial_\alpha r) = (df/dr)(x_\alpha/r)$. If $f = K/r$, we find $\partial_\alpha f(r) = (-K/r^2)(x_\alpha/r) = -Kx_\alpha/r^3$.

P1.8 a) If $V = K(\mathbf{p}.\mathbf{r})/r^3$, $\mathbf{E} = -\nabla V = (K/r^3)[3(\mathbf{p}.\mathbf{r})\,\mathbf{r}/r^2 - \mathbf{p}]$. If $\mathbf{p} = p\mathbf{e}_z$, $V = Kpz/r^3$ and $\mathbf{E} = (Kp/r^3)[3z\mathbf{r}/r^2 - \mathbf{e}_z]$. **b)** If $\mathbf{A} = k(\mathcal{M}\times\mathbf{r})/r^3$, we find $\mathbf{B} = (k/r^3)[3(\mathcal{M}.\mathbf{r})\mathbf{r}/r^2 - \mathcal{M}]$. Using spherical coordinates, if $\mathcal{M} = \mathcal{M}\mathbf{e}_z$, we find $\mathbf{A} = k\mathcal{M} \sin\theta\, \mathbf{e}_\varphi/r^2$ and $\mathbf{B} = (k\,\mathcal{M}/r^3)[3 \cos\theta\, \mathbf{e}_r - \mathbf{e}_z]$.

P1.9 As $\nabla \times \mathbf{E} = 0$, $\mathbf{E}$ is the gradient of the scalar function $f = 3x^2 - 5xz - 4y^2 + C$.

P1.10 As $\nabla.\mathbf{B} = 0$, we may write $\mathbf{B} = \nabla \times \mathbf{A}$, where $\mathbf{A} = -\tfrac{1}{2}By\mathbf{e}_x + \tfrac{1}{2}Bx\,\mathbf{e}_y + \nabla f$.

P1.12 a) $\nabla.(k\mathbf{r}) = 3k$ and $\nabla.[\mathbf{r}f(\mathbf{r})] = 3f + \mathbf{r}.\nabla f$.

P1.16 a) Verify that $\nabla_1 = \nabla_r$ and $\nabla_2 = -\nabla_r$. **b)** $\mathbf{F}_{2\to1} = -\mathbf{F}_{1\to2} = (dU/dr)(\mathbf{r}/r)$.

P1.18 $N_e = N_p = 1.37 \times 0^{24}$ and $N_n = 1.65 \times 10^{24}$. $Q = 2.2 \times 10^5$ C. $t = 2.2 \times 10^4$ s $\cong 6.1$ h.

Chapter 2

P2.1 a) $d\mathbf{F} = 2\pi K_o qq_s R^2 \sin\theta\, d\theta\, (z - R\cos\theta)\,[R^2 + z^2 - 2zR\cos\theta]^{-3/2}\,\mathbf{e}_z$
$\mathbf{F} = 2\pi K_o qq_s (R/2z^2)\,\mathbf{e}_z\{2R - |z - R| + (z^2 - R^2)/|z - R|\}$. If $z < R$, $\mathbf{F} = 0$ and, if $z > R$, $\mathbf{F} = (K_o qQ/z^2)\,\mathbf{e}_z$. **b)** $F_{z<R} = K_o qQz/R^3$ and $F_{z>R} = K_o qQ/z^2$

P2.2 $\mathbf{F} = 2K_o qq'\,y(y^2 + a^2)^{-3/2}\,\mathbf{e}_y$. $W = 2K_o qq'\,(y^2 + a^2)^{-1/2}$ independently of the path.

P2.4 a) $V(x) = 90\{1/|x| - 2/|x'|\}$ and $E(x) = 90\{1/x|x| - 2/x'|x'|\}$, where $x' = x - 0.05$.
b) $V(x, y) = 90(1/r - 2/d)$ and $\mathbf{E} = 90[x/r^3 - 2(x - 0.05)/d^3]\mathbf{e}_x + 90y[1/r^3 - 2/d^3]\mathbf{e}_y$, where $r = (x^2 + y^2)^{1/2}$ and $d = [(x - 0.05)^2 + y^2]^{1/2}$. **c)** $V(x,y) \rightarrow -90/r$ and $\mathbf{E} \rightarrow -90\mathbf{r}/r^3$.

P2.6 $V_M = V_o - Ex$ and $W =$ 40 eV independently of the path.

P2.7 a) $q_{v,1} = 0$ and $q_{v,2} = -4\varepsilon_o b$. $\mathbf{E}_1 = -2ax\,\mathbf{e}_x + 2ay\,\mathbf{e}_y$ and $\mathbf{E}_2 = -2bx\,\mathbf{e}_x - 2by\,\mathbf{e}_y$.
b) $\nabla \times \mathbf{E}_1 = 2\mathbf{e}_z$ and $\nabla \times \mathbf{E}_2 = 0$, $\mathbf{E}_2$ may be an electrostatic field with a potential $V = yx - (3/2)x^2 - (3/2)y^2 + C$ and charge density $q_v = 6\varepsilon_o$.

P2.8 a) If $x < -d/2$, $E_1 = -q_v d/2\varepsilon_o$ and $V_1 = q_v dx/2\varepsilon_o + q_v d^2/8\varepsilon_o + B$. If $-d/2 < x < d/2$, $E_2(x) = q_v x/\varepsilon_o$ and $V_2 = -q_v x^2/2\varepsilon_o + B$. If $x > d/2$, $E_3 = q_v d/2\varepsilon_o$ and $V_3 = -q_v x\,d/2\varepsilon_o + q_v d^2/8\varepsilon_o + B$. **c)** The total energy is $U_T = \frac{1}{2}mv^2 + (qq_v\,d/2\varepsilon_o)(d/4 + x_o)$ and it remains constant. If $qq_v > 0$ and $U_T > 0$, the particle crosses the plate. Let $U_o = -qq_v\,d^2/8\varepsilon_o$. If $U_o < U_T < 0$, the particle penetrates to $x_1 = (-\varepsilon_o mv^2/qq_v - d^2/4 - dx_o)^{1/2}$. If $U_T < U_o$, the particle can reach the point $x_1 = x_o + \varepsilon_o mv^2/qq_v d$ in front of the plate. If $qq_v < 0$, the particle crosses the plate and oscillates between the points $-x_1$ and x_1, where $x_1 = -\varepsilon_o mv^2/qq_v d - x_o$. **d)** For $x < -d/2$, $V_1 = (q_v d/2\varepsilon_o)x + q_v d^2/8\varepsilon_o - q_s d/2\varepsilon_o + B$ and $E_1 = -q_v d/2\varepsilon_o$. For $-d/2 < x < d/2$, $V_2 = -q_v x^2/2\varepsilon_o + q_s x/\varepsilon_o + B$ and $E_2 = q_v x/\varepsilon_o - q_s/\varepsilon_o$. For $x > d/2$, $V_3 = -(q_v\,d/2\varepsilon_o)x + q_v\,d^2/8\varepsilon_o + q_s\,d/2\varepsilon_o + B$ and $E_3 = q_v\,d/2\varepsilon_o$.

P2.9 a) R is the radius of the atom and $q_o = 15Ze/8\pi R^3$.
b) $q(r) = -(Zer^3/2R^5)(5R^2 - 3r^2)$ for $r < R$. **c)** $E(r) = K_o Ze/r^2 - \frac{1}{2}(K_o Ze/R^5)(5R^2 r - 3r^3)$ for $r < R$ and $E(r) = 0$ for $r > R$. $V(r) = K_o Ze/r + (K_o Ze/8R^5)(10R^2 r^2 - 3r^4 - 15R^4)$ for $r < R$ and $V(r) = 0$ for $r > R$.

P2.12 b) $\mathbf{p} = q\mathbf{r}$ in the case of a single charge, $\mathbf{p} = -q\,\overrightarrow{BA}$ in the case of two charges of Figure 2.16a and $\mathbf{p} = 0$ in the case of four charges of Figure 2.16b. **c)** Setting $\mathbf{p} = q\mathbf{r}$, $\langle\mathbf{E}\rangle = -K_o\mathbf{p}/a^3$ if q is inside the sphere and $\langle\mathbf{E}\rangle = -K_o\mathbf{p}/r^3$ if q is outside the sphere. This result also holds in the case of several charges.

P2.13 $p \approx 9.0 \times 10^{-28}$ C.m.

P2.14 a) $U_E = -K_o\,q(\mathbf{R}.\mathbf{p})/R^3$, where $\mathbf{R} = \mathbf{r} - \mathbf{r}'$. $\mathbf{F}_{q\rightarrow p} = K_o\,q[\mathbf{p}/R^3 - 3\mathbf{R}(\mathbf{p}.\mathbf{R})/R^5]$ and $\mathbf{\Gamma}_{q\rightarrow p} = -K_o q\,\mathbf{R} \times \mathbf{p}/R^3 = \mathbf{p} \times \mathbf{E}$.
b) $\mathbf{F}_{p'\rightarrow p} = 3(K_o/R^5)\,[\mathbf{p}\,(\mathbf{p}'.\mathbf{R}) + \mathbf{p}'(\mathbf{p}.\mathbf{R}) + \mathbf{R}\,(\mathbf{p}.\mathbf{p}') - 5\,\mathbf{R}(\mathbf{R}.\mathbf{p}')(\mathbf{R}.\mathbf{p})/R^2]$ and $\mathbf{\Gamma}_{p'\rightarrow p} = K_o[(\mathbf{p}' \times \mathbf{p})/R^3 - 3(\mathbf{R}.\mathbf{p}')\,(\mathbf{R} \times \mathbf{p})/R^5]$

P2.15 $U_E = -qV(\mathbf{r} - \mathbf{d}/2) + qV(\mathbf{r} + \mathbf{d}/2) - K_o\,q^2/d \cong -q\mathbf{d}.\mathbf{E} - K_o\,q^2/d$ and $W = -\mathbf{p}.\mathbf{E}$.

P2.17 a) $V(z) = K_o q/(R^2 + z^2)^{½}$, $E_z(0, 0, z) = -\partial_z V(0, 0, z) = K_o qz/(R^2 + z^2)^{3/2}$. The field has a minimum $E_{min} = -2K_o q/3^{3/2}R^2$ for $z = -R/\sqrt{2}$ and a maximum $E_{max} = 2K_o q/3^{3/2}R^2$ for $z = R/\sqrt{2}$. **b)** $F = -K_o eqz/(R^2 + z^2)^{3/2}$. For $|z| >> R$, $F \to -K_o eq/z|z|$. For $z = 0$, $F = 0$. If the electron may move only on Oz, this is a stable equilibrium position. Near $z = 0$, $F \cong -K_o eqz/R^3$, the oscillation frequency is $\tilde{\nu} = (1/2\pi)(K_o eq/mR^3)^{½}$. **c)** If the charge is not uniform, V remains the same but the field is not in the direction Oz. However, the component of **E** in the direction Oz remains the same.

P2.19 $\mathbf{E} = (q_s/2\varepsilon_o)\,[z/|z| - z(R^2 + z^2)^{-½}]\mathbf{e}_z$ and $V = (q_s/2\varepsilon_o)\,[(R^2 + z^2)^{1/2} - |z|]$.
If $|z| << R$, $V \to (q_s/2\varepsilon_o)\,[R - |z|]$ and $\mathbf{E} \to (q_s/2\varepsilon_o)\,[z/|z|]\,\mathbf{e}_z$.

P2.20 a) For $-d/2 < z < d/2$, $E = q_s/\varepsilon_o - q'_s/\varepsilon_o$ and $V = z(q'_s/\varepsilon_o - q_s/\varepsilon_o)$.
For $z > d/2$, $E = q_s/\varepsilon_o + q'_s/\varepsilon_o$ and $V = z(-q_s/\varepsilon_o - q'_s/\varepsilon_o) + q'_s d/\varepsilon_o$.
For $z < -d/2$, $E = -q_s/\varepsilon_o - q'_s/\varepsilon_o$ and $V = z(q_s/\varepsilon_o + q'_s/\varepsilon_o) + q_s d/\varepsilon_o$.
b) The invariance in translations perpendicular to Oz implies that V and **E** do not depend on x and y, hence $V = V(z)$ and $\mathbf{E} = -\partial_z V\,\mathbf{e}_z = E(z)\,\mathbf{e}_z$.

P2.21 Because of the symmetries, **E** is normal to the surface. Gauss law and the continuity of V give: $E_1 = -q_v d/2\varepsilon_o$, $E_2 = q_v d/2\varepsilon_o$, $E_3(z) = q_v z/\varepsilon_o$, $V_1 = q_v d(z + d/4)/2\varepsilon_o$, $V_2 = q_v d(d/4 - z)/2\varepsilon_o$ and $V_3 = -q_v z^2/2\varepsilon_o$. If $d \to 0$, in such a way that $q_s = q_v d$ remains constant, $E_1 = -q_s/2\varepsilon_o$, $E_2 = q_s/2\varepsilon_o$, $V_1 = q_s z/2\varepsilon_o$ and $V_2 = -q_s/2\varepsilon_o$.

P2.22 a) $\mathbf{E}_{in} = 0$ and $\mathbf{E}_{ex}(\rho) = (q_s R/\rho\varepsilon_o)\,\mathbf{e}_\rho$. **c)** $\mathbf{E}_{in}(\rho) = (q_v\rho/2\varepsilon_o)\,\mathbf{e}_\rho$ and $\mathbf{E}_{ex}(\rho) = (q_v/2\varepsilon_o)(R^2/\rho)\,\mathbf{e}_\rho$. **d)** $V_{in}(\rho) = (q_v/4\varepsilon_o)(R^2 - \rho^2) + C$ and $V_{ex}(\rho) = (q_v/2\varepsilon_o)\,R^2 \ln(R/\rho) + C$.

P2.23 a) $\mathbf{E}_{in} = K_o(Q\,r/R^3)\,\mathbf{e}_z$, $V_{in} = ½K_oQ(3/R - r^2/R^3)$, $\mathbf{E}_{ex} = K_o(Q/r^2)\,\mathbf{e}_z$ and $V_{ex} = K_oQ/r$. **b)** $\mathbf{E}_{in} = K_o(Q\,r/R^3 + q/r^2)\,\mathbf{e}_r$, $V_{in} = ½K_oQ(3/R - r^2/R^3) + K_o q/r$, $\mathbf{E}_{ex} = [K_o(Q + q)/r^2]\,\mathbf{e}_r$ and $V_{ex} = K_o(Q + q)/r$. In the case of the atom, $\mathbf{E}_{in} = K_oZe(1/r^2 - r/R^3)\mathbf{e}_r$, $\mathbf{E}_{ex} = 0$, $V_{in} = (K_oZe)(1/r + r^2/2R^3 - 3/2R)$ and $V_{ex} = 0$.

P2.24 a) For $r > R_1$, $E_1(r) = (q_v/3\varepsilon_o r^2)(R_1{}^3 - R_2{}^3)\,\mathbf{e}_r$ and $V_1(r) = (q_v/3\varepsilon_o r)(R_1{}^3 - R_2{}^3)$.
For $R_2 < r < R_1$, $E_2(r) = (q_v/3\varepsilon_o r^2)(r^3 - R_2{}^3)\,\mathbf{e}_r$ and $V_2(r) = (q_v/6\varepsilon_o)(3R_1{}^2 - r^2 - 2R_2{}^3/r)$.
For $r < R_2$, $E_3 = 0$ and $V_3 = (q_v/2\varepsilon_o)(R_1{}^2 - R_2{}^2)$. **b)** Setting $r' = (r^2 + d^2 - 2Rd\cos\theta)^{½}$, we find $V_1 = (q_v/3\varepsilon_o)(R_1{}^3/r - R_2{}^3/r')$, $V_2 = (q_v/6\varepsilon_o)(3R_1{}^2 - r^2 - 2R_2{}^3/r')$, $V_3 = (q_v/6\varepsilon_o)(3R_1{}^2 - r^2 - 3R_2{}^2 + r'^2)$, $\mathbf{E}_1 = (q_v/3\varepsilon_o)[(R_1{}^3/r^2 - R_2{}^3 r/r'^3)\,\mathbf{e}_r + (R_2{}^3 d/r'^3)\,\mathbf{e}_z]$, $\mathbf{E}_2 = (q_v/3\varepsilon_o)[(r - rR_2{}^3/r'^3)\,\mathbf{e}_r + (R_2{}^3 d/r'^3)\,\mathbf{e}_z]$ and $\mathbf{E}_3 = (q_v d/3\varepsilon_o)\,\mathbf{e}_z$.

P2.26 $C^2/N.m$, $U_E = -4.50 + 9.0\,[(100x + 40)^2]^{-½} - 18.0\,[(100x - 6.0)^2]^{-½}$.
$F = 900(100x + 40)/|100x + 40|^3 - 1800(100x - 6.0)/|100x - 6.0|^3 = -46.5$ N.

P2.27 a) $\mathbf{F} = K_o qq'\,\mathbf{e}_x/(x'^2 - L^2/4)$. **b)** $\mathbf{F} = K_o(qq'/LL')\ln(A^-/A^+)$, where $A^\pm = 4D^2 - (L \pm L')^2$. If $D >> L$ and $D >> L'$, we find $\mathbf{F} \to K_o\,(qq'/d^2)\,\mathbf{e}_x$.

P2.28 $V = 3yx - 5y^2 + zy + C$. We find $\int_O^M d\mathbf{r}'.\mathbf{E}(\mathbf{r}') = 14 = V(M) - V(O)$.

P2.29 $U_E \approx (K_o q^2/d)(1 - 2N \ln 2) \approx -2N\,(K_o\,q^2/d) \ln 2$.

P2.30 b) $U_E = (3/5)\ K_oQ^2/R$.

P2.31 c) $R_e = (3/5)\ K_oe^2/mc^2 = 1.69 \times 10^{-15}$ m .

P2.32 The potentials and the fields are:
Inside (1): $V_{(1)} = (K_oQ_1/2R_1^3)(3R_1^2 - r_1^2) + K_oQ_2/r_2$ and $\mathbf{E}_{(1)} = K_oQ_1\mathbf{r}_1/R_1^3 + K_oQ_2\mathbf{r}_2/r_2^3$.
Inside (2): $V_{(2)} = K_oQ_1/r_1 + (K_oQ_2/2R_2)(3 - r_2^2/R_2^2)$ and $\mathbf{E}_{(2)} = K_oQ_1\mathbf{r}_1/r_1^3 + K_oQ_2\mathbf{r}_2/R_2^3$.
Outside the balls: $V_{(3)} = K_o\,Q_1/r_1 + K_o\,Q_2/r_2$ and $\mathbf{E}_{(3)} = K_oQ_1\mathbf{r}_1/r_1^3 + K_oQ_2\mathbf{r}_2/r_2^3$.
b) $U_{E(1)} = (3/5)\ K_oQ_1^2/R_1$, $U_{E(2)} = (3/5)\ K_oQ_2^2/R_2$ and $U_{int} = W = K_oQ_1Q_2/d$.
c) $F_{1\to2} = K_oQ_1Q_2/z^2$. **d)** $q_s = q_v\,d\cos\theta$.

P2.34 a) $U_E = (3/5)\ K_o\,Z(Z-1)e^2/R = 1.56 \times 10^{-10}$ J. **b)** $U_i - U_f = 0.58 \times 10^{-10}$ J

Chapter 3

P3.1 a) $q_s = q_vd$. $\mathbf{E}_{in} = 0$ for $x > d$, $E(x) = (q_s/\varepsilon_o)(1 - x/d)$ for $0 < x < d$) and $E_{ex} = q_s/\varepsilon_o$ for $x < 0$. **b)** $q_s = A/\delta$, $E_{in} = (q_s/\varepsilon_o)\exp(-\delta x)$ and $E_{ex} = q_s/\varepsilon_o$. **c)** $Q = R^2E/K_o = 4.4 \times 10^{-4}\ R^2$ (in coulombs), $V = RE = 4 \times 10^6\ R$ (in volts), $V_{5\ cm} = 0.20$ MV and $V_{1\ m} = 4$ MV.
$d = 3.8 \times 0^{-15}$ m.

P3.2 a) For $r > R_3$, $E_{ex} = K_oq/r^2$ and $V_{ex} = K_oq/r$. For $R_1 < r < R_2$, $E_{in} = K_oq/r^2$ and $V_{in} = K_oq/r + K_oq(1/R_3 - 1/R_2)$. $V_{ball} = 420$ V and $V_{shell} = 150$ V. The ball acts on a charge q' placed outside the shell but q' does not act on the ball. **b)** Let q_1, q_2 and q_3 be the charge of the ball, of the internal surface of the shell and the external surface of the shell. Then, $q_2 = -q_1$, $E_{ex} = K_oq_3/r^2$ and $V_{ex} = K_oq_3/r$, $E_{in} = K_oq_1/r^2$ and $V_{in} = K_oq_1/r + A$.
The given potentials and the continuity equations give:
$q_1 = R_1R_2(V_2 - V_1)/K_o(R_1 - R_2) = -1.85$ nC, $q_3 = R_3V_2/K_o = 6.67$ nC
$V_{cav} = K_oq_1/r + A$ and $V_{ex} = K_oq_3/r$, where $A = (V_1R_1 - V_2R_2)/(R_1 - R_2) = 267$ V.

P3.3 a) $E = 101$ V/m pointing downward and $\Delta V = 182$ V. There will be no current because the body is equipotential. **b)** The plate has a charge of -0.9 nC on its upper face and 0.9 nC on its lower face. The galvanometer indicates 0.9 nC.

P3.4 $E_{wire} = 2.17 \times 10^6$ V/m, $E_{shell} = 2.17 \times 10^4$ V/m, $q_L = 1.21 \times 10^{-8}$ C/m, $C_L = 12.1$ pF/m and $r = 0.217$ mm.

P3.5 $Q_1 = 500$ µC, $Q_2 = 1000$ µC, $U_{E1} = 2.5 \times 10^{-2}$ J, $U_{E2} = 5.0 \times 10^{-2}$ J. The potentials, the charges and the energies are not modified. If one connects the plates of opposite polarities, we find $Q'_1 = 166.7$ µC, $Q'_2 = 333.3$ µC, $V' = 33.3$ V, $U'_{E1} = U_{E1}/9$ and $U'_{E2} = U_{E2}/9$. Thus, 8/9 of the energy is dissipated as Joule heat.

P3.6 a) $C = 2.21$ nF, $Q_1 = 30$ µC and $Q_2 = 20$ µC.

P3.7 b) In the case of a cylindrical capacitor $U_E = ½(Q^2/2\pi\varepsilon_oL)\ln(R_2/R_1)$. **c)** The stored energy between R_1 and r is $U_E(r) = ½(Q^2/2\pi\varepsilon_oL)\ln(r/R_1)$. It is equal to $½U_E$ if $r = \sqrt{R_1R_2}$.

P3.8 $U_{E,o} = ½K_oQ^2/R_1$ and $U_E = ½K_oq^2/R_1 + ½K_o(Q-q)^2/R_2$. The minimum of U_E corresponds to $q_{min} = QR_1/(R_1+R_2)$, a potential $V_{min} = K_oQ/(R_1+R_2)$ and an energy $U_{E,\,min} = ½K_oQ^2/(R_1+R_2)$. There is some energy loss. $q_{min} = 4.95\times10^{-2}$ µC, $V_{min} = 44.55$ kV, $U_{E,o} = 11.25$ J and $U_{E,\,min} = 0.111$ J.

P3.9 a) $\delta C = -\varepsilon_o S\,\delta x/x^2$, $\delta U_E(x) = (Q_o^2/2\varepsilon_o S)\,\delta x$, hence $F = -(Q_o^2/2\varepsilon_o S)$. **b)** $\delta U_E(x) = -½\,\varepsilon_o SV_o^2\,\delta x/x^2$, $\delta U_{bat} = -\varepsilon_o SV_o^2\,\delta x/x^2$, $dW' = dU_E - dU_{bat}$, hence $F = -½\,\varepsilon_o SV_o^2/x^2$.

P3.10 a) $d = 0.125$ mm. **b)** $C = \varepsilon_o S/h = 35$ nF, $E = q/\varepsilon_o S$, $q = \varepsilon_o SE_c = 141.5$ C, $V = 4\times10^9$ V. $U_E = 2.83\times10^{11}$ J. **c)** $Q = 0.45$ C and $U_E = 1.01$ kJ and P = 2.05×10^5 W.

P3.11 a) $C' = \varepsilon_o S/(d-d') = Cd/(d-d')$ independently of the position of the plate and its inclination. **b)** $U_E = ½Q^2d(d-d')/\varepsilon_o L[L(d-d')+xd')]$, $F = -dU_E/dx|_{Q\,=\,\text{constant}}$ or $dU_E/dx|_{V\,=\,\text{constant}}$, hence $F = ½\,Q^2\,dd'(d-d')/\varepsilon_o L(Ld-d'L+xd')^2$.

P3.12 $\Gamma = (½\,Q^2/C^2)(\partial C/\partial\theta) \approx -\varepsilon_o L^3V^2/4d^2$.

P3.13 $v_d = 0.15$ mm/s.

P3.14 a) $R = (\rho/2\pi L)\ln(R_2/R_1)$. **b)** $R = (\rho/4\pi)(1/R_1 - 1/R_1)$.

P3.15 At 220 V, the lost power is $P_J = 8.26\times10^8$ W (all the energy is lost). At 22 kV, the lost power is $P_J = 82.6$ kW. At 220 kV, the lost power is $P_J = 826$ W.

P3.17 a) $\mathbf{j} = E(\sigma_{13}\,\mathbf{e}_1 + \sigma_{23}\,\mathbf{e}_2 + \sigma_{33}\,\mathbf{e}_3)$, **b)** $\mathbf{E} = j(\sigma^{-1}{}_{13}\,\mathbf{e}_1 + \sigma^{-1}{}_{23}\,\mathbf{e}_2 + \sigma^{-1}{}_{33}\,\mathbf{e}_p)$, $V_x = a\sigma^{-1}{}_{13}j$, $V_y = b\sigma^{-1}{}_{23}j$ and $V_z = c\sigma^{-1}{}_{33}j$, where $\sigma^{-1}{}_{ij}$ is the inverse matrix of σ_{ij}.

P3.18 $\underline{I}_{CD} = \mathcal{E}(\underline{Z}_1\underline{Z}_3 - \underline{Z}_2\underline{Z}_4)/[\,\underline{Z}_1\underline{Z}_2(\underline{Z}_3+\underline{Z}_4) + \underline{Z}_3\underline{Z}_4(\underline{Z}_1+\underline{Z}_2) + \underline{z}(\underline{Z}_1+\underline{Z}_2)(\underline{Z}_3+\underline{Z}_4)]$.

Chapter 4

P4.1 a) The field and charge densities are $E_{out} = q_s/\varepsilon_o$, $E_{in} = q_s/\varepsilon$, $q_s = \varepsilon V/[b+\varepsilon_r(d-b)]$, hence $C = \varepsilon S/[b+\varepsilon_r(d-b)]$ and $q_s' = V(\varepsilon-\varepsilon_o)/[b+\varepsilon_r(d-b)]$. **b)** $C_o = 0.885$ nF and $Q_o = 0.310$ µC, $C = 1.107$ nF, $\varepsilon_r = 1.25$, $q_s = 3.10$ µC/m² and $q_s' = 0.620$ µC/m².

P4.3 $Q = 2\pi\varepsilon_o LV/\alpha$, where $\alpha = [\ln(R_4R_2/R_3R_1) + \varepsilon_r^{-1}\ln(R_3/R_4)]$, $\mathbf{D} = (\varepsilon_o V/\alpha\rho)\,\mathbf{e}_\rho$,

$E(\rho) = V/\alpha\rho$ outside the dielectric and $E(\rho) = V/\varepsilon_r\alpha\rho$ inside the dielectric $\mathbf{P}(\rho) = \varepsilon_o(1-\varepsilon_r^{-1})\,(V/\alpha\rho)\mathbf{e}_\rho$ and $q'_v = -\nabla.\mathbf{P} = 0$.

P4.4 $\mathbf{D}(\mathbf{r}) = (q/4\pi)(\mathbf{r}/r^3)$ everywhere, $\mathbf{E}_{in} = Kq\mathbf{r}/r^3$ and $\mathbf{E}_{ex} = K_oq\mathbf{r}/r^3$. $\mathbf{P}_{ex}(\mathbf{r}) = 0$ and $\mathbf{P}_{in} = (q/4\pi)(1-\varepsilon_o/\varepsilon)(\mathbf{r}/r^3)$. The bound charge densities are $q_s' = (q/4\pi R^2)(1-\varepsilon_o/\varepsilon)$ and $q'_v(\mathbf{r}) = 0$. The total bound charge is $Q' = 4\pi R^2q_s' = q(1-\varepsilon_o/\varepsilon)$.

P4.5 $q'_s = P\cos\theta$, $q' = \pm\pi PR^2$, $x_{barycenters} = \pm(2/3)R$, $\mathbf{p} = (4/3)\pi R^3P\,\mathbf{e}_z = \mathbf{P}\mathcal{V}$. **b)** $V^{ex}(\mathbf{r}) = R^3(\mathbf{P}.\mathbf{r})/3\varepsilon_o r^3$ and $V^{in}(\mathbf{r}) = (\mathbf{P}.\mathbf{r})/3\varepsilon_o$. $\mathbf{E}^{ex} = (R^3/3\varepsilon_o)[3(\mathbf{P}.\mathbf{r})\mathbf{r}/r^5 - \mathbf{P}/r^3]$, $\mathbf{E}^{in} = \mathbf{r}/3\varepsilon_o$, $\mathbf{D}^{ex} = (R^3/3)\,[3\,(\mathbf{P}.\mathbf{r})\,\mathbf{r}/r^5 - \mathbf{P}/r^3]$ and $\mathbf{D}^{in} = 2\mathbf{P}/3$. We find on the sphere $(R = r)$: $\mathbf{E}^{ex}{}_{//} = \mathbf{E}^{in}{}_{//} = (1/3\varepsilon_o)[(\mathbf{P}.\mathbf{r})\mathbf{r}/r^2 - \mathbf{P}]$ and $\mathbf{D}^{ex}{}_\perp = \mathbf{D}^{in}{}_\perp = (2/3r^2)(\mathbf{P}.\mathbf{r})\,\mathbf{r}$.

P4.6 $\mathbf{E}_{in} = (P/2\varepsilon_o)[(z+\frac{1}{2}h)/R_+ - (z-\frac{1}{2}h)/R_- - 2]$, $\mathbf{D}_{in}(z) = \frac{1}{2}P[(z+\frac{1}{2}h)/R_+ - (z-\frac{1}{2}h)/R_-]$
$\mathbf{E}_{ex} = (P/2\varepsilon_o)[(z+\frac{1}{2}h)/R_+ - (z-\frac{1}{2}h)/R_-]$, $\mathbf{D}_{ex}(z) = \frac{1}{2}P[(z+\frac{1}{2}h)/R_+ - (z-\frac{1}{2}h)/R_-]$
where $R_\pm = (R^2 + z^2 + h^2/4 \pm zh)^{1/2}$. If $h << R$ and $|z| >> R$, $\mathbf{D}_{ex} = \varepsilon_o\mathbf{E}_{ex} = PR^2h/2z^3$ and, if $h << R$, $\mathbf{E}_{in} \cong -P/\varepsilon_o$ and $\mathbf{D}_{in} \cong Ph/2\varepsilon_o R$.

P4.7 $\mathbf{e}_n = (-1/3)^n\mathbf{E}_o$ and $\mathbf{p}_n = \varepsilon_o\mathbf{e}_{n-1}$; thus, $\mathbf{E} = \mathbf{E}_o/(1+\chi_e/3)$ and $\mathbf{P} = 3\varepsilon_o\mathbf{E}_o(\varepsilon_r-1)/(2+\varepsilon_r)$.

P4.9 $U_E = \frac{1}{2}C_oV_o^2 = Q_o^2/2C_o$ and $U'_E = Q_o^2/2\varepsilon_r C_o = U_E/\varepsilon_r$. The variation of the energy comes from the work of the force $F' = -Q_o^2 d(\varepsilon - \varepsilon_o)/2L'[\varepsilon_o(L-x) + \varepsilon x)]^2$.

P4.10 $U_E = \frac{1}{2}CV^2$, where $C = \varepsilon_o L[(L-x)/d + \varepsilon x/(\varepsilon d - \varepsilon d' + \varepsilon_o d')]$,
$F = \frac{1}{2}V^2\, \partial_x C = \frac{1}{2}V^2\varepsilon_o L d'(\varepsilon - \varepsilon_o)/d(\varepsilon d - \varepsilon d' + \varepsilon_o d')]$.

Chapter 5

P5.2 $V = (q_L/2\pi\varepsilon_o)\ln(2h/R)$ and $C_L = (2\pi\varepsilon_o)/\ln(2h/R)$. In the case of two conductors in a horizontal plane, $C'_L = \pi\varepsilon_o/\ln(d/R)$ and in the case of two conductors in a vertical plane, $C''_L = (2\pi\varepsilon_o)/\ln[(hd^2)(h+d)^2/R^2(h+d/2)^3]$. If $d << h$, $C''_L = (\pi\varepsilon_o)/\ln(d/R)$

P5.3 $V = V_o\ln(\rho/r)/\ln(R/r)$ and $\mathbf{E} = -[V_o/\rho\ln(R/r)]\,\mathbf{e}_\rho$

P5.5 If the potential of the electrode is V = Constant, we must have
$V(x,y,z) = (4V_o/\pi)\,\Sigma_{0\le p\le\infty}(2p+1)^{-1}\exp[-(2p+1)\pi y/d]\sin[(2p+1)\pi z/d]$.

Chapter 6

P6.1 $v = 1.85 \times 10^{-5}$ m/s, $F_M = 5.92 \times 10^{-24}$ N, $E_H = 37$ μV/m, $V_H = 3.7$ μV and $q_s = 3.27 \times 10^{-16}$ C/m². If the conductor has the same number of positive and negative charge carriers, there will be no Hall voltage.

P6.2 $\mathbf{F} = 2IRB\,\mathbf{e}_y$.

P6.4 b) $\mathbf{A} = -\frac{1}{2}yB\,\mathbf{e}_x + \frac{1}{2}xB\,\mathbf{e}_y + \nabla f$. We find $\Phi = \int_{\mathcal{C}} d\mathbf{r}.\mathbf{A} = \pi r^2 B$.

P6.5 a) $\mathbf{B}(z) = [\mu_o INa^2/2(a^2+z^2)^{3/2}]\mathbf{e}_z$, $\mathbf{B}_O = (\mu_o IN/2a)\,\mathbf{e}_z$. If $z>>R$, $\mathbf{B} = (\mu_o INa^2/2z^3)\mathbf{e}_z$. The field of a magnetic moment $\mathcal{M} = \pi NIa^2\,\mathbf{e}_z$ is $\mathbf{B}_{\mathcal{M}}(\mathbf{r}) = (\mu_o/4\pi r^5)[3(\mathcal{M}.\mathbf{r})\mathbf{r} - r^2\mathcal{M}]$.
$\mathbf{B}$ differs from $B_{\mathcal{M}}$ by less than 1% if $z = 12.2\,a$. The circulation of B is $\mu_o I$.
b) $\mathbf{B} = \frac{1}{2}\mu_o NIa^2(1/R_+^3 + 1/R_-^3)\,\mathbf{e}_z$, where $R_\pm = [(D/2 \pm z)^2 + a^2]^{1/2}$.

P6.8 $\mathbf{B} = (\mu_o qN/a^2)\{(a^2+z^2)^{1/2} + z^2(a^2+z^2)^{-1/2} - 2|z|^{1/2}\}\,\mathbf{e}_z \to (\mu_o qNa^2/4z^3$.
$\mathbf{B}(\mathbf{r}) = (\mu_o/4\pi r^5)\,[3(\mathcal{M}.\mathbf{r})\,\mathbf{r} - r^2\mathcal{M}]$. If $\mathbf{r}$ is in the direction of $\mathcal{M}$, $\mathbf{B} = (\mu_o/2\pi r^3)\,\mathcal{M}$, where $\mathcal{M} = \frac{1}{2}\pi qNa^2 = \frac{1}{4}qa^2\omega$. **b)** $\mathbf{B} = (\mu_o qNR^2/3z^3)\,\mathbf{e}_z = (\mu_o/2\pi r^3)\,\mathcal{M}$, where $\mathcal{M} = (2\pi qNR^2/3)\,\mathbf{e}_z = (q\omega R^2/3)\,\mathbf{e}_z$.

P6.10 a) $\mathbf{B}(\mathbf{r}) = (\mu_o IL/2\pi\rho D)\,\mathbf{e}_\varphi$ and $\mathbf{A}(\mathbf{r}) = (\mu_o I/4\pi)\ln[(D+L)/(D-L)]\,\mathbf{e}_z$, where we use cylindrical coordinates and $D = (\rho^2 + L^2)^{1/2}$. In the case of the square circuit $\mathbf{A} = 0$ and

$\mathbf{B} = (2\mu_o IL^2/\pi\rho^2 D)\,\mathbf{e}_z$, where $D = (z^2 + 2L^2)^{1/2}$. If $z >> L$, we find $\mathbf{B} = \mu_o\mathcal{M}/4\pi|z|^3$, where $\mathcal{M} = 4L^2I$. **b)** $B^{(ex)}(\rho) = (\mu_o I/2\pi\rho)\mathbf{e}_\varphi$ and $B^{(in)}(\rho) = (\mu_o I\rho/2\pi a^2)\mathbf{e}_\varphi$. We may write $\mathbf{A}^{(ex)} = [-(\mu_o I/2\pi)\ln\rho]\,\mathbf{e}_z + \nabla f$ and $\mathbf{A}^{(in)} = [\mu_o I(a^2-\rho^2)/4\pi a^2 - (\mu_o I/2\pi)\ln a]\,\mathbf{e}_z + \nabla f$. If $a \to 0$, $\mathbf{A}^{(in)}$ diverges. **c)** Set $\mathbf{j} = j\mathbf{e}_x$, we find $A_x^{(in)} = -\frac{1}{2}\mu_o j_s z^2/d$ and $B_y^{(in)} = -\mu_o j_s\, z/d$
$\mathbf{A}^{(ex)} = [-\frac{1}{2}\mu_o j_s z + (1/8)\mu_o j_s d]\,\mathbf{e}_x$ and $\mathbf{B}^{(ex)} = -\frac{1}{2}\mu_o j_s\,\mathbf{e}_y$ (for $z > d/2$)
$\mathbf{A}^{(ex)} = [-\frac{1}{2}\mu_o j_s z + (1/8)\mu_o j_s d]\,\mathbf{e}_x$ and $B^{(ex)} = \frac{1}{2}\mu_o j_s\,\mathbf{e}_y$ (for $z < -d/2$)

P6.11 For $r < r_1$, $\mathbf{B} = (\mu_o Ir/2\pi r_1^2)\,\mathbf{e}_\varphi$. For $r_1 < r < r_2$, $\mathbf{B} = (\mu_o I/2\pi r)\,\mathbf{e}_\varphi$. For $r_2 < r < r_3$, $\mathbf{B} = [\mu_o I(r_3^{\,2} - r^2)/2\pi r\,(r_3^{\,2} - r_2^2)]\,\mathbf{e}_\varphi$. For $r > r_3$, $\mathbf{B} = 0$

P6.12 a) $\mathbf{B}(\rho) = (\mu_o I\rho/2\pi a^2)\,\mathbf{e}_\rho$, $B(a) = (\mu_o I/2\pi a) = 4\times 10^{-4}$ T. **b)** $v_d = 4.74$ µm/s. **c)** $\mathbf{F}_M = -eBv_d\,\mathbf{e}_\rho$. **d)** $\mathbf{E} = -(\mu_o I^2\rho/2\pi^2 a^4 en_e)\,\mathbf{e}_\rho$, $q_v = -I^2/\pi^2 c^2 a^4 en_e$, where $n_e \approx 8.4\times 10^{28}$ is the number of free electrons per m^3. Note that $q_v/n_e e = 2.5\times 10^{-28}$.

P6.13 a) $\mathbf{E} = N_v q\rho\mathbf{e}_\rho/2\varepsilon_o$, $\mathbf{B} = \frac{1}{2}\mu_o\rho N_v qv\mathbf{e}_\varphi$, $\mathbf{F}_m = -\frac{1}{2}\mu_o\rho N_v q^2v^2\mathbf{e}_\rho$ and $\mathbf{F}_e = N_v q^2\rho\mathbf{e}_\rho/2\varepsilon_o$, hence $\mathbf{F}_m = -(v^2/c^2)\mathbf{F}_e$. The total force is $\mathbf{F} = \mathbf{F}_e + \mathbf{F}_m = (qI\rho/2\pi\varepsilon_o R^2 v)(1-v^2/c^2)\,\mathbf{e}_\rho$, where $I = \pi R^2 N_v qv$ is the beam intensity. **b)** The equation of the radial motion of the particle at the periphery is $\ddot{\rho} = C^2/2\rho$. Integrating, we find $\dot{\rho} = C\,[\ln(\rho/\rho_o)]^{1/2}$.

P6.15 b) $\int_{\mathcal{C}} d\mathbf{r}.\mathbf{A} = \Phi_B$, **c)** $\mathbf{A}^{(in)} = \frac{1}{2}\mu_o nI\rho\,\mathbf{e}_\varphi$ et $A^{(ex)}{}_z = -(\mu_o I/2\pi)\ln\rho + C$.

P6.19 $\mathbf{A}^{(in)} = \frac{1}{4}\mu_o j(a^2-\rho^2)\,\mathbf{e}_z$, $\mathbf{B}^{(in)} = (\frac{1}{2}\mu_o j\rho)\,\mathbf{e}_\varphi$, $\mathbf{A}^{(ex)} = \frac{1}{2}\mu_o ja^2\ln(a/\rho)\,\mathbf{e}_z$ and $\mathbf{B}^{(ex)} = (\frac{1}{2}\mu_o ja^2/\rho)\,\mathbf{e}_\varphi$.

P6.20 $A_\varphi^{(cond)} = -(1/3)\mu_o j\rho^2 + \frac{1}{2}\mu_o jb\rho - (1/6\rho)\mu_o jb^3$, $\mathbf{B}^{(cond)} = (-\mu_o j\rho + \mu_o jb)\,\mathbf{e}_z$. $A_\varphi^{(in)} = \frac{1}{2}\rho\mu_o j(b-a) + (1/6\rho)\mu_o j(a^3 - b^3)$, $\mathbf{B}^{(in)} = \mu_o j(b-a)\,\mathbf{e}_z$, $A_\varphi^{(ex)} = 0$ and $\mathbf{B}^{(ex)} = 0$.

P6.23 $\mathbf{F} = \pi R^2 IB_o p\mathbf{e}_x$. **b)** Laplace's force method gives $\mathbf{F} = -\pi pIR^2B_o\mathbf{e}_z$, while the energy method gives $\mathbf{F} = 0$ since $\nabla.\mathbf{B} = pB_o$; hence $\mathbf{B}$ cannot be a magnetic field.

P6.24 $\mathbf{F}_{wire\to e} = (\mu_o Iev/2\pi\rho)\mathbf{e}_\rho = 4.8\times 10^{-19}\,\mathbf{e}_z$ (in N), $\mathbf{B}_e = (\mu_o ev\rho/4\pi)\,\mathbf{e}_\varphi/[\rho^2 + (z-z_e)^2]^{3/2}$, $\mathbf{F}_{e\to wire} = -(\mu_o Iev/2\pi\rho)\,\mathbf{e}_\rho = -\mathbf{F}_{wire\to e}$.

P6.25 $\mathbf{B} = (\mu_o/2\pi D)[(I_1 - I_2)\sin\alpha\,\mathbf{e}_x + (I_1 + I_2)\cos\alpha\,\mathbf{e}_y]$, $B = 3.29\times 10^{-5}$ T. $\mathbf{B}_1 = (\mu_o/2\pi d)I_1\,\mathbf{e}_x$. The exerted force on the unit length of circuit (2) is $\mathbf{F}_{1\to 2} = (\mu_o I_1 I_2/2\pi d)\,\mathbf{e}_y$. It is attractive with a magnitude $F_L = 1.2\times 10^{-3}$ N/m.

P6.26 a) $\mathbf{B}(\mathbf{r}) = (\mu_o\mathcal{M}/2\pi z^3)\,\mathbf{e}_z$, $\mathbf{\Gamma} = (\mu_o\mathcal{M}^2/2\pi z^3)\sin\theta\,\mathbf{e}_x$. The stable equilibrium position is $\theta = 0$. **b)** $U_M = -(\mu_o\mathcal{M}^2/2\pi z^3)\cos\theta$. $W = 1.6\times 10^{-22}$ J, $F_z = -(3\mu_o\mathcal{M}^2/2\pi z^4) = -2.4\times 10^{-12}$ N. **c)** $T = (2\pi r/e)(mr/K_o)^{1/2}$, $I = (e^2/2\pi r)(mr/K_o)^{1/2}$ and $\mathcal{M}_o = el/2m$, where $l = mrv$ is the orbital angular momentum, $\mathbf{B}(O) = (\mu_o el/4\pi mr^3)\,\mathbf{e}_z$ and $U_M = -(\mu_o e/4\pi mr^3)\,\mathcal{M}_p.\mathbf{l}$.

P6.27 $\mathbf{B}_2 = -\mu_o\,\mathbf{n}_{12}\times\mathbf{j}_s$.

Chapter 7

P7.5 The internal fields for $|z| << h$ and the external fields for $|z| >> h$ are respectively $\mathbf{B}^{(in)} = \frac{1}{2}\mu_o(R_1^2 - R_2^2)(h^2+z^2)/(h^2-z^2)^2]\mathbf{M}$ and $\mathbf{H}^{(in)} = \frac{1}{2}(R_1^2 - R_2^2)(h^2+z^2)/(h^2-z^2)^2]\mathbf{M}$, $\mathbf{B}^{(ex)} = [\mu_o h|z|(R_2^2 - R_1^2)/(h^2 - z^2)^2]\mathbf{M}$ and $\mathbf{H}^{(ex)} = [h|z|(R_2^2 - R_1^2)/(h^2 - z^2)^2]\mathbf{M}$.

P7.6 a) For $R > |z|$, $\mathbf{B}^{(in)} = (2/3)\mu_o M\,\mathbf{e}_z$ and for $R < |z|$, $\mathbf{B}^{(ex)} = (\mu_o \mathcal{M}/2\pi|z|^3)\,\mathbf{e}_z$.

P7.8 a) For $\rho < R_1$, $\mathbf{B}_{wire} = \mu_o\mathbf{H}_{wire} = (\mu_o I\rho/2\pi R_1^2)\mathbf{e}_\varphi$ and $\mathbf{M}_{wire} = 0$.
For $R_1 < \rho < R_2$, $\mathbf{B}_{cavity} = \mu_o\mathbf{H}_{cavity} = (\mu_o I/2\pi\rho)\mathbf{e}_\varphi$ and $\mathbf{M}_{cavity} = 0$.
For $R_2 < r < R_3$, $\mathbf{B}_{medium} = \mu\mathbf{H}_{medium} = (\mu I/2\pi\rho)\mathbf{e}_\varphi$ and $\mathbf{M}_{medium} = (I/2\pi\rho)(\mu/\mu_o - 1)\mathbf{e}_\varphi$.
For $R_3 < \rho$, $\mathbf{H}_{exterior} = (I/2\pi\rho)\mathbf{e}_\varphi$, $\mathbf{B}_{exterior} = (\mu_o I/2\pi\rho)\mathbf{e}_\varphi$ and $\mathbf{M}_{exterior} = 0$.
b) $\mathbf{j}' = 0$, $\mathbf{j}'_{s2} = (I/2\pi R_2)(\mu/\mu_o - 1)\,\mathbf{e}_z$ and $\mathbf{j}'_{s3} = -(I/2\pi R_3)(\mu/\mu_o - 1)\,\mathbf{e}_z$.

P7.9 $\chi_M = -1.58 \times 10^{-9}$.

P7.10 a) $B^{in}(0) = 5.03\times10^{-3}$ T, $H^{in}(0) = 1.96 \times 10^5$ A.m^{-1}, $B(3\text{ cm}) = 3.17\times10^{-3}$ T and $H(3\text{ cm}) = 2.52\times10^3$ A.m^{-1}. The magnetic moment of the disk is $\mathcal{M} = 3.1416$ A.m^2 and its field on the axis is $B_{\mathcal{M}} = 6.28\times10^{-7}/z^3$ (in teslas). The B field differs from $B_{\mathcal{M}}$ by 5% for $z \approx 27$ cm. **b)** $z \approx 4.71$ cm.

P7.11 a) $\mathcal{M} = lSM$.
$\mathbf{F}_{1\to2} = (3\mu_o/4\pi r^5)\{(\mathcal{M}_2.\mathbf{r})\,\mathcal{M}_1 + (\mathcal{M}_1.\mathbf{r})\,\mathcal{M}_2 - (5/r^2)(\mathcal{M}_1.\mathbf{r})(\mathcal{M}_2.\mathbf{r})\,\mathbf{r} + (\mathcal{M}_1.\mathcal{M}_2)\,\mathbf{r}\}$.

P7.13 a) $\chi_m = 2.771 \times 10^{-3}$, $\mathcal{M} = 5.51 \times 10^{-3}$ A.m^2. **b)** $\mathcal{M} = 7.61$ A.m^2 and $\Gamma = 7.61 \times 10^{-2}$ N.m.

P7.14 a) $N_a = 8.39\times10^{28}$ atoms/m^3, $N_e = ZN = 2.18\times10^{30}$ electrons/m^3, $M_{sat} = 2.03\times10^7$ A/m. $M/M_{sat} = 10\%$, i.e. 2.6 electrons per atom. **b)** $\mathcal{M}_{1\text{ kg}} = 256$ A.m^2 and $F_{1\text{ kg}} = 5\,130$ N.

P7.15 a) $\Phi = 2.5 \times 10^{-4}$ Wb, **b)** $B = 1.48 \times 10^{-2}$ T.

Chapter 8

P8.1 $\mathcal{E} = -\mathcal{S}k$, $I = -\mathcal{S}k/R$, $q(t) = -\mathcal{S}kt/R$ and $U_J(t) = \mathcal{S}^2k^2t/R$.

P8.2 $\mathbf{A} = \frac{1}{2}\mu_o nI\rho\,\mathbf{e}_\varphi$ for $\rho < a$ and $\mathbf{A} = \mu_o nI(a^2/2\rho)\,\mathbf{e}_\varphi$ for $\rho < a$.
$\mathbf{E} = -\frac{1}{2}\mu_o n\,\partial_t I\,\rho\,\mathbf{e}_\varphi$ for $\rho < a$ and $\mathbf{E} = -\mu_o n\,\partial_t I\,(a^2/2\rho)\,\mathbf{e}_\varphi$ for $\rho > a$.
The circulation of $\mathbf{E}$ over the circuit of radius a is $\mathcal{E} = -\pi\mu_o na^2\,\partial_t I$.

P8.3 a) $E_\varphi = -\frac{1}{2}\mu n\rho\,\partial_t I$. The induced e.m.f. is $\mathcal{E} = L\,\partial_t I$, where $L = \pi\mu n^2r^2h$ is the inductance. **b)** $\mathbf{A} = \frac{1}{2}B\rho\,\mathbf{e}_\varphi$.

P8.4 a) $\mathcal{E} = v_oDB = 0.2$ V, $I = \mathcal{E}/R = 50$ mA. The exerted force is $F' = v_oD^2B^2/R = 5\times10^{-3}$ N. The power is $P' = v_oD^2B^2/R = 0.01$ W. **b)** $x = IDBt^2/2m$. The e.m.f. must be $\mathcal{E} = IR + ID^2B^2t/m$. The conservation of energy gives the equation $\mathcal{E}I = RI^2 + \partial_t U_M$.
c) $x = (DB\mathcal{E}/Rm)[t + (Rm/D^2B^2)\,e^{-t/\tau} - Rm/D^2B^2]$. The velocity reaches the limit value

$v_{\lim} = DB\mathcal{E}/Rm$. Then, $I = (\mathcal{E}/R)(1 - D^2B^2/Rm)$. **d)** $v = v_o[1 - e^{-t/\tau}]$ and $x = v_o[t + (Rm/D^2B^2)\,e^{-t/\tau} - Rm/D^2B^2]$. $I = (Dv_oB/R)\,[e^{-t/\tau} - 1]$

P8.5 a) $V_A - V_O = \frac{1}{2}(\boldsymbol{\omega}.\mathbf{B})L^2$. **b)** $V_A - V_O = \frac{1}{2}\omega BR^2 = 0.94$ V.

P8.7 a) The induced current in the small loop opposes the variation of Φ through this loop. Thus, it is in the direction of I if I decreases. **b)** If the small loop is approached toward the large loop, a current is induced in the opposite direction to I.
c) $M = \pi\mu_o r^2R^2/2(R^2+h^2)^{3/2}$.

P8.8 a) $L = \mu_o\pi a^2n^2[(a^2 + 4h^2)^{1/2} - a]$. $L = 18.8$ mH and $L_\infty = 2\mu_o\pi a^2n^2h = 19.7$ mH.

P8.9 $\mathcal{E} = 800$V, $U_J = 20.67$ J, $\Delta U_M = -240$ J. $U_g = -219.33$ J (the generator takes back energy).

P8.10 a) $L\ddot{Q} + R\dot{Q} + Q/C = 0$. The solution is $Q = (\omega_o/\tilde{\omega})Q_o\,e^{-\beta t}\cos(\tilde{\omega}t+\varphi)$, where $\omega_o = 1/\sqrt{LC} = 70.71$ rad/s, $\beta = R/2L = 50$ rad/s, $\tilde{\omega} = \sqrt{\omega_o{}^2 - \beta^2} = 50$ rad/s and $\varphi = \arctan(-\beta/\tilde{\omega}) = -\pi/4$. **b)** $I = -(\omega_o{}^2/\tilde{\omega})Q_o\,e^{-\beta t}\sin(\tilde{\omega}t)$,
$U_E = (Q_o{}^2/2C)\,(\omega_o/\tilde{\omega})^2\,e^{-2\beta t}\cos^2(\tilde{\omega}t + \varphi)$, $U_M = (Q_o{}^2/2C)\,(\omega_o/\tilde{\omega})^2\,e^{-2\beta t}\sin^2(\tilde{\omega}t)$,
$U = U_E + U_M$. **c)** $U_J = U(0) - U(t)$. **d)** $\tau = 20$ ms, $f_q = 0.7071$. $t = 9.9$ ms.

P8.11 $\langle\mathcal{E}\rangle = -0.625$ V, $\langle I\rangle = 125$ mA, $Q = 25$ mC, $\Delta U_M = 0$ and W $= 1.56\times10^{-2}$ J.

P8.14 a) $L_i = (\mu aN_i{}^2/\pi)\ln[(R + a)/(R - a)]$ and $M_{12} = (\mu aN_1N_2/\pi)\ln[(R+a)/(R-a)]$.
b) $I_1R_1 + L_1\partial_tI_1 + M\partial_tI_2 = \mathcal{E}_1$ and $I_2R_2 + L_2\partial_tI_2 + M\partial_tI_1 = 0$,
$P = R_1I_1{}^2 + R_2I_2{}^2 + \partial_t(\frac{1}{2}L_1I_1{}^2 + \frac{1}{2}L_2I_2{}^2) + \partial_t(MI_1I_2)$.
c) $I_1 = (\mathcal{E}_1/D)[R_2{}^2 + \omega^2L_2{}^2]\cos(\omega t - \phi_1)$ and $I_2 = \omega M(\mathcal{E}_1/D)\cos(\omega t - \alpha - \pi/2)$,
where $D = [(R_1R_2)^2 + \omega^2(L_1R_2 + L_2R_1)^2]^{1/2}$, $\alpha = \text{Arctan}(\omega L_1/R_1 + \omega L_2/R_2)$
and $\phi_1 = \text{Arctan}\{R_2\omega L_1/[R_1R_2 + \omega^2L_2(L_1 + L_2R_1/R_2)]\}$.

P8.15 a) $L = 19.74$ mH, $U_M = 0.987$ J and $t = 2.47$ ms. **b)** $q_b = -59.22$ μC independently on the coil radius and the time of variation of the intensity.
c) If $I = I_m\sin(\omega t)$, $B = \mu_o(N_s/h)\,I_m\sin(\omega t)$ and $I_b = -\pi\mu_oa^2(N_bN_s/hR)\,\omega I_m\cos(\omega t)$.

P8.16 a) $U_M = \pi\mu\,N^2R^2I^2/2h$. **b)** $M = \pi\mu n_1n_2xR_2{}^2$ and $F_{12} = I_1I_2\,\partial_xM = \pi\mu\,n_1n_2R_2{}^2$.

P8.17 $d\mathbf{F} = \frac{1}{2}\,dS(B^2/\mu_o)\,\mathbf{e}_p$.

P8.18 a) $\Gamma_M = -\pi NB_or^2I\sin\theta$. **b)** Neglecting the self-induction, $\Gamma' = \pi NB_or^2I\sin\theta$; then, $I = (\pi\omega NB_or^2/R)\sin(\omega t)$. The conservation of energy requires that $dW = dU_J = R(\pi\omega NB_or^2/R)^2\sin^2(\omega t)\,dt$. Taking into account the self-induction, the circuit equation becomes $RI + L\partial_tI = \pi\omega NB_or^2\cos(\omega t)$. The solution of this equation is $I = A\sin(\omega t - \alpha)$, where $A = \pi\omega NB_or^2/(R^2 + L^2\omega^2)^{1/2}$ and $\alpha = \text{Arctan}\,(L\omega/R)$. The conservation of energy requires that $dW = dU_M + dU_J$, which is verified.

P8.19 $\mathbf{j} = \sigma\mathbf{E} = -\frac{1}{2}\sigma\rho\omega B_o\cos\omega t\,\mathbf{e}_\varphi$ and $\langle P_J\rangle = (\pi R^4h/16)\sigma\omega^2B_o{}^2 = 372$ W.

P8.20 a) $\mathbf{F} = \frac{1}{2}\, e\rho <\partial_t B> + eB\partial_t\rho)\,\mathbf{e}_\varphi - eB\rho\, \partial_\tau\varphi\, \mathbf{e}_\rho$, hence the equations of motion $eB\rho\, \partial_t\varphi = m\,(v^2/\rho)$ and $\partial_t v = (e/m)\,(\frac{1}{2}\,\rho <\partial_t B> + B\partial_t\rho)$.

Chapter 9

P9.2 a) The equation $\nabla.\mathbf{E} = q_v/\varepsilon_o$ requires that $q_v = 4 \times 10^3\varepsilon_o$. **b)** The Maxwell's equations (with $q_v = 0$ and $\mathbf{j} = 0$) are verified if $\rho^{-1}\,\partial_\rho(\rho E_\rho) = 0$ and $-\partial_\rho E_z\,\mathbf{e}_\varphi = -b\mathbf{e}_z$, hence $b = 0$, $E_\rho = \alpha/\rho$ and $E_z = \beta$, where α and β are arbitrary constants.

P9.4 a) The symmetries require that $\mathbf{A} = A(r)\,\mathbf{r}/r$ or 0 by making a gauge transformation with $f = -\int dr\, A(r)$, hence $\mathbf{B} = \nabla\times\mathbf{A} = 0$. Also, $\mathbf{E} = E(r)\,\mathbf{r}/r$, where $E(r) = -\partial_r V$ and Gauss law give $E_{ex}(\mathbf{r}) = (1/4\pi\varepsilon)\, q\mathbf{r}/r^3$, hence $V_{ex} = (1/4\pi\varepsilon)\, q/r$. **b)** The symmetries require that $\mathbf{A} = A(\rho, t)\,\mathbf{e}_z$ and $\mathbf{E} = E(\rho, t)\,\mathbf{e}_\rho$. Setting $t' = t - (\rho^2 + u^2)^{\frac{1}{2}}/v$, the retarded potentials are $V = 0$ and $\mathbf{A}(\mathbf{r}, t) = (\mu/2\pi)\,\mathbf{e}_z \int_0^L du\, I(t')/(\rho^2 + u^2)^{\frac{1}{2}}$, hence the fields

$\mathbf{B} = \nabla \times \mathbf{A} = (\mu\rho/2\pi)\mathbf{e}_\varphi \int_0^L du\, [I'(t')/v(\rho^2+u^2) + I(t')/(\rho^2+u^2)^{3/2}]$ and

$\mathbf{E} = -\,\partial_t\mathbf{A} = -(\mu/2\pi)\mathbf{e}_z \int_0^L du\;\; I'(t')/(\rho^2+u^2)^{\frac{1}{2}}$. **c)** If $I = I_o$ (constant),

$\mathbf{A}(\mathbf{r}, t) = -(\mu I_o/2\pi)\mathbf{e}_z \ln\rho + C^{te}$, $\mathbf{E} = -\,\partial_t\mathbf{A} = 0$ and $\mathbf{B} = \nabla \times \mathbf{A} = -\partial_\rho A_z\,\mathbf{e}_\varphi = (\mu I/2\pi\rho)$.
If $I = I_o$ starting at $t = 0$, i.e. $I = I_o\theta(t)$, setting $\xi = (t^2\, v^2 - \rho^2)^{\frac{1}{2}}$, we may write
$V = 0$, $\mathbf{A}(\mathbf{r}, t) = (\mu I_o/2\pi)\,\theta(vt-\rho)\,\mathbf{e}_z \ln[(\xi + tv)/\rho]$,
$\mathbf{E} = -\,\partial_t\mathbf{A} = -(\mu v I_o/2\pi\xi)\,\theta(vt-\rho)\,\mathbf{e}_z$ and $\mathbf{B} = \nabla \times \mathbf{A} = (\mu t v I_o/2\pi\rho\xi)\,\theta(vt - \rho)\,\mathbf{e}_z$.

P9.5 a) Gauss law and Ampère's law give $E = N\mathcal{S}/2\pi\varepsilon_o\rho$ and $B = \mu_o N v\mathcal{S}/2\pi\rho$.
b) We find $q'_v(\mathbf{r}') = q_v(\mathbf{r})$ and $\mathbf{j}'_v(\mathbf{r}') = \mathbf{j}(\mathbf{r}) - q_v(\mathbf{r})\mathbf{v}_o$, $\mathbf{B}' = (\mu_o N v\mathcal{S}/2\pi\rho)\,\mathbf{e}_\varphi$ and $\mathbf{E}' = E(1-v^2/c^2)\mathbf{e}_\rho$. $\mathbf{E}'$ does not verify Gauss law, which gives $\mathbf{E}' = (N\mathcal{S}/2\pi\varepsilon_o\rho)\,\mathbf{e}_\rho$.

P9.7 a) $\mathbf{B}_{MQ\text{-}P}(\mathbf{r},t) \cong \mathbf{B}_P(\mathbf{r},t) = \mu n I(t)\mathbf{e}_z$, $V_{MQ\text{-}P}(\mathbf{r},t) \cong V_P(\mathbf{r},t) = 0$ and
$\mathbf{A}_{MQ\text{-}P}(\mathbf{r},t) \cong \mathbf{A}_P(\mathbf{r},t) = \frac{1}{2}\mu n I\rho\mathbf{e}_\varphi$, while $\mathbf{E}_{MQ\text{-}P} \cong -\nabla V_P - \partial_t\mathbf{A}_P = -\frac{1}{2}\mu n\rho\, \partial_t I\,\mathbf{e}_\varphi$.
b) $\mathbf{E}_{EQ\text{-}P} \cong \mathbf{E}_P(\mathbf{r},t) = [q(t)/\pi\varepsilon R^2]\,\mathbf{e}_z$ and $V_{EQ\text{-}P} \cong V_P(\mathbf{r},t) = q(t)\,(d - z)/\pi\varepsilon R^2$, while $\mathbf{B}_{EQ\text{-}P}$ is calculated as in the permanent regime but with the current density $\mathbf{j}_T = \mathbf{j} + \varepsilon\,\partial_t\mathbf{E} = (\partial_t q/\pi R^2)\mathbf{e}_z$, hence $\mathbf{B}_{EQ\text{-}P} = (\mu\rho/2\pi R^2)\,\partial_t q\,\mathbf{e}_\varphi$ and $\mathbf{A}_{EQ\text{-}P} = -(\mu\rho^2/4\pi R^2)\,\partial_t q\,\mathbf{e}_z + \nabla f$.
c) We find $\mathbf{j} = \mathbf{j}_o + \mathbf{j}'_o\, e^{-t/\tau_c}$, where $\mathbf{j}_o$ and $\mathbf{j}'_o$ are constant vectors, $\mathbf{B} = \frac{1}{2}\mu\rho j_o\,\mathbf{e}_\varphi$ and $\mathbf{A} = -(\frac{1}{4}\mu\rho^2 j_o)\mathbf{e}_z$, $\mathbf{E} = \mathbf{j}/\sigma$ and $V = -jz/\sigma + C$. In the MQ-P approximation $\mathbf{j}'_o \ll \mathbf{j}_o$.

P9.8 $V_{MQ\text{-}P}(\mathbf{r},t) = q/4\pi\varepsilon R$, $\mathbf{A}_{MQ\text{-}P}(\mathbf{r},t) = \mu q\,\mathbf{v}_q(t)/4\pi R = -\,(\mu q a\omega/4\pi R)\,\sin(\omega t)\,\mathbf{e}_z$,
$\mathbf{B}_{MQ\text{-}P}(\mathbf{r},t) = \mu q\,\mathbf{v}_q(t)\times\mathbf{R}\,/4\pi R^3 = -(\mu q a\omega/4\pi R^3)\,\sin(\omega t)\,\mathbf{e}_z\times\mathbf{R}$., where $\mathbf{R} = \mathbf{r} - \mathbf{r}_q$,
$\mathbf{E}_{MQ\text{-}P} \cong -\nabla V_{QP} - \partial_t\,\mathbf{A}_{QP} = q\mathbf{R}/4\pi\varepsilon R^3 + (\mu q a\omega^2\mathbf{e}_z\,/4\pi R^3)[R^2\cos(\omega t) - aR_z\sin^2(\omega t)]$.

P9.9 $q_v = -Q'(t - r/v_e)/4\pi r^2 v_e$ and $\mathbf{j} = -[Q'(t - r/v_e)/4\pi r^2]\,\mathbf{e}_r$. The potentials are:
For $r < v_e t$: $V_{MQ\text{-}P}(\mathbf{r}, t) = (K_o/r)\,Q(t - r/v_e) - (K_o/v_e)\int_r^{v_e t} dr'\, Q'(t - r'/v_e)/r'$

and $\mathbf{A}_{MQ\text{-}P}(\mathbf{r}, t) = -(\mu_o/12\pi r^2)\,\mathbf{e}_r\{\int_0^r dr'\, r'\, Q'(t - r'/v) + r^3 \int_r^{v_e t} dr'\, Q'(t - r'/v_e)/r'^2\,\}$

For $r > v_e t$: $V_{MQ\text{-}P}(\mathbf{r}, t) = 0$ and $\mathbf{A}_{MQ\text{-}P}(\mathbf{r}, t) = -(\mu/12\pi r^2)\,\mathbf{e}_r \int_0^{v_e t} dr'\, r' Q'(t - r'/v_e)$

P9.10 In the case of the solenoid, for instance, $U_{EM,v} \cong (1/8v^2)\mu\rho^2 n^2 \dot{I}^2 + \frac{1}{2}\mu n^2 I^2$, $\partial_t U_{EM,v} \cong \mu n^2 \dot{I}\,[\rho^2 \ddot{I}/4v^2 + I]$, $\mathbf{S} \cong -\frac{1}{2}\rho\mu n^2 I \dot{I}\,\mathbf{e}_\rho$ and $\nabla.\mathbf{S} = \rho^{-1}\partial_\rho(\rho S) \cong -\mu n^2 I \dot{I}$. Verify that $\nabla.\mathbf{S} + \partial_t U_{EM,v} \cong 0$. For the cylinder verify that $\nabla.\mathbf{S} + \partial_t U_{EM,v} + \mathbf{j}.\mathbf{E} = 0$.

P9.11 $\mathbf{S} = -(\rho j_o j/2\sigma)\mathbf{e}_\rho$, $P_{entering} = 2\pi aLS = \pi La^2 j j_o/\sigma$, where $j_o = j + \tau_c \partial_t j$, $P_J = (j^2/\sigma)\pi a^2 L$ and $U_{EM,v} = 2\pi L[\varepsilon j^2 a^2/4\sigma^2 + (\mu a^4/32)j_o^2]$. The conservation of energy requires that $P_{entering} = P_J + \partial_t U_{EM}$, which is verified.

P9.12 The non-zero elements of the Maxwell's tensor are: $\tau_{11} = -(\mu_o a^2 j^2/8)\cos(2\varphi)$, $\tau_{22} = (\mu a^2 j^2/8)\cos(2\varphi)$ and $\tau_{12} = \tau_{21} = -(\mu a^2 j^2/8)\sin(2\varphi)$. The force acting on $d\mathcal{S}$ is $f_\alpha = d\mathcal{S}\,\Sigma_\beta\, n_\beta\, \tau_{\alpha\beta}$, where $\mathbf{n}$ is the unit vector normal to $d\mathcal{S}$, hence $\mathbf{f} = d\mathcal{S}\,(B^2/2\mu)\,\mathbf{n}$.

Chapter 10

P10.2 a) $x_{max} = (\Delta\omega/\Delta k)t - n\pi/\Delta k$, where $\Delta\omega = \frac{1}{2}(\omega_2 - \omega_1)$ and $\Delta k = \frac{1}{2}(k_2 - k_1)$.

P10.3 d) $v_{(p)} = c/(A + B/\lambda^2) = 1.85\times10^8$m/s and $v_{(g)} = c(A - B/\lambda^2)/(A + B/\lambda^2)^2 = 1.77\times10^8$ m/s.

P10.4 $\partial^2_{xx} f + k^2 f = 0$, where $k = \omega/v$. Its general solution is $f = A\sin(kx + \phi)$. The modes are $f_n = A_n \sin(k_n x)$, where $\omega_n = n\pi v/L$ and $\lambda_n = 2\pi/k_n = 2L/n$.

P10.5 a) $\mathbf{E} = A\cos(\omega t - kz)\,\mathbf{e}_x + B\cos(\omega t - kz)\,\mathbf{e}_y$ (linear polarization).
b) $\mathbf{E} = A\cos(\omega t - kz)\,\mathbf{e}_x + A\cos(\omega t - kz - \pi/2)\,\mathbf{e}_y$ (left-circular polarization).
c) $\mathbf{E} = A\cos(\omega t - kx)\,\mathbf{e}_y + A\cos(\omega t - kx + \pi/2)\,\mathbf{e}_z$ (right-circular polarization).
d) $\mathbf{E} = vA\,[\cos(\omega t - kz - \pi/2)\,\mathbf{e}_x + \cos(\omega t - kz - \pi/2 - \pi/6)\,\mathbf{e}_y]$ (left-elliptic polarization).
e) $\mathbf{E} = vA[\cos(\omega t - ky + \pi)\,\mathbf{e}_z + \cos(\omega t - ky + \pi - \pi/2)\mathbf{e}_x]$ (left-circular polarization).
f) $\mathbf{E} = vA\,[\cos(\omega t + kx + \pi/2)\,\mathbf{e}_z + \cos(\omega t + kx + \pi/2 + \pi/2)\,\mathbf{e}_y\,]$ (right-circular polarization).

P10.6 Let $\xi = \omega t - kx\sin\theta - kz\cos\theta + \phi$ and $\mathbf{e}_1 = \cos\theta\,\mathbf{e}_x - \sin\theta\,\mathbf{e}_z$. **a)** $\mathbf{E} = E\mathbf{e}_1\cos\xi$. **b)** $\mathbf{E} = E\mathbf{e}_y\cos\xi$. **c)** $\mathbf{E} = E\cos\xi\,\mathbf{e}_1 - E\sin\xi\,\mathbf{e}_y$. **d)** $\mathbf{E} = E\cos\xi\,\mathbf{e}_1 + E\sin\xi\,\mathbf{e}_y$.

P10.7 b) $E'_x = [E_x^2\cos^2\alpha + E_y^2\sin^2\alpha + E_x E_y\sin 2\alpha\cos\phi]^{1/2}$.
$E'_y = [E_x^2\sin^2\alpha + E_x^2\cos^2\alpha - E_x E_y\sin 2\alpha\cos\phi]^{1/2}$.
$\cos\phi' = [(E_y^2 - E_x^2)\sin 2\alpha + 2E_x E_y\cos 2\alpha\cos\phi]/2E'_x E'_y$.
$\sin\phi' = (E_x E_y/E'_x E'_y)\sin\phi$. d) $\tan 2\alpha = 2E_x E_y\cos\phi/(E_x^2 - E_y^2)$.

P10.8 b) $I_{11} = \frac{1}{2}(s_o + s_3)$, $I_{12} = \frac{1}{2}(s_1 - is_2)$, $I_{21} = \frac{1}{2}(s_1 + is_2)$, $I_{22} = \frac{1}{2}(s_o - s_3)$.
$\det I = \frac{1}{4}s_o^2(1 - s^2/s_o)$, where $s^2 = s_1^2 + s_2^2 + s_3^2$ and $\operatorname{tr} I = s_o$.
$I'_{11} = \frac{1}{2}(s_o + s_2)$, $I_{12} = -\frac{1}{2}(s_3 + is_2)$, $I_{21} = \frac{1}{2}(-s_1 + is_2)$, $I_{22} = \frac{1}{2}(s_o - s_2)$.
$I''_{++} = \frac{1}{2}(s_o + s_2)$, $I''_{+-} = \frac{1}{2}(s_3 - is_1)$, $I''_{-+} = \frac{1}{2}(s_3 + is_1)$, $I''_{--} = \frac{1}{2}(s_o - s_2)$.
b) If the wave is completely non-polarized, $s_1 = s_2 = s_3 = 0$, thus $P = 0$.

P10.10 a) $<U_{EM}> = <U_{EM,v\,1}> + <U_{EM,v\,2}>$ and $\mathcal{I} = \mathcal{I}_1 + \mathcal{I}_2$ if $\mathbf{E}_1$ and $\mathbf{E}_2$ are polarized in two perpendicular directions or non-polarized. c) The transmitted intensity is half the incident intensity.

P10.11 a) $\mathbf{E} = E_m \cos\xi\,[-\cos\theta\,\mathbf{e}_y + \sin\theta\,\mathbf{e}_z]$ and $\mathbf{B} = (nE_m/c)\cos\xi\,\mathbf{e}_x$, where we have set $\xi = \omega t - ky\sin\theta + kz\cos\theta + \phi$ and $k = n\omega/c$. **b)** $U_{EM,v} = n^2\varepsilon_o E_m^2 \cos^2\xi$, $\mathbf{S} = (n/\mu_o c)E_m^2 \cos^2\xi\,\mathbf{e}$, $\mathcal{I} = (n/2\mu_o c)E_m^2$ and $<dP/d\mathcal{S}> = \mathcal{I}\cos\theta$, $<P_s> = \mathcal{I}\cos\theta$.

P10.12 $f = 0.33 \times 10^{-2}$ N and $v = 0.29$ m/s.

P10.13 $E_m = (2\mu_o c\mathcal{I})^{1/2} = 8.7$ kV.m^{-1}, $P = 4\pi r^2\mathcal{I} = 1.26$ MW, $r = 2.6 \times 10^{-8}$ m.

P10.14 a) $\delta\mathbf{p}/\delta t = (\mathcal{S}\mathcal{I}/c)\cos\theta\,[\sin\theta\,\mathbf{e}_y - (\cos\theta + \tfrac{1}{2})\mathbf{e}_z]$.
b) $\delta\mathbf{p}/\delta t = (\mathcal{S}\mathcal{I}/c)\cos\theta\,\{\sin\theta\,\mathbf{e}_y - (\cos\theta + \tfrac{1}{2} - \alpha/2)\,\mathbf{e}_z\}$ and $p_r = (\mathcal{I}/c)\cos\theta\,(\cos\theta + \tfrac{1}{2} - \alpha/2)\,\mathbf{e}_z$. **c)** $f = 1.0 \times 10^{-5}$ N and $p_r = 1.0 \times 10^{-3}$ Pa.

P10.15 a) $E_m = 1.00$ kV/m. **b)** $P = 3.74 \times 10^{26}$ W and $m = P/c^2 = 4.15 \times 10^9$ kg/s. **c)** $\mathcal{S} = 2.5$ m^2. **d)** $f_r/F_G = 3P/16GM_S\pi r_o m_v c = 5.6{\times}10^{-7}/r_o$. We find $f_r > F_G$ if $r_o < 0.56$ µm.

P10.16 a) $\mathcal{I}(r) = P/\Omega r^2$. **b)** $P_i = 0.428\,\mathcal{S}PR^4/\lambda^2D^4$.

P10.17 We have in this case $\mathbf{E} = E\cos(\omega t - kz)\mathbf{e}_x$ and $\mathbf{B} = (E/v)\cos(\omega t - kz)\mathbf{e}_y$, where $k = \omega/v$. The Maxwell's tensor has only one non-zero component $\tau_{33} = -U_{EM,v} = -\varepsilon E^2\cos^2(\omega t - kz)$. The force is $\mathbf{F} = \mathcal{S}U_{EM,v}\cos\theta\,\mathbf{e}_z$.

P10.18 a) $v_g = cn/[1 + (\omega_p/\omega_o)^2(n^2 - 1)^2]$.

P10.21 a) $\mathbf{B} = (i/\omega)\nabla \times \mathbf{F}\,e^{i\omega t}$, $\mathbf{j} = \{(i/\omega\mu_o)[\nabla(\nabla.\mathbf{F}) - \Delta\mathbf{F}] + (\varepsilon_o/\mu_o)\mathbf{F}\}\,e^{i\omega t}$ and $q_v = \varepsilon_o\nabla.\mathbf{F}\,e^{i\omega t}$. **b)** $j_d/j = \omega\varepsilon_o/\sigma$.

P10.22 If $\mathbf{E} = E_m e^{-\delta z}\cos\xi\,\mathbf{e}_x$ and $\mathbf{B} = (E_m/v)\,e^{-\delta z}\cos(\xi - \alpha)\,\mathbf{e}_y$, where $\xi = \omega t - kz$, $\mathbf{S} = (E_m^2/\mu v)(1+Q^2)^{1/4}e^{-2\delta z}\cos\xi\cos(\xi - \alpha)\,\mathbf{e}_z$ and $dP_J = \sigma\mathcal{S}\,dz\,e^{-2\delta z}E_m^2\cos^2\xi$.

P10.23 b) $\mathbf{S} = -(j^2r/2\sigma)\,\mathbf{e}_\rho$.

P10.24 a) $\mathbf{E} = E_o\exp(-\eta z)\exp(-ikz)\,\mathbf{e}_x$ and $\mathbf{B} = -(iE_o/\omega)(\eta + ik)\exp(-\eta z)\exp(-ikz)\,\mathbf{e}_y$, where $\eta^2 = \tfrac{1}{2}(\omega/c)^2[(1 + \omega_o^2/\omega^2)^{1/2} + 1]$, $k = \sigma\omega\mu_o/2\eta$ and $\omega_o = \sigma/\varepsilon_o$. **b)** The charge density is negligible if $\omega >> \omega_o$; then, $\eta \approx (\omega/c)$ and $k = \tfrac{1}{2}\sigma\mu_o c$.

P10.25 b) $\mathbf{j}_s \cong (-i\sigma/p)\,\underline{E}_m e^{i\omega t}\,\mathbf{e}_x$, $\mathbf{E}^{ex} \cong 0$ and $\mathbf{B}^{ex} = i\mu_o(\sigma/p)\underline{E}_m e^{i\omega t}\,\mathbf{e}_y$.

P10.26 a) $N = 5.86 \times 10^{28}$ electrons/m^3, $\tilde{\nu}_p = 2.17 \times 10^{15}$ Hz.
b) $\delta = (\omega_p - \omega^2)^{1/2}/v = 4.4 \times 10^7$ m^{-1}, $l = 0.023$ µ.
c) $x_{0.5} = 7.9 \times 10^{-3}$ µ, $x_{0.01} = 0.052$ µ, $\lambda < 2\pi c/\omega_p = 0.138$ µ (X-rays).

P10.28 b) If $E = E_m e^{i\omega t}$, the equation of motion is $m\dot{v} = -bv - eE$, hence $v = -emE/(b + im\omega)$ and $j = -eNv = Nme^2E/(b + im\omega)$. The complex conductivity is $\underline{\sigma} = Nme^2/(b + im\omega)$. **c)** $<P_v> = ½\, bNe^2E_m^2/(b^2 + m^2\,\omega^2)$.
d) $k^2 = (\omega/v^2)(m\omega^2 - m\omega_p^2 - ib\omega)/(m\omega - ib)$.

P10.29 a) $U_{EM,v} = \varepsilon E_m^2 \cos^2\xi$ and $\mathbf{S} = (E_m^2/\mu v)\cos^2\xi\ \mathbf{e}_z = U_{EM,v}\, v\mathbf{e}_z$, where $\xi = \omega t - kz$. **b)** $E_\gamma = h\omega/2\pi$, $p_\gamma = h\omega/2\pi c$, $<N_\gamma> = \pi\varepsilon E_m^2/h\omega$. $N_S = c <N_\gamma>$, $P_s = p_r = ½\, \varepsilon E_m^2$.

P10.30 $\lambda = 121.5023$ nm, $U_K = 1.6349 \times 10^{-18}$ J and $U'_K = 2.1799 \times 10^{-18}$ J.

Chapter 11

P11.1 a) $f_R = \tan^2(\theta - \theta'')(1 + \tan^2\theta \tan^2\theta'')/(\tan\theta + \tan\theta'')^2$
$f_T = 2\tan\theta\tan\theta''\ [2 + \tan^2(\theta-\theta'')]/(\tan\theta + \tan\theta'')^2$. The conservation of energy requires that $f_R + f_T = 1$ and it is verified. **b)** $\theta'' = 28.13°$ and $\mathcal{R}_{//} = -0.0920$, $\mathcal{T}_{//} = 0.7280$, $f_{R\,//} = 8.46 \times 10^{-3}$ and $f_{T\,//} = 0.9915$. $\mathcal{R}_\perp = -0.303$, $\mathcal{T}_\perp = 0.69$, $f_R = 9.199\times10^{-2}$, $f_T = 0.9081$. $\mathcal{P}' = -0.832$ and $\mathcal{P}'' = 0.044$

P11.2 b) $\theta'' = 40°.6$. $\mathcal{R}_{//} = 0.066$, $\mathcal{T}_{//} = 0.702$, $\mathcal{R}_\perp = -0.337$, $\mathcal{T}_\perp = 0.6$, $\alpha' = 79°$ and $\alpha'' = 40°$.

P11.3 a) $\mathcal{I}' = \mathcal{R}^2\mathcal{I}$, $\mathcal{I}'' = \mathcal{T}^2(Z_1/Z_2)\mathcal{I}$. The energy conservation on the surface requires that $(1 - \mathcal{R}^2)\cos\theta' = \mathcal{T}^2(Z_1/Z_2)\cos\theta''$, which is verified by $\mathcal{R}_{//}$ and $\mathcal{T}_{//}$ and by $\mathcal{R}_\perp$ and $\mathcal{T}_\perp$, thus in the case of any wave. **b)** $I_3 = 16n^2 I_1/(1+n)^4 I_1 = 0.9216\, I_1$. After four plates, the intensity becomes $I_f = (0.9216)^4\, I_1 = 0.7214\, I_1$.

P11.4 $f_T = 4n^2\cos^2\theta\cos^2\theta''/[(n^2-1)^2\sin^2\alpha + 4\cos^2\theta(n^2 - \sin^2\theta)]$, where $\alpha = (n\omega L/c)\cos\theta''$. We find $\alpha = 4.7124 \times 10^4$ rad, $\sin\alpha = -0,589$ and $f_T = 0,94$.

P11.5 a) $\underline{\mathbf{E}} = A_{//}\, e^{i(\omega t - \mathbf{k.r})}\, \mathbf{e}_{//} + A_\perp\, e^{i(\omega t - \mathbf{k.r} + \phi)}\, \mathbf{e}_\perp$ such that $\phi = 0$ ou π if the wave is polarized linearly, $A_{//} = A_\perp$ and $\phi = \pm\,\pi/2$ if the wave is polarized circularly and the amplitudes or ϕ varying at random in the case of a non-polarized wave.

b) $f_R = \tan^2(\theta - \theta'')\dfrac{1 + \tan^2\theta\,\tan^2\theta''}{(\tan\theta + \tan\theta'')^2}$, $f_T = \dfrac{2\tan\theta\,\tan\theta''}{(\tan\theta + \tan\theta'')^2}[2 + \tan^2(\theta - \theta'')]$,

$$\mathcal{P}' = \frac{2n(1-n^2)\cos\theta''\cos\theta(\cos^2\theta - \cos^2\theta'')}{(1-n^2)^2\cos^2\theta''\cos^2\theta + n^2(\cos^2\theta'' - \cos^2\theta)^2}$$

$$\mathcal{P}'' = \frac{(n^2-1)(\cos^2\theta - \cos^2\theta'')}{(n^2+1)(\cos^2\theta + \cos^2\theta'') + 4n\cos\theta\cos\theta''}.$$

$\mathcal{P}' = 1$ and $\mathcal{P}'' = -0.04$ at Brewster incidence. $\mathcal{P}' = 0.44$ and $\mathcal{P}'' = -0.0097$ at 30°.

P11.6 We set $\Delta_1 = [(n^2 - \sin^2\theta]^{1/2}$ and $\Delta_2 = [(1 - n^2 \sin^2\theta]^{1/2}$.
a) $\mathcal{T}'_\perp(\theta)/\mathcal{T} = [\Delta_1 - \cos\theta]^4/(n^2-1)^2$, $\mathcal{T}'_\perp(0) = 0.04\,\mathcal{T}$, $\mathcal{T}'_\perp(\pi/2) = \mathcal{T}$, $\mathcal{T}'_\perp(\pi/4) = 0.092\,\mathcal{T}$.
b) $\mathcal{T}'_{//}(\theta)/\mathcal{T} = [\Delta_1 - n^2\cos\theta]^2/[\Delta_1 + n^2\cos\theta]^2$, $\mathcal{T}'_{//}(0)/\mathcal{T} = 0.04$, $\mathcal{T}'_{//}(\pi/2)/\mathcal{T} = 1$, $\mathcal{T}'_{//}(\theta_B)/\mathcal{T} = 0$, $\mathcal{T}'_{//}(\pi/4)/\mathcal{T} = 8.47 \times 10^{-3}$. **c)** $\mathcal{T}'_\perp/\mathcal{T} = (n\cos\theta - \Delta_2)^2/(n\cos\theta + \Delta_2)^2$, $\mathcal{T}'_{//}/\mathcal{T} = (n\Delta_2 - \cos\theta)^2/(n\Delta_2 + \cos\theta)^2$. $\mathcal{T}'_\perp/\mathcal{T}$ increases from $(n-1)^2/(n+1)^2 = 0.04$ for $\theta = 0$ to 1 for $\theta = i_L = 41.8°$. $\mathcal{T}'_{//}/\mathcal{T} = \mathcal{R}_{//}^2$ decreases from $(n-1)^2/(n+1)^2 = 0.04$ for $\theta = 0$, vanishes for $\theta = \theta'_B = \text{Arctan}(1/n)$ and then increases to 1 for $\theta = i_L$.

P11.7 a) $i_L \approx 41.8°$, $\delta = 5.927\ \mu^{-1}$, $l \approx 1\ \mu\text{m}$. **b)** Setting $\eta = \omega\cos\theta/\delta cn$, one finds $\mathcal{R} = -(1/\Delta)(1 + \eta^2)\,\text{sh}(L\delta)$ and $\mathcal{T} = 2i\eta/\Delta$,, where $\Delta = (1 - \eta^2)\,\text{sh}(L\delta) + 2i\,\text{ch}(L\delta)$. Numerical values: $|\mathcal{R}| = 1 - 1.31 \times 10^{-5}$ and $|\mathcal{T}| = 5.13 \times 10^{-3}$.

P11.8 b) For $\lambda = 500$ nm, $|\mathcal{R}| \cong 0.968$ and $\phi = 3.11$ rad. For $\lambda = 100$ nm $|\mathcal{R}| \cong 0.930$ and $\phi = 3.07$ rad. The thickness of the film must exceed 3.0 nm.

P11.9 $\mathbf{E} = \mathbf{E}_m\, e^{i(\omega t - \mathbf{q}.\mathbf{r})}$ and $\mathbf{B} = \mathbf{B}_m\, e^{i(\omega t - \mathbf{q}.\mathbf{r})}$ with the conditions $\mu_o\underline{\varepsilon}\omega\mathbf{E}_m = -\mathbf{q} \times \mathbf{B}_m$, $\omega\mathbf{B}_m = \mathbf{q} \times \mathbf{E}_m$, $\mathbf{q}.\mathbf{E}_m = 0$, $\mathbf{q}.\mathbf{B}_m = 0$ and the dispersion relation $q^2 = \omega^2\mu_o\underline{\varepsilon}$ i.e. $q^2 = \omega^2\mu_o(\varepsilon_2 - i\sigma_c/\omega)$. **b)** It is possible to introduce a complex index, but the ratio $\sin\theta''/\sin\theta$ is not constant and it is not possible to write Snell's law.

P11.11 b) $\lambda = 476 \pm 12$ nm. The dark spectral lines correspond to $\lambda = 600$ nm and $\lambda = 429$ nm. **c)** $I = 4I_o\,[1 + \cos(2\pi\, x/i)\cos(\pi\, x\, \delta\lambda/i\lambda)]$, where I_o is the intensity of each source. $\mathcal{C} = |\cos(\pi\, x\, \delta\lambda/i\lambda)|$, where $i = \lambda D/d$.

P11.13 b) If the antennas are in phase the order 0 maximum is in the normal direction and the order 1 at 30°A. The half-width of the principal maximums is $\lambda/6d = 4.8°$. The sweeping is obtained if the phase shift is $\phi(t) = -2\pi(d/\lambda)\sin(\omega t)$.

P11.14 a) x(first minimum) = 3 mm, x(first secondary maximum) = 4.5 mm. **b)** The principal maximums coincide at $x = 0$, the first minimum for $\lambda = 0.5\ \mu$ is at $x' = 2.5$ mm and its first secondary maximum is at $x' = 3.75$ mm. **c)** The first minimum is at $x = 57.74 \pm 0.46$ cm.

P11.15 $I = I_o\,\mathcal{F}_d(\Phi/2)\,F(\Phi/2)$, where $\Phi = 2\pi(d/\lambda)\sin\theta$, $\mathcal{F}_d(x) = (\sin x)^2/x^2$ and $F(x) = [1 - x^2/\pi^2]^{-2}$. The zeros of I are at $\Phi/2 = n\pi$. The intensity of the principal maximum is increased and that of the secondary maximums is reduced.

P11.17 c) The radius of the water droplets or the dust particles is $R \approx 7\ \mu$m.

P11.18 a) $E_1 = E_m \sin[\omega(t - r_1/c)]$, $E_2 = E_m \sin[\omega(t - r_2/c) + \phi]$, $E = 2E_m\cos(\pi\Delta r/\lambda - \phi/2)\sin[\omega(t - r_1/2c - r_2/2c) + \phi/2]$, where $\Delta r = r_2 - r_1 \approx d\sin\theta$. $\mathcal{T}$ is maximal if $\sin\theta_{max} = (p + \phi/2\pi)\lambda/d$ and minimal if $\sin\theta_{min} = (p + \phi/2\pi + 1/2)\lambda/d$. To modify the direction of the maximums, one has only to modify the phase ϕ.
b) If $d = \lambda/2$ and $\phi = 0$, $\mathcal{T}(\theta) = 4\mathcal{T}_o\cos^2(\tfrac{1}{2}\pi\sin\theta)$; hence $\mathcal{T}(0) = 4\mathcal{T}_o$, $\mathcal{T}(30°) = 2\mathcal{T}_o$, $\mathcal{T}(60°) = 0.175\,\mathcal{T}_o$ and $\mathcal{T}(90°) = 0$. **c)** If $d = \lambda/2$ and $\phi = 30°$, $\mathcal{T}$ is maximal for $\theta_{max} = 9.6°$.

d) With N antennas, $\mathcal{I} = N^2\mathcal{I}_o \sin^2(N\phi_T/2)/N^2 \sin^2(\phi_T/2)$ where $\phi_T = \phi - 2\pi(d/\lambda)\sin\theta$. The width of the principal maximum is $\delta\theta \approx 2\lambda/Nd$.

P11.19 a) The maximums are at $\theta = 0, \pm 2.334\times10^{-3}, \pm 4.668\times10^{-3}, \ldots$ in degrees. **b)** The angular width of the central maximum is about $\delta\theta = 0.107°$, the principal maximums are at $\theta = 0, \pm 1.72°$, etc.

P11.20 a) $I = I_o\,\mathcal{F}_i(\theta)\,\mathcal{F}_d(\Phi/2)$, where $\mathcal{F}_i(\theta) = [1 + 2\cos\alpha]^2$ and $\mathcal{F}_d(x) = (\sin x)^2/x^2$ with $\alpha = 2\pi(a/\lambda)\sin\theta$ and $x = \Phi/2 = \pi(d/\lambda)\sin\theta$. **b)** $I' = I_o\,\mathcal{F}'_i\,\mathcal{F}_d(\theta)$, where $\mathcal{F}'_i = 1 + 4\cos^2\alpha + 4\cos\alpha\cos\phi$.

P11.21 a) $d = 2.50$ µm. **b)** $\lambda' = 484$ nm. **c)** $N > 3650/p$, where p is the order. **d)** 411.0 nm $< \lambda <$ 450.8 nm.

P11.22 d) $\theta = 0$, 10.24°, 20.82°, 32.22°, 45.30°, 62.69°.

P11.23 a) $F = N_i c$. $\theta = \pi - 2\,\mathrm{Arcsin}(b/R)$, **b)** $dN_s = \frac{1}{4}R^2F\,d\Omega$. $\sigma(\Omega) = \frac{1}{4}R^2$, $\sigma = \pi R^2$, **c)** $\mu = 7.85\times10^{-3}$ m^{-1}.

Chapter 12

P12.1 a) V and I verify the equations of the line if $k = \omega/v$, where $v = \sqrt{C_l L_l}$ and $Z = \sqrt{L_l/C_l}$. **b)** $\underline{V} = \underline{V}_m\,e^{i(\omega t - kx)}$, $\underline{V}' = \underline{V}'_m\,e^{i(\omega t + k'x)}$, $\underline{V}'' = \underline{V}''_m\,e^{i(\omega t - k''x)}$, hence $\underline{I} = \underline{I}_m\,e^{i(\omega t - kx)}$, $\underline{I}' = -\underline{I}'_m\,e^{i(\omega t + k'x)}$, $\underline{I}'' = \underline{I}''_m\,e^{i(\omega t - k''x)}$, where $\underline{I}_m = \underline{V}_m/Z_1$, $\mathcal{R} = \underline{V}'_m/\underline{V}_m = e^{-2ikD}(Z_2 - Z_1)/(Z_2 + Z_1)$, $\underline{I}'_m/\underline{I}_m = -e^{-2ikD}(Z_2 - Z_1)/(Z_2 + Z_1)$, $k = k' = \omega L_{l1}/Z_1$ and $k'' = \omega L_{l2}/Z_2$. $\mathcal{T} = \underline{V}''_m/\underline{V}_m = 2\,e^{i(k''-k)D}\,Z_2/(Z_2 + Z_1)$, $\underline{I}''_m/\underline{I}_m = 2\,e^{i(k''-k)D}\,Z_1/(Z_2 + Z_1)$.
c) $\underline{I}(D,t) + \underline{I}'(D,t) = 0$, $\mathcal{R} \rightarrow \underline{V}'_m/\underline{V}_m = e^{-2ikD}$, $I'_m/I_m = -\,e^{-2ikD}$.
d) $V_m\,e^{-ikD} + V'_m\,e^{ikD} = 0$, $\mathcal{R} = \underline{V}'_m/\underline{V}_m = -e^{-2ikD}$, $I'_m/I_m = e^{-2ikD}$, **e)** $\underline{Z}_c = Z$.

P12.2 a) $Z = (L_l/C_l)^{1/2} = 96.5\ \Omega$. **b)** $\underline{V}(L,t) = Z_c\underline{I}(L,t)$, $\mathcal{R} = \underline{V}'_m/\underline{V}_m = (Z_c - Z)/(Z_c + Z)\,e^{-i\phi}$
c) $V(x,t) = V_m[\cos(\omega t - kx) + \mathcal{R}_o\cos(\omega t + kx - \phi)]$,
$I(x,t) = (V_m/Z)[\cos(\omega t - kx) - \mathcal{R}_o\cos(\omega t + kx - \phi)]$, $\langle P_g\rangle = 2V_m^2 Z_c/(Z_c + Z)^2 = P_c$ where $P_c = V(L,t)I(L,t)$. **d)** The wave is totally reflected if $Z_c = 0$. No wave is reflected if $Z_c = Z$. The mean supplied power is maximal if $Z_c = Z$.

P12.3 c) There is no dispersion if $kL_lG_l = r_lC_l$ (Heaviside's condition). The attenuation coefficient is then $\eta = G_l(L_l/C_l)^{1/2} = ZG_l$ independently on the frequency.

P12.4 b) If the line is short-circuited ($\underline{Z}_c = 0$), $\omega_n = \pi nv/D$,
$V_n = A\cos(\omega_n t + \phi)\cos(n\pi z/D)$ and $I_n = (A/Z_l)\sin(\omega_n t + \phi)\sin(n\pi z/D)$.

If the line is open ($\underline{Z}_c = \infty$), $\omega_n = \pi(n + \frac{1}{2})v/D$, $V_n = A\cos[\omega_n t + \phi]\cos[(n + \frac{1}{2})\pi z/D]$ and $I_n = (A/Z_l)\sin[\omega_n t + \phi]\sin[(n + \frac{1}{2})\pi z/D]$.

P12.5 a) $\mathbf{E}_T = 2E_m \sin(kz) \sin(\omega t)\, \mathbf{e}_x$ and $\mathbf{B}_T = 2(E_m/c) \cos(kz) \cos(\omega t)\, \mathbf{e}_y$. The nodal planes of $\mathbf{E}_T$ are at $z_p = p\pi/k = p\pi c/\omega = p\lambda/2$.
b) $U_{EM,v} = 2\varepsilon_o E_m^2[\sin^2(kz) \sin^2(\omega t) + \cos^2(kz) \cos^2(\omega t)]$ and $<U_{EM,v}> = \varepsilon_o E_m^2$.
$\mathbf{S} = (4E_m^2/\mu c) \sin(kz) \sin(\omega t) \cos(kz) \cos(\omega t)\, \mathbf{e}_z$ and $<\mathbf{S}> = 0$.
c) $q_s = 0$, $\mathbf{j}_s = (2E_m/c\mu_o) \cos(\omega t)\, \mathbf{e}_x$ and $p_r = 2(\varepsilon_o E_m^2) \cos^2(\omega t)$.
d) $N_\gamma = \pi c\varepsilon_o E_m^2/h\omega$. The radiation pressure is equal to the momentum transfer per unit time and per unit area, hence $p_r = \Delta P = \varepsilon_o E^2$ and $\Delta U = 0$.
e) $\mathbf{E} = E_m[\cos(\omega t - kz)\, \mathbf{e}_x + E \cos(\omega t - kz + \pi/2)\, \mathbf{e}_y]$
$\mathbf{B} = (E_m/c)[-\cos(\omega t - kz + \pi/2)\, \mathbf{e}_x + \cos(\omega t - kz)\, \mathbf{e}_y]$
$\mathbf{E}' = -E_m [\cos(\omega t + kz)\, \mathbf{e}_x + \cos(\omega t + kz + \pi/2)\, \mathbf{e}_y]$
$\mathbf{B}' = (E_m/c)[-\cos(\omega t + kz + \pi/2)\, \mathbf{e}_x + \cos(\omega t + kz)\, \mathbf{e}_y]$
$\mathbf{E}_T = 2E_m \sin(kz)[\sin(\omega t)\mathbf{e}_x + \cos(\omega t)\mathbf{e}_y]$, $\mathbf{B}_T = (2E_m/c) \cos(kz)[\sin(\omega t)\mathbf{e}_x + \cos(\omega t)\mathbf{e}_y]$
$\mathbf{S} = 0$ and $p_r = 2\varepsilon_o E_m^2$. **f)** $\tilde{\nu}_p = pc/2d$ and $k_p = p\pi c/d$. The fields in the mode p are

$\mathbf{E}_T = 2E_p \sin(p\pi cz/d) \sin(p\pi ct/d + \phi_p)\, \mathbf{e}_x$, $\mathbf{B}_T = 2(E_p/c) \cos(p\pi cz/d) \cos(p\pi ct/d + \phi_p)\, \mathbf{e}_y$.

P12.6 a) $\mathbf{B} = (A/v) \sin(\omega t) \sin(kz)\mathbf{e}_y$, $v = \omega/k$. **b)** $U_{E,v} = \frac{1}{2}\varepsilon A^2 \cos^2(\omega t)\cos^2(kz)$,
$U_{M,v} = \frac{1}{2}\varepsilon A^2 \sin^2(\omega t) \sin^2(kz)$, $\mathbf{S} = (A^2/4\mu v) \sin(2\omega t) \sin(2kz)\, \mathbf{e}_z$.
d) U_{EM} (half wave zone) $= (1/16)\, \varepsilon A^2\lambda \mathcal{S}$.

P12.7 a) If $\mathbf{E}^\pm = E_o [\cos(\omega t)\, \mathbf{e}_x + \cos(\omega t \pm \pi/2)\, \mathbf{e}_y] \sin(kz)$, $(\mathbf{E}^\pm)^2 = E_o^2 \sin^2(kz)$. The tip of $\mathbf{E}^\pm$ moves on a circle of radius $E_o|\sin(kz)|$. **b)** $\mathbf{E}^\pm_i = u_o [\cos(\omega t - kz)\mathbf{e}_x + \cos(\omega t - kz \pm \pi/2)\, \mathbf{e}_y]$ and $\mathbf{E}^\pm_r = E_o [\cos(\omega t + kz + \pi)\, \mathbf{e}_x + \cos(\omega t + kz + \pi \pm \pi/2)\mathbf{e}_y]$, hence $\mathbf{E}^\pm = \mathbf{E}^\pm_i + \mathbf{E}^\pm_r = 2E_o [\cos(\omega t - \pi/2)\mathbf{e}_x + \cos(\omega t - \pi/2 \pm \pi/2)\mathbf{e}_y] \sin(kz)$.
c) The angular momentum transfer per unit time of the incident wave and that of the reflected wave are $\mathbf{L}_{EM,v}{}^\pm_i = \mp (U_{EM}{}^\pm{}_{,v\,i}/\omega)\mathbf{e}_z$ and $\mathbf{L}_{EM,v}{}^\pm_r = \mp (U_{EM}{}^\pm{}_{,v\,r}/\omega)(-\mathbf{e}_z)$
Their sum $\mathbf{L}_{EM,v}{}^\pm_i + \mathbf{L}_{EM,v}{}^\pm_r$ is equal to zero at each point.

P12.8 a) $\mathbf{E} = E_m \cos(\omega t - \mathbf{k.r})\mathbf{e}_y$, $\mathbf{B} = -(E_m/c) \cos(\omega t - \mathbf{k.r})[\sin\theta\, \mathbf{e}_x + \cos\theta\, \mathbf{e}_z]$,
$\mathbf{k} = (\omega/c)(-\cos\theta\, \mathbf{e}_x + \sin\theta\, \mathbf{e}_z)$, $\mathbf{E}' = -E_m \cos(\omega t - \mathbf{k}'.\mathbf{r})\, \mathbf{e}_y$,
$\mathbf{B}' = (E_m/c) \cos(\omega t - \mathbf{k}'.\mathbf{r})[\sin\theta\, \mathbf{e}_x - \cos\theta\, \mathbf{e}_z]$, $\mathbf{k}' = (\omega/c)(\cos\theta\, \mathbf{e}_x + \sin\theta\, \mathbf{e}_z)$
b) Setting $\xi = kx\cos\theta$ and $\eta = kx \sin\theta$, $\mathbf{E}_T = -2E_m \sin\xi \sin(\omega t - kz\sin\theta)\, \mathbf{e}_y$
$\mathbf{B}_T = (2E_m/c)[\sin\theta \sin\xi \sin(\omega t - kz\sin\theta)\, \mathbf{e}_x - \cos\theta \cos\xi \cos(\omega t - kz\sin\theta)\, \mathbf{e}_z]$
$v_p = c\sin\theta$. The nodes of $\mathbf{E}$ are at $x_m = m\lambda/2\cos\theta$, where m is an integer.
c) $U_{em} = 2\varepsilon_o E_m^2\{\cos^2\theta \cos^2\xi + [\sin^2\theta - \cos(2\xi)] \sin^2(\omega t - kz\sin\theta)\}$
$\mathbf{S} = (E_m^2/c\mu_o)[\cos\theta \sin(2\eta) \sin(2\omega t - 2kz\sin\theta)\, \mathbf{e}_x + 4\sin\theta \sin^2\xi \sin^2(\omega t - kz\sin\theta)\mathbf{e}_z]$
$\mathcal{I} = (2E_m^2/c\mu_o) \sin\theta \sin^2\xi$. **d)** $\cos\theta_n = n\pi c/\omega d$, $\omega_n = n\pi c/d$, $k_n = n\pi/d$
e) $q_s = 0$, $\mathbf{j}_s(x = 0) = 2\varepsilon_o cE_m \cos\theta_n \cos(\omega_n t - k_n z\sin\theta_n)\, \mathbf{e}_y$
and $\mathbf{j}_s(x = d) = (-1)^{n+1}(2\varepsilon_o cE_m) \cos\theta_n \cos(\omega_n t - k_n z\sin\theta_n)\, \mathbf{e}_y$
$\Pi_r = 2\varepsilon_o E_m^2 \cos^2\theta_n \cos^2(\omega t - kz\sin\theta_n)$.

P12.10 a) $\tilde{\nu}_{m,n} = \frac{1}{2}c(m^2/a^2 + n^2/b^2)^{1/2}$. $\tilde{\nu}_{1,0} = 7.5$ GHz, $k = (\omega/c)\, \gamma_{1,0}$,

where $\gamma_{1,0} = (1 - \omega_{10}{}^2/\omega^2)^{1/2}$. The fields are $\mathbf{E}(\text{TE}) = E_o \sin(\pi x/a)\, e^{i(\omega t - kz)}\, \mathbf{e}_y$

and $\mathbf{B}(\mathrm{TE}) = (E_o/c)\gamma_{1,0}[-\sin(\pi x/a)\,\mathbf{e}_x + i(\pi/ak)\cos(\pi x/a)\,\mathbf{e}_z]\ e^{i(\omega t - kz)}$
$\tilde{\nu}_{1,0} = 7.5$, $\tilde{\nu}_{2,0} = \tilde{\nu}_{0,1} = 15.0$, $\tilde{\nu}_{1,1} = 16.8$, $\tilde{\nu}_{2,1} = 21.2$, $\tilde{\nu}_{3,0} = 22.5$ (in GHz).

b) $\omega^2 = \omega_{1,0}^2 + c^2k^2 \Rightarrow v_p = \omega/k = c(1 - \tilde{\nu}_{1,0}^2/\tilde{\nu}^2)^{-½} = 4.54\times10^8$ m/s,

$v_g = d\omega/dk = c(1 - \tilde{\nu}_{1,0}^2/\tilde{\nu}^2)^{½} = 1.98\times10^8$ m/s.

P12.11 $\omega_{p,q} = (\pi v/a)(p^2+q^2)^{½}$, $a_{min} = \lambda/2n = 0.1667$ µm. $v_{(p)} = \omega/k = v/\gamma_{p,q}$, $v_{(g)} = v\gamma_{p,q}$ $\Rightarrow \gamma_{0,1} = \gamma_{1,0} = 0.866$, $v_{(p)} = 2.31\times10^8$ m/s, $v_{(g)} = 1.73\times10^8$ m/s.

P12.12 a) $\tilde{\nu}_{1,0,0} = \tilde{\nu}_{0,1,0} = \tilde{\nu}_{0,0,1} = v/2a$ (degeneracy 3), $\tilde{\nu}_{1,1,0} = \tilde{\nu}_{1,0,1} = \tilde{\nu}_{0,1,1} = v/2^{½}a$ (degeneracy 3) and $\tilde{\nu}_{1,1,1} = 3^{½}v/2a$ (non-degenerate). **b)** The lowest cut-off frequencies are $\tilde{\nu}_{1,1}\,(TE) = 0.878$ GHz, $\tilde{\nu}_{0,1}\,(TM) = 1.15$ GHz and $\tilde{\nu}_{2,1}\,(TE) = 1.46$ GHz. The lowest frequencies of the cavity are $\tilde{\nu}_{1,1,1}(\mathrm{TE}) = 3.14$ GHz, $\tilde{\nu}_{0,1,1}(\mathrm{TM}) = 3.91$ GHz and $\tilde{\nu}_{1,1,2}(\mathrm{TE}) = 4.08$ GHz.

P12.13 $\tilde{\nu}_{m,j}(TM) = (c/2\pi R)\,x_{m,j}$, where $x_{0,1} = 2.40$, $x_{1,1} = 3.83$, $x_{2,1} = 5.14$, $x_{0,2} = 5.52$. $\tilde{\nu}'_{m,j}(TE) = (c/2\pi R)x'_{m,j}$, where $x'_{1,1} = 1.841$, $x'_{2,1} = 3.054$, $x'_{0,1} = 3.832$, $x'_{1,2} = 5.331$. For $R = 1.5$ cm, $\tilde{\nu}'_{1,1}(TE) = 5.856$, $\tilde{\nu}_{0,1}(TM) = 7.650$ and $\tilde{\nu}'_{2,1}(TE) = 9.714$ (in GHz). The bandwidth is $\tilde{\nu}_{0,1}(TM) - \tilde{\nu}'_{1,1}(TE) = 1.794$ GHz. This corresponds to 1.8×10^5 telephone calls or 287 TV channels.

Chapter 13

P13.1 a) $v' = (c^2 + v_o^2)^{½}$ and $\tan\theta' = -c/v_o$, hence $\tan\alpha = v_o/c \cong 10^{-4} \cong \alpha$. **b)** The transformation law of the group velocity gives $v'_{gx} = v_{gx} = -c$, $v'_{gy} = v_{gy} = 0$, $v'_{gz} = v_{gz} - v_o = -v_o$.

P13.3 $z'_1 = \gamma(z_1 - v_ot_1) = 0$, $t'_1 = \gamma(t_1 - v_oz_1/c^2) = 0$, $z'_2 = \gamma(z_2 - v_ot_2)$, $t'_2 = \gamma(t_2 - v_oz_2/c^2)$.

Event (2) precedes event (1) if $t'_2 < 0$, thus $t_2 < v_o z_2/c^2$. As $v_o < c$, event (2) will never precede (1) if $c > z_2/t_2$. This is the condition on events related by causality.

P13.6 $A'_1 = A_1$, $A'_2 = A_2$, $A'_3 = \gamma(A_3 - \beta A_4)$, $A'_4 = \gamma(\beta A_3 - A_4)$ etc.

P13.7 $\partial_x = \partial_{x'}$, $\partial_y = \partial_{y'}$, $\partial_z = \gamma\partial_{z'} - (\gamma v_o/c^2)\partial_{t'}$, $\partial_t = -\gamma v_o\partial_{z'} + \gamma\partial_{t'}$
$\Rightarrow \partial_\mu\partial_\mu f(x_\rho) = \{\partial_{x'}\partial_{x'} + \partial_{y'}\partial_{y'} + \partial_{z'}\partial_{z'} - (1/c^2)\partial_{t'}\partial_{t'}\}f(x'_\rho)$

P13.8 $D < cT/\pi\alpha = 4.14\times10^{22}$ m, i.e. 4.4×10^6 light-years.

P13.9 a) $P = 0.294$ MeV/c $= 1.578\times10^{-22}$ kg.m/s, $E = 0.589$ MeV $= 0.947\times10^{-13}$ J and $U_K = 0.079$ MeV $= 0.129\times10^{-14}$ J, compared with the non-relativistic values $p = 1.366\times10^{-22}$ kg.m/s and $U_K = 0.102\times10^{-14}$J. **b)** $v_{rel} = 0.8\,c$ compared to the non-relativistic value c.

P13.10 $(\overline{\mathbf{P}}.\overline{\mathbf{P}}) = \mathbf{P}^2 - E^2/c^2 = 0^2 - (mc^2)^2 \Rightarrow E^2 = c^2\mathbf{P}^2 + m^2c^4$.

P13.11 a) $mc^2 = 8.20 \times 10^{-14}$ J = 0.511 MeV. **b)** $v_{rel} = 0.776\ c$ and $v_{cl} = 1.08\ c$ **c)** $P_{rel} = 3.36 \times 10^{-22}$ kg.m/s. $p_{cl} = 2.95 \times 10^{-22}$ kg.m/s. **d)** E(at rest) = 1.02 MeV = 1.64×10^{-13} J and E(in motion) = 1.08 MeV = 1.74×10^{-13} J.

P13.12 b) In light propagates in the same direction as the motion of the medium, $v = (v' + v_o)/(1+\beta v'/c) \approx (v' + v_o)(1-\beta v'/c) \approx (c/n + v_o)(1-\beta/n) \approx c/n + v_o(1-1/n^2)$.

P13.13 $v_o = 1.52 \times 10^8$ m/s.

P13.14 $v = 2.70 \times 10^5$ km.s^{-1}, $D = 1.4{\times}10^{10}$ light-years = 1.3×10^{23} km.

P13.15 $\mathbf{B}' = 0$ and $\mathbf{E}' = (q_s'/\varepsilon_o)\mathbf{e}_x$. Transforming to the observer's frame, we get $\mathbf{E} = \gamma\mathbf{E}' = \gamma(q_s'/\varepsilon_o)\mathbf{e}_x$ and $\mathbf{B} = (\gamma/c)(\beta \times \mathbf{E}') = \gamma(q_s'/\varepsilon_o c^2)v_o\mathbf{e}_y$.. Directly in S, we get $\mathbf{E} = (q_s/\varepsilon_o)\mathbf{e}_x$ and $\mathbf{B} = \mu_o j_s\mathbf{e}_y$ with the relation $q_s = \gamma q_s'$ and $\mathbf{j}_s = q_s = \gamma q_s'\,\mathbf{v}_o$.

P13.16 a) $\mathbf{v} = v\mathbf{e}_x$ and $\mathbf{v}' = (v/\gamma)\mathbf{e}_x$. $\mathbf{f} = -e\mathbf{E} = -eE(\cos\alpha\,\mathbf{e}_y + \sin\alpha\,\mathbf{e}_z)$ and $\mathbf{f}' = [\,\mathbf{f} + (\gamma-1)(\mathbf{v}_o.\mathbf{f})\,\mathbf{v}_o/v_o^2 - \gamma\mathbf{v}_o(\mathbf{f}.\mathbf{v})/c^2]/\gamma[1-(\mathbf{v}_o.\mathbf{v})/c^2] = -eE\,(\cos\alpha\,\mathbf{e}_y/\gamma + \sin\alpha\,\mathbf{e}_z)$ **b)** $\mathbf{E}' = E(\gamma\cos\alpha\,\mathbf{e}_y + \sin\alpha\,\mathbf{e}_z)$, $\mathbf{B}' = (\gamma/c)E\beta\cos\alpha\,\mathbf{e}_x$, hence $\mathbf{f}' = -e[\,\mathbf{E}' + \mathbf{v}' \times \mathbf{B}']$.

P13.18 b) $\mathbf{B}' = 0$ if $\mathbf{v}_o = c_o^2(\mathbf{E} \times \mathbf{B})/E^2 + k\mathbf{E}$, where k is arbitrary. This is possible if $v_o < c$, i.e. $cB < E$. **c)** $\mathbf{E}' = 0$ if $\mathbf{v}_o = (\mathbf{E} \times \mathbf{B})/B^2 + k\mathbf{B}$, possible if $v_o < c$, i.e. $cB > E$.

P13.19 a) $F_M/F_E = \beta^2 \sin\theta < \beta^2 < 1$.
b) $\mathbf{F}_{1\to 2} = (K_o q_1 q_2/c^2 r_1'^3)\,[c^2\mathbf{r}_{12} - (\mathbf{v}_2.\mathbf{v}_1)\mathbf{r}_{12} + \mathbf{v}_1(\mathbf{v}_2.\mathbf{r}_{12})]$, where $\mathbf{r}_{12} = \mathbf{r}_2 - \mathbf{r}_1$ and $r'_1 = \gamma_1[\mathbf{r}_{12}{}^2 - \beta_1{}^2\,\mathbf{r}_{12}{}^2 + (\mathbf{r}_{12}.\boldsymbol{\beta}_1)^2]^{1/2}$, hence $\mathbf{F}_{21} \neq -\mathbf{F}_{12}$. **c)** Evaluating the force exerted by the element dz' of the conductor containing the conduction charges $-eN_L dz'$ of velocity $\mathbf{v}_1$ and the charge $eN_L dz'$ of the ions at rest and integrating over the conductor, we get the expected result $\mathbf{F} = (\mu_o qI/2\pi\rho)(\mathbf{v}_2 \times \mathbf{e}_\varphi)$, where $I = -eN_L v_1$.

P13.20 $\mathbf{B}' = 0$, $\mathbf{A}' = 0$, $V'(\rho', \varphi', z') = -\,2K_o\,q'_L \ln(\rho')$, $\mathbf{E}'(\rho', \varphi', z') = 2(K_o\,q'_L/\rho')\,\mathbf{e}_\rho$
$\mathbf{A} = (\gamma/c^2)\mathbf{v}_o V' = -(\mu_o/2\pi)(\gamma q'_L \mathbf{v}_o)\ln(\rho)$, $V = \gamma V' = -(\gamma q'_L/2\pi\varepsilon_o)\ln(\rho)$
$\mathbf{E} = \gamma\mathbf{E}' + (1-\gamma)(\mathbf{E}'.\mathbf{e})\mathbf{e} = (\gamma q'_L/2\pi\varepsilon_o\rho)\,\mathbf{e}_\rho$, $\mathbf{B} = (\gamma/c)(\boldsymbol{\beta} \times \mathbf{E}') = (\mu_o/2\pi)(\gamma q'_L v_o/\rho)\,\mathbf{e}_\varphi$.
Note that $\gamma q'_L = q_L$(because of length contraction) and the current intensity is $I = q_L v_o$, hence $\mathbf{B} = (\mu_o I/2\pi\rho)\,\mathbf{e}_\varphi$ as given by Ampère's law.

P13.21 Use the relationships $\nabla = \nabla' + (\gamma-1)(\boldsymbol{\beta}.\nabla')\boldsymbol{\beta}/\beta^2 + (\gamma/c)\boldsymbol{\beta}\partial_{t'}$ and $\partial_t = \gamma(\partial_{t'} + c\boldsymbol{\beta}.\nabla')$.

P13.23 Write Maxwell's equations in the frame S', use the transformation laws of the fields, that of charge and current densities and the relationships
$\nabla = \nabla' + (\gamma-1)(\boldsymbol{\beta}.\nabla')\boldsymbol{\beta}/\beta^2 + (\gamma/c)\boldsymbol{\beta}\partial_{t'}$ and $\partial_t = \gamma\,(\partial_{t'} + c\boldsymbol{\beta}.\nabla')$

Chapter 14

P14.1 a) $\ddot{x} = -\,a$, where $a = eE/m$, $\ddot{y} = 0$ and $\ddot{z} = 0$, $\mathbf{v} = (-at + v_o)\mathbf{e}_x$ and $\mathbf{r} = (-\tfrac{1}{2}at^2 + v_o t)\mathbf{e}_x$. **b)** $\mathbf{a} = -e\mathbf{E}/m = -1.76{\times}10^{16}$ m/s^2 $\mathbf{e}_x$, $v = -\,at$ and $x = -\tfrac{1}{2}at^2$. It reaches the speed $-c/10$ at $t = 1.71 \times 10^{-9}$ s after travelling a distance of 2.56 cm. **c)** $x = 2.85 \times 10^{-5}$ m and $t = 5.69 \times 10^{-11}$ s.

P14.2 $r_i \approx 1$ cm.

P14.3 The limit velocity is $v_l = [qE + (4/3)\pi r^3 g(\mu - \mu')]/6\pi\eta r$, $q \approx -5e$.

P14.4 a) $y_S = -0.267$ mm, $v = 2.57\times10^7$ m/s, $\theta = -3.05°$ and $U_K = 3.0085 \times 10^{-16}$ J. **b)** $y = 1.07$ cm.

P14.5 $E_{cyl} = -30.7$ kV/m, $F_{cyl} = 4.91\times10^{-15}$ N, $F_{rod} = 2.45 \times 10^{-13}$ N and $a_{rod} = 2.45\times10^{-3}$ m/s^2.

P14.6 a) $v = 8.38 \times 10^7$ m/s and **b)** $v = 8.15 \times 10^7$ m/s.

P14.7 $\mathbf{F} = -e\mathbf{E} = -K_o Ze^2\mathbf{r}/r^3 + K_o(Z'-1)e^2\mathbf{r}/R^3$ and $U_E = -(3/10)(K_o Z'e^2/R)[5Z-2Z']$. As $5Z > 2Z'$, we deduce that U_E is always negative. The atom is stable.

P14.8 a) $\mathbf{F} = -K_o e^2\mathbf{r}/R^3$. The general motion is elliptic with a frequency of 2.53×10^{15} Hz. **b)** According to Bohr's model, $f = (F/mR)^{½}/2\pi = (e/2\pi R)(K_o Z/mR)^{½} = 2.53 \times 10^{15}$ Hz compared to 5×10^{14} for the visible wavelength $\lambda = 0.6$ μm.

P14.9 $v = 1.58\times10^7$ m/s. The shortest distance is $d = 4.55 \times 10^{-14}$ m, $E = 5.49 \times 10^{19}$ V/m and $V = 2.5 \times 10^6$ V. The maximum acceleration is $a = 2.66 \times 10^{27}$ m/s^2.

P14.10 a) $L = b(2mU_K)^{½}$, $\eta = [1 + (8\pi\varepsilon_o bU_K ZZ'e^2)^2]^{½}$ and $C = -ZZ'/8\pi\varepsilon_o b^2 U_K$.

P14.11 a) $v = 1.237 \times 10^8$ m/s. **b)** The beam is deviated toward the south by a distance of $x = 2.53$ mm.

P14.12 a) Using the expressions, $P = mv/(1-v^2/c^2)^{½} = eBR = 2.563\times10^{-19}$ kg.m/s, we get $v = cP/(P^2 + m^2c^2)^{½} = 0.4548$ c and $W = (m^2c^4 + c^2P^2)^{½} = 1.688594 \times 10^{-10}$ J, thus $U_K = 1.84976 \times 10^{-11}$ J. $\omega_c = c^2|q|B/W = 1.70533 \times 10^8$ rad/s and $V = 114.98$ MV. **b)** The electron would have a speed $v_e = 0.99975\ c = 2.9972 \times 10^8$ m/s.

P14.13 $\delta\tilde{\nu} = \pm½\, eBr\,\tilde{\nu}/mv = \pm\, eB/4\pi m$.

P14.15 As the kinetic energy is comparable to the rest energy ($mc^2 \cong 0.5$ MeV), we must use the relativistic expressions, which give $v = 1.237\times10^8$ m/s. The component of **v** in the direction of **B** is $v_z = 6.1886\times10^7$ m/s and its normal component is $v_\perp = 1.0715 \times 10^7$ m/s. The trajectory is a helix of radius 30.5 μm and the angular velocity is $\omega = 3.52\times10^{11}$ rad/s in the right-hand direction. The pitch is $a = 1.106$ mm.

P14.16 $\tilde{\nu} = 7.62$ MHz, $U_K = (eBR_o)^2/2m = 2.99$ MeV and $N = 74.8$ turns. The cyclotron cannot be used to accelerate electrons because they become extreme relativistic.

P14.17 **B** must have a component $-E/v_y$ in the direction Ox.

P14.18 $m = 1.9915 \times 10^{-26}$ kg, $R(C^{12}) = 16.625$ cm and $R(C^{13}) = 17.306$ cm. The lines are separated by a distance of 1.36 cm.

Chapter 15

P15.1 Use the charge density $q_v(\mathbf{r}', t') = q(t')\,\delta^3(\mathbf{r}'-\frac{1}{2}\mathbf{d}) - q(t')\,\delta^3(\mathbf{r}'+\frac{1}{2}\mathbf{d})$ with $d << r$.

P15.3 c) Seting $I_m = \omega Q_m$, $k = \omega/c$, we find $\mathrm{A}(\mathbf{r},t) = (\mu_o d\, I_m/4\pi r)\sin(\omega t')\mathbf{e}_z$, where $t' \equiv t - r/c$ .Thus, $V(\mathbf{r}, t) = (K_o z d\, I_m/\omega r^3)\,[-\cos(\omega t') + kr\sin(\omega t')]$.
$\mathbf{E} = (K_o d I_m/\omega r^3)\{\,2[kr\sin(\omega t') - \cos(\omega t')]\cos\theta\,\mathbf{e}_r$
$+ [k^2r^2\cos(\omega t') + kr\sin(\omega t') - \cos(\omega t')]\sin\theta\,\mathbf{e}_\theta\}$.
$\mathbf{B} = (\mu_o d/4\pi r^2)\, I_m\,[\sin(\omega t') + kr\cos(\omega t')]\sin\theta\,\mathbf{e}_\phi$. At large distance,
$\mathbf{S} \approx (\mu_o d^2/16\pi^2 c r^2)\,\omega^2 I_m^2\cos^2(\omega t')\sin^2\theta\,\mathbf{e}_r$, $P = (\pi/3\varepsilon_o c)(d/\lambda)^2 I_m^2$, $R = 789\,(d/\lambda)^2$.

P15.4 Setting $t' = t - \tau$, where $\tau = r/c$ is the propagation time, we find
a) $\mathcal{M}(t) = \mathcal{S}\, I(t)\,\mathbf{e}_z$, $\mathbf{A}(\mathbf{r},t) = (\mu_o/4\pi r^3)\,[\,\mathcal{M}(t')\times\mathbf{r} + \tau\,\mathcal{M}'(t')\times\mathbf{r}\,]_{ret}$, $V(\mathbf{r},t) = 0$.
b) $\mathbf{E}(\mathbf{r},t) = -(\mu_o/4\pi r^3)\,[(\mathcal{M}'\times\mathbf{r}) + \tau\,(\mathcal{M}''\times\mathbf{r})]_{ret}$.
$\mathbf{B}(\mathbf{r},t) = (\mu_o/4\pi r^3)\,[\,3(\mathcal{M}.\mathbf{e}_r)\mathbf{e}_r - \mathcal{M} + 3\tau(\mathcal{M}'.\mathbf{e}_r)\,\mathbf{e}_r - \tau\,\mathcal{M}' - \tau^2\,\mathcal{M}'' + \tau^2\,(\mathcal{M}''.\mathbf{e}_r)\,\mathbf{e}_r\,]$.
c) If $I = I_m e^{i\omega t}$, setting $\mathcal{M}_m = I_m\,\mathcal{S}$ and $k = \omega/c$, we find
$\mathbf{A}(\mathbf{r},t) = (\mu_o/4\pi r^2)(1+ikr)\,\mathcal{M}_m e^{i(\omega t-rk)}\sin\theta\,\mathbf{e}_\varphi \to i(\mu_o/4\pi r)\,k\,\mathcal{M}_m e^{i(\omega t-rk)}\sin\theta\,\mathbf{e}_\varphi$,
$\mathbf{E}(\mathbf{r},t) = (\mu_o/4\pi r^2)\,\omega(-i+kr)\,\mathcal{M}_m e^{i(\omega t-rk)}\sin\theta\,\mathbf{e}_\varphi \to (\mu_o/4\pi r)\omega k\,\mathcal{M}_m e^{i(\omega t-rk)}\sin\theta\,\mathbf{e}_\varphi$,
$\mathbf{B}(\mathbf{r},t) = (\mu_o/4\pi r^3)\{(2+2ikr)\cos\theta\,\mathbf{e}_r + (1+ikr-k^2r^2)\sin\theta\,\mathbf{e}_\theta\}\,\mathcal{M}_m e^{i(\omega t-rk)}$
$\to -(\mu_o/4\pi r)\,k^2\,\mathcal{M}_m e^{i(\omega t-rk)}\sin\theta\;\mathbf{e}_\theta$.
d) $\langle dP\rangle = (c\mu_o/32\pi^2)\,k^4\,\mathcal{M}_m^2\sin^2\theta\,d\Omega$, $\langle P\rangle = (4/3)\,\pi^3\,\mu_o c\,\mathcal{M}_m^2/\lambda^4$.

P15.5 a) $L = 1.5$ m. $P_{(DE)} = 3.63$ kW. **b)** $P_{(DM)} = 1.23$ kW.

P15.6 The retarded potential at $\mathbf{r}$ and t is $V(\mathbf{r},t) = (Q_o/4\pi\varepsilon_o r)\cos[\omega(t - |\mathbf{r}|/c)]$.

P15.10 a) $\mathbf{p} = q z_m\cos(\omega t)\,\mathbf{e}_z$, $P_{(ray)} = (\mu q^2/12\pi c)\,\omega^4 z_m^2$. The power that is emitted in the directions making less than 45° with the Oxy plane is 79%.

P15.13 a) $\mathbf{S} = \mathbf{E}\times\mathbf{B}/\mu$

$$= \frac{q^2}{16\pi^2\varepsilon v g^6 R^4}\{\,\mathbf{e}R^2[\alpha^2 g^2 - (\mathbf{e}.\boldsymbol{\alpha})^2(1-\beta^2) + 2g\,(\mathbf{e}.\boldsymbol{\alpha})(\boldsymbol{\alpha}.\boldsymbol{\beta})] + v^2(1-\beta^2)^2[g\boldsymbol{\beta} - \mathbf{e}(\mathbf{e}.\boldsymbol{\beta} - \beta^2)]$$
$$+ gvR(1-\beta^2)[\,\boldsymbol{\beta}(\mathbf{e}.\boldsymbol{\alpha}) + \boldsymbol{\alpha}g] + \mathbf{e}\,vR(1-\beta^2)[(\mathbf{e}.\boldsymbol{\alpha})(2\beta^2 - 1 - \mathbf{e}.\boldsymbol{\beta}) + 2g(\boldsymbol{\alpha}.\boldsymbol{\beta})]\}_{ret}.$$

b) $$\mathbf{S} \approx \frac{q^2}{16\pi^2\varepsilon v g^6 R^2}\,[\alpha^2 g^2 - (\mathbf{e}.\boldsymbol{\alpha})^2(1-\beta^2) + 2g(\mathbf{e}.\boldsymbol{\alpha})(\boldsymbol{\alpha}.\boldsymbol{\beta})]\,\mathbf{e}\,\Big|_{ret}.$$

P15.14 a) $P_{(rad)} = \mu_o\,q^2a^2/6\pi c$. **b)** Writing $P_{(rad)} = -fv$, we get the braking force $\mathbf{f} = -(\mu_o\,q^2a^2/6\pi c)(\mathbf{v}/v^2)$. The equation of motion $ma = f$ has the solution $v = v_o\exp(-t/\tau)$, where $\tau = \mu_o q_2/6\pi mc$. In the case of the electron, $\tau = 6.26\times10^{-24}$ s.
c) $a = U_K/mL$, $U_{(ray)} = 2(\tau/t)\,U_K = 2.4\times10^{-32}$ J $= 1.5\times10^{-13}$ eV.

Appendix A

Mathematical Review

In this appendix, we designate the natural or Napierian logarithm as $\ln(x)$, and the hyperbolic functions as $\sinh(x)$, $\cosh(x)$, and $\tanh(x)$. The inverse functions are designated by $\sinh^{-1}(x)$, $\cosh^{-1}(x)$, $\tanh^{-1}(x)$, $\sin^{-1}(x)$, $\cos^{-1}(x)$, and $\tan^{-1}(x)$, instead of Arcsin x, etc. The unit of angles is the radian. To simplify the notations, the partial derivatives (or derivatives) are designated by $\partial_x f$ for $\partial f/\partial x$, $\partial^2{}_{xy} f$ for $\partial^2 f/\partial x\, \partial y$, etc.

A.1. Taylor series

Taylor series about $x = 0$ and $x = a$ are, respectively,

$$f(x) = f(0) + \partial_x f|_{x=0}\, x/1! + \partial^2{}_x f|_{x=0}\, x^2/2! + \partial^3{}_x f|_{x=0}\, x^3/3! + \ldots$$

$$f(x) = f(a) + \partial_x f|_{x=a}\, (x-a)/1! + \partial^2{}_x f|_{x=a}\, (x-a)^2/2! + \partial^3{}_x f|_{x=a}\, (x-a)^3/3! + \ldots$$

Examples:

$$(1+x)^n = 1 + n\,x + n(n-1)\,x^2/2! + n(n-1)(n-2)\,x^3/3! + \ldots \qquad (|x| < 1)$$

$$(x+y)^n = x^n + n\,x^{n-1}y + n(n-1)\,x^{n-2}y^2/2! + n(n-1)(n-2)\,x^{n-3}y^3/3! + \ldots \qquad (|y| < |x|)$$

A.2. Logarithmic, exponential, hyperbolic and trigonometric functions

$y = e^x = 1 + x/1! + x^2/2! + x^3/3! + \ldots,$ $\qquad \ln(1+x) = x - x^2/2! + x^3/3! - \ldots \quad (x^2 < 1)$

$\sinh(x) = \tfrac{1}{2}(e^x - e^{-x}) = x/1! + x^3/3! + x^5/5! \ldots,$ $\qquad \cosh(x) = \tfrac{1}{2}(e^x + e^{-x}) = 1 + x^2/2! + x^4/4!$

$\tanh(x) = \sinh(x)/\text{coch}(x) = x - x^3/3 + 2x^5/15...,$ $\cosh^2(x) - \sinh^2(x) = 1$

$\sinh(x \pm y) = \sinh x \cosh y \pm \cosh x \sinh y,$ $\cosh(x \pm y) = \cosh x \cosh y \pm \sinh x \sinh y$

$\cosh(2x) = 2\cosh^2 x - 1 = 2\sinh^2 x + 1,$ $\sinh(2x) = 2\sinh x \cosh x$

$\sin x = x/1! - x3/3! + x^5/5!\ ...,$ $\cos x = 1 - x^2/2! + x^4/4!\ ...$

$\cos x = \sin(\pi/2 - x) = -\cos(\pi - x),$ $\sin x = \cos(x - \pi/2) = \sin(\pi - x)$

$\tan x = \sin x/\cos x = x + x^3/3 + 2x^5/15 + ...|_{|x|<\pi/2},$ $\cos^2(x) + \sin^2(x) = 1$

$\sin(x \pm y) = \sin x \cos y \pm \cos x \sin y,$ $\cos(x \pm y) = \cos x \cos y \mp \sin x \sin y$

$\sin(2x) = 2\sin x \cos x,$ $\cos(2x) = 2\cos^2 x - 1 = 1 - 2\sin^2 x$

$\cos x + \cos y = 2\cos \tfrac{1}{2}(x+y) \cos \tfrac{1}{2}(x-y),$ $\cos x - \cos y = 2\sin \tfrac{1}{2}(x+y) \sin \tfrac{1}{2}(y-x)$

$\sin x + \sin y = 2\sin \tfrac{1}{2}(x+y) \cos \tfrac{1}{2}(x-y),$ $\sin x - \sin y = 2\sin \tfrac{1}{2}(x-y) \cos \tfrac{1}{2}(x+y)$

$\tan(x \pm y) = (\tan x \pm \tan y)/(1 \mp \tan x.\tan y),$ $\tan x \pm \tan y = \sin(x \pm y)/\cos x\ .\ \cos y$

$$\sin^{-1}(x) = \frac{\pi}{2} - \cos^{-1}(x) = \frac{1}{1}x + \frac{1}{2.3}x^3 + \frac{1.3}{2.4.5}x^5 + .. = \Sigma_{k\geq 0} \frac{(2k)!\, x^{2k+1}}{(2k+1)\times 2^{2k}(k!)^2} \quad (x^2 < 1)$$

$\tan^{-1}(x) = x - x^3/3 + x^5/5 - x^7/7 + ...|_{\text{if } |x|<1}$ or $\pi/2 - 1/x + 1/3x^3 - 1/5x^5 + ...\ |_{\text{if } |x| \geq 1}$

A.3. Integrals

The *indefinite integral* $f(x)$ is its primitive: $\int dx\, f(x) = F(x)$, where $dF/dx \equiv f$. *The definite integral* $\int_a^b dx\, f(x)$ is the area under the curve representing the function $f(x)$ between the points $x = a$ and $x = b$. It may be shown that $\int_a^b dx\, f(x) = F(b) - F(a)$. The *average value* of a function $f(x)$ between the points $x = a$ and $x = b$ is

$$<f> = \frac{1}{b-a} \int_a^b dx\, f(x) = \frac{F(b) - F(a)}{b-a}.$$

Thus, the area under the curve $f(x)$ between a and b is the same as that of the rectangle of sides $(b - a)$ and $<f>$.

$\partial_x \int dx\, f(x) = f(x),$ $\int dx\, \partial_x f \cdot g(x) = f(x)\, g(x) - \int dx\, \partial_x g\ .\ f(x)$

$\int dx\, x^{-1} = \ln(x)$ $\int dx\, x^n = x^{n+1}/(n+1) \qquad (n \neq -1),$

$\int dx\, \partial_x f/f = \ln f(x),$ $\partial_x \int_{a(x)}^{b(x)} dt\, f(t) = f[b(x)]\, \partial_x b - f[a(x)]\partial_x a$

$\int dx\ e^{ax} = e^{ax}/a,$ $\int dx\ x\,e^{ax} = (ax-1)\,e^{ax}/a^2$

$\int dx \sinh x = \cosh x,$ $\int dx \cosh x = \sinh x$

$\int dx \tanh x = \ln(\cosh x),$ $\int dx/\tanh x = \ln(\sinh x)$

$\int dx \sin x = -\cos x,$ $\int dx \cos x = \sin x$

$\int dx \tan x = -\ln(\cos x),$ $\int dx/\tan x = \ln(\sin x)$

$\int dx \sin^2 x = \frac{1}{2}x - \frac{1}{4}\sin 2x,$ $\int dx \cos^2 x = \frac{1}{2}x + \frac{1}{4}\sin 2x$

$\int dx\,(a + b\cos x)^{-1} = 2(a^2-b^2)^{-1/2}\tan^{-1}[f/(a+b)]$ (where $f = |a^2-b^2|^{1/2}\tan(x/2)$ if $a^2 > b^2$)
or $(b^2-a^2)^{-1/2}\ln[(f+a+b)/(f-a-b)]$ (where $f = |b^2-a^2|^{1/2}\tan(x/2)$ if $a^2 < b^2$)

$\int dx\,(a^2+x^2)^{-1} = (1/a)\tan^{-1}(x/a)$ $\int dx(a^2-x^2)^{-1} = (1/2a)\ln[(a+x)/(a-x)]$

$\int dx(x^2 \pm a^2)^{-1/2} = \ln[x + (x^2 \pm a^2)^{1/2}]$ $\int dx(a^2-x^2)^{-1/2} = \sin^{-1}(x/a) = \pi/2 - \cos^{-1}(x/a)$

Integrals involving $v = a + bx + cx^2$ (with $w = b + 2cx$ and $\Delta = b^2 - 4ac$)

$\int dx\ w^{-1/2} = (1/c)\,w^{1/2}$ $\int dx\ w^{-3/2} = -(1/c)\,w^{-1/2}$

$\int dx/v = \Delta^{-1/2}\ln[(w-\Delta^{1/2})/(w+\Delta^{1/2})]|_{\text{if }\Delta>0},$ or $-2/w|_{\text{if }\Delta=0}$ or $= 2|\Delta|^{-1/2}\tan^{-1}(w|\Delta|^{-1/2})|_{\text{if }\Delta<0}$

$\int \frac{dx}{\sqrt{v}} = c^{-1/2}\ln(w + 2\sqrt{cv}\,)|_{\text{if } c>0},$ or $= c^{-1/2}\sinh^{-1}[w|\Delta|^{-1/2}]|_{\text{if } c>0 \text{ and } \Delta<0},$
or $= -|c|^{-1/2}\sin^{-1}(w\Delta^{-1/2})\,|_{\text{if } c<0 \text{ and } \Delta>0},$ or $= c^{-1/2}\ln w\,|_{\text{if } c>0 \text{ and } \Delta=0}$

$\int dx\ v^{-3/2} = -2w/\Delta v^{1/2},$ $\int dx\ xv^{-3/2} = 2(2a + bx)/\Delta v^{1/2}$

A.4. Complex numbers

A *complex number* (represented by an underlined symbol) is the association of two real numbers:

$$\underline{z} = x + \mathrm{i}y, \qquad \text{where} \quad \mathrm{i}^2 = -1 .$$

x and y are the *real part* and the *imaginary part*, respectively, of $\underline{z}$:

$$x = \mathcal{Re}\,\underline{z}, \qquad \text{and} \qquad y = \mathcal{Im}\,\underline{z} .$$

The *complex conjugate* of $\underline{z} \equiv x + \mathrm{i}y$ is $\underline{z}^* \equiv x - \mathrm{i}y$. A function of $\underline{z}$ is defined by the same Taylor series as for a real variable. For instance,

$$e^{i\phi} = 1 + \frac{1}{1!}(i\phi) + \frac{1}{2!}(i\phi)^2 + \frac{1}{3!}(i\phi)^3 + \frac{1}{4!}(i\phi)^4 \ldots$$
$$= 1 - \frac{1}{2!}\phi^2 + \frac{1}{4!}\phi^4 \ldots + i(\frac{1}{1!}\phi - \frac{1}{3!}\phi^3 + \ldots) = \cos\phi + i\sin\phi \text{ (Euler equation)}.$$

We deduce that

$$\sinh(x) = -i\sin(ix), \qquad \cosh(x) = \cos(ix), \qquad \tanh(x) = -i\tan(ix)$$
$$\sinh(ix) = i\sin(x), \qquad \cosh(ix) = \cos(x), \qquad \tanh(ix) = i\tan(x).$$

A complex number $\underline{z} = x + iy$ may be represented by a point of coordinates x and y in the (x, y) (Argand diagram, Figure A.1). We may use the polar coordinates ρ and ϕ for this point; then,

$$\underline{z} = x + iy = \rho\cos\phi + i\rho\sin\phi = \rho e^{i\phi}$$

with the relations

$$x = \rho\cos\phi, \qquad y = \rho\sin\phi, \qquad \tan\phi = y/x, \qquad \rho = |\underline{z}| = (x^2 + y^2)^{1/2}.$$

ρ is the *modulus* of $\underline{z}$ and ϕ is its *argument* or its *phase* (determined up to 2π).

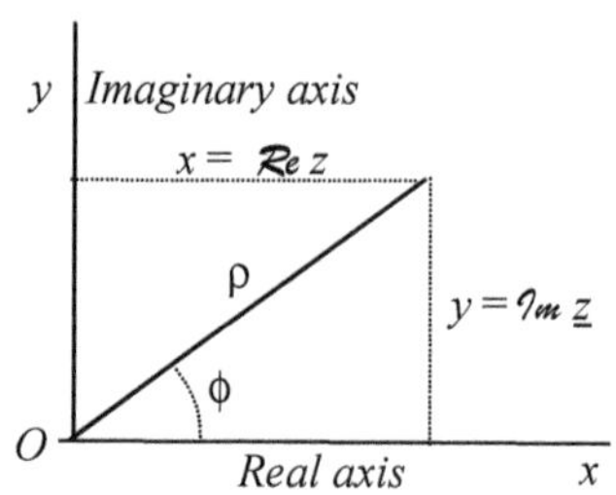

Figure A.1. *Argand diagram for complex numbers*

The sum of two complex numbers is

$$\underline{z}_1 + \underline{z}_2 = (x_1 + iy_1) + (x_2 + iy_2) = (x_1 + x_2) + i(y_1 + y_2).$$

The product and the ratio of two complex numbers may easily be evaluated using the exponential form

$$\underline{z}_1\underline{z}_2 = (\rho_1 e^{i\phi_1})(\rho_2 e^{i\phi_2}) = \rho_1\rho_1 e^{i(\phi_1+\phi_2)}, \qquad \frac{\underline{z}_1}{\underline{z}_2} = \frac{\rho_1 e^{i\phi_1}}{\rho_2 e^{i\phi_2}} = \frac{\rho_1}{\rho_2}e^{i(\phi_1-\phi_2)}.$$

If the algebraic form is used, we find

$$\underline{z}_1\,\underline{z}_2 = (x_1 + \mathrm{i}y_1)(x_2 + \mathrm{i}y_2) = x_1 x_2 - y_1 y_2 + \mathrm{i}(x_1 y_2 + x_2 y_1),$$

$$\frac{\underline{z}_1}{\underline{z}_2} = \frac{x_1 + \mathrm{i}y_1}{x_2 + \mathrm{i}y_2} = \frac{x_1 + \mathrm{i}y_1}{x_2 + \mathrm{i}y_2}\,\frac{x_2 - \mathrm{i}y_2}{x_2 - \mathrm{i}y_2} = \frac{x_1x_2 + y_1y_2 + \mathrm{i}(x_2y_1 - x_1y_2)}{x_2^2 + y_2^2}.$$

A.5. Functions of several variables

a) Consider the function of several variables $f(x, y, z)$. The *partial derivative* $\partial_x f \equiv \partial f/\partial x$, for instance, corresponds to the variation of x, while the other variables are maintained constant. Higher-order derivatives are defined in the same way. These derivatives are independent of the order of differentiation. We have, for instance, $\partial_x\partial_y f = \partial_y\partial_x f$. Thus, it is not necessary to note this order and we may write $\partial^2_{x,y} f \equiv \partial^2 f/\partial x\,\partial y$, etc.

The variation of f if x varies by dx, while the other variables remain constant, is $df\,|_{y,z} = \partial_x f\;dx$. If all the variables vary, the corresponding total variation of f is $df = dx\;\partial_x f + dy\;\partial_y f + dz\;\partial_z f$. Conversely, a differential expression of the form $P(x, y, z)\,dx + Q(x, y, z)\,dy + R(x, y, z)\,dz$ is the total differential of a scalar function $f(x, y, z)$ if P, Q and R are the partial derivatives of f, that is, $P = \partial_x f$, $Q = \partial_y f$ and $R = \partial_z f$. For this, it is necessary and sufficient that $\partial_y P = \partial_x Q$, $\partial_z P = \partial_x R$, etc., then, $df = P(x, y, z)\,dx + Q(x, y, z)\,dy + R(x, y, z)\,dz$.

b) Consider the finite integral $I = \int_a^b dx\,f(x)$. Making a change of variable $x = x(u)$, we obtain $I = \int_{a'}^{b'} du\;(\partial_u x)\,f[x(u)]$, where a' and b' are the values of u that correspond to a and b, i.e. $x(a') = a$ and $x(b') = b$. This result can be generalized to multiple integrals. For instance, the double integral $I = \iint_{\mathcal{D}} dx\;dy\;f(x, y)$ over a domain $\mathcal{D}$, becomes in the change of variables $x = x(u, v)$ and $y = y(u, v)$ a double integral over a domain $\mathcal{D}'$ in (u, v)

$$I = \iint_{\mathcal{D}'} du\;dv\;\frac{D(x,y)}{D(u,v)}\;f[x(u, v),y(u, v)], \qquad \text{where } \frac{D(x,y)}{D(u,v)} \equiv \begin{bmatrix} \partial_u x & \partial_v x \\ \partial_u y & \partial_v y \end{bmatrix}.$$

$D(x, y)/D(u, v)$ is the *Jacobian* of the transformation.

c) Let $I(t) = \int_{a(t)}^{b(t)} dx\;f(x, t)$. The derivative of I with respect to t is

$$\partial_t I = f(b, t)\partial_t b - f(a, t)\;\partial_t a + \int_{a(t)}^{b(t)} dx\;\partial_t f.$$

A.6. Vector analysis in Cartesian coordinates

To write summations over coordinates, we designate the Cartesian coordinates by $x_1 \equiv x$, $x_2 \equiv y$, and $x_3 \equiv z$ and, for instance, the partial derivative with respect to x_1 by $\partial_1 f \equiv \partial_x f \equiv \partial f/\partial x$. The unit vectors of the axes are $\mathbf{e}_x$, $\mathbf{e}_y$, and $\mathbf{e}_z$; then, for instance

$$\Sigma_\alpha \partial_\alpha f\, \mathbf{e}_\alpha \equiv (\partial f/\partial x_1)\mathbf{e}_1 + (\partial f/\partial x_2)\mathbf{e}_2 + (\partial f/\partial x_3)\mathbf{e}_3 \equiv (\partial f/\partial x)\mathbf{e}_x + (\partial f/\partial y)\mathbf{e}_y + (\partial f/\partial z)\mathbf{e}_z.$$

A vector $\mathbf{V}$ may be specified by its Cartesian components such that

$$\mathbf{V} = V_x\, \mathbf{e}_x + V_y\, \mathbf{e}_y + V_z\, \mathbf{e}_z = \Sigma_\alpha V_\alpha\, \mathbf{e}_\alpha .$$

The sum of two vectors $\mathbf{V}$ and $\mathbf{W}$ is

$$\mathbf{V} + \mathbf{W} = (V_x + W_x)\, \mathbf{e}_x + (V_y + W_y)\, \mathbf{e}_y + (V_z + W_z)\, \mathbf{e}_z = \Sigma_\alpha (V_\alpha + W_\alpha)\mathbf{e}_\alpha .$$

If $\mathbf{V}$ and $\mathbf{W}$ have magnitudes V and W, and form an angle θ, their *scalar product* is

$$\mathbf{V.W} = VW \cos\alpha = V_x W_x + V_y W_y + V_z W_z = \Sigma_\alpha V_\alpha W_\alpha .$$

The *cross products* of the basis vectors are $\mathbf{e}_y \times \mathbf{e}_z = \mathbf{e}_x$, $\mathbf{e}_z \times \mathbf{e}_x = \mathbf{e}_y$, and $\mathbf{e}_x \times \mathbf{e}_y = \mathbf{e}_z$. Therefore, the cross product (also called the *vector product*) of $\mathbf{V}$ and $\mathbf{W}$ is

$$\mathbf{V} \times \mathbf{W} = (V_x\, \mathbf{e}_x + V_y\, \mathbf{e}_y + V_z\, \mathbf{e}_z) \times (W_x\, \mathbf{e}_x + W_y\, \mathbf{e}_y + W_z\, \mathbf{e}_z) = \begin{bmatrix} \mathbf{e}_x & \mathbf{e}_y & \mathbf{e}_z \\ V_x & V_y & V_z \\ W_x & W_y & W_z \end{bmatrix}$$

$$= (V_y W_z - V_z W_y)\, \mathbf{e}_x + (V_z W_x - V_x W_z)\, \mathbf{e}_y + (V_x W_y - V_y W_x)\, \mathbf{e}_z .$$

The *double cross product* of three vectors is

$$\mathbf{U} \times (\mathbf{V} \times \mathbf{W}) = (\mathbf{U.W})\, \mathbf{V} - (\mathbf{U.V})\, \mathbf{W}.$$

The vector differential operator $\boldsymbol{\nabla}$ (called *del* or *nabla*) is

$$\boldsymbol{\nabla} = \Sigma_\alpha \mathbf{e}_\alpha \partial_\alpha \equiv \mathbf{e}_x \partial_x + \mathbf{e}_y \partial_y + \mathbf{e}_z \partial_z .$$

The *gradient* of a scalar field f is a vector field

$$\boldsymbol{\nabla} f \equiv \Sigma_\alpha \partial_\alpha f\, \mathbf{e}_\alpha \equiv \partial_x f\, \mathbf{e}_x + \partial_y f\, \mathbf{e}_y + \partial_z f\, \mathbf{e}_z .$$

It verifies the properties

$$df = d\mathbf{r}.\boldsymbol{\nabla} f, \quad \boldsymbol{\nabla}(fg) = f\boldsymbol{\nabla} g + g\boldsymbol{\nabla} f .$$

The *divergence* of a vector field **V** is the scalar field

$$\nabla.\mathbf{V} = \Sigma_\alpha\, \partial_\alpha V_\alpha \equiv \partial_x V_x + \partial_y V_y + \partial_z V_z\,.$$

The *curl* of a vector field **V** is the vector field

$$\nabla \times \mathbf{V} = (\partial_y V_z - \partial_z V_y)\,\mathbf{e}_x + (\partial_z V_x - \partial_x V_z)\,\mathbf{e}_y + (\partial_x V_y - \partial_y V_x)\,\mathbf{e}_z\,.$$

The *Laplacian* is the scalar operator

$$\Delta = \nabla^2 = \partial^2_{xx} + \partial^2_{yy} + \partial^2_{zz}\,.$$

Here are some useful vector analysis relationships:

$$\nabla \times (\nabla f) = 0\,, \qquad \nabla.(\nabla f) = \Delta f\,,$$

$$\nabla.(f\mathbf{V}) = f\,\nabla.\mathbf{V} + \nabla f.\mathbf{V}\,, \qquad \nabla.(\mathbf{V}\times\mathbf{U}) = \mathbf{U}.(\nabla\times\mathbf{V}) - \mathbf{V}.(\nabla\times\mathbf{U})\,,$$

$$\Delta(fg) = f\,\Delta g + 2\,\nabla f.\nabla g + g\,\Delta f\,, \qquad \nabla.(\nabla\times\mathbf{V}) = 0\,,$$

$$\nabla\times(\nabla\times\mathbf{V}) = \nabla\,(\nabla.\mathbf{V}) - \Delta\mathbf{V}\,, \qquad \nabla\times(f\mathbf{V}) = \nabla f\times\mathbf{V} + f\,\nabla\times\mathbf{V})\,,$$

The *circulation* of a vector field **V** on a path $\mathcal{C}$ is

$$\int_{\mathcal{C}} d\mathbf{r}.\mathbf{V} = \int_{\mathcal{C}} \Sigma_\alpha\, dx_\alpha\, V_\alpha \equiv \int_{\mathcal{C}} (dx\, V_x + dy\, V_y + dz\, V_z)\,.$$

The *flux* of a vector field **V** through a surface $\mathcal{S}$ (of orthogonal unit vector **n**) is

$$\iint_{\mathcal{S}} d\mathcal{S}\; \mathbf{n}.\mathbf{V}\,.$$

A.7. Two theorems

The two following theorems are very useful in vector analysis:

THEOREM 1. The necessary and sufficient condition for a vector field **E** to be the gradient of a scalar field f is that its curl $\nabla \times \mathbf{E}$ is equal to zero.

Indeed, if **E** is the gradient of a function f, we have $E_\alpha = \partial_\alpha f$. Thus, the components of the curl of **E** are $(\nabla\times\mathbf{E})_\alpha = \partial_\beta E_\gamma - \partial_\gamma E_\beta = \partial_\beta(\partial_\gamma f) - \partial_\gamma(\partial_\beta f) = 0$, since the differentiations commute. Conversely, if $\nabla \times \mathbf{E} = 0$, we have $\partial_\alpha E_\beta - \partial_\beta E_\alpha = 0$. By taking $\alpha = 1$, for instance, we find $\partial_1 E_\beta = \partial_\beta E_1$. Integrating with respect to x and designating by f the primitive of E_1 with respect to x, we find

$$E_\beta = \int dx\, \partial_\beta E_1 = \partial_\beta \int dx\, E_1 \equiv \partial_\beta f, \qquad \text{where} \quad f \equiv \int dx\, E_1 .$$

We note that f is defined up to the addition of a constant.

THEOREM 2. The necessary and sufficient condition that a vector field **B** be the curl of a vector field **A** is that its divergence $\nabla.\,\mathbf{B}$ is equal to zero.

Indeed, if $\mathbf{B} = \nabla \times \mathbf{A}$, we have $\nabla.\mathbf{B} = \nabla.(\nabla \times \mathbf{A}) = 0$. Conversely, for any vector **B**, we may write, for instance, $B_1 = -\partial_3 F$ and $B_2 = \partial_3 G$. Then, the equation $\nabla.\mathbf{B} \equiv \partial_1 B_1 + \partial_2 B_2 + \partial_3 B_3 = 0$ gives $B_3 = -\partial_1 F + \partial_2 G$. These expressions of B_1, B_2, and B_3 are equivalent to the vector relation

$$\mathbf{B} = \nabla \times \mathbf{A}, \qquad \text{where } \mathbf{A} = G\,\mathbf{e}_1 + F\,\mathbf{e}_2 .$$

We note that, if **A** is replaced by $\mathbf{A}' = \mathbf{A} + \nabla f$, the field **B** is not modified for any f. Thus, **A** is determined up to the addition of the gradient of an arbitrary function.

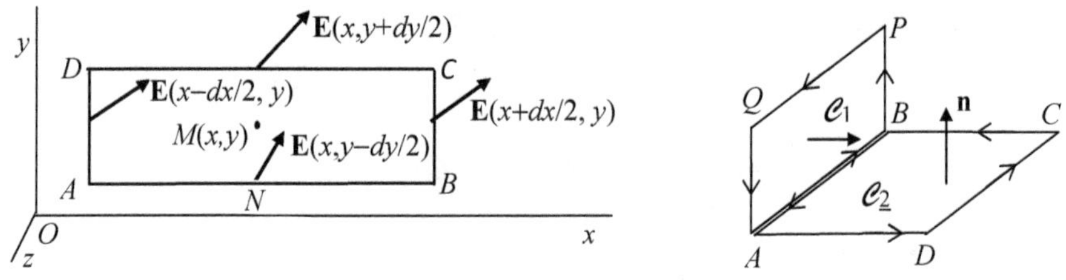

a) (b)

Figure A.2. *Stokes's theorem: a) case of an infinitesimal plane rectangle, b) generalization*

A.8. Stokes's theorem

This theorem allows us to write the circulation of a vector **V** on a closed path $\mathcal{C}$ as the flux of $\nabla \times \mathbf{V}$ through the surface $\mathcal{S}$ bounded by $\mathcal{C}$ and such that $\mathcal{C}$ is orientated according to the right-hand rule around the normal to $\mathcal{S}$. Consider first a small rectangular contour of center $M(x, y)$ and sides dx and dy (Figure A.2a). The circulation of the vector **E** over its perimeter is

$$dC = \int d\mathbf{r}.\mathbf{E}(\mathbf{r}) = \overrightarrow{AB}.\mathbf{E}(x,y-\tfrac{1}{2}dy) + \overrightarrow{BC}.\mathbf{E}(x+\tfrac{1}{2}dx, y) + \overrightarrow{CD}.\mathbf{E}(x, y+\tfrac{1}{2}dy) + \overrightarrow{DA}.\mathbf{E}(x-\tfrac{1}{2}dx, y)$$

where, over AB for instance, we took the field at the middle point of this segment $N(x, y - \tfrac{1}{2}dy)$ up to the first order in dx and dy. Thus, we may write

$$dC = dx\, E_x(x, y - \tfrac{1}{2}dy) + dy\, E_y(x+\tfrac{1}{2}dx, y) - dx\, E_x(x, y + \tfrac{1}{2}dy) - dy E_y(x-\tfrac{1}{2}dx, y)$$

As $E_x(x, y - \frac{1}{2}dy) \approx E_x(x, y) - \frac{1}{2}dy\, \partial_y E_x(x, y)$, for instance, we may write

$$dC = dx\, dy\, [\partial_x E_y - \partial_y E_x] = (\nabla \times \mathbf{E}).\mathbf{n}\, d\mathcal{S},$$

where $d\mathcal{S} = dx\, dy$ is the area of the rectangle and $\mathbf{n} = \mathbf{e}_3$ is the unit vector that is normal to $d\mathcal{S}$.

This result may be generalized to any finite surface $\mathcal{S}$ bounded by a closed path $\mathcal{C}$. Indeed, this surface may always be considered as a juxtaposition of small surfaces bounded by infinitesimal rectangles C_i parallel to the planes of coordinates. Consider two such rectangles *ABCD* and *BAEF* having a common side *AB* (Figure A.2b). The circulation over *AB* for the first rectangle and the circulation over *BA* for the second rectangle cancel, since they are in opposite directions. Thus, the sum of the circulations over these rectangles is equal to the circulation over the external path *CDAEFBC* and the sum of the circulations over all the C_i is equal to the circulation over the path $\mathcal{C}$. By choosing **n** according to the right-hand rule and the orientation of $\mathcal{C}$, the circulation over each rectangle C_i is equal to the flux of $(\nabla \times \mathbf{E})$. Thus, the total circulation over $\mathcal{C}$ is equal to the flux of $(\nabla \times \mathbf{E})$ through the surface $\mathcal{S}$ bounded by $\mathcal{C}$. This result is known as *Stokes's theorem*

$$\int_C d\mathbf{r}.\mathbf{E}(\mathbf{r}) = \iint_{\mathcal{S}} d\mathcal{S}\, (\nabla \times \mathbf{E}).\, \mathbf{n}.$$

In particular, if $(\nabla \times \mathbf{E}) = 0$, this theorem implies that the circulation of **E** over any closed path is equal to zero and, consequently, the circulation of **E** between any two points A and B depends on these points and not on the path. Then, by the theorem (1) of section A.7, there exists a potential V such that

$$\mathbf{E} = -\nabla V, \qquad \text{and} \qquad \int_A^B d\mathbf{r}.\mathbf{E} = V(A) - V(B)\,.$$

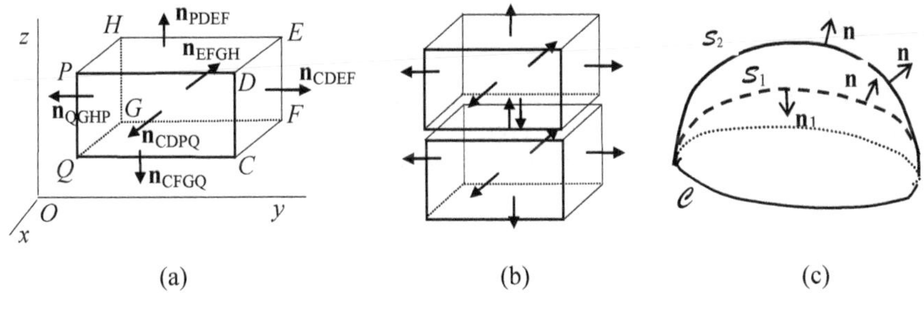

Figure A.3 *a) Gauss-Ostragradsky's theorem for an infinitesimal parallelepiped; b) flux through two adjacent parallelepipeds; c) conservation of the flux through the surfaces bounded by the same closed path* $\mathcal{C}$ *if* $\nabla.\mathbf{B} = 0$

A.9. Gauss-Ostrogradsky's Theorem

This theorem allows the flux of any vector field **B** through any closed surface $\mathcal{S}$ to be written as the integral of $\nabla.\mathbf{B}$ over the volume $\mathcal{V}$ enclosed by $\mathcal{S}$. Consider first an infinitesimal rectangular parallelepiped of center $M(x, y, z)$ and sides dx, dy, and dz (Figure A.3a). Taking the normal unit vector **n** pointing outward, the fluxes through the opposite faces *PQCD* and *EFGH* are

$$d\Phi_{PQCD} = \mathbf{B}(x + \tfrac{1}{2}dx, y, z).\mathbf{n}_{PQCD}\, dy\, dz = B_1(x + \tfrac{1}{2}dx, y, z)\, dy\, dz$$

$$d\Phi_{EFGH} = \mathbf{B}(x - \tfrac{1}{2}dx, y, z).\mathbf{n}_{EFGH}\, dy\, dz = -\, B_1(x - \tfrac{1}{2}dx, y, z)\, dy\, dz\ .$$

Keeping only the first order in dx, dy, and dz, we find for the sum of these fluxes

$$d\Phi_{PQCD} + d\Phi_{EFGH} = [B_x(x+\tfrac{1}{2}dx, y, z) - B_x(x-\tfrac{1}{2}dx, y, z)\ ]dy\, dz = \partial_x B_x(x, y, z)\, dx\, dy\, dz.$$

Calculating the fluxes through the other two pairs of opposite faces in the same way, as $dx\, dy\, dz$ is the volume $d\mathcal{V}$ of the parallelepiped, we find the total outward flux

$$d\Phi = (\partial_x B_x + \partial_y B_y + \partial_z B_z)\, dx\, dy\, dz = (\nabla.\, \mathbf{B})\, d\mathcal{V}.$$

This result may be generalized to any finite volume $\mathcal{V}$ enclosed by a surface $\mathcal{S}$. We may always consider $\mathcal{V}$ as a juxtaposition of infinitesimal parallelepipeds $d\mathcal{V}_i$. If two parallelepipeds have a common face (Figure A.3b), the normal unit vector outgoing from one of them is the opposite of the normal unit vector outgoing from the other and the fluxes of **B** through this common face cancel in the sum $\Sigma_i\, d\Phi_i$. As each $d\Phi_i$ is equal to $(\nabla.\mathbf{B})\, d\mathcal{V}_i$, by summing over all the parallelepipeds, we find that the outward flux through the external surface $\mathcal{S}$ is equal to the volume integral of the divergence of **B**. This result is known as *Gauss-Ostragradsky's theorem*

$$\iint_{\mathcal{S}} d\mathcal{S}\, \mathbf{n}.\mathbf{B} = \iiint_{\mathcal{V}} d\mathcal{V}\, (\nabla.\mathbf{B})\ .$$

In particular, if $\nabla.\mathbf{B} = 0$, this theorem implies that the flux of **B** through any closed surface is equal to zero and, consequently, the flux of **B** through two surfaces $\mathcal{S}_1$ and $\mathcal{S}_2$ bounded by the same contour $\mathcal{C}$ are equal (Figure A.3c). Then, by the theorem (2) of section A.7, there is a *vector potential* **A** such that

$$\mathbf{B} = \nabla \times \mathbf{A}\ , \qquad \text{and} \qquad \iint_{\mathcal{S}} d\mathcal{S}\, \mathbf{n}.\mathbf{B} = \iint_{\mathcal{S}} d\mathcal{S}\, \mathbf{n}.(\nabla\times\mathbf{A}) = \int_{\mathcal{C}} d\mathbf{r}.\mathbf{A}\ .$$

A.10. Vector analysis in curvilinear coordinates

Cylindrical coordinates

If a physical system has rotational symmetry around the z axis, it is more convenient to use cylindrical coordinates (ρ, φ, z) (Figure A.4a). The cylindrical coordinates of a point M are related to its Cartesian coordinates by the equations:

$$\rho = \sqrt{x^2 + y^2}, \qquad \tan\varphi = y/x, \qquad x = \rho\cos\varphi, \qquad y = \rho\sin\varphi\,.$$

Note that the whole space corresponds to $0 < \rho < \infty$, $0 < \varphi < 2\pi$ and $-\infty < z < \infty$. The position of a point M is written as

$$\mathbf{r} = \overrightarrow{OM} = \rho\cos\varphi\,\mathbf{e}_x + \rho\sin\varphi\,\mathbf{e}_y + z\,\mathbf{e}_z\,.$$

The unit vectors tangent to the coordinate curves are related to the Cartesian vectors $\mathbf{e}_\alpha$ by the equations

$$\mathbf{e}_\rho = \cos\varphi\,\mathbf{e}_x + \sin\varphi\,\mathbf{e}_y, \qquad \text{and} \qquad \mathbf{e}_\varphi = -\text{sinx}\,\varphi\,\mathbf{e}_x + \cos\varphi\,\mathbf{e}_y$$

$$\text{or} \quad \mathbf{e}_x = \cos\varphi\,\mathbf{e}_\rho - \sin\varphi\,\mathbf{e}_\varphi, \qquad \text{and} \qquad \mathbf{e}_y = \sin\varphi\,\mathbf{e}_\rho + \cos\varphi\,\mathbf{e}_\varphi\,.$$

The displacement vector $d\mathbf{r}$ generated by the variations $d\rho$, $d\varphi$, and dz is given by

$$d\mathbf{r} = d\rho\,\mathbf{e}_\rho + \rho\,d\varphi\,\mathbf{e}_\varphi + dz\,\mathbf{e}_z\,.$$

A vector field $\mathbf{A}$ is specified by its cylindrical components A_ρ, A_φ and A_z or by its Cartesian components A_x, A_y and A_z according to

$$\mathbf{A} = A_x\,\mathbf{e}_x + A_y\,\mathbf{e}_y + A_z\,\mathbf{e}_z = A_\rho\,\mathbf{e}_\rho + A_\varphi\,\mathbf{e}_\varphi + A_z\mathbf{e}_z\,.$$

Thus, the components are related by the equations

$$A_\rho = A_x\cos\varphi + A_y\sin\varphi\,, \qquad \text{and} \qquad A_\varphi = -A_x\sin\varphi + A_y\cos\varphi$$

and conversely,

$$A_x = A_\rho\cos\varphi - A_\varphi\sin\varphi\,, \qquad \text{and} \qquad A_y = A_\rho\sin\varphi + A_\varphi\cos\varphi\,.$$

The symmetry about Oz, requires that the components A_ρ, A_φ, and A_z are independent of φ.

The elements of area $ds_{\alpha\beta}$ generated by two displacements $d\mathbf{r}_\alpha$ and $d\mathbf{r}_\beta$ along the lines of coordinates α and β and the element of volume $d\mathcal{V}$ are

$$ds_{23} = \rho\, d\varphi\, dz, \qquad ds_{31} = d\rho\, dz, \qquad ds_{12} = \rho\, d\rho\, d\varphi, \qquad d\mathcal{V} = \rho\, d\rho\, d\varphi\, dz\,.$$

The gradient of a scalar field *f*, the divergence of a vector field **A**, and the curl of **A**, as well as the Laplacian of *f* and **A** are given by:

$$\nabla f = \partial_\rho f\ \mathbf{e}_\rho + \rho^{-1}\,\partial_\varphi f\ \mathbf{e}_\varphi + \partial_z f\ \mathbf{e}_z$$

$$\nabla . \mathbf{A} = \rho^{-1}\,\partial_\rho(\rho A_\rho) + \rho^{-1}\,\partial_\varphi A_\varphi + \partial_z A_z$$

$$\nabla\times \mathbf{A} = [\,\rho^{-1}\,\partial_\varphi A_z - \partial_z A_\varphi]\,\mathbf{e}_\rho + [\partial_z A_\rho - \partial_\rho A_z]\,\mathbf{e}_\varphi + \rho^{-1}[\partial_\rho(\rho A_\varphi) - \partial_\varphi A_\rho)]\,\mathbf{e}_z$$

$$\Delta f = \rho^{-1}\,\partial_\rho(\rho\,\partial_\rho f) + \rho^{-2}\,\partial^2{}_{\varphi\varphi} f + \partial^2{}_{zz} f$$

$$\begin{aligned}\Delta\mathbf{A} = {} & [\partial^2{}_{\rho\rho}\, A_\rho - \rho^{-2} A_\rho + \rho^{-1}\,\partial_\rho A_\rho + \rho^{-2}\,\partial^2{}_{\varphi\varphi} A_\rho - 2\,\rho^{-2}\,\partial_\varphi A_\varphi + \partial^2{}_{zz} A_\rho\,]\,\mathbf{e}_\rho \\ & + [\,\rho^{-1}\,\partial^2{}_{\rho\rho}(\rho A_\varphi) - \rho^{-2}\,\partial_\rho(\rho A_\varphi) + 2\,\rho^{-2}\,\partial_\varphi A_\rho + \rho^{-2}\,\partial^2{}_{\varphi\varphi} A_\varphi + \partial^2{}_{zz} A_\varphi\,]\,\mathbf{e}_\varphi \\ & + [\partial^2{}_{\rho\rho} A_z + \rho^{-1}\,\partial_\rho A_z + \rho^{-2}\,\partial^2{}_{\varphi\varphi} A_z + \partial^2{}_{zz} A_z]\,\mathbf{e}_z\end{aligned}$$

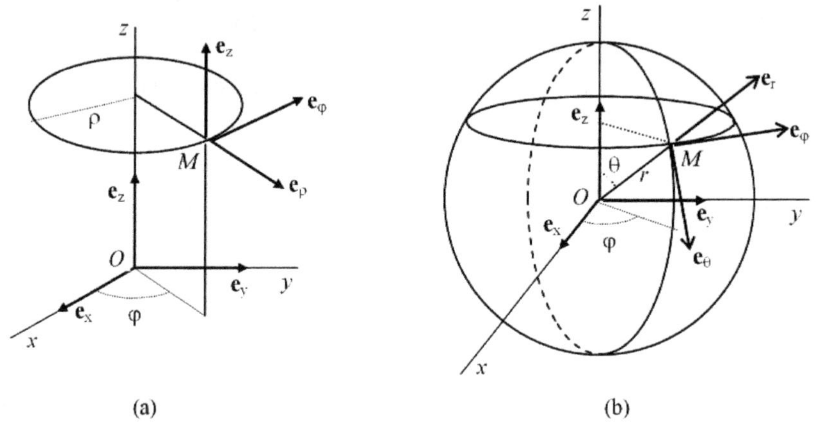

Figure A.4. *a) Cylindrical coordinates, b) spherical coordinates*

Spherical coordinates

If a physical system has rotational symmetry around a point *O*, it is convenient to use spherical coordinates *r*, θ and φ (Figure A.4b). The whole space corresponds to $0 < r < \infty$, $0 < \theta < \pi$ and $0 < \varphi < 2\pi$. The spherical coordinates of a point *M* are related to its Cartesian coordinates by the equations

$$r = \sqrt{x^2+y^2+z^2}, \qquad \tan\theta = \sqrt{x^2+y^2}/z,$$

$$\cos\varphi = x/\sqrt{x^2+y^2}, \qquad \sin\varphi = y/\sqrt{x^2+y^2}, \qquad \tan\varphi = y/x$$

$$x = r\sin\theta\cos\varphi, \qquad y = r\sin\theta\sin\varphi, \qquad z = r\cos\theta$$

The position of a point M is written as

$$\mathbf{r} = \overrightarrow{OM} = r\sin\theta\cos\varphi\,\mathbf{e}_x + r\sin\theta\sin\varphi\,\mathbf{e}_y + r\cos\theta\,\mathbf{e}_z.$$

The unit vectors tangent to the coordinate curves are related to the Cartesian vectors $\mathbf{e}_\alpha$ by the equations

$$\mathbf{e}_r = \sin\theta\cos\varphi\,\mathbf{e}_x + \sin\theta\sin\varphi\,\mathbf{e}_y + \cos\theta\,\mathbf{e}_z,$$

$$\mathbf{e}_\theta = \cos\theta\cos\varphi\,\mathbf{e}_x + \cos\theta\sin\varphi\,\mathbf{e}_y - \sin\theta\,\mathbf{e}_z,$$

$$\mathbf{e}_\varphi = -\sin\varphi\,\mathbf{e}_x + \cos\varphi\,\mathbf{e}_y.$$

and conversely,

$$\mathbf{e}_x = \sin\theta\cos\varphi\,\mathbf{e}_r + \cos\theta\cos\varphi\,\mathbf{e}_\theta - \sin\theta\,\mathbf{e}_\varphi,$$

$$\mathbf{e}_y = \sin\theta\sin\varphi\,\mathbf{e}_r + \cos\theta\sin\varphi\,\mathbf{e}_\theta + \cos\theta\,\mathbf{e}_\varphi,$$

$$\mathbf{e}_z = \cos\theta\,\mathbf{e}_r - \sin\theta\,\mathbf{e}_\theta.$$

The displacement vector $d\mathbf{r}$ generated by the variations dr, $d\theta$, and $d\varphi$ is

$$d\mathbf{r} = dr\,\mathbf{e}_r + r\,d\theta\,\mathbf{e}_\theta + r\sin\theta\,d\varphi\,\mathbf{e}_\varphi.$$

A vector field $\mathbf{A}$ is specified by its spherical components A_r, A_θ, and A_φ or its Cartesian components A_x, A_y, and A_z according to

$$\mathbf{A} = A_x\,\mathbf{e}_x + A_y\,\mathbf{e}_y + A_z\,\mathbf{e}_z = A_r\,\mathbf{e}_r + A_\theta\,\mathbf{e}_\theta + A_\varphi\,\mathbf{e}_\varphi.$$

Thus, the components are related by the equations

$$A_r = -A_x\sin\theta\cos\varphi + A_y\sin\theta\sin\varphi + A_z\cos\theta,$$

$$A_\theta = A_x\sin\theta\sin\varphi + A_y\sin\theta\cos\varphi - A_z\sin\theta, \qquad A_\varphi = -A_x\sin\theta + A_y\cos\theta$$

and conversely,

$$A_x = A_r\sin\theta\cos\varphi + A_\theta\cos\theta\cos\varphi - A_\varphi\sin\varphi,$$

$$A_y = A_r \sin\theta \sin\varphi + A_\theta \cos\theta \sin\varphi + A_\varphi \cos\varphi, \qquad A_z = A_r \cos\theta - A_\theta \sin\theta.$$

The rotational symmetry about O requires that the components A_r, A_θ, and A_φ are independent of θ and φ.

The elements of area $ds_{\alpha\beta}$ generated by two displacements $d\mathbf{r}_\alpha$ and $d\mathbf{r}_\beta$ along the lines of coordinates α and β and the element of volume $d\mathcal{V}$ are

$$ds_{23} = r^2 \sin\theta\, d\theta\, d\varphi, \quad ds_{31} = r \sin\theta\, dr\, d\varphi, \quad ds_{12} = r\, dr\, d\theta, \quad d\mathcal{V} = r^2 \sin\theta\, d\theta\, d\varphi$$

The gradient of a scalar field f, the divergence of a vector field $\mathbf{A}$ and the curl of $\mathbf{A}$, as well as the Laplacian of f and $\mathbf{A}$ are given by:

$$\nabla f = \partial_r f\, \mathbf{e}_r + r^{-1}\, \partial_\theta f\, \mathbf{e}_\theta + (r \sin\theta)^{-1}\, \partial_\varphi f\, \mathbf{e}_\varphi$$

$$\nabla.\mathbf{A} = r^{-2}\, \partial_r(r^2 A_r) + (r \sin\theta)^{-1} \partial_\theta(\sin\theta\, A_\theta) + (r \sin\theta)^{-1}\, \partial_\varphi A_\varphi$$

$$\nabla\times\mathbf{A} = (r \sin\theta)^{-1}[\partial_\theta(\sin\theta\, A_\varphi) - \partial_\varphi A_\theta]\, \mathbf{e}_r + (r \sin\theta)^{-1}[\, \partial_\varphi A_r - \sin\theta\, \partial_r(r A_\varphi)]\, \mathbf{e}_\theta + r^{-1}\, [\partial_r(r A_\theta) - \partial_\theta A_r]\, \mathbf{e}_\varphi$$

$$\Delta f = r^{-2}\, \partial_r\, (r^2\, \partial_r f) + (r^2 \sin\theta)^{-1}\, \partial_\theta\, (\sin\theta\, \partial_\theta f) + (r \sin\theta)^{-2} \partial^2_{\varphi\varphi} f$$

$$\begin{aligned}\Delta\mathbf{A} = r^{-2}\{&[r^2\, \partial^2_{rr} A_r + 2r\, \partial_r A_r - 2A_r + \partial^2_{\theta\theta} A_r - 2\, \partial_\theta A_\theta + (\tan\theta)^{-1}(\partial_\theta A_r - 2A_\theta) \\ &-2\, (\sin\theta)^{-1}\partial_\varphi A_\varphi + (\sin\theta)^{-2}\, \partial^2_{\varphi\varphi} A_r]\, \mathbf{e}_r \\ &+ [r^2\, \partial^2_{rr} A_\theta + 2r\, \partial_r A_\theta + \partial^2_{\theta\theta} A_\theta + 2\, \partial_\theta A_r \\ &\quad + (\tan\theta)^{-1}\partial_\theta A_\theta - 2\cos\theta\, (\sin\theta)^{-2}\, \partial_\varphi A_\varphi + (\sin\theta)^{-2}\, (\partial^2_{\varphi\varphi} A_\theta - A_\theta)\,]\, \mathbf{e}_\theta \\ &+ [r^2\, \partial^2_{rr} A_\varphi + 2r\, \partial_r A_\varphi - 2A_\varphi + \partial^2_{\theta\theta} A_\varphi + (\sin\theta)^{-1}(\partial^2_{\varphi\theta} A_r + 2\partial_\varphi A_r - r\partial^2_{r\varphi} A_r) \\ &\quad + 3\, (\tan\theta)^{-1}\, \partial_\theta A_\varphi + \partial^2_{\varphi\varphi}\, A_\varphi + A_\varphi\,]\, \mathbf{e}_\varphi\}\end{aligned}$$

A.11. Dirac delta function

A narrow rectangular function $u_d(z)$ (Figure A.5a) of width d and area under the curve equal to 1 may be written as

$$u_d(z) = 1/d \qquad (\text{if } z_o - d/2 < z < z_o + d/2)\ ,$$

$$u_d(z) = 0 \qquad (\text{if } z < z_o - d/2 \ \text{ or } \ z > z_o + d/2)\ .$$

In the limit $d \to 0$ (Figure A.5b), the function u_d becomes very narrow and $u_d(z_o) \to \infty$. This is a representation of the *Dirac delta function* $\delta(z - z_o)$, centered at z_o and such that

$$\delta(z - z_o) = \infty \quad \text{if } z = z_o\ , \qquad \text{and} \qquad \delta(z - z_o) = 0 \quad \text{if } z \neq z_o,$$

$$\int_{-\infty}^{\infty} dz\ \delta(z - z_o) = 1.$$

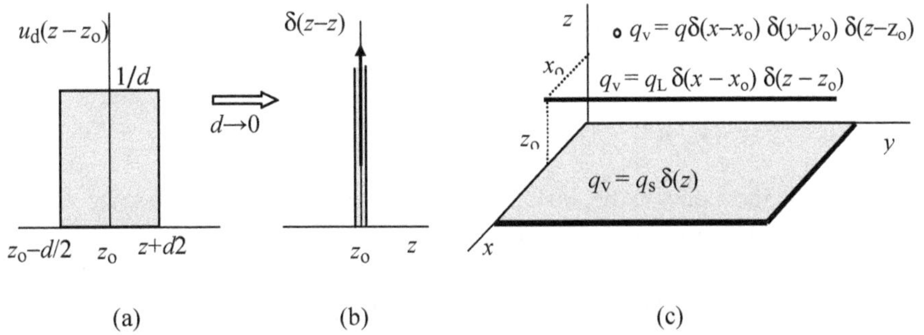

Figure A.5. *a) Rectangular function of width d and unit area under the curve, b) its limit as d→0 is the Dirac delta function, c) representation of a charged surface, charged line and point-charge as volume charge densities proportional respectively to a one-dimensional, two-dimensional, and three-dimensional Dirac delta functions*

Let $f(z)$ be a regular function at z_o and $F(z)$ its primitive. Evaluating the infinite integral of the product $f(z)\,u_d(z)$, we find in the limit $d \to 0$

$$\lim_{d\to0}\int_{-\infty}^{\infty} dz\, u_d(z)f(z) = \lim_{d\to0}\int_{z_o-d/2}^{z_o+d/2} dz\,\frac{f(z)}{d} = \lim_{d\to0}\frac{1}{d}\,[F(z_o+½d) - F(z_o-½d)] = \left.\frac{dF}{dz}\right|_{z_o} = f(z_o)$$

Thus, the integral of the product $u_d(z)f(z)$ gives $f(z_o)$. As in the limit $d\to 0$, $u_d(z)$ becomes the Dirac delta function, we may write the very useful relation

$$\int_{-\infty}^{\infty} dz\,\delta(z-z_o)\,f(z) = f(z_o)\,.$$

We note that, if the integration is restricted to an interval $[z_1, z_2]$, we obtain $f(z_o)$ if $z_1 < z_o < z_2$, and 0 if z_o is outside the interval $[z_1, z_2]$.

A rectangular function $u_d(t)$, where t is time, may represent an electric signal of duration d. Similarly, the function $q_s\,u_d(z-z_o)$ may represent the volume charge density of a plate $z_o - d/2 < z < z_o + d/2$ of charge q_s per unit area. In the limit $d \to 0$, the function $q_s\,\delta(z-z_o)$ represents the volume charge density concentrated on a charged plane $z = z_o$ with a surface charge density q_s (Figure A.5c). By a similar argument, a linear charge of density q_L on an axis parallel to Oy and having the coordinates $x = x_o$ and $z = z_o$ may be represented by a volume charge density $q_v(\mathbf{r}) = q_L\,\delta(x-x_o)\,\delta(z-z_o)$. Also, a point-charge q at a point x_o, y_o, and z_o is represented by the volume charge density $q_v(\mathbf{r}) = q\,\delta^3(\mathbf{r}-\mathbf{r}_o)$, where the three-dimensional Dirac delta function $\delta^3(\mathbf{r}-\mathbf{r}_o)$ is a short-hand notation for the product of three Dirac functions $q\,\delta(x - x_o)\,\delta(y - y_o)\,\delta(z - z_o)$. It has the important property that

$$\iiint_{\mathcal{V}} d\mathcal{V}\,\delta^3(\mathbf{r} - \mathbf{r}_o)\,f(\mathbf{r}) = f(\mathbf{r}_o)$$

provided that the point $\mathbf{r}_o$ be inside the volume $\mathcal{V}$. In the case of several point charges, the total charge density is $q_v(\mathbf{r}) = \Sigma_j\, q_j\, \delta^3(\mathbf{r} - \mathbf{r}_j)$.

As an application, using this expression of q_v and the expression [2.5] of the field of a continuous volume charge distribution, for the point charges we find

$$\mathbf{E}(\mathbf{r}) = K_o \iiint_{\mathcal{V}} d\mathcal{V}'\,\Sigma_j\, q_j\, \delta^3(\mathbf{r}' - \mathbf{r}_j)(\mathbf{r} - \mathbf{r}')/|\,\mathbf{r} - \mathbf{r}'|^3 = K_o\,\Sigma_j\, q_j(\mathbf{r} - \mathbf{r}_j)/|\,\mathbf{r} - \mathbf{r}_j|^3\,.$$

It may be shown that the Dirac delta functions verify the two useful relations:

$$\delta[g(x)] = \Sigma_j\, \delta(x - x_i)\,/\,|g'(x_j)|, \qquad \Delta(1/|\mathbf{r} - \mathbf{r}'|) = -\,4\pi\,\delta^3(\mathbf{r} - \mathbf{r}')\,,$$

where the x_j are the roots of the equation $g(x) = 0$ and $g'(x)$ is the derivative of $g(x)$.

Appendix B

Units in Physics

B.1. Multiples and submultiples of units

The multiples of units and their submultiples by powers of 10 are designated as in Table B.1.

10^n	**Prefixe**	**Example**	10^n	**Prefixe**	**Example**
10^{18}	exa- (E)	exajoule, EJ	10^{-1}	deci- (d)	decibel, dB
10^{15}	peta- (P)	petasecond, Ps	10^{-2}	centi- (c)	centimeter, cm
10^{12}	tera- (T)	terahertz, THz	10^{-3}	milli- (m)	millimeter, mm
10^{9}	giga- (G)	gigavolt, GV	10^{-6}	micro- (μ)	microgram, μg
10^{6}	miga- (M)	megawatt, MW	10^{-9}	nano- (n)	nanometer, nm
10^{3}	kilo- (K)	kilogram, kg	10^{-12}	pico- (p)	picofarad, pF
10^{2}	hecto- (H)	(rarely used)	10^{-15}	femto- (f)	femtometer, fm
10^{1}	deca- (Da)	(rarely used)	10^{-18}	atto- (atto)	attocoulomb, aC

Table B.1. *Multiples and submultiples*

B.2. Fundamental and derived SI units

The units of all physical quantities are defined in terms of six *fundamental* (or *basic*) *units*, which are chosen by convention: *length*, *time*, *mass*, *current intensity*, *temperature*, and *luminous intensity*. The units of the other physical quantities (the so-called *derived units*) are defined in terms of the fundamental units by using the dimensional homogeneity of physical laws. The International System of Units (SI) used in the book has the *meter* (m), *second* (s), *kilogram* (kg), *ampere* (A), *kelvin* (K), and *candela* (cd) as fundamental units. However, some branches of science and engineering continue to use the CGS system (based on the *centimeter*, *gram*, and *second*) for mechanical quantities. As the whole circumference is divided into 2π *radians*, the unit of angles is the radian. We also use the degree (°), *minute* (′), and *second* (″), and sometimes *revolution* (2π radians).

B.3. Mechanical units

Quantity (dimension)	Unit (SI)	Unit (CGS)	Remarks
Length (L)	meter (m)	centimeter (cm)	1 m = 100 cm, 1 μm = 10^{-6} m, 1 F (fermi) = 10^{-15} m
Time (T)	second (s)	second (s)	1 year = 3.155812×10^7 s
Mass (M)	kilogram (kg)	gram (g)	1 kg = 1,000 g, 1 u = $1.660\,056\,55 \times 10^{-27}$ kg, 1 MeV/c^2 = $1.782\,68 \times 10^{-30}$ kg
Area (L^2)	m^2	cm^2	1 m^2 = 10^4 cm^2
Volume (L^3)	m^3	cm^3	1 m^3 = 10^6 cm^3, 1 liter = 10^{-3} m^3
Frequency (T^{-1})	hertz (Hz) = 1 s^{-1}		
Velocity (LT^{-1})	m/s	cm/s	1 km/h = 0.2778 m/s
Acceleration (LT^{-2})	m/s^2	cm/s^2	1 m/s^2 = 100 cm/s^2
Angular velocity (T^{-1})	rad/s	rad/s	1 tour/s = 6.283 18 rad/s
Angular acceleration (T^{-2})	rad/s^2	rad/s^2	1 tour/s^2 = 6.283 18 rad/s^2
Force (MLT^{-2})	newton (N)	dyne	1 N = 10^5 dyne
Moment of force (ML^2T^{-2})	N.m	dyne.cm	1 N.m = 10^7 dyne.cm
Momentum (MLT^{-1})	kg.m/s	g.cm/s	1 kg.m/s = 10^5 g.cm/s
Angular momentum (ML^2T^{-1})	kg.m^2/s	g.cm^2/s	1 kg.m^2/s = 10^7 g.cm^2/s
Moment of inertia (ML^2)	kg.m^2	g.cm^2	1 kg.m^2 = 10^7 g.cm^2
Mass density (ML^{-3})	kg/m^3	g/cm^3	1 kg/m^3 = 10^{-3} g/cm^3
Pressure ($ML^{-1}T^{-2}$)*	pascal (N/m^2)	barye = dyne/cm^2	1 N/m^2 = 10 dyne/cm^2 1 atm = $1.013\,25 \times 10^5$ N/m^2 = 76 cm Hg 1 cm Hg = 1.316×10^{-2} 1 atm = 1,333.22 N/m^2 1 bar = 10^5 N/m^2
Work, energy (ML^2T^{-2})	joule (J)	erg	1 J = 10^7 erg, 1 calorie = 4.186 J, 1 eV = $1.602\,189\,2 \times 10^{-19}$ J
Power (ML^2T^{-3})	watt(W)	erg/s	1 W = 10^7 ergs, 1 horse-power = 745.7 W
Surface tension (MT^{-2})	kg/s^2	g/s^2	1 kg/s^2 = 1,000 g/s^2
Molar mass (M mole^{-1})	kg/k mole	g/mole	1 g/mole = 1 kg/k mole
Specific volume (L^3M^{-1})	m^3/kg	cm^3/g	1 m^3/kg = 10^3 cm^3/g

Table B.2. *Mechanical units*

B.4. SI electromagnetic units

Quantity	Dimensions	Unit (abbreviation)	Remarks
Current intensity	I	ampere (A)	
Electric charge	$Q = TI$	coulomb (C) = 1A.s	1 has – h = 3,600 C
Potential, emf	$\Phi = L^2MT^{-3}I^{-1}$	volt (V)	
Electric field	$E = LMT^{-3}I^{-1}$	V/m	
Capacitance	$C = L^{-2}M^{-1}T^4I^2$	farad (F)	
Displacement field	$D = L^{-2}TI$	C/m^2	
Electric displacement flux	$\Phi_D = TI$	C	
Electric dipole moment	$p_e = LTI$	C.m	
Polarization density	$P = L^{-2}TI$	C/m^2	
Electric permittivity	$\varepsilon = L^{-3}M^{-1}T^4I^2$	F/m	
Resistance	$R = L^2MT^{-3}I^{-2}$	ohm (Ω)	
Resistivity	$\rho = L^3MT^{-3}I^{-2}$	Ω.m	
Electric conductivity	$\sigma = L^{-3}M^{-1}T^3I^2$	$\Omega^{-1}m^{-1}$	
Magnetic flux	$\Phi_B = L^2MT^{-2}I^{-1}$	weber (Wb)	1 maxwell = 10^{-8} Wb
Magnetic induction field	$B = MT^{-2}I^{-1}$	tesla (T) = Wb/m^2	1 gauss = 10^{-4} T
Magnetic field strength	$H = L^{-1}I$	A/m	1 oersted =$10^3/4\pi$ A/m
Magnetic moment	$M = L^2I$	A/m^2	
Intensity of magnetization	$M = L^{-1}I$	A.m	A.m
Inductance	$L = L^2MT^{-2}I^{-2}$	henry (H)	
Magnetic permeability	$\mu = LMT^{-2}I^{-2}$	H/m	

Table B.3. *Electromagnetic units*

B.5. CGS electromagnetic units

The CGS unit for a mechanical quantity is related to that of the corresponding SI unit by a conversion factor that is always a power of 10 obtained from the *dimensional relation*. For example, the CGS unit of force ($F = MLT^{-2}$) is the dyne, while the SI unit is the newton, hence 1 dyne = 1 g.cm/s^2 = (10^{-3} kg) × (10^{-2} m)/s^2 =

10^{-5} N. This is not the case for electromagnetic quantities, as the equations for physical laws are not exactly the same in the SI and the CGS. Furthermore, within CGS, there are several choices of electromagnetic units called *electrostatic units* (ESU), *electromagnetic units* (EMU), *Heaviside-Lorentz units* and *Gaussian units*. This arises because electromagnetism needs another fundamental unit besides the mechanical units of length, mass and time. The choice of this unit depends on the values assigned to the constants ε_o and μ_o (that are related by the reation $\varepsilon_o\mu_o = 1/c^2$). In the SI, this supplementary unit is the ampere (A) defined from Ampère force law (exerted between two thin and parallel wires) $F_A = (\mu_o/2\pi)(II'L/d)$ with $\mu_o = 4\pi\times10^{-7}$. The first CGS electromagnetic systems attempted to define the new electromagnetic unit in terms of mechanical units. The ESU system defines the unit of charge from Coulomb law written as $F_C = qq'/r^2$. The MSU system defines the unit of intensity from Ampère force law written as $F_A = 2(II'L/d)$. The Gaussian system writes both laws as $F_C = qq'/r^2$ and $F_A = 2(II'L/c^2d)$.

The electromagnetic equations in the Gaussian system may be obtained from the corresponding SI equations by using the following rules: do not change the charge q (and consequently, the charge densities, the dipole moment **p**, the polarization **P**, the current and the current densites), the electric field **E** (and consequently the potential V), replace ε_o by $1/4\pi$ (and consequently μ_o by $4\pi/c^2$) and make the following substitutions:

$$\mathbf{D} \rightarrow \mathbf{D}/4\pi,\ \chi_E \rightarrow 4\pi\chi_E,\ \mathbf{B} \rightarrow \mathbf{B}/c,\ \mathbf{H} \rightarrow c\mathbf{H}/4\pi,\ \mathcal{M} \rightarrow c\mathcal{M},\ \mathbf{M} \rightarrow c\mathbf{M},\ \chi_M \rightarrow 4\pi\chi_M$$

The electric displacement and the magnetic field are defined as $\mathbf{D} = \mathbf{E} + 4\pi\mathbf{P}$ and $\mathbf{H} = \mathbf{B} - 4\pi\mathbf{M}$. The relative permittivity and permeability become $\varepsilon_r = 1 + 4\pi\chi_E$ and $\mu_r = 1 + 4\pi\chi_M$, the vector potential **A** becomes $\mathbf{A}/c$, the gyromagnetic ratio g becomes g/c while the inductance and the mutual inductance become L/c^2 and M_{ij}/c^2. As an appliction, Maxwell equations in the Gaussian system are written as

$$\nabla.\mathbf{D} = 4\pi q_v,\quad \nabla\times\mathbf{E} + (1/c)\partial_t\mathbf{B} = 0,\quad \nabla.\mathbf{B} = 0 \ \text{ and } \ \nabla\times\mathbf{H} = (4\pi/c)\,\mathbf{j} + (1/c)\,\partial_t\mathbf{D}$$

The electromagnetic units in the ESU CGS are usually denoted by their SI name with an attached prefix "stat", or with a separate abbreviation "esu", for instance, statampere (statA) or esu current, statcoulomb (statC) or esu charge, etc. Those in EMU CGS system are denoted with prefix "ab" or with a separate abbreviation "emu". The units which have proper names are the *franklin* (Fr) for the ESU charge, the *biot* (Bi) for the EMU current, the *gauss* (G) for the EMU and Gaussian unit of magnetic induction **B**, the *oersted* (Oe) for the EMU CGS and Gaussian unit of magnetic field **H**.

Quantity	SI unit	Its equivalent in ESU Units	Its equivalent in EMU Units	Its equivalent in Gaussian Units
Current intensity	ampere (A) =	(10^{-1} c) statA	(10^{-1}) abA	($10^{-1}c$) Fr·s^{-1}
Electric charge	coulomb (C) =	(10^{-1} c) statC	(10^{-1}) abC	($10^{-1}c$) Fr
Potential, emf	volt (V) =	(10^{8} c^{-1}) statV	(10^{8}) abV	($10^{8}c^{-1}$) statV
Electric field	V/m =	($10^{6}c^{-1}$) statV/cm	(10^{6}) abV/cm	($10^{6}c^{-1}$) statV/cm
Capacitance	farad (F) =	($10^{-9}c^{2}$) cm	(10^{-9}) abF	($10^{-9}c^{2}$) cm
Resistance	ohm (Ω) =	($10^{9}c^{-2}$) s/cm	(10^{9}) abΩ	($10^{9}c^{-2}$) s/cm
Resistivity	Ω.m =	($10^{11}c^{-2}$) s	(10^{11}) abΩ·cm	($10^{11}c^{-2}$) s
Magnetic flux	weber (Wb) =	($10^{8}c^{-1}$) statT.cm²	(10^{8}) Mw	(10^{8}) G.cm²
Magnetic induction field	tesla (T) =	(10^{4} c^{-1}) statT	(10^{4}) G	(10^{4}) G
Magnetic field strength	A/m =	(4π $10^{-3}c$) statA/cm	(4π 10^{-3}) Oe	(4π 10^{-3}) Oe
Magnetic moment	A/m^{2} =	= (10^{3} c) statA.cm²	(10^{3}) abA.cm²	(10^{3}) erg/G
Inductance	henry (H) =	= (10^{9} c^{-2}) cm^{-1}.s^{2}	(10^{9}) abH	(10^{9} c^{-2}) cm^{-1}.s^{2}

Table B.4 *CGS electromagnetic units*

Appendix C

Some Physical Constants

C.1. Mechanical and thermodynamic constants

Speed of light in a vacuum (exact value)	$c = 2.997\ 924\ 58 \times 10^8$ m/s
Gravitational constant	$G = 6.672\ 59\ (85) \times 10^{-11}$ N.m^2/kg^2
Standard acceleration of free fall	$g = 9.806\ 65$ m/s^2
Sidereal year	365.2564 days
Standard pressure	$p_0 = 1$ atm $= 1.013\ 25 \times 10^5$ N/m^2
Standard temperature	$T_0 = 273.16$ K
Mechanical equivalent of the calorie	J = 4.1855 (4) J/Cal
Mass density of dry air (STP)	$\rho_0 = 1.2929$ kg/m^3
Speed of sound in air (STP)	$V_s = 331.36$ m/s
Molar volume of ideal gases (STP)	$V_m = 2.241\ 383\ (70) \times 10^{-2}$ m^3/mole
Avogadro's number	$N_A = 6.022\ 136\ 7\ (36) \times 10^{23}$ mole^{-1}
Boltzmann's constant	$k = 1.380\ 658\ (12) \times 10^{-23}$ J/K
Gas constant	$R = 8.314\ 510\ (70)$ J/K.mole
Acoustic impedance of air (STP)	$Z_a = 428$ N.s/m^3
Threshold of hearing sound intensity	$I_0 = 10^{-12}$ W/m^2
Stefan-Boltzmann's constant	$\sigma = 5.670\ 51\ (7) \times 10^{-8}$ J/m^2.s.K^4
Mass density of mercury	13,595.5 kg/m^3

Table C.1. *Mechanical and thermodynamic quantities. Standard g means at sea level and at latitude of* 45°. STP *means standard temperature* (0 °C) *and pressure* (760 mm Hg)

C.2. Electromagnetic and atomic constants

Elementary charge	$e = 1.602\ 189\ 2\ (46) \times 10^{-19}$ C
Electron-volt	eV = $1.602\ 189\ 2\ (46) \times 10^{-19}$ J
Planck's constant	$h = 6.626\ 075\ (40) \times 10^{-34}$ J.s $= 4.135\ 669\ (11)\ 10^{-15}$ eV.s
Reduced Planck's constant	$\hbar = h/2\pi = 1.054\ 572\ 66\ (63) \times 10^{-34}$ J.s $= 6.582\ 0728\ (17) \times 10^{-16}$ eV.s
Electron mass	$m_e = 9.109\ 389\ 7\ (54) \times 10^{-31}$ kg $= 0.510\ 999\ 06\ (15)$ MeV/c^2
Proton mass	$m_p = 1.672\ 623\ (10) \times 10^{-27}$ kg $= 938.272\ 3\ (27)$ MeV/c^2
Neutron mass	$m_n = 1.674\ 928\ 6\ (10) \times 10^{-27}$ kg $= 939.565\ 63\ (28)$ MeV/c^2
Atomic mass unit	$u = \mathrm{m}(C^{12})/12 = 931.494\ 32\ (28)$ MeV/c^2 $= 1.660\ 540\ 2\ (10) \times 10^{-27}$ kg
Bohr radius	$a_o = 4\pi\varepsilon_o\hbar^2/m_e e^2$ $= 0.529\ 177\ 249\ (24) \times 10^{-10}$ m
Bohr magneton	$\mu_B = e\hbar/2m_e = 9.2741 \times 10^{-24}$ A.m^2
Electric permittivity of vacuum	$\varepsilon_o = 8.854\ 187\ 817 \times 10^{-12}$ A^{2}s^4/kg m^3 $1/4\pi\varepsilon_o = 8.987\ 551\ 787 \times 10^9$ N.m^2.C^{-2}
Magnetic permeability of vacuum	$\mu_o = 4\pi \times 10^{-7}$ kg m/A^{2}s^2

Table C.2. *Electromagnetic and atomic constants*

Further Reading

[BEC 08] BECHERRAWY T., *Vibrations, ondes et optique, tome 3: ondes électromagnétiques*, Hermès, 2008.

[BOR 99] BORN M. and WOLF E., *Principles of Optics* (7th edition), Cambridge University Press, 1999.

[COR 06] CORDIER D., *Electromagnétisme*, Volumes 1 and 2, Dunod, 2006.

[DUF 90] DUFFIN W. J., *Electricity and Magnetism* (4th edition), McGraw-Hill Inc., 1990

[DUR 69] DURNEY C.C. and JOHNSON C.H., *Introduction to Modern Electromagnetics*, McGraw-Hill Inc., 1969.

[EIN 10] EINSTEIN A., *Relativity: The Special and General Theory*, Rober Publishing Company, 2010.

[FEY 89] FEYNMAN R.P., LEIGHTON R.B. and SANDS M., *The Feynman Lectures on Physics, Vol. II: the Electromagnetic Field,* Addison-Wesley, 1989.

[FIT 07] FITZPATRICK R., *Maxwell's Equations and the Principles of Electromagnetism*, Infinity Science Press, 2007.

[FLE 08] FLEISCH D., *A Student's Guide to Maxwell's Equations*, Cambridge University Press, 2008.

[GRA 90] GRANT I.S. and PHILIPS W.R., *Elecromagnetism* (2nd edition), Wiley-Blackwell, 1990.

[GRE 98] GREINER W., *Classical Elecrodynamics*, Springer-Verlag, 1998.

[GRI 98] GRIFFITHS D. J., *Introduction to Electrodynamics* (3rd edition), Prentice Hall, 1998.

[HEL 12] HELRICH C. S., *The Classical Theory of Fields, Electromagnetism*, Springer-Verlag, 2012.

[HOF 99] HOFFMAN B., *Relativity and its Roots*, Dover Publications Inc., 1999.

[JAC 98] JACKSON J.D., *Classical Electrodynamics* (3rd edition), Wiley, 1998.

[JIN 02] JIN J., *The Finite Element Method in Electromagnetics* (2nd edition), Wiley-Blackwell, 2002.

[JON 64] JONES D.S., *The Theory of Electromagnetism*, Elsevier, 1964.

[LAN 87] LANDAU L.D. and LISSCHITZ E.M., *Classical Theory of Fields* (4th edition), Butterworth-Heinemann, 1987.

[LAN 84] LANDAU L.D. and LIFSCHITZ E.M., *Electrodynamics of Continuous Media* (2nd edition), Butterworth-Heinemann, 1984.

[LOR 62] LORRAIN P. and CORSON D., *Electromagnetic Fields and Waves*, Freeman, 1962.

[LOR 79] LORRAIN P. and CORSON D., *Electromagnetism, Principles and Applications* (2nd edition), Freeman, 1979.

[MOL 06] MOLITON A., *Basic Electromagnetism and Materials*, Springer-Verlag, 2006.

[MOR 53] MORSE P.M. and FESHBACH H., *Methods in Theoretical Physics*, McGraw Hill Inc., 1953.

[PAN 05] PANOFSKY W. K. H. and PHILLIPS M., *Classical Electricity and Magnetism* (2nd edition), Dover Publications Inc., 2005.

[PAU 86] PAULI W., *Theory of Relativity,* Dover Publications Inc., 1986

[POL 01] POLLACK G. and STUMP D., *Electromagnetism*, Addison-Wesley, 2001.

[PUR 86] PURCELL E., *Electricity and Magnetism, Berkeley Physics Course Volume 2* (2nd edition), McGraw Hill Inc., 1986.

[REI 08] REITZ J.R., MILFORD F.J. and CHRISTY R.W., *Foundations of Electromagnetic Theory* (4th edition), Addison Wesley, 2008.

[RIN 66] RINDLER W., *Special Relativity* (2nd edition), Interscience, 1966.

[ROS 97] ROSSER G., *Interpretation of Classical Electromagnetism*, Springer-Verlag, 1997.

[SAD 10] SADIKU, M.N.O., *Elements of Electromagnetics* (5th edition), Oxford University Press, 2010.

[SCH 88] SCHWARTZ M., *Principles of Electrodynamics*, Dover Publications Inc., 1988.

[SLA 69] SLATER J.C . and FRANK N.H., *Electromagnetism*, Dover Publications Inc., 1969.

[STR 08] STRATTON J.A., *Electromagnetic Theory*, Read Books, 2008.

Index

F

G, H

N, O, P

Q, R

www.ingramcontent.com/pod-product-compliance
Lightning Source LLC
LaVergne TN
LVHW080030301125
826312LV00003B/36

* 9 7 8 1 8 4 8 2 1 3 5 5 5 *